Lecture Notes in Artificial Intelligence 9706

Subseries of Lecture Notes in Computer Science

More information about this series at http://www.springer.com/series/1244

Nicola Olivetti · Ashish Tiwari (Eds.)

Automated Reasoning

8th International Joint Conference, IJCAR 2016
Coimbra, Portugal, June 27 – July 2, 2016
Proceedings

Springer

Editors
Nicola Olivetti
Aix-Marseille University
Marseille
France

Ashish Tiwari
SRI International
Menlo Park, CA
USA

ISSN 0302-9743 ISSN 1611-3349 (electronic)
Lecture Notes in Artificial Intelligence
ISBN 978-3-319-40228-4 ISBN 978-3-319-40229-1 (eBook)
DOI 10.1007/978-3-319-40229-1

Library of Congress Control Number: 2016940352

LNCS Sublibrary: SL7 – Artificial Intelligence

Printed on acid-free paper

This Springer imprint is published by Springer Nature
The registered company is Springer International Publishing AG Switzerland

Preface

This volume contains the proceedings of the 8th International Joint Conference on Automated Reasoning, IJCAR 2016, held in Coimbra (Portugal) during June 27 – July 2, 2016. IJCAR is the premier international conference covering all topics in automated reasoning, including foundations, implementations, and applications. The 2016 edition of the conference was a merger of three leading events in automated reasoning: International Conference on Automated Deduction (CADE), International Symposium on Frontiers of Combining Systems (FroCoS) and International Conference on Analytic Tableaux and Related Methods (TABLEAUX). Previous IJCAR conferences were held at Siena (Italy) in 2001, Cork (Ireland) in 2004, Seattle (USA) in 2006, Sydney (Australia) in 2008, Edinburgh (UK) in 2010, Manchester (UK) in 2012, and Vienna (Austria) in 2014.

The IJCAR 2016 program consisted of presentations of original research papers and invited talks. Original papers were divided into two categories: regular papers and system desriptions. There were 79 submissions, consisting of 65 regular papers and 14 systems descriptions. Each paper was carefully reviewed by at least three reviewers. All reviewers were either members of the Program Committee (PC) or experts in the area chosen by the PC members. After reviewing and discussing the submissions, the PC accepted 26 regular papers and nine system descriptions.

The program also included four invited talks of the highest scientific value given by Arnon Avron (Tel Aviv University), Gilles Barthe (IMDEA Madrid), Sumit Gulwani, (MSR, Redmond) and André Platzer (CMU, Pittsburgh). The abstracts of the invited talks are included in the present proceedings.

The peer-reviewed research papers are organized in the proceedings in the following sections: Satisfiability of Boolean Formulas, Satisfiability Modulo Theory, Rewriting, Arithmetic Reasoning and Mechanized Mathematics, First-Order Logic and Proof Theory, First-Order Theorem Proving, Higher-Order Theorem Proving, Modal and Temporal Logics, Non-Classical Logics, and Verification. The wide range of sections reflect the variety of topics covered in IJCAR 2016 and witness the maturity of the area of automated reasoning.

During the conference, the International Conference on Automated Deduction (CADE) Herbrand Award for Distinguished Contributions to Automated Reasoning was presented to Zohar Manna and Richard Waldinger. The Best Paper Award was conferred to Jasmin Christian Blanchette (Inria, France), Mathias Fleury (MPI, Germany), and Christoph Weidenbach (MPI, Germany) for their paper titled "A Verified SAT Solver Framework with Learn, Forget, Restart, and Incrementality." Several students received the Woody Bledsoe Travel Awards, named after the late Woody Bledsoe, and funded by CADE Inc. to support student participation.

Several people helped make IJCAR 2016 a success. We want to express our gratitude to the conference chair, Pedro Quaresma, and to the local Organizing Committee who made IJCAR 2016 possible: Sandra Marques Pinto (publicity chair), Reinhard

Kahle (workshop chair), Nuno Baeta, Carlos Caleiro, Nelma Moreira, João Rasga, and Vanda Santos. We thank all the members of the PC for their active participation in the process of evaluating and selecting papers for publication, and during the selection of the invited speakers. We also thank the external reviewers for their precious contribution. The combined expertise of the PC members and the external reviewers ensured that the papers accepted for publication were of the highest scientific quality. We whole-heartedly thank all the authors for submitting their work to IJCAR 2016. On behalf of the PC, we thank the invited speakers for their contribution. We also acknowledge the contributions of the workshop and competition organizers. We extend our thanks to Andrei Voronkov and the EasyChair development team for providing their conference management platform.

We finally thank the University of Coimbra, the hosting institution, and all sponsors for their contribution to the success of the event.

April 2016

Nicola Olivetti
Ashish Tiwari

Organization

IJCAR 2016 was organized by the Department of Mathematics of the Faculty of Sciences and Technology of the University of Coimbra.

Program Committee Chairs

Nicola Olivetti	LSIS, University of Aix-Marseille, France
Ashish Tiwari	SRI International, USA

Program Committee

Franz Baader	TU Dresden, Germany
Peter Baumgartner	NICTA, The Australian National University, Australia
Maria Paola Bonacina	Università degli Studi di Verona, Italy
Agata Ciabattoni	TU Wien, Austria
Leonardo de Moura	Microsoft Research, USA
Hans De Nivelle	University of Wroclaw, Poland
Stephanie Delaune	LSV, CNRS, ENS Cachan, France
Stéphane Demri	LSV, CNRS, ENS Cachan, France
Clare Dixon	University of Liverpool, UK
Christian Fermüller	TU Wien, Austria
Didier Galmiche	Université de Lorraine - LORIA, France
Silvio Ghilardi	Università degli Studi di Milano, Italy
Jürgen Giesl	RWTH Aachen University, Germany
Birte Glimm	Universität Ulm, Germany
Rajeev Goré	The Australian National University, Australia
Reiner Hähnle	Technical University of Darmstadt, Germany
Stefan Hetzl	TU Wien, Austria
Dejan Jovanović	SRI International, USA
Reinhard Kahle	CENTRIA Universidade Nova de Lisboa, Portugal
Deepak Kapur	University of New Mexico, USA
Jordi Levy	IIIA CSIC, Bellaterra, Catalonia, Spain
Carsten Lutz	University of Bremen, Germany
Christopher Lynch	Clarkson University, USA
George Metcalfe	University of Bern, Switzerland
Aart Middeldorp	University of Innsbruck, Austria
Dale Miller	Inria and LIX/Ecole Polytechnique, France
Sara Negri	University of Helsinki, Finland
Nicola Olivetti	LSIS, Aix-Marseille University, France
Jens Otten	University of Potsdam, Germany
Lawrence Paulson	University of Cambridge, UK
Nicolas Peltier	CNRS LIG, Grenoble, France

Andrei Popescu	Middlesex University, London, UK
Christophe Ringeissen	LORIA-Inria Nancy, France
Philipp Ruemmer	Uppsala University, Sweden
Masahiko Sakai	Nagoya University, Japan
Renate A. Schmidt	University of Manchester, UK
Roberto Sebastiani	University of Trento, Italy
Martina Seidl	Johannes Kepler University Linz, Austria
Viorica Sofronie-Stokkermans	Max Planck Institute for Informatics, Germany
Ashish Tiwari	SRI International, USA
Josef Urban	Radboud University, Nijmegen, The Netherlands
Christoph Weidenbach	Max Planck Institute for Informatics, Germany

Local Organizing Committee

Conference Chair

Pedro Quaresma	University of Coimbra, Portugal

Publicity Chair

Sandra Marques Pinto	University of Coimbra, Portugal

Workshops Chair

Reinhard Kahle	New University of Lisbon, Portugal

Local Organization

Nuno Baeta	Polytechnic Institute of Coimbra, Portugal
Carlos Caleiro	IST, University of Lisbon, Portugal
Nelma Moreira	University of Porto, Portugal
João Rasga	CMAF-CIO, University of Lisbon, Portugal
Vanda Santos	CISUC, University of Coimbra, Portugal

Additional Reviewers

E. Abraham
B. Afshari
S. Ahmetaj
P. Backeman
A. Bate
M. Bender
H. Bensaid
C. Benzmüller
M. Bilkova
J.C. Blanchette
F. Blanqui
M. Brenner
T. Brock-Nannestad
M. Bromberger
J. Brotherston
C. Brown
R. Bubel
R. Chadha
K. Chaudhuri
Chung-Kil Hur
C. Dragoi
B. Dutertre
G. Ebner
M. Echenim
S. Enqvist
J.C. Espírito Santo
M. Färber
B. Felgenhauer
M. Ferrari
A.E. Flores Montoya

P. Fontaine
P. Fournier
F. Frohn
J. Giráldez-Cru
N. Gorogiannis
A. Griggio
C. Haase
K. Hashimoto
J. Hensel
M. Hentschel
J. Hölzl
Z. Hou
U. Hustadt
J. Ilin
D. Jiang
S. Joosten
C. Kaliszyk
M. Kaminski
Y. Kazakov
W. Keller
M. Kerber
E. Kieronski
T. King
I. Konnov
T. Kotek
C. Kupke
T. Kutsia
R. Kuznets
P. Lammich
M. Lange
D. Larchey-Wendling
A. Leitsch
B. Lellmann
S. Lucas
A. Marshall
D. Mery
P.J. Meyer
J. Nagele
C. Nalon
N. Nishida
M. Ogawa
E. Orlandelli
D. Petrisan
E. Pimentel
G. Primiero
R. Ramanayake
G. Reis
A. Reynolds
R. Rowe
D. Rydeheard
A. Sangnier
M. Schiller
T. Schneider
S. Schulz
M. Suda
G. Sutcliffe
G. Sutre
R. Thiemann
T.K. Tran
P. Trentin
V. Van Oostrom
L. Vigneron
M. Villaret
M. Volpe
J. Von Plato
J. Vyskocil
U. Waldmann
H. Wansing
F. Wiedijk
T. Wies
B. Woltzenlogel Paleo
N. Zhang
A. Zeljić
D. Zufferey

IJCAR Steering Committee

Franz Baader	TU Dresden, Germany
Maria Paola Bonacina	Università degli Studi di Verona, Italy
Christian Fermüller	TU Wien, Austria
Stefan Hetzl	TU Wien, Austria
Nicola Olivetti	LSIS, University of Aix-Marseille, France
Jens Otten	University of Potsdam, Germany
Ashish Tiwari	SRI International, USA

Sponsors

University of Coimbra
CISUC, Centre for Informatics and Systems of the University of Coimbra
CMA.FCT.UNL Centre for Mathematics and Applications, FCT/UNL
CMUC, Centre for Mathematics, University of Coimbra
CMUP, Centre for Mathematics, University of Porto
IT, Instituto de Telecomunicações
FCT, Portuguese Foundation for Science and Technology
CMC, Câmara Municipal de Coimbra
Coimbra City Hall

Abstracts of Invited Talks

A Logical Framework for Developing and Mechanizing Set Theories

Arnon Avron

School of Computer Science, Tel Aviv University, 69978 Tel Aviv, Israel
aa@cs.tau.ac.il

Abstract. We describe a framework for formalizing mathematics which is based on the usual set theoretical foundations of mathematics. Its most important feature is that it reflects real mathematical practice in making an extensive use of statically defined abstract set terms, in the same way they are used in ordinary mathematical discourse. We also show how large portions of scientifically applicable mathematics can be developed in this framework in a straightforward way, using just rather weak set theories which are predicatively acceptable. The key property of those theories is that every object which is used in it is defined by some closed term of the theory. This allows for a very concrete, computationally-oriented interpretation. However, the development is not committed to such interpretation, and can easily be extended for handling stronger set theories, including *ZFC* itself.

Verification of Differential Private Computations

Gilles Barthe

IMDEA Software Institute, Madrid, Spain

Differential privacy [3, 4], is a statistical notion of privacy which achieves compelling trade-offs between input privacy and accuracy (of outputs). Differential privacy is also an attractive target for verification: despite their apparent simplicity, recently proposed algorithms have intricate privacy and accuracy proofs. We present two program logics for reasoning about privacy and accuracy properties of probabilistic computations. Our first program logic [2] is used for proving accuracy bounds and captures reasoning about the union bound, a simple but effective tool from probablility theory. Our second program logic [1] is used for proving privacy and captures fine-grained reasoning about probabilistic couplings [6, 8], a powerful tool for studying Markov chains. We illustrate the strengths of our program logics with novel and elegant proofs of challenging examples from differential privacy. Finally, we discuss the relationship between our approach and general-purpose frameworks for the verification of probabilistic programs, such as PPDL [5] and pGCL [7].

References

1. Barthe, G., Gaboardi, M., Grégoire, B., Hsu, J., Strub, P.: Proving differential privacy via probabilistic couplings. In: Proceedings of LICS 2016 (2016). http://arxiv.org/abs/1601.05047
2. Barthe, G., Gaboardi, M., Grégoire, B., Hsu, J., Strub, P.: A program logic for union bounds. CoRR, abs/1602.05681 (2016). http://arxiv.org/abs/1602.05681
3. Dwork, C.: Differential privacy. In: Bugliesi, M., Preneel, B., Sassone, V., Wegener, I. (eds.) ICALP 2006. LNCS, vol. 4052, pp. 1–12. Springer, Berlin (2006)
4. Dwork, C., McSherry, F., Nissim, K., Smith, A.: Calibrating noise to sensitivity in private data analysis. In: IACR Theory of Cryptography Conference (TCC), New York, New York, pp. 265–284 (2006). http://dx.doi.org/10.1007/11681878_14
5. Kozen, D.: A probabilistic PDL. J. Comput. Syst. Sci. **30**(2), 162–178 (1985). http://dx.doi.org/10.1016/0022-0000(85)90012-1. Preliminary version at STOC 1983
6. Lindvall, T.: Lectures on the Coupling Method. Courier Corporation (2002)
7. Morgan, C., McIver, A., Seidel, K.: Probabilistic predicate transformers. ACM Trans. Program. Lang. Syst. **18**(3), 325–353 (1996). doi: 10.1145/229542.229547
8. Thorisson, H.: Coupling, Stationarity, and Regeneration. Springer, New York (2000)

Programming by Examples: Applications, Algorithms, and Ambiguity Resolution

Sumit Gulwani

Microsoft Corporation, Redmond, WA, USA
sumitg@microsoft.com

Abstract. 99 % of computer end users do not know programming, and struggle with repetitive tasks. Programming by Examples (PBE) can revolutionize this landscape by enabling users to synthesize intended programs from example based specifications. A key technical challenge in PBE is to search for programs that are consistent with the examples provided by the user. Our efficient search methodology is based on two key ideas: (i) Restriction of the search space to an appropriate domain-specific language that offers balanced expressivity and readability (ii) A divide-and-conquer based deductive search paradigm that inductively reduces the problem of synthesizing a program of a certain kind that satisfies a given specification into sub-problems that refer to sub-programs or sub-specifications. Another challenge in PBE is to resolve the ambiguity in the example based specification. We will discuss two complementary approaches: (a) machine learning based ranking techniques that can pick an intended program from among those that satisfy the specification, and (b) active-learning based user interaction models. The above concepts will be illustrated using FlashFill, FlashExtract, and FlashRelate—PBE technologies for data manipulation domains. These technologies, which have been released inside various Microsoft products, are useful for data scientists who spend 80 % of their time wrangling with data. The Microsoft PROSE SDK allows easy construction of such technologies.

Logic and Proofs for Cyber-Physical Systems

André Platzer

Computer Science Department, Carnegie Mellon University, Pittsburgh, USA
aplatzer@cs.cmu.edu

Abstract. *Cyber-physical systems* (CPS) combine cyber aspects such as communication and computer control with physical aspects such as movement in space, which arise frequently in many safety-critical application domains, including aviation, automotive, railway, and robotics. But how can we ensure that these systems are guaranteed to meet their design goals, e.g., that an aircraft will not crash into another one?

This paper highlights some of the most fascinating aspects of cyberphysical systems and their dynamical systems models, such as hybrid systems that combine discrete transitions and continuous evolution along differential equations. Because of the impact that they can have on the real world, CPSs deserve proof as safety evidence.

Multi-dynamical systems understand complex systems as a combination of multiple elementary dynamical aspects, which makes them natural mathematical models for CPS, since they tame their complexity by compositionality. The family of *differential dynamic logics* achieves this compositionality by providing compositional logics, programming languages, and reasoning principles for CPS. Differential dynamic logics, as implemented in the theorem prover KeYmaera X, have been instrumental in verifying many applications, including the Airborne Collision Avoidance System ACAS X, the European Train Control System ETCS, automotive systems, mobile robot navigation, and a surgical robot system for skullbase surgery. This combination of strong theoretical foundations with practical theorem proving challenges and relevant applications makes *Logic for CPS* an ideal area for compelling and rewarding research.

Contents

Rewriting

Arithmetic Reasoning and Mechanizing Mathematics

First-Order Logic and Proof Theory

First-Order Theorem Proving

Verification

Invited Talks

A Logical Framework for Developing and Mechanizing Set Theories

Arnon Avron(✉)

School of Computer Science, Tel Aviv University, 69978 Tel Aviv, Israel
aa@cs.tau.ac.il

Abstract. We describe a framework for formalizing mathematics which is based on the usual set theoretical foundations of mathematics. Its most important feature is that it reflects real mathematical practice in making an extensive use of statically defined abstract set terms, in the same way they are used in ordinary mathematical discourse. We also show how large portions of scientifically applicable mathematics can be developed in this framework in a straightforward way, using just rather weak set theories which are predicatively acceptable. The key property of those theories is that every object which is used in it is defined by some closed term of the theory. This allows for a very concrete, computationally-oriented interpretation. However, the development is not committed to such interpretation, and can easily be extended for handling stronger set theories, including ZFC itself.

Set theory is almost universally accepted as the foundational theory in which the whole of mathematics can be developed. As such, it is the most natural framework for MKM (Mathematical Knowledge Management). Moreover: as is emphasized and demonstrated in [7], set theory also has a great computational potential. However, in order to be used for these tasks it is necessary to overcome the following serious gaps that exist between the "official" formulations of set theory (like ZFC) and actual mathematical practice:

- Unlike the language used in real mathematical practice, the language(s) used in official formalizations of set theories are rather poor and inconvenient.
- ZFC treats all the mathematical objects on a par, and so hid the computational significance of many of them.
- Core mathematics practically deals only with a fraction of the set-theoretical "universe" of ZFC. Therefore easier to mechanize systems, corresponding to universes which are better suited for computations, should do.

The goal of this paper is to present a unified, type-free, user-friendly framework (originally developed in [2,3]) for formalizations of axiomatic set theories of different strength, from rudimentary set theory to full ZFC. Our framework makes it possible to employ in a natural way all the usual set notations and constructs as found in textbooks on naive or axiomatic set theory (and *only* such notations). Another important feature of this framework is that its set of

N. Olivetti and A. Tiwari (Eds.): IJCAR 2016, LNAI 9706, pp. 3–8, 2016.
DOI: 10.1007/978-3-319-40229-1_1

closed terms suffices for denoting every concrete set (including infinite ones!) that might be needed in applications, as well as for *computations* with sets.

Our basic assumption is that the sets which are interesting from a computational point of view are those which can be *defined* by abstract terms the form $\{x \mid \varphi\}$, using formulas in some, intuitively meaningful, formal language. Now the use of such terms is also indispensible for any user-friendly treatment of set theories. Therefore they are used in all textbooks on first-order set theories, as well as in several computerized systems. However, whenever they are intended to denote *sets* (rather than classes) they are introduced (at least partially) in a *dynamic* way, with different semantic justification each time. In contrast, what abstract set terms may be used in our framework is *statically* defined in a precise, purely *syntactic* way, using the mechanism of *safety relations.*

A safety relation is a syntactic relation between formulas and sets of variables, which provide a common generalization of the notions of domain-independence (in database theory), absoluteness (in set theory), and decidability (in formal arithmetics). Intuitively, φ is *safe* with respect to $\{y_1, ..., y_k\}$ (where $Fv(\varphi) = \{x_1, ..., x_n, y_1, ..., y_k\}$ and $k > 0$) if for every "accepted" sets $a_1, ..., a_n$, the collection $\{\langle y_1, ..., y_k\rangle \mid \varphi(a_1, ..., a_n, y_1, ..., y_k)\}$ is also an "accepted" set, which can be constructed from $a_1, ..., a_n$. Safety with respect to the empty set intuitively means "definiteness", and should be thought of as a generalization of decidability and of absoluteness. The differences between set theories is mainly reduced in our framework to different interpretations of the vague notions of "acceptable", "can be constructed", and "definite".

1 Outline of the Formal Framework

1.1 Logics

We allow the use of four different types of logics in our framework. The basic two are classical first-order logic and intuitionistic first-order logic. However, in our opinion the first-order level is not sufficient for handling infinity in a satisfactory way, while second-order logic is too strong. In [1] it was argued that *TC*-logic (also called ancestral logic — *AL*) which allows the use of a transitive closure operation *TC* provides a better framework for the formalization of mathematics. This suggestion (again in two versions: classical and intuitionistic) seems particularly promising for the present project, since with *TC* the difference between set theories with infinity and those without it can again be reduced to differences in the underlying syntactic safety relations.

1.2 Languages

A language L for a set theory S should be based in our framework on some first-order signature σ which includes $\in$ and $=$, and it is introduced by a simultaneous recursive definition of its terms, formulas, and the safety relation $\succ$ that underlies it. The clauses for the terms and formulas of such L always include the usual ones, together with the following additional clauses:

- If x is a variable, φ is a formula, and $\varphi \succ \{x\}$, then $\{x \mid \varphi\}$ is a term.
- If the underlying logic is a TC-logic then $(TC_{x,y}\varphi)(t, s)$ is a formula whenever φ is a formula, x, y are distinct variables, and t, s are terms.

The clauses defining the safety relation $\succ$ of L should include the set of syntactic conditions given below (which generalize those used for d.i. in database theory).

1. $\varphi \succ \emptyset$ if φ is atomic.
2. $\varphi \succ \{x\}$ if $\varphi \in \{x = t, t = x, x \in x, x \in t\}$, and $x \notin Fv(t)$.
3. $\neg\varphi \succ \emptyset$ if $\varphi \succ \emptyset$.
4. $\varphi \vee \psi \succ X$ if $\varphi \succ X$ and $\psi \succ X$.
5. $\varphi \wedge \psi \succ X \cup Y$ if $\varphi \succ X$, $\psi \succ Y$ and $Y \cap Fv(\varphi) = \emptyset$.
6. $\exists y\varphi \succ X - \{y\}$ if $y \in X$ and $\varphi \succ X$.
7. $\forall x(\varphi \rightarrow \psi) \succ \emptyset$ if $\varphi \succ \{x\}$ and $\psi \succ \emptyset$[1]

More clauses may then be added, depending on the theory S. In particular, if TC-logic is used as the underlying logic then the following clause will also be included: $(TC_{x,y}\varphi)(x, y) \succ X$ if $\varphi \succ X$, and $\{x, y\} \cap X \neq \emptyset$.

Definition 1. Given an underlying logic $\mathcal{L}$ and a first-order signature σ which includes $\in$ and $=$, the language $L^{\mathcal{L}}_{\sigma}$ is the minimal language which satisfies all the above conditions.

The basic language used in our framework will be $RSL = L^{FOL}_{\sigma_{ZF}}$, where FOL denotes (classical or intuitionistic) first-order logic, and $\sigma_{ZF} = \{\in, =\}$. Already in this language (and in its extensions) we can introduce as *abbreviations* most of the standard notations for sets used in mathematics, like: $\emptyset$, $\{t_1, \ldots, t_n\}$, $\langle t, s\rangle$, $\{x \in t \mid \varphi\}$ in case $\varphi \succ \emptyset$, $\{t \mid x \in s\}$, $s \times t$, $s \cap t$, $s \cup t$, $\bigcup t$, $\bigcap t$, $\iota x\varphi$ (in case $\varphi \succ \{x\}$), and $\lambda x \in s.t$. An exact characterization (proved in [4]) of the expressive power of RSL can be given in terms of the well-known class of rudimentary set functions (see [8]): For any n-ary rudimentary function F there exists a formula φ such that $Fv(\varphi) = \{y, x_1, \ldots, x_n\}$, $\varphi \succ_{RSL} \{y\}$ and $F(x_1, \ldots, x_n) = \{y \mid \varphi\}$. Conversely, if $Fv(\varphi) = \{y_1, \ldots, y_k, x_1, \ldots, x_n\}$, and $\varphi \succ_{RSL} \{y_1, \ldots, y_k\}$, then there exists a rudimentary function F s.t. $F(x_1, \ldots, x_n) = \{\langle y_1, \ldots, y_k\rangle \mid \varphi\}$.

1.3 The Basic Axioms and Systems

The main part of every Theory T in our framework consists of the following axiom schemas (our version of the "ideal calculus" [10]):

Extensionality: $\forall z(z \in x \leftrightarrow z \in y) \rightarrow x = y$
Comprehension$_L$: $\forall x(x \in \{x \mid \varphi\} \leftrightarrow \varphi)$

Given a signature σ and a logic $\mathcal{L}$, we denote by $RST^{\mathcal{L}}_{\sigma}$ the theory in $L^{\mathcal{L}}_{\sigma}$ whose axioms are the basic ones listed above. Note that the strength of $RST^{\mathcal{L}}_{\sigma}$ depends on the set of terms available in $L^{\mathcal{L}}_{\sigma}$, and so on the safety relation used in $L^{\mathcal{L}}_{\sigma}$. Now the most important feature of $RST^{\mathcal{L}}_{\sigma}$ is that its two main axioms directly lead (and are equivalent) to the following *set-theoretical* reduction rules:

[1] In the classical case this condition is derivable from the others.

(β) $\vdash_{RST^{\mathcal{L}}_{\sigma}} t \in \{x \mid \varphi\} \leftrightarrow \varphi\{t/x\}$ (provided t is free for x in φ).
(η) $\vdash_{RST^{\mathcal{L}}_{\sigma}} \{x \mid x \in t\} = t$ (provided $\{x \mid x \in t\}$ is a term, i.e. $x \notin Fv(t)$).

It is easy to see that the usual reduction rules of the typed λ-calculus follow from these reduction rules. In particular: $\vdash_{RST^{\mathcal{L}}_{\sigma}} a \in s \rightarrow (\lambda x \in s.t)(a) = t\{a/x\}$.

1.4 Extensions by Definitions

It was argued in [12] that the language of ZFC with definitions and partial functions provides the most promising "bedrock semantics for communicating and sharing mathematical knowledge". Regularly expanding the language employed is indeed an essential part of every mathematical research and its presentation. There are two principles that govern this process in our framework. First, its static nature demands that conservatively expanding the language of a given theory should be reduced to the use of *abbreviations*. Second, since the introduction of new predicates and function symbols creates new atomic formulas and terms, one should be careful that the above conditions concerning the underlying safety relation $\succ$ are preserved. Thus only formulas φ such that $\varphi \succ \emptyset$ can be used for defining new predicated symbols. Now in the set-theoretical context it is more convenient to write $t \in X$ (instead of $X(t)$) when X is a defined unary predicate symbol[2], viewing X as a *class*. Thus we allow the use of class terms of the form $\{x \hat{|} \varphi\}$, provided that $\varphi \succ \emptyset$. The treatment of such terms is done in the standard way, as described, e.g., in [13]. New function symbols, corresponding to global operations (like the "rudimentary functions"), can then be introduced in the form $\lambda x \in X.t$, where X is a class term. See [6] for details.

2 Handling the Axioms of ZF and ZFC

The definability of $\{t, s\}$ and of $\bigcup t$ means that $\succ_{RSL}$ suffices for the axioms of pairing and union. Next we turn to the comprehension axioms that remain valid if we limit ourselves to hereditarily finite sets. It can be shown ([3]) that each of them can be captured (in a modular way) by adding to the definition of $\succ_{RSL}$ a certain syntactic condition. The separation axiom, for example, is available whenever $\varphi \succ \emptyset$ (where φ is the separating formula and $\succ$ is the safety relation used), and this is already quite strong. However, to capture the full power of this schema we need to add the condition that $\varphi \succ \emptyset$ for *every* formula φ (implying that we see any formula of the language as defining a "definite" property). Similarly, the replacement schema is available whenever the corresponding function is explicitely definable (in the form $\lambda x \in s.t$), but a more complicated condition corresponds to the full schema. As for the non-predicative powerset axiom, the simplest way to get it is to enrich the language with the binary relation $\subseteq$, add an axiom connecting it with $\in$, and then add to the definition of the safety relation the simple condition: $x \subseteq t \succ \{x\}$ if $x \notin Fv(t)$.

[2] The use of binary predicates etc. can be reduced, of course, to the use of unary ones.

Next we turn to the axiom of Infinity — the only comprehension axiom that necessarily takes us out of the realm of finite sets. As long as we stick to first-order languages, it seems impossible to incorporate it into our systems by just imposing new simple syntactic conditions on the safety relation. Instead, the best way to capture it is to add to the basic signature a new constant HF (interpreted as the collection $\mathcal{HF}$ of hereditarily finite sets) together with the obvious counterparts of *Peano's axioms*. On the other hand, if a TC-logic is used as the underlying logic then we get the infinity axiom for free, since the set ω of the finite ordinals is definable by a safe formula in this extended language: $\omega = \{y \mid \exists x.x = \emptyset \wedge (TC_{x,y} y = \{z \mid z = x \vee z \in x\})(x, y)\}$.

The regularity axiom can best be incorporated into our framework in the form of ϵ-induction. Finally, the most natural way to handle the axiom of choice in that framework is to further extend its set of terms by allowing the use of Hilbert's ε symbol (together with its usual characterizing axiom, which is equivalent to the axiom of global choice).

3 Predicative Theories and Computational Universes

Let $\mathcal{T}$ be a theory formulated within our framework. From the Platonist point of view, its set of closed terms $\mathcal{D}(\mathcal{T})$ induces some subset $\mathcal{S}(\mathcal{T})$ of the universe V of sets. (The identity of $\mathcal{S}(\mathcal{T})$ depends only on the *language* of $\mathcal{T}$ and on the interpretations of the symbols in its signature other than $\in$ and $=$). $\mathcal{D}(\mathcal{T})$ also determines some subset $\mathcal{M}(\mathcal{T})$ of any transitive model $\mathcal{M}$ of $\mathcal{T}$. We call a theory $\mathcal{T}$ *predicative* if the set $\mathcal{S}(\mathcal{T})$ it induces is a "universe" in the sense that it is a transitive model of $\mathcal{T}$, and in addition the identity of $\mathcal{S}(\mathcal{T})$ is *absolute* in the sense that $\mathcal{M}(\mathcal{T}) = \mathcal{S}(\mathcal{T})$ for any transitive model $\mathcal{M}$ of $\mathcal{T}$ (implying that $\mathcal{S}(\mathcal{T})$ is actually a *minimal* transitive model of $\mathcal{T}$). We call a transitive set *a computational universe* if it is $\mathcal{S}(\mathcal{T})$ for some predicative theory $\mathcal{T}$. In [4,5] it is shown that some theories which naturally arise in our framework are predicative (and so their minimal models are computational). This includes:

- *RST*: This is the theory $RST^{FOL}_{\sigma_{ZF}}$ (which can be shown to be equivalent to Gandy's basic set theory [11]). Its minimal model $\mathcal{S}(RST)$ is identical to $\mathcal{HF}$ (the collection of hereditarily finite sets), which is J_1 in Jensen's hierarchy.
- RST_{HF}: This is $RST^{FOL}_{\sigma_{ZF}\cup\{HF\}}$ extended with Peano's axioms for HF. Its minimal model is J_2.
- *PZF*: This is $RST^{\mathcal{TCL}}_{\sigma_{ZF}}$, where $\mathcal{TCL}$ is some reasonable TC-logic. Its minimal model is $J_{\omega^\omega} = L_{\omega^\omega}$.

In a series of papers (e.g. [9]), Feferman showed that predicative mathematics is sufficient for the formalization of the scientifically applicable mathematics. However, Feferman's systems have the drawbacks of not using the standard set-theoretical framework, and their languages and basic concepts are rather complicated in comparison to ZFC. The predicative theories of our framework seem therefore to be a better choice. This thesis has been pursued in [5,6]. [6] is devoted to the system RST_{HF}, which is the *minimal* system that meets all the basic predicative principles (in particular, it allows the introduction of the natural numbers as a complete set). It is shown there how to develop large portions of applicable mathematics within this minimal theory and its minimal universe J_2. Not surprisingly, the restriction to this minimal framework has its price: the development of mathematics within it involves a lot of coding, as well as treating even the real line as a proper class. In contrast, in [5] the development is done in a way which is very close to mathematical practice, using stronger, but still strictly first-order, predicative theories. The next step of this project will examine the use of PZF. PZF seems rather promising in this respect, since its minimal model, J_{ω^ω}, allows a natural interpretation of cumulative type theory, in which J_ω, J_{ω^2}, J_{ω^3},... are taken as the major types. Thus the real numbers can be taken to be those that are available in J_ω (which is far beyond what is available in J_2), and $\mathbb{R}$ itself will be an ordinary object of 'type' J_{ω^2}.

References

1. Avron, A.: Transitive closure and the mechanization of mathematics. In: Kamareddine, F.D. (ed.) Thirty Five Years of Automating Mathematics. Applied Logic Series, pp. 149–171. Springer, Heidelberg (2003)
2. Avron, A.: Formalizing set theory as it is actually used. In: Asperti, A., Bancerek, G., Trybulec, A. (eds.) MKM 2004. LNCS, vol. 3119, pp. 32–43. Springer, Heidelberg (2004)
3. Avron, A.: A framework for formalizing set theories based on the use of static set terms. In: Avron, A., Dershowitz, N., Rabinovich, A. (eds.) Pillars of Computer Science. LNCS, vol. 4800, pp. 87–106. Springer, Heidelberg (2008)
4. Avron, A.: A new approach to predicative set theory. In: Schindler, R. (ed.) Ways of Proof Theory. Onto Series in Mathematical Logic, pp. 31–63. onto Verlag (2010)
5. Avron, A., Cohen, L.: Formalizing scientifically applicable mathematics in a definitional framework. J. Formalized Reasoning **9**(1), 53–70 (2016)
6. Avron, A., Cohen, L.: A minimal framework for applicable mathematics (to appear, 2016)
7. Cantone, D., Omodeo, E., Policriti, A.: Set Theory for Computing: From Decision Procedures to Declarative Programming With Sets. Springer, New York (2001)
8. Devlin, K.: Constructibility. Perspectives in Mathematical Logic. Springer, Heidelberg (1984)
9. Feferman, S.: Weyl vindicated: Das kontinuum 70 years later. Termi e prospettive della logica e della filosofia della scienza contemporanee, vol. 1 (1988)
10. Fraenkel, A.A., Bar-Hillel, Y., Levy, A.: Foundations of Set Theory. Elsevier, Amsterdam (1973)
11. Gandy, R.O.: Set-theoretic functions for elementary syntax. In: Proceedings of the Symposium in Pure Mathematics, vol. 13, pp. 103–126 (1974)
12. Kieffer, S., Avigad, J., Friedman, H.: A language for mathematical language management. Stud. Logic Grammar Rhetoric **18**, 51–66 (2009)
13. Levy, A.: Basic Set Theory. Perspectives in Mathematical Logic. Springer, Heidelberg (1979)

Programming by Examples: Applications, Algorithms, and Ambiguity Resolution

Sumit Gulwani(✉)

Microsoft Corporation, Redmond, WA, USA
sumitg@microsoft.com

Abstract. 99 % of computer end users do not know programming, and struggle with repetitive tasks. Programming by Examples (PBE) can revolutionize this landscape by enabling users to synthesize intended programs from example based specifications. A key technical challenge in PBE is to search for programs that are consistent with the examples provided by the user. Our efficient search methodology is based on two key ideas: (i) Restriction of the search space to an appropriate domain-specific language that offers balanced expressivity and readability (ii) A divide-and-conquer based deductive search paradigm that inductively reduces the problem of synthesizing a program of a certain kind that satisfies a given specification into sub-problems that refer to sub-programs or sub-specifications. Another challenge in PBE is to resolve the ambiguity in the example based specification. We will discuss two complementary approaches: (a) machine learning based ranking techniques that can pick an intended program from among those that satisfy the specification, and (b) active-learning based user interaction models. The above concepts will be illustrated using FlashFill, FlashExtract, and FlashRelate—PBE technologies for data manipulation domains. These technologies, which have been released inside various Microsoft products, are useful for data scientists who spend 80 % of their time wrangling with data. The Microsoft PROSE SDK allows easy construction of such technologies.

1 Introduction

Program Synthesis [4] is the task of synthesizing a program that satisfies a given specification. The traditional view of program synthesis has been to synthesize programs from logical specifications that relate the inputs and outputs of the program. Programming by Examples (PBE) [6] is a sub-field of program synthesis, where the specification consists of input-output examples, or more generally, output properties over given input states. PBE has emerged as a favorable paradigm for two reasons: (i) the example-based specification in PBE makes it more tractable than general program synthesis. (ii) Example-based specifications are much easier for the users to provide in many scenarios.

2 Applications

PBE has been applied to various domains [3,15], and some recent applications include parsing [14], refactoring [17], and query construction [20]. However, the

N. Olivetti and A. Tiwari (Eds.): IJCAR 2016, LNAI 9706, pp. 9–14, 2016.
DOI: 10.1007/978-3-319-40229-1_2

killer application of PBE today is in the broad space of *data wrangling*, which refers to the tedious process of converting data from one form to another. The data wrangling pipelines includes tasks related to extraction, transformation, and formatting.

Extraction: A first step in a data wrangling pipeline is often that of ingesting or extracting tabular data from semi-structured formats such as text/log files, web pages, and XML/JSON documents. These documents offer their creators great flexibility in storing and organizing hierarchical data by combining presentation/formatting with the underlying data. However, this makes it extremely hard to extract the relevant data. The FlashExtract technology allows extracting structured (tabular or hierarchical) data out of semi-structured documents from examples [12]. For each field in the output data schema, the user provides positive/negative instances of that field and FlashExtract generates a program to extract all instances of that field. The FlashExtract technology ships as the *ConvertFrom-String* cmdlet in Powershell in Windows 10, wherein the user provides examples of the strings to be extracted by inserting tags around them in test. The FlashExtract technology also ships in Azure OMS (Operations Management Suite), where it enables extracting *custom fields* from log files.

Transformation: The *Flash Fill* feature, released in Excel 2013 and beyond, is a PBE technology for automating syntactic string transformations of the kind such as converting "FirstName LastName" into "LastName, FirstName" [5]. PBE can also facilitate more sophisticated string transformations that require lookup into other tables [21]. PBE is also a very natural fit for automating transformations of other data types such as numbers [22] and dates [24].

Formatting: Another useful application of PBE is in the space of formatting data tables. This can be useful in converting semi-structured tables found commonly in spreadsheets into proper relational tables [2], or for re-pivoting the underlying hierarchical data that has been locked into a two-dimensional tabular format [10]. PBE can also be useful in automating repetitive formatting in a powerpoint slide deck such as converting all red colored text into green, or switching the direction of all horizontal arrows [19].

3 Algorithms

Our methodology for designing and developing PBE algorithms involves three key insights: domain-specific languages, deductive search, and a framework that provides rich reusable machinery.

Domain-specific Language: A key idea in program synthesis is to restrict the search space to an underlying domain-specific language (DSL) [1,7]. The DSL should be expressive enough to represent a wide variety of tasks in the underlying task domain, but also restricted enough to allow efficient search. We have designed many functional domain-specific languages for this purpose, each of which is characterized by a set of operators and a syntactic restriction on how those operators can be composed with each other (as opposed to allowing all possible type-safe composition of those operators) [6].

Deductive Search: A simple search strategy is to enumerate all programs in order of increasing size [27]. Another commonly used search strategy is to reduce the search problem to constraint solving via an appropriate reduction and then leverage off-the-shelf SAT/SMT constraint solvers [8,25,26]. None of these search strategies work effectively for our domains: the underlying DSLs are too big for an enumerative strategy to scale, and involve operators that are too sophisticated for existing constraint solvers to reason about.

Our synthesis algorithms employ a novel deductive search methodology [18] that is based on standard algorithmic paradigm of divide-and-conquer. The key idea is to recursively reduce the problem of synthesizing a program expression e of a certain kind and that satisfies a certain specification ψ to simpler sub-problems (where the search is either over sub-expressions of e or over sub-specifications of ψ), followed by appropriately combining those results. The *reduction logic* for reducing a synthesis problem to simpler synthesis problems depends on the nature of the involved expression e and the inductive specification ψ. In contrast to enumerative search, this search methodology is top-down—it fixes the top-part of an expression and then searches for its sub-expressions. Enumerative search is usually bottom-up—it enumerates smaller sub-expressions before enumerating larger expressions.

Framework: Developing a synthesis algorithm for a specific domain is an expensive process: The design of the algorithm requires domain-specific insights. A robust implementation requires non-trivial engineering. Furthermore any extensions or modifications to the underlying DSL are not easy.

The divide-and-conquer strategy underneath the various synthesis algorithms can be refactored out inside a framework. Furthermore, since the reduction logic depends on the logical properties of the top-level operator, these properties can be captured modularly by the framework for re-use inside synthesizers for others DSLs that use that operator. Our PROSE framework [18] builds over these ideas and has facilitated development of industrial-strength PBE implementations for various domains.

4 Ambiguity Resolution

Examples are an ambiguous form of specification; there are often many programs that are consistent with the specification provided by a user. A challenge is to identify an intended program that has the desired behavior on the various inputs that the user cares about. Tessa Lau presented a critical discussion of PBE systems in 2009 noting that PBE systems are not yet widespread due to lack of usability and confidence in such systems [11]. We present two complementary techniques for increasing usability and confidence of a PBE system.

Ranking: Our synthesis algorithms generate the set of all/most programs in the underlying DSL that are consistent with the specification provided by the user. We rank these programs and pick the top-ranked program. Ranking is a function of both program features and data features. Program features typically capture simplicity and size of a program. Data features are over the data that is generated by the program when executed on various inputs. Weights over these features can be learned using machine learning techniques in an offline manner [23].

User Interaction models: In case the ranking does not pick an intended program, or even otherwise, we need appropriate user interaction models that can provide the equivalent of debugging experience in standard programming environments. We can allow the user to navigate between all programs synthesized by the underlying synthesizer (in an efficient manner) and to pick an intended program [16]. Another complementary technique can be to ask questions to the user as in active learning. These questions can be generated based on the differences in the results produced by executing the multiple synthesized programs on the available inputs [16].

5 Conclusion and Future Work

The programming languages research community has traditionally catered to the needs of professional programmers in the continuously evolving technical industry. The widespread access to computational devices has brought a new opportunity, that of enabling non-programmers to create small programs for automating their repetitive tasks. PBE becomes a very valuable paradigm in this setting.

It is interesting to compare PBE with Machine learning (ML) since both involve example-based training and prediction on new unseen data. PBE learns from very few examples, while ML typically requires large amount of training data. The models generated by PBE are human-readable and editable programs unlike many black-box models produced by ML. On the other hand, ML is better suited for fuzzy/noisy tasks.

There are many interesting future directions. The next generation of programming experience shall be built around *multi-modal specifications* that are natural and easy for the user to provide. While this article has focused on example-based specifications, natural language-based specifications can complement example-based specifications and might even be a better fit for various class of tasks such as spreadsheet queries [9] and smartphone scripts [13]. Furthermore, the specifications may be provided iteratively, implying the need for incremental synthesis algorithms. Another interesting future direction is to build systems that learn user preferences based on past user interactions across different programming sessions. (For instance, the underlying ranking can be dynamically updated). This can pave the way for personalization and learning across users.

References

1. Alur, R., Bodik, R., Juniwal, G., Martin, M.M., Raghothaman, M., Seshia, S.A., Singh, R., Solar-Lezama, A., Torlak, E., Udupa, A.: Syntax-guided synthesis. In: FMCAD (2013)
2. Barowy, D.W., Gulwani, S., Hart, T., Zorn, B.G.: FlashRelate: extracting relational data from semi-structured spreadsheets using examples. In: PLDI (2015)
3. Cypher, A. (ed.): Watch What I Do: Programming by Demonstration. MIT Press, Cambridge (1993)
4. Gulwani, S.: Dimensions in program synthesis. In: PPDP (2010)
5. Gulwani, S.: Automating string processing in spreadsheets using input-output examples. In: POPL (2011)
6. Gulwani, S.: Programming by examples (and its applications in data wrangling). In: Esparza, J., Grumberg, O., Sickert, S. (eds.) Verification and Synthesis of Correct and Secure Systems. IOS Press (2016)
7. Gulwani, S., Harris, W., Singh, R.: Spreadsheet data manipulation using examples. Commun. ACM (2012)
8. Gulwani, S., Jha, S., Tiwari, A., Venkatesan, R.: Synthesis of loop-free programs. In: PLDI (2011)
9. Gulwani, S., Marron, M.: NLyze: interactive programming by natural language for spreadsheet data analysis and manipulation. In: SIGMOD (2014)
10. Harris, W.R., Gulwani, S.: Spreadsheet table transformations from examples. In: PLDI (2011)
11. Lau, T.: Why PBD systems fail: Lessons learned for usable AI. In: CHI 2008 Workshop on Usable AI (2008)
12. Le, V., Gulwani, S.: FlashExtract: a framework for data extraction by examples. In: PLDI (2014)
13. Le, V., Gulwani, S., Smartsynth, Z.: Synthesizing smartphone automation scripts from natural language. In: MobiSys (2013)
14. Leung, A., Sarracino, J., Lerner, S.: Interactive parser synthesis by example. In: PLDI (2015)
15. Lieberman, H.: Your Wish Is My Command: Programming by Example. Morgan Kaufmann, San Francisco (2001)
16. Mayer, M., Soares, G., Grechkin, M., Le, V., Marron, M., Polozov, O., Singh, R., Zorn, B., Gulwani, S.: User interaction models for disambiguation in programming by example. In: UIST (2015)
17. Meng, N., Kim, M., McKinley, K.S.: LASE: locating and applying systematic edits by learning from examples. In: ICSE (2013)
18. Polozov, O., Gulwani, S.: FlashMeta: a framework for inductive program synthesis. In: OOPSLA (2015). https://microsoft.github.io/prose/
19. Raza, M., Gulwani, S., Milic-Frayling, N.: Programming by example using least general generalizations. In: AAAI (2014)
20. Shen, Y., Chakrabarti, K., Chaudhuri, S., Ding, B., Novik, L.: Discovering queries based on example tuples. In: SIGMOD (2014)
21. Singh, R., Gulwani, S.: Learning semantic string transformations from examples. PVLDB **5**, 740–751 (2012)
22. Singh, R., Gulwani, S.: Synthesizing number transformations from input-output examples. In: Madhusudan, P., Seshia, S.A. (eds.) CAV 2012. LNCS, vol. 7358, pp. 634–651. Springer, Heidelberg (2012)

23. Singh, R., Gulwani, S.: Predicting a correct program in programming by example. In: Kroening, D., Păsăreanu, C.S. (eds.) CAV 2015. LNCS, vol. 9206, pp. 398–414. Springer, Heidelberg (2015)
24. Singh, R., Gulwani, S.: Transforming spreadsheet data types using examples. In: POPL (2016)
25. Solar-Lezama, A.: Program Synthesis by Sketching. Ph.D. thesis, UC Berkeley (2008)
26. Srivastava, S., Gulwani, S., Foster, J.S.: From program verification to program synthesis. In: POPL (2010)
27. Udupa, A., Raghavan, A., Deshmukh, J.V., Mador-Haim, S., Martin, M.M.K., Alur, R.: TRANSIT: specifying protocols with concolic snippets. In: PLDI (2013)

Logic & Proofs for Cyber-Physical Systems

André Platzer(✉)

Computer Science Department, Carnegie Mellon University, Pittsburgh, USA
aplatzer@cs.cmu.edu

Abstract. *Cyber-physical systems* (CPS) combine cyber aspects such as communication and computer control with physical aspects such as movement in space, which arise frequently in many safety-critical application domains, including aviation, automotive, railway, and robotics. But how can we ensure that these systems are guaranteed to meet their design goals, e.g., that an aircraft will not crash into another one?

This paper highlights some of the most fascinating aspects of cyber-physical systems and their dynamical systems models, such as hybrid systems that combine discrete transitions and continuous evolution along differential equations. Because of the impact that they can have on the real world, CPSs deserve proof as safety evidence.

Multi-dynamical systems understand complex systems as a combination of multiple elementary dynamical aspects, which makes them natural mathematical models for CPS, since they tame their complexity by compositionality. The family of *differential dynamic logics* achieves this compositionality by providing compositional logics, programming languages, and reasoning principles for CPS. Differential dynamic logics, as implemented in the theorem prover KeYmaera X, have been instrumental in verifying many applications, including the Airborne Collision Avoidance System ACAS X, the European Train Control System ETCS, automotive systems, mobile robot navigation, and a surgical robot system for skull-base surgery. This combination of strong theoretical foundations with practical theorem proving challenges and relevant applications makes *Logic for CPS* an ideal area for compelling and rewarding research.

1 Logical Foundations of Cyber-Physical Systems

Can we trust a computer to control physical processes? That depends on how it has been programmed and what will happen if it malfunctions. When a lot is at stake, computers need to be *guaranteed* to interact correctly with the physical world. So, we need ways of analyzing, designing, and guaranteeing the behavior of such systems. Providing these ways is an *intellectual grand challenge* with substantial scientific, economical, societal, and educational impact. Its solution is the key to enabling computer assistance that we can bet our lives on.

This paper focuses on illustrating important principles of cyber-physical systems here. Technical surveys can be found in the literature, e.g., [2,7,8,12,20,32,41,42].
This material is based upon work supported by the National Science Foundation under NSF CAREER Award CNS-1054246.

N. Olivetti and A. Tiwari (Eds.): IJCAR 2016, LNAI 9706, pp. 15–21, 2016.
DOI: 10.1007/978-3-319-40229-1_3

1.1 Cyber-Physical Systems

Computer control has been suggested to remedy inefficiencies, reliability issues, or defects for virtually all physical systems. But computer control only helps our society if we can ensure that it works correctly. As has been argued on numerous occasions [1–8,11,12,17,18,20,21,23–27,38,41–44], we must, thus, *verify* the correctness of these systems, as testing may miss bugs. This problem is confounded, because the behavior of the system under one circumstance can radically differ from the behavior under another, especially when complex computer decisions for different objectives interact. It is crucial to prove the absence of bugs so that we are confident to bet our lives on the system functioning correctly, since that is what we do every time we get into an airplane or car.

Systems like these are called *cyber-physical systems* (*CPS*). They combine cyber capabilities (communication, computation and control) with physical capabilities (sensing and actuation) to solve problems *that neither part could solve alone*. While CPS are widely appreciated for their broad range of application domains (e.g., automotive, aerospace, medical, transportation, civil engineering, materials, chemistry, energy), the goal of the *Logical Foundations of CPS* is to identify the *common foundational core* that constitutes the true essence of CPS and their proof principles to serve as the simultaneous mathematical basis for all those applications. The foundations of digital computer science have revolutionized how systems are designed and our whole society works. We need even stronger foundations when software reaches out into our physical world.

1.2 Multi-dynamical Systems

The first crucial insight for CPS foundations is the multi-dynamical systems principle [32] of understanding complex systems as a combination of multiple elementary dynamical aspects. Mathematically, CPS are *multi-dynamical systems* [32], i.e. systems characterized by multiple facets of dynamical systems, schematically summarized in Fig. 1. CPS involve computer control decisions and are, thus, *discrete*. CPS are *continuous*, because they evolve along differential equations of motion or other physical processes. CPS are *uncertain*, because their behavior is subject to choices coming from environmental variability or intentional uncertainties that simplify their model. This uncertainty can manifest in different ways. Uncertainties make CPS *stochastic* when good information about the distribution of choices is available. Uncertainties make CPS *nondeterministic* when no commitment about the resolution of choices is made. Uncertainties make CPS *adversarial* when they involve multiple agents with potentially conflicting goals or even active competition in a game. Verifying that CPS work correctly requires dealing with all of these dynamical features—and sometimes even more—at the same time.

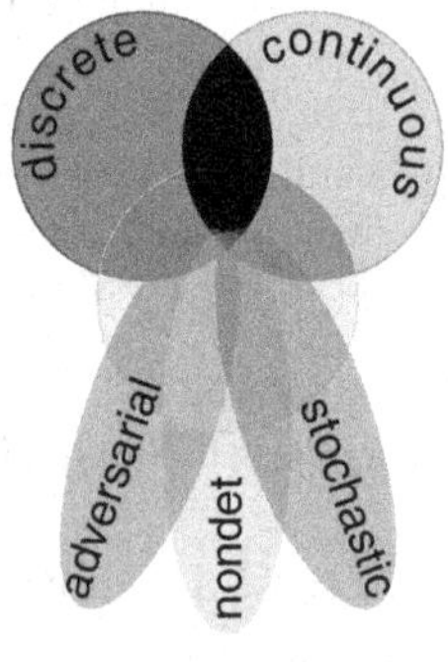

Fig. 1. Dynamical aspects of CPS

1.3 CPS Proofs

Multi-dynamical systems study complex CPS as a combination of multiple elementary dynamical aspects. This approach helps to tame the complexity of CPS by understanding that their complexity just comes from combining lots of simple dynamical effects with one another. The overall system is quite complex, but each of its pieces is better-behaved, since it only has one dynamics. What miracle translates this *descriptive simplification* of a CPS in terms of a combination of multiple dynamical aspects into an *analytic simplification* in terms of multiple dynamical systems that can be considered side-by-side?

The key to this mystery is to integrate the CPS dynamics all within a single, compositional logic [32]. Since compositionality is an intrinsic feature starting from the very semantics of logic [9,10,13,14,37,39,40], logics naturally reason compositionally, too. With suitable generalizations of logics to embrace multi-dynamical systems [27–31,34,35], this compositionality generalizes to CPS. Verification works by constructing a proof in such a logic. The whole proof verifies a complex CPS. Yet, each proof step only reasons separately about one dynamical aspect at a time using, e.g., local dynamics of differential equations, the theory of real-closed fields, symbolic logic, differential form computations [35], fixpoint theory [34], and so on, each captured in a separate, modular axiom or proof rule.

1.4 Theory

This logical view on CPS has already made it possible to develop rich theories of *hybrid systems* that combine discrete change and continuous differential equations [27,31,35], theories of *distributed hybrid systems* that combine distributed systems with hybrid systems [30], theories of *hybrid games* that combine discrete, continuous, and adversarial dynamics [34], all of which are sound and relatively complete, but was also used for *stochastic hybrid systems* [29]. The approach was instrumental in formulating and proving the first [27] and second [31] *completeness theorem* for hybrid systems, which characterize and align the discrete and continuous challenges of hybrid systems, and reveal their fundamental symmetry. The theory of hybrid systems forms a *proof-theoretical bridge* aligning the theory of continuous systems with the theory of discrete systems. Proof theory was essential in the study of *provability of properties of differential equations* and *differential cut elimination* [33], which turn out to generalize ideas from Lie's results on Lie groups [19] but also relate to Gentzen's cut elimination theorem in classical logic [10]. Logic was equally crucial for the development of *differential ghosts* that create extra dimensions [33] as proof-theoretical analogues of dark matter, whose existence was speculated to balance out energy invariants in astrophysics [16].

As a logical rendition of Lie's ideas, *differential invariants* [28,33,35] enable induction principles for differential equations characterizing the rate of change of truth of a formula in the direction of the dynamics; see Fig. 2. Intuitively, F always remains true after following the differential equation $x' = f(x)$ within the domain H (conclusion), if F started out true initially (conclusion's assumption), and if, within H, the differential F' of F (which characterizes the infinitesimal change of F as a function of x') holds after assigning the right-hand side $f(x)$ of the differential equation to its left-hand side x' (premise). Differential invariants lift the high descriptive power of differential equations to a high analytic power, so that their properties can be proved even if the equations cannot be solved. Solutions ruin the descriptive power even *if* the differential equations can be solved, so that differential invariants are advantageous regardless.

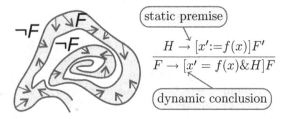

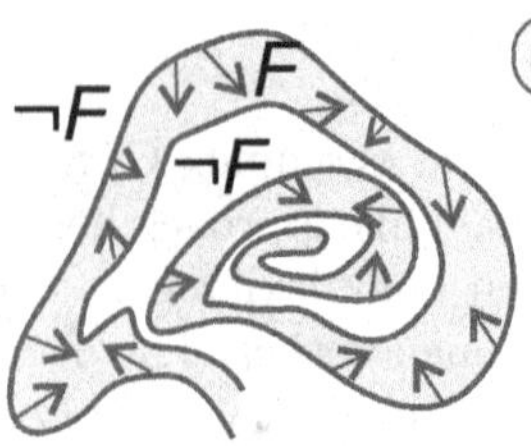

static premise

$$\frac{H \rightarrow [x' := f(x)]F'}{F \rightarrow [x' = f(x) \& H]F}$$

dynamic conclusion

Fig. 2. (left) Differential invariant F **(right)** Proof rule for invariance of F along differential equation $x' = f(x)$ in evolution domain H

1.5 Applications

Logical Foundations of CPS play an increasingly important role in practical applications by way of their implementations in the theorem prover KeYmaera and its clean-slate successor[1] KeYmaera X. This includes finding and fixing [36] flaws in an air traffic conflict resolution maneuver, verifying and identifying issues in the *Next-generation Airborne Collision Avoidance System* ACAS X [15], verifying the *European train control system* ETCS, car control systems, mobile ground robot navigation, and finding and fixing bugs in a skull-base surgical robot system. Logic also identified a way of correctly relating proof in a model to truth in reality [22], which is an inevitable challenge for CPS.

Finally, multi-dynamical systems impact education in the *Foundations of Cyber-Physical Systems* course that is breaking with the myth that cyber-physical systems are too challenging to be taught at the undergraduate level. The compositionality principles of logic and multi-dynamical systems considerably tame the educational complexity of CPS by making it possible to focus on one aspect at a time without losing the ability to combine the understanding attained for each aspect. The rich variety of systems that the students verified for their final course projects[2] indicates that this approach effectively conveys the principles for a successful separation of concerns for CPS.

[1] http://www.keymaeraX.org/.

[2] The students' self-defined 3-week course projects and their presentations to a panel of experts from industry in the CPS V&V Grand Prix are available from the course web pages http://lfcps.org/course/fcps.html.

1.6 Summary

Logical foundations make a big difference for cyber-physical systems, certainly in understanding the basic principles of CPS, but also in real applications like the Next-generation Airborne Collision Avoidance System. Lessons from centuries of logic and foundations research can have a huge impact on advancing CPS. Yet, conversely, the questions that CPS pose can have an equally significant impact on advancing logic. Cyber-physical systems serve as a catalytic integrator for other sciences, because they benefit from combining numerous exciting areas of logic, mathematics, computer science, and control theory that previously seemed unrelated. The mix of enabling strong analytic foundations with the need for practical advances of rigorous reasoning and the significance of its applications, as well as its fruitful interactions with many other sciences, make cyber-physical systems an ideal field for compelling and rewarding research that has only just begun. Numerous wonders remain yet to be discovered.

References

1. Alur, R.: Formal verification of hybrid systems. In: Chakraborty, S., Jerraya, A., Baruah, S.K., Fischmeister, S. (eds.) EMSOFT, pp. 273–278. ACM (2011)
2. Alur, R.: Principles of Cyber-Physical Systems. MIT Press, Cambridge (2015)
3. Alur, R., Courcoubetis, C., Halbwachs, N., Henzinger, T.A., Ho, P.H., Nicollin, X., Olivero, A., Sifakis, J., Yovine, S.: The algorithmic analysis of hybrid systems. Theoret. Comput. Sci. **138**(1), 3–34 (1995)
4. Alur, R., Henzinger, T., Lafferriere, G., Pappas, G.J.: Discrete abstractions of hybrid systems. Proc. IEEE **88**(7), 971–984 (2000)
5. Branicky, M.S.: General hybrid dynamical systems: modeling, analysis, and control. In: Alur, R., Sontag, E.D., Henzinger, T.A. (eds.) HS 1995. LNCS, vol. 1066, pp. 186–200. Springer, Heidelberg (1996)
6. Clarke, E.M., Emerson, E.A., Sifakis, J.: Model checking: algorithmic verification and debugging. Commun. ACM **52**(11), 74–84 (2009)
7. Davoren, J.M., Nerode, A.: Logics for hybrid systems. IEEE **88**(7), 985–1010 (2000)
8. Doyen, L., Frehse, G., Pappas, G.J., Platzer, A.: Verification of hybrid systems. In: Clarke, E.M., Henzinger, T.A., Veith, H. (eds.) Handbook of Model Checking, Chap. 28. Springer, Heidelberg (2017)
9. Frege, G.: Begriffsschrift, eine der arithmetischen nachgebildete Formelsprache des reinen Denkens. Verlag von Louis Nebert, Halle (1879)
10. Gentzen, G.: Untersuchungen über das logische Schließen. I. Math. Zeit. **39**(2), 176–210 (1935)
11. Henzinger, T.A., Sifakis, J.: The discipline of embedded systems design. Computer **40**(10), 32–40 (2007)
12. Henzinger, T.A.: The theory of hybrid automata. In: LICS, pp. 278–292. IEEE Computer Society, Los Alamitos (1996)
13. Hilbert, D.: Die Grundlagen der Mathematik. Abhandlungen aus dem Seminar der Hamburgischen Universität **6**(1), 65–85 (1928)
14. Hoare, C.A.R.: An axiomatic basis for computer programming. Commun. ACM **12**(10), 576–580 (1969)

15. Jeannin, J.-B., Ghorbal, K., Kouskoulas, Y., Gardner, R., Schmidt, A., Zawadzki, E., Platzer, A.: A formally verified hybrid system for the next-generation airborne collision avoidance system. In: Baier, C., Tinelli, C. (eds.) TACAS 2015. LNCS, vol. 9035, pp. 21–36. Springer, Heidelberg (2015)
16. Kapteyn, J.C.: First attempt at a theory of the arrangement and motion of the sidereal system. Astrophys. J. **55**, 302 (1922)
17. Larsen, K.G.: Verification and performance analysis for embedded systems. In: Chin, W., Qin, S. (eds.) TASE 2009, Third IEEE International Symposium on Theoretical Aspects of Software Engineering, 29–31 July 2009, pp. 3–4. IEEE Computer Society, Tianjin, China (2009)
18. Lee, E.A., Seshia, S.A.: Introduction to Embedded Systems - A Cyber-Physical Systems Approach. Lulu Press, Raleigh (2013). Lulu.com
19. Lie, S.: Vorlesungen über continuierliche Gruppen mit geometrischen und anderen Anwendungen. Teubner, Leipzig (1893)
20. Lunze, J., Lamnabhi-Lagarrigue, F. (eds.): Handbook of Hybrid Systems Control: Theory, Tools, Applications. Cambridge University Press, Cambridge (2009)
21. Maler, O.: Control from computer science. Ann. Rev. Control **26**(2), 175–187 (2002)
22. Mitsch, S., Platzer, A.: ModelPlex: verified runtime validation of verified cyber-physical system models. In: Bonakdarpour, B., Smolka, S.A. (eds.) RV 2014. LNCS, vol. 8734, pp. 199–214. Springer, Heidelberg (2014)
23. Nerode, A.: Logic and control. In: Cooper, S.B., Löwe, B., Sorbi, A. (eds.) CiE 2007. LNCS, vol. 4497, pp. 585–597. Springer, Heidelberg (2007)
24. Nerode, A., Kohn, W.: Models for hybrid systems: automata, topologies, controllability, observability. In: Grossman, R.L., Ravn, A.P., Rischel, H., Nerode, A. (eds.) HS 1991 and HS 1992. LNCS, vol. 736, pp. 317–356. Springer, Heidelberg (1993)
25. NITRD CPS Senior Steering Group: CPS vision statement. NITRD (2012)
26. Pappas, G.J.: Wireless control networks: modeling, synthesis, robustness, security. In: Caccamo, M., Frazzoli, E., Grosu, R. (eds.) Proceedings of the 14th ACM International Conference on Hybrid Systems: Computation and Control, HSCC 2011, April 12–14, 2011, pp. 1–2. ACM, Chicago (2011)
27. Platzer, A.: Differential dynamic logic for hybrid systems. J. Autom. Reas. **41**(2), 143–189 (2008)
28. Platzer, A.: Differential-algebraic dynamic logic for differential-algebraic programs. J. Log. Comput. **20**(1), 309–352 (2010)
29. Platzer, A.: Stochastic differential dynamic logic for stochastic hybrid programs. In: Bjørner, N., Sofronie-Stokkermans, V. (eds.) CADE 2011. LNCS, vol. 6803, pp. 446–460. Springer, Heidelberg (2011)
30. Platzer, A.: Quantified differential dynamic logic for distributed hybrid systems. In: Dawar, A., Veith, H. (eds.) CSL 2010. LNCS, vol. 6247, pp. 469–483. Springer, Heidelberg (2010)
31. Platzer, A.: The complete proof theory of hybrid systems. In: LICS, pp. 541–550. IEEE (2012)
32. Platzer, A.: Logics of dynamical systems. In: LICS, pp. 13–24. IEEE (2012)
33. Platzer, A.: The structure of differential invariants and differential cut elimination. Log. Meth. Comput. Sci. **8**(4), 1–38 (2012)
34. Platzer, A.: Differential game logic. ACM Trans. Comput. Log. **17**(1), 1: 1–1: 51 (2015)
35. Platzer, A.: A uniform substitution calculus for differential dynamic logic. In: Felty, A., Middeldorp, A. (eds.) CADE. LNCS, vol. 9195, pp. 467–481. Springer, Heidelberg (2015)

36. Platzer, A., Clarke, E.M.: Formal verification of curved flight collision avoidance maneuvers: a case study. In: Cavalcanti, A., Dams, D.R. (eds.) FM 2009. LNCS, vol. 5850, pp. 547–562. Springer, Heidelberg (2009)
37. Pratt, V.R.: Semantical considerations on Floyd-Hoare logic. In: FOCS, pp. 109–121. IEEE (1976)
38. President's Council of Advisors on Science and Technology: Leadership under challenge: Information technology R&D in a competitive world. An Assessment of the Federal Networking and Information Technology R&D, Program, August 2007
39. Scott, D., Strachey, C.: Toward a mathematical semantics for computer languages? Technical report, PRG-6, Oxford Programming Research Group (1971)
40. Smullyan, R.M.: First-Order Logic. Dover, Mineola (1968)
41. Tabuada, P.: Verification and Control of Hybrid Systems: A Symbolic Approach. Springer, New York (2009)
42. Tiwari, A.: Abstractions for hybrid systems. Form. Meth. Syst. Des. **32**(1), 57–83 (2008)
43. Tiwari, A.: Logic in software, dynamical and biological systems. In: LICS, pp. 9–10. IEEE Computer Society (2011)
44. Wing, J.M.: Five deep questions in computing. Commun. ACM **51**(1), 58–60 (2008)

Satisfiability of Boolean Formulas

A Verified SAT Solver Framework with Learn, Forget, Restart, and Incrementality

Jasmin Christian Blanchette[1,2], Mathias Fleury[2(✉)], and Christoph Weidenbach[2]

[1] Inria Nancy – Grand Est & LORIA, Villers-lès-Nancy, France
jasmin.blanchette@inria.fr
[2] Max-Planck-Institut für Informatik, Saarbrücken, Germany
mathias.fleury@mpi-inf.mpg.de

Abstract. We developed a formal framework for CDCL (conflict-driven clause learning) in Isabelle/HOL. Through a chain of refinements, an abstract CDCL calculus is connected to a SAT solver expressed in a functional programming language, with total correctness guarantees. The framework offers a convenient way to prove metatheorems and experiment with variants. Compared with earlier SAT solver verifications, the main novelties are the inclusion of rules for forget, restart, and incremental solving and the application of refinement.

1 Introduction

Researchers in automated reasoning spend a significant portion of their work time specifying logical calculi and proving metatheorems about them. These proofs are typically carried out with pen and paper, which is error-prone and can be tedious. As proof assistants are becoming easier to use, it makes sense to employ them.

In this spirit, we started an effort, called IsaFoL (Isabelle Formalization of Logic), that aims at developing libraries and methodology for formalizing modern research in the field, using the Isabelle/HOL proof assistant [7]. Our initial emphasis is on established results about propositional and first-order logic. In particular, we are formalizing large parts of Weidenbach's forthcoming textbook, tentatively called *Automated Reasoning—The Art of Generic Problem Solving*. Our inspiration for formalizing logic is the IsaFoR project, which focuses on term rewriting [40].

The objective of formalization work is not to eliminate paper proofs, but to complement them with rich formal companions. Formalizations help catch mistakes, whether superficial or deep, in specifications and theorems; they make it easy to experiment with changes or variants of concepts; and they help clarify concepts left vague on paper.

This paper presents our formalization of CDCL from *Automated Reasoning* on propositional satisfiability (SAT), developed via a refinement of Nieuwenhuis,

N. Olivetti and A. Tiwari (Eds.): IJCAR 2016, LNAI 9706, pp. 25–44, 2016.
DOI: 10.1007/978-3-319-40229-1_4

Oliveras, and Tinelli's account of CDCL [29]. CDCL is the algorithm implemented in modern SAT solvers. We start with a family of abstract DPLL [11] and CDCL [2,18,28,39] transition systems (Sect. 3). Some of the calculi include rules for learning and forgetting clauses and for restarting the search. All calculi are proved sound and complete, as well as terminating under a reasonable strategy. The abstract CDCL calculus is refined into the more concrete calculus presented in *Automated Reasoning* and recently published [42] (Sect. 4). The latter specifies a criterion for learning clauses representing first unit implication points (1UIPs) [2], with the guarantee that learned clauses are not redundant and hence derived at most once. The calculus also supports incremental solving. This concrete calculus is refined further, as a certified functional program extracted using Isabelle's code generator (Sect. 5).

Any formalization effort is a case study in the use of a proof assistant. Beyond the code generator, we depended heavily on the following features of Isabelle:

- *Isar* [43] is a textual proof format inspired by the pioneering Mizar system [27]. It makes it possible to write structured, readable proofs—a requisite for any formalization that aims at clarifying an informal proof.
- *Locales* [1,19] parameterize theories over operations and assumptions, encouraging a modular style of development. They are useful to express hierarchies of related concepts and to reduce the number of parameters and assumptions that must be threaded through a formal development.
- *Sledgehammer* integrates superposition provers and SMT (satisfiability modulo theories) solvers in Isabelle to discharge proof obligations. The SMT solvers, and one of the superposition provers [41], are built around a SAT solver, resulting in a situation where SAT solvers are employed to prove their own metatheory.

Our work is related to other verifications of SAT solvers, typically with the aim of increasing their trustworthiness (Sect. 6). This goal has lost some of its significance with the emergence of formats for certificates that are easy to generate, even in highly optimized solvers, and that can be processed efficiently by verified checkers [17]. In contrast, our focus is on formalizing the metatheory of CDCL, to study and connect the various members of the family. The main novelties of our framework are the inclusion of rules for forget, restart, and incremental solving and the application of refinement to transfer results. The framework is available online as part of the IsaFoL repository [13].

2 Isabelle

Isabelle [31,32] is a generic proof assistant that supports many object logics. The metalogic is an intuitionistic fragment of higher-order logic (HOL) [10]. The types are built from type variables $'a, 'b, \ldots$ and n-ary type constructors, normally written in postfix notation (e.g., $'a$ *list*). The infix type constructor $'a \Rightarrow 'b$ is interpreted as the (total) function space from $'a$ to $'b$. Function applications are written in a curried style (e.g., $\mathsf{f}\ x\ y$). Anonymous functions $x \mapsto y_x$

are written $\lambda x.\ y_x$. The judgment $t :: \tau$ indicates that term t has type τ. Propositions are simply terms of type *prop*. Symbols belonging to the signature are uniformly called *constants*, even if they are functions or predicates. The metalogical operators include universal quantification $\bigwedge :: ('a \Rightarrow prop) \Rightarrow prop$ and implication $\Longrightarrow :: prop \Rightarrow prop \Rightarrow prop$. The notation $\bigwedge x.\ p_x$ is syntactic sugar for $\bigwedge (\lambda x.\ p_x)$ and similarly for other binder notations.

Isabelle/HOL is the instantiation of Isabelle with HOL, an object logic for classical HOL extended with rank-1 (top-level) polymorphism and Haskell-style type classes. It axiomatizes a type *bool* of Booleans as well as its own set of logical symbols ($\forall$, $\exists$, False, True, $\neg$, $\wedge$, $\vee$, $\longrightarrow$, $\longleftrightarrow$, $=$). The object logic is embedded in the metalogic via a constant Trueprop :: $bool \Rightarrow prop$, which is normally not printed. The distinction between the two logical levels is important operationally but not semantically.

Isabelle adheres to the tradition initiated in the 1970s by the LCF system [14]: All inferences are derived by a small trusted kernel; types and functions are defined rather than axiomatized to guard against inconsistencies. High-level specification mechanisms let us define important classes of types and functions, notably inductive predicates and recursive functions. Internally, the system synthesizes appropriate low-level definitions.

Isabelle developments are organized as collections of theory files, or modules, that build on one another. Each file consists of definitions, lemmas, and proofs expressed in Isar, Isabelle's input language. Proofs are specified either as a sequence of tactics that manipulate the proof state directly or in a declarative, natural deduction format. Our formalization almost exclusively employs the more readable declarative style.

The Sledgehammer tool [4,34] integrates automatic theorem provers in Isabelle/HOL, including CVC4, E, LEO-II, Satallax, SPASS, Vampire, veriT, and Z3. Upon invocation, it heuristically selects relevant lemmas from the thousands available in loaded libraries, translates them along with the current proof obligation to SMT-LIB or TPTP, and invokes the automatic provers. In case of success, the machine-generated proof is translated to an Isar proof that can be inserted into the formal development.

Isabelle locales are a convenient mechanism for structuring large proofs. A locale fixes types, constants, and assumptions within a specified scope. For example:

$$\texttt{locale X} = \texttt{fixes c} :: \tau_{'a}\ \texttt{assumes}\ A_{'a,\mathsf{c}}$$

The definition of locale X implicitly fixes a type $'a$, explicitly fixes a constant c whose type $\tau_{'a}$ may depend on $'a$, and states an assumption $A_{'a,\mathsf{c}} :: prop$ over $'a$ and c. Definitions made within the locale may depend on $'a$ and c, and lemmas proved within the locale may additionally depend on $A_{'a,\mathsf{c}}$. A single locale can introduce several types, constants, and assumptions. Seen from the outside, the lemmas proved in X are polymorphic in type variable $'a$, universally quantified over c, and conditional on $A_{'a,\mathsf{c}}$.

Locales support inheritance, union, and embedding. To embed Y into X, or make Y a *sublocale* of X, we must recast an instance of Y into an instance of

X, by providing, in the context of Y, definitions of the types and constants of X together with proofs of X's assumptions. The command `sublocale` Y $\subseteq$ X t emits the proof obligation $A_{\upsilon,t}$, where υ and $t :: \tau_\upsilon$ may depend on types and constants from Y. After the proof, all the lemmas proved in X become available in Y, with $'a$ and $\mathsf{c} :: \tau_{'a}$ instantiated with υ and $t :: \tau_\upsilon$.

3 Abstract CDCL

The abstract CDCL (conflict-driven clause learning) calculus by Nieuwenhuis et al. [29] forms the first layer of our refinement chain. Our formalization relies on basic Isabelle libraries for lists and multisets and on custom libraries for propositional logic. Properties such as partial correctness and termination are inherited by subsequent layers.

3.1 Propositional Logic

We represent raw and annotated literals by freely generated datatypes parameterized by the types $'v$ (propositional variable), $'lvl$ (decision level), and $'cls$ (clause):

```
datatype 'v literal =          datatype ('v, 'lvl, 'cls) ann_literal =
  Pos 'v                         Decided ('v literal) 'lvl
| Neg 'v                       | Propagated ('v literal) 'cls
```

The syntax is similar to that of Standard ML and other typed functional programming languages. For example, *literal* has two constructors, Pos and Neg, of type $'v \Rightarrow 'v$ *literal*. Informally, we write A, $\neg A$, and $L^\dagger$ for positive, negative, and decided literals, and $-L$ for the negation of a literal, with $-(\neg A) = A$. The simpler calculi do not use $'lvl$ or $'cls$; they take $'lvl = 'cls = unit$, a singleton type whose unique value is denoted by ().

A $'v$ *clause* is a (finite) multiset over $'v$ *literal*. Clauses themselves are often stored in multisets of clauses. To ease reading, we write clauses using logical symbols (e.g., $\bot$, L, and $C \vee D$ for $\emptyset$, $\{L\}$, and $C \uplus D$). Given a set I of literals, $I \vDash C$ is true if and only if C and I share a literal. This is lifted to (multi)sets of clauses: $I \vDash N \longleftrightarrow \forall C{\in}N.\ I \vDash C$. A set is satisfiable if there exists a (consistent) set of literals I such that $I \vDash N$. Finally, $N \vDash N' \longleftrightarrow \forall I.\ I \vDash N \longrightarrow I \vDash N'$.

3.2 DPLL with Backjumping

Nieuwenhuis et al. present CDCL as a set of transition rules on states. A state is a pair (M, N), where M is the *trail* and N is the set of clauses to satisfy. The trail is a list of annotated literals that represents the partial model under construction. In accordance with Isabelle conventions for lists, the trail grows on the left: Adding a literal L to M results in the new trail $L \cdot M$, where the list constructor $\cdot$ has type $'a \Rightarrow 'a$ *list* $\Rightarrow 'a$ *list*. The concatenation of two lists is written $M @ M'$. To lighten the notation, we often build lists from elements and other lists by simple juxtaposition, writing MLM' for $M @ L \cdot M'$.

The core of the CDCL calculus is defined as a transition relation DPLL+BJ, an extension of classical DPLL (Davis–Putnam–Logemann–Loveland) [11] with nonchronological backtracking, or *backjumping*. We write $S \Longrightarrow_{\mathsf{DPLL+BJ}} S'$ for DPLL+BJ S S'. The DPLL+BJ calculus consists of three rules, starting from an initial state (ϵ, N):

Propagate $(M, N) \Longrightarrow_{\mathsf{DPLL+BJ}} (LM, N)$
  if N contains a clause $C \vee L$ such that $M \vDash \neg C$ and L is undefined in M (i.e., neither $M \vDash L$ nor $M \vDash -L$)

Backjump $(M'L^{\dagger}M, N) \Longrightarrow_{\mathsf{DPLL+BJ}} (L'M, N)$
  if N contains a conflicting clause C (i.e., $M'L^{\dagger}M \vDash \neg C$) and there exists a clause $C' \vee L'$ such that $N \vDash C' \vee L'$, $M \vDash \neg C'$, and L' is undefined in M but occurs in N or in $M'L^{\dagger}$

Decide $(M, N) \Longrightarrow_{\mathsf{DPLL+BJ}} (L^{\dagger}M, N)$
  if the atom of L belongs to N and is undefined in M

The Backjump rule is more general than necessary for capturing DPLL, where it suffices to swap the leftmost decision literal. In this form, the rule can also represent CDCL backjumping, if $C' \vee L'$ is a new clause derived from N.

A natural representation of such rules in Isabelle is as an inductive predicate. Isabelle's `inductive` command lets us specify the transition rules as introduction rules. From this specification, it produces elimination rules to perform a case analysis on a hypothesis of the form DPLL+BJ S S'. In the interest of modularity, we formalized the rules individually as their own predicates and combined them to obtain DPLL+BJ:

```
inductive DPLL+BJ :: 'st ⇒ 'st ⇒ bool where
  decide S S' ⟹ DPLL+BJ S S'
| propagate S S' ⟹ DPLL+BJ S S'
| backjump S S' ⟹ DPLL+BJ S S'
```

The predicate operates on states (M, N) of type *'st*. To allow for refinements, this type is kept as a parameter of the calculus, using a locale that abstracts over it and that provides basic operations to manipulate states:

```
locale dpll_state =
  fixes
    trail :: 'st ⇒ ('v, unit, unit) ann_literal list and
    clauses :: 'st ⇒ 'v clause multiset and
    prepend_trail :: ('v, unit, unit) ann_literal ⇒ 'st ⇒ 'st and ... and
    remove_clause :: 'v clause ⇒ 'st ⇒ 'st
  assumes
    ⋀S L. trail (prepend_trail L S) = L · trail S and ... and
    ⋀S C. clauses (remove_cls C S) = remove_mset C (clauses S)
```

The predicates corresponding to the individual calculus rules are phrased in terms of such an abstract state. For example:

$\texttt{inductive}$ decide :: $'st \Rightarrow 'st \Rightarrow bool$ $\texttt{where}$
  undefined_lit L (trail S) $\Longrightarrow$ atm_of $L \in$ atms_of (clauses S) $\Longrightarrow$
  $S' \sim$ prepend_trail (Decided L ()) $S \Longrightarrow$ decide S S'

States are compared extensionally: $S \sim S'$ is true if the two states have identical trails and clause sets, ignoring other fields. This flexibility is necessary to allow refinements with more sophisticated data structures.

In addition, each rule is defined in its own locale, parameterized by additional side conditions. Complex calculi are built by inheriting and instantiating locales providing the desired rules. Following a common idiom, the DPLL+BJ calculus is distributed over two locales: The first locale, DPLL+BJ_ops, defines the DPLL+BJ calculus; the second locale, DPLL+BJ, extends it with an assumption expressing a structural invariant over DPLL+BJ that is instantiated when proving concrete properties later. This cannot be achieved with a single locale, because definitions may not precede assumptions.

Theorem 1 (Termination [13, **wf_dpll_bj**]**).** *The relation* DPLL+BJ *is well founded.*

Termination is proved by exhibiting a well-founded relation $\prec$ such that $S' \prec S$ whenever $S \Longrightarrow_{\mathsf{DPLL+BJ}} S'$. Let $S = (M, N)$ and $S' = (M', N')$ with the decompositions

$$M = M_n L_n^\dagger \cdots M_1 L_1^\dagger M_0 \qquad M' = M'_{n'} L'^\dagger_{n'} \cdots M'_1 L'^\dagger_1 M'_0$$

where $M_0, \ldots, M_n, M'_0, \ldots, M'_{n'}$ contain no decision literals. Let V be the number of distinct variables occurring in the initial clause set N. Now, let $\nu\, M = V - |M|$, indicating the number of unassigned variables in the trail M. Nieuwenhuis et al. define $\prec$ such that $S' \prec S$ if (1) there exists $i \leq n, n'$ for which $[\nu\, M'_0, \ldots, \nu\, M'_{i-1}] = [\nu\, M_0, \ldots, \nu\, M_{i-1}]$ and $\nu\, M'_i < \nu\, M_i$ or (2) $[\nu\, M_0, \ldots, \nu\, M_n]$ is a strict prefix of $[\nu\, M'_0, \ldots, \nu\, M'_{n'}]$. This order is not to be confused with the lexicographic order—we have $[0] \prec \epsilon$ by condition (2), whereas $\epsilon <_{\mathsf{lex}} [0]$. Yet the authors justify well-foundedness by appealing to the well-foundedness of $<_{\mathsf{lex}}$ on bounded lists over finite alphabets. In our proof, we clarify and simplify matters by mapping states to lists $\big[|M_0|, \ldots, |M_n|\big]$, without appealing to ν. Using the standard lexicographic ordering, states become *larger* with each transition:

Propagate	$[k_1, \ldots, k_n] <_{\mathsf{lex}} [k_1, \ldots, k_n + 1]$	
Backjump	$[k_1, \ldots, k_n] <_{\mathsf{lex}} [k_1, \ldots, k_j + 1]$	with $j \leq n$
Decide	$[k_1, \ldots, k_n] <_{\mathsf{lex}} [k_1, \ldots, k_n, 0]$	

The lists corresponding to possible states are $\prec$-bounded by the list consisting of V occurrences of V, thereby delimiting a finite domain $D = \{[k_1, \ldots, k_n] \mid k_1, \ldots, k_n, n \leq V\}$. We take $\prec$ to be the restriction of $>_{\mathsf{lex}}$ to D. A variant of this approach is to encode lists into a measure $\mu_V\, M = \sum_{i=0}^{n} |M_i|\, V^{n-i}$ and let $S' \prec S \leftrightarrow \mu_V\, M' > \mu_V\, M$, building on the well-foundedness of $>$ over bounded sets of integers.

A *final* state is a state from which no transitions are possible. Given a relation $\Longrightarrow$, we write $\Longrightarrow^{*!}$ for the right-restriction of its reflexive transitive closure to final states.

Theorem 2 (Partial Correctness [13, **full_dpll_backjump_final_state_from_init_state**]**).** *If* $(\epsilon, N) \Longrightarrow^{*!}_{\mathsf{DPLL+BJ}} (M, N)$*, then* N *is satisfiable if and only if* $M \vDash N$.

We first prove structural invariants on arbitrary states (M', N) reachable from (ϵ, N), namely: (1) each variable occurs at most once in M'; (2) if $M' = M_2 L M_1$ where L is propagated, then $M_1, N \vDash L$. From these invariants, together with the constraint that (M, N) is a final state, it is easy to prove the conclusion.

3.3 Classical DPLL

The locale machinery allows us to derive a classical DPLL [11] calculus from DPLL with backjumping. This is achieved through a DPLL locale that restricts the Backjump rule so that it performs only chronological backtracking:

Backtrack $(M' L^{\dagger} M, N) \Longrightarrow_{\mathsf{DPLL}} (-L \cdot M, N)$
  if there exists a conflicting clause and M' contains no decided literals

Lemma 3 (Backtracking [13, **backtrack_is_backjump**]**).** *Backtracking is a special case of backjumping.*

The Backjump rule depends on a conflict clause C and a clause $C' \vee L'$ that justifies the propagation of L'. The conflict clause is specified by Backtrack. As for $C' \vee L'$, given a trail $M' L^{\dagger} M$ decomposable as $M_n L^{\dagger} M_{n-1} L^{\dagger}_{n-1} \cdots M_1 L^{\dagger}_1 M_0$ where $M_0, \ldots, M_n$ contain no decided literals, we can take $C' = -L_1 \vee \cdots \vee -L_{n-1}$.

Consequently, the inclusion DPLL $\subseteq$ DPLL+BJ holds. In Isabelle, this is expressed as a locale instantiation: DPLL is made a sublocale of DPLL+BJ, with a side condition restricting the application of the Backjump rule. The partial correctness and termination theorems are inherited from the base locale. DPLL instantiates the abstract state type *'st* with a concrete type of pairs. By discharging the locale assumptions emerging with the `sublocale` command, we also verify that these assumptions are consistent. Roughly:

```
locale DPLL =
begin
  type_synonym 'v state = ('v, unit, unit) ann_literal list × 'v clause multiset
  inductive backtrack :: 'v state ⇒ 'v state ⇒ bool where ...
end

sublocale DPLL ⊆ dpll_state fst snd (λL (M, N). (L · M, N)) ...
sublocale DPLL ⊆ DPLL+BJ_ops ... (λC L S S'. DPLL.backtrack S S') ...
sublocale DPLL ⊆ DPLL+BJ ...
```

If a conflict cannot be resolved by backtracking, we would like to have the option of stopping even if some variables are undefined. A state (M, N) is *conclusive* if $M \models N$ or if N contains a conflicting clause and M contains no decided literals. For DPLL, all final states are conclusive, but not all conclusive states are final.

Theorem 4 (Partial Correctness [13, dpll_conclusive_state_correctness]). *If $(\epsilon, N) \Longrightarrow^*_{\mathsf{DPLL}} (M, N)$ and (M, N) is a conclusive state, N is satisfiable if and only if $M \models N$.*

The theorem does not require stopping at the first conclusive state. In an implementation, testing $M \models N$ can be expensive, so a solver might continue for a while. In the worst case, it will stop in a final state—which exists by Theorem 1.

3.4 The CDCL Calculus

The abstract CDCL calculus extends DPLL+BJ with a pair of rules for learning new lemmas and forgetting old ones:

Learn $(M, N) \Longrightarrow_{\mathsf{CDCL\_NOT}} (M, N \uplus \{C\})$ if $N \models C$ and each atom of C is in N or M

Forget $(M, N \uplus \{C\}) \Longrightarrow_{\mathsf{CDCL\_NOT}} (M, N)$ if $N \models C$

In practice, the Learn rule is normally applied to clauses built exclusively from atoms in M, because the learned clause is false in M. This property eventually guarantees that the learned clause is not redundant (e.g., it is not already contained in N).

We call this calculus CDCL_NOT after Nieuwenhuis, Oliveras, and Tinelli. Because of the locale parameters, it is strictly speaking a family of calculi. In general, CDCL_NOT does not terminate, because it is possible to learn and forget the same clause infinitely often. But for some instantiations of the parameters with suitable restrictions on Learn and Forget, the calculus always terminates. In particular, DPLL+BJ always terminates.

Theorem 5 (Termination [13, wf_cdcl$_{\mathsf{NOT}}$_no_learn_and_forget_infinite_chain]). *Let C be an instance of the CDCL_NOT calculus (i.e., $\mathsf{C} \subseteq$ CDCL_NOT). If C admits no infinite chains consisting exclusively of Learn and Forget transitions, then C is well founded.*

In many SAT solvers, the only clauses that are ever learned are the ones used for backtracking. If we restrict the learning so that it is always done immediately before backjumping, we can be sure that some progress will be made between a Learn and the next Learn or Forget. This idea is captured by the following combined rule:

Learn+Backjump $(M' L^\dagger M, N) \Longrightarrow_{\mathsf{CDCL\_NOT\_merge}} (L' M, N \uplus \{C' \vee L'\})$
if C, $L^\dagger$, L', M, M', N satisfy Backjump's side conditions

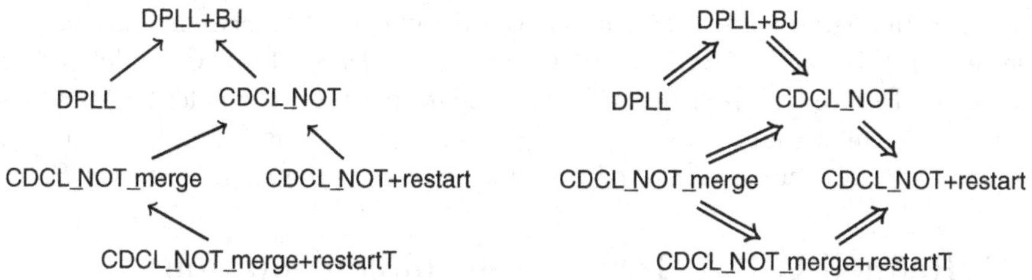

(a) Syntactic dependencies (b) Refinements

Fig. 1. Connections between the abstract calculi

The calculus variant that performs this rule instead of Learn or Backjump is called CDCL_NOT_merge. Because a single Learn+Backjump transition corresponds to two transitions in CDCL_NOT, the inclusion CDCL_NOT_merge $\subseteq$ CDCL_NOT does not hold. Instead, we have CDCL_NOT_merge $\subseteq$ CDCL_NOT$^+$, which is proved by simulation.

3.5 Restarts

Modern SAT solvers rely on a dynamic decision literal heuristic. They periodically restart the proof search to apply the effects of a changed heuristic. This helps the calculus focus on a part of the initial clauses where it can make progress. Upon a restart, some learned clauses may be removed, and the trail is reset to ϵ. Since our calculus has a Forget rule, our Restart rule needs only to clear the trail. Adding Restart to CDCL_NOT yields CDCL_NOT+restart. However, this calculus does not terminate, because Restart can be applied infinitely often.

A working strategy is to gradually increase the number of transitions between successive restarts. This is formalized via a locale parameterized by a base calculus C and an unbounded function $f :: \mathbb{N} \Rightarrow \mathbb{N}$. Nieuwenhuis et al. require f to be strictly increasing, but unboundedness is sufficient.

The extended calculus C+restartT is defined by the two rules

Restart $(S, n) \Longrightarrow_{C\text{+restartT}} ((\epsilon, N'), n + 1)$ if $S \Longrightarrow_C^m (M', N')$ and $m \geq f\ n$
Finish $(S, n) \Longrightarrow_{C\text{+restartT}} (S',\ n + 1)$ if $S \Longrightarrow_C^{*!} S'$

The T in restartT reminds us that we count the number of *transitions*; in Sect. 4.4, we will review an alternative strategy based on the number of conflicts or learned clauses. Termination relies on a measure μ_V associated with C that may not increase from restart to restart: If $S \Longrightarrow_C^* S' \Longrightarrow_{\text{restartT}} S''$, then $\mu_V\ S'' \leq \mu_V\ S$. The measure may depend on V, the number of variables occurring in the problem. We instantiated the locale parameter C with CDCL_NOT_merge and f with the Luby sequence $(1, 1, 2, 1, 1, 2, 4, \ldots)$ [23], with the restriction that no clause containing duplicate literals is ever learned, thereby bounding the number of learnable clauses and hence the number of transitions taken by C.

Figure 1(a) summarizes the syntactic dependencies between the calculi reviewed in this section. An arrow $C \longrightarrow B$ indicates that C is defined in terms of B. Figure 1(b) presents the refinements between the calculi. An arrow $C \Longrightarrow B$ indicates that we proved $C \subseteq B^*$ or some stronger result—either by locale embedding (`sublocale`) or by simulating C's behavior in terms of B.

4 A Refined CDCL Towards an Implementation

The CDCL_NOT calculus captures the essence of modern SAT solvers without imposing a policy on when to apply specific rules. In particular, the Backjump rule depends on a clause $C' \vee L'$ to justify the propagation of a literal, but does not specify a procedure for coming up with this clause. For *Automated Reasoning*, Weidenbach developed a calculus that is more specific in this respect, and closer to existing implementations, while keeping many aspects unspecified [42]. This calculus, CDCL_W, is also formalized in Isabelle and connected to CDCL_NOT.

4.1 The New CDCL Calculus

The CDCL_W calculus operates on states (M, N, U, k, D), where M is the trail; N and U are the sets of initial and learned clauses, respectively; k is the decision level (i.e., the number of decision literals in M); D is a conflict clause, or the distinguished clause $\top$ if no conflict has been detected. In M, each decision literal is annotated with a level (Decided L k or L^k), and each propagated literal is annotated with the clause that caused its propagation (Propagated L C or L^C). The level of a propagated literal L is the level of the closest decision literal that follows it in the trail, or 0 if no such literal exists. The level of a clause is the highest level of any of its literals (0 for $\bot$). The calculus assumes that N contains no duplicate literals and never produces clauses containing duplicates.

The calculus starts in a state $(\epsilon, N, \emptyset, 0, \top)$. The following rules apply as long as no conflict has been detected:

Propagate $(M, N, U, k, \top) \Longrightarrow_{\mathsf{CDCL\_W}} (L^{C \vee L} M, N, U, k, \top)$
  if $C \vee L \in N \uplus U$, $M \vDash \neg C$, and L is undefined in M
Conflict $(M, N, U, k, \top) \Longrightarrow_{\mathsf{CDCL\_W}} (M, N, U, k, D)$ if $D \in N \uplus U$ and $M \vDash \neg D$
Decide $(M, N, U, k, \top) \Longrightarrow_{\mathsf{CDCL\_W}} (L^{k+1} M, N, U, k+1, \top)$
  if L is undefined in M and occurs in N
Restart $(M, N, U, k, \top) \Longrightarrow_{\mathsf{CDCL\_W}} (\epsilon, N, U, 0, \top)$ if $M \nvDash N$
Forget $(M, N, U \uplus \{C\}, k, \top) \Longrightarrow_{\mathsf{CDCL\_W}} (M, N, U, k, \top)$
  if $M \nvDash N$ and M contains no literal L^C

Once a conflict clause is detected and stored in the state, the following rules collaborate to reduce it and backtrack, exploring a first unique implication point [2]:

Skip $(L^C M, N, U, k, D) \Longrightarrow_{\mathsf{CDCL\_W}} (M, N, U, k, D)$
  if $D \notin \{\bot, \top\}$ and $-L$ does not occur in D
Resolve $(L^{C \vee L} M, N, U, k, D \vee -L) \Longrightarrow_{\mathsf{CDCL\_W}} (M, N, U, k, C \cup D)$ if D is of level k

Backtrack
$(M'K^{i+1}M, N, U, k, D \vee L) \Longrightarrow_{\mathsf{CDCL\_W}} (L^{D \vee L}M, N, U \uplus \{D \vee L\}, i, \top)$
if L is of level k and D is of level i

(In Resolve, $C \cup D$ is the same as $C \vee D$, except that it avoids duplicating literals present in both C and D.) In combination, these three rules can be simulated by the combined learning and nonchronological backjump rule Learn+Backjump from CDCL_NOT_merge.

Several structural invariants hold on all states reachable from an initial state, including the following: The trail is consistent; the k decided literals in the trail are annotated with levels k to 1; and the clause annotating a propagated literal of the trail is contained in $N \uplus U$. Some of the invariants were not mentioned in the textbook (e.g., whenever L^C occurs in the trail, L is a literal of C); formalization helped develop a better understanding of the data structure and clarify the book.

Like CDCL_NOT, CDCL_W has a notion of conclusive state. A state (M, N, U, k, D) is *conclusive* if $D = \top$ and $M \vDash N$ or if $D = \bot$ and N is unsatisfiable. The calculus always terminates but, without suitable strategy, it can stop in an inconclusive state. Consider this derivation: $(\epsilon, \{A, B\}, \emptyset, 0, \top) \Longrightarrow_{\mathsf{Decide}} (\neg A^1, \{A, B\}, \emptyset, 1, \top) \Longrightarrow_{\mathsf{Decide}} (\neg B^2\, \neg A^1, \{A, B\}, \emptyset, 2, \top) \Longrightarrow_{\mathsf{Conflict}} (\neg B^2\, \neg A^1, \{A, B\}, \emptyset, 2, A)$. The conflict cannot be processed by Skip or Resolve. The calculus is blocked.

4.2 A Reasonable Strategy

To prove correctness, we assume a *reasonable* strategy: Propagate and Conflict are preferred over Decide; Restart and Forget are not applied. (We will lift the restriction on Restart and Forget in Sect. 4.4.) The resulting calculus, CDCL_W+stgy, refines CDCL_W with the assumption that derivations are produced by a reasonable strategy. This assumption is enough to ensure that the calculus can backjump after detecting a conflict clause other than $\bot$. The crucial invariant is the existence of a literal with the highest level in any conflict, so that Resolve can be applied.

Theorem 6 (Partial Correctness [13, full_cdcl$_\mathsf{W}$_stgy_final_state_conclusive_from_init_state]). *If* $(\epsilon, N, \emptyset, 0, \top) \Longrightarrow^{*!}_{\mathsf{CDCL\_W+stgy}} S'$ *and* N *contains no clauses with duplicate literals,* S' *is a conclusive state.*

Once a conflict clause has been stored in the state, the clause is first reduced by a chain of Skip and Resolve transitions. Then, there are two scenarios: (1) the conflict is solved by a Backtrack, at which point the calculus may resume propagating and deciding literals; (2) the reduced conflict is $\bot$, meaning that N is unsatisfiable—i.e., for unsatisfiable clause sets, the calculus generates a resolution refutation.

The CDCL_W+stgy calculus is designed to have respectable complexity bounds. One of the reasons for this is that the same clause cannot be learned twice:

Theorem 7 (Relearning [13, **cdcl$_W$_stgy_distinct_mset_clauses**]**).** *Let* $(\epsilon, N, \emptyset, 0, \top) \Longrightarrow^*_{\mathsf{CDCL\_W+stgy}} (M, N, U, k, D)$. *No* Backtrack *transition is possible from the latter state causing the addition of a clause from* $N \uplus U$ *to* U.

The formalization of this theorem posed some challenges. The informal proof in *Automated Reasoning* is as follows (with slightly adapted notations):

> *Proof.* By contradiction. Assume CDCL learns the same clause twice, i.e., it reaches a state $(M, N, U, k, D \vee L)$ where Backtrack is applicable and $D \vee L \in N \uplus U$. More precisely, the state has the form $(K_n \cdots K_2 K_1^k M_2 K^{i+1} M_1, N, U, k, D \vee L)$ where the K_i, $i{>}1$ are propagated literals that do not occur complemented in D, as for otherwise D cannot be of level i. Furthermore, one of the K_i is the complement of L. But now, because $D \vee L$ is false in $K_n \cdots K_2 K_1^k M_2 K^{i+1} M_1$ and $D \vee L \in N \uplus U$ instead of deciding K_1^k the literal L should be propagated by a reasonable strategy. A contradiction. Note that none of the K_i can be annotated with $D \vee L$. □

Many details are missing. To find the contradiction, we must show that there exists a state in the derivation with the trail $M_2 K^{i+1} M_1$, and such that $D \vee L \in N \uplus U$. The textbook does not explain why such a state is guaranteed to exist. Moreover, inductive reasoning is hidden under the ellipsis notation $(K_n \cdots K_2)$. Such a high level proof might be suitable for humans, but the details are needed in Isabelle, and Sledgehammer alone cannot fill in such large gaps, especially if induction is needed. The full formal proof is over 700 lines long and was among the most difficult proofs we carried out.

Using this theorem and assuming that only backjumping has a cost, we get a complexity of $O(3^V)$, where V is the number of different propositional variables. If Conflict is always preferred over Propagate, the learned clause in never redundant in the sense of ordered resolution [42], yielding a complexity bound of $O(2^V)$. Formalizing this is planned for future work.

In *Automated Reasoning*, and in our formalization, Theorem 7 is also used to establish the termination of CDCL_W+stgy. However, the argument for the termination of CDCL_NOT also applies to CDCL_W irrespective of the strategy, a stronger result. To lift this result, we must show that CDCL_W refines CDCL_NOT.

4.3 Connection with Abstract CDCL

It is interesting to show that CDCL_W refines CDCL_NOT_merge, to establish beyond doubt that CDCL_W is a CDCL calculus and to lift the termination proof and any other general results about CDCL_NOT_merge. The states are easy to connect: We interpret a CDCL_W tuple (M, N, U, k, C) as a CDCL_NOT pair (M, N).

The main difficulty is to relate the low-level conflict-related CDCL_W rules to their high-level counterparts. Our solution is to introduce an intermediate calculus, called CDCL_W_merge, that combines consecutive low-level transitions into

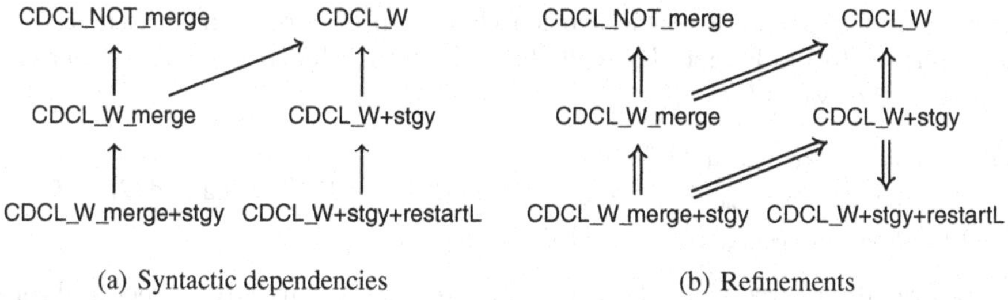

Fig. 2. Connections involving the refined calculi

a single transition. This calculus refines both CDCL_W and CDCL_NOT_merge and is sufficiently similar to CDCL_W so that we can transfer termination and other properties from CDCL_NOT_merge to CDCL_W through it.

Whenever the CDCL_W calculus performs a low-level sequence of transitions of the form Conflict (Skip | Resolve)* Backtrack$^?$, the CDCL_W_merge calculus performs a single transition of a new rule that subsumes all four low-level rules:

Reduce+Maybe_Backtrack $S \Longrightarrow_{\mathsf{CDCL\_W\_merge}} S''$
if $S \Longrightarrow_{\mathsf{Conflict}} S' \Longrightarrow^{*!}_{\mathsf{Skip}\,|\,\mathsf{Resolve}\,|\,\mathsf{Backtrack}} S''$

When simulating CDCL_W_merge in terms of CDCL_NOT, two interesting scenarios arise. In the first case, Reduce+Maybe_Backtrack's behavior comprises a backtrack. The rule can then be simulated using CDCL_NOT_merge's Learn+Backjump rule. The second scenario arises when the conflict clause is reduced to $\bot$, leading to a conclusive final state. Then, Reduce+Maybe_Backtrack has no counterpart in CDCL_NOT_merge. More formally, the two calculi are related as follows: If $S \Longrightarrow_{\mathsf{CDCL\_W\_merge}} S'$, either $S \Longrightarrow_{\mathsf{CDCL\_NOT\_merge}} S'$ or S is a conclusive state. Since CDCL_NOT_merge is well founded, so is CDCL_W_merge. This implies that CDCL_W without Restart terminates.

Since CDCL_W_merge is mostly a rephrasing of CDCL_W, it makes sense to restrict CDCL_W_merge to a *reasonable* strategy that prefers Propagate and Reduce+Maybe_Backtrack over Decide, yielding CDCL_W_merge+stgy. The two strategy-restricted calculi have the same end-to-end behavior:

$$S \Longrightarrow^{*!}_{\mathsf{CDCL\_W\_merge+stgy}} S' \longleftrightarrow S \Longrightarrow^{*!}_{\mathsf{CDCL\_W+stgy}} S'$$

4.4 A Strategy with Restart and Forget

We could use the same strategy for restarts as in Sect. 3.5, but we prefer to exploit Theorem 7, which asserts that no relearning is possible. Since only finitely many different duplicate-free clauses can ever be learned, it is sufficient to increase the number of learned clauses between two restarts to obtain termination. This criterion is the norm in existing implementations. The lower bound on the number of learned clauses is given by an unbounded function $f :: \mathbb{N} \Rightarrow \mathbb{N}$. In addition, we allow an arbitrary subset of the learned clauses to be forgotten upon a restart but otherwise forbid Forget. The calculus C+restartL that realizes these ideas is defined by the two rules

Restart $(S, n) \Longrightarrow_{C\text{+restartL}} (S''', n+1)$
  if $S \Longrightarrow_{C}^{*} S' \Longrightarrow_{\mathsf{Restart}} S'' \Longrightarrow_{\mathsf{Forget}}^{*} S'''$ and $|\mathsf{learned}\ S'| - |\mathsf{learned}\ S| \geq f\ n$
Finish $(S, n) \Longrightarrow_{C\text{+restartL}} (S', n+1)$ if $S \Longrightarrow_{C}^{*!} S'$

We formally proved that CDCL_W+stgy+restartL is partially correct and terminating. Figure 2 summarizes the situation, following the conventions of Fig. 1.

4.5 Incremental Solving

SMT solvers combine a SAT solver with theory solvers (e.g., for uninterpreted functions and linear arithmetic). The main loop runs the SAT solver on a set of clauses. If the SAT solver answers "unsatisfiable," the SMT solver is done; otherwise, the main loop asks the theory solvers to provide further, theory-motivated clauses to exclude the current candidate model and force the SAT solver to search for another one. This design crucially relies on incremental SAT solving: the possibility of adding new clauses to the clause set C of a conclusive satisfiable state and of continuing from there.

As a step towards formalizing SMT, we designed a calculus CDCL_W+stgy+incr that provides incremental solving on top of CDCL_W+stgy:

Add_Nonconflict$_C$ $(M, N, U, k, \top) \Longrightarrow_{\mathsf{CDCL\_W+stgy+incr}} S'$
  if $M \not\models \neg C$ and $(M, N \uplus \{C\}, U, k, \top) \Longrightarrow_{\mathsf{CDCL\_W+stgy}}^{*!} S'$
Add_Conflict$_C$ $(M'LM, N, U, k, \top) \Longrightarrow_{\mathsf{CDCL\_W+stgy+incr}} S'$
  if $LM \models \neg C$, $-L \in C$, M' contains no literal of C, L is of level i in LM, and $(LM, N \uplus \{C\}, U, i, C) \Longrightarrow_{\mathsf{CDCL\_W+stgy}}^{*!} S'$

We first run the CDCL_W+stgy calculus on a set of clauses N, as usual. If N is satisfiable, we can add a nonempty, duplicate-free clause C to the set of clauses and apply one of the two above rules. These rules adjust the state and relaunch CDCL_W+stgy.

Theorem 8 (Partial Correctness [13, **incremental_conclusive_state**]**).** *If S is a conclusive state and $S \Longrightarrow_{\mathsf{CDCL\_W+stgy+incr}} S'$, then S' is a conclusive state.*

The key is to prove that the structural invariants that hold for CDCL_W+stgy still hold after adding the new clause to the state. The proof is easy because we can reuse the invariants we have already proved about CDCL_W+stgy.

5 An Implementation of CDCL

The previous sections presented variants of DPLL and CDCL as parameterized transition systems, formalized using locales and inductive predicates. The final link in our refinement chain is a deterministic SAT solver that implements

CDCL_W+stgy, expressed as a functional program in Isabelle. When implementing a calculus, we must make many decisions regarding the data structures and the order of rule applications. We choose to represent states by tuples (M, N, U, k, D), where propositional variables are coded as natural numbers and multisets as lists.[1] Each transition rule in CDCL_W+stgy is implemented by a corresponding function. For example, the function that implements the Propagate rule is given below:

$$
\begin{array}{l}
\textbf{definition}\ \mathsf{do\_propagate\_step} :: {}'v\ \mathit{solver\_state} \Rightarrow {}'v\ \mathit{solver\_state}\ \textbf{where} \\
\quad \mathsf{do\_propagate\_step}\ S\ = \\
\quad (\text{case}\ S\ \text{of} \\
\qquad (M, N, U, k, \top) \Rightarrow \\
\qquad (\text{case}\ \mathsf{find\_first\_unit\_propagation}\ M\ (N\ @\ U)\ \text{of} \\
\qquad\quad \mathsf{Some}\ (L, C) \Rightarrow (\mathsf{Propagated}\ L\ C \cdot M, N, U, k, \top) \\
\qquad\ \mid \mathsf{None} \Rightarrow S) \\
\quad \mid S \Rightarrow S)
\end{array}
$$

The main loop invokes the functions for the rules, looking for conflicts before propagating literals. It is a recursive program, specified using the `function` command [21]. For Isabelle to accept the recursive definition of the main loop as a terminating program, we must discharge a proof obligation stating that its call graph is well founded. This is a priori unprovable: The solver is not guaranteed to terminate if starting in an arbitrary state. To work around this, we restrict the input by introducing a subset type that contains a strong enough structural invariant, including the duplicate-freedom of all the lists in the data structure. With the invariant in place, it is easy to show that the call graph is included in CDCL_W+stgy, allowing us to reuse its termination argument. The partial correctness theorem can then be lifted, meaning that the SAT solver is a decision procedure for propositional logic.

The final step is to extract running code. Using Isabelle's code generator [15], we can translate the program into Haskell, OCaml, Scala, or Standard ML code. The resulting program is syntactically analogous to the source program in Isabelle (including its dependencies) and uses the target language's facilities for datatypes and recursive functions with pattern matching. Invariants on subset types are ignored; when invoking the solver from outside Isabelle, the caller is responsible for ensuring that the input satisfies the invariant. The entire program is about 700 lines long in OCaml. It is not efficient, due to its extensive reliance on lists, but it satisfies the need for a proof of concept.

6 Discussion and Related Work

Our formalization consists of about 28 000 lines of Isabelle text. It was done over a period of 10 months almost entirely by Fleury, who also taught himself

[1] We have started formalizing the two-watched-literal optimization [28] but have yet to connect it with our SAT solver implementation. The `README.md` file in our repository is frequently updated to mention the latest developments [13].

Isabelle during that time. It covers nearly all of the metatheoretical material of Sects. 2.6 to 2.11 of *Automated Reasoning* and Sect. 2 of Nieuwenhuis et al., including normal form transformations and ground unordered resolution [12].

It is difficult to quantify the cost of formalization as opposed to paper proofs. For a sketchy paper proof, formalization may take an arbitrarily long time; indeed, Weidenbach's nine-line proof of Theorem 7 took 700 lines of Isabelle. In contrast, given a very detailed paper proof, one can obtain a formalization in less time than it took to write the paper proof [44]. A common hurdle to formalization is often the lack of suitable libraries. For CDCL, we spent considerable time adding definitions, lemmas, and automation hints to Isabelle's multiset library but otherwise did not need any special libraries. We also found that organizing the proof at a high level—especially locale engineering—is more challenging than discharging proof obligations.

Given the varied level of formality of the proofs in the draft of *Automated Reasoning*, it is unlikely that Fleury will ever catch up with Weidenbach. But the insights arising from formalization have already enriched the textbook in many ways. The most damning mistake was in the proof of the resolution calculus without reductions, where the completeness theorem was stated with "$N \Longrightarrow^* \{\bot\}$" instead of "$N \Longrightarrow^* N'$ and $\bot \in N'$." For CDCL, the main issues were that key invariants were omitted and some proofs were too sketchy to be accessible to the intended audience of the book.

For discharging proof obligations, we relied extensively on Sledgehammer, including its facility for generating detailed Isar proofs [3] and the SMT-based *smt* tactic [9]. We found the SMT solver CVC4 particularly useful, corroborating earlier empirical evaluations [36]. In contrast, the counterexample generators Nitpick and Quickcheck [5] were seldom of any use. We often discovered flawed conjectures by seeing Sledgehammer fail to solve an easy-looking problem. As one example among many, we lost perhaps one hour working from the hypothesis that converting a set to a multiset and back is the identity: $\mathsf{set\_mset}\ (\mathsf{mset\_set}\ A) = A$. Because Isabelle multisets are finite, the property does not hold for infinite sets A; yet Nitpick and Quickcheck fail to find a counterexample, because they try only finite sets as values for A.

Formalizing logic in a proof assistant is an enticing, if somewhat self-referential, prospect. Shankar's proof of Gödel's first incompleteness theorem [37], Harrison's formalization of basic first-order model theory [16], and Margetson and Ridge's formalized completeness and cut elimination theorems [24] are among the first results in this area. Recently, SAT solvers have been formalized in proof assistants. Marić [25,26] verified a CDCL-based SAT solver in Isabelle/HOL, including two watched literals, as a purely functional program. The solver is monolithic, which complicates extensions. In addition, he formalized the abstract CDCL calculus by Nieuwenhuis et al. Marić's methodology is quite different from ours, without the use of refinements, inductive predicates, locales, or even Sledgehammer. In his Ph.D. thesis, Lescuyer [22] presents the formalization of the CDCL calculus and the core of an SMT solver in Coq. He also developed a reflexive DPLL-based SAT solver for Coq, which can be used as a tactic in the proof assistant. Another formalization of a CDCL-based SAT solver, including termination but excluding two watched literals, is by Shankar and Vaucher in PVS [38]. Most of this work was done by Vaucher during a two-month internship, an impressive achievement. Finally, Oe et al. [33] verified an imperative and fairly efficient CDCL-based SAT solver, expressed using the Guru language for verified programming. Optimized data structures are used, including for two watched literals and conflict analysis. However, termination is not guaranteed, and model soundness is achieved through a run-time check and not proved.

7 Conclusion

The advantages of computer-checked metatheory are well known from programming language research, where papers are often accompanied by formalizations and proof assistants are used in the classroom [30,35]. This paper, like its predecessors [6,8], reported on some steps we have taken to apply these methods to automated reasoning. Compared with other application areas of proof assistants, the proof obligations are manageable, and little background theory is required.

We presented a formal framework for DPLL and CDCL in Isabelle/HOL, covering the ground between an abstract calculus and a certified SAT solver. Our framework paves the way for further formalization of metatheoretical results. We intend to keep following *Automated Reasoning*, including its generalization of ordered ground resolution with CDCL, culminating with a formalization of the full superposition calculus and extensions. Thereby, we aim at demonstrating that interactive theorem proving is mature enough to be of use to practitioners in automated reasoning, and we hope to help them by developing the necessary libraries and methodology.

The CDCL algorithm, and its implementation in highly efficient SAT solvers, is one of the jewels of computer science. To quote Knuth [20, p. iv], "The story of satisfiability is the tale of a triumph of software engineering blended with rich doses of beautiful mathematics." What fascinates us about CDCL is not only *how* or *how well* it works, but also *why* it works so well. Knuth's remark is accurate, but it is not the whole story.

Acknowledgment. Stephan Merz made this work possible. Dmitriy Traytel remotely cosupervised Fleury's M.Sc. thesis and provided ample advice on using Isabelle (as opposed to developing it). Andrei Popescu gave us his permission to reuse, in a slightly adapted form, the succinct description of locales he cowrote on a different occasion [6]. Simon Cruanes, Anders Schlichtkrull, Mark Summerfield, and Dmitriy Traytel suggested textual improvements.

References

1. Ballarin, C.: Locales: a module system for mathematical theories. J. Autom. Reasoning **52**(2), 123–153 (2014)
2. Biere, A., Heule, M., van Maaren, H., Walsh, T. (eds.): Handbook of Satisfiability. Frontiers in Artificial Intelligence and Applications, vol. 185. IOS Press (2009)
3. Blanchette, J.C., Böhme, S., Fleury, M., Smolka, S.J., Steckermeier, A.: Semi-intelligible Isar proofs from machine-generated proofs. J. Autom. Reasoning **55**(2), 155–200 (2016)
4. Blanchette, J.C., Böhme, S., Paulson, L.C.: Extending Sledgehammer with SMT solvers. J. Autom. Reasoning **51**(1), 109–128 (2013)
5. Blanchette, J.C., Bulwahn, L., Nipkow, T.: Automatic proof and disproof in Isabelle/HOL. In: Tinelli, C., Sofronie-Stokkermans, V. (eds.) FroCoS 2011. LNCS, vol. 6989, pp. 12–27. Springer, Heidelberg (2011)
6. Blanchette, J.C., Popescu, A.: Mechanizing the metatheory of Sledgehammer. In: Fontaine, P., Ringeissen, C., Schmidt, R.A. (eds.) FroCoS 2013. LNCS, vol. 8152, pp. 245–260. Springer, Heidelberg (2013)
7. Blanchette, J.C., Fleury, M., Schlichtkrull, A., Traytel, D.: IsaFoL: Isabelle formalization of logic. https://bitbucket.org/jasmin_blanchette/isafol
8. Blanchette, J.C., Popescu, A., Traytel, D.: Unified classical logic completeness. In: Demri, S., Kapur, D., Weidenbach, C. (eds.) IJCAR 2014. LNCS, vol. 8562, pp. 46–60. Springer, Heidelberg (2014)
9. Böhme, S., Weber, T.: Fast LCF-style proof reconstruction for Z3. In: Kaufmann, M., Paulson, L.C. (eds.) ITP 2010. LNCS, vol. 6172, pp. 179–194. Springer, Heidelberg (2010)
10. Church, A.: A formulation of the simple theory of types. J. Symb. Logic **5**(2), 56–68 (1940)
11. Davis, M., Logemann, G., Loveland, D.W.: A machine program for theorem-proving. Commun. ACM **5**(7), 394–397 (1962)
12. Fleury, M.: Formalisation of ground inference systems in a proof assistant. https://www.mpi-inf.mpg.de/fileadmin/inf/rg1/Documents/fleury_master_thesis.pdf
13. Fleury, M., Blanchette, J.C.: Formalization of Weidenbach's Automated Reasoning—The Art of Generic Problem Solving. https://bitbucket.org/jasmin_blanchette/isafol/src/master/Weidenbach_Book/README.md
14. Gordon, M.J.C., Milner, R., Wadsworth, C.P.: Edinburgh LCF: A Mechanised Logic of Computation. LNCS, vol. 78. Springer, Heidelberg (1979)
15. Haftmann, F., Nipkow, T.: Code generation via higher-order rewrite systems. In: Blume, M., Kobayashi, N., Vidal, G. (eds.) FLOPS 2010. LNCS, vol. 6009, pp. 103–117. Springer, Heidelberg (2010)
16. Harrison, J.V.: Formalizing basic first order model theory. In: Newey, M., Grundy, J. (eds.) TPHOLs 1998. LNCS, vol. 1479, pp. 153–170. Springer, Heidelberg (1998)
17. Heule, M.J., Hunt Jr., W.A., Wetzler, N.: Bridging the gap between easy generation and efficient verification of unsatisfiability proofs. Softw. Test. Verif. Reliab. **24**(8), 593–607 (2014)

18. Bayardo Jr., R.J., Schrag, R.: Using CSP look-back techniques to solve exceptionally hard SAT instances. In: Freuder, E.C. (ed.) CP 1996. LNCS, vol. 1118, pp. 46–60. Springer, Heidelberg (1996)
19. Kammüller, F., Wenzel, M., Paulson, L.C.: Locales—a sectioning concept for Isabelle. In: Bertot, Y., Dowek, G., Hirschowitz, A., Paulin, C., Théry, L. (eds.) TPHOLs 1999. LNCS, vol. 1690, pp. 149–166. Springer, Heidelberg (1999)
20. Knuth, D.E.: The Art of Computer Programming, Volume 4, Fascicle 6: Satisfiability. Addison-Wesley, Reading (2015)
21. Krauss, A.: Partial recursive functions in higher-order logic. In: Furbach, U., Shankar, N. (eds.) IJCAR 2006. LNCS (LNAI), vol. 4130, pp. 589–603. Springer, Heidelberg (2006)
22. Lescuyer, S.: Formalizing and implementing a reflexive tactic for automated deduction in Coq. Ph.D. thesis (2011)
23. Luby, M., Sinclair, A., Zuckerman, D.: Optimal speedup of Las Vegas algorithms. Inf. Process. Lett. **47**(4), 173–180 (1993)
24. Margetson, J., Ridge, T.: Completeness theorem, vol. 2004. Formal proof development. http://afp.sf.net/entries/Completeness.shtml
25. Marić, F.: Formal verification of modern SAT solvers. Archive of Formal Proofs (2008). Formal proof development. http://afp.sf.net/entries/SATSolverVerification.shtml
26. Marić, F.: Formal verification of a modern SAT solver by shallow embedding into Isabelle/HOL. Theoret. Comput. Sci. **411**(50), 4333–4356 (2010)
27. Matuszewski, R., Rudnicki, P.: Mizar: the first 30 years. Mechanized Math. Appl. **4**(1), 3–24 (2005)
28. Moskewicz, M.W., Madigan, C.F., Zhao, Y., Zhang, L., Malik, S.: Chaff: engineering an efficient SAT solver. In: DAC 2001, pp. 530–535. ACM (2001)
29. Nieuwenhuis, R., Oliveras, A., Tinelli, C.: Solving SAT and SAT modulo theories: from an abstract Davis-Putnam-Logemann-Loveland procedure to DPLL(T). J. ACM **53**(6), 937–977 (2006)
30. Nipkow, T.: Teaching semantics with a proof assistant: no more LSD trip proofs. In: Rybalchenko, A., Kuncak, V. (eds.) VMCAI 2012. LNCS, vol. 7148, pp. 24–38. Springer, Heidelberg (2012)
31. Nipkow, T., Klein, G.: Concrete Semantics: With Isabelle/HOL. Springer, New York (2014)
32. Nipkow, T., Paulson, L.C., Wenzel, M.: Isabelle/HOL: A Proof Assistant for Higher-Order Logic. LNCS, vol. 2283. Springer, Heidelberg (2002)
33. Oe, D., Stump, A., Oliver, C., Clancy, K.: `versat`: a verified modern SAT solver. In: Kuncak, V., Rybalchenko, A. (eds.) VMCAI 2012. LNCS, vol. 7148, pp. 363–378. Springer, Heidelberg (2012)
34. Paulson, L.C., Blanchette, J.C.: Three years of experience with Sledgehammer, a practical link between automatic and interactive theorem provers. In: Sutcliffe, G., Schulz, S., Ternovska, E. (eds.) IWIL-2010. EPiC, vol. 2, pp. 1–11. EasyChair (2012)
35. Pierce, B.C.: Lambda, the ultimate TA: using a proof assistant to teach programming language foundations. In: Hutton, G., Tolmach, A.P. (eds.) ICFP 2009, pp. 121–122. ACM (2009)

36. Reynolds, A., Tinelli, C., de Moura, L.: Finding conflicting instances of quantified formulas in SMT. In: Claessen, K., Kuncak, V. (eds.) FMCAD 2014, pp. 195–202. IEEE Computer Society Press (2014)
37. Shankar, N.: Metamathematics, Machines, and Gödel's Proof. Cambridge Tracts in Theoretical Computer Science, vol. 38. Cambridge University Press, Cambridge (1994)
38. Shankar, N., Vaucher, M.: The mechanical verification of a DPLL-based satisfiability solver. Electron. Notes Theoret. Comput. Sci. **269**, 3–17 (2011)
39. Marques-Silva, J.P., Sakallah, K.A.: GRASP—A new search algorithm for satisfiability. In: ICCAD 1996, pp. 220–227. IEEE Computer Society Press (1996)
40. Sternagel, C., Thiemann, R.: An Isabelle/HOL formalization of rewriting for certified termination analysis. http://cl-informatik.uibk.ac.at/software/ceta/
41. Voronkov, A.: AVATAR: the architecture for first-order theorem provers. In: Biere, A., Bloem, R. (eds.) CAV 2014. LNCS, vol. 8559, pp. 696–710. Springer, Heidelberg (2014)
42. Weidenbach, C.: Automated reasoning building blocks. In: Meyer, R., Platzer, A., Wehrheim, H. (eds.) Olderog-Festschrift. LNCS, vol. 9360, pp. 172–188. Springer, Heidelberg (2015). doi:10.1007/978-3-319-23506-6_12
43. Wenzel, M.: Isabelle/Isar—A generic framework for human-readable proof documents. In: Matuszewski, R., Zalewska, A. (eds.) From Insight to Proof: Festschrift in Honour of Andrzej Trybulec, Studies in Logic, Grammar, and Rhetoric, vol. 10(23). University of Białystok (2007)
44. Woodcock, J., Banach, R.: The verification grand challenge. J. Uni. Comput. Sci. **13**(5), 661–668 (2007)

Super-Blocked Clauses

Benjamin Kiesl[1(✉)], Martina Seidl[2], Hans Tompits[1], and Armin Biere[2]

[1] Institute for Information Systems,
Vienna University of Technology, Vienna, Austria
kiesl@kr.tuwien.ac.at
[2] Institute for Formal Models and Verification, JKU Linz, Linz, Austria

Abstract. In theory and practice of modern SAT solving, clause-elimination procedures are essential for simplifying formulas in conjunctive normal form (CNF). Such procedures identify redundant clauses and faithfully remove them, either before solving in a preprocessing phase or during solving, resulting in a considerable speed up of the SAT solver. A wide number of effective clause-elimination procedures is based on the clause-redundancy property called *blocked clauses*. For checking if a clause C is blocked in a formula F, only those clauses of F that are resolvable with C have to be considered. Hence, the blocked-clauses redundancy property can be said to be *local*. In this paper, we argue that the established definitions of blocked clauses are not in their most general form. We introduce more powerful generalizations, called *set-blocked clauses* and *super-blocked clauses*, respectively. Both can still be checked locally, and for the latter it can even be shown that it is the most general local redundancy property. Furthermore, we relate these new notions to existing clause-redundancy properties and give a detailed complexity analysis.

1 Introduction

Over the last two decades, we have seen enormous progress in the performance of SAT solvers, i.e., tools for solving the satisfiability problem of propositional logic (SAT) [1]. As a consequence, SAT solvers have become attractive reasoning engines in many user domains like the verification of hardware and software [2] as well as in the backends of other reasoning tools like SMT solvers [3] or even first-order theorem provers [4]. In such applications, however, SAT solvers often reach their limits, motivating the quest for more efficient SAT techniques.

Clause-elimination procedures which simplify formulas in conjunctive normal form (CNF) play a crucial role regarding the performance of modern SAT solvers [5–12]. Either before solving ("preprocessing") or during solving ("inprocessing"), such procedures identify redundant clauses and remove them without changing the satisfiability or unsatisfiability of the formula [6,7].

This work has been supported by the Austrian Science Fund (FWF) under projects W1255-N23 and S11408-N23.

N. Olivetti and A. Tiwari (Eds.): IJCAR 2016, LNAI 9706, pp. 45–61, 2016.
DOI: 10.1007/978-3-319-40229-1_5

An important redundancy property is that of *blocked clauses* [13,14]. Informally, a clause C is blocked in a CNF-formula F if it contains a literal l such that all possible resolvents of C on l with clauses from F are tautologies. As only the resolution environment of a clause C and not the whole formula F has to be considered to check whether C is blocked, the blocked-clauses condition is said to be a *local* redundancy property.

Blocked clauses have not only shown to be important for speeding up the solving process [8,14], but they also yield the basis for blocked-clause decomposition which splits a CNF into two parts such that blocked-clause elimination can solve it. Blocked-clause decomposition [9] is successfully used for gate extraction, for efficiently finding backbone variables, and for the detection of implied binary equivalences [10,11]. The winner of the SATRace 2015 competition, the solver abcdSAT [12], uses blocked-clause decomposition as core technology.

These success stories motivate us to have a closer look at local redundancy properties in general, and at blocked clauses in particular. We show in this paper that the established definitions of local clause redundancy properties like blocked clauses are not in their most general form and introduce more powerful generalizations, called *set-blocked clauses* and *super-blocked clauses*. Both can still be checked locally and for the latter we show that it is actually the most general local redundancy property. Furthermore, we relate these new notions to existing clause redundancy properties and give a detailed complexity analysis.

Our paper is structured as follows. After introducing the necessary preliminaries in Sect. 2, we present some observations on blocked clauses in Sect. 3. In Sect. 4, we introduce the notion of *semantic blocking* and show that it is the most general local redundancy property. After this, the syntax-based notions of *set-blocking* and *super-blocking* are introduced in Sect. 5, where we also relate the different redundancy properties to each other and show that super-blocking coincides with semantic blocking. In Sect. 6, we give a detailed complexity analysis and in Sect. 7, we outline the relationship to existing redundancy properties before concluding with an outlook to future work in Sect. 8.

2 Preliminaries

We consider propositional formulas in *conjunctive normal form* (CNF) which are defined as follows. A *literal* is either a Boolean variable x (a *positive literal*) or the negation $\neg x$ of a variable x (a *negative literal*). For a literal l, we define $\bar{l} = \neg x$ if $l = x$ and $\bar{l} = x$ if $l = \neg x$. Accordingly, for a set L of literals, we define $\bar{L} = \{\bar{l} \mid l \in L\}$. A *clause* is a disjunction of literals. A *formula* is a conjunction of clauses. A clause can be seen as a set of literals and a formula as a set of clauses. A *tautology* is a clause that contains both l and $\bar{l}$ for some literal l. For a literal, clause, or formula F, $var(F)$ denotes the variables in F. For convenience, we treat $var(F)$ as a variable if F is a literal, and as a set of variables otherwise.

An *assignment* over a set V of variables is a function that assigns to every variable in V either 1 or 0. If for an assignment τ and a formula F, the domain of τ coincides with $var(F)$, then τ is said to be an assignment *of* F. Given an assignment τ and a literal l, τ_l is the assignment obtained from τ by interchanging ("flipping") the truth value of l, i.e., by defining $\tau_l(v) = 1 - \tau(v)$ if $v = var(l)$ and $\tau_l(v) = \tau(v)$ otherwise.

A literal l is *satisfied* by an assignment τ if l is positive and $\tau(var(l)) = 1$ or if it is negative and $\tau(var(l)) = 0$. A clause is satisfied by an assignment τ if it contains a literal that is satisfied by τ. Finally, a formula is satisfied by an assignment τ if all of its clauses are satisfied by τ. A formula is *satisfiable* if there exists an assignment that satisfies it. Two formulas are *logically equivalent* if they are satisfied by the same assignments. Two formulas F and F' are *satisfiability equivalent* if F is satisfiable if and only if F' is satisfiable.

Given two clauses C_1 and C_2 with literal $l \in C_1$ and $\bar{l} \in C_2$, the clause $C = (C_1 \setminus \{l\}) \cup (C_2 \setminus \{\bar{l}\})$ is called the *resolvent* of C_1 and C_2 on l. Given a formula F and a clause C, the *resolution environment*, $env_F(C)$, of C in F is the set of all clauses in F that can be resolved with C:

$$env_F(C) = \{C' \in F \mid \exists l \in C' \text{ such that } \bar{l} \in C\}.$$

The variables in $var(C)$ are referred to as *local variables* and the variables in $var(env_F(C)) \setminus var(C)$ are the *external variables*, denoted by $ext_F(C)$.

Next, we formally introduce the redundancy of clauses. Intuitively, a clause C is redundant w.r.t. a formula F if neither its addition to F nor its removal from F changes the satisfiability or unsatisfiability of F.

Definition 1. *A clause C is* redundant *w.r.t. a formula F if $F \setminus \{C\}$ and $F \cup \{C\}$ are satisfiability equivalent. A* redundancy property *is a set of pairs (F, C) where C is redundant w.r.t. F. Finally, for two redundancy properties $\mathcal{P}_1$ and $\mathcal{P}_2$, $\mathcal{P}_1$ is* more general *than $\mathcal{P}_2$ if $\mathcal{P}_2 \subseteq \mathcal{P}_1$. Accordingly, $\mathcal{P}_1$ is* strictly more general *than $\mathcal{P}_2$ if $\mathcal{P}_2 \subset \mathcal{P}_1$.*

As an example, consider the formula $F = \{(a \vee b), (\neg a \vee \neg b)\}$. The clause $C = (\neg a \vee \neg b)$ is redundant w.r.t. F since $F \setminus \{C\}$ and $F \cup \{C\}$ are satisfiability equivalent (although they are not logically equivalent). Furthermore, the set $\{(F, C) \mid F$ is a formula and C is a tautology$\}$ is a redundancy property since for every formula F and every tautology C, $F \setminus \{C\}$ is satisfiability equivalent to $F \cup \{C\}$.

Also note that C is *not* redundant w.r.t. F if and only if $F \setminus \{C\}$ is satisfiable and $F \cup \{C\}$ is unsatisfiable. Redundancy properties as defined above yield not only the basis for clause-elimination but also for clause-addition procedures [7].

3 Observations on Blocked Clauses

In the following, we recapitulate the notion of blocked clauses due to Heule et al. [6] which we will refer to as *literal-blocked clauses* in the rest of the paper. Motivated by the examples given in this section, we will generalize this notion of blocking to more powerful redundancy properties.

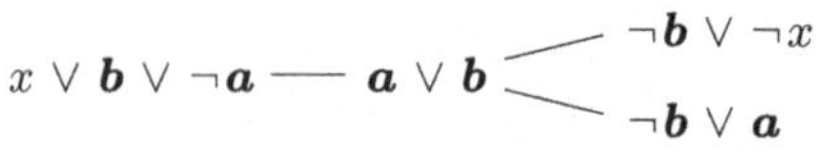

Fig. 1. The clause $(a \vee b)$ from Example 3 and its resolution environment.

Definition 2. *Given a formula F, a clause C, and a literal $l \in C$, l* blocks *C in F if for each clause $C' \in F$ with $\bar{l} \in C'$, $C \cup (C' \setminus \{\bar{l}\})$ is a tautology. A clause C is* literal-blocked *in F if there exists a literal that blocks C in F. By* BC *we denote the set $\{(F, C) \mid C$ is literal-blocked in $F\}$.*

Example 1. Consider the formula $F = \{(\neg a \vee c), (\neg b \vee \neg a)\}$ and the clause $C = (a \vee b)$. The literal b blocks C in F since the only clause in F that contains $\neg b$ is the clause $C' = (\neg b \vee \neg a)$, and $C \cup (C' \setminus \{\bar{l}\}) = (a \vee b \vee \neg a)$ is a tautology.

Proposition 1. BC *is a redundancy property.*

Proposition 1 paraphrases results from [6] and actually follows from results in this paper (cf. Proposition 6 and Corollary 9). Intuitively, if an assignment τ satisfies $F \setminus \{C\}$ but falsifies C which is blocked by literal l, then τ_l satisfies C. The condition that l blocks C thereby guarantees that τ_l does not falsify any other clauses in F. Hence, τ_l satisfies $F \cup \{C\}$ and thus $F \setminus \{C\}$ and $F \cup \{C\}$ are satisfiability equivalent. Next, we illustrate how a satisfying assignment of $F \cup \{C\}$ can be obtained from one of $F \setminus \{C\}$ [6]. This approach is used when blocked clauses have been removed from a formula during pre- or inprocessing.

Example 2. Consider again the formula $F = \{(\neg a \vee c), (\neg b \vee \neg a)\}$ and the clause $C = (a \vee b)$ from Example 1. We already know that b blocks C in F. So let τ be the assignment that falsifies the variables a, b, and c. Clearly, τ satisfies F but falsifies C. Now, the assignment τ_b, obtained from τ by flipping the truth value of b, satisfies not only C but also all clauses of F: The only clause that could have been falsified by flipping the truth value of b is $(\neg b \vee \neg a)$, but since $\neg a$ is still satisfied by τ_b we get that τ_b satisfies $F \cup \{C\}$. □

Literal-blocked clauses generalize many other redundancy properties like *pure literal* or *tautology* [6]. One of their particularly important properties is that for testing if some clause C is literal-blocked in a formula F it suffices to consider only those clauses of F that can be resolved with C, i.e., the clauses in the resolution environment, $env_F(C)$, of C. This raises the question whether there exist redundant clauses which can be identified by considering only their resolution environment, but which are not literal-blocked. This is indeed the case:

Example 3. Let $C = (a \vee b)$ and F an arbitrary formula with the resolution environment $env_F(C) = \{(x \vee b \vee \neg a), (\neg b \vee \neg x), (\neg b \vee a)\}$ (see Fig. 1). The clause C is not literal-blocked in F but redundant: Suppose that there exists an assignment τ that satisfies F but falsifies C. Then, τ must satisfy either x or $\neg x$. If $\tau(x) = 1$, then C can be satisfied by flipping the truth value of a, resulting in assignment $\tau' = \tau_a$. Thereby, $\tau'(x) = 1$ guarantees that the clause $(x \vee b \vee \neg a)$ stays satisfied. In contrast, if $\tau(x) = 0$, we can satisfy C by the assignment τ'', obtained from τ by flipping the truth values of both a and b: Then, $\tau''(b) = 1$ guarantees that $(x \vee b \vee \neg a)$ stays satisfied whereas $\tau''(x) = 0$ and $\tau''(a) = 1$ guarantee that both $(\neg b \vee \neg x)$ and $(\neg b \vee a)$ stay satisfied. Since flipping the truth values of literals in C does not affect the truth of clauses outside the resolution environment, $env_F(C)$, we obtain in both cases a satisfying assignment of F. □

4 A Semantic Notion of Blocking

In the examples of the preceding section, when arguing that a clause C is redundant w.r.t. some formula F, we showed that every assignment τ that satisfies $F \setminus \{C\}$, but falsifies C, can be turned into a satisfying assignment τ' of $F \cup \{C\}$ by flipping the truth values of certain literals in C. Since this flipping only affects the truth of clauses in the resolution environment, $env_F(C)$, of C, it suffices to make sure that τ' satisfies $env_F(C)$ in order to guarantee that it satisfies $F \cup \{C\}$. This naturally leads to the following semantic notion of blocking:

Definition 3. *A clause C is* semantically blocked *in a formula F if, for every satisfying assignment τ of $env_F(C)$, there exists a satisfying assignment τ' of $env_F(C) \cup \{C\}$ such that $\tau(v) = \tau'(v)$ for all $v \notin var(C)$. By* $\mathsf{SEM}_{\mathsf{BC}}$ *we denote the set $\{(F, C) \mid C$ is semantically blocked in $F\}$.*

Note that clause C in Example 3 is semantically blocked in F. Note also that if the resolution environment, $env_F(C)$, of a clause C is not satisfiable, then C is semantically blocked.

Theorem 2. $\mathsf{SEM}_{\mathsf{BC}}$ *is a redundancy property.*

Proof. Let F be a formula and C a clause that is semantically blocked in F. We show that $F \cup \{C\}$ is satisfiable if $F \setminus \{C\}$ is satisfiable. Suppose that there exists a satisfying assignment τ of $F \setminus \{C\}$. We proceed by a case distinction.

Case 1: C contains a literal l with $var(l) \notin var(F \setminus \{C\})$. Then, τ can be easily extended to a satisfying assignment τ' of $F \cup \{C\}$ that satisfies l.

Case 2: $var(C) \subseteq var(F \setminus \{C\})$. In this case, τ is an assignment of $F \cup \{C\}$. Suppose that τ falsifies C. It follows that C is not a tautology and so it does not contain a literal l such that $\bar{l} \in C$, hence $C \notin env_F(C)$. Thus, $env_F(C) \subseteq F \setminus \{C\}$ and so τ satisfies $env_F(C)$. Since C is semantically blocked in F, there exists a satisfying assignment τ' of $env_F(C) \cup \{C\}$ such that $\tau(v) = \tau'(v)$ for all $v \notin var(C)$. Now, since $\tau'(v)$ differs from τ only on variables in $var(C)$, the only clauses in F that could possibly be falsified by τ' are those with a literal $\bar{l}$ such that $l \in C$. But those are exactly the clauses in $env_F(C)$, so τ' satisfies $F \cup \{C\}$.

Hence, C is redundant w.r.t. F and thus $\mathsf{SEM}_{\mathsf{BC}}$ is a redundancy property. □

If a clause C is redundant w.r.t. some formula F and this redundancy can be identified by considering only its resolution environment in F, then we expect C to be redundant w.r.t. every formula F' in which C has the same resolution environment as in F. This leads us to the notion of *local redundancy properties.*

Definition 4. *A redundancy property $\mathcal{P}$ is* local *if, for any two formulas F, F' and every clause C with $env_F(C) = env_{F'}(C)$, either $\{(F,C), (F',C)\} \subseteq \mathcal{P}$ or $\{(F,C), (F',C)\} \cap \mathcal{P} = \emptyset$.*

Theorem 3. $\mathsf{SEM_{BC}}$ *is a local redundancy property.*

Preparatory for showing that $\mathsf{SEM_{BC}}$ is actually the most general local redundancy property (cf. Theorem 5 below), we first prove the following lemma.

Lemma 4. *Let F be a formula and C a clause that is not semantically blocked in F. Then, there exists a formula F' with $env_{F'}(C) = env_F(C)$ such that C is not redundant w.r.t. F'.*

Proof. Let F be a formula and C a clause that is not semantically blocked in F, i.e., there exists a satisfying assignment τ of $env_F(C)$ but there does not exist a satisfying assignment τ' of $env_F(C) \cup \{C\}$ such that $\tau(v) = \tau'(v)$ for all $v \notin var(C)$. We define the set T of (unit) clauses as follows:

$$T = \{(v) \mid v \notin var(C), \tau(v) = 1\} \cup \{(\neg v) \mid v \notin var(C), \tau(v) = 0\}.$$

We furthermore define $F' = env_F(C) \cup \{C\} \cup T$. Clearly, since C can be falsified and since the clauses in T contain only literals with variables that do not occur in C, we get that neither C nor any clause of T contains a literal $\bar{l}$ with $l \in C$. We thus have that $env_{F'}(C) = env_F(C)$.

Now observe the following: The assignment τ satisfies $env_F(C)$ and, clearly, also T, hence $F' \setminus \{C\}$ is satisfiable. Furthermore, by the construction of T, every assignment that satisfies F' must agree with τ on all variables $v \notin var(C)$. Now, since there does not exist a satisfying assignment τ' of $env_F(C) \cup \{C\}$ such that $\tau(v) = \tau'(v)$ for all $v \notin var(C)$, it follows that $F' \cup \{C\} = F'$ is unsatisfiable. Therefore, $F' \setminus \{C\}$ and $F' \cup \{C\}$ are not satisfiability equivalent and thus C is not redundant w.r.t. F'. □

Theorem 5. $\mathsf{SEM_{BC}}$ *is the most general local redundancy property.*

Proof. Suppose there exists a local redundancy property $\mathcal{P}$ that is strictly more general than $\mathsf{SEM_{BC}}$. Then, there exists some pair (F, C) such that $(F, C) \in \mathcal{P}$ but $(F, C) \notin \mathsf{SEM_{BC}}$. Now, since $(F, C) \notin \mathsf{SEM_{BC}}$ it follows by Lemma 4 that there exists a formula F' with $env_{F'}(C) = env_F(C)$ such that C is not redundant w.r.t. F'. But since $\mathcal{P}$ is local and $env_{F'}(C) = env_F(C)$, it follows that $(F', C) \in \mathcal{P}$, hence $\mathcal{P}$ is not a redundancy property, a contradiction. □

5 Super-Blocked Clauses

In the following, we introduce syntax-based notions of blocking which strictly generalize the original notion of literal-blocking as given in Definition 2. We will first introduce the notion of set-blocking which is already a strict generalization of literal-blocking. This notion will then be further generalized to the so-called notion of super-blocking which, as we will prove, coincides with the notion of semantic blocking given in Definition 3.

Definition 5. *Let F be a formula and C a clause. A non-empty set $L \subseteq C$* blocks *C in F if, for each clause $C' \in F$ with $C' \cap \bar{L} \neq \emptyset$, $(C \setminus L) \cup \bar{L} \cup C'$ is a tautology. We say that a clause is* set-blocked *in F if there exists a set that blocks it. We write $\mathsf{SET}_{\mathsf{BC}}$ to refer to $\{(F, C) \mid C \text{ is set-blocked in} F\}$.*

Example 4. Let $C = (a \vee b)$ and $F = \{(\neg a \vee b), (\neg b \vee a)\}$. Then, C is set-blocked by $L = \{a, b\}$ but not literal-blocked in F. □

Given an assignment τ that satisfies $F \setminus \{C\}$ but falsifies C, the existence of a blocking set L guarantees that a satisfying assignment τ' of $F \cup \{C\}$ can be obtained from τ by flipping the truth values of the literals in L. Since $(C \setminus L) \cup \bar{L} \cup C'$ is a tautology for every C' in the resolution environment of C, it holds that (i) C' itself is a tautology and thus satisfied by τ', or (ii) C' contains a literal of L which is satisfied by τ' since its truth value is flipped, or (iii) C' contains a literal l which is satisfied since $\bar{l} \in C$ is falsified by τ and the truth value of l is not flipped. Hence, τ' satisfies $F \cup \{C\}$.

Proposition 6. *Set-blocking is strictly more general than literal-blocking, i.e., it holds that* $\mathsf{BC} \subset \mathsf{SET}_{\mathsf{BC}}$.

Proof. Example 4 shows that $\mathsf{BC} \neq \mathsf{SET}_{\mathsf{BC}}$. It remains to show that $\mathsf{BC} \subseteq \mathsf{SET}_{\mathsf{BC}}$. Let F be a formula and C a literal-blocked clause in F. We distinguish two cases:

CASE 1: C is a tautology. Then, $l, \bar{l} \in C$ for some literal l. Let $L = \{l, \bar{l}\}$. It follows that $(C \setminus L) \cup \bar{L} \cup C'$ is a tautology for every C' with $C' \cap \bar{L} \neq \emptyset$.

CASE 2: C is not a tautology. Since C is literal-blocked, there exists some literal $l \in C$ such that for every clause $C' \in F$ with $\bar{l} \in C'$, $C \cup (C' \setminus \{\bar{l}\})$ is a tautology. Let $L = \{l\}$ and let $C' \in F$ with $C' \cap \bar{L} \neq \emptyset$. Then, as C' contains $\bar{l}$, $C \cup (C' \setminus \{\bar{l}\})$ is a tautology. Since C is not a tautology, C' contains some literal $l' \neq \bar{l}$ such that $\bar{l'} \in C \cup (C' \setminus \{\bar{l}\})$. Now, since $l' \neq \bar{l}$ we have that $\bar{l'} \neq l$ and thus $\bar{l'} \in (C \setminus \{l\}) \cup C'$. But then, $(C \setminus L) \cup \bar{L} \cup C'$ is a tautology.

Thus, C is set-blocked in F and therefore $\mathsf{BC} \subseteq \mathsf{SET}_{\mathsf{BC}}$. □

We already argued slightly informally why set-blocked clauses are redundant. However, the fact that $\mathsf{SET}_{\mathsf{BC}}$ is a redundancy property follows directly from the properties of super-blocked clauses, which we introduce next. In the following, for a formula F and an assignment τ, we denote by $F|\tau$ the set of clauses obtained from F by removing all clauses that are satisfied by τ. Recall that the external variables, $ext_F(C)$, are those that are contained in $env_F(C)$ but not in C.

Definition 6. *A clause C is* super-blocked *in a formula F if, for every assignment τ over the external variables, $ext_F(C)$, C is set-blocked in $F|\tau$. We write* $\mathsf{SUP_{BC}}$ *for the set $\{(F, C) \mid C$ is super-blocked in $F\}$.*

For instance, the clause C in Example 3 is not set-blocked but super-blocked in F since it is set-blocked in $F|\tau$ and $F|\tau'$ for $\tau(x) = 1$ and $\tau'(x) = 0$. Again, the idea is that from an assignment τ that satisfies $F \setminus \{C\}$ but falsifies C, a satisfying assignment τ' of $F \cup \{C\}$ can be obtained by flipping the truth values of certain literals of C. However, for making sure that the flipping does not falsify any clauses C' in the resolution environment of C, also the truth values of literals $l \in C'$ with $var(l) \in ext_F(C)$ are considered. This is in contrast to set-blocking, where only the truth values of literals whose variables are contained in $var(C)$ are considered. Finally, note that if a clause is set-blocked in F, then it is also set-blocked in every $F' \subseteq F$ and thus in every $F|\tau$. Hence we get:

Proposition 7. *Super-blocking is strictly more general than set-blocking, i.e., it holds that* $\mathsf{SET_{BC}} \subset \mathsf{SUP_{BC}}$.

Theorem 8. *A clause is super-blocked in a formula F if and only if it is semantically blocked in F, i.e., it holds that* $\mathsf{SUP_{BC}} = \mathsf{SEM_{BC}}$.

Proof. For the "only if" direction, let F be a formula, C a clause that is super-blocked in F, and τ a satisfying assignment of $env_F(C)$. If τ satisfies C, or C contains a literal l with $var(l) \notin var(F)$ (implying that τ can be straightforwardly extended to a satisfying assignment of C), then it trivially follows that C is semantically blocked in F. Assume thus that $var(C) \subseteq var(F)$ and that τ does not satisfy C. Furthermore, let τ_E be obtained from τ by restricting it to the external variables $ext_F(C)$. Since C is super-blocked in F, there exists a non-empty set $L \subseteq C$ that blocks C in $F|\tau_E$. Consider the following assignment:

$$\tau'(v) = \begin{cases} 0 & \text{if} \neg v \in L, \\ 1 & \text{if} v \in L, \\ \tau(v) & \text{otherwise.} \end{cases}$$

Since τ falsifies C there is no literal l with $l, \bar{l} \in L$, hence τ' is well-defined. Clearly, τ' satisfies C and $\tau'(v) = \tau(v)$ for all $v \notin var(C)$. It remains to show that τ' satisfies $env_F(C)$. Since τ' differs from τ only on the truth values of variables in $var(L)$, τ' can only falsify clauses containing a literal $\bar{l}$ with $l \in L$. Let C' be such a clause. We proceed by a case distinction.

CASE 1: C' contains an external literal l (i.e., $var(l) \in ext_F(C)$) that is satisfied by τ. Then, since $var(l) \notin var(C)$ and thus $l \notin L$, it follows that τ' agrees with τ on the truth value of l and thus l is satisfied by τ'.

CASE 2: C' does not contain an external literal that is satisfied by τ. In this case, C' is contained in $F|\tau_E$ and thus, since L set-blocks C in $F|\tau_E$, we have that $(C \setminus L) \cup \bar{L} \cup C'$ is a tautology. If C' is a tautology, then it is easily satisfied by τ', so assume that it is not a tautology. Clearly, since C is not a tautology, we have that $(C \setminus L) \cup \bar{L}$ is not a tautology, hence there are two literals $l, \bar{l}$ such that $l \in C'$ and $\bar{l}$ is in $C \setminus L$ or in $\bar{L}$. If $\bar{l} \in C \setminus L$, then τ' agrees with τ on $\bar{l}$, hence $\bar{l}$ is falsified by τ' and thus l is satisfied by τ'. In contrast, if $\bar{l} \in \bar{L}$, then $l \in L$ and thus l is satisfied by τ'. In both cases τ' satisfies l and thus C'.

For the "if" direction, let F be a formula and C a clause that is not super-blocked in F, i.e., there exists an assignment τ_E over the external variables, $ext_F(C)$, such that C is not set-blocked in $F|\tau_E$. Then, let

$$\tau(v) = \begin{cases} 1 & \text{if} \neg v \in C, \\ 0 & \text{if} v \in C, \\ \tau_E(v) & \text{otherwise.} \end{cases}$$

Clearly, τ is well-defined since C cannot be a tautology, for otherwise it would be set-blocked in $F|\tau_E$. Furthermore, τ falsifies C and since (by definition) every clause $C' \in env_F(C)$ contains a literal $\bar{l}$ such that $l \in C$ it satisfies $env_F(C)$.

Now let τ' be a satisfying assignment of C such that $\tau'(v) = \tau(v)$ for all $v \notin var(C)$. As τ' satisfies C, it is obtained from τ by flipping the truth values of some literals $L \subseteq C$. We show that τ' does not satisfy $env_F(C)$. Clearly, τ' agrees with τ_E over the external variables $ext_F(C)$ and since C is not set-blocked in $F|\tau_E$, there exists a clause $C' \in F|\tau_E$ with $C' \cap \bar{L} \neq \emptyset$ such that $(C \setminus L) \cup \bar{L} \cup C'$ is not a tautology and neither τ_E nor τ' satisfy any external literal in C'.

Let $l \in C'$ be a (local) literal with $var(l) \in var(C)$. Since $(C \setminus L) \cup \bar{L} \cup C'$ is not a tautology it follows that $\bar{l} \notin C \setminus L$ and $\bar{l} \notin \bar{L}$. Since $var(l) \in var(C)$ we get that $l \in C \setminus L$ or $l \in \bar{L}$. In both cases, l is not satisfied by τ'. Thus, no literal in C' is satisfied by τ' and consequently τ' does not satisfy $C' \in env_F(C)$, which then allows to conclude that C is not semantically blocked in F. □

Corollary 9. $\mathsf{SET_{BC}}$ *is a (local) redundancy property.*

6 Complexity Analysis

In this section, we analyze the complexity of testing whether a clause is set-blocked or super-blocked. We further consider the complexity of testing restricted variants of set-blocking and super-blocking which gives rise to a whole family of blocking notions. Note that all complexity results are w.r.t. the size of a clause and its resolution environment.

Definition 7. *The* set-blocking problem *is the following decision problem: Given a pair (F, C), where F is a set of clauses and C a clause such that every $C' \in F$ contains a literal $\bar{l}$ with $l \in C$, is C set-blocked in F?*

Theorem 10. *The set-blocking problem is* NP*-complete.*

Proof. We first show NP-membership followed by NP-hardness.

NP-MEMBERSHIP: For a non-empty set $L \subseteq C$, it can be checked in polynomial time whether $(C \setminus L) \cup \bar{L} \cup C'$ is a tautology for every C' with $C' \cap \bar{L} \neq \emptyset$. The following is thus an NP-procedure: Guess a non-empty set $L \subseteq C$ and check if it blocks C in F.

NP-HARDNESS (*Proof Sketch*): We give a reduction from SAT by defining the following reduction function on input formula F which is w.l.o.g. in CNF:

$$f(F) = (F', C), \text{ with } C = (u \vee x_1 \vee x_1' \vee \cdots \vee x_n \vee x_n'),$$

where $var(F) = \{x_1, \ldots, x_n\}$ and $u, x_1', \ldots, x_n'$ are new variables that do not occur in F. Furthermore, F' is obtained from F by

- replacing every clause $D \in F$ by a clause $t(D)$ obtained from D by adding $\neg u$ and replacing every negative literal $\neg x_i$ by the positive literal x_i', and
- adding the clauses $(\neg x_i \vee \neg x_i'), (\neg x_i \vee u), (\neg x_i' \vee u)$ for every $x_i \in var(F)$.

The intuition behind the construction of F' and C is as follows. By including u in C and adding $\neg u$ to every $t(D)$ with $D \in F$, we guarantee that all clauses in F' contain a literal l with $\bar{l} \in C$. This makes (F', C) a valid instance of the set-blocking problem. The main idea, however, is, that blocking-sets L of C in F' correspond to satisfying assignments τ of F.

An assignment τ, obtained from a blocking set L by defining $\tau(x_i) = 1$ if $x_i \in L$ and $\tau(x_i) = 0$ otherwise, satisfies F because of the following:

1. Since all $C' = t(D)$ with $D \in F$, as well as C, contain—apart from $\neg u$—only positive literals, $(C \setminus L) \cup \bar{L} \cup C'$ is only a tautology if L contains a literal of C'. Now, the clauses $(\neg x_i \vee u), (\neg x_i' \vee u)$ force u to be contained in L and thus L must contain a literal $l \neq \neg u$ of every $t(D)$ with $D \in F$.
2. The reason why negative literals $\neg x_i$ are replaced by positive literals x_i' is as follows: If C were of the form $(u \vee x_1 \vee \neg x_1 \vee \cdots \vee x_n \vee \neg x_n)$, it would be trivially blocked by every set L containing two complementary literals $x_i, \neg x_i$. Hence, satisfying assignments would not correspond to blocking sets.
3. The clauses $(\neg x_i \vee \neg x_i')$ guarantee that x_i and x_i' cannot both be contained in L. Since L contains a literal of every $t(D)$, it is thus guaranteed that τ satisfies every $D \in F$: If L contains a positive literal $x_i \in t(D)$, then $x_i \in D$ is satisfied by τ. If L contains a negative literal $x_i' \in t(D)$, then $x_i \notin L$, hence $\tau(x_i) = 0$ and thus $\neg x_i \in D$ is satisfied by τ.

Similarly, one can show that every set L, obtained from a satisfying assignment τ of F by defining $L = \{u\} \cup \{x_i \mid \tau(x_i) = 1\} \cup \{x_i' \mid \tau(x_i) = 0\}$, blocks C in F'. □

We next analyze the complexity of testing whether a clause is super-blocked. To do so, we define the following problem:

Definition 8. *The* super-blocking problem *is the following decision problem: Given a pair (F, C), where F is a set of clauses and C a clause such that every $C' \in F$ contains a literal $\bar{l}$ with $l \in C$, is C super-blocked in F?*

Theorem 11. *The super-blocking problem is Π_2^P-complete.*

Proof. Again, we first show Π_2^P-membership followed by Π_2^P-hardness.

Π_2^P-MEMBERSHIP: The following is a Σ_2^P-procedure for testing whether C is *not* super-blocked in F: Guess an assignment τ over the external variables, $ext_F(C)$, and ask an NP-oracle whether C is set-blocked in $F|\tau$. If the oracle answers *no*, then return *yes*, otherwise return *no*.

Π_2^P-HARDNESS (*Proof Sketch*): We give a reduction from $\forall\exists$-SAT to the super-blocking problem. Let $\phi = \forall X \exists Y F$ be an instance of $\forall\exists$-SAT and assume w.l.o.g. that F is in CNF. We define the reduction function

$$f(\phi) = (F', C), \text{ with } C = (u \vee y_1 \vee y_1' \vee \cdots \vee y_n \vee y_n'),$$

where $Y = \{y_1, \ldots, y_n\}$ and $u, y_1', \ldots, y_n'$ are new variables not occurring in ϕ. Furthermore, F' is obtained from F by

- replacing every clause $D \in F$ by a clause $t(D)$ which is obtained from D by adding $\neg u$ and replacing every negative literal $\neg y_i$ by the positive literal y_i' for $y_i \in Y$; and by
- adding the clauses $(\neg y_i \vee \neg y_i'), (\neg y_i \vee u), (\neg y_i' \vee u)$ for every $y_i \in Y$.

As super-blocking coincides with semantic blocking, we show that ϕ is satisfiable if and only if C is semantically blocked in F'.

The reduction is similar to the one used for proving Theorem 10. Here, however, only the existentially quantified variables of ϕ are encoded into C, hence all $x_i \in X$ are external variables.

For the "only if" direction, we assume that ϕ is satisfiable and that we are given some arbitrary satisfying assignment τ of F'. By restricting τ to the variables in X we can then obtain an assignment σ_X over the variables in X. Since ϕ is satisfiable, there exists an assignment σ_Y over the variables in Y such that $\sigma_X \cup \sigma_Y$ satisfies F. From this we can in turn obtain a satisfying assignment τ' of $F' \cup \{C\}$ by defining $\tau'(x_i) = \sigma_X$ for $x_i \in X$, $\tau'(y_i) = \sigma_Y(y_i)$ and $\tau'(y_i') = 1 - \sigma_Y(y_i)$ for $y_i \in Y$, and finally $\tau'(u) = 1$. Since τ' differs from τ only on variables in $var(C)$, C is semantically blocked in F'.

Likewise, for showing the "if" direction, we assume that C is semantically blocked in F' and that we are given some arbitrary assignment σ_X over the variables in X. The crucial observation is then that for σ_X we can construct an assignment τ that satisfies F', by defining $\tau(x_i) = \sigma_X(x_i)$ for all $x_i \in X$ and $\tau(v) = 0$ for all $v \in C$. The assignment τ satisfies F' since every $C' \in F'$ contains a literal $\bar{l}$ with $l \in C$. Then, since C is semantically blocked in F', there exists a satisfying assignment τ' of $F' \cup \{C\}$ that corresponds with σ_X over X. Since $(\neg y_i \vee u)$ and $(\neg y_i' \vee u)$ are in F' for every $y_i \in Y$, it is also guaranteed that u must be satisfied by τ' and thus τ' satisfies a literal $l \neq \neg u$ in every $t(D)$ with $D \in F$. Finally, an assignment σ_Y over the variables in Y can be obtained by defining $\sigma_Y(y_i) = 1$ if and only if $\tau'(y_i) = 1$. Then, $\sigma_X \cup \sigma_Y$ is a satisfying assignment of F. □

We have already seen that the set-blocking problem is NP-complete in the general case. However, a restricted variant of set-blocking is obtained by only allowing blocking sets whose size is bounded by a constant. Then, the resulting problem of testing whether a clause C is blocked by some non-empty set $L \subseteq C$, whose size is at most k for $k \in \mathbb{N}^+$, turns out to be polynomial: For a finite set C and $k \in \mathbb{N}^+$, there are only polynomially many non-empty subsets $L \subseteq C$ with $|L| \leq k$. To see this, observe (by basic combinatorics) that the exact number of such subsets is given by the following sum which reduces to a polynomial with degree at most k:

$$\sum_{i=1}^{k} \binom{|C|}{i}.$$

Hence, the number of non-empty subsets $L \subseteq C$ with $|L| \leq k$ is polynomial in the size of C. This line of argumentation is actually very common. For the sake of completeness, however, we provide the following example:

Example 5. Let $|C| = n$ and $k = 3$ (with $k \leq n$). Then, the number of non-empty subsets $L \subseteq C$ with $|L| \leq k$ is given by the polynomial $\sum_{i=1}^{3} \binom{n}{i} = \frac{1}{6}n^3 + \frac{5}{6}n$ of degree $k = 3$. □

Now, as there are only polynomially many potential blocking sets and since it can be checked in polynomial time whether a given set $L \subseteq C$ blocks C in F (as argued in the proof of Theorem 10), it can be checked in polynomial time whether for some clause C there exists a blocking set L of size at most k.

Since the definition of super-blocking is based on the definition of set-blocking, one can also consider the complexity of restricted versions of super-blocking where the size of the according blocking sets is bounded by a constant. We thus define an infinite number of decision problems (one for every $k \in \mathbb{N}^+$) as follows:

Definition 9. *For any $k \in \mathbb{N}^+$, the* k-super-blocking problem *is the following decision problem: Given a pair (F, C), where F is a set of clauses and C a clause such that every $C' \in F$ contains a literal $\bar{l}$ with $l \in C$, does it hold that, for every assignment τ over the external variables $ext_F(C)$, there exists a non-empty set $L \subseteq C$ with $|L| \leq k$ that blocks C in $F|\tau$?*

Theorem 12. *The k-super-blocking problem is in* co-NP *for all $k \in \mathbb{N}^+$.*

Proof. Consider the statement that has to be tested for the complement of the k-super-blocking problem:

> There exists an assignment τ over the external variables $ext_F(C)$ such that no non-empty subset of C with $|C| \leq k$ blocks C in $F|\tau$.

Since it can be checked in polynomial time whether a given set $L \subseteq C$ blocks C in $F|\tau$, the following is an NP-procedure:

> Guess an assignment τ over the external variables $ext_F(C)$ and check for every non-empty subset of C (with $|C| \leq k$) whether it blocks C in $F|\tau$. If there is one, return *no*, otherwise return *yes*.

Hence, for every integer $k \in \mathbb{N}^+$, the k-super-blocking problem is in co-NP. □

Hardness for the complexity class co-NP can be shown already for $k = 1$.

Theorem 13. *The 1-super-blocking problem is* co-NP*-hard.*

Proof. By a reduction from the unsatisfiability problem of propositional logic. Let $F = \{C_1, \ldots, C_n\}$ be a formula in CNF and define the reduction function

$$f(F) = (F', C), \text{ with } C = (u_1 \vee \cdots \vee u_n),$$

where $u_1, \ldots, u_n$ are new variables that do not occur in F, and $F' = \bigcup_{i=1}^n F_i$ with $F_i = \{(\neg u_i \vee \bar{l}) \mid l \in C_i\}$. Clearly, (F', C) is a valid instance of the 1-super-blocking problem and $var(F) = ext_{F'}(C)$. We show that F is unsatisfiable if and only if, for every assignment τ over $ext_{F'}(C)$, there exists a $u_i \in C$ such that $\{u_i\}$ set-blocks C in $F'|\tau$.

For the "only if" direction, assume that F is unsatisfiable and let τ be an assignment over $ext_{F'}(C)$. Since $var(F) = ext_{F'}(C)$ it follows that there exists a clause C_i in F that is falsified by τ. But then, since every clause in F_i contains a literal $\bar{l}$ with $l \in C_i$, it follows that F_i is satisfied by τ. Hence, $F_i \cap F'|\tau = \emptyset$ and thus, since $\neg u_i$ only occurs in F_i, $\{u_i\}$ trivially set-blocks C in F'.

For the "if" direction, assume that for every τ over $ext_{F'}(C)$, there exists a $u_i \in C$ such that $\{u_i\}$ set-blocks C in $F'|\tau$. Since $var(F) = ext_{F'}(C)$ it follows that for every assignment τ of F and every clause $(\neg u_i \vee \bar{l}) \in F'|\tau$ (with $l \in C_i$), $T = (C \setminus \{u_i\}) \cup \{\neg u_i\} \cup \{\neg u_i, \bar{l}\}$ is a tautology. But since T cannot contain complementary literals it must be the case that $(\neg u_i \vee \bar{l}) \notin F'|\tau$ which implies that every $l \in C_i$ is falsified by τ. It follows that F is unsatisfiable. □

Corollary 14. *The k-super-blocking problem is* co-NP*-complete for all $k \in \mathbb{N}^+$.*

The notions of set-blocking and super-blocking, together with the corresponding restrictions discussed in this section, give rise to a whole family of blocking notions which differ in both generality and complexity. We conclude the following: (i) Considering the assignments over external variables (as is the case for super-blocking) leads to co-NP-hardness. (ii) If blocking sets of arbitrary size are considered, the (sub-)problem of checking whether there exists a blocking set is NP-hard. (iii) If the size of blocking sets is bounded by a constant k, the (sub-)problem of testing whether there exists a blocking set turns out to be polynomial. (iv) The problem of testing whether a clause is super-blocked in the most general sense, where the size of blocking sets is not bounded by a constant, is Π_2^P-complete. Hence, we can summarize the following complexity results:

	$\|L\|$ is unrestricted	$\|L\| \leq k$ for $k \in \mathbb{N}^+$
Super-blocking	Π_2^P-complete	co-NP-complete
Set-blocking	NP-complete	P

Note that the cardinality $|L|$ of blocking sets is of course bounded by the length of the clauses, thus we can restrict $|L| \leq |C|$. This is particularly interesting for formula instances with (uniform) constant or maximal clause length.

Finally, we conclude the discussion by returning to the starting point of this paper: literal-blocked clauses. Obviously, we can write the definition for set-blocking with $|L| \leq 1$ as follows: A set $\{l\} \subseteq C$ *blocks* a clause C in a formula F if for each clause $C' \in F$ with $\bar{l} \in C'$, $(C \setminus \{l\}) \cup C'$ is a tautology. (Note that we write $(C \setminus \{l\}) \cup C'$ instead of $(C \setminus \{l\}) \cup \{\bar{l}\} \cup C'$ since $\bar{l}$ is anyhow required to be contained in C'.) This is very similar to the original definition of literal-blocked clauses which requires $C \cup (C' \setminus \{l\})$ to be a tautology.

7 Comparison with Other Redundancy Properties

In the following, we consider several local and non-local redundancy properties as presented in [7] and relate them to the previously discussed local redundancy properties. From the three basic redundancy properties *tautology* (T), *subsumption* (S), and *literal-blocked clauses* (BC), extended redundancy properties are derived as follows.

Given a formula F and a clause C, $\mathsf{ALA}(F, C)$ is the *unique* clause obtained from C by repeating *asymmetric literal addition*, as defined in the following, until a fixed point is reached: If $l_1, \ldots, l_k \in C$ and there is a clause $(l_1 \vee \cdots \vee l_k \vee l) \in F \setminus \{C\}$ for some literal l, let $C := C \cup \{\bar{l}\}$. The special case where $k = 1$ is called *hidden literal addition* (HLA). Due to space limitations, we will not consider HLA separately. Given a formula F and a clause C, $(F, C) \in \mathsf{AT}$ (resp., AS or ABC) if $(F, \mathsf{ALA}(F, C)) \in \mathsf{T}$ (resp., S or BC).

Finally, we introduce the redundancy properties prefixed with R [7]. Given a formula F and a clause C, $(F, C) \in \mathsf{R}\mathcal{P}$ if either (i) $(F, C) \in \mathcal{P}$ or (ii) there is a literal l in C such that for each clause $C' \in F$ with $\bar{l} \in C'$, $(F, C \cup C' \setminus \{\bar{l}\}) \in \mathcal{P}$. Examples are RT, RS, and RAT. Especially RAT is extremely powerful, because it captures all known SAT solving techniques including preprocessing, inprocessing, and clause learning [7,15].

These notions of redundancy lead to the hierarchy depicted in Fig. 2 which we extend with the previously introduced set-blocked and super-blocked clauses. We discuss the incomparability with redundancy properties based on T in detail; incomparability with subsumption-based properties works analogously.

Proposition 15. $\mathsf{AT} \not\subseteq \mathsf{SET}_{\mathsf{BC}}$ *and* $\mathsf{SET}_{\mathsf{BC}} \not\subseteq \mathsf{AT}$.

Proof. Let $C = (a \vee b \vee c)$ and $F = \{(\neg a \vee x), (\neg b \vee x), (\neg c \vee x), (a \vee b)\}$. Since $\neg b \in \mathsf{ALA}(F, C)$, it follows that $(F, C) \in \mathsf{AT}$. Now, assume that C is set-blocked by some set $L \subseteq C$, i.e., for every C' with $C' \cap \bar{L} \neq \emptyset$, $(C \setminus L) \cup \bar{L} \cup C'$ is a tautology. Since $L \subseteq C$ is non-empty, $(\neg v \vee x) \cap \bar{L} \neq \emptyset$ for at least one $(\neg v \vee x)$ with $v \in \{a, b, c\}$. Let therefore C' be such a $(\neg v \vee x)$. Then, $v \notin (C \setminus L)$ and $v \notin \bar{L}$. Hence, $(C \setminus L) \cup \bar{L} \cup C'$ is not a tautology and thus C is not set-blocked by L, a contradiction. We conclude that $(F, C) \notin \mathsf{SET}_{\mathsf{BC}}$.

Finally, let $F = \emptyset$ and $C = (a)$. Then, $(F, C) \in \mathsf{SET}_{\mathsf{BC}}$, but $(F, C) \notin \mathsf{AT}$. □

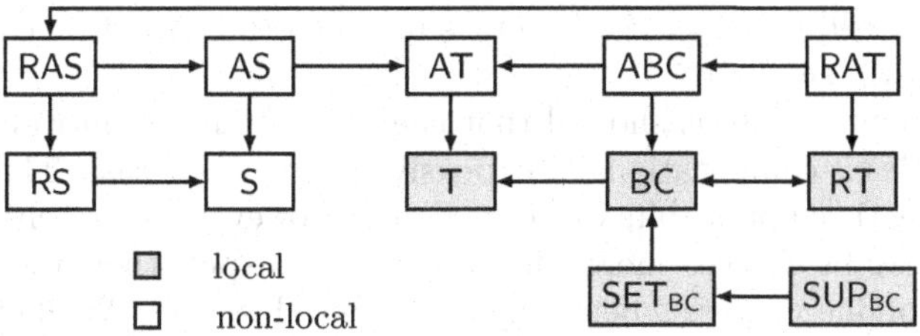

Fig. 2. Hierarchy of redundancy properties [7] extended with novel local redundancies. For redundancy properties $\mathcal{P}_1$ and $\mathcal{P}_2$, an arrow from $\mathcal{P}_1$ to $\mathcal{P}_2$ denotes that $\mathcal{P}_2 \subseteq \mathcal{P}_1$.

Proposition 16. $\mathsf{AT} \not\subseteq \mathsf{SUP_{BC}}$ *and* $\mathsf{SUP_{BC}} \not\subseteq \mathsf{AT}$.

Proof. Consider again the clause $C = (a \vee b \vee c)$ and the formula $F = \{(\neg a \vee x), (\neg b \vee x), (\neg c \vee x), (a \vee b)\}$ from the proof of Proposition 15, and observe that $ext_F(C) = \{x\}$. Here, for the assignment τ that falsifies the external variable x, $F|\tau = F$ and since C is not set-blocked in F (as shown in the proof of Proposition 15), it is not set-blocked in $F|\tau$, hence $(F, C) \notin \mathsf{SUP_{BC}}$.

To see that $\mathsf{SUP_{BC}} \not\subseteq \mathsf{AT}$, let $F = \emptyset$ and $C = (a)$. Then, since $(F, C) \in \mathsf{SET_{BC}}$ and $\mathsf{SET_{BC}} \subset \mathsf{SUP_{BC}}$, we get that $(F, C) \in \mathsf{SUP_{BC}}$ but $(F, C) \notin \mathsf{AT}$. □

From Proposition 16 together with the fact that $\mathsf{AT} \subset \mathsf{RAT}$ we get:

Corollary 17. $\mathsf{RAT} \not\subseteq \mathsf{SUP_{BC}}$.

Proposition 18. $\mathsf{SET_{BC}} \not\subseteq \mathsf{RAT}$.

Proof. Consider the clause $C = (a \vee b)$ and the formula $F = \{(a \vee b), (\neg a \vee b), (a \vee \neg b)\}$. Clearly, C is set-blocked by $L = \{a, b\}$ in F and thus $(F, C) \in \mathsf{SET_{BC}}$.

Now, for the literal a there is only the clause $C' = (\neg a \vee b)$ that contains $\neg a$ and $C \cup C' \setminus \{\neg a\} = (a \vee b)$. Furthermore, for the literal b there is only the clause $C'' = (a \vee \neg b)$ that contains $\neg b$ and here again we get that $C \cup C'' \setminus \{\neg b\} = (a \vee b)$. Since $\mathsf{ALA}(F \setminus \{C\}, (a \vee b)) = (a \vee b)$ is not a tautology, $(F, C) \notin \mathsf{RAT}$. □

Corollary 19. RAT *is incomparable with* $\mathsf{SET_{BC}}$ *and* $\mathsf{SUP_{BC}}$.

8 Conclusion

Previous research and recent SAT competitions have clearly revealed the power of solving techniques based on the redundancy property of literal-blocked clauses. One reason for the success of this redundancy property is that it is local in the sense that it can be efficiently checked by considering only the resolution environment of a clause [8,12,14]. In this paper, we showed that there are even more general local redundancy properties like set-blocked clauses ($\mathsf{SET_{BC}}$) and super-blocked clauses ($\mathsf{SUP_{BC}}$). Local redundancy properties are particularly appealing in the context of real-world verification, where problem encodings into SAT often lead to very large formulas in which the resolution environments of clauses are still small.

Our complexity analysis showed that checking the newly introduced redundancy properties is computationally expensive in the worst case. This seemingly limits their practical applicability at first glance. However, we presented bounded variants that can be checked more efficiently and we expect them to considerably improve the solving process when added to our SAT solvers. While the focus of this paper lies on the theoretical investigation of local redundancy properties, thereby contributing to gaining a deeper understanding of blocked clauses, a practical evaluation is subject to future work.

Another direction for future work is lifting the new redundancy properties to QSAT, the satisfiability problem of quantified Boolean formulas (QBF). There, literal-blocked clauses have been shown to be even more effective than in SAT solving [6,16] and we expect that this also holds for quantified variants of $\mathsf{SET}_{\mathsf{BC}}$ and $\mathsf{SUP}_{\mathsf{BC}}$.

References

1. Biere, A., Heule, M., van Maaren, H., Walsh, T. (eds.): Handbook of Satisfiability. IOS Press, Amsterdam (2009)
2. Vizel, Y., Weissenbacher, G., Malik, S.: Boolean satisfiability solvers and their applications in model checking. Proc. IEEE **103**(11), 2021–2035 (2015)
3. Barrett, C.W., Sebastiani, R., Seshia, S.A., Tinelli, C.: Satisfiability modulo theories. In: Handbook of Satisfiability, pp. 825–885. IOS Press (2009)
4. Reger, G., Suda, M., Voronkov, A.: Playing with AVATAR. In: Felty, P.A., Middeldorp, A. (eds.) Automated Deduction - CADE-25. LNCS, vol. 9195, pp. 399–415. Springer, Switzerland (2015)
5. Järvisalo, M., Biere, A., Heule, M.: Simulating circuit-level simplifications on CNF. J. Autom. Reasoning **49**(4), 583–619 (2012)
6. Heule, M., Järvisalo, M., Lonsing, F., Seidl, M., Biere, A.: Clause elimination for SAT and QSAT. J. Artif. Intell. Res. **53**, 127–168 (2015)
7. Järvisalo, M., Heule, M.J.H., Biere, A.: Inprocessing rules. In: Gramlich, B., Miller, D., Sattler, U. (eds.) IJCAR 2012. LNCS, vol. 7364, pp. 355–370. Springer, Heidelberg (2012)
8. Manthey, N., Philipp, T., Wernhard, C.: Soundness of inprocessing in clause sharing SAT solvers. In: Järvisalo, M., Van Gelder, A. (eds.) SAT 2013. LNCS, vol. 7962, pp. 22–39. Springer, Heidelberg (2013)
9. Heule, M.J.H., Biere, A.: Blocked clause decomposition. In: McMillan, K., Middeldorp, A., Voronkov, A. (eds.) LPAR-19 2013. LNCS, vol. 8312, pp. 423–438. Springer, Heidelberg (2013)
10. Iser, M., Manthey, N., Sinz, C.: Recognition of nested gates in CNF formulas. In: Heule, M., Weaver, S. (eds.) SAT 2015. LNCS, vol. 9340, pp. 255–271. Springer, Heidelberg (2015). doi:10.1007/978-3-319-24318-4_19
11. Balyo, T., Fröhlich, A., Heule, M.J.H., Biere, A.: Everything you always wanted to know about blocked sets (but were afraid to ask). In: Sinz, C., Egly, U. (eds.) SAT 2014. LNCS, vol. 8561, pp. 317–332. Springer, Heidelberg (2014)
12. Chen, J.: Fast blocked clause decomposition with high quality (2015). CoRR abs/1507.00459

13. Kullmann, O.: On a generalization of extended resolution. Discrete Appl. Math. **96–97**, 149–176 (1999)
14. Järvisalo, M., Biere, A., Heule, M.: Blocked clause elimination. In: Esparza, J., Majumdar, R. (eds.) TACAS 2010. LNCS, vol. 6015, pp. 129–144. Springer, Heidelberg (2010)
15. Heule, M.J.H., Hunt Jr., W.A., Wetzler, N.: Verifying refutations with extended resolution. In: Bonacina, M.P. (ed.) CADE 2013. LNCS, vol. 7898, pp. 345–359. Springer, Heidelberg (2013)
16. Lonsing, F., Bacchus, F., Biere, A., Egly, U., Seidl, M.: Enhancing search-based QBF solving by dynamic blocked clause elimination. In: Davis, M., Fehnker, A., McIver, A., Voronkov, A. (eds.) LPAR-20 2015. LNCS, vol. 9450, pp. 418–433. Springer, Heidelberg (2015). doi:10.1007/978-3-662-48899-7_29

Satisfiability Modulo Theory

Counting Constraints in Flat Array Fragments

Francesco Alberti[1], Silvio Ghilardi[2(✉)], and Elena Pagani[2]

[1] Fondazione Centro San Raffaele, Milano, Italy
[2] Università Degli Studi di Milano, Milano, Italy
silvio.ghilardi@unimi.it

Abstract. We identify a fragment of Presburger arithmetic enriched with free function symbols and cardinality constraints for interpreted sets, which is amenable to automated analysis. We establish decidability and complexity results for such a fragment and we implement our algorithms. The experiments run in discharging proof obligations coming from invariant checking and bounded model-checking benchmarks show the practical feasibility of our decision procedure.

1 Introduction

Enriching logic formalisms with counting capabilities is an important task in view of the needs of many application areas, ranging from database theory to formal verification. Such enrichments have been designed both in the description logics area and in the area of Satisfiability Modulo Theories (SMT), where some of the most important recent achievements were decidability and complexity bounds for BAPA [14] - the enrichment of Presburger arithmetic with the ability of talking about finite sets and their cardinalities. As pointed out in [15], BAPA constraints can be used for program analysis and verification by expressing data structure invariants, simulations between program fragments or termination conditions. The analysis of BAPA constraints was successfully extended also to formalisms encompassing multisets [18] as well as direct/inverse images along relations and functions [23].

A limitation of BAPA and its extensions lies in the fact that only uninterpreted symbols (for sets, relations, functions, etc.) are allowed. On the other hand, it is well-known that a different logical formalism, namely unary counting quantifiers, can be used in order to reason about the cardinality of definable (i.e. of interpreted) sets. Unary counting quantifiers can be added to Presburger arithmetic without compromising decidability, see [19], however they might be quite problematic if combined in an unlimited way with free function symbols. In this paper, we investigate the extension of Presburger arithmetic including both counting quantifiers and uninterpreted function symbols, and we isolate fragments where we can achieve decidability and in some cases also relatively good complexity bounds. The key ingredient to isolate such fragments is the notion of flatness: roughly, in a flat formula, subterms of the kind $a(t)$ (where a is a free function symbol) can occur only if t is a variable. By itself, this naif

N. Olivetti and A. Tiwari (Eds.): IJCAR 2016, LNAI 9706, pp. 65–81, 2016.
DOI: 10.1007/978-3-319-40229-1_6

flatness requirement is useless (any formula can match it to the price of introducing extra quantified variables); in order to make it effective, further syntactic restrictions should be incorporated in it, as witnessed in [2]. This is what we are going to do in this paper, where suitable notions of 'flat' and 'simple flat' formulæ are introduced in the rich context of Presburger arithmetic enriched with free function symbols and with unary counting quantifiers (we use free function symbols to model arrays, see below).

The fragments we design are all obviously more expressive than BAPA, but they do not come from pure logic motivations, on the contrary they are suggested by an emerging application area, namely the area of verification of fault-tolerant distributed systems. Such systems (see [8] for a good account) are modeled as partially synchronous systems, where a finite number of identical processes operate in lock-step (in each *round* they send messages, receive messages, and update their local state depending on the local state at the beginning of the round and the received messages). Messages can be lost, processes may omit to perform some tasks or also behave in a malicious way; for these reasons, the fact that some actions are enabled or not, and the correctness of the algorithms themselves, are subject to threshold conditions saying for instance that some qualified majority of processes are in a certain status or behave in a non-faulty way. Verifications tasks thus have to handle cardinality constraints of the kind studied in this paper (the reader interested in full formalization examples can directly go to Sect. 5).

The paper is organized as follows: we first present basic syntax (Sect. 2), then decidability (Sect. 3) and complexity (Sect. 4) results; experiments with our prototypical implementation are supplied in Sect. 5, and Sect. 6 concludes the work.

2 Preliminaries

We work within Presburger arithmetic enriched with free function symbols and cardinality constraints. This is a rather expressive logic, whose syntax is summarized in Fig. 1. Terms and formulæ are interpreted in the natural way over the domain of integers $\mathbb{Z}$; as a consequence, satisfiability of a formula ϕ means that it is possible to assign values to parameters, free variables and array-ids so as to make ϕ true in $\mathbb{Z}$ (validity of ϕ means that $\neg\phi$ is not satisfiable, equivalence of ϕ and ψ means that $\phi \leftrightarrow \psi$ is valid, etc.). We nevertheless implicitly assume few constraints (to be explained below) about our intended semantics.

To denote integer numbers, we have (besides variables and numerals) also parameters: the latter denote unspecified integers. Among parameters, we always include a specific parameter (named N) identifying the dimension of the system - alias the length of our arrays: in other words, it is assumed that for all array identifiers $a \in Arr$, the value $a(x)$ is conventional (say, zero) outside the interval $[0, N) = \{n \in \mathbb{Z} \mid 0 \leq n < N\}$. Although binary free function symbols are quite useful in some applications, in this paper we prefer not to deal with them. The operator $\sharp\{x \mid \phi\}$ *indicates the cardinality of the finite set formed by the* $x \in [0, N)$ *such that* $\phi(x)$ *holds.*

$0, 1, \ldots$		$\in \mathbb{Z}$	numerals (numeric constants)
$x, y, z, \ldots$		$\in Var$	individual variables
$M, N, \ldots$		$\in Par$	parameters (free constants)
$a, b, \ldots$		$\in Arr$	array ids (free unary function symbols)
$t, u, \ldots$	$::=$	$n \mid M \mid x \mid t + t \mid -t \mid a(t) \mid \sharp\{x \mid \phi\}$	terms
$A, B, \ldots$	$::=$	$t < t \mid t = t \mid t \equiv_n t$	atoms
$\phi, \psi, \ldots$	$::=$	$A \mid \phi \wedge \phi \mid \neg\phi \mid \exists x\, \phi$	formulae

Fig. 1. Syntax

Notice that the cardinality constraint operator $\sharp\{x \mid -\}$, as well the quantifier $\exists x$, bind the variable x; below, we indicate with $\psi(\underline{x})$ (resp. $t(\underline{x})$) the fact that the formula ψ (the term t) has free individual variables included in the list $\underline{x}$. When we speak of a substitution, we always mean 'substitution without capture', meaning that, when we replace the free occurrences of a variable x with a term u in a formula ϕ or in a term t, the term u should not contain free variables that might be located inside the scope of a binder for them once the substitution is performed; the result of the substitution is denoted with $\phi(u/x)$ and $t(u/x)$.

The logic of Fig. 1 is far from being tractable, because even the combination of free function symbols and Presburger arithmetic lands in a highly undecidable class [10]. We are looking for a mild fragment, nevertheless sufficiently expressive for our intended applications. These applications mostly come from verification tasks, like bounded model checking or invariant checking. Our aim is to design a decidable fragment (so as to be able not only to produce certifications, but also to find bugs) with some minimal closure properties; from this point of view, notice that for bounded model checking closure under conjunctions is sufficient, but for invariant checking we need also closure under negations in order to discharge entailments.

2.1 Flat Formulæ

We now introduce some useful subclasses of the formulæ built up according to the grammar of Fig. 1 (all subclasses are closed under Boolean operations):

- *Arithmetic formulæ* : these are built up from the grammar of Fig. 1 without using neither array-ids nor cardinality constraint operators; we use letter $\alpha, \beta, \ldots$ for arithmetic formulæ. Recall that, according to the well-known quantifier elimination result, arithmetic formulæ are equivalent to *quantifier-free* arithmetic formulæ.
- *Constraint formulæ* : these are built up from the grammar of Fig. 1 without using array-ids.
- *Basic formulæ* : these are obtained from an arithmetic formula by simultaneously replacing some free variables by terms of the kind $a(y)$, *where* y *is a variable and* a *an array-id.* When we need to display full information,

we may use the notation $\alpha(\underline{y}, \mathbf{a}(\underline{y}))$ to indicate basic formulæ. By this notation, we mean that $\underline{y} = y_1, \ldots, y_n$ are variables, $\mathbf{a} = a_1, \ldots, a_s$ are array-ids and that $\alpha(\underline{y}, \mathbf{a}(\underline{y}))$ is obtained from an arithmetic formula $\alpha(\underline{x}, \underline{z})$ (where $\underline{z} = z_{11}, \ldots, z_{sn}$) by replacing z_{ij} with $a_i(y_j)$ ($i = 1, \ldots, s$ and $j = 1, \ldots, n$).
- *Flat formulæ* : these are recursively defined as follows (i) basic formulæ are flat formulæ; (ii) if ϕ is a flat formula, β is a basic formula, z and x are variables, then $\phi(\sharp\,\{x \mid \beta\}\,/\,z)$ is a flat formula. Thus in flat formulæ all dereferenced indexes are either free or the ones defining the comprehension.[1]

Example 1. The formulæ $a(y) + a(z) \leq z$ and $z = \sharp\,\{x \,|\, \sharp\,\{x' \,|\, a(x') < 1\} = a(x)\}$ are flat (the former also basic) whereas $z = \sharp\,\{x \,|\, \sharp\,\{x' \,|\, a(x') < x\} = a(x)\}$ is not such (the binder $\sharp\,\{x \,|\, \cdots$ captures a free occurrence of x in $\sharp\,\{x' \,|\, a(x') < x\}$).

The following result is proved in [19] (see also the Appendix of [1]):

Theorem 1. *For every constraint formula one can compute an arithmetic formula equivalent to it.*

3 Satisfiability for Flat formulæ

We shall show that flat formulæ are decidable for satisfiability. In fact, we shall show decidability of the slightly larger class covered by the following

Definition 1. *Extended flat formulæ (briefly, E-flat formulæ) are formulæ of the kind*

$$\exists \underline{z}.\ \alpha\ \wedge\ \sharp\{x \mid \beta_1\} = z_1\ \wedge \cdots \wedge\ \sharp\{x \mid \beta_K\} = z_K \tag{1}$$

where $\underline{z} = z_1, \ldots, z_K$ and $\alpha, \beta_1, \ldots, \beta_K$ are basic formulæ and x does not occur in α.

Notice that α and the β_j in (1) above may contain further free variables $\underline{y}$ (besides $\underline{z}$) as well as the terms $\mathbf{a}(\underline{y})$ and $\mathbf{a}(\underline{z})$; the β_j may contain occurrences of x and of $\mathbf{a}(x)$.

That flat formulæ are also E-flat can be seen as follows: due to the fact that our substitutions avoid captures, we can use equivalences like $\phi(t/z) \leftrightarrow \exists z\,(t = z \wedge \phi)$ in order to abstract out the terms $t := \sharp\,\{x \,|\, \alpha\}$ occurring in the recursive construction of a flat formula ϕ. By repeating this linear time transformation, we end up in a formula of the kind (1). However, not all E-flat formulæ are flat because the dependency graph associated to (1) might not be acyclic (the graph we are talking about has the z_j as nodes and has an arc $z_j \to z_i$ when z_i occurs in β_j). The above conversion of a flat formula into a formula of the form (1) on the other hand produces an E-flat formula whose associated graph is acyclic.

We prove a technical lemma showing how we can manipulate E-flat formulæ without loss of generality. Formulae $\varphi_1, \ldots, \varphi_K$ are said to be a partition iff the

[1] If we want to emphasize the way the basic formula β is built up, following the above conventions, we may write it as $\beta(x, \underline{y}, \mathbf{a}(x), \mathbf{a}(\underline{y}))$; here, supposing that $\mathbf{a}$ is $a_1, \ldots, a_s$, since x is a singleton, the tuple $\mathbf{a}(x)$ is $a_1(x), \ldots, a_s(x)$.

formulæ $\bigvee_{l=1}^{K} \varphi_l$ and $\neg(\varphi_l \wedge \varphi_h)$ (for $h \neq l$) are valid. Recall that the existential closure of a formula is the sentence obtained by prefixing it with a string of existential quantifiers binding all variables having a free occurrence in it.

Lemma 1. *The existential closure of an E-flat formula is equivalent to a sentence of the kind*

$$\exists \underline{z}\, \exists \underline{y}.\ \alpha(\underline{y}, \underline{z}) \wedge \sharp\{x \mid \beta_1(x, \mathbf{a}(x), \underline{y}, \underline{z})\} = z_1 \wedge \cdots \wedge \sharp\{x \mid \beta_K(x, \mathbf{a}(x), \underline{y}, \underline{z})\} = z_K \tag{2}$$

where $\underline{y}$ and $\underline{z} := z_1, \ldots, z_K$ are variables, α is arithmetical, and the formulæ $\beta_1, \ldots, \beta_K$ are basic and form a partition.

Proof. The differences between (the matrices of) (2) and (1) are twofold: first in (2), the β_l form a partition and, second, in (1) the terms $a_s(y_i)$ and $a_s(z_h)$ (for $a_s \in \mathbf{a}$ and $y_i \in \underline{y}$, $z_h \in \underline{z}$) may occur in α and in the β_l.

We may disregard the $a_s(z_h)$ without loss of generality, because we can include them in the $a_s(y_i)$: to this aim, it is sufficient to take a fresh y, to add the conjunct $y = z_h$ to α and to replace everywhere $a_s(z_h)$ by $a_s(y)$. In order to eliminate also a term like $a_s(y_i)$, we make a guess and distinguish the case where $y_i \geq N$ and the case where $y_i < N$ (formally, 'making a guess' means to replace (1) with a disjunction - the two disjuncts being obtained by adding to α the case description). According to the semantics conventions we made in Sect. 2, the first case is trivial because we can just replace $a_s(y_i)$ by 0. In the other case, we first take a fresh variable u and apply the equivalence $\gamma(\ldots a_s(y_j) \ldots) \leftrightarrow \exists u\, (a_s(y_j) = u \wedge \gamma(\ldots u \ldots))$ (here γ is the whole (1)); then we replace $a_s(y_j) = u$ by the equivalent formula $\sharp\{x \mid x = y_j \wedge a_s[x] = u\} = 1$ and finally the latter by $\exists u'\, (u' = 1 \wedge \sharp\, \{x \mid x = y_j \wedge a_s[x] = u\} = u')$ (the result has the desired shape once we move the new existential quantifiers in front).

After this, we still need to modify the β_l so that they form a partition (this further step produces an exponential blow-up). Let $\psi(\underline{y})$ be the matrix of a formula of the kind (2), where the β_l are not a partition. Let us set $\underline{K} := \{1, \ldots, K\}$ and let us consider further variables $\underline{u} = \langle u_\sigma \rangle_\sigma$, for $\sigma \in 2^{\underline{K}}$. Then it is clear that the existential closure of ψ is equivalent to the formula obtained by prefixing the existential quantifiers $\exists \underline{u}\, \exists \underline{z}$ to the formula

$$\left(\alpha \wedge \bigwedge_{l=1}^{K} z_l = \sum_{\sigma \in 2^{\underline{K}},\ \sigma(l)=1} u_\sigma \right) \wedge \bigwedge_{\sigma \in 2^{\underline{K}}} \sharp\{x \mid \beta_\sigma\} = u_\sigma \tag{3}$$

where $\beta_\sigma ;= \bigwedge_{l=1}^{K} \epsilon_{\sigma(l)} \beta_l$ (here $\epsilon_{\sigma(l)}$ is '$\neg$' if $\sigma(l) = 0$, it is a blank space otherwise). ⊣

Theorem 2. *Satisfiability of E-flat formulæ is decidable.*

Proof. We reduce satisfiability of (2) to satisfiability of a constraint formula (4) which is decidable by Theorem 1. The matrix of (2) has free variables $\underline{z}, \underline{y}$ and these are inherited by the equi-satisfiable formula (4), but the latter contains extra free variables $z_S, z_{l,S}$: variables z_S count new Venn regions, whereas variables $z_{l,S}$ counts how many elements are taken from z_S to contribute to the old Venn region counted by z_l. In detail, we show that (2) is equisatisfiable with

$$\alpha \wedge \bigwedge_{S \in \wp(\underline{K})} \left(z_S = \sharp\{x \mid \bigwedge_{l \in S} \exists \underline{u}\, \beta_l(x, \underline{u}, \underline{y}, \underline{z}) \wedge \bigwedge_{l \notin S} \forall \underline{u} \neg \beta_l(x, \underline{u}, \underline{y}, \underline{z})\} \right) \wedge$$
$$\wedge \bigwedge_{S \in \wp(\underline{K})} \left(z_S = \sum_{l \in S} z_{l,S} \right) \wedge \bigwedge_{l=1}^{K} \left(z_l = \sum_{S \in \wp(\underline{K}), l \in S} z_{l,S} \right) \wedge \bigwedge_{l \in S \in \wp(\underline{K})} z_{l,S} \geq 0 \tag{4}$$

(according to our notations, the basic formulæ $\beta_l(x, \mathbf{a}(x), \underline{y}, \underline{z})$ from (2) were supposed to be built up from the arithmetic formulæ $\beta_l(x, \underline{u}, \underline{y}, \underline{z})$ by replacing the variables $\underline{u} = u_1, \ldots, u_s$ with the terms $\mathbf{a}(x) = a_1 x), \ldots, a_s(x))$.

Suppose that (4) *is satisfiable.* Then there is an assignment V to the free variables occurring in it so that (4) is true in the standard structure of the integers (for simplicity, we use the same name for a free variable and for the integer assigned to it by V). If $\mathbf{a} = a_1, \ldots, a_s$, we need to define $a_s(i)$ for all s and for all $i \in [0, N)$. For every $l = 1, \ldots, K$ this must be done in such a way that there are exactly z_l integer numbers taken from $[0, N)$ satisfying $\beta_l(x, \mathbf{a}(x), \underline{y}, \underline{z})$. The interval $[0, N)$ can be partioned by associating with each $i \in [0, N)$ the set $i_S = \{l \in \underline{K} \mid \exists \underline{u}\, \beta_l(i, \underline{u}, \underline{y}, \underline{z})$ holds under $V\}$. For every $S \in \wp(\underline{K})$ the number of the i such that $i_S = S$ is z_S; for every $l \in S$, pick $z_{l,S}$ among them and, for these selected i, let the s-tuple $\mathbf{a}(i)$ be equal to an s-tuple $\underline{y}$ such that $\beta_l(i, \underline{u}, \underline{y}, \underline{z})$ holds (for this tuple $\underline{y}$, since the β_l are a partition, $\beta_h(i, \underline{u}, \underline{y}, \underline{z})$ does not hold, if $h \neq l$). Since $z_S = \sum_{l \in S} z_{l,S}$ and since $\sum_S z_S$ is equal to the length of the interval $[0, N)$ (because the formulæ $\bigwedge_{l \in S} \exists \underline{u}\, \beta_l \wedge \bigwedge_{l \notin S} \forall \underline{u} \neg \beta_l$ are a partition), the definition of the $\mathbf{a}$ is complete. The formula (2) is true by construction.

On the other hand *suppose that* (2) *is satisfiable* under an assignment V; we need to find $V(z_S)$, $V(z_{l,S})$ (we again indicate them simply as $z_S, z_{l,S}$) so that (4) is true. For z_S there is no choice, since $z_S = \sharp\{i \mid \bigwedge_{l \in S} \exists \underline{u}\, \beta_l(i, \underline{u}, \underline{y}, \underline{z}) \wedge \bigwedge_{l \notin S} \forall \underline{u} \neg \beta_l(i, \underline{u}, \underline{y}, \underline{z})\}$ must be true; for $z_{l,S}$, we take it to be the cardinality of the set of the i such that $\beta_l(i, \mathbf{a}(i), \underline{y}, \underline{z})$ holds under V and $S = \{h \in \underline{K} \mid \exists \underline{u}\ \beta_h(i, \underline{u}, \underline{y}, \underline{z})$ holds under $V\}$. In this way, for every S, the equality $z_S = \sum_{l \in S} z_{l,S}$ holds and for every l, the equality $z_l = \sum_{S \in \wp(\underline{K}), l \in S} z_{l,S}$ holds too. Thus the formula (2) becomes true under our extended V. ⊣

Example 2. Let us test the satisfiability of

$$N > 3 \wedge z_2 \equiv_5 1 \wedge z_1 = \sharp\{x \mid x + |a(x)| < 3\} \wedge z_2 = \sharp\{x \mid x + |a(x)| \geq 3\} \tag{5}$$

We have $K = 2$ and let us put $S_1 := \{1, 2\}, S_2 := \emptyset, S_3 := \{1\}, S_4 := \{2\}$. Since the absolute value is a positive number, when writing down (4), we easily realize that we must have $z_{S_1} = \sharp\{0, 1, 2\} = 3, z_{S_2} = z_{S_3} = 0, z_{S_4} = \sharp\{3, \ldots, N-1\} = N - 3$. Thus (5) is satisfiable iff there are $z_{1S_1}, z_{2S_1}, z_{2S_4} \geq 0$ such that

$$N > 3 \wedge z_2 \equiv_5 1 \wedge z_{1S_1} + z_{2S_1} = 3 \wedge z_{2S_4} = N - 3 \wedge z_1 = z_{1S_1} \wedge z_2 = z_{2S_1} + z_{2S_4}$$

which is in fact the case (but notice that an additional conjunct like $N \equiv_5 0$ would make (5) unsatisfiable).

4 A More Tractable Subcase

We saw that satisfiability of flat formulæ is decidable, but the complexity of the decision procedure is very high: Lemma 1 introduces an exponential blow-up and other exponential blow-ups are introduced by Theorem 2 and by the decision procedure (via quantifier elimination) from [19]. Of course, all this might be subject to dramatic optimizations (to be investigated by future reseach); in this paper we show that there is a much milder (and still practically useful) fragment.

Definition 2. *Simple flat formulæ are recursively defined as follows: (i) basic formulæ are simple flat formulæ; (ii) if ϕ is a simple flat formula, $\beta(\mathbf{a}(x), \mathbf{a}(\underline{y}), \underline{y})$ is a basic formula and x, z are variables, then $\phi(\sharp\{x \mid \beta\} / z)$ is a simple flat formula.*

As an example of a simple flat formula consider the following one

$$a'(y) = z \wedge \sharp\{x \mid a'(x) = a(x)\} \geq N{-}1 \wedge (\sharp\{x \mid a'(x) = a(x)\} < N \to a(y) \neq z)$$

expressing that $a' = write(a, y, z)$ (i.e. that the array a' is obtained from a by over-writing z in the entry y).

Definition 3. *Simple E-flat formulæ are formulæ of the kind*

$$\begin{aligned} \exists \underline{z}.\ \alpha(\mathbf{a}(\underline{y}), \mathbf{a}(\underline{z}), \underline{y}, \underline{z}) \ \wedge\ \sharp\{x \mid \beta_1(\mathbf{a}(x), \mathbf{a}(\underline{y}), \mathbf{a}(\underline{z}), \underline{y}, \underline{z})\} = z_1 \ \wedge \cdots \\ \cdots \wedge\ \sharp\{x \mid \beta_K(\mathbf{a}(x), \mathbf{a}(\underline{y}), \mathbf{a}(\underline{z}), \underline{y}, \underline{z}))\} = z_K \end{aligned} \tag{6}$$

where α and the β_i are basic.

It is easily seen that (once again) simple flat formulæ are closed under Boolean combinations and that simple flat formulæ are simple E-flat formulæ (the converse is not true, for ciclicity of the dependence graph of the z_i's in (6)).

The difference between simple and non-simple flat/E-flat formulæ is that in simple formulæ *the abstraction variable cannot occur outside the read of an array symbol* (in other words, the β, β_i from the above definition are of the kind $\beta_i(\mathbf{a}(x), \mathbf{a}(\underline{y}), \mathbf{a}(\underline{z}), \underline{y}, \underline{z})$ and not of the kind $\beta_i(\mathbf{a}(x), \mathbf{a}(\underline{y}), \mathbf{a}(\underline{z}), x, \underline{y}, \underline{z})$). This restriction has an important semantic effect, namely that formulæ (6) are equi-satisfiable to formulæ which are *permutation-invariant*, in the following sense. The truth value of an arithmetical formula or of a formula like $z = \sharp\{x \,|\, \alpha(\mathbf{a}(x), \underline{y})\}$ is not affected by a permutation of the values of the $\mathbf{a}(x)$ for $x \in [0, N)$, because x does not occur free in α (permuting the values of the $\mathbf{a}(x)$ may on the contrary change the value of a flat non-simple sentence like $z = \sharp\{x \,|\, a(x) \leq x\}$). This 'permutation invariance' will be exploited in the argument proving the correctness of decision procedure of Theorem 3 below. Formulae (6) themselves are not permutation-invariant because of subterms $\mathbf{a}(\underline{z}), \mathbf{a}(\underline{y})$, so we first show how to eliminate them up to satisfiability:

Lemma 2. *Simple E-flat formulæ are equi-satisfiable to disjunctions of permutation-invariant formulæ of the kind*

$$\exists \underline{z}.\ \alpha(\underline{y}, \underline{z}) \wedge \sharp\{x \mid \beta_1(\mathbf{a}(x), \underline{y}, \underline{z})\} = z_1 \wedge \cdots \wedge \sharp\{x \mid \beta_K(\mathbf{a}(x), \underline{y}, \underline{z}))\} = z_K \qquad (7)$$

Proof. Let us take a formula like (6): we convert it to an equi-satisfiabe disjunction of formulæ of the kind (7). The task is to eliminate terms $\mathbf{a}(\underline{z})$, $\mathbf{a}(\underline{y})$ by a series of guessings (each guessing will form the content of a disjunct). Notice that we can apply the procedure of Lemma 1 to eliminate the $\mathbf{a}(\underline{z})$, but for the $\mathbf{a}(\underline{y})$ we must operate differently (the method used in Lemma 1 introduced non-simple abstraction terms).

Let us suppose that $\underline{y} := y_1, \ldots, y_m$ and that, after a first guess, α contains the conjunct $y_j < N$ for each $j = 1, \ldots, m$ (if it contains $y_j \geq N$, we replace $a_s(y_j)$ by 0); after a second series of guesses, we can suppose also that α contains the conjuncts $y_{j_1} \neq y_{j_2}$ for $j_1 \neq j_2$ (if it contains $y_{j_1} = y_{j_2}$, we replace y_{j_1} by y_{j_2} everywhere, making y_{j_1} to disappear from the whole formula). In the next step, (i) we introduce for every $a \in \mathbf{a}$ and for every $j = 1, \ldots, m$ a fresh variable u_{aj}, (ii) we replace everywhere $a(y_j)$ by u_{aj} and (iii) we conjoin to α the equalities $a(y_j) = u_{aj}$. In this way we get a formula of the following kind

$$\exists \underline{z}.\ \bigwedge_{a \in \mathbf{a}, y_j \in \underline{y}} a(y_j) = u_{aj} \wedge \alpha(\underline{y}, \underline{u}, \underline{z}) \ \wedge \bigwedge_{l=1}^{K} \sharp\{x \mid \beta_l(\mathbf{a}(x), \underline{y}, \underline{u}, \underline{z})\} = z_l \qquad (8)$$

where $\underline{u}$ is the tuple formed by the u_{aj} (varying a and j). We now make another series of guesses and conjoin to α either $u_{aj} = u_{a'j'}$ or $u_{aj} \neq u_{a'j'}$ for $(a, j) \neq (a', j')$. Whenever $u_{aj} = u_{a'j'}$ is conjoined, u_{aj} is replaced by $u_{a'j'}$ everywhere, so that u_{aj} disappears completely. The resulting formula still has the form (8), but now the map $(a, j) \mapsto u_{aj}$ is not injective anymore (otherwise said, u_{aj} now indicates the element from the tuple $\underline{u}$ associated with the pair (a, j) and we might have that the same u_{aj} is associated with different pairs (a, j)).

Starting from (8) modified in this way, let us define now the equivalence relation among the y_j that holds between y_j and $y_{j'}$ whenever for all $a \in \mathbf{a}$ there is $u_a \in \underline{u}$ such that α contains the equalities $a(y_j) = u_a$ and $a(y'_j) = u_a$. Each equivalence class E is uniquely identified by the corresponding function f_E from $\mathbf{a}$ into $\underline{u}$ (it is the function that for each $y_j \in E$ maps $a \in \mathbf{a}$ to the $u_a \in \underline{u}$ such that α contains the equality $a(y_j) = u_a$ as a conjunct). Let $E_1, \ldots, E_r$ be the equivalence classes and let $n_1, \ldots, n_r$ be their cardinalities. We claim that (8) is equisatisfiable to

$$\begin{aligned}\exists \underline{z}.\ \alpha(\underline{y}, \underline{u}, \underline{z}) \ \wedge \bigwedge_{q=1}^{r} \sharp\{x \mid \bigwedge_{a \in \mathbf{a}} a(x) = f_{E_q}(a)\} \geq n_q \ \wedge \\ \wedge \bigwedge_{l=1}^{K} \sharp\{x \mid \beta_l(\mathbf{a}(x), \underline{y}, \underline{u}, \underline{z})\} = z_l\end{aligned} \qquad (9)$$

In fact, satisfiability of (8) trivially implies the satisfiability of the formula (9); vice versa, since (9) is permutation-invariant, if it is satisfiable we can modify any assignment satisfying it via a simultaneous permutation of the values of the $a \in \mathbf{a}$ so as to produce an assignment satisfying (8).

We now need just the trivial observation that the inequalities $\sharp\{x \mid \bigwedge_{a\in\mathbf{a}} a(x) = f_{E_q}(a)\} \geq n_q$ can be replaced by the formulæ $\sharp\{x \mid \bigwedge_{a\in\mathbf{a}} a(x) = f_{E_q}(a)\} = z'_q \ \wedge \ z'_q \geq n_q$ (for fresh z'_q) in order to match the syntactic shape of (7). ⊣

We can freely assume that quantifiers do not occur in simple flat formulæ: this is without loss of generality because such formulæ are built up from arithmetic and basic formulæ.[2]

Theorem 3. *Satisfiability of simple flat formulæ can be decided in NP (and thus it is an NP-complete problem).*

Proof. First, by applying the procedure of the previous Lemma we can reduce to the problem of checking the satisfiability of formulæ of the kind

$$\alpha(\underline{y},\underline{z}) \ \wedge \ \sharp\{x \mid \beta_1(\mathbf{a}(x),\underline{y},\underline{z})\} = z_1 \ \wedge \cdots \wedge \ \sharp\{x \mid \beta_K(\mathbf{a}(x),\underline{y},\underline{z})\} = z_K \quad (10)$$

where $\alpha, \beta_1, \ldots, \beta_K$ are basic (notice also that each formula in the output of the procedure of the previous Lemma comes from a polynomial guess).

Suppose that $A_1(\mathbf{a}(x),\underline{y},\underline{z}), \ldots, A_L(\mathbf{a}(x),\underline{y},\underline{z})$ are the atoms occurring in $\beta_1, \ldots, \beta_K$. For a Boolean assignment σ to these atoms, we indicate with $[\![\beta_j]\!]^\sigma$ the Boolean value (0 or 1) the formula β_l has under such assignment. We first claim that (10) is satisfiable iff there exists *a set of assignments* Σ such that the formula

$$\alpha(\underline{y},\underline{z}) \ \wedge \ \bigwedge_{\sigma\in\Sigma} \exists \underline{u} \left(\bigwedge_{j=1}^{L} \epsilon_{\sigma(A_j)} A_j(\underline{u},\underline{y},\underline{z}) \right) \wedge \begin{bmatrix} z_1 \\ z_2 \\ \vdots \\ z_K \end{bmatrix} = \sum_{\sigma\in\Sigma} v_\sigma \begin{bmatrix} [\![\beta_1]\!]^\sigma \\ [\![\beta_2]\!]^\sigma \\ \vdots \\ [\![\beta_K]\!]^\sigma \end{bmatrix} \wedge \quad (11)$$
$$\wedge \sum_{\sigma\in\Sigma} v_\sigma = N \wedge \bigwedge_{\sigma\in\Sigma} v_\sigma > 0$$

is satisfiable (we introduced extra fresh variables v_σ, for $\sigma \in \Sigma$; notation $\epsilon_{\sigma(A_j)}$ is the same as in the proof of Lemma 1). In fact, on one side, if (10) is satisfiable under V, we can take as Σ the set of assigments for which $\bigwedge_{j=1}^{L} \epsilon_{\sigma(A_j)} A_j(\mathbf{a}(i),\underline{y},\underline{z})$ is true under V for some $i \in [0,N)$ and for v_σ the cardinality of the set of the $i \in [0,N)$ for which $\bigwedge_{j=1}^{L} \epsilon_{\sigma(A_j)} A_j(\mathbf{a}(i),\underline{y},\underline{z})$ holds. This choice makes (11) true. Vice versa, if (11) is true under V, in order to define the value of the tuple $\mathbf{a}(i)$ (for $i \in [0,N)$), pick for every $\sigma \in \Sigma$ some $\underline{u}_\sigma$ such that $\bigwedge_{j=1}^{L} \epsilon_{\sigma(A_j)} A_j(\underline{u}_\sigma,\underline{y},\underline{z})$ holds; then, supposing $\Sigma = \{\sigma_1, \ldots, \sigma_h\}$, let

[2] By the quantifier-elimination result for Presburger arithmetic, it is well-known that arithmetic formulæ are equivalent to quantifier-free ones. The same is true for basic formulæ because they are obtained from arithmetic formulae by substitutions without capture.

$\mathbf{a}(i)$ be equal to $\underline{u}_{\sigma_1}$ for $i \in [0, v_{\sigma_1})$, to $\underline{u}_{\sigma_2}$ for $i \in [v_{\sigma_1}, v_{\sigma_2})$, etc. Since we have that $\sum_{\sigma \in \Sigma} v_\sigma = N$, the definition of the interpretation of the $\mathbf{a}$ is complete (any other permutation of the values $\mathbf{a}(x)$ inside $[0, N)$ would fit as well). In this way, formula (10) turns out to be true.

We so established that our original formula is satisfiable iff there is some Σ such that (11) is satisfiable; the only problem we still have to face is that Σ might be exponentially large. To reduce to a polynomial Σ, we use the same technique as in [16]. In fact, if (11) is satisfiable, then the column vector $(z_1, \ldots, z_K, N)^T$ is a linear combination with positive integer coefficients of the 0/1-vectors $([\![\beta_1]\!]^\sigma, \cdots, [\![\beta_K]\!]^\sigma, 1)^T$ and it is known from [9] that, if this is the case, the same result can be achieved by assuming that at most $2K' \log_2(4K')$ of the v_σ are nonzero (we put $K' := K + 1$). Thus polynomially many Σ are sufficient and for such Σ, a satisfying assignment for the existential Presburger formula (11) is a polynomial certificate. ⊣

4.1 Some Heuristics

We discuss here some useful heuristics for the satisfiability algorithm for simple flat formulæ (most of these heuristics have been implemented in our prototype).

1.- The satisfiability test involves all formulæ (11) for each set of assignments Σ having cardinality *at most* $M = \lceil 2K' \log_2(4K') \rceil$ (actually, one can improve this bound, see [16]). If we replace in (11), for every σ, the conjunct $v_\sigma > 0$ by $v_\sigma \geq 0$ and the conjunct[3] $\exists \underline{u}\, (\bigwedge_{j=1}^{L} \epsilon_{\sigma(A_j)} A_j(\underline{u}, \underline{y}, \underline{z}))$ by $v_\sigma > 0 \to \exists \underline{u}\, (\bigwedge_{j=1}^{L} \epsilon_{\sigma(A_j)} A_j(\underline{u}, \underline{y}, \underline{z}))$, we can limit ourselves to the Σ having cardinality *equal to* M. This trick is useful if, for some reason, we prefer to go through any sufficient set of assignments (like the set of all assignments supplied by some Boolean propagation, see below).

2.- There is no need to consider assignments σ over the set of the atoms A_j occurring in the $\beta_1, \ldots, \beta_K$: any set of formulæ generating the $\beta_1, \ldots, \beta_K$ by Boolean combinations fits our purposes. As a consequence, the choice of these 'atoms' is subject to case-by-case evaluations.

3.- Universally quantified formulæ of the kind $\forall x\, (0 \leq x \wedge x < N \to \beta)$ can be turned into flat formulæ by rewriting them as $N = \sharp\{x \mid \beta\}$ (and in fact such universally quantified formulæ often occur in our benchmarks suite). These formulæ contribute to (10) via the conjuncts of the kind $z_i = N \wedge \sharp\{x \mid \beta_i(\mathbf{a}(x), \underline{y}, \underline{z})\} = z_i$. It is quite useful to consider the $\{\beta_{i_1}, \ldots, \beta_{i_L}\}$ arising in this way as atoms (in the sense of point 2 above) and restrict to the assignments σ such that $\sigma(\beta_{i_1}) = \cdots = \sigma(\beta_{i_L}) = 1$.

4.- Boolean propagation is a quite effective strategy to prune useless assignments. In our context, as soon as a partial assignment σ is produced inside the assignments enumeration subroutine, an SMT solver is invoked to test the satisfiability of $\alpha(\underline{y}, \underline{z}) \wedge \bigwedge_{j \in dom(\sigma)} \epsilon_{\sigma(A_j)} A_j(\underline{u}, \underline{y}, \underline{z})$. Since this is implied by a

[3] These conjuncts (varying $\sigma \in \Sigma$) are needed in (11) to ensure that the assignments we are using can coexist in a model.

(skolemized) conjunct of (11), if the test is negative the current partial assignment is discarded and next partial assignment (obtained by complementing the value of the last assigned literal) is taken instead.

5 Examples and Experiments

We implemented a prototype ArCa-Sat[4] producing out of simple E-flat formulæ (10) the proof obbligations (11) (written in SMT-LIB2 format), exploiting the heuristics explained in Sect. 4.1. To experiment the feasibility of our approach for concrete verification problems, we also implemented a (beta) version of a tool called ArCa producing out of the specification of a parametric distributed system and of a safety-like problem, some E-flat simple formulæ whose unsatisfiability formalizes invariant-checking and bounded-model checking problems. A script executing in sequence ArCa, ArCa-Sat and z3 can then solve such problems by reporting a 'sat/unsat' answer.

A system is specified via a pair of flat (simple) formulæ $\iota(\underline{p})$ and $\tau(\underline{p}, \underline{p}')$ and a safety problem via a further formula $\upsilon(\underline{p})$ (here the $\underline{p}$ are parameters and array-ids, the $\underline{p}'$ are renamed copies of the $\underline{p}$). A bounded model checking problem is the problem of checking whether the formula

$$\iota(\underline{p}_0) \wedge \tau(\underline{p}_0, \underline{p}_1) \wedge \cdots \wedge \tau(\underline{p}_n, \underline{p}_{n+1}) \wedge \upsilon(\underline{p}_{n+1})$$

is satisfiable for a fixed n. An invariant-cheking problem, given also a formula $\phi(\underline{p})$, is the problem of checking whether the three formulæ

$$\iota(\underline{p}) \wedge \neg\phi(\underline{p}), \quad \phi(\underline{p}) \wedge \tau(\underline{p}, \underline{p}') \wedge \neg\phi(\underline{p}'), \quad \phi(\underline{p}) \wedge \upsilon(\underline{p})$$

are unsatisfiable. Notice that since all our algorithms terminate and are sound and complete, the above problems are always solved by the above tool combination (if enough computation resources are available). Thus, our technique is able *both to make safety certifications and to find bugs.*

To validate our technique, in the following we describe in detail the formalization of the send-receive broadcast primitive (SRBP) in [21]. SRBP is used as a basis to synchronize clocks in systems where processes may fail in sending and/or receiving messages. Periodically, processes broadcast the virtual time to be adopted by all, as a *(session s)* message. Processes that accept this message set s as their current time. SRBP aims at guaranteeing the following properties:

Correctness: if at least $f + 1$ correct processes broadcast the message *(session s)*, all correct processes accept the message.

Unforgeability: if no correct process broadcasts *(session s)*, no correct process accepts the message.

Relay: if a correct process accepts *(session s)*, all correct processes accept it.

[4] ArCa stands for *Array with Cardinalities.*

where $f < N/2$ is the number of processes failing during an algorithm run, with N the number of processes in the system. Algorithm 1 shows the pseudo-code.

We model SRBP as follows: $IT(x)$ is the initial state of a process x; it is s when x broadcasts a *(init, session s)* message, and 0 otherwise. $SE(x) = s$ indicates that x has broadcast its own echo. $AC(x) = s$ indicates that x has accepted *(session s)*. Let pc be the program counter, r the round number, and G a flag indicating whether one round has been executed. We indicate with $F(x) = 1$ the fact that x is faulty, and $F(x) = 0$ otherwise. Finally, $CI(x)$ and $CE(x)$ are the number of respectively inits and echoes received. In the following, $\forall x$ means $\forall x \in [0, N)$. Some sentences are conjoined to all our proof obligations, namely: $\#\{x|F(x) = 0\} + \#\{x|F(x) = 1\} = N \wedge \#\{x|F(x) = 1\} < N/2$. For the Correctness property, we write ι_c as follows:

$$\iota_c := pc = 1 \wedge r = 0 \wedge G = 0 \wedge s \neq 0 \wedge \tag{12}$$
$$\#\{x|IT(x) = 0\} + \#\{x|IT(x) = s\} = N \wedge \tag{13}$$
$$\#\{x|F(x) = 0 \wedge IT(x) = s\} \geq (\#\{x|F(x) = 1\} + 1) \wedge \tag{14}$$
$$\forall x.SE(x) = 0 \wedge AC(x) = 0 \wedge CI(x) = 0 \wedge CE(x) = 0 \tag{15}$$

where we impose that the number of correct processes broadcasting the init message is at least the number of faulty processes, f, plus 1. It is worth to notice that – from the above definition – our tool produces a specification that is checked for any $N \in \mathbb{N}$ number of processes. The constraints on IT allow to verify all admissible assignments of 0 or s to the variables. Similarly for $F(x)$.

The algorithm safety is verified by checking that the bad properties cannot be reached from the initial state. For Correctness, we set $\upsilon_c := pc = 1 \wedge G = 1 \wedge \#\{x|F(x) = 0 \wedge AC(x) = 0\} > 0$, that is, Correctness is not satisfied if – after one round – some correct process exists that has yet to accept. The algorithm evolution is described by two transitions: τ_1 and τ_2. The former allows to choose the number of both inits and echoes received by each process. The latter describes the actions in Algorithm 1.

$$\begin{aligned}
\tau_1 := \; & pc = 1 \wedge pc' = 2 \wedge r' = r \wedge G' = G \wedge s' = s \wedge \exists K1, K2, K3, K4. \\
& K1 = \#\{x|F(x) = 0 \wedge IT(x) = s\} \wedge K2 = \#\{x|F(x) = 0 \wedge SE(x) = s\} \wedge \\
& K3 = \#\{x|F(x) = 1 \wedge IT(x) = s\} \wedge K4 = \#\{x|F(x) = 1 \wedge SE(x) = s\} \wedge \\
& \forall x.F(x) = 0 \Rightarrow (CI'(x) \geq K1 \wedge CI'(x) \leq (K1 + K3) \wedge CE'(x) \geq K2 \wedge \\
& CE'(x) \leq (K2 + K4)) \wedge \\
& \forall x.F(x) = 1 \Rightarrow (CI'(x) \geq 0 \wedge CI'(x) \leq (K1 + K3) \wedge CE'(x) \geq 0 \wedge \\
& CE'(x) \leq (K2 + K4)) \wedge \\
& \forall x.IT'(x) = IT(x) \wedge SE'(x) = SE(x) \wedge AC'(x) = AC(x) \\
\tau_2 := \; & pc = 2 \wedge pc' = 1 \wedge r' = (r + 1) \wedge s' = s \wedge G' = 1 \wedge \\
& \forall x.(CI(x) \geq \#\{x|F(x) = 1\} + 1 \Rightarrow SE'(x) = s \wedge AC'(x) = s) \wedge \\
& \forall x.(CI(x) < \#\{x|F(x) = 1\} + 1 \wedge CE(x) \geq 1 \Rightarrow SE'(x) = s \wedge AC'(x) = s) \wedge \\
& \forall x.(CI(x) < \#\{x|F(x) = 1\} + 1 \wedge CE(x) < 1 \Rightarrow SE'(x) = 0 \wedge AC'(x) = 0) \wedge \\
& \forall x.IT'(x) = IT(x) \wedge CI'(x) = CI(x) \wedge CE'(x) = CE(x)
\end{aligned}$$

Algorithm 1. Pseudo-code for the send-receive broadcast primitive.

Initialization:
To broadcast a *(session s)* message, a correct process sends *(init, session s)* to all.
End Initialization
for each correct process:
1. **if** received *(init, session s)* from at least $f+1$ distinct processes **or**
2. received *(echo, session s)* from any process **then**
3. accept *(session s)*;
4. send *(echo, session s)* to all;
5. **endif**

end for

The same two transitions are used to verify both the Unforgeability and the Relay properties, for which however we have to change the initial and final formula. For Unforgeability, (13) in ι changes as $... \wedge \#\{x|F(x) = 0 \wedge IT(x) = 0\} = \#\{x|F(x) = 0\} \wedge ...$; while $\upsilon_u := pc = 1 \wedge G = 1 \wedge \#\{x|F(x) = 0 \wedge AC(x) = s\} > 0$. In ι_u we say that all non-faulty processes have $IT(x) = 0$. Unforgeability is not satisfied if some correct process accepts. For Relay, we use:

$$\begin{aligned}\iota_r := {} & pc = 1 \wedge r = 0 \wedge s \neq 0 \wedge G = 0 \wedge \\ & \#\{x|F(x) = 0 \wedge AC(x) = s \wedge SE(x) = s\} = 1 \wedge \\ & \#\{x|AC(x) = 0 \wedge SE(x) = 0\} = (N-1) \wedge \#\{x|AC(x) = s \wedge SE(x) = s\} = 1 \wedge \\ & \forall x.IT(x) = 0 \wedge CI(x) = 0 \wedge CE(x) = 0\end{aligned}$$

while $\upsilon_r = \upsilon_c$. In this case, we start the system in the worst condition: by the hypothesis, we just know that one correct process has accepted. Upon acceptance, by the pseudo-code, it must have sent an echo. All the other processes are initialized in an idle state. We also produce an unsafe model of Correctness: we modify ι_c by imposing that just f correct processes broadcast the init message.

In Table 1, we report the results of validating these and other models with our tool. In the first column, the considered algorithm is indicated. The second column indicates the property to be verified; the third column reports the conditions of verification. In the fourth column, we indicate whether we consider either a bounded model checking (BMC) or an invariant-checking (IC) problem. The fifth column supplies the obtained results (for BMC problems, 'safe' means of course 'safe up to the analyzed bound'). The sixth column shows the time jointly spent by ARCA, ARCA-SAT and Z3 for the verification, considering for BMC the sum of the times spent for every traces of length up to 10. We used a PC equipped with Intel Core i7 processor and operating system Linux Ubuntu 14.04 64 bits. We focused on BMC problems as they produce longer formulas thus stressing more the tools. Specifically, following the example above, we modeled:

Table 1. Evaluated algorithms and experimental results.

Algorithm	Property	Condition	Problem	Outcome	Time (s.)
SRBP [21]	Correctness	$\geq (f+1)$ init's	BMC	safe	0.82
SRBP [21]	Correctness	$\leq f$ init's	BMC	unsafe	2.21
SRBP [21]	Unforgeability	$\geq (f+1)$ init's	BMC	safe	0.85
SRBP [21]	Relay	$\geq (f+1)$ init's	BMC	safe	1.93
BBP [22]	Correctness	$N > 3f$	BMC	safe	6.17
BBP [22]	Unforgeability	$N > 3f$	BMC	safe	0.25
BBP [22]	Unforgeability	$N \geq 3f$	BMC	unsafe	0.25
BBP [22]	Relay	$N > 3f$	BMC	safe	1.01
OT [4]	Agreement	threshold $>2N/3$	IC	safe	4.20
OT [4]	Agreement	threshold $>2N/3$	BMC	safe	278.95
OT [4]	Agreement	threshold $\leq 2N/3$	BMC	unsafe	17.75
OT [4]	Irrevocability	threshold $>2N/3$	BMC	safe	8.72
OT [4]	Irrevocability	threshold $\leq 2N/3$	BMC	unsafe	9.51
OT [4]	Weak Validity	threshold $>2N/3$	BMC	safe	0.45
OT [4]	Weak Validity	threshold $\leq 2N/3$	BMC	unsafe	0.59
UV [5]	Agreement	$\mathcal{P}_{nosplit}$ violated	BMC	unsafe	4.18
UV [5]	Irrevocability	$\mathcal{P}_{nosplit}$ violated	BMC	unsafe	2.04
UV [5]	Integrity	-	BMC	safe	1.02
$U_{T,E,\alpha}$ [3]	Integrity	$\alpha = 0 \wedge \mathcal{P}_{safe}$	BMC	safe	1.16
$U_{T,E,\alpha}$ [3]	Integrity	$\alpha = 0 \wedge \neg\mathcal{P}_{safe}$	BMC	unsafe	0.83
$U_{T,E,\alpha}$ [3]	Integrity	$\alpha = 1 \wedge \mathcal{P}_{safe}$	BMC	safe	5.20
$U_{T,E,\alpha}$ [3]	Integrity	$\alpha = 1 \wedge \neg\mathcal{P}_{safe}$	BMC	unsafe	4.93
$U_{T,E,\alpha}$ [3]	Agreement	$\alpha = 0 \wedge \mathcal{P}_{safe}$	BMC	safe	59.80
$U_{T,E,\alpha}$ [3]	Agreement	$\alpha = 0 \wedge \neg\mathcal{P}_{safe}$	BMC	unsafe	7.78
$U_{T,E,\alpha}$ [3]	Agreement	$\alpha = 1 \wedge \mathcal{P}_{safe}$	BMC	safe	179.67
$U_{T,E,\alpha}$ [3]	Agreement	$\alpha = 1 \wedge \neg\mathcal{P}_{safe}$	BMC	unsafe	31.94
MESI [17]	cache coherence	-	IC	safe	0.11
MOESI [20]	cache coherence	-	IC	safe	0.08
Dekker [6]	mutual exclusion	-	IC	safe	2.05

- the byzantine broadcast primitive (BBP) [22] used to simulate authenticated broadcast in the presence of malicious failures of the processes,
- the one-third algorithm (OT) [4] for consensus in the presence of benign transmission failures,
- the Uniform Voting (UV) algorithm [5] for consensus in the presence of benign transmission failures,

- the $U_{T,E,\alpha}$ algorithm [3] for consensus in the presence of malicious transmission failures,
- the MESI [17] and MOESI [20] algorithms for cache coherence,
- the Dekker's algorithm [6] for mutual exclusion.

All the models, together with our tools to verify them, are available at http://users.mat.unimi.it/users/ghilardi/arca (for the z3 solver see http://rise4fun.com/z3).

As far as the processing times are concerned, we observed that on average z3 accounts for around 68% of the processing time, while ArCa and ArCa-Sat together account for the remaining 32%. Indeed, the SMT tests performed by ArCa-Sat are lightweight – as they only prune assignments – yet effective, as they succeed in reducing the number of assignments of at least one order of magnitude.

6 Conclusions, Related and Further Work

We identified two fragments of the rich syntax of Fig. 1 and we showed their decidability (for the second fragment we showed also a tight complexity bound). Since our fragments are closed under Boolean connectives, it is possible to use them not only in bounded model checking (where they can both give certifications and find bugs), but also in order to decide whether an invariant holds or not. We implemented our algorithm for the weaker fragment and used it in some experiments. As far as we know, this is the first implementation of a *complete* algorithm for a fragment of arithmetic with arrays and counting capabilities for interpreted sets. In future, we plan to extend both our tool ArCa and our results in order to deal with more complex verification problems.

Since one of the major intended applications concerns fault-tolerant distributed systems, we briefly review and compare here some recent work in the area. Papers [11–13] represent a very interesting and effective research line, where cardinality constraints are not directly handled but abstracted away using interval abstract domains and counters. As a result, a remarkable amount of algorithms are certified, although the method might suffer of some lack of expressiveness for more complex examples. On the contrary, paper [4] directly handles cardinality constraints for interpreted sets; nontrivial invariant properties are synthesized and checked, based on Horn constraint solving technology. At the level of decision procedures, some incomplete inference schemata are employed (completeness is nevertheless showed for array updates against difference bounds constraints). Paper [7] introduces a very expressive logic, specifically tailored to handle consensus problems (whence the name 'consensus logic' CL). Such logic employs arrays with values into power set types, hence it is situated in a higher order logic context. Despite this, our flat fragment is not fully included into CL, because we allow arithmetic constraints on the sort of indexes and also mixed constraints between indexes and data: in fact, we have a unique sort for indexes and data, leading to the possibility of writing typically non permutation-invariant formulæ like $\sharp\{x \mid a(x)+x = N\} = z$. As pointed out in [2], this mono-sorted approach is useful in the analysis of programs, when pointers to the memory (modeled as an array) are stored into array variables. From the point of view of deduction, the paper [7] uses an incomplete algorithm in order to certify invariants. A smaller decidable fragment (identified via several syntactic restrictions) is introduced in the final part of the paper; the sketch of the decidability proof supplied for this smaller fragment uses bounds for minimal solutions of Presburger formulæ as well as Venn regions decompositions in order to build models where all nodes in the same Venn region share the same value for their function symbols.

References

1. Alberti, F., Ghilardi, S., Pagani, E.: Counting constraints in flat array fragments (2016). CoRR, abs/1602.00458
2. Alberti, F., Ghilardi, S., Sharygina, N.: Decision procedures for flat array properties. In: Ábrahám, E., Havelund, K. (eds.) TACAS 2014 (ETAPS). LNCS, vol. 8413, pp. 15–30. Springer, Heidelberg (2014)
3. Biely, M., Charron-Bost, B., Gaillard, A., Hutle, M., Schiper, A., Widder, J.: Tolerating corrupted communication. In: Proceedings of PODC, pp. 244–253 (2007)
4. Bjørner, N., von Gleissenthall, K., Rybalchenko, A.: Synthesizing cardinality invariants for parameterized systems (2015). https://www7.in.tum.de/~gleissen/papers/sharpie.pdf
5. Charron-Bost, B., Schiper, A.: The heard-of model: computing in distributed systems with benign faults. Distrib. Comput. **22**, 49–71 (2009)
6. Dijkstra, E.W.: Cooperating sequential processes. In: Genuys, F. (ed.) Programming Languages, pp. 43–112. Academic Press, New York (1968)
7. Drăgoi, C., Henzinger, T.A., Veith, H., Widder, J., Zufferey, D.: A logic-based framework for verifying consensus algorithms. In: McMillan, K.L., Rival, X. (eds.) VMCAI 2014. LNCS, vol. 8318, pp. 161–181. Springer, Heidelberg (2014)
8. Dragoi, C., Henzinger, T.A., Zufferey, D.: The need for language support for fault-tolerant distributed systems. In: Proceedings of SNAPL (2015)
9. Eisenbrand, F., Shmonin, G.: Carathéodory bounds for integer cones. Oper. Res. Lett. **34**(5), 564–568 (2006)
10. Halpern, J.Y.: Presburger arithmetic with unary predicates is Π^1_1 complete. J. Symbol. Logic **56**(2), 637–642 (1991)
11. John, A., Konnov, I., Schmid, U., Veith, H., Widder, J.: Parameterized model checking of fault-tolerant distributed algorithms by abstraction. In: Proceedings of FMCAD, pp. 201–209, August 2013
12. Konnov, I., Veith, H., Widder, J.: On the completeness of bounded model checking for threshold-based distributed algorithms: reachability. In: Baldan, P., Gorla, D. (eds.) CONCUR 2014. LNCS, vol. 8704, pp. 125–140. Springer, Heidelberg (2014)
13. Konnov, I., Veith, H., Widder, J.: SMT and POR beat counter abstraction: parameterized model checking of threshold-based distributed algorithms. In: Kroening, D., Păsăreanu, C.S. (eds.) CAV 2015. LNCS, vol. 9206, pp. 85–102. Springer, Heidelberg (2015)
14. Kuncak, V., Nguyen, H.H., Rinard, M.: An algorithm for deciding BAPA: boolean algebra with Presburger arithmetic. In: Nieuwenhuis, R. (ed.) CADE 2005. LNCS (LNAI), vol. 3632, pp. 260–277. Springer, Heidelberg (2005)

15. Kuncak, V., Nguyen, H.H., Rinard, M.: Deciding boolean algebra with Presburger arithmetic. J. Autom. Reasoning **36**(3), 213–239 (2006)
16. Kuncak, V., Rinard, M.: Towards efficient satisfiability checking for boolean algebra with Presburger arithmetic. In: Pfenning, F. (ed.) CADE 2007. LNCS (LNAI), vol. 4603, pp. 215–230. Springer, Heidelberg (2007)
17. Papamarcos, M.S., Patel, J.H.: A low-overhead coherence solution for multiprocessors with private cache memories. In: Proceedings of ISCA, p. 348 (1984)
18. Piskac, R., Kuncak, V.: Decision procedures for multisets with cardinality constraints. In: Logozzo, F., Peled, D.A., Zuck, L.D. (eds.) VMCAI 2008. LNCS, vol. 4905, pp. 218–232. Springer, Heidelberg (2008)
19. Schweikhart, N.: Arithmetic, first-order logic, and counting quantifiers. ACM TOCL **6**, 1–35 (2004)
20. Solihin, Y.: Fundamentals of Parallel Computer Architecture Multichip and Multicore Systems. Solihin Publishing & Consulting LLC (2008)
21. Srikanth, T.K., Toueg, S.: Optimal clock synchronization. J. ACM **34**(3), 626–645 (1987)
22. Srikanth, T.K., Toueg, S.: Simulating authenticated broadcasts to derive simple fault-tolerant algorithms. Distrib. Comput. **2**(2), 80–94 (1987)
23. Yessenov, K., Piskac, R., Kuncak, V.: Collections, cardinalities, and relations. In: Barthe, G., Hermenegildo, M. (eds.) VMCAI 2010. LNCS, vol. 5944, pp. 380–395. Springer, Heidelberg (2010)

A New Decision Procedure for Finite Sets and Cardinality Constraints in SMT

Kshitij Bansal[1], Andrew Reynolds[2], Clark Barrett[1(✉)], and Cesare Tinelli[2]

[1] Department of Computer Science, New York University, New York, USA
barrett@cs.nyu.edu

[2] Department of Computer Science, The University of Iowa, Iowa City, USA

Abstract. We consider the problem of deciding the theory of finite sets with cardinality constraints using a satisfiability modulo theories solver. Sets are a common high-level data structure used in programming; thus, such a theory is useful for modeling program constructs directly. More importantly, sets are a basic construct of mathematics and thus natural to use when formalizing the properties of computational systems. We develop a calculus describing a modular combination of a procedure for reasoning about membership constraints with a procedure for reasoning about cardinality constraints. Cardinality reasoning involves tracking how different sets overlap. For efficiency, we avoid considering Venn regions directly, as done previous work. Instead, we develop a novel technique wherein potentially overlapping regions are considered incrementally as needed. We use a graph to track the interaction among the different regions. Initial experimental results demonstrate that the new technique is competitive with previous techniques and scales much better on certain classes of problems.

1 Introduction

Satisfiability modulo theories (SMT) solvers are at the heart of many verification tools. One of the reasons for their popularity is that fast, dedicated decision procedures for fragments of first-order logic are extremely useful for reasoning about constructs common in hardware and software verification. In particular, they provide a good balance between speed and expressiveness. Common fragments include theories such as bitvectors, arithmetic, and arrays, which are useful both for modeling basic constructs as well as for performing general reasoning.

As the use of SMT solvers has spread, there has been a corresponding demand for SMT solvers to support additional useful theories. Although it is possible to encode finitely axiomatizable theories using quantifiers, the performance and robustness gap between a custom decision procedure and an encoding using quantifiers can be quite significant.

In this paper, we present a new decision procedure for a fragment of set theory. Our main motivation is that sets are a common abstraction used in programming. As with other SMT theories like the theories of arrays and bitvectors,

This work was partially supported by NSF grants 1228765, 1228768, and 1320583.

N. Olivetti and A. Tiwari (Eds.): IJCAR 2016, LNAI 9706, pp. 82–98, 2016.
DOI: 10.1007/978-3-319-40229-1_7

we expect the theory of sets to be useful in modeling a variety of program constructs. Sets are also used directly in high-level programming languages like SETL and in specification languages like Alloy, B and Z. More generally, sets are a basic construct in mathematics and come up quite naturally when trying to express properties of systems.

While the full language of set theory is undecidable, many interesting fragments are known to be decidable. We present a calculus which can handle basic set operations, such as membership, union, intersection, and difference, and which can also reason efficiently about set cardinalities. The calculus is also designed for easy integration into the DPLL(T) framework [12].

1.1 Related Work

In the SMT community, the desire to support a theory of finite sets with cardinality goes back at least to a 2009 proposal [9]. However, the focus there is on formalizing the semantics and representation of the theory within the context of the SMT-LIB language, rather than on a decision procedure for deciding it.

There is an existing stream of research on exploring decidable fragments of set theory (often referred to in the literature as syllogistics) [5]. One such subfragment is MLSS, more precisely, the ground set-theoretic fragment with basic Boolean set operators (union, intersection, set difference), singleton operator and membership predicate. A tableau-based procedure for this fragment was presented in [6], and the part of our calculus covering this same fragment builds on that work. In [7], an extension of the theory of arrays is presented, which can be used to encode the MLSS fragment. However, this approach cannot be used to encode cardinality constraints.

In this paper, we consider the MLSS fragment extended with set cardinality operations. The decidability of this fragment was established in [14]. The procedure given there involves making an up-front guess that is exponential in the number of set variables, making it non-incremental and highly impractical. That said, the focus of [14] is on establishing decidability and not on providing an efficient procedure.

Another logical fragment that is closely related is the Boolean Algebra and Presburger Arithmetic (BAPA) fragment, for which several algorithms have been proposed [10,11,13]. Though BAPA doesn't have the membership predicate or the singleton operator in its language, [13, Sect. 4] shows how one can generalize the algorithm for such reasoning. Intuitively, singleton sets can be simulated by imposing a cardinality constraint $\mathsf{card}(X) = 1$. Similarly, a membership constraint, say $x \sqsubseteq S$, is encoded by introducing a singleton set, say X, and then using the subset operation: $X \sqsubseteq S$.

This reduction can lead to significant inefficiencies, however. Consider the following simple example: $x \sqsubseteq S_1 \sqcup (S_2 \sqcup (\ldots \sqcup (S_{99} \sqcup S_{100})))$. It is easy to see that the constraint is satisfiable. In our calculus, a straightforward repeated application of one of the rules for set unions can determine this. On the other hand, in a reduction to BAPA, the membership reasoning is reduced to reasoning about cardinalities of different sets. For example, the algorithm in [13] will reduce the problem to arithmetic constraints involving variables for 2^{101} Venn regions derived from $S_1, S_2, \ldots, S_{100}$, and the singleton set introduced for x.

The broader point is that reasoning about cardinalities of Venn regions is the main bottleneck for this fragment. As we show in our calculus, it is possible to avoid using Venn regions for membership predicates by instead reasoning about them directly. For explicit cardinality constraints, our calculus minimizes the number of Venn regions that need to be considered by reasoning about only a limited number of relevant regions that are introduced lazily.

1.2 Formal Preliminaries

We work in the context of many-sorted first-order logic with equality. We assume the reader is familiar with the following notions: signature, term, literal, formula, free variable, interpretation, and satisfiability of a formula in an interpretation (see, e.g., [3] for more details). Let Σ be a many-sorted signature. We will use $\approx$ as the (infix) logical symbol for equality—which has type $\sigma \times \sigma$ for all sorts σ in Σ and is always interpreted as the identity relation. We write $s \not\approx t$ as an abbreviation of $\neg\, s \approx t$. If e is a term or a formula, we denote by $\mathcal{V}(e)$ the set of e's free variables, extending the notation to tuples and sets of terms or formulas as expected.

If φ is a Σ-formula and $\mathcal{I}$ a Σ-interpretation, we write $\mathcal{I} \models \varphi$ if $\mathcal{I}$ satisfies φ. If t is a term, we denote by $t^{\mathcal{I}}$ the value of t in $\mathcal{I}$. A *theory* is a pair $T = (\Sigma, \mathbf{I})$, where Σ is a signature and $\mathbf{I}$ is a class of Σ-interpretations that is closed under variable reassignment (i.e., every Σ-interpretation that differs from one in $\mathbf{I}$ only in how it interprets the variables is also in $\mathbf{I}$). $\mathbf{I}$ is also referred to as the *models* of T. A Σ-formula φ is *satisfiable* (resp., *unsatisfiable*) *in* T if it is satisfied by some (resp., no) interpretation in $\mathbf{I}$. A set Γ of Σ-formulas *entails in* T a Σ-formula φ, written $\Gamma \models_T \varphi$, if every interpretation in $\mathbf{I}$ that satisfies all formulas in Γ satisfies φ as well. We write $\models_T \varphi$ as an abbreviation for $\emptyset \models_T \varphi$. We write $\Gamma \models \varphi$ to denote that Γ entails φ in the class of all Σ-interpretations. The set Γ is *satisfiable in* T if $\Gamma \not\models_T \bot$ where $\bot$ is the universally false atom. Two Σ-formulas are *equisatisfiable in* T if for every model $\mathcal{A}$ of T that satisfies one, there is a model of T that satisfies the other and differs from $\mathcal{A}$ at most over the free variables not shared by the two formulas. When convenient, we will tacitly treat a finite set of formulas as the conjunction of its elements and vice versa.

2 A Theory of Finite Sets with Cardinality

We consider a typed theory $\mathfrak{T}_S$ of finite sets with cardinality. In a more general logical setting, this theory would be equipped with a parametric set type, with a type parameter for the set's elements, and a corresponding collection of polymorphic set operations.[1] For simplicity here, we will describe instead a many-sorted theory of sets of sort Set whose elements are all of sort Element.

[1] In fact, this is the setting supported in our implementation in CVC4.

Constant and function symbols:

n : Card for all $n \in \mathbb{N}$		- : Card $\rightarrow$ Card	+ : Card $\times$ Card $\rightarrow$ Card
$\emptyset$: Set	card$(\cdot)$: Set $\rightarrow$ Card	$\{\cdot\}$: Element $\rightarrow$ Set	$\sqcup, \sqcap, \setminus$: Set $\times$ Set $\rightarrow$ Set

Predicate symbols:

< : Card $\times$ Card	>= : Card $\times$ Card	$\sqsubseteq$: Set $\times$ Set	$\sqsubset$: Element $\times$ Set

Fig. 1. The signature of $\mathfrak{T}_S$.

The theory $\mathfrak{T}_S$ can be combined with any other theory $\mathfrak{T}$ in a standard way, i.e., Nelson-Oppen-style, by identifying the Element sort with a sort σ in $\mathfrak{T}$, with the restriction that σ must be interpreted in $\mathfrak{T}$ as an infinite set.[2] Note that we limit our language to consider only *flat* sets (i.e. no sets of sets). However, this can be simulated by combining $\mathfrak{T}$ with itself using the mechanism just mentioned. The theory $\mathfrak{T}_S$ has also a sort Card for terms denoting set cardinalities. Since we consider only finite sets, all cardinalities will be natural numbers.

Atomic formulas in $\mathfrak{T}_S$ are built over a signature with these three sorts, and an infinite set of variables for each sort. Modulo isomorphism, $\mathfrak{T}_S$ is the theory of a single many-sorted structure, and its models differ in essence only on how they interpret the variables. Each model of $\mathfrak{T}_S$ interprets Element as some *countably infinite* set E, Set as the set of *finite* subsets of E, and Card as $\mathbb{N}$. The signature of $\mathfrak{T}_S$ has the following predicate and function symbols, summarized in Fig. 1: the usual symbols of *linear* integer arithmetic, the usual set composition operators, an empty set ($\emptyset$) and a singleton set ($\{\cdot\}$) constructor, and a cardinality operator (card$(\cdot)$), all interpreted as expected. The signature includes also symbols for the cardinality comparison (<), subset ($\sqsubseteq$) and membership ($\sqsubset$) predicates.

We call *set term* any term of sort Set or of the form card(s), and *cardinality term* any term of sort Card with no occurrences of card$(\cdot)$. A *set constraint* is an atomic formula of the form $s \approx t$, $s \sqsubseteq t$, $e \sqsubset t$ or their negation, with s and t set terms and e a term of sort Element. A *cardinality constraint* is a [dis]equality $[\neg]c \approx d$ or an inequality $c <d$ or $c >= d$ where c and d are cardinality terms. An *element constraint* is a [dis]equality $[\neg]x \approx y$ where x and y are variables of sort Element. A $\mathfrak{T}_S$*-constraint* is a set, cardinality or element constraint.

We will use x, y for variables of sort Element; S, T, U for variables of sort Set; s, t, u, v for terms of sort Set; and c with subscripts for variables of sort Card. Given $\mathcal{C}$, a set of constraints, Vars$(\mathcal{C})$ (respectively, Terms$(\mathcal{C})$) denotes the set of variables (respectively, terms) in $\mathcal{C}$. For notational convenience, we fix an injective mapping from terms of sort Set to variables of sort Card that allows us to associate to each such term s a unique cardinality variable c_s.

We are interested in checking the satisfiability in $\mathfrak{T}_S$ of finite sets of $\mathfrak{T}_S$-constraints. While this problem is decidable, it has high worst-case time complex-

[2] An extension that allows σ to be interpreted as finite by relying on polite combination [8] is planned as future work.

ity [14]. So our efforts are in the direction of producing a solver for $\mathfrak{T}_S$-constraints that is efficient in practice, in addition to being correct and terminating. Our solver relies on the modular combination of a solver for set constraints and an off-the-shelf solver for linear integer arithmetic, which handles arithmetic constraints over set cardinalities.

3 A Calculus for the Theory

In this section, we describe a tableaux-style calculus capturing the essence of our combined solver for $\mathfrak{T}_S$. As we describe in the next section, that calculus admits a proof procedure that decides the satisfiability of $\mathfrak{T}_S$-constraints.

For simplicity, we consider as input to the calculus only conjunctions $\mathcal{C}$ of constraints whose set constraints are in *flat form*. These are (well-sorted) set constraints of the form $S \approx T$, $S \not\approx T$, $S \approx \emptyset$, $S \approx \{x\}$, $S \approx T \sqcup U$, $S \approx T \sqcap U$, $S \approx T \setminus U$, $x \sqsubseteq S$, $x \not\sqsubseteq S$, or $c_S \approx \mathsf{card}(S)$, where S, T, U, c_S, and x are variables of the expected sort. We also assume that any set variable S of $\mathcal{C}$ appears in at most one union, intersection or set difference term. Thanks to common satisfiability-preserving transformations,[3] all of these assumptions can be made without loss of generality.

The calculus is described as a set of derivation rules which modify a *state* data structure. A state is either the special state unsat or a tuple of the form $\langle \mathcal{S}, \mathcal{M}, \mathcal{A}, \mathcal{G} \rangle$, where $\mathcal{S}$ is a set of set constraints, $\mathcal{M}$ is a set of element constraints, $\mathcal{A}$ is a set of cardinality constraints, and $\mathcal{G}$ is a directed graph over set terms with nodes $V(\mathcal{G})$ and edges $E(\mathcal{G})$. Since cardinality constraints can be processed by a standard arithmetic solver, and element constraints by a simple equality solver,[4] we present and discuss only rules that deal with set constraints.

The derivation rules are provided in Fig. 2 through 9 in *guarded assignment form*. In such form, the premises of a rule refer to the current state and the conclusion describes how each state component is changed, if at all, by the rule's application. A derivation rule *applies* to a state σ if all the conditions in the rule's premises hold for σ *and* the resulting state is different from σ. In the rules, we write S, t as an abbreviation for $S \cup \{t\}$. Rules with two or more conclusions separated by the symbol $\|$ are non-deterministic branching rules.

The rules are such that it is possible to generate a closed tableau (or *derivation tree*) from an initial state $\langle \mathcal{S}_0, \mathcal{M}_0, \mathcal{A}_0, \mathcal{G}_0 \rangle$, where $\mathcal{G}_0$ is an empty graph, if and only if the conjunction of all the constraints in $\mathcal{S}_0 \cup \mathcal{M}_0 \cup \mathcal{A}_0$ is unsatisfiable in $\mathfrak{T}_S$. Broadly speaking, the derivation rules can be divided into three categories. First are those that reason about membership constraints (of form $x \sqsubseteq S$). These rules only update the components $\mathcal{S}$ and $\mathcal{M}$ of the current state, although their premises may depend on other parts of the state, in particular, the nodes of the graph $\mathcal{G}$. Second are rules that handle constraints of the form

---

3 Including replacing constraints of the form $s \sqsubseteq t$ with $s \approx (s \sqcap t)$.

4 Recall that $\mathfrak{T}_S$ has no terms of sort $\mathsf{Element}$ besides variables.

Union Down I
$$\frac{x \not\sqsubseteq s \sqcup t \in \mathcal{S}^*}{\mathcal{S} := \mathcal{S} \lhd (x \not\sqsubseteq s) \lhd (x \not\sqsubseteq t)}$$

Union Down II
$$\frac{x \sqsubseteq s \sqcup t \in \mathcal{S}^* \qquad \{u, v\} = \{s, t\} \qquad x \not\sqsubseteq u \in \mathcal{S}^*}{\mathcal{S} := \mathcal{S} \lhd (x \sqsubseteq v)}$$

Union Up I
$$\frac{x \not\sqsubseteq s \in \mathcal{S}^* \qquad x \not\sqsubseteq t \in \mathcal{S}^* \qquad s \sqcup t \in \mathcal{T}}{\mathcal{S} := \mathcal{S} \lhd (x \not\sqsubseteq s \sqcup t)}$$

Union Up II
$$\frac{x \sqsubseteq u \in \mathcal{S}^* \qquad u \in \{s, t\} \qquad s \sqcup t \in \mathcal{T}}{\mathcal{S} := \mathcal{S} \lhd (x \sqsubseteq s \sqcup t)}$$

Inter Down I
$$\frac{x \sqsubseteq s \sqcap t \in \mathcal{S}^*}{\mathcal{S} := \mathcal{S} \lhd (x \sqsubseteq s) \lhd (x \sqsubseteq t)}$$

Inter Down II
$$\frac{x \not\sqsubseteq s \sqcap t \in \mathcal{S}^* \qquad \{u, v\} = \{s, t\} \qquad x \sqsubseteq u \in \mathcal{S}^*}{\mathcal{S} := \mathcal{S} \lhd (x \not\sqsubseteq v)}$$

Inter Up I
$$\frac{x \sqsubseteq s \in \mathcal{S}^* \qquad x \sqsubseteq t \in \mathcal{S}^* \qquad s \sqcap t \in \mathcal{T}}{\mathcal{S} := \mathcal{S} \lhd (x \sqsubseteq s \sqcap t)}$$

Inter Up II
$$\frac{x \not\sqsubseteq u \in \mathcal{S}^* \qquad u \in \{s, t\} \qquad s \sqcap t \in \mathcal{T}}{\mathcal{S} := \mathcal{S} \lhd (x \not\sqsubseteq s \sqcap t)}$$

Union split
$$\frac{x \sqsubseteq s \sqcup t \in \mathcal{S} \qquad x \sqsubseteq s, x \sqsubseteq t \notin \mathcal{S}^*}{\mathcal{S} := \mathcal{S} \lhd (x \sqsubseteq s) \ \| \ \mathcal{S} := \mathcal{S} \lhd (x \sqsubseteq t)}$$

Inter split
$$\frac{s \sqcap t \in \mathcal{T} \qquad \{u, v\} = \{s, t\} \qquad x \sqsubseteq u \in \mathcal{S}^* \qquad x \sqsubseteq v, x \not\sqsubseteq v \notin \mathcal{S}^*}{\mathcal{S} := \mathcal{S} \lhd (x \sqsubseteq v) \ \| \ \mathcal{S} := \mathcal{S} \lhd (x \not\sqsubseteq v)}$$

Fig. 2. Union and intersection rules.

$c_S \approx \mathsf{card}(S)$. The graph incrementally built by the calculus is central to satisfying these constraints. Third are rules for propagating element and cardinality constraints respectively to $\mathcal{M}$ and $\mathcal{A}$.

3.1 Set Reasoning Rules

Figures 2 and 3 focus on sets without cardinality. They are based on the MLSS decision procedure by Cantone and Zarba [6], though with some key differences. First, the rules operate over a set $\mathcal{T}$ of Set terms which may be larger than just the terms in $\mathcal{S}$. This generalization is required because of additional terms that may be introduced when reasoning about cardinalities. Second, the reasoning is done modulo equality. A final, technical difference is that we work with sets of ur-elements rather than untyped sets.

These rules rely on the following additional notation. Given a set $\mathcal{C}$ of constraints, let $\text{Terms}_{\mathsf{Sort}}(\mathcal{C})$ refer to terms of sort Sort in $\mathcal{C}$, with $\text{Terms}(\mathcal{C})$ denoting all terms in $\mathcal{C}$. We define the binary relation $\approx^*_{\mathcal{C}} \subseteq \text{Terms}(\mathcal{C}) \times \text{Terms}(\mathcal{C})$ to be the reflexive, symmetric, and transitive closure of the relation on terms induced by equality constraints in $\mathcal{C}$. Now, we define the following closures:

$$\begin{aligned}\mathcal{M}^* &= \{x \approx y \mid x \approx^*_{\mathcal{M}} y\} \cup \{x \not\approx y \mid \exists x', y'.\ x \approx^*_{\mathcal{M}} x',\ y \approx^*_{\mathcal{M}} y',\ x' \not\approx y' \in \mathcal{M}\}\\ \mathcal{S}^* &= \mathcal{S} \cup \{x \sqsubseteq s \mid \exists x', s'.\ x \approx^*_{\mathcal{M}} x',\ s \approx^*_{\mathcal{S}} s',\ x' \sqsubseteq s' \in \mathcal{S}\}\\ &\quad \cup \{x \not\sqsubseteq s \mid \exists x', s'.\ x \approx^*_{\mathcal{M}} x',\ s \approx^*_{\mathcal{S}} s',\ x' \not\sqsubseteq s' \in \mathcal{S}\}\end{aligned}$$

where x, y, x', y' in $\text{Terms}_{\mathsf{Element}}(\mathcal{M} \cup \mathcal{S})$, and s, s' in $\text{Terms}_{\mathsf{Set}}(\mathcal{S})$. Next, we define a left-associative operator $\triangleleft$. Intuitively, given a set of constraints $\mathcal{C}$ and a literal l, $\mathcal{C} \triangleleft (l)$ adds l to $\mathcal{C}$ only if l is not in $\mathcal{C}$'s closure. More precisely,

$$\mathcal{C} \triangleleft (l) = \begin{cases} \mathcal{C} & \text{if } l \in \mathcal{C}^* \\ \mathcal{C} \cup \{l\} & \text{otherwise.} \end{cases} \tag{1}$$

Finally, the set of *relevant* terms for these rules is denoted by $\mathcal{T}$ and consists of terms from $\mathcal{S}$ and $\mathcal{G}$: $\mathcal{T} = \text{Terms}(\mathcal{S}) \cup V(\mathcal{G})$.

Figure 2 shows the rules for reasoning about membership in unions and intersections. Each rule covers one case in which a new membership (or non-membership) constraint can be deduced. The justification for these rules is straightforward based on the semantics of the set operations. Due to space limitations, we do not show the rules that process set difference constraints. However, they are analogous to those given for union and intersection constraints. Figure 3 shows rules for singletons, disequalities, and contradictions. Note in particular that the Set Disequality rule introduces a variable y, denoting an element that is in one set but not in the other.

Singleton
$$\frac{\{x\} \in \mathcal{T}}{\mathcal{S} := \mathcal{S} \triangleleft (x \sqsubseteq \{x\})}$$

Single Member
$$\frac{x \sqsubseteq \{y\} \in \mathcal{S}^*}{\mathcal{M} := \mathcal{M} \triangleleft (x \approx y)}$$

Single Non-member
$$\frac{x \not\sqsubseteq \{y\} \in \mathcal{S}^*}{\mathcal{M} := \mathcal{M} \triangleleft (x \not\approx y)}$$

Set Disequality
$$\frac{s \not\approx t \in \mathcal{S}^* \quad \nexists x \in \text{Terms}(\mathcal{S}) \text{ such that } x \sqsubseteq s \in \mathcal{S}^* \text{ and } x \not\sqsubseteq t \in \mathcal{S}^* \quad \nexists x \in \text{Terms}(\mathcal{S}) \text{ such that } x \not\sqsubseteq s \in \mathcal{S}^* \text{ and } x \sqsubseteq t \in \mathcal{S}^*}{\mathcal{S} := \mathcal{S} \triangleleft (y \sqsubseteq s) \triangleleft (y \not\sqsubseteq t) \quad \| \quad \mathcal{S} := \mathcal{S} \triangleleft (y \not\sqsubseteq s) \triangleleft (y \sqsubseteq t)}$$

Eq Unsat
$$\frac{(x \not\approx x) \in \mathcal{M}^*}{\mathsf{unsat}}$$

Set Unsat
$$\frac{(x \sqsubseteq s) \in \mathcal{S}^* \qquad (x \not\sqsubseteq s) \in \mathcal{S}^*}{\mathsf{unsat}}$$

Empty Unsat
$$\frac{(x \sqsubseteq \emptyset) \in \mathcal{S}^*}{\mathsf{unsat}}$$

Fig. 3. Singleton, disequality and contradiction rules. Here, y is a fresh variable.

Example 1. Let $\mathcal{S} = \{S \approx A \sqcup B, S \approx C \sqcap D, x \sqsubseteq C, x \not\sqsubseteq D, y \not\sqsubseteq S, y \sqsubseteq D\}$. Using the rules in Fig. 2, we can directly deduce the additional constraints: $x \not\sqsubseteq C \sqcap D$ (by Inter Up II), $x \not\sqsubseteq A$, $x \not\sqsubseteq B$, $y \not\sqsubseteq A$, $y \not\sqsubseteq B$ (by Union Down I), and $y \not\sqsubseteq C$ (by Inter Down II). This gives a complete picture, modulo equality, of exactly which sets contain x and y. □

3.2 Cardinality of Sets

The next set of rules is based on two observations: (*i*) the cardinality of two sets, and that of their union, intersection and set difference are inter-related; (*ii*) if two set terms are asserted to be equal, their cardinalities must match. Figure 4 shows the Venn regions for two sets, T and U. Notice the following relationships: T is a disjoint union of $T \setminus U$ and $T \sqcap U$; $T \sqcup U$ is a disjoint union of $T \setminus U$ and $T \sqcap U$ and $U \setminus T$; and U is a disjoint union of $T \sqcap U$ and $U \setminus T$. Knowing that the sets are *disjoint* is important; it allows us to infer the constraints:

$$\begin{aligned} \mathsf{card}(T) &\approx \mathsf{card}(T \setminus U) + \mathsf{card}(T \sqcap U) \\ \mathsf{card}(T \sqcup U) &\approx \mathsf{card}(T \setminus U) + \mathsf{card}(T \sqcap U) + \mathsf{card}(U \setminus T) \\ \mathsf{card}(U) &\approx \mathsf{card}(U \setminus T) + \mathsf{card}(T \sqcap U). \end{aligned}$$

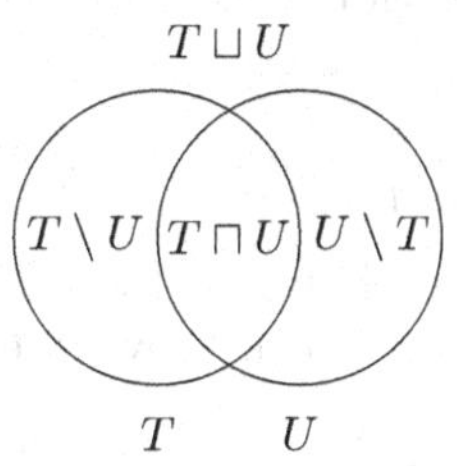

Fig. 4. Venn regions for T and U.

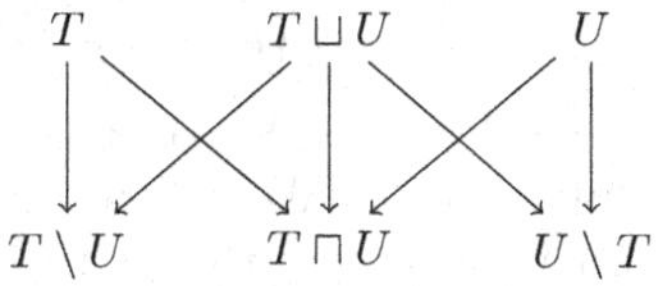

Fig. 5. The same structure as a graph.

We can represent these same relationships using a graph. The nodes of the graph are set terms, and each node has the property that it is the disjoint union of its children in the graph. The graph for the regions in Fig. 4 is shown in Fig. 5. We ensure that the graph contains all nodes whose cardinality is implicitly or explicitly constrained by the current state. Set terms with implicit cardinality constraints include (*i*) union, intersection, and set difference terms appearing in $\mathcal{S}$, for which one of the operands is already in the graph; and (*ii*) terms occurring in an equality whose other member is already in the graph. A careful analysis[5] reveals that we can actually avoid adding intersection terms $t \sqcap u$ unless both t and u are already in the graph, and set difference terms $t \setminus u$ unless t is already in the graph.

The rules in Fig. 6 make use of a function add which takes a graph $\mathcal{G}$ and a term s and returns the graph $\mathcal{G}'$ defined as follows:

1. For $s = T$ or $s = \emptyset$ or $s = \{x\}$:
   $V(\mathcal{G}') = V(\mathcal{G}) \cup \{s\}$
   $E(\mathcal{G}') = E(\mathcal{G})$
2. For $s = T \sqcap U$ or $s = T \setminus U$:
   $V(\mathcal{G}') = V_2 = V(\mathcal{G}) \cup \{T, U, T \setminus U, T \sqcap U, U \setminus T\}$

[5] See completeness proof in [1, Chap. 2] for further details.

$E(\mathcal{G}') = E_e = E(\mathcal{G}) \cup \{(T, T \setminus U), (T, T \sqcap U), (U, T \sqcap U), (U, U \setminus T)\}$

3. For $s = T \sqcup U$ and V_2 and E_2 as above:
   $V(\mathcal{G}') = V_2 \cup \{T \sqcup U\}$
   $E(\mathcal{G}') = E_2 \cup \{(T \sqcup U, T \setminus U), (T \sqcup U, T \sqcap U), (T \sqcup U, U \setminus T)\}$

Recall that, by assumption, each set variable participates in at most one union, intersection, or set difference. This ensures that edges from a set variable node are added only once, maintaining the invariant that its children in the graph are disjoint. Terms with explicit constraints on their cardinality are added to the graph by Introduce Card. Terms that have implicit constraints on their cardinality, specifically, singletons and the empty set, are added by rules Introduce Singleton and Introduce Empty Set.

If two nodes s and t in the graph are asserted to be equal (that is, $s \approx t \in \mathcal{S}$ or $t \approx s \in \mathcal{S}$), we can ensure they have the same cardinality by systematically modifying the graph. Let $\mathcal{L}(n)$ denote the set of leaf nodes for the subtree rooted at node n which are not known to be empty. Formally,

$$\mathcal{L}(n) = \{n' \in \text{Leaves}(n) \mid n' \approx \emptyset \notin \mathcal{S}^*\}, \tag{2}$$

where Leaves $(v) = \{w \in V(\mathcal{G}) \mid C(w) = \emptyset, w \text{ is reachable from } v\}$ and $C(w)$ denotes the children of w. We call two nodes n and n' *merged* if they have the same set of nonempty leaves, that is if $\mathcal{L}(n) = \mathcal{L}(n')$.

Introduce Eq Right
$$\frac{S \approx t \in \mathcal{S} \qquad S \in V(\mathcal{G}) \qquad t \notin V(\mathcal{G})}{\mathcal{G} := \text{add}(\mathcal{G}, t)}$$

Introduce Union
$$\frac{S \approx T \sqcup U \in \mathcal{S} \qquad T \sqcup U \notin V(\mathcal{G}) \qquad T \in V(\mathcal{G}) \text{ or } U \in V(\mathcal{G})}{\mathcal{G} := \text{add}(\mathcal{G}, T \sqcup U)}$$

Introduce Eq Left
$$\frac{S \approx t \in \mathcal{S} \qquad S \notin V(\mathcal{G}) \qquad t \in V(\mathcal{G})}{\mathcal{G} := \text{add}(\mathcal{G}, S)}$$

Introduce Inter
$$\frac{S \approx T \sqcap U \in \mathcal{S} \qquad T \sqcap U \notin V(\mathcal{G}) \qquad T \in V(\mathcal{G}) \qquad U \in V(\mathcal{G})}{\mathcal{G} := \text{add}(\mathcal{G}, T \sqcap U)}$$

Introduce Card
$$\frac{c_s \approx \mathsf{card}(S) \in \mathcal{S}}{\mathcal{G} := \text{add}(\mathcal{G}, S)}$$

Introduce Singleton
$$\frac{\{x\} \in \text{Terms}(\mathcal{S})}{\mathcal{G} := \text{add}(\mathcal{G}, \{x\})}$$

Introduce Empty Set
$$\frac{}{\mathcal{G} := \text{add}(\mathcal{G}, \emptyset)}$$

Fig. 6. Graph introduction rules.

The rules in Fig. 7 ensure that for all equalities over set terms, the corresponding nodes in the graph are merged. Consider an equality $s \approx t$. Rule Merge Equality I handles the case when either $\mathcal{L}(s)$ or $\mathcal{L}(t)$ is a proper subset of the other by constraining the extra leaves in the superset to be empty. Rule Merge Equality II handles the remaining case where neither is a subset of the other.

Merge Equality I

$$\frac{s \approx t \in \mathcal{S} \qquad s, t, \emptyset \in V(\mathcal{G}) \qquad \{u, v\} = \{s, t\} \qquad \mathcal{L}(u) \subsetneq \mathcal{L}(v)}{\mathcal{S} := \{s' \approx \emptyset \mid s' \in \mathcal{L}(v) \setminus \mathcal{L}(u)\} \cup \mathcal{S}}$$

Merge Equality II

$$\frac{s \approx t \in \mathcal{S} \qquad s, t \in V(\mathcal{G}) \qquad \mathcal{L}(s) \nsubseteq \mathcal{L}(t) \qquad \mathcal{L}(t) \nsubseteq \mathcal{L}(s)}{\mathcal{G} := \text{merge}(\mathcal{G}, s, t)}$$

Fig. 7. Merge rules.

The graph $\mathcal{G}' = \text{merge}(\mathcal{G}, s, t)$ is defined as follows, where $L_1 = \mathcal{L}(s) \setminus \mathcal{L}(t)$ and $L_2 = \mathcal{L}(t) \setminus \mathcal{L}(s)$:

$$V(\mathcal{G}') = V(\mathcal{G}) \cup \{l_1 \sqcap l_2 \mid l_1 \in L_1, l_2 \in L_2\}$$
$$E(\mathcal{G}') = E(\mathcal{G}) \cup \{(l_1, l_1 \sqcap l_2), (l_2, l_1 \sqcap l_2) \mid l_1 \in L_1, l_2 \in L_2\}$$

We denote by $\hat{\mathcal{G}}$ the collection of all of the following arithmetic constraints imposed by graph $\mathcal{G}$:

1. For each set term $s \in V(\mathcal{G})$, its corresponding cardinality is the sum of the corresponding non-empty leaf nodes: $\left\{c_s \approx \sum_{t \in \mathcal{L}(s)} c_t \;\middle|\; s \in V(\mathcal{G})\right\}$.
2. Each cardinality is non-negative: $\{c_s \texttt{>=} 0 \mid s \in V(\mathcal{G})\}$.
3. A singleton set has cardinality 1: $\{c_s \approx 1 \mid s \in V(\mathcal{G}),\ s = \{x\}\}$.
4. The empty set has cardinality 0: $\{c_s \approx 0 \mid s \in V(\mathcal{G}),\ s = \emptyset\}$.

Rule Arithmetic contradiction, shown in Fig. 8 makes use of the arithmetic solver to check whether the constraints in $\hat{\mathcal{G}}$ are inconsistent with the input constraints. Also shown is rule Guess Empty Set which can be used to guess if a leaf node is empty. This is useful to apply early on, to reduce the impact of merge operations on the size of the graph. Here and in Fig. 9, Leaves $(\mathcal{G}) = \{v \in V(\mathcal{G}) \mid C(v) = \emptyset\}$.

Arithmetic contradiction

$$\frac{\mathcal{A} \cup \hat{\mathcal{G}} \models_{\mathcal{T}_A} \bot}{\textsf{unsat}}$$

Guess Empty Set

$$\frac{t \in \text{Leaves}(\mathcal{G})}{\mathcal{S} := \mathcal{S} \triangleleft (t \approx \emptyset) \quad || \quad \mathcal{S} := \mathcal{S} \triangleleft (t \not\approx \emptyset)}$$

Fig. 8. Additional graph rules.

3.3 Cardinality and Membership Interaction

The rules in Fig. 9 propagate consequences of set membership constraints to the sets $\mathcal{M}$ and $\mathcal{A}$. Let $\mathcal{E}$ denote the set of equalities in $\mathcal{M}$, and let $[x]_{\mathcal{E}}$ denote the equivalence class of x with respect to $\mathcal{E}$. Then for a Set term t, $t_{\mathcal{S}} = \{[x]_{\mathcal{E}} \mid x \sqsubseteq t \in \mathcal{S}^*\}$, the set of equivalence classes of elements known to be in t. The notation $\mathcal{A} \Rrightarrow c_t \geq n$ means that $c_t \texttt{>=} k \in \mathcal{A}$ for some concrete constant $k \geq n$.

$$\textsc{Members Arrangement}\quad \frac{t \in \text{Leaves}(\mathcal{G}) \quad \mathcal{A} \not\Rightarrow c_t \geq |t_S| \quad [x]_\mathcal{E}, [y]_\mathcal{E} \in t_S \quad [x]_\mathcal{E} \neq [y]_\mathcal{E} \quad x \not\approx y \notin \mathcal{M}^*}{\mathcal{M} := \mathcal{M} \triangleleft (x \approx y) \quad \| \quad \mathcal{M} := \mathcal{M} \triangleleft (x \not\approx y)}$$

$$\textsc{Guess Lower Bound}\quad \frac{t \in \text{Leaves}(\mathcal{G}) \qquad \mathcal{A} \not\Rightarrow c_t \geq |t_S| \qquad c_t < |t_S| \notin \mathcal{A}}{\mathcal{A} := c_t >= |t_S|, \mathcal{A} \quad \| \quad \mathcal{A} := c_t < |t_S|, \mathcal{A}}$$

$$\textsc{Propagate Minsize}\quad \frac{x_1 \sqsubseteq s, \ldots, x_n \sqsubseteq s \in \mathcal{S}^* \quad \mathcal{A} \not\Rightarrow c_s \geq n \quad x_i \not\approx x_j \in \mathcal{M}^* \text{ for all } 1 \leq i < j \leq n}{\mathcal{A} := c_s >= n, \mathcal{A}}$$

Fig. 9. Cardinality and membership interaction rules.

Rule Members Arrangement is used to decide which elements of a set should be equal or disequal. Once applied to completion, Rule Propagate Minsize can then be used to determine a lower bound for the cardinality of that set. Rule Guess Lower Bound can be used to short-circuit this process by guessing a conservative lower bound based on the number of distinct equivalence classes of elements known to be members of a set. If this does not lead to a contradiction, a model can be found without resorting to extensive use of Members Arrangement.

Example 2. Consider again the constraints from Example 1, but now augmented with cardinality constraints $\{c_S \approx \mathsf{card}(S),\ c_C \approx \mathsf{card}(C),\ c_D \approx \mathsf{card}(D)\}$ and arithmetic constraints $\{c_S >= 4,\ c_C + c_D < 10\}$. Using the rules in Fig. 6, the following nodes get added to the graph: S, C, D (by Introduce Card), $A \sqcup B$, $C \sqcap D$ (by Introduce Eq Right). $A \sqcup B$ is added with children $A \setminus B$, $A \sqcap B$, and $B \setminus A$; and by adding $C \sqcap D$, we also get $C \setminus D$ and $D \setminus C$, with the corresponding edges from C and D. Now, using two applications of Merge Equality II, we force the sets S, $A \sqcup B$ and $C \sqcap D$ to have the same set of 3 leaves, labeled $S \sqcap (A \setminus B) \sqcap (C \sqcap D)$, $S \sqcap (A \sqcap B) \sqcap (C \sqcap D)$, and $S \sqcap (B \setminus A) \sqcap (C \sqcap D)$. Let us call these nodes l_1, l_2, and l_3 for convenience. Let us also designate $l_4 = C \setminus D$ and $l_5 = D \setminus C$. Notice that the induced arithmetic constraints now include $c_S \approx c_{l_1} + c_{l_2} + c_{l_3}$, $c_C \approx c_{l_1} + c_{l_2} + c_{l_3} + c_{l_4}$, and $c_D \approx c_{l_1} + c_{l_2} + c_{l_3} + c_{l_5}$. With the addition of $C \setminus D$ and $D \setminus C$ to the graph, these are also added to $\mathcal{T}$. We can then deduce $x \sqsubseteq C \setminus D$ and $y \sqsubseteq D \setminus C$ using the (not shown) rules for propagation over set difference. Finally, we can use Propagate Minsize to deduce $c_{l_4} >= 1$ and $c_{l_5} >= 1$. It is now not hard to see that using pure arithmetic reasoning, we can deduce that $c_C + c_D >= 10$ which leads to unsat using Arithmetic contradiction. □

4 Calculus Correctness

Our calculus is terminating and sound for any derivation strategy, that is, regardless of how the rules are applied. It is also refutation complete for any *fair* strategy, defined as a strategy that does not delay indefinitely the application of an applicable derivation rule. For space reasons, we only outline the proof arguments here. Complete proofs are given in [1].

We group the derivation rules of the calculus in the following subsets.

$\mathcal{R}_1$: membership predicate reasoning rules, from Figs. 2 and 3.
$\mathcal{R}_2$: graph rules to reason about cardinality, from Figs. 6, 7 and 8.
$\mathcal{R}_3$: rules from Fig. 9 other than Rule GUESS LOWER BOUND.
$\mathcal{R}_4$: Rule GUESS LOWER BOUND.

The rules are used to construct derivation trees. A *derivation tree* is a tree over states, where the root is a state of the form $\langle \mathcal{S}_0, \mathcal{M}_0, \mathcal{A}_0, (\emptyset, \emptyset)\rangle$, (and $\mathcal{S}_0$, $\mathcal{M}_0$, $\mathcal{A}_0$ obey the input constraints mentioned at the beginning of Sect. 3), and where the children of each non-root node are obtained by applying one of the derivation rules of the calculus to that node. A branch of a derivation tree is *closed* if it ends with unsat; it is *saturated* with respect to a set $\mathcal{R}$ of rules if it is not closed and no rules in $\mathcal{R}$ apply to its leaf. A derivation tree is *closed* if all of its branches are closed. A derivation tree *derives* from a derivation tree T if it is obtained from T by the application of exactly one of the derivation rules to one of T's leaves.

Let S be a set of $\mathfrak{T}_S$-constraints. A *derivation (of S)* is a sequence $(T_i)_{0\leq i\leq\kappa}$ of derivation trees, with κ finite or countably infinite, such that T_{i+1} derives from T_i for all i, and T_0 is a one-node tree whose root is a state $\langle \mathcal{S}_0, \mathcal{M}_0, \mathcal{A}_0, (\emptyset, \emptyset)\rangle$ where $\mathcal{S}_0 \cup \mathcal{M}_0 \cup \mathcal{A}_0$ is $\mathfrak{T}_S$-equisatisfiable with S. A *refutation (of S)* is a (finite) derivation of S that ends with a closed tree.

4.1 Termination

Proposition 1 (Termination). *Every derivation in the calculus is finite.*

Proof (Sketch). It is enough to show that every application of a derivation rule to a state produces smaller states with respect to a well-founded relation $\succ$ over states other than unsat. For simplicity, we ignore the rule GUESS LOWER BOUND, although the proof could be extended to that rule as well. To define $\succ$ we first define the following functions, each of which maps a state $\sigma = \langle \mathcal{S}, \mathcal{M}, \mathcal{A}, \mathcal{G}\rangle$ to a natural number (from $\mathbb{N}$).

- $f_1(\sigma)$: number of equalities $t_1 \approx t_2$ in $\mathcal{S}$ such that either $t_1 \notin V(\mathcal{G})$, $t_2 \notin V(\mathcal{G})$, or $\mathcal{L}(t_1) \neq \mathcal{L}(t_2)$.
- $f_2(\sigma)$: cardinality of $(\mathrm{Terms}_{\mathsf{Set}}(\mathcal{S}) \cup \{\emptyset\}) \setminus V(\mathcal{G})$.
- $f_3(\sigma)$: cardinality of $\{t \in \mathrm{Leaves}\,(\mathcal{G}) \mid t \approx \emptyset \notin \mathcal{S}^*, t \not\approx \emptyset \notin \mathcal{S}^*\}$.
- $f_4(\sigma)$: number of disequalities $t_1 \not\approx t_2$ in $\mathcal{S}$ such that the premise of SET DISEQUALITY holds.
- $f_5(\sigma)$: cardinality of $T = \mathrm{Terms}_{\mathsf{Set}}(\mathcal{S}) \cup \{\emptyset\} \cup V(\mathcal{G})$.
- $f_6(\sigma)$: cardinality of $\mathrm{Terms}_{\mathsf{Element}}(\mathcal{S} \cup \mathcal{M})$.
- $f_7(\sigma)$: $2 \cdot f_6(\sigma)^2$ minus the cardinality of $\mathcal{M}^*$.[6]
- $f_8(\sigma)$: $2 \cdot f_5(\sigma)^2 + 2 \cdot f_5(\sigma) \cdot f_6(\sigma)$ minus the cardinality of $\mathcal{S}^*$.[7]
- $f_9(\sigma)$: cardinality of $T \setminus \{t \in \mathrm{Leaves}\,(\mathcal{G}) \mid \mathcal{A} \not\Rightarrow c_t \geq |t_S|\}$.

Let $(\mathbb{N}^9, >^9_{\mathsf{lex}})$ be the 9-fold lexicographic product of $(\mathbb{N}, >)$. We define $\succ$ as the relation such that $\sigma \succ \sigma'$ iff $(f_1(\sigma), \ldots, f_9(\sigma)) >^9_{\mathsf{lex}} (f_1(\sigma'), \ldots, f_9(\sigma'))$. □

[6] Note that the cardinality of $\mathcal{M}^*$ is at most $2 \cdot (f_6(\sigma))^2$.
[7] One can show that this value is non-negative.

4.2 Completeness

We develop the proof in stages, proving properties about different subsets of rules. We start with a proposition about the rule set $\mathcal{R}_1$.

Proposition 2. *Let $\langle \mathcal{S}, \mathcal{M}, \mathcal{A}, \mathcal{G} \rangle$ be a state to which none of rules in $\mathcal{R}_1$ apply. There is a model $\mathfrak{S}$ of $\mathfrak{T}_S$ that satisfies the constraints $\mathcal{S}$ and $\mathcal{M}$ and has the following properties.*

1. *For all $x, y \in \mathrm{Vars}(\mathcal{M}) \cup \mathrm{Vars}(\mathcal{S})$ of sort* Element*, $x^{\mathfrak{S}} = y^{\mathfrak{S}}$ if and only if $x \approx y \in \mathcal{M}^*$.*
2. *For all $S \in \mathrm{Vars}(\mathcal{S})$ of sort* Set*, $S^{\mathfrak{S}} = \{x^{\mathfrak{S}} \mid x \sqsubseteq S \in \mathcal{S}^*\}$.*
3. *For all $c_S \in \mathrm{Vars}(\mathcal{S})$ of sort* Card*, $c_S^{\mathfrak{S}} = |S^{\mathfrak{S}}|$.*

For the next two results, let $\langle \mathcal{S}, \mathcal{M}, \mathcal{A}, \mathcal{G} \rangle$ be the leaf of a branch saturated with respect to rules $\mathcal{R}_1 \cup \mathcal{R}_2 \cup \mathcal{R}_3$ in a derivation tree. The first result is about the effects of the rules in $\mathcal{R}_2$. The second is about the rules in $\mathcal{R}_3$.

Proposition 3. *For every $s \in V(\mathcal{G})$ the following holds.*

1. *If $s \approx t \in \mathcal{S}$ or $t \approx s \in \mathcal{S}$ for some t, then $\mathcal{L}(s) = \mathcal{L}(t)$.*
2. *If $s = T \sqcup U$, then $\mathcal{L}(T \sqcup U) = \mathcal{L}(T) \cup \mathcal{L}(U)$.*
3. *If $s = T \sqcap U$, then $\mathcal{L}(T \sqcap U) = \mathcal{L}(T) \cap \mathcal{L}(U)$.*
4. *If $s = T \setminus U$, then $\mathcal{L}(T \setminus U) = \mathcal{L}(T) \setminus \mathcal{L}(U)$.*
5. *For all distinct $t, u \in \mathrm{Leaves}(s)$, $\models_{\mathfrak{T}_S} t \sqcap u \approx \emptyset$.*
6. *$\{t \approx u \mid t \approx u \in \mathcal{S}^*\} \models_{\mathfrak{T}_S} s \approx \bigsqcup_{t \in \mathcal{L}(s)} t$.*

Proposition 4. *Let $\mathfrak{S}$ be an interpretation as the one specified in Proposition 2 and let $\mathfrak{A}$ be any model of $\mathfrak{T}_S$ satisfying $\mathcal{A}$. Then, for all $t \in \mathcal{L}(\mathcal{G})$, $c_t^{\mathfrak{A}} \geq |t^{\mathfrak{S}}|$.*

Completeness is a direct consequence of the following result.

Proposition 5. *Let $\mathcal{D}$ be a derivation tree with root $\langle \mathcal{S}_0, \mathcal{M}_0, \mathcal{A}_0, (\emptyset, \emptyset) \rangle$. If $\mathcal{D}$ has a branch saturated with respect to rules $\mathcal{R}_1 \cup \mathcal{R}_2 \cup \mathcal{R}_3$, then there exists a model $\mathfrak{I}$ of $\mathfrak{T}_S$ that satisfies $\mathcal{S}_0 \cup \mathcal{M}_0 \cup \mathcal{A}_0$.*

Proof (Sketch). We build the model of the leaf nodes in the graph by modifying as needed the model obtained from Proposition 2. We add additional elements to these sets to make the cardinalities match the model satisfying the arithmetic constraints and the constraints induced by the graph. Propositions 3 and 4 ensure that it is always possible to do so without violating the set constraints. □

Proposition 6 (Completeness). *Under any fair derivation strategy, every derivation of a set S of $\mathfrak{T}_S$-unsatisfiable constraints extends to a refutation.*

Proof. Contrapositively, suppose that S has a derivation **D** that cannot be extended to a refutation. By Proposition 1, **D** must be extensible to one that ends with a tree with a saturated branch. By Proposition 5, S is satisfiable in $\mathfrak{T}_S$. □

4.3 Soundness

We start by showing that every rule preserves constraint satisfiability.

Lemma 1. *For every rule of the calculus, the premise state is satisfied by a model $\mathfrak{I}_p$ of $\mathfrak{T}_S$ iff one of its conclusion configurations is satisfied by a model $\mathfrak{I}_c$ of $\mathfrak{T}_S$ where $\mathfrak{I}_p$ and $\mathfrak{I}_c$ agree on the variables shared by the two states.*

Proof (Sketch). Soundness of the rules in Figs. 2 and 3 follows trivially from the semantics of set operators and the definition of $\mathcal{S}^*$. Soundness of MERGE EQUALITY I follows from properties of the graph (see Proposition 3, in particular the property that leaf terms are disjoint). The rules in Fig. 6 and rule MERGE EQUALITY II do not modify the constraints, but we need them to establish properties of the graph. Soundness of the induced graph constraints in ARITHMETIC CONTRADICTION follows from Proposition 3 (in particular properties 3 and 3). Soundness of PROPAGATE MINSIZE follows from the semantics of cardinality. Soundness of GUESS EMPTY SET, MEMBERS ARRANGEMENT and GUESS LOWER BOUND is trivial. □

Proposition 7 (Soundness). *Every set of $\mathfrak{T}_S$-constraints that has a refutation is $\mathfrak{T}_S$-unsatisfiable.*

Proof (Sketch). Given Lemma 1, one can show by structural induction on derivation trees that the root of any closed derivation tree is $\mathfrak{T}_S$-unsatisfiable. The claim then follows from the fact that every refutation of a set S of $\mathfrak{T}_S$-constraints starts with a state $\mathfrak{T}_S$-equisatisfiable with S. □

5 Evaluation

We have implemented a decision procedure based on the calculus above in the SMT solver CVC4 [2]. We describe a high-level, non-deterministic version of it here, followed by an initial evaluation on benchmarks from program analysis.

5.1 Derivation Strategy

The decision procedure can be thought of as a specific strategy for applying the rules given in Sect. 3, divided into the sets $\mathcal{R}_1$, ..., $\mathcal{R}_4$ introduced in Sect. 4.

Our derivation strategy can be summarized as follows. We start with the derivation from the initial state $\langle \mathcal{S}_0, \mathcal{M}_0, \mathcal{A}_0, \mathcal{G}_0 \rangle$ with $\mathcal{G}_0$ the empty graph, as described in Sect. 3, and apply the steps listed below, in the given order. The steps are described as rules being applied to a *current* branch. Initially, the current branch is the only branch in tree. On application of a rule with more than one conclusion, we select one of the branches (say, the left branch) as the current branch.

1. If a rule that derives `unsat` is applicable to the current branch, we apply one and close the branch. We then pick another open branch as the current branch and repeat Step 1. If no open branch exists, we stop and output `unsat`.

2. If a *propagation* rule (those with one conclusion) in $\mathcal{R}_1$ is applicable, apply one and go to Step 1.
3. If a *split* rule (those with more than one conclusion) in $\mathcal{R}_1$ is applicable, apply one and go to Step 1.
4. If Guess Empty Set rule is applicable, apply it and go to Step 1.
5. If an introduce or merge rule in $\mathcal{R}_2$ is applicable, apply it and go to Step 1.
6. If any of the remaining rules is applicable, apply one and go to Step 1.
7. At this point, the current branch is saturated. Stop and output sat.

Note that if there are no constraints involving the cardinality operator, then steps 1 to 3 above are sufficient for completeness.

file	output	time (s.)	# V	# L
vc1	unsat	0.00	3	3
vc2a	unsat	0.01	17	8
vc2b	sat	0.01	15	7
vc2	unsat	0.00	8	5
vc3a	unsat	0.00	6	0
vc3b	sat	0.01	17	8
vc3	unsat	0.00	6	0
vc4b	sat	0.22	45	16
vc4	unsat	0.07	57	18
vc5b	sat	1.71	71	22
vc5	unsat	0.36	68	21
vc6a	unsat	0.02	34	14
vc6b	sat	0.14	31	13
vc6c	sat	0.06	34	14
vc6	sat	0.02	38	18

(a) Jahob

file	output	time (s.)	# V	# L
vc1	1 sat/4 unsat	0.02	12	6
vc2	1 sat/3 unsat	0.07	39	23
vc3	2 sat/2 unsat	0.09	54	21
vc4	1 sat/3 unsat	0.02	0	0
vc5	2 sat/2 unsat	0.08	27	13
vc6	1 sat/3 unsat	0.01	0	0
vc7	2 sat/4 unsat	0.34	56	33
vc8	1 sat/3 unsat	0.01	0	0
vc9	2 sat/2 unsat	0.09	39	19
vc10	2 sat/2 unsat	0.32	94	32

(b) Leon

Fig. 10. Results on program verification benchmarks.

5.2 Experimental Evaluation

We evaluated our procedure on benchmarks obtained from verification of programs. The experiments were run on a machine with 3.40GHz Intel i7 CPU with a memory limit of 3 GB and timeout of 300 seconds. We used a development version of CVC4 for this evaluation.[8] Benchmarks are available on our website[9].

The first set of benchmarks consists of single query benchmarks obtained from verifying programs manipulating pointer-based data structures. These were generated by the Jahob system, and have been used to evaluate earlier work on decision procedures for finite sets and cardinality [10,11,13]. The results from running CVC4 on these benchmarks are provided in Fig. 10a. The output reported

[8] Git commit c833e17 at https://github.com/CVC4/CVC4/commit/c833e176 .

[9] http://cs.nyu.edu/~kshitij/setscard/.

by CVC4 is in the second column. The third column shows the solving time. The fourth and fifth columns give the maximum number of vertices (# V) and leaves[10] (# L) in the graph at any point during the run of the algorithm. Keeping the number of leaves low is important to avoid a blowup from the MERGE EQUALITY II rule.

Although we have not rerun the algorithms from [10,11,13], we report here the experimental results as stated in the respective papers. As the experiments were run on different machines the comparison is only indicative, but it does suggest that our algorithm has comparable performance.

In [11], the procedure from [10] is reported to solve 12 of the 15 benchmarks with a timeout of 100 seconds, while the novel procedure in [11] is reported to solve 11 of the 15 benchmarks with the same timeout. The best-performing previous algorithm ([13]) can solve all 15 benchmarks in under a second.[11] As another point of comparison, we tested the algorithm from [13] on a benchmark of the type mentioned in Sect. 1.1: a single constraint of the form $x \sqsubseteq A_1 \sqcup \ldots \sqcup A_{21}$. As expected, the algorithm failed (it ran out of memory after 85 seconds). In contrast, CVC4 solves this problem instantaneously.

Finally, another important difference compared to earlier work is that our implementation is completely integrated in an actively developed and maintained solver, CVC4.[12] To highlight the usefulness of an implementation in a full-featured SMT solver, we did a second evaluation on a set of incremental (i.e., multiple-query) benchmarks obtained from the Leon verification system [4]. These contain a mix of membership and cardinality constraints together with the theories of datatypes and bitvectors. The results of this evaluation are shown in Fig. 10b. The output column reports the number of sat and unsat queries in each benchmark. CVC4 successfully solves all of the queries in these benchmarks in under one second. To the best of our knowledge, no other SMT solver can handle this combination of theories.

6 Conclusion

We presented a new decision procedure for deciding finite sets with cardinality constraints and proved its correctness. A novel feature of the procedure is that it can reason directly and efficiently about both membership constraints and cardinality constraints. We have implemented the procedure in the CVC4 SMT solver, and demonstrated the feasibility as well as some advantages of our approach. We hope this work will enable the use of sets and cardinality in many new

[10] The # L statistic is updated only when explicitly computed, so the numbers are approximate. For the same reason, # L is 0 on certain benchmarks even though # V is not. This is because CVC4 was able to report unsat before the need for computing the set of leaves arose.

[11] Note that [13] includes a second set of benchmarks, but we were unable to evaluate our algorithm on these, as they were only made available in a non-standard format and were missing crucial datatype declarations.

[12] One reason we were unable to do a more thorough comparison with previous work is that those implementations are no longer being maintained.

applications. We also expect to use it to drive the development of a standard theory of sets under the SMT-LIB initiative.

Acknowledgements. We thank the reviewers for their valuable and constructive suggestions. We thank Viktor Kuncak and Etienne Kneuss for valuable scientific discussions and for providing the Leon benchmarks. We thank Philippe Suter for his help running the algorithm from [13].

References

1. Bansal, K.: Decision Procedures for Finite Sets with Cardinality and Local Theory Extensions. Ph.D. thesis, New York University, January 2016
2. Barrett, C., Conway, C.L., Deters, M., Hadarean, L., Jovanović, D., King, T., Reynolds, A., Tinelli, C.: CVC4. In: Gopalakrishnan, G., Qadeer, S. (eds.) CAV 2011. LNCS, vol. 6806, pp. 171–177. Springer, Heidelberg (2011)
3. Barrett, C., Sebastiani, R., Seshia, S., Tinelli, C.: Satisfiability modulo theories. In: Biere, A., Heule, M.J.H., van Maaren, H., Walsh, T. (eds.) Handbook of Satisfiability, vol. 185, pp. 825–885, chap. 26. IOS Press, February 2009
4. Blanc, R.W., Kneuss, E., Kuncak, V., Suter, P.: An overview of the Leon verification system: verification by translation to recursive functions. In: Scala Workshop (2013)
5. Cantone, D., Omodeo, E.G., Policriti, A.: Set Theory for Computing: From Decision Procedures to Logic Programming with Sets. Monographs in Computer Science. Springer, Heidelberg (2001)
6. Cantone, D., Zarba, C.G.: A new fast tableau-based decision procedure for an unquantified fragment of set theory. In: Caferra, R., Salzer, G. (eds.) FTP 1998. LNCS (LNAI), vol. 1761, pp. 126–136. Springer, Heidelberg (2000)
7. De Moura, L., Bjørner, N.: Generalized, efficient array decision procedures. In: Formal Methods in Computer-Aided Design (FMCAD 2009), pp. 45–52. IEEE (2009)
8. Jovanović, D., Barrett, C.: Polite theories revisited. In: Fermüller, C.G., Voronkov, A. (eds.) LPAR-17. LNCS, vol. 6397, pp. 402–416. Springer, Heidelberg (2010)
9. Kröning, D., Rümmer, P., Weissenbacher, G.: A proposal for a theory of finite sets, lists, and maps for the SMT-LIB standard. In: Proceedings of the $7^t h$ International Workshop on Satisfiability Modulo Theories (SMT 2009), August 2009
10. Kuncak, V., Nguyen, H.H., Rinard, M.: Deciding Boolean algebra with Presburger arithmetic. J. Autom. Reasoning **36**(3), 213–239 (2006)
11. Kuncak, V., Rinard, M.: Towards efficient satisfiability checking for Boolean algebra with Presburger arithmetic. In: Pfenning, F. (ed.) CADE 2007. LNCS (LNAI), vol. 4603, pp. 215–230. Springer, Heidelberg (2007)
12. Nieuwenhuis, R., Oliveras, A., Tinelli, C.: Solving SAT and SAT Modulo theories: from an abstract Davis-Putnam-Logemann-Loveland procedure to DPLL(T). J. ACM **53**(6), 937–977 (2006)
13. Suter, P., Steiger, R., Kuncak, V.: Sets with cardinality constraints in satisfiability modulo theories. In: Jhala, R., Schmidt, D. (eds.) VMCAI 2011. LNCS, vol. 6538, pp. 403–418. Springer, Heidelberg (2011)
14. Zarba, C.G.: Combining sets with integers. In: Armando, A. (ed.) FroCos 2002. LNCS (LNAI), vol. 2309, pp. 103–116. Springer, Heidelberg (2002)

Congruence Closure in Intensional Type Theory

Daniel Selsam[1] and Leonardo de Moura[2(✉)]

[1] Stanford University, Stanford, USA
dselsam@stanford.edu
[2] Microsoft Research, Redmond, USA
leonardo@microsoft.com

Abstract. Congruence closure procedures are used extensively in automated reasoning and are a core component of most satisfiability modulo theories solvers. However, no known congruence closure algorithms can support any of the expressive logics based on intensional type theory (ITT), which form the basis of many interactive theorem provers. The main source of expressiveness in these logics is dependent types, and yet existing congruence closure procedures found in interactive theorem provers based on ITT do not handle dependent types at all and only work on the simply-typed subsets of the logics. Here we present an efficient and proof-producing congruence closure procedure that applies to every function in ITT no matter how many dependencies exist among its arguments, and that only relies on the commonly assumed *uniqueness of identity proofs* axiom. We demonstrate its usefulness by solving interesting verification problems involving functions with dependent types.

1 Introduction

Congruence closure procedures are used extensively in automated reasoning, since almost all proofs in both program verification and formalized mathematics require reasoning about equalities [23]. The algorithm constitutes a fundamental component of most satisfiability modulo theories (SMT) solvers [4,20]; it is often distinguished as the "core theory solver", and is responsible for communicating literal assignments to the underlying SAT solver and equalities to the other "satellite solvers" [10,20]. However, no known congruence closure algorithms can support any of the expressive logics based on intensional type theory (ITT). Yet despite the lack of an algorithm for congruence closure, the benefits that ITTs confer in terms of expressiveness, elegance, and trustworthiness have proved substantial enough that different flavors of ITT form the basis of many interactive theorem provers, such as Coq [8], Lean [21], and Matita [2], and also several emerging programming languages, such as Agda [5], Epigram [16], and Idris [6]. Many of the most striking successes in both certified programming and formalized mathematics have been in variants of ITT, such as the development of a fully-certified compiler for most of the C language [14] and the formalization of the odd-order theorem [11].

There are currently two main workarounds for the lack of a congruence closure algorithm for ITT, and for the lack of robust theorem proving tools for ITT

N. Olivetti and A. Tiwari (Eds.): IJCAR 2016, LNAI 9706, pp. 99–115, 2016.
DOI: 10.1007/978-3-319-40229-1_8

more generally. One option is to rely much more on manual proving. Although many impressive projects have been formalized with little to no automation, this approach is not very attractive since the cost of manual proving can be tremendous. We believe that as long as extensive manual proving is a central part of writing certified software or formalizing mathematics, these will remain niche activities for the rare expert. The other option is to relinquish the use of dependent types whenever manual reasoning becomes too burdensome so that more traditional automation can be used. Note that the Coq system even has a tactic `congruence` that performs congruence closure, but it does not handle dependent types at all and only works on the simply-typed subset of the language. This sacrifice may be appropriate in certain contexts, but losing all the benefits of dependent types makes this an unsatisfactory solution in general.

Given the limitations of these two workarounds, it would be preferable to perform congruence closure and other types of automated reasoning directly in the richer language of ITT. Unfortunately, equality and congruence are both surprisingly subtle in ITT, and as we will see, the theorem that could justify using the standard congruence closure procedure for functions with dependent types is not provable in the core logic, nor does it follow from any of the axioms commonly assumed in existing systems. In this paper, we introduce a new notion of congruence that applies to every function in ITT no matter how many dependencies exist among its arguments, along with a simple and efficient extension of the standard congruence closure procedure to fully automate reasoning about this more general notion of congruence. Our procedure is applicable to a wide variety of projects since it only relies on the *uniqueness of identity proofs* axiom, which is built into the logic of many systems including Agda, Idris, and Lean, and which is commonly assumed in the others. We hope our procedure helps make it possible for users to have the best of both worlds: to reap all the benefits of dependent types while still enjoying all the power of traditional automation.

2 Preliminaries

We assume the term language is a dependent λ-calculus in which terms are described by the following grammar:

$$t, s ::= x \mid c \mid \texttt{Type} \mid t\ s \mid \lambda x : s, t \mid \Pi x : s, t$$

where x is a variable and c is a constant. To simplify the presentation, we omit type universes at sort `Type`. It is not relevant to this paper whether the universe hierarchy is cumulative or not, nor whether there is a distinguished sort `Prop` (the sort of all propositions). The term Π`x:A, B` denotes the type of functions `f` that map any element `a:A` to an element of `B[a/x]`. When `x` appears in `B` we say that `f` is *dependently-typed*; otherwise we write Π`x:A, B` as `A` $\rightarrow$ `B` to denote the usual non-dependent function space. When `B` is a proposition, Π`x:A, B` can be read as the universally quantified formula $\forall$`x:A, B`, or as the logical implication `A` $\Rightarrow$ `B` if `x` does not appear in `B`. The term `f a` denotes a function application, and the lambda abstraction λ`x:A, t` denotes a function that given an element `a` of type

A produces t[a/x]. As usual in Type Theory, a *context* Γ is a sequence of *typing assumptions* a:A and (local) definitions c:A := t, where t has type A and c does not occur in t. We often omit the type A and simply write c := t to save space when no confusion arises. Similarly, an *environment* Δ is a sequence of (global) definitions f:A := t. We use $type(\Delta, \Gamma, t)$ to denote the type of t with respect to Δ and Γ, and $type(t)$ when no confusion arises. Given an environment Δ and a context Γ, every term reduces to a normal form by the standard $\beta\delta\eta\iota\zeta$-reduction rules. For this paper we will assume a fixed environment Δ that contains all definitions and theorems that we present. As usual, we write Π(a:A)(b:B),C as a shorthand for Πa:A,(Πb:B,C). We use a similar shorthand for λ-terms.

2.1 Equality

One of the reasons that congruence is subtle in ITT is that equality itself is subtle in ITT. The single notion of equality in most other logics splits into at least three different yet related notions in ITT.

Definitional equality. The first notion of equality in ITT is *definitional equality.* We write a ≡ b to mean that a and b are equal by definition, which is the case whenever a and b reduce to the same normal form. For example, if we define a function f : ℕ → ℕ := λ n : ℕ, 0 in the environment Δ, then the terms 0 and f 0 both reduce to the same normal form 0 and so are equal by definition. On the other hand, (λ n m: ℕ, n + m) is not definitionally equal to (λ n m: ℕ, m + n), since they are both in normal form and these normal forms are not the same. Note that definitional equality is a judgment at the meta-level, and the theory itself cannot refer to it; in particular, it is not possible to assume or negate a definitional equality.

Homogeneous propositional equality. The second notion of equality in ITT is *homogeneous propositional equality*, which we will usually shorten to *homogeneous equality* since "propositional" is implied. Unlike definitional equality which is a judgment at the meta-level, homogeneous equality can be assumed, negated, and proved inside the logic itself. There is a constant eq : Π (A : Type), A → A → Type in Δ such that, for any type A and elements a b : A, the expression eq A a b represents the proposition that a and b are "equal". Note that we call this homogeneous equality because the types of a and b must be definitionally equal to even *state* the proposition that a and b are equal. We write a =A b as shorthand for eq A a b, or a = b if the type A is clear from context. We say a term t of type a = b is a *proof* for a = b.

The meaning of homogeneous equality is given by the introduction and elimination rules for eq, which state how to prove that two elements are equal and what one can do with such a proof respectively. The introduction rule for eq is the dependent function refl : Π (A : Type) (a : A), a = a, which says that every element of type A is equal to itself. We call refl the reflexivity axiom, and write refl a whenever the type A is clear from context. Note that if a b : A are definitionally equal, then refl a is a proof for a = b. The elimination principle (also known as the recursor) for the type eq is the dependent function erec:

erec : Π (A : Type) (a : A) (C : A → Type), C a → Π (b : A), a = b → C b

This principle states that if a property C holds for an element a, and a = b for some b, then we can conclude that C must hold of b as well. We say C is the *motive*, and we write (erec C p e) instead of (erec A a C p b e) since A, a and b can be inferred easily from e : a = b. Note that by setting C to be the identity function id : Type → Type, erec can be used to change the type of a term to an equal type; that is, given a term a : A and a proof e : A = B, the term (erec id a e) has type B. We call this a *cast*, and say that we *cast* a to have type B. Note that it is straightforward to use erec and refl to prove that eq is symmetric and transitive and hence an equivalence relation.

Heterogeneous propositional equality. As we saw above, homogeneous equality suffers from a peculiar limitation: it is not even possible to form the proposition a = b unless the types of a and b are definitionally equal. The further one strays from the familiar confines of simple type theory, the more severe this handicap becomes. For example, a common use of dependent types is to include the length of a list inside its type in order to make out-of-bounds errors impossible. The resulting type is often called a *vector* and has type vector : Π (A : Type), ℕ → Type. It is easy to define an append function on vectors:

app : Π (A : Type) (n m : ℕ), vector A n → vector A m → vector A (n + m)

However, we cannot even state the proposition that app is associative using homogeneous equality, since the type vector A (n + (m + k)) is not definitionally equal to the type vector A ((n + m) + k), only propositionally equal. The same issue arises when reasoning about vectors in mathematics. For example, we cannot even state the proposition that concatenating zero-vectors of different lengths m and n over the real numbers $\mathbb{R}$ is commutative, since the type $\mathbb{R}^{m+n}$ is not definitionally equal to the type $\mathbb{R}^{n+m}$. In both cases, we could use erec to cast one of the two terms to have the type of the other, but this approach would quickly become unwieldy as the number of dependencies increased, and moreover every procedure that reasoned about equality would need to do so modulo casts.

Thus there is a need for a third notion of equality in ITT, *heterogeneous propositional equality*, which we will usually shorten to *heterogeneous equality* since "propositional" is implied. There is a constant heq : Π (A : Type) (B : Type), A → B → Type that behaves like eq except that its arguments may have different types.[1] We write a == b as shorthand for heq A B a b. Heterogeneous equality has an introduction rule hrefl : Π (A : Type) (a : A), a == a analogous to refl, and it is straightforward to show that heq is an equivalence relation by proving the following theorems:

hsymm : Π (A B : Type) (a : A) (b : B), a == b → b == a
htrans : Π (A B C : Type) (a : A) (b : B) (c : C), a == b → b == c → a == c

---

[1] There are many equivalent ways of defining heq. One popular way is "John Major equality" [15]. Additional formulations and formal proofs of equivalence can be found at http://leanprover.github.io/ijcar16/congr.lean.

Unfortunately, the flexibility of `heq` does not come without a cost: as we discuss in Sect. 3, `heq` turns out to be weaker than `eq` in subtle ways and does not permit as simple a notion of congruence.

Converting from heterogeneous equality to homogeneous equality. It is straightforward to convert a proof of homogeneous equality `p : a = b` into one of heterogeneous equality using the lemma

```
lemma ofeq (A : Type) (a b : A) : a = b → a == b
```

However, we must assume an axiom in order to prove the reverse direction

```
ofheq (A : Type) (a b : A) : a == b → a = b
```

The statement is equivalent to the *uniqueness of identity proofs* (UIP) principle [26], to Streicher's *Axiom K* [26], and to a few other variants as well. Although these axioms are not part of the core logic of ITT, they have been found to be consistent with ITT by means of a meta-theoretic argument [18], and are built into the logic of many systems including Agda, Idris, and Lean. They also follow from various stronger axioms that are commonly assumed, such as *proof irrelevance* and *excluded middle.* In Coq, UIP or an axiom that implies it is often assumed when heterogeneous equality is used, including in the CompCert project [14]. Our approach is built upon being able to recover homogeneous equalities from heterogeneous equalities between two terms of the same type and so makes heavy use of `ofheq`.

3 Congruence

Congruence over homogeneous equality. It is straightforward to prove the following lemma using `erec`:

```
lemma congr : Π (A B : Type) (f g : A → B) (a b : A), f = g → a = b → f a = g b
```

and thus prove that `eq` is indeed a congruence relation for simply-typed functions. Thus the standard congruence closure algorithm can be applied to the simply-typed subset of ITT without much complication. In particular, we have the familiar property that `f a` and `g b` can be proved equal if and only if either an equality `f a = g b` has been asserted, or if `f` can be proved equal to `g` and `a` can be proved equal to `b`.

Congruence over heterogeneous equality. Unfortunately, once we introduce functions with dependent types, we must switch to `heq` and lose the familiar property discussed above that `eq` satisfies for simply-typed functions. Ideally we would like the following congruence lemma for heterogeneous equality:

```
hcongr_ideal : Π (A A' : Type) (B : A → Type) (B' : A' → Type)
  (f : Π (a : A), B a) (f' : Π (a' : A'), B' a') (a : A) (a' : A'),
  f == f' → a == a' → f a == f' a'
```

Unfortunately, this theorem is not provable in ITT [1], even when we assume UIP. The issue is that we need to establish that `B = B'` as well, and this fact does not follow from (Π (a : A), B a) = (Π (a' : A'), B' a'). Assuming **hcongr_ideal** as an axiom is not a satisfactory solution because it would limit the applicability of our approach, since as far as we know it is not assumed in any existing interactive theorem provers based on ITT.

However, for any given n, it is straightforward to prove the following congruence lemma using only **erec**, **ofheq** and **hrefl**[2]:

```
lemma hcongr_n
  (A₁: Type)
  (A₂: A → Type)
  ...
  (A_n: Π a₁ ... a_{n−2}, A_{n−1} a₁ ... a_{n−2} → Type)
  (B: Π a₁ ... a_{n−1}, A_n a₁ ... a_{n−1} → Type) :
  Π (f g: Π a₁ ... a_n, B a₁ ... a_n), f = g →
  Π (a₁ b₁: A₁), a₁ == b₁ →
  Π (a₂: A₂ a₁) (b₂: A₂ b₁), a₂ == b₂ →
  ...
  Π (a_n: A_n a₁ ... a_{n−1}) (b_n : A_n b₁ ... b_{n−1}), a_n == b_n →
  f a₁ ... a_n == g b₁ ... b_n
```

The lemmas $\texttt{hcongr}_n$ are weaker than `hcongr_ideal` because they require the outermost functions `f` and `g` to have the same type. Although we no longer have the property that `f == g` and `a == b` implies `f a == g b`, we show in the next section how to extend the congruence closure algorithm to deal with the additional restriction imposed by $\texttt{hcongr}_n$.

When using $\texttt{hcongr}_n$ lemmas, we omit the parameters $\texttt{A}_i$, `B`, $\texttt{a}_i$ and $\texttt{b}_i$ since they can be inferred from the parameters with types `f = g` and $\texttt{a}_i == \texttt{b}_i$. Note that even if some arguments of an n-ary function `f` do not depend on all previous ones, it is still straightforward to find parameters $\texttt{A}_i$ and `B` that do depend on all previous arguments and so fit the theorem, and yet become definitionally equal to the types of the actual arguments of `f` once applied to the preceding arguments. We remark that we avoid this issue in our implementation by synthesizing custom congruence theorems for every function we encounter.

4 Congruence Closure

We now have all the necessary ingredients to describe a very general congruence closure procedure for ITT. Our procedure is based on the one proposed by Nieuwenhuis and Oliveras [24] for first-order logic, which is efficient, is proof producing, and is used by many SMT solvers. We assume the input to our congruence closure procedure is of the form $\Gamma \vdash$ `a == b`, where Γ is a context and `a == b` is the goal. Note that a goal of the form `a = b` can be converted into

[2] The formal statements and proofs for small values of n can be found at http://leanprover.github.io/ijcar16/congr.lean, along with formal proofs of all other lemmas described in this paper.

a == b before we start our procedure, since when a and b have the same type, any proof for a == b can be converted into a proof for a = b using ofheq. Similarly, any hypothesis of the form e: a = b can be replaced with e: a == b using ofeq. As in abstract congruence closure [3,13], we introduce new variables c to name all proper subterms of every term appearing on either side of an equality, both to simplify the presentation and to obtain the efficiency of DAG-based implementations.[3] For example, we encode f N a == f N b using the local definitions (c_1 := f N) (c_2 := c_1 a) (c_3 := c_1 b) and the equality c_2 == c_3. We remark that c_2 == c_3 is definitionally equal to f N a == f N b by ζ-reduction. Here is an example problem instance for our procedure:

(N: Type) (a b: N) (f: Π A: Type, A → A) (c_1 := f N)
(c_2 := c_1 a) (c_3 := c_1 b) (e: a == b) ⊢ c_2 == c_3

The term (hcongr$_2$ (refl f) (hrefl N) e) is a proof for the goal c_2 == c_3.

As in most congruence closure procedures, ours maintains a union-find data structure that partitions the set of terms into a number of disjoint subsets such that if a and b are in the same subset (denoted a ≈ b) then the procedure can generate a proof that a == b. Each subset is an *equivalence class*. The union-find data structure computes the equivalence closure of the relation == by merging the equivalence classes of a and b whenever e: a == b is asserted. However, the union-find data structure alone does not know anything about congruence, and in particular it will not automatically propagate the assertion a == b to other terms that contain a or b; for example, it would not merge the equivalence classes of c := f a and d := f b. Thus, additional machinery is required to find and propagate new equivalences implied by the rules of congruence.

We say that two terms are *congruent* if they can be proved to be equivalent using a congruence rule given the current partition of the union-find data structure. We also say two local definitions c := f a and d := g b are congruent whenever f a and g b are congruent. We remark that congruence closure algorithms can be parameterized by the structure of the congruence rules they propagate. In our case, we use the family of hcongr$_n$ lemmas as congruence rules.

We now describe our congruence closure procedure in full, although the overall structure is similar to the one presented in [24]. The key differences are in how we determine whether two terms are congruent, how we build formal proofs of congruence using hcongr$_n$, and what local definitions we need to visit after merging two equivalence classes to ensure that all new congruences are detected. The basic data structures in our procedure are

- *repr*: a mapping from variables to variables, where *repr*[x] is the representative for the equivalence class x is in. We say variable x is a *representative* if and only if *repr*[x] is x.
- *next*: a mapping from variables to variables that induces a circular list for each equivalence class, where *next*[x] is the next element in the equivalence class x is in.

[3] To simplify the presentation further, we ignore the possibility that any of these subterms themselves include partial applications of equality.

- *pr*: a mapping from variables to pairs consisting of a variable and a proof, where if $pr[x]$ is (y, p), then p is a proof for $x == y$ or $y == x$. We use $target[x]$ to denote $pr[x].1$. This structure implements the *proof forests* described in [24].
- *size*: a mapping from representatives to natural numbers, where for each representative x, $size[x]$ is the number of elements in the equivalence class represented by x.
- *pending*: a list of local definitions and typing assumptions to be processed. It is initialized with the context Γ.
- *congrtable*: a set of local definitions such that given a local definition E, the function *lookup*(E) returns a local definition in *congrtable* congruent to E if one exists.
- *uselists*: a mapping from representatives to sets of local definitions, such that local definition D is in $uselists[x]$ if D might become congruent to another definition if the equivalence class of x were merged with another equivalence class.

Our procedure maintains the following invariants for the data structures described above.

1. $repr[next[x]] \equiv repr[repr[x]] \equiv repr[x]$
2. If $repr[x] \equiv repr[y]$, then $next^k[x] \equiv y$ for some k.
3. $target^k[x] \equiv repr[x]$ for some k. That is, we can view $target^k[x]$ as a "path" from x to $repr[x]$. Moreover, the proofs in *pr* can be used to build a proof from x to any element along this path.
4. Let s be $size[repr[x]]$, then $next^s[x] \equiv x$. That is, *next* does indeed induce a set of disjoint circular lists, one for each equivalence class.

Whenever a new congruence proof for `c` $==$ `d` is inferred by our procedure, we add the auxiliary local definition `e`: `c` $==$ `d` $:=$ `p` to *pending*, where `e` is a fresh variable, and `p` is a proof for `c` $==$ `d`. The proof `p` is always an application of the lemma $\texttt{hcongr}_n$ for some n. We say `e : c` $==$ `d` and `e: c` $==$ `d` $:=$ `p` are *equality proofs* for `c` $==$ `d`. Given an equality proof E, the functions *lhs*(E) and *rhs*(E) return the left and right hand sides of the proved equality. Given a local definition E of the form `c` $:=$ `f a`, the function *var*(E) returns `c`, and *app*(E) the pair `(f, a)`. We say a variable `c` is a local definition when Γ contains the definition `c` $:=$ `f a`, and the auxiliary partial function *def*(`c`) returns this local definition.

Implementing congrtable. In order to implement the congruence closure procedure efficiently, the congruence rules must admit a data structure *congrtable* that takes a local definition and quickly returns a local definition in the table that it is congruent to if one exists. It is easy to implement such a data structure with a Boolean procedure CONGRUENT(D, E) that determines if two local definitions are congruent, along with a congruence-respecting hash function. Although the family of $\texttt{hcongr}_n$ lemmas does not satisfy the property that `f a` and `g b` are congruent whenever `f` $\approx$ `g` and `a` $\approx$ `b`, we still have a straightforward criterion for determining whether two terms are congruent.

Proposition 1. *Consider the terms f `a` and g `b`. If* `a` $\approx$ `b`*, then f `a` and g `b` are congruent provided either:*

1. *f and g are homogeneously equal;*
2. *f and g are congruent.*

Proof. First note that in both cases, we can generate a proof that `a` $==$ `b` since we have assumed that `a` $\approx$ `b`. In the first case, if `f` and `g` are homogeneously equal, then no matter how many partial applications they contain, we can apply $\texttt{hcongr}_1$ to the proof of homogeneous equality and the proof that `a` $==$ `b`. In the second case, if `f` and `g` are congruent, it means that we can generate proofs of all the preconditions of $\texttt{hcongr}_k$ for some k, and the only additional precondition to $\texttt{hcongr}_{k+1}$ is a proof that `a` $==$ `b`, which we can generate as well.

```
1: procedure CONGRUENT(D, E)
2:     (f, a) ← app(D); (g, b) ← app(E)
3:     return a ≈ b and
4:             [(f ≈ g and type(f) ≡ type(g)) or
5:             (f and g are local definitions and CONGRUENT(def(f), def(g)))]
6: procedure CONGRHASH(D)
7:     given: h, a hash function on terms
8:     (f, a) ← app(D)
9:     return hashcombine(h(repr[f]), h(repr[a]))
```

Fig. 1. Implementing *congrtable*

Proposition 1 suggests a simple recursive procedure to detect when two terms are congruent, which we present in Fig. 1. The procedure CONGRUENT(D, E), where D and E are local definitions of the form `c` $:=$ `f a` and `d` $:=$ `g b`, returns `true` if a proof for `c` $==$ `d` can be constructed using an $\texttt{hcongr}_n$ lemma for some n. Note that although the congruence lemmas $\texttt{hcongr}_n$ are themselves n-ary, it is not sufficient to view the two terms being compared for congruence as applications of n-ary functions. We must compare each pair of partial applications for homogeneous equality as well (line 4), since two terms with n arguments each might be congruent using $\texttt{hcongr}_m$ for any m such that $m \leq n$. For example, `f a1 c` and `g b1 c` are congruent by $\texttt{hcongr}_2$ if `f` $=$ `g` and `a1` $==$ `b1`, and yet are only congruent by $\texttt{hcongr}_1$ if all we know is `f a1` $=$ `g b1`. It is even possible for two terms to be congruent that do not have the same number of arguments. For example, `f` $=$ `g a` implies that `f b` and `g a b` are congruent by $\texttt{hcongr}_1$.

Proposition 1 also suggests a simple way to hash local definitions that respects congruence. Given a hash function on terms, the procedure CONGRHASH(D) hashes a local definition of the form `c` $:=$ `f a` by simply combining the hashes of the representatives of `f` and `a`. This hash function respects congruence because if `c` $:=$ `f a` and `d` $:=$ `g b` are congruent, it is a necessary (though not sufficient) condition that `f` $\approx$ `g` and `a` $\approx$ `b`.

```
procedure CC(Γ ⊢ a == b)
    pending ← Γ
    while pending is not empty do
        remove next E from pending
        if E is an equality proof then PROCESSEQ(E)
        else INITIALIZE(E)
    if repr[a] ≡ repr[b] then return MKPR(a, b)
    else fail
```

Fig. 2. Congruence closure procedure

The procedure. Fig. 2 contains the main procedure CC. It initializes *pending* with the input context Γ. Variables in typing assumptions and local definitions are processed using INITIALIZE (Fig. 3), and equality proofs are processed using PROCESSEQ (Fig. 4).

```
procedure INITIALIZE(E)
    c ← var(E)
    repr[c] ← c; next[c] ← c; size[c] ← 1; uselists[c] ← ∅
    pr[c] ← (c, 'hrefl c')
    if E is a local definition then
        INITUSELIST(E, E)
        if D = lookup(E) then
            d ← var(D); e ← make fresh variable
            add (e : d == c := MKCONGR(D, E, [])) to pending and Γ
        else add E to congrtable
procedure INITUSELIST(E, P)
    (f, a) ← app(E)
    add P to uselists[f] and uselists[a]
    if f is a local definition then INITUSELIST(def(f), P)
```

Fig. 3. Initialization procedure

The INITIALIZE(E) procedure invokes INITUSELIST(E, E) whenever E is a local definition $c := f\ a$. The second argument at INITUSELIST(E, P) represents the *parent* local definition that must be included in the *uselists*. We must ensure that for every local definition D that could be inspected during a call to CONGRUENT(E_1, E_2) for some E_2, we add $var(E_1)$ to the *uselist* of $var(D)$ when initializing E_1. Thus the recursion in INITUSELIST must mirror the recursion in CONGRUENT conservatively, and always recurse whenever CONGRUENT might recurse. For example, assume the input context Γ contains

$$(\mathtt{A\!:\ Type})\ (\mathtt{a\ b\ d\!:\ A})\ (\mathtt{g} : \mathtt{A} \to \mathtt{A} \to \mathtt{A})\ (\mathtt{f} : \mathtt{A} \to \mathtt{A})\ (\mathtt{c}_1 := \mathtt{g\ a})\ (\mathtt{c}_2 := \mathtt{c}_1\ \mathtt{b})\ (\mathtt{c}_3 := \mathtt{f\ d}).$$

When INITIALIZE($\mathtt{c}_2$:= $\mathtt{c}_1$ b) is invoked, $\mathtt{c}_2 := \mathtt{c}_1\ \mathtt{d}$ is added to the *uselists* of $\mathtt{c}_1$, b, g and a. By a slight abuse of notation, we write '**hrefl** a' to represent in the pseudocode the expression that creates the `hrefl`-application using as argument the term stored in the program variable a.

The procedure PROCESSEQ is used to process equality proofs $a == b$. If a and b are already in the same equivalence class, it does nothing. Otherwise, it first removes every element in $uselists[repr[a]]$ from $congrtable$ (procedure REMOVEUSES). Then, it merges the equivalence classes of a and b so that for every a' in the equivalence class of a, $repr[a']$ is set to $repr[b]$. This operation can be implemented efficiently using the $next$ data structure. As in [24], the procedure also reorients the path from a to $repr[a]$ induced by pr (procedure FLIPPROOFS) to make sure invariant 3 is still satisfied and *locally irredundant transitivity proofs* [22] can be generated. It then reinserts the elements removed by REMOVEUSES into $congrtable$ (procedure REINSERTUSES); if any are found to be congruent to an existing term in a different partition, it proves equivalence using the congruence lemma $hcongr_n$ (procedure MKCONGR) and puts the new proof onto the queue. Finally, PROCESSEQ updates $next$, $uselists$ and $size$ data structures.

```
 1: procedure PROCESSEQ(E)
 2:     a ← lhs(E);  b ← rhs(E)
 3:     if repr[a] ≡ repr[b] then return
 4:     if size(repr[a]) > size(repr[b]) then swap(a, b)
 5:     rₐ ← repr[a];  r_b ← repr[b]
 6:     REMOVEUSES(rₐ); FLIPPROOFS(a)
 7:     for all a' s.t. repr[a'] ≡ rₐ do repr[a'] ← r_b
 8:     pr[a] ← (b, E)
 9:     REINSERTUSES(rₐ)
10:     swap(next[rₐ], next[r_b])
11:     move uselists[rₐ] to uselists[r_b]; size[r_b] ← size[r_b] + size[rₐ]
12: procedure FLIPPROOFS(a)
13:     if repr[a] ≡ a then return
14:     (b, p) ← pr[a]; FLIPPROOFS(b); pr[b] ← (a, p)
15: procedure REMOVEUSES(a)
16:     for all E in uselists[a] do remove E from congrtable
17: procedure REINSERTUSES(a)
18:     for all E in uselists[a] do
19:         if D = lookup(E) then
20:             d ← var(D); e ← var(E); p ← make fresh variable
21:             add (p : d == e := MKCONGR(D, E, [])) to pending and Γ
22:         else add E to congrtable
```

Fig. 4. Process equality procedure

Figure 5 contains a simple recursive procedure MKCONGR to construct the proof that two congruent local definitions are equal. The procedure takes as input two local definitions D and E of the form $c := f\ a$ and $d := g\ b$ such that

CONGRUENT(D, E), along with a possibly empty list of equality proofs es for `a1 == b1`, ..., `an == bn`, and returns a proof for `f a a1 ... an == g b b1 ... bn`. The two cases in the MKCONGR procedure mirror the two cases of the CONGRUENT procedure. If the types of `f` and `g` are definitionally equal we construct an instance of the lemma $\texttt{hcongr}_{|es|+1}$. The procedure MKPR(`a`, `b`) (Fig. 5) creates a proof for `a == b` if `a` and `b` are in the same equivalence class by finding the common element $target^n[\texttt{a}] \equiv target^m[\texttt{b}]$ in the "paths" from `a` and `b` to the equivalence class representative. Note that, if CONGRUENT(D, E) is true, then MKCONGR(D, E, []) is a proof for `c == d`.

procedure MKCONGR(D, E, es)
    assumption: CONGRUENT(D, E)
    $(f, a) \leftarrow app(D)$; $(g, b) \leftarrow app(E)$; $e_{ab} \leftarrow$ MKPR(a, b)
    if $type(f) \equiv type(g)$ **then**
        $n \leftarrow len(es)$; $e_{fg} \leftarrow$ MKPR(f, g)
        return '$\texttt{hcongr}_{n+1}$ (`ofheq` e_{fg}) e_{ab} es'
    else return MKCONGR($def(f)$, $def(g)$, $[es, e_{ab}]$)
procedure MKPR(a, b)
    if $a \equiv b$ **then return** '`hrefl` a'
    let n and m be the smallest values s.t. $target^n[a] \equiv target^m[b]$
    $e_a \leftarrow$ MKTRANS(a, n); $e_b \leftarrow$ MKTRANS(b, m); **return** '`htrans` e_a (`hsymm` e_b)'
procedure MKTRANS(a, n)
    if $n = 0$ **then return** '`hrefl` a'
    $(b, e_{ab}) \leftarrow pr[a]$; $e \leftarrow$ MKTRANS(b, $n - 1$)
    if $lhs(e_{ab}) \equiv a$ **and** $rhs(e_{ab}) \equiv b$ **then return** '`htrans` e_{ab} e'
    else return '`htrans` (`hsymm` e_{ab}) e'

Fig. 5. Transitive proof generation procedure

Finally, we remark that the main loop of CC maintains the following two invariants.

Theorem 1. *If a and b are in the same equivalence class (i.e., $a \approx b$), then* MKPR(a, b) *returns a correct proof that $a == b$.*

Theorem 2. *If $type(f) \equiv type(g)$, $f \approx g$, $a_1 \approx b_1$, $\ldots a_n \approx b_n$, $c \equiv f\ a_1 \ldots a_n$ and $d \equiv g\ b_1 \ldots b_n$, then $c \approx d$.*

Extensions. There are many standard extensions to the congruence closure procedure that are straightforward to support in our framework, such as tracking disequalities to find contradictions and propagating injectivity and disjointness for inductive datatype constructors [17]. Here we present a simple extension for propagating equalities among elements of *subsingleton* types that is especially important when proving theorems in ITT. We say a type `A:Type` is a subsingleton if it has at most one element; that is, if for all `(a b:A)`, we have that `a = b`. Subsingletons are used extensively in practice, and are especially ubiquitous when *proof irrelevance* is assumed, in which case every proposition is a subsingleton.

One common use of dependent types is to extend functions to take extra arguments that represent proofs that certain preconditions hold. For example, the logarithm function only makes sense for positive real numbers, and we can make it impossible to even call it on a non-positive number by requiring a proof of positivity as a second argument: $c := f\ a$. The second argument is a proposition and hence is a subsingleton when we assume *proof irrelevance*. Consider the following goal: $(\mathtt{a}\ \mathtt{b} : \mathbb{R})\ (\mathtt{Ha} : \mathtt{a} > 0)\ (\mathtt{Hb} : \mathtt{b} > 0)\ (\mathtt{e} : \mathtt{a} = \mathtt{b}) \vdash$ $\mathtt{safe\_log}\ \mathtt{a}\ \mathtt{Ha} = \mathtt{safe\_log}\ \mathtt{b}\ \mathtt{Hb}$. The core procedure we presented above would not be able to prove this theorem on its own because it would never discover that $\mathtt{Ha} == \mathtt{Hb}$. We show how to extend the procedure to automatically propagate facts of this kind.

We assume we have an oracle $issub(\Gamma, A)$ that returns true for subsingleton types for which we have a proof $\Pi \mathtt{a}\ \mathtt{b}{:}\mathtt{A},\ \mathtt{a} = \mathtt{b}$. Many proof assistants implement an efficient (and incomplete) *issub* using *type classes* [7,19], but it is beyond the scope of this paper to describe this mechanism. Given a subsingleton type $\mathtt{A}$ with proof $\mathtt{sse}_A$, we can prove

$$\mathtt{hsse}_A\colon \Pi\ (\mathtt{C}{:}\mathtt{Type})\ (\mathtt{c}{:}\mathtt{C})\ (\mathtt{a}{:}\mathtt{A}),\ \mathtt{C} == \mathtt{A} \to \mathtt{c} == \mathtt{a},$$

which we can use as an additional propagation rule in the congruence closure procedure. The idea is to merge the equivalence classes of $\mathtt{a}{:}\mathtt{A}$ and $\mathtt{c}{:}\mathtt{C}$ whenever $\mathtt{A}$ is a subsingleton and $\mathtt{C} \approx \mathtt{A}$. First, we add a mapping *subrep* from subsingleton types to their representatives. Then, we include the following additional code in INITIALIZE:

$C \leftarrow type(c)$; $A \leftarrow repr[C]$
if $issub(\Gamma, A)$ **then**
  if $a = subrep[A]$ **then**
    $p \leftarrow$ MKPR(C, A); $e \leftarrow$ make fresh variable
    add $(e : c == a := \mathtt{hsse}_A\ C\ p\ c\ a)$ to *pending* and Γ
  else $subrep[A] \leftarrow c$

Finally, at PROCESSEQ whenever we merge the equivalence classes of subsingleton types A and C, we also propagate the equality $subrep[A] == subrep[C]$.

With this extension, our procedure can prove $\mathtt{safe\_log}\ \mathtt{a}\ \mathtt{Ha} = \mathtt{safe\_log}\ \mathtt{b}\ \mathtt{Hb}$ in the example above, since the terms $\mathtt{a} > 0$ and $\mathtt{b} > 0$ are both subsingleton types with representative elements $\mathtt{Ha}$ and $\mathtt{Hb}$ respectively, and when their equivalence classes are merged, the subsingleton extension propagates the fact that their representative elements are equal, i.e. that $\mathtt{Ha} == \mathtt{Hb}$.

5 Applications

We have implemented our congruence closure procedure for Lean[4] along with many of the standard extensions as part of a long-term effort to build a robust theorem prover for ITT. Although congruence closure can be useful on its own, its

[4] https://github.com/leanprover/lean/blob/master/src/library/blast/congruence_closure.cpp.

power is greatly enhanced when it is combined with a procedure for automatically instantiating lemmas so that the user does not need to manually collect all the ground facts that the congruence closure procedure will need. We use an approach called *e-matching* [10] to instantiate lemmas that makes use of the equivalences represented by the state of the congruence closure procedure when deciding what to instantiate, though the details of e-matching are beyond the scope of this paper. The combination of congruence closure and e-matching is already very powerful, as we demonstrate in the following two examples, the first from software verification and the second from formal mathematics. The complete list of examples we have used to test our procedure can be found at http://leanprover.github.io/ijcar16/examples.

Vectors (indexed lists). As we mentioned in Sect. 2.1, a common use of dependent types is to include the length of a list inside its type in order to make out-of-bounds errors impossible. The constructors of `vector` mirror those of `list`:

$$\texttt{nil} : \Pi\ \{\texttt{A} : \texttt{Type}\},\ \texttt{vector A}\ 0$$
$$\texttt{cons} : \Pi\ \{\texttt{A} : \texttt{Type}\}\ \{\texttt{n} : \mathbb{N}\},\ \texttt{A} \to \texttt{vector A n} \to \texttt{vector A}\ (\texttt{succ n})$$

where `succ` is the successor function on natural numbers, and where curly braces indicate that a parameter should be inferred from context. We use the notation `[x]` to denote the one-element `vector` containing only `x`, i.e. `cons x nil`, and `x::v` to denote `cons x v`. It is easy to define append and reverse on `vector`:

$$\texttt{app} : \Pi\ \{\texttt{A} : \texttt{Type}\}\ \{\texttt{n}_1\ \texttt{n}_2 : \mathbb{N}\},\ \texttt{vector A n}_1 \to \texttt{vector A n}_2 \to \texttt{vector A}\ (\texttt{n}_1 + \texttt{n}_2)$$
$$\texttt{rev} : \Pi\ \{\texttt{n} : \mathbb{N}\},\ \texttt{vector A n} \to \texttt{vector A n}$$

When trying to prove the basic property $\texttt{rev}\ (\texttt{app}\ \texttt{v}_1\ \texttt{v}_2) == \texttt{app}\ (\texttt{rev}\ \texttt{v}_2)\ (\texttt{rev}\ \texttt{v}_1)$ about these two functions, we reach the following goal:

$$(\texttt{A} : \texttt{Type})\ (\texttt{n}_1\ \texttt{n}_2 : \mathbb{N})\ (\texttt{x}_1\ \texttt{x}_2 : \texttt{A})\ (\texttt{v}_1 : \texttt{vector A n}_1)\ (\texttt{v}_2 : \texttt{vector A n}_2)$$
$$(\texttt{IH} : \texttt{rev}\ (\texttt{app}\ \texttt{v}_1\ (\texttt{x}_2\texttt{::}\texttt{v}_2)) == \texttt{app}\ (\texttt{rev}\ (\texttt{x}_2\texttt{::}\texttt{v}_2))\ (\texttt{rev}\ \texttt{v}_1))$$
$$\vdash \texttt{rev}\ (\texttt{app}\ (\texttt{x}_1\texttt{::}\texttt{v}_1)\ (\texttt{x}_2\texttt{::}\texttt{v}_2)) == \texttt{app}\ (\texttt{rev}\ (\texttt{x}_2\texttt{::}\texttt{v}_2))\ (\texttt{rev}\ (\texttt{x}_1\texttt{::}\texttt{v}_1))$$

Given basic lemmas about how to push `app` and `rev` in over `cons`, a lemma stating the associativity of `app`, and a few basic lemmas about natural numbers, our congruence closure procedure together with the e-matcher can solve this goal. Once the e-matcher establishes the following ground facts:

$$\texttt{H}_1 : \texttt{rev}\ (\texttt{x}_1\texttt{::}\texttt{v}_1) == \texttt{app}\ (\texttt{rev}\ \texttt{v}_1)\ [\texttt{x}_1]$$
$$\texttt{H}_2 : \texttt{app}\ (\texttt{x}_1\texttt{::}\texttt{v}_1)\ (\texttt{x}_2\texttt{::}\texttt{v}_2) == \texttt{x}_1\texttt{::}(\texttt{app}\ \texttt{v}_1\ (\texttt{x}_2\texttt{::}\texttt{v}_2))$$
$$\texttt{H}_3 : \texttt{rev}\ (\texttt{x}_1\texttt{::}(\texttt{app}\ \texttt{v}_1\ (\texttt{x}_2\texttt{::}\texttt{v}_2))) == \texttt{app}\ (\texttt{rev}\ (\texttt{app}\ \texttt{v}_1\ (\texttt{x}_2\texttt{::}\texttt{v}_2)))\ [\texttt{x}_1]$$
$$\texttt{H}_4 : \texttt{app}\ (\texttt{app}\ (\texttt{rev}\ (\texttt{x}_2\texttt{::}\texttt{v}_2))\ (\texttt{rev}\ \texttt{v}_1))\ [\texttt{x}_1] == \texttt{app}\ (\texttt{rev}\ (\texttt{x}_2\texttt{::}\texttt{v}_2))\ (\texttt{app}\ (\texttt{rev}\ \texttt{v}_1)\ [\texttt{x}_1])$$

as well as a few basic facts about the natural numbers, the result follows by congruence.

Safe arithmetic. As we mentioned in Sect. 4, another common use of dependent types is to extend functions to take extra arguments that represent proofs that certain preconditions hold. For example, we can define safe versions of the logarithm function and the inverse function as follows:

$$\texttt{safe\_log} : \Pi\ (\texttt{x} : \mathbb{R}),\ \texttt{x} > 0 \to \mathbb{R} \qquad \texttt{safe\_inv} : \Pi\ (\texttt{x} : \mathbb{R}),\ \texttt{x} \neq 0 \to \mathbb{R}$$

Although it would be prohibitively cumbersome to prove the preconditions manually at every invocation, we can relegate this task to the theorem prover, so that `log x` means `safe_log x p` and $\texttt{y}^{-1}$ means `safe_inv y q`, where `p` and `q` are proved automatically. Given basic lemmas about arithmetic identities, our congruence closure procedure together with the e-matcher can solve many complex equational goals like the following, despite the presence of embedded proofs:

$$\forall\ (\texttt{x y z w} : \mathbb{R}),\ \texttt{x} > 0 \to \texttt{y} > 0 \to \texttt{z} > 0 \to \texttt{w} > 0 \to \texttt{x} * \texttt{y} = \texttt{exp z} + \texttt{w} \to$$
$$\texttt{log}\ (2 * \texttt{w} * \texttt{exp z} + \texttt{w}^2 + \texttt{exp}\ (2 * \texttt{z}))\ /\ -2 = \texttt{log y}^{-1} - \texttt{log x}$$

6 Related Work

Corbineau [9] presents a congruence closure procedure for the simply-typed subset of ITT and a corresponding implementation for Coq as the tactic `congruence`. The procedure uses homogeneous equality and does not support dependent types at all. Hur [12] presents a library of tactics for reasoning over a different variant of heterogeneous equality in Coq, for which the user must manually separate the parts of the type that are allowed to vary between heterogeneously equal terms from those that must remain the same. The main tactic provided is `Hrewritec`, which tries to rewrite with a heterogeneous equality by converting it to a cast-equality, rewriting with that, and then generalizing the proof that the types are equal. There does not seem to be any general notion of congruence akin to our family of $\texttt{hcongr}_n$ lemmas.

Sjöberg and Weirich [25] propose using congruence closure during type checking for a new dependent type theory in which definitional equality is determined by the congruence closure relation instead of by the standard forms of reduction. Their type theory is not compatible with any of the standard flavors of ITT such as the calculus of inductive constructions, and so their procedure cannot be used to prove theorems in systems such as Coq and Lean. The congruence rules they use are also not as general as ours, since they require the two functions being applied to be the same, whereas $\texttt{hcongr}_n$ allows them to differ as long as they are homogeneously equal. As a result, given $\texttt{x} = \texttt{y}$, they cannot conclude $\texttt{f x} = \texttt{g y}$ from $\texttt{f} = \texttt{g}$, let alone $\texttt{f a x} = \texttt{g y}$ from $\texttt{f a} = \texttt{g}$. Moreover, they do not discuss why or whether the natural binary congruence rule (i.e. `hcongr_ideal`) would be unsound in their type theory, nor why their congruence rule needs to be n-ary.

7 Conclusion

We have presented a very general notion of congruence for ITT based on heterogeneous equality that applies to all dependently typed functions. We also presented a congruence closure procedure that can propagate the associated congruence rules efficiently and so automatically prove a large and important set of goals. Just as congruence closure procedures (along with DPLL) form the foundation of modern SMT solvers, we hope that our congruence closure procedure can form the foundation of a robust theorem prover for intensional type theory. We are building such a theorem prover for Lean, and it can already solve many interesting problems.

Acknowledgments. We would like to thank David Dill, Jeremy Avigad, Robert Lewis, Nikhil Swany, Floris van Doorn and Georges Gonthier for providing valuable feedback on early drafts.

References

1. Private communication with Jeremy Avigad and Floris van Doorn
2. Asperti, A., Ricciotti, W., Sacerdoti Coen, C., Tassi, E.: The Matita interactive theorem prover. In: Bjørner, N., Sofronie-Stokkermans, V. (eds.) CADE 2011. LNCS, vol. 6803, pp. 64–69. Springer, Heidelberg (2011)
3. Bachmair, L., Tiwari, A., Vigneron, L.: Abstract congruence closure. J. Autom. Reason. **31**(2), 129–168 (2003)
4. Barrett, C., Conway, C.L., Deters, M., Hadarean, L., Jovanović, D., King, T., Reynolds, A., Tinelli, C.: CVC4. In: Gopalakrishnan, G., Qadeer, S. (eds.) CAV 2011. LNCS, vol. 6806, pp. 171–177. Springer, Heidelberg (2011)
5. Bove, A., Dybjer, P., Norell, U.: A brief overview of Agda – a functional language with dependent types. In: Berghofer, S., Nipkow, T., Urban, C., Wenzel, M. (eds.) TPHOLs 2009. LNCS, vol. 5674, pp. 73–78. Springer, Heidelberg (2009)
6. Brady, E.: Idris, a general-purpose dependently typed programming language: design and implementation. J. Funct. Program. **23**(05), 552–593 (2013)
7. Castéran, P., Sozeau, M.: A gentle introduction to type classes and relations in Coq. Technical report. Citeseer (2012)
8. Coq Development Team: The Coq proof assistant reference manual: Version 8.5. INRIA (2015–2016)
9. Corbineau, P.: Autour de la clôture de congruence avec Coq. Master's Thesis, Université Paris-Sud (2001)
10. Detlefs, D., Nelson, G., Saxe, J.B.: Simplify: a theorem prover for program checking. J. ACM **52**(3), 365–473 (2005)
11. Gonthier, G., Asperti, A., Avigad, J., Bertot, Y., Cohen, C., Garillot, F., Le Roux, S., Mahboubi, A., O'Connor, R., Ould Biha, S., Pasca, I., Rideau, L., Solovyev, A., Tassi, E., Théry, L.: A machine-checked proof of the odd order theorem. In: Blazy, S., Paulin-Mohring, C., Pichardie, D. (eds.) ITP 2013. LNCS, vol. 7998, pp. 163–179. Springer, Heidelberg (2013)
12. Hur, C.K.: Heq: a Coq library for heterogeneous equality (2010)
13. Kapur, D.: Shostak's congruence closure as completion. In: Comon, H. (ed.) RTA 1997. LNCS, vol. 1232, pp. 23–37. Springer, Heidelberg (1997)
14. Leroy, X.: Formal verification of a realistic compiler. Commun. ACM **52**(7), 107–115 (2009)
15. McBride, C.: Elimination with a motive. In: Callaghan, P., Luo, Z., McKinna, J., Pollack, R. (eds.) TYPES 2000. LNCS, vol. 2277, pp. 197–216. Springer, Heidelberg (2002)
16. McBride, C.: Epigram: practical programming with dependent types. In: Vene, V., Uustalu, T. (eds.) AFP 2004. LNCS, vol. 3622, pp. 130–170. Springer, Heidelberg (2005)
17. McBride, C., Goguen, H.H., McKinna, J.: A few constructions on constructors. In: Filliâtre, J.-C., Paulin-Mohring, C., Werner, B. (eds.) TYPES 2004. LNCS, vol. 3839, pp. 186–200. Springer, Heidelberg (2006)
18. Miquel, A., Werner, B.: The not so simple proof-irrelevant model of CC. In: Geuvers, H., Wiedijk, F. (eds.) TYPES 2002. LNCS, vol. 2646, pp. 240–258. Springer, Heidelberg (2003)

19. de Moura, L., Avigad, J., Kong, S., Roux, C.: Elaboration in dependent type theory. Technical report (2015). http://arXiv.org/abs/1505.04324
20. de Moura, L., Bjørner, N.S.: Z3: an efficient SMT solver. In: Ramakrishnan, C.R., Rehof, J. (eds.) TACAS 2008. LNCS, vol. 4963, pp. 337–340. Springer, Heidelberg (2008)
21. de Moura, L., Kong, S., Avigad, J., Van Doorn, F., von Raumer, J.: The Lean theorem prover (system description). In: Felty, A.P., Middeldorp, A. (eds.) CADE-25. LNAI, vol. 9195, pp. 378–388. Springer, Heidelberg (2015)
22. de Moura, L., Rueß, H., Shankar, N.: Justifying equality. Electron. Notes Theoret. Comput. Sci. **125**(3), 69–85 (2005)
23. Nelson, G., Oppen, D.C.: Fast decision procedures based on congruence closure. J. ACM (JACM) **27**(2), 356–364 (1980)
24. Nieuwenhuis, R., Oliveras, A.: Proof-producing congruence closure. In: Giesl, J. (ed.) RTA 2005. LNCS, vol. 3467, pp. 453–468. Springer, Heidelberg (2005)
25. Sjöberg, V., Weirich, S.: Programming up to congruence. In: POPL 2015, NY, USA, pp. 369–382. ACM, New York (2015)
26. Streicher, T.: Investigations into Intensional Type Theory, Habilitations-schrift, Ludwig-Maximilians-Universität München (1993)

Fast Cube Tests for LIA Constraint Solving

Martin Bromberger[1,2(✉)] and Christoph Weidenbach[1]

[1] Max Planck Institute for Informatics, Saarbrücken, Germany
{mbromber,weidenb}@mpi-inf.mpg.de
[2] Graduate School of Computer Science, Saarbrücken, Germany

Abstract. We present two tests that solve linear integer arithmetic constraints. These tests are sound and efficiently find solutions for a large number of problems. While many complete methods search along the problem surface for a solution, these tests use cubes to explore the interior of the problems. The tests are especially efficient for constraints with a large number of integer solutions, e.g., those with infinite lattice width. Inside the SMT-LIB benchmarks, we have found almost one thousand problem instances with infinite lattice width, and we have shown the advantage of our cube tests on these instances by comparing our implementation of the cube test with several state-of-the-art SMT solvers. Our implementation is not only several orders of magnitudes faster, but it also solves all instances, which most SMT solvers do not. Finally, we discovered an additional application for our cube tests: the extraction of equalities implied by a system of linear arithmetic inequalities. This extraction is useful both as a preprocessing step for linear integer constraint solving as well as for the combination of theories by the Nelson-Oppen method.

Keywords: Linear arithmetic · SMT · Integer arithmetic · Constraint solving

1 Introduction

Finding an integer solution for a polyhedron that is defined by a system of linear inequalities $Ax \leq b$ is a well-known NP-complete problem [18]. Systems of linear inequalities have many real-world applications so that this problem has been investigated in different research areas, e.g., in optimization via *(mixed) integer linear programming* (MILP) [15] and in constraint solving via *satisfiability modulo theories* (SMT) [2,4,7,12].

It is standard for commercial MILP implementations to integrate preprocessing techniques, heuristics, and specialized tests [15]. Although these techniques are not complete, they are much more efficient on their designated target systems of linear inequalities than a complete algorithm alone. Since there exist specialized techniques for many classes of real-world problems representable as polyhedra, commercial MILP solvers are efficient on many real-world inputs—even though the problem, in general, is NP-complete.

N. Olivetti and A. Tiwari (Eds.): IJCAR 2016, LNAI 9706, pp. 116–132, 2016.
DOI: 10.1007/978-3-319-40229-1_9

The constraint solving community is still in the process of developing the same variety in specialized tests as the MILP community. The biggest challenge is to adopt the tests from the MILP community so that they still fit the input systems relevant for constraint solving. For example, SMT theory solvers have to solve a large number of incrementally connected, small systems of linear inequalities. Exploiting this incremental connection is key for making SMT theory solvers efficient [11]. In contrast, MILP solvers typically target one large system. The same holds for their specialized tests, which are not well suited to exploit incremental connections.

In this paper, we present two tests tailored toward SMT solvers: the *largest cube test* and the *unit cube test*. The idea is to find hypercubes that are contained inside the input polyhedron and guarantee the existence of an integer solution. Due to computational complexity, we will restrict ourselves to only those hypercubes that are parallel to the coordinate axes. The largest cube test finds a hypercube with maximum edge length contained in the input polyhedron, determines its real valued center, and rounds it to a potential integer solution. The unit cube test determines if a polyhedron contains a hypercube with edge length one, which is the minimal edge length that guarantees an integer solution.

Most SMT linear integer arithmetic theory solvers are based on a branch-and-bound algorithm on top of the simplex algorithm. They search for a solution at the surface of a polyhedron. However, our tests search in the interior of the polyhedron. This gives them an advantage on polyhedra with a large number of integer solutions, e.g., polyhedra with infinite lattice width [16]. Since the only difference between the input polyhedron $Ax \leq b$ and the associated unit cube polyhedron $Ax \leq b'$ are the row bounds, our unit cube test is especially easy to implement and integrate into SMT theory solvers.

SMT theory solvers are designed to efficiently exchange bounds [9]. This efficient exchange is the main reason why SMT theory solvers exploit the incremental connection between the different polyhedra so well. Our unit cube test also requires only an exchange of bounds. After applying the test, we can easily recover the original polyhedron by reverting to the original bounds. In doing so, the unit cube test conserves the incremental connection between the different original polyhedra. We make a similar observation about the largest cube test.

A variant of the linear program for the unit cube test first appeared in 1969 as a subroutine in a heuristic by Hillier for MILP optimization [13]. While Hillier was aware of the unit cube test, he applied it only to cones, a special class of polyhedra. His work never mentioned applications beyond cones, nor did he prove any structural properties connected to hypercubes. As mentioned before, the main advantage of the cube tests is that they compute interior point candidates. The same can be done using an interior point method [17] instead of the simplex algorithm. Therefore, Hillier's heuristic tailored for MILP optimization lost popularity as soon as interior point methods became efficient in practice. Nonetheless, our cube tests remain relevant for SMT theory solvers because there are no competitive incremental interior point methods.

Also, Bobot et al. discuss relations between hypercubes, called ∞-norm balls, and polyhedra [2]. In their paper, they detail the same relation between polyhedra with infinite lattice width and hypercubes that we discovered. Their work also includes a linear optimization program that detects polyhedra with infinite lattice width and positive linear combinations between inequalities. Our largest cube test can detect all of the above because it is, with some minor changes, the dual of the linear optimization program of Bobot et al. However, our tests are a lot closer to the original polyhedron and are, therefore, easier to construct, and the tests produce sample points as well. Via rounding, our tests use these sample points to compute an actual integer solution as proof. Moreover, our cube tests also find solutions for polyhedra with finite lattice width.

Our contributions are as follows: we define the linear cube transformation (Corollary 3) that allows us to efficiently compute whether a polyhedron $Ax \leq b$ contains a hypercube of edge length e by solely changing the bounds b in Sect. 3. Based on this transformation, we develop in Sect. 4 two tests: the largest cube test and the unit cube test. For polyhedra with infinite lattice width, both tests always succeed (Lemma 5). Inside the SMT-LIB benchmarks, there are almost one thousand problem instances with infinite lattice width, and we show the advantage of our cube tests on these instances by comparing our implementation of the cube test with several state-of-the-art SMT solvers in Sect. 5. Our implementation is not only several orders of magnitudes faster, but it also solves all instances, which most SMT solvers do not (Fig. 7). It is more robust than the test suggested by Bobot et al. [2] (Fig. 7). Eventually, we introduce in Sect. 6 an additional application for our cube tests: the extraction of equalities implied by a system of linear arithmetic inequalities. The paper ends with a discussion on possible directions for future research, Sect. 7.

2 Preliminaries

While the difference between matrices, vectors, and their components is always clear in context, we generally use upper case letters for matrices (e.g., A), lower case letters for vectors (e.g., x), and lower case letters with an index i or j (e.g., b_i, x_j) as components of the associated vector at position i or j, respectively. The only exceptions are the row vectors $a_i^T = (a_{i1}, \dots, a_{in})$ of a matrix $A = (a_1, \dots, a_m)^T$, which already contain an index i that indicates the row's position inside A. In order to save space, we write vectors only implicitly as columns via the transpose $(\ )^T$ operator, which turns all rows $(b_1, \dots, b_m)$ into columns $(b_1, \dots, b_m)^T$ and vice versa. We will also abbreviate $(\dots, 0, \dots)^T$ as $\mathbf{0}$.

In this paper, we treat *polyhedra* and their definitions through a *system of inequalities* $Ax \leq b$ as interchangeable. For such a system of inequalities, the row coefficients are given by $A = (a_1, \dots, a_m)^T \in \mathbb{Q}^{m \times n}$, the inequality bounds are given by $b = (b_1, \dots, b_m)^T \in \mathbb{Q}^m$, and the variables are given by $x = (x_1, \dots, x_n)^T$.

We denote by $P_b^A = \{x \in \mathbb{R}^n : Ax \leq b\}$ the *set of real solutions* to the system of inequalities $Ax \leq b$ and, therefore, the points inside the polyhedron.

Similarly, we denote by $C_e^n(z) = \{x \in \mathbb{R}^n : \forall j \in 1, \ldots, n.\ |x_j - z_j| \leq \frac{e}{2}\}$ the set of points contained in the *n-dimensional hypercube* $C_e^n(z)$ that is parallel to the coordinate axes, has *edge length* $e \in \mathbb{R}_{\geq 0}$, and has *center* $z \in \mathbb{R}^n$. For the remainder of this paper, we will consider only hypercubes that are parallel to the coordinate axes. For simplicity, we call these restricted hypercubes *cubes.* Similar to polyhedra, we will use the set of points $C_e^n(z)$ interchangeably with the cube defined by the set.

Besides cubes and polyhedra, we use multiple p-norms $\|.\|_p$ in this paper [10]. These p-norms are defined as functions ($\|.\|_p : \mathbb{R}^n \to \mathbb{R}$) for $p \geq 1$ such that $\|x\|_p = (|x_1|^p + \ldots + |x_n|^p)^{1/p}$. A special p-norm is the *maximum norm.* It is defined by the limit of $\|.\|_p$ for $p \to \infty$: $\|x\|_\infty = \max\{|x_1|, \ldots, |x_n|\}$. If we compare the maximum norm and the definition of $C_e^n(z)$, we see that cubes and p-norms are related: $\left(\|x - z\|_\infty \leq \frac{e}{2}\right) \iff \left(\forall j \in 1, \ldots, n.\ |x_j - z_j| \leq \frac{e}{2}\right)$.

Using p-norms, we define a *closest integer* for a point x as a point $x' \in \mathbb{Z}^n$ with minimal distance $\|x - x'\|_p$ for all p-norms. We also define the operators $\lceil x_j \rfloor$ and $\lceil x \rfloor$ such that they describe a *closest integer* for x_j and x, respectively. Formally, this means that $\lceil x \rfloor = (\lceil x_1 \rfloor, \ldots, \lceil x_n \rfloor)^T$ and

$$\lceil x_j \rfloor = \begin{cases} \lfloor x_j \rfloor & \text{if } x_j - \lfloor x_j \rfloor < 0.5\,, \\ \lceil x_j \rceil & \text{if } x_j - \lfloor x_j \rfloor \geq 0.5\,. \end{cases}$$

This definition of $\lceil x \rfloor$ is also known as *simple rounding.*

Lemma 1. *For $x \in \mathbb{R}^n$, $\lceil x \rfloor$ is a closest integer to x:*

$$\forall p \geq 1.\ \forall x' \in \mathbb{Z}^n.\ \|x - \lceil x \rfloor\|_p \leq \|x - x'\|_p\,.$$

Proof. We first look at the one-dimensional case, where $\|x_j\|_p$ simplifies to $|x_j|$:

$$\forall p \geq 1.\ \forall x'_j \in \mathbb{Z}.\ |x_j - \lceil x_j \rfloor| \leq |x_j - x'_j|\,.$$

For $\lceil x_j \rfloor, x'_j \in \mathbb{Z}$, there exists $z_j \in \mathbb{Z}$ such that $x'_j = \lceil x_j \rfloor - z_j$. For $x_j \in \mathbb{R}$, there exists a $d_j \in [-0.5, 0.5]$ such that $d_j := x_j - \lceil x_j \rfloor$. The inequality trivially holds for $z_j = 0$:

$$|x_j - x'_j| = |x_j - \lceil x_j \rfloor + z_j| = |x_j - \lceil x_j \rfloor|\,.$$

Via the triangle inequality, for the remaining $z_j \neq 0$ we get :

$$|x_j - x'_j| = |x_j - \lceil x_j \rfloor + z_j| = |d_j + z_j| \geq |z_j| - |d_j|\,.$$

Since $z_j \neq 0$, and $d_j \in [-0.5, 0.5]$ imply $|z_j| \geq 1$, and $|d_j| \leq 0.5$, respectively, we get:

$$|x_j - x'_j| \geq |z_j| - |d_j| \geq 1 - |d_j| \geq 0.5 \geq |d_j| = |x_j - \lceil x_j \rfloor|\,.$$

The multidimensional case follows from the p-norms' monotonicity [10], i.e., if $|x_j - \lceil x_j \rfloor| \leq |x_j - x'_j|$ for all $j \in \{1, \ldots, n\}$, then $\|x - \lceil x \rfloor\|_p \leq \|x - x'\|_p$. □

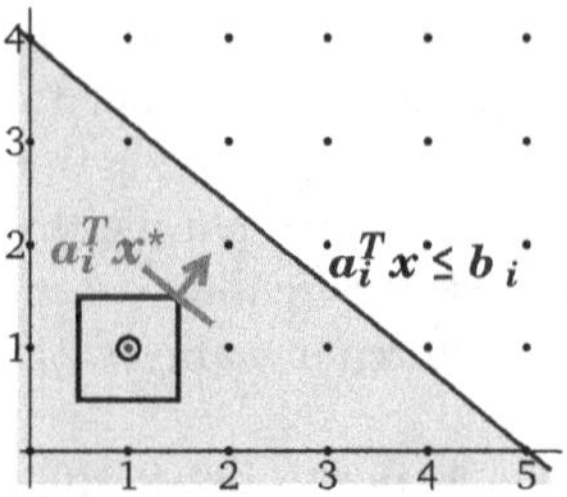

Fig. 1. A square (two-dimensional cube) fitting into an inequality $a_i^T x \le b_i$ and the cube's maximum $a_i^T x^*$ for the objective $a_i^T x$

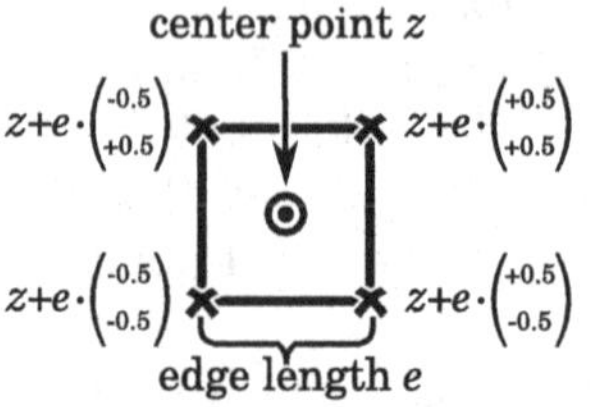

Fig. 2. The vertices of an arbitrary square parallel to the coordinate axes (two-dimensional cube with edge length e and center z)

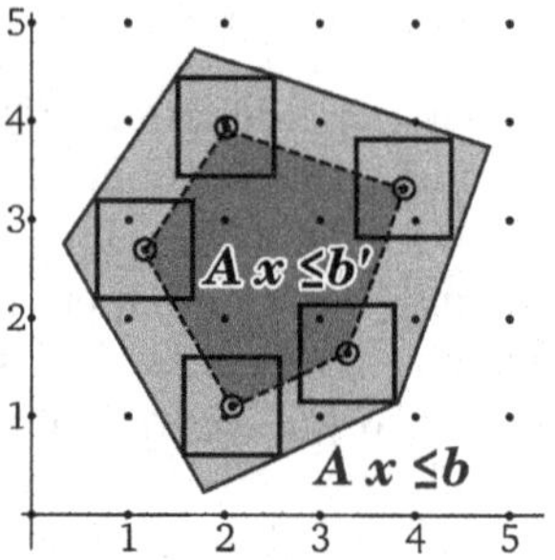

Fig. 3. The transformed polyhedron $Ax \le b'$ for edge length 1 together with the original polyhedron $Ax \le b$

3 Fitting Cubes into Polyhedra

We say that a cube $C_e^n(z)$ *fits* into a polyhedron defined by $Ax \le b$ if all points inside the cube $C_e^n(z)$ are solutions of $Ax \le b$, or formally: $C_e^n(z) \subseteq P_b^A$. In order to compute this, we transform the polyhedron $Ax \le b$ into another polyhedron $Ax \le b'$. For this new polyhedron, we merely have to test whether the cube's center point z is a solution ($z \in P_{b'}^A$) in order to also determine whether the cube $C_e^n(z)$ fits into the original polyhedron ($C_e^n(z) \subseteq P_b^A$). This is a simple test that requires only evaluation. We call this entire transformation the *linear cube transformation.*

We start explaining the linear cube transformation by looking at the case where the polyhedron is defined by a single inequality $a_i^T x \le b_i$. A cube $C_e^n(z)$ fits into the inequality $a_i^T x \le b_i$ if all points inside the cube $C_e^n(z)$ are solutions of $a_i^T x \le b_i$, or formally: $\forall x \in C_e^n(z).\ a_i^T x \le b_i$.

We can think of $a_i^T x$ as an objective function that we want to maximize and see b_i as a guard for the maximum objective of any solution in the cube. Thus, we can express the universal quantifier in the above equation as an optimization problem (see Fig. 1): $\max\{a_i^T x : x \in C_e^n(z)\} \le b_i$. This also means that all points in $x \in C_e^n(z)$ satisfy the inequality $a_i^T x \le b_i$ if a point $x^* \in C_e^n(z)$ with maximum value $a_i^T x^* = \max\{a_i^T x : x \in C_e^n(z)\}$ for the objective function $a_i^T x$ satisfies the inequality $a_i^T x^* \le b_i$. We can formalize the above optimization problem as a linear program:

$$\begin{array}{ll} \text{maximize} & a_i^T x \\ \text{subject to} & z_j - \frac{e}{2} \le x_j \le z_j + \frac{e}{2} \ \text{ for } j = 1, \ldots, n\,. \end{array}$$

However, for the case of cubes, there is an even easier way to determine the maximum objective value. Since every cube is a bounded polyhedron, one of the points with maximum objective value is a vertex $x^v \in C_e^n(z)$. A vertex x^v of the cube $C_e^n(z)$ is one of the points with maximum distance to the center z (see Fig. 2), or formally: $x^v = \left(z_1 \pm \frac{e}{2}, \ldots, z_n \pm \frac{e}{2}\right)^T$. If we insert the above equation into the objective function $a_i^T x$, we get:

$$a_i^T \left(z_1 \pm \frac{e}{2}, \ldots, z_n \pm \frac{e}{2}\right)^T = a_i^T z + \frac{e}{2} \sum\nolimits_{j=1}^{n} \pm a_{ij} \,,$$

which in turn is maximal if we choose x^v such that $\pm a_{ij}$ is always positive:

$$a_i^T x^v = a_i^T z + \frac{e}{2} \sum\nolimits_{j=1}^{n} |a_{ij}| = a_i^T z + \frac{e}{2} \left\|a_i\right\|_1 .$$

Hence, we transform the inequality $a_i^T x \leq b_i$ into $a_i^T x \leq b_i - \frac{e}{2} \left\|a_i\right\|_1$, and $C_e^n(z)$ fits into $a_i^T x \leq b_i$ if $a_i^T z \leq b_i - \frac{e}{2} \left\|a_i\right\|_1$.

Corollary 2. *Let $C_e^n(z)$ be a cube and $a_i^T x \leq b_i$ be an inequality. All $x \in C_e^n(z)$ fulfill $a_i^T x \leq b_i$ if and only if $a_i^T z \leq b_i - \frac{e}{2} \left\|a_i\right\|_1$.*

Next, we look at the case where multiple inequalities $a_i^T x \leq b_i$ (for $i = 1, \ldots, m$) define the polyhedron $Ax \leq b$. Since P_b^A is the intersection of all $P_{b_i}^{a_i}$, the cube fits into $Ax \leq b$ if and only if it fits into all inequalities $a_i^T x \leq b_i$, respectively:

$$\forall i \in \{1, \ldots, m\}.\ \forall x \in C_e^n(z).\ a_i^T x \leq b_i \,.$$

We can express this by m optimization problems:

$$\forall i \in \{1, \ldots, m\}.\ \max\{a_i^T x : x \in C_e^n(z)\} \leq b_i$$

and, after applying Corollary 2, by the following m inequalities:

$$\forall i \in \{1, \ldots, m\}.\ a_i^T z \leq b_i - \frac{e}{2} \left\|a_i\right\|_1 .$$

Hence, the linear cube transformation transforms the polyhedron $Ax \leq b$ into the polyhedron $Ax \leq b'$, where $b'_i = b_i - \frac{e}{2} \left\|a_i\right\|_1$, and $C_e^n(z)$ fits into $Ax \leq b$ if $Az \leq b'$.

Corollary 3. *Let $C_e^n(z)$ be a cube and $Ax \leq b$ be a polyhedron. $C_e^n(z) \subseteq P_b^A$ if and only if $Az \leq b'$, where $b'_i = b_i - \frac{e}{2} \left\|a_i\right\|_1$.*

Until now, we have discussed how to use the linear cube transformation to determine if one cube $C_e^n(z)$ with fixed center point z fits into a polyhedron $Ax \leq b$. A generalization of this problem determines whether a polyhedron $Ax \leq b$ contains a cube of edge length e at all. Actually, a closer look at the transformed polyhedron $Ax \leq b'$ reveals that the linear cube transformation ($b'_i = b_i - \frac{e}{2} \left\|a_i\right\|_1$) is dependent only on the edge length e of the cube. Therefore, the solutions $P_{b'}^A$ of the transformed polyhedron $Ax \leq b'$ are exactly all center points of cubes with edge length e that fit into the original polyhedron $Ax \leq b$ (see Fig. 3). By determining the satisfiability of the transformed polyhedron $Ax \leq b'$, we can now also determine whether a polyhedron $Ax \leq b$ contains a cube of edge length e at all. If we choose a suitable algorithm, e.g., the simplex algorithm, then we even get the center point z of a cube $C_e^n(z)$ that fits into $Ax \leq b$. This observation is the foundation for the cube tests that we will present in Sect. 4.

4 Fast Cube Tests

In contrast to arbitrary polyhedra, determining whether a cube $C_e^n(z)$ contains an integer point is easy. Because of the cubes symmetry, it is enough to test whether it contains a closest integer point $\lceil z \rfloor$ to the center z.

Lemma 4. *A cube $C_e^n(z)$ contains an integer point if and only if it contains a closest integer point $\lceil z \rfloor$ to the center z.*

Proof. The implication from left to right follows directly from Lemma 1 and from the relation between the maximum norm and cubes. The implication from right to left is obvious. □

Note that every point $z \in \mathbb{R}^n$ is also a cube $C_0^n(z)$ of edge length 0. In order to be efficient, our tests will look only at cubes with special properties. In the case of the largest cube test, we check for an integer solution in one of the largest cubes fitting into the polyhedron $Ax \leq b$. In the case of the unit cube test, we look for a cube of edge length one, which always guarantees an integer solution. Due to these restrictions, both tests are not complete but very fast to compute.

4.1 Largest Cube Test

A well-known test, implemented in most ILP solvers, is *simple rounding*. For simple rounding, the ILP solver computes a real solution x for a set of inequalities, *rounds* it to a closest integer $\lceil x \rfloor$, and determines whether this point is an integer solution. Not all types of real solutions are good candidates for this test to be successful. Especially *surface points*, such as *vertices*, the usual output of the simplex algorithm, are not good candidates for rounding. For many polyhedra, *center and interior points* z are a better choice because all integer points adjacent to z are solutions, including a closest integer point $\lceil z \rfloor$.

To calculate a real center point with the simplex algorithm, we use the linear cube transformation (Sect. 3). The center point will be the center point of a largest cube that fits into the polyhedron $Ax \leq b$ (see Fig. 4). We determine the center z of this largest cube and the associated edge length e with the following LP:

$$\begin{array}{ll} \text{maximize} & x_e \\ \text{subject to} & Ax + a'\frac{x_e}{2} \leq b, \text{ where } a'_i = \|a_i\|_1 \\ & x_e \geq 0\,. \end{array}$$

This linear program employs the linear cube transformation from Sect. 3. The only generalization is a variable x_e for the edge length instead of a constant value e. Additionally, this linear program maximizes the edge length as an optimization goal.

If the resulting maximum edge length is unbounded, the original polyhedron contains cubes of arbitrary edge length (see Fig. 5) and, thus, infinitely many integer solutions. Since the linear program contains all solutions of the original polyhedron (see $x_e = 0$), the original polyhedron is empty if and only if the

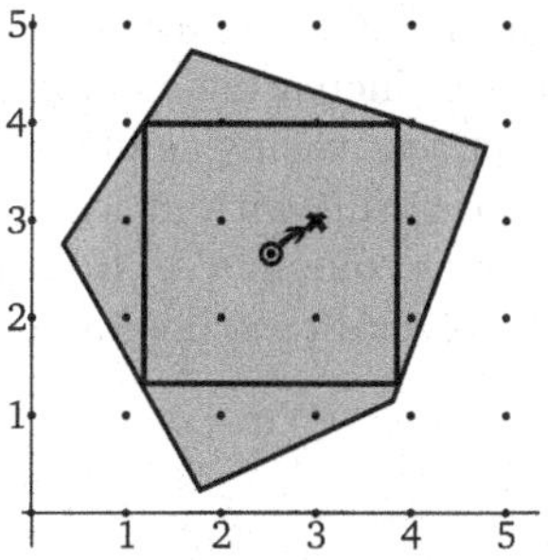

Fig. 4. The largest cube inside a polyhedron, its center point, and a closest integer point to the center

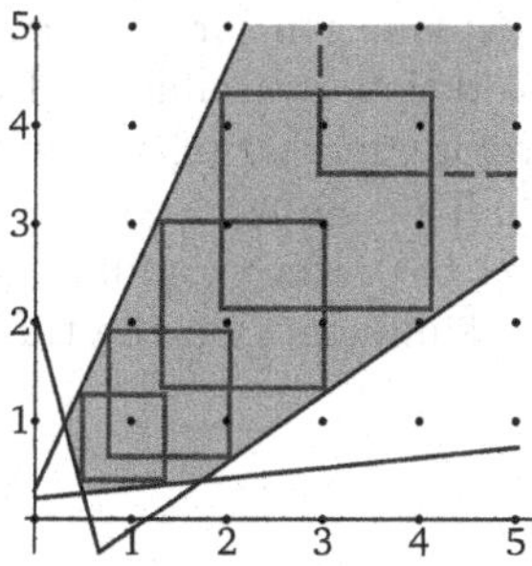

Fig. 5. An infinite lattice width polyhedron, containing cubes for every edge length $e > 0$.

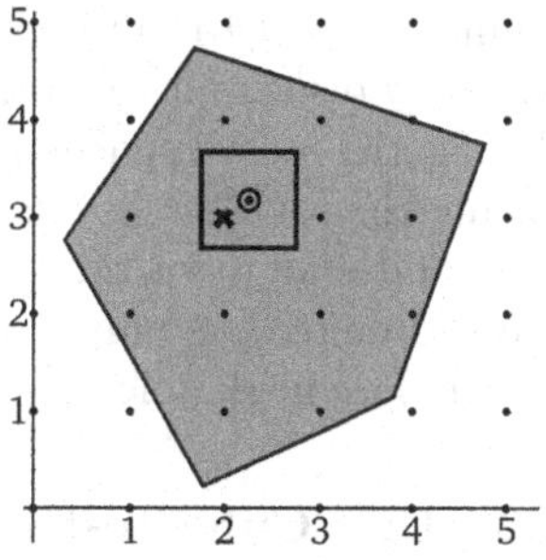

Fig. 6. A unit cube inside a polyhedron, its center point, and a closest integer point to the center

above linear program is infeasible. If the maximum edge length is a finite value e, we use the resulting assignment z for the variables x as a center point and $C_e^n(z)$ is a largest cube that fits into the polyhedron. From the center point, we round to a closest integer point $\lceil z \rfloor$ and determine if it fits into the original polyhedron. If this is the case, we are done because we have found an integer solution for $Ax \leq b$. Otherwise, the largest cube test does not know whether or not $Ax \leq b$ has an integer solution. An example for the latter case, are the following inequalities: $3x_1 - x_2 \leq 0$, $-2x_1 - x_2 \leq -2$, and $-2x_1 + x_2 \leq 1$. These inequalities have exactly one integer solution $(1,3)^T$, but the largest cube contained by the inequalities has edge length $e = \frac{3}{17}$ and center point $(\frac{3}{17}, \frac{3}{2})^T$, which rounds to $(0,2)^T$.

Instead of a cube, it is also possible to use a ball to compute a center point. The result is the *Chebyshev center* [3], i.e., the center of a largest ball that fits into the polyhedron:

$$\begin{array}{ll} \text{maximize} & x_r \\ \text{subject to} & Ax + a'x_r \leq b, \text{ where } a'_i = \|a_i\|_2 \\ & x_r \geq 0\,. \end{array}$$

However, the coefficients a'_i are then defined via the 2-norm $\|a_i\|_2 = \sqrt{\sum_{j=1}^{n} a_{ij}^2}$ and are, therefore, potentially irrational. As theory solvers in the SMT context use exact rational arithmetic, the Chebyshev center is not straightforward to integrate.

The largest cube test also upholds the incremental advantages of the dual simplex algorithm proposed by Dutertre and de Moura [9]. The only difference is the extra column $a'\frac{x_e}{2}$, which the theory solver can internally create while it is notified of all potential arithmetic literals. Adding this column from the start does not influence the correctness of the solution because $x_e \geq 0$ guarantees that the largest cube test is satisfiable exactly when the original inequalities $Ax \leq b$ are satisfiable. Even for explanations of unsatisfiability, it suffices to remove the bound $x_e \geq 0$ to obtain an explanation for the original inequalities $Ax \leq b$. The only disadvantage is the additional variable x_e. However, increasing x_e only shrinks the search space. Therefore, increasing x_e can never resolve any conflicts during the satisfiability search. The simplex solver recognizes this with at least one additional pivot that sets x_e to 0. Hence, adding the extra column $a' \frac{x_e}{2}$ from the beginning has only constant influence on the theory solver's run-time, and is therefore negligible.

4.2 Unit Cube Test

Most SMT theory solvers implement a simplex algorithm that is specialized towards feasibility and not towards optimization [1,6,9,12]. Therefore, a test based on optimization, such as the largest cube test, does not fit well with existing implementations. As an alternative, we have developed a second test based on cubes that does not need optimization.

We avoid optimization by fixing the edge length e to the value 1 for all the cubes $C_e^n(z)$ we consider (see Fig. 6). We do so because cubes $C_1^n(z)$ of edge length 1 are the smallest cubes to always guarantee an integer solution, completely independent of the center point z. A cube with edge length 1 is also called a *unit cube*. To prove this guarantee, we first fix $e = 1$ in the definition of cubes, $C_1^n(z) = \left\{x \in \mathbb{R}^n : \forall j \in 1, \ldots, n.\ |x_j - z_j| \leq \frac{1}{2}\right\}$, and look at the following property for the rounding operator $\lceil . \rfloor$: $\forall z_j \in \mathbb{R}. |\lceil z_j \rfloor - z_j| \leq \frac{1}{2}$. We see that any unit cube contains a closest integer $\lceil z \rfloor$ to its center point z. Furthermore, 1 is the smallest edge length that guarantees an integer solution for a cube with center point $z = (\ldots, \frac{1}{2}, \ldots)^T$. Thus, 1 is the smallest value that we can fix as an edge length to guarantee an integer solution for all cubes $C_1^n(z)$.

Our second test tries to find a unit cube that fits into the polyhedron $Ax \leq b$ and, thereby, a guarantee for an integer solution for $Ax \leq b$. Again, we employ the linear cube transformation from Sect. 3 and obtain the linear program:

$$Az \leq b', \text{ where } b'_i = b_i - \tfrac{1}{2} \left\|a_i\right\|_1 .$$

In addition to being a linear program without an optimization objective, we only have to change the row bounds b'_i of the original inequalities. In the dual simplex algorithm proposed by Dutertre and de Moura [9] and implemented in many SMT theory solvers [1,6,9,12], such a change of bounds is already part of the framework so that integrating the unit cube test into theory solvers is possible with only minor adjustments to the existing implementation. Since our unit cube test requires only an exchange of bounds, we can easily return to the original polyhedron by reverting the bounds. In doing so, the unit cube test upholds the incremental connection between the different original polyhedra.

5 Experiments

While our tests are useful for many types of polyhedra, the motivation for our tests stems from a special type of polyhedra, so-called *infinite lattice width* polyhedra [16]. A polyhedron $Ax \leq b$ has *infinite lattice width* if for every objective

$c \in \mathbb{R}^n \setminus \{\mathbf{0}\}$, either its maximum or minimum objective value is unbounded, or formally:

$$\forall c \in \mathbb{R}^n \setminus \{\mathbf{0}\}.\ \sup\left\{c^T x \mid x \in P_b^A\right\} = \infty \text{ or } \inf\left\{c^T x \mid x \in P_b^A\right\} = -\infty .$$

Polyhedra with infinite lattice width seem trivial at first glance because their interior expands arbitrarily far in all directions (see Fig. 5). Therefore, a polyhedron with infinite lattice width contains an infinite number of integer solutions [16]. Nonetheless, many SMT theory solvers have proven to be inefficient on those polyhedra because they use a branch-and-bound approach with an underlying simplex solver [9]. Although such an approach will terminate inside finite a priori bounds [18], it does not explore the infinite interior, but rather directs the search along the solutions suggested by the simplex solver: the vertices of the polyhedron. Thus, the SMT theory solvers concentrate their search on a bounded part of the polyhedron. This bounded part contains only a finite number of integer solutions, whereas the complete interior contains infinitely many integer solutions. The advantage of our cube tests is that they actually exploit the infinite interior because polyhedra with infinite lattice width contain cubes for every edge length (see Fig. 5). Our tests are always successful on polyhedra with infinite lattice width and usually need only a small number of pivoting steps before finding a solution.

Lemma 5. *Let $Ax \leq b$ be a polyhedron. Let $a' \in \mathbb{Z}^m$ be a vector such that its components are $a'_i = \|a_i\|_1$. Then, the following two statements are equivalent:*
(1) $Ax \leq b$ contains a cube $C_e^n(z)$ for every $e \in \mathbb{R}_{\geq 0}$, and
(2) $Ax \leq b$ has infinite lattice width.
Or formally:
(1) $\forall e \in \mathbb{R}_{\geq 0}.\ \exists x \in \mathbb{R}^n.\ Ax \leq b - \frac{e}{2} \cdot a'$,
(2) $\forall c \in \mathbb{R}^n \setminus \{\mathbf{0}\}.\ \sup\left\{c^T x \mid x \in P_b^A\right\} = \infty$ *or* $\inf\left\{c^T x \mid x \in P_b^A\right\} = -\infty$.

Proof. (1) ⇒ (2): We first assume that $Ax \leq b$ contains a cube $C_e^n(z)$ for every $e \in \mathbb{R}_{\geq 0}$. Note that the center point z depends on the edge length e. Furthermore, we define the function:

$$\operatorname{width}(c, S) = \left(\sup\left\{c^T x \mid x \in S)\right\} + \sup\left\{-c^T x \mid x \in S)\right\}\right) \tag{1}$$

for every vector $c \in \mathbb{R}^n \setminus \{\mathbf{0}\}$ and for every set of points $S \subseteq \mathbb{R}^n$. Then, we prove that:

$$\lim{}_{e \to \infty} \operatorname{width}(c, C_e^n(.)) \to \infty .$$

In Sect. 3, we have shown that:

$$\sup\left\{c^T x \mid x \in C_e^n(z)\right\} = c^T z + \frac{e}{2} \cdot \|c\|_1 \text{ , and} \tag{2}$$

$$\sup\left\{-c^T x \mid x \in C_e^n(z)\right\} = -c^T z + \frac{e}{2} \cdot \|c\|_1 . \tag{3}$$

Therefore, $\operatorname{width}(c, C_e^n(z)) = e \cdot \|c\|_1$, which is independent of z. After inserting (2) and (3) into (1), we get:

$$\lim_{e \to \infty} \operatorname{width}(c, C_e^n(.)) = \lim_{e \to \infty} e \cdot \|c\|_1 \to \infty .$$

Since $Ax \leq b$ contains cubes $C_e^n(z)$ for all $e \in \mathbb{R}$, it holds for all $e \in \mathbb{R}$ that

$$\text{width}(c, P_b^A) \geq \text{width}(c, C_e^n(.)),$$

and, thus, $\text{width}(c, P_b^A) = \infty$. Since P_b^A is also convex, it must hold that:

$$\sup\left\{c^T x \mid x \in P_b^A\right\} = \infty \text{ or } \inf\left\{c^T x \mid x \in P_b^A\right\} = -\infty.$$

(2) ⇒ (1): By contradiction. Assume that $Ax \leq b$ has infinite lattice width but that there exists an $e \in \mathbb{R}_{\geq 0}$ such that $Ax \leq b$ contains no cube $C_e^n(z)$ of edge length e. By Corollary 3, $Ax \leq b$ contains no cube $C_e^n(z)$ of edge length e implies that $Ax \leq b - \frac{e}{2} \cdot a'$ is unsatisfiable. By Farkas Lemma [3], $Ax \leq b - \frac{e}{2} \cdot a'$ is unsatisfiable implies that there exists a $y \in \mathbb{R}^m$ such that: (a) $y_i \geq 0$ for all $i \in \{1, \ldots, m\}$, (b) $y_k > 0$ for at least one $k \in \{1, \ldots, m\}$, (c) $y^T A = 0$, and (d) $0 > y^T b - \frac{e}{2} \cdot y^T a'$. Because of (b), we can transform the equality (c) into the following form:

$$a_k = -\sum\nolimits_{i=1, i \neq k}^{m} \left(\frac{y_i}{y_k} a_i\right). \tag{4}$$

By multiplying (4) with an $x \in P_b^A$, we get: $a_k^T x = -\sum\nolimits_{i=1, i \neq k}^{m} \left(\frac{y_i}{y_k} a_i^T x\right)$. Since $a_i^T x \leq b_i$ and $y_i \geq 0$, we get a finite lower bound for $a_k^T x$:

$$a_k^T x = -\sum\nolimits_{i=1, i \neq k}^{m} \left(\frac{y_i}{y_k} a_i^T x\right) \geq -\sum\nolimits_{i=1, i \neq k}^{m} \left(\frac{y_i}{y_k} b_i\right).$$

Thus, the upper bound $\sup\left\{a_k^T x \mid x \in P_b^A\right\} \leq b_k < \infty$ and the lower bound $\inf\left\{a_k^T x \mid x \in P_b^A\right\} \geq -\sum\nolimits_{i=1, i \neq k}^{m} \left(\frac{y_i}{y_k} b_i\right) > -\infty$ are finite, which contradicts the assumption that $Ax \leq b$ has infinite lattice width. □

We have found instances of polyhedra with the infinite lattice width property in some classes of the SMT-LIB benchmarks. These instances are 229 of the 233 *dillig* benchmarks designed by Dillig et al. [7], 503 of the 591 *CAV-2009* benchmarks also by Dillig et al. [7], 229 of the 233 *slacks* benchmarks which are the dillig benchmarks extended with slack variables [14], and 19 of the 37 *prime-cone* benchmarks, that is, "a group of crafted benchmarks encoding a tight n-dimensional cone around the point whose coordinates are the first n prime numbers" [14]. The remaining problems (4 from dillig, 88 from CAV-2009, 4 from slacks, and 18 from prime-cone) do not fulfill the infinite lattice width property because they are either tightly bounded or unsatisfiable. For our experiments, we look only at the instances of those benchmark classes that actually fulfill the infinite lattice width property.

Using these benchmark instances, we have confirmed our theoretical assumptions (Lemma 5) in practice. We integrated the unit cube test into our own branch-and-bound solver *SPASS-IQ*[1] and ran it on the infinite lattice width instances; once with the unit cube test turned on (*SPASS-IQ-0.1+uc*) and once

[1] http://www.spass-prover.org/spass-iq.

Benchmark Name	CAV-2009		DILLIG		PRIME-CONE		SLACKS		ROTATE	
#Instances	503		229		19		229		229	
Solvers:	**solved**	**time**	**solved**	**time**	**solved**	**time**	**solved**	**time**	**solved**	**time**
SPASS-IQ-0.1+uc	**503**	22	**229**	9	**19**	0.4	**229**	**26**	**229**	**9**
SPASS-IQ-0.1	**503**	713	**229**	218	**19**	0.4	197	95	**229**	214
ctrl-ergo	**503**	**12**	**229**	**5**	**19**	0.4	**229**	46	24	6760
cvc4-1.4	467	12903	206	4146	18	3	152	4061	208	6964
mathsat5-3.9	**503**	6409	225	2314	**19**	3.5	181	4577	**229**	1513
yices-2.4.2	472	11461	213	2563	**19**	**0.1**	147	5767	180	10171
z3-4.4.0	466	764	213	525	**19**	0.2	158	383	213	528

Fig. 7. Experimental results

with the test turned off (*SPASS-IQ-0.1*). For every problem, SPASS-IQ-0.1+uc applies the unit cube test exactly once. This application happens before we start the branch-and-bound approach. We also compared our solver with some of the state-of-the-art SMT solvers currently available for linear integer arithmetic: *cvc4-1.4* [1], *mathsat5-3.9* [5], *yices2.4.2* [8], and *z3-4.4.0* [6]. As mentioned before, all these solvers employ a branch-and-bound approach with an underlying dual simplex solver [9].

The solvers had to solve each problem in under 10 min. For the experiments, we used a Debian Linux server with 32 Intel Xeon E5-4640 (2.4 GHz) processors and 512 GB RAM. Figure 7 lists the results of the different solvers (column one) on the different benchmark classes (row one). Row two lists the number of benchmark instances we considered for our experiments. For each combination of benchmark class and solver, we have listed the number of instances the solver could solve in the given time as well as the total time (in seconds) of the instances solved (columns labelled with "solved" and "time", respectively).

Our solver that employs the unit cube test solves all instances with the application of the unit cube test and is 25 times faster than our solver without the test. The SMT theory solvers in their standard setting were not able to solve all instances within the allotted time. Moreover, our unit cube test was over 100 times faster than any state-of-the-art SMT solver.

We also compared our test with the *ctrl-ergo* solver, which includes a subroutine that is essentially the dual to our largest cube test [2]. As expected, both approaches are comparable for infinite lattice width polyhedra. In order to also compare the two approaches on benchmarks without infinite lattice width, we created the *rotate* benchmarks by adding the same four inequalities to all infinite width instances of the dillig benchmarks. These four inequalities essentially describe a square bounding the variables x_0 and x_1 in an interval $[-u, u]$. For a large enough choice of u (e.g., $u = 2^{10}$), the square is so large that the benchmarks are still satisfiable and not absolutely trivial for branch-and-bound solvers. To add a challenge, we rotated the square by a small factor $1/r$, which resulted in the following four inequalities:

$$-b \cdot r \cdot r + r \leq b \cdot r \cdot x_0 - x_1 \leq b \cdot r \cdot r - r\,, \text{ and}$$
$$-b \cdot r \cdot r + r \leq x_0 + b \cdot r \cdot x_1 \leq b \cdot r \cdot r - r\,.$$

These changes have nearly no influence on SPASS-IQ, and two SMT solvers even benefit from the proposed changes. However, the *rotate* benchmarks are very hard for ctrl-ergo because its subroutine detects only infinite lattice width. Without infinite lattice width, ctrl-ergo starts its search from the boundaries of the polyhedron instead of looking at the polyhedron's interior. We can even control the number of iterations (r^2) ctrl-ergo spends on the parts of the boundary without any integer solutions if we choose r accordingly (e.g., $r = 2^{10}$). In contrast, we use our cube tests to also extract interior points for rounding. This difference makes our tests much more stable under consideration of small changes to the polyhedron.

There exist alternative methods for solving linear integer constraints that do not rely on a branch-and-bound approach [4,14]. These have not yet matured enough to be competitive with our tests or state-of-the-art SMT theory solvers.

Most problems in the linear integer arithmetic SMT-LIB benchmarks with finite lattice width can be solved without using any actual integer arithmetic technique. A standard simplex solver for the reals typically finds a real solution for such a problem that is also an integer solution. Applying the unit cube test on these trivial problem classes is a waste of time, worst case it doubles the eventual solution time. For these examples it is beneficial to first compute a general real solution and to check it for integer satisfiability before applying the unit cube test. This has the additional benefit that real unsatisfiable problems are also filtered out before applying the unit cube test. Also, the unit cube test is almost guaranteed to fail on problems containing boolean variables, i.e., variables that are either 0 or 1, unless they are absolutely trivial and describe a unit cube themselves. Whenever the problem contains a boolean variable, it is often beneficial to skip the unit cube test.

6 Further Cube Test Applications

Equalities are the greatest challenge for the applicability of our cube tests. A polyhedron contains an equality $a_E^T x = b_E$ if $a_E^T x = b_E$ holds for all $x \in P_b^A$. An equality contained in $Ax \leq b$ is *explicit* if $Ax \leq b$ includes the inequalities $a_E^T x \leq b_E$ and $-a_E^T x \leq -b_E$. Otherwise, the equality is *implicit*. Polyhedra containing equalities have only surface points and, therefore, neither an interior nor a center. Thus, a largest cube has edge length zero and is just a point in the original polyhedron. Similar problems occur if we allow not only inequalities but also other types of constraints, such as negated equalities ($a_i^T x \neq b_i$), divisibility constraints ($d \mid a_i^T x + b_i$, i.e., $d \in \mathbb{Z}$ divides $a_i^T x + b_i$), and negated divisibility constraints ($d \nmid a_i^T x + b_i$). In this section, we propose additional transformations and strategies that are useful for resolving the aforementioned challenges and are also applicable even beyond our tests.

First of all, we can transform any divisibility constraint and negated divisibility constraint into an equality by introducing additional variables. For divisibility constraints $d \mid a_i^T x + b_i$, this transformation is known as the diophantine representation: $\exists q \in \mathbb{Z}.\ dq - a_i^T x = b_i$. For negated divisibility constraints $d \nmid a_i^T x + b_i$, there exists a similar transformation: $\exists q \in \mathbb{Z}. \exists r \in \mathbb{Z}.\ dq + r - a_i^T x = b_i \ \wedge\ 1 \leq r \leq d-1$. Both of these transformations describe the formal definition of dividing $a_i^T x + b_i$ by d: $a_i^T x + b_i = dq + r$, where q is the quotient of the division and r the remainder. Since the divisibility constraint enforces that d divides $a_i^T x + b_i$, the remainder r must be zero. Likewise, the negated divisibility constraint enforces that d does not divide $a_i^T x + b_i$. Therefore, the remainder r lies between 1 and $d - 1$. These transformations are useful beyond our tests because they can be used to integrate (negated) divisibility constraints into the simplex algorithm. The only disadvantage is that we have to introduce additional variables q (and r) for every (negated) divisibility constraint.

Next, we eliminate all equalities from $Ax \leq b$. We do so by taking an equality $a_i^T x = b_i$ contained in $Ax \leq b$ and replacing a variable x_k in $Ax \leq b$ by substituting with $x_k := \frac{1}{a_{ik}}(b_i - \sum_{j=1, j\neq k}^{n} a_{ij}x_j)$, where $a_{ik} > 0$. Naturally, replacing x_k in $Ax \leq b$ creates a new system of inequalities $A'x' \leq b'$, where $A' \in \mathbb{Q}^{(m-1)\times(n-1)}$, $b' \in \mathbb{Q}^{(m-1)}$, and $x' = (x_1, \ldots, x_{k-1}, x_{k+1}, \ldots, x_n)^T$. We iteratively repeat this approach until our system of inequalities $A^I x^I \leq b^I$ contains no more equalities. As a by-product, we get a system of equalities $A^E x = b^E$ consisting of all equalities we have found. The two systems of constraints $A^I x^I \leq b^I$ and $A^E x = b^E$ together are equivalent to $Ax \leq b$, but $A^I x^I \leq b^I$ contains no equalities while $A^E x = b^E$ contains (at least implicitly) all equalities of $Ax \leq b$. We can now completely eliminate the equalities $A^E x = b^E$ from $Ax \leq b$ by combining this approach with a *diophantine equation handler* [12]. The result is a new system of inequalities that contains no equalities and has an integer solution if and only if $Ax \leq b$ has one.

Extracting equalities has further applications; for instance, the derivation of equalities is needed for the combination of theories by the *Nelson-Oppen method.* We can even check whether an arbitrary equality $a_E^T x = b_E$ is an equality of $Ax \leq b$ by transforming the equalities $A^E x = b^E$ into a substitution and applying this substitution to $a_E^T x = b_E$. The equality $a_E^T x = b_E$ is only contained in $Ax \leq b$ if $a_E^T x = b_E$ simplifies to $0 = 0$ after the substitution.

However, we are still missing one step in our elimination approach: how do we efficiently find an equality $a_i^T x = b_i$ contained in $Ax \leq b$ so that we can substitute with it? The answer are cubes, presented in the below lemma.

Lemma 6. *Let $Ax \leq b$ be a polyhedron. Then, exactly one of the following statements is true:*
(1) $Ax \leq b$ contains an equality $a_E^T x = b_E$ with $a_E \neq \mathbf{0}$, or
(2) $Ax \leq b$ contains a cube with edge length $e > 0$.

Proof. This proof is a case distinction over the sign of x_e for the following slightly simplified version of the largest cube test:

$$\begin{aligned} &\text{maximize} \ \ x_e \\ &\text{subject to} \ \ Ax + a'x_e \leq b, \ \text{where } a'_i = \tfrac{1}{2}\left\|a_i\right\|_1 . \end{aligned} \tag{5}$$

If the maximum objective value is positive, $Ax \leq b$ contains a cube with edge length $e > 0$. Therefore, we have to prove that $Ax \leq b$ contains no equality

$a_E^T x = b_E$ with $a_E \neq 0$, which we will do by contradiction. Assume $Ax \leq b$ contains an equality $a_E^T x = b_E$ with $a_E \neq 0$. Then, by transitivity of the subset relation, the polyhedron consisting of the inequalities $a_E^T x \leq b_E$ and $-a_E^T x \leq -b_E$ must also contain a cube of edge length e. However, applying the transformation from Corollary 3 to this new polyhedron results in two contradicting inequalities: $a_E^T x \leq b_E - \|a_E\|_1 \cdot \frac{e}{2}$ and $-a_E^T x \leq -b_E - \|a_E\|_1 \cdot \frac{e}{2}$. Thus, (1) and (2) cannot hold at the same time.

If the maximum objective value is zero, then $Ax \leq b$ is satisfiable but contains no cube with edge length $e > 0$. Therefore, we have to prove that $Ax \leq b$ contains an equality $a_E^T x = b_E$ with $a_E \neq 0$. Consider the dual linear program of (5):

$$\begin{array}{ll} \text{minimize} & y^T b \\ \text{subject to} & y^T A = 0\,, \\ & y^T a' = 1\,, \text{ where } a_i' = \frac{1}{2}\|a_i\|_1\,, \\ & y \geq 0\,. \end{array} \tag{6}$$

Due to strong duality, the objectives of the dual and primal linear programs are equal. Therefore, there exists a $y \in \mathbb{R}^m$ that has objective $y^T b = 0$ and that satisfies the dual (6). Since $y^T a' = 1$ and $a_i' \geq 0$ and $y_i \geq 0$ holds, there exists a $k \in \{1, \ldots, m\}$ such that $y_k > 0$. By multiplying $y^T A = 0$ with an $x \in P_b^A$ and isolating $a_k^T x$, we get: $a_k^T x = -\sum_{i=1, i\neq k}^m \left(\frac{y_i}{y_k} a_i^T x\right)$. Using $y_i \geq 0$, and our original inequalities $a_i^T x \leq b_i$, we get a finite lower bound for $a_k^T x$:

$$a_k^T x = -\sum\nolimits_{i=1, i\neq k}^m \left(\frac{y_i}{y_k} a_i^T x\right) \geq -\sum\nolimits_{i=1, i\neq k}^m \left(\frac{y_i}{y_k} b_i\right)\,.$$

Now, we reformulate $y^T b = 0$ analogously and get: $b_k = -\sum_{i=1, i\neq k}^m \left(\frac{y_i}{y_k} b_i\right)$. Thus, $a_k^T x = b_k$ is an equality contained in the original inequalities $Ax \leq b$.

If the maximum objective value is negative, $Ax \leq b$ is unsatisfiable and contains no cube with edge length $e > 0$. Since P_b^A is now empty, $Ax \leq b$ contains all equalities. □

By Lemma 6 a polyhedron contains a cube with a positive edge length $e > 0$, or an equality. Since e is arbitrarily small, the factor $\frac{e}{2}\|a_i\|_1$ is also arbitrarily small and $a_i^T x + \frac{e}{2}\|a_i\|_1 \leq b_i$ converges to $a_i^T x < b_i$. Therefore, $Ax \leq b$ contains an equality if and only if $Ax < b$ is unsatisfiable. We can solve this system of strict inequalities with the dual simplex algorithm by Dutertre and de Moura [9]. In case $Ax < b$ is unsatisfiable, the algorithm returns an explanation, i.e., a minimal set C of unsatisfiable constraints $a_i^T x < b_i$ from $Ax < b$. If $Ax \leq b$ itself was satisfiable, then we can extract equalities from this explanation: every $a_i^T x < b_i \in C$ implies that $Ax \leq b$ contains the equality $a_i^T x = b_i$.

Finally, we have two ways of handling negated equalities $a_i^T x \neq b_i$. Either we split our set of constraints into two sets of constraints, replacing $a_i^T x \neq b_i$ in the first one with $a_i^T x \leq b_i - 1$ and in the second one with $-a_i^T x \leq -b_i - 1$; or, we ignore all negated equalities during the calculation of the tests themselves and use the negated equalities only to verify the integer solutions returned by the tests.

7 Conclusion

We have presented two tests based on cubes: the largest cube test and the unit cube test. Our tests can be integrated into SMT theory solvers without sacrificing the advantages SMT solvers gain from the incremental structure of subsequent subproblems. Furthermore, our experiments have shown that these tests increase efficiency on certain polyhedra such that previously hard sets of constraints become trivial. We have even shown that major obstacles to our tests, for example equalities, can be handled through generally useful preprocessing steps. Moreover, these preprocessing steps led to an additional application for our tests: finding equalities.

Our future research will investigate further applications of our tests. We expect that we can use cubes not only for the detection of equalities, but also for the detection of (un)bounded directions. We can likely use the largest cube test as a selection strategy for branching by always choosing the branch containing the largest cube. This is in all likelihood a beneficial strategy since the largest cube is a good heuristic for the branch with the most space for integer solutions.

References

1. Barrett, C., Conway, C.L., Deters, M., Hadarean, L., Jovanović, D., King, T., Reynolds, A., Tinelli, C.: CVC4. In: Gopalakrishnan, G., Qadeer, S. (eds.) CAV 2011. LNCS, vol. 6806, pp. 171–177. Springer, Heidelberg (2011)
2. Bobot, F., Conchon, S., Contejean, E., Iguernelala, M., Mahboubi, A., Mebsout, A., Melquiond, G.: A simplex-based extension of Fourier-Motzkin for solving linear integer arithmetic. In: Gramlich, B., Miller, D., Sattler, U. (eds.) IJCAR 2012. LNCS, vol. 7364, pp. 67–81. Springer, Heidelberg (2012)
3. Boyd, S., Vandenberghe, L.: Convex Optimization. Cambridge University Press, New York (2004)
4. Bromberger, M., Sturm, T., Weidenbach, C.: Linear integer arithmetic revisited. In: Felty, A.P., Middeldorp, A. (eds.) Automated Deduction - CADE-25. LNCS, vol. 9195, pp. 623–637. Springer, Heidelberg (2015)
5. Cimatti, A., Griggio, A., Schaafsma, B.J., Sebastiani, R.: The MathSAT5 SMT solver. In: Piterman, N., Smolka, S.A. (eds.) TACAS 2013 (ETAPS 2013). LNCS, vol. 7795, pp. 93–107. Springer, Heidelberg (2013)
6. de Moura, L., Bjørner, N.S.: Z3: an efficient SMT solver. In: Ramakrishnan, C.R., Rehof, J. (eds.) TACAS 2008. LNCS, vol. 4963, pp. 337–340. Springer, Heidelberg (2008)
7. Dillig, I., Dillig, T., Aiken, A.: Cuts from proofs: a complete and practical technique for solving linear inequalities over integers. In: Bouajjani, A., Maler, O. (eds.) CAV 2009. LNCS, vol. 5643, pp. 233–247. Springer, Heidelberg (2009)
8. Dutertre, B.: Yices 2.2. In: Biere, A., Bloem, R. (eds.) CAV 2014. LNCS, vol. 8559, pp. 737–744. Springer, Heidelberg (2014)
9. Dutertre, B., de Moura, L.: A fast linear-arithmetic solver for DPLL(T). In: Ball, T., Jones, R.B. (eds.) CAV 2006. LNCS, vol. 4144, pp. 81–94. Springer, Heidelberg (2006)
10. Ehrgott, M.: Scalarization techniques. In: Multicriteria Optimization, pp. 97–126. Springer, Heidelberg (2005)

11. Faure, G., Nieuwenhuis, R., Oliveras, A., Rodríguez-Carbonell, E.: SAT modulo the theory of linear arithmetic: exact, inexact and commercial solvers. In: Kleine Büning, H., Zhao, X. (eds.) SAT 2008. LNCS, vol. 4996, pp. 77–90. Springer, Heidelberg (2008)
12. Griggio, A.: A practical approach to satisfiability modulo linear integer arithmetic. JSAT **8**((1/2)), 1–27 (2012)
13. Hillier, F.S.: Efficient heuristic procedures for integer linear programming with an interior. Oper. Res. **17**(4), 600–637 (1969)
14. Jovanović, D., de Moura, L.: Cutting to the chase. JAR **51**(1), 79–108 (2013)
15. Jünger, M., Liebling, T.M., Naddef, D., Nemhauser, G.L., Pulleyblank, W.R., Reinelt, G., Rinaldi, G., Wolsey, L.A. (eds.): 50 Years of Integer Programming 1958–2008: From the Early Years to the State-of-the-Art. Springer, Heidelberg (2010)
16. Kannan, R., Lovász, L.: Covering minima and lattice point free convex bodies. In: Nori, K. (ed.) FSTTCS. LNCS, vol. 241, pp. 193–213. Springer, Heidelberg (1986)
17. Karmarkar, N.: A new polynomial-time algorithm for linear programming. Combinatorica **4**(4), 373–396 (1984)
18. Papadimitriou, C.H.: On the complexity of integer programming. J. ACM **28**(4), 765–768 (1981)

Model Finding for Recursive Functions in SMT

Andrew Reynolds[1(✉)], Jasmin Christian Blanchette[2,3], Simon Cruanes[2], and Cesare Tinelli[1]

[1] Department of Computer Science, The University of Iowa, Iowa City, USA
andrew.j.reynolds@gmail.com
[2] Inria Nancy – Grand Est & LORIA, Villers-lès-Nancy, France
[3] Max-Planck-Institut für Informatik, Saarbrücken, Germany

Abstract. SMT solvers have recently been extended with techniques for finding models of universally quantified formulas in some restricted fragments of first-order logic. This paper introduces a translation that reduces axioms specifying a large class of recursive functions, including terminating functions, to universally quantified formulas for which these techniques are applicable. An evaluation confirms that the approach improves the performance of existing solvers on benchmarks from three sources. The translation is implemented as a preprocessor in the CVC4 solver and in a new higher-order model finder called Nunchaku.

1 Introduction

Many solvers based on SMT (satisfiability modulo theories) can reason about quantified formulas using incomplete instantiation-based methods [15,31]. These methods work well for proving the unsatisfiability of an input set of formulas, but they are of little help for finding models of them when they are satisfiable. Often, a single universal quantifier in one of the axioms of a problem is enough to prevent the discovery of models.

In the past few years, techniques have been developed to find models for quantified formulas in SMT. Ge and de Moura [19] introduced an instantiation-based procedure for formulas in the *essentially uninterpreted* fragment. This fragment is limited to universally quantified formulas where all variables occur as direct subterms of uninterpreted functions, as in $\forall x : \mathsf{Int}.\ \mathsf{f}(x) \approx \mathsf{g}(x) + 5$. Other syntactic criteria extend this fragment slightly, including some cases when variables occur as arguments of arithmetic predicate symbols. Subsequently, Reynolds et al. [32,33] introduced techniques for finding finite models for quantified formulas over uninterpreted types and types having a fixed finite interpretation. These techniques can find a model for a formula such as $\forall x, y : \tau.\ x \approx y \vee \neg\, \mathsf{f}(x) \approx \mathsf{f}(y)$, where τ is an uninterpreted type.

Unfortunately, none of these fragments can accommodate the vast majority of quantified formulas that correspond to recursive function definitions. The essentially uninterpreted fragment does not allow the argument of a recursive function to be used inside a complex term on the right-hand side of the definition, whereas the finite model finding techniques are not applicable in the presence

N. Olivetti and A. Tiwari (Eds.): IJCAR 2016, LNAI 9706, pp. 133–151, 2016.
DOI: 10.1007/978-3-319-40229-1_10

of functions over infinite domains such as the integers or algebraic datatypes. A simple example where both approaches fail is

$$\forall x : \mathsf{Int}.\ \mathsf{p}(x) \approx \mathsf{ite}\big(x \leq 0,\ 1,\ 2 * \mathsf{p}(x-1)\big)$$

where ite is the 'if–then–else' operator. This state of affairs is unsatisfactory, given the frequency of recursive definitions in practice.

We present a method for translating formulas involving recursive function definitions to formulas where finite model finding techniques can be applied. The definitions must meet a semantic criterion to be admissible (Sect. 2). This criterion is general enough to include well-founded (terminating) recursive function definitions and restrictive enough to exclude inconsistent equations such as $\forall x : \mathsf{Int}.\ \mathsf{f}(x) \approx \mathsf{f}(x) + 1$.

We define a translation for a class of formulas involving admissible recursive function definitions (Sect. 3). A recursive equation $\forall x : \tau.\ \mathsf{f}(x) \approx t$ is translated to $\forall a : \alpha_\tau.\ \mathsf{f}(\gamma_\mathsf{f}(a)) \approx t[\gamma_\mathsf{f}(a)/x]$, where α_τ is an "abstract" uninterpreted type and $\gamma_\mathsf{f} : \alpha_\tau \to \tau$ is an uninterpreted function from α_τ to the corresponding concrete type τ. Additional constraints ensure that the abstract values that are relevant to the formula's satisfiability exist. The translation preserves satisfiability and, for admissible definitions, unsatisfiability, and makes finite model finding possible for problems in this class.

The approach is implemented as a preprocessor in the SMT solver CVC4 [2] and in a new higher-order model finder called Nunchaku (Sect. 4). We evaluated the two implementations on benchmarks from IsaPlanner [22], Leon [6], and Isabelle/HOL, to demonstrate that this translation improves the effectiveness of two SMT solvers, CVC4 and Z3 [17], in finding countermodels to verification conditions (Sect. 5). Unlike earlier work (Sect. 6), our approach relies on off-the-shelf SMT solvers.

An earlier version of this paper was presented at the SMT 2015 workshop in San Francisco [30]. This paper extends the workshop paper with proof sketches, an expanded implementation section covering Nunchaku and relevant CVC4 optimizations, and the evaluation on Isabelle benchmarks produced by Nunchaku.

2 Preliminaries

Our setting is a monomorphic (or many-sorted) first-order logic like the one defined by SMT-LIB [3]. A *signature* Σ consists of a set Σ^{ty} of first-order types (or sorts) and a set Σ^{f} of function symbols over these types. We assume that signatures always contain a Boolean type Bool and constants $\top, \bot$: Bool for truth and falsity, an infix equality predicate $\approx : \tau \times \tau \to \mathsf{Bool}$ for each $\tau \in \Sigma^{\mathrm{ty}}$, standard Boolean connectives ($\neg$, $\wedge$, $\vee$, etc.), and an if–then–else function symbol $\mathsf{ite} : \mathsf{Bool} \times \tau \times \tau \to \tau$ for each $\tau \in \Sigma^{\mathrm{ty}}$. We fix an infinite set Σ^{v}_τ of *variables of type* τ for each $\tau \in \Sigma^{\mathrm{ty}}$ and define Σ^{v} as $\bigcup_{\tau \in \Sigma^{\mathrm{ty}}} \Sigma^{\mathrm{v}}_\tau$. (Well-typed) Σ-terms are built as usual over functions symbols in Σ and variables in Σ^{v}. Formulas are terms of type Bool. We write t^τ to denote terms of type τ and $\mathcal{T}(t)$ to denote the set of subterms of t. Given a term u, a variable tuple $\overline{x} = (x_1^{\tau_1}, \ldots, x_n^{\tau_n})$ and a term tuple $\overline{t} = (t_1^{\tau_1}, \ldots, t_n^{\tau_n})$, we write $u[\overline{t}/\overline{x}]$ to denote the result of simultaneously replacing all occurrences of x_i with t_i in u, for each $i = 1, \ldots, n$. When convenient, we will treat a finite set of formulas as the conjunction of its elements.

A Σ-*interpretation* $\mathcal{I}$ maps each type $\tau \in \Sigma^{\mathrm{ty}}$ to a nonempty set $\tau^{\mathcal{I}}$, the *domain* of τ in $\mathcal{I}$, each function symbol $\mathsf{f} : \tau_1 \times \cdots \times \tau_n \to \tau$ in Σ^{f} to a total function $\mathsf{f}^{\mathcal{I}} : \tau_1^{\mathcal{I}} \times \cdots \times \tau_n^{\mathcal{I}} \to \tau^{\mathcal{I}}$, and each variable $x : \tau$ of Σ^{v} to an element of $\tau^{\mathcal{I}}$. A *theory* is a pair $T = (\Sigma, \mathbf{I})$ where Σ is a signature and $\mathbf{I}$ is a class of Σ-interpretations, the *models* of T, closed under variable reassignment (i.e., for every $\mathcal{I} \in \mathbf{I}$, every Σ-interpretation that differs from $\mathcal{I}$ only on the variables of Σ^{v} is also in $\mathbf{I}$). A Σ-formula φ is T-*satisfiable* if it is satisfied by some interpretation in $\mathbf{I}$, which we call a T-*model of* φ. A formula φ T-*entails* ψ, written $\varphi \models_T \psi$, if all interpretations in $\mathbf{I}$ that satisfy φ also satisfy ψ. Two formulas φ and ψ are T-*equivalent* if each T-entails the other. We call T a *Herbrand* theory if for every quantifier-free Σ-formula ψ over the variables $\overline{x}$, $\{\psi[\overline{t}/\overline{x}] \mid t \, \text{ground}$ $\Sigma\text{-term}\} \models_T \forall \overline{x}.\, \psi$.

If $T_0 = (\Sigma_0, \mathbf{I}_0)$ is a theory and Σ is a signature with $\Sigma_0^{\mathrm{f}} \subseteq \Sigma^{\mathrm{f}}$ and $\Sigma_0^{\mathrm{ty}} \subseteq \Sigma^{\mathrm{ty}}$, the *extension of* T_0 *to* Σ is the theory $T = (\Sigma, \mathbf{I})$ where $\mathbf{I}$ is the set of all Σ-interpretations $\mathcal{I}$ whose Σ_0-reduct is a model of T_0. Note that T is a *conservative* extension of T_0, in the sense that a Σ_0-formula is T_0-satisfiable if and only it is T-satisfiable. We refer to the symbols of Σ that are not in Σ_0 as *uninterpreted.*

For the rest of the paper, we fix a theory $T = (\Sigma, \mathbf{I})$ with uninterpreted symbols, constructed as above, and assume it is a Herbrand theory. While this is an actual restriction, it can be shown that the theories typically considered in SMT are Herbrand.

Unconventionally, we consider *annotated quantified formulas* of the form $\forall_{\mathsf{f}}\, \overline{x}.\, \varphi$, where $\mathsf{f} \in \Sigma^{\mathrm{f}}$ is uninterpreted. Their semantics is the same as for standard quantified formulas $\forall \overline{x}.\, \varphi$. Given $\mathsf{f} : \tau_1 \times \cdots \times \tau_n \to \tau$, a formula $\forall_{\mathsf{f}}\, \overline{x}.\, \varphi$ is a *function definition* (*for* f) if $\overline{x}$ is a tuple of variables $x_1^{\tau_1}, \ldots, x_n^{\tau_n}$ and φ is a quantifier-free formula T-equivalent to $\mathsf{f}(\overline{x}) \approx t$ for some term t of type τ. We will consider only annotated quantified formulas that are function definitions. We write $\exists \overline{x}.\, \varphi$ as an abbreviation for $\neg\, \forall \overline{x}.\, \neg\, \varphi$.

Definition 1. A formula φ is in *definitional form with respect to* $\{\mathsf{f}_1, \ldots, \mathsf{f}_n\} \subseteq \Sigma^{\mathrm{f}}$ if it is of the form $(\forall_{\mathsf{f}_1} \overline{x}_1.\, \varphi_1) \wedge \cdots \wedge (\forall_{\mathsf{f}_n} \overline{x}_n.\, \varphi_n) \wedge \varphi_0$, where $\mathsf{f}_1, \ldots, \mathsf{f}_n$ are distinct function symbols, $\forall_{\mathsf{f}_i} \overline{x}_i.\, \varphi_i$ is a function definition for $i = 1, \ldots, n$, and φ_0 contains no function definitions. We call φ_0 the *goal* of φ.

In the signature Σ, we distinguish a subset $\Sigma^{\mathrm{dfn}} \subseteq \Sigma^{\mathrm{f}}$ of *defined* uninterpreted function symbols. We consider Σ-formulas that are in definitional form with respect to Σ^{dfn}.

Definition 2. Given a set of function definitions $\Delta = \{\forall_{\mathsf{f}_1} \overline{x}_1.\, \varphi_1, \ldots, \forall_{\mathsf{f}_n} \overline{x}_n.\, \varphi_n\}$, a ground formula ψ is *closed under function expansion with respect to* Δ if

$$\psi \models_T \bigwedge\nolimits_{i=1}^{n} \{\varphi_i[\overline{t}/\overline{x}_i] \mid \mathsf{f}_i(\overline{t}) \in \mathcal{T}(\psi)\}$$

The set Δ is *admissible* if for every T-satisfiable formula ψ closed under function expansion with respect to Δ, the formula $\psi \wedge \Delta$ is also T-satisfiable.

In Definition 1, notice that the goal φ_0 is a formula possibly containing quantifiers. Given an admissible set of function definitions Δ, we may establish a model exists for $\Delta \wedge \varphi_0$ if we are able to find a ground formula ψ_0 that entails φ_0, and subsequently extend ψ_0 to a T-satisfiable ground formula ψ that is closed under function expansion with respect to Δ. We may obtain ψ from ψ_0 by conjoining to it ground formulas (typically, conjunctions of ground literals) that entail instances of function definitions from Δ until the criterion in Definition 2 is met.

Admissibility is a semantic criterion that must be satisfied for each function definition before applying our translation, described in Sect. 3. It is useful to connect it to the standard notion of *well-founded* function definitions, often called *terminating* definitions. In such definitions, all recursive calls are decreasing with respect to a well-founded relation, which must be supplied by the user or inferred automatically using a termination prover. This ensures that the function is uniquely defined at all points.

First-order logic has no built-in notion of computation or termination. To ensure that a function specification is well founded, it is sufficient to require that the defined function be terminating when seen as a functional program, under *some* evaluation order. For example, the definition $\forall x : \mathsf{Int}.\ \mathsf{p}(x) \approx \mathsf{ite}\big(x \leq 0,\ 1,\ 2 * \mathsf{p}(x-1)\big)$, where the theory T is integer arithmetic extended with $\mathsf{p} : \mathsf{Int} \to \mathsf{Int}$, can be shown to be well founded under a strategy that evaluates the condition of an ite before evaluating the relevant branch, ignoring the other branch. Logically, such dependencies can be captured by congruence rules. Krauss developed these ideas in the general context of higher-order logic [24, Sect. 2], where theories such as integer arithmetic can be axiomatized.

Theorem 3. *If Δ is a set of well-founded function definitions for $\Sigma^{\mathrm{dfn}} = \{\mathsf{f}_1, \ldots, \mathsf{f}_n\}$, then it is admissible.*

Proof Sketch. Let ψ be a T-satisfiable formula closed under function expansion with respect to Δ. We show that $\psi \wedge \Delta$ is also T-satisfiable. Let $\mathscr{I}$ be a T-model of ψ, and let $\mathscr{I}_0$ be the reduct of $\mathscr{I}$ to the function symbols in $\Sigma^{\mathrm{f}} \setminus \Sigma^{\mathrm{dfn}}$. Because well-founded definitions uniquely characterize the interpretation of the functions they define, there exists a model $\mathscr{I}'$ of T that extends $\mathscr{I}_0$ such that $\mathscr{I}' \models \Delta$. Since ψ is closed under function expansion, it already constrains the functions in Σ^{dfn} recursively as far as is necessary for interpreting ψ. Thus, any point v for which $\mathsf{f}_i^{\mathscr{I}}(v)$ is needed for interpreting ψ will have its expected value according to its definition and hence coincide with $\mathscr{I}'$. And since $\psi^{\mathscr{I}}$ does not depend on the interpretation at the other points, $\mathscr{I}'$ is, like $\mathscr{I}$, a T-model of ψ. Since $\mathscr{I}' \models \Delta$ by assumption, we have $\mathscr{I}' \models \psi \bigwedge \Delta$ as desired. □

Another useful class of function definitions is that of *productive* corecursive functions. Corecursive functions are functions to a coalgebraic datatype. These functions can be ill founded without being inconsistent. Intuitively, productive corecursive functions are functions that progressively reveal parts of their potentially infinite output [1,38]. For instance, given a type of infinite streams of integers constructed by $\mathsf{scons} : \mathsf{int} \times \mathsf{stream} \to \mathsf{stream}$, the function defined by $\forall_{\mathsf{e}}\, x.\ \mathsf{e}(x) \approx \mathsf{scons}(x,\ \mathsf{e}(x+1))$ falls within this class: Each call to e produces one constructor before entering the next call. Like terminating recursion, productive corecursion totally specifies the functions it defines, and the proof of Theorem 3 can be adapted to cover that case.

Theorem 4. *If* Δ *is a set of productive function definitions for* Σ^{dfn}*, then it is admissible.*

It is even possible to mix recursion and corecursion in the same function [11] while preserving totality and admissibility. Beyond totality, an admissible set can contain underspecified functions such as $\forall_{\mathsf{f}}\, x : \mathsf{Int}.\ \mathsf{f}(x) \approx \mathsf{f}(x)$ or $\forall_{\mathsf{g}}\, x.\ \mathsf{g}(x) \approx \mathsf{g}(x+1)$. The latter is problematic operationally, because in general the closure of a formula ψ that depends on some term $\mathsf{f}(\mathsf{a})$ is an infinite set: $\{\psi\} \cup \{\mathsf{g}(\mathsf{a}+k) \approx \mathsf{g}(\mathsf{a}+k+1) \mid k \geq 0\}$. A similar issue arises with corecursive definitions specifying infinite *acyclic* objects, such as the e stream introduced above. Nonetheless, admissibility is still useful if a goal formula does not refer to g or e, because it tells us that we can safely ignore their definitions. We conjecture that it is safe to ignore all tail-recursive calls (i.e., calls that occupy the right-hand side of the definition, potentially under some ite branch) when establishing well-foundedness or productivity, without affecting admissibility.

An example of an inadmissible set is $\{\forall_{\mathsf{f}}\, x : \mathsf{Int}.\ \mathsf{f}(x) \approx \mathsf{f}(x) + 1\}$, where T is integer arithmetic extended to a set of uninterpreted symbols $\{\mathsf{f}, \mathsf{g} : \mathsf{Int} \to \mathsf{Int}, \ldots\}$. The set is inadmissible because the formula $\top$ is closed under function expansion with respect to this set (trivially, since f does not occur in $\top$), and yet there is no model of T satisfying f's definition. A more subtle example is $\{\forall_{\mathsf{f}}\, x : \mathsf{Int}.\ \mathsf{f}(x) \approx \mathsf{f}(x),\ \forall_{\mathsf{g}}\, x : \mathsf{Int}.\ \mathsf{g}(x) \approx \mathsf{g}(x) + \mathsf{f}(x)\}$. While this set has a model where f and g are interpreted as the constant function 0, it is not admissible since $\mathsf{f}(0) \approx 1$ is closed under function expansion but there is no interpretation satisfying both $\mathsf{f}(0) \approx 1$ and g's definition.

3 The Translation

For the rest of the section, let φ be a Σ-formula in definitional form with respect to Σ^{dfn} whose definitions are admissible. We present a method that constructs an extended signature $\mathcal{E}(\Sigma)$ and an $\mathcal{E}(\Sigma)$-formula φ' that is T'-satisfiable if and only if φ is T-satisfiable, where T' is the extension of T to $\mathcal{E}(\Sigma)$—i.e., φ and φ' are *equisatisfiable* (*in* T'). Since T' is a conservative extension of T, for simplicity we will refer to it also as T from now on. The idea behind this construction is to use an uninterpreted type α_{f} to abstract the set of *relevant* input tuples for each defined function f, and restrict the quantification of f's definition to a single variable of this type. Informally, the relevant input tuples $\bar{t}$ of a function f are the ones for which the interpretation of $\mathsf{f}(\bar{t})$ is relevant to the satisfiability of φ.

$$
\begin{array}{l}
\mathcal{A}_0(t^\tau, p) = \\
\quad \text{if } \tau = \mathsf{Bool} \text{ and } t = \mathsf{b}(t_1, \ldots, t_n) \text{ then} \\
\quad\quad \text{let } (t'_i, \chi_i) = \mathcal{A}_0(t_i, \mathsf{pol}(\mathsf{b}, i, p)) \text{ for } i = 1, \ldots, n \text{ in} \\
\quad\quad \text{let } \chi = \chi_1 \wedge \cdots \wedge \chi_n \text{ in} \\
\quad\quad \text{if } p = \mathsf{pos} \text{ then } \big(\mathsf{b}(t'_1, \ldots, t'_n)) \wedge \chi, \top\big) \\
\quad\quad \text{else if } p = \mathsf{neg} \text{ then } \big(\mathsf{b}(t'_1, \ldots, t'_n) \vee \neg\chi, \top\big) \\
\quad\quad \text{else } \big(\mathsf{b}(t'_1, \ldots, t'_n), \chi\big) \\
\quad \text{else if } t = \forall_{\mathsf{f}}\, \overline{x}.\ u \text{ then} \\
\quad\quad \text{let } (u', \chi) = \mathcal{A}_0(u, p) \text{ in } \big(\forall a : \alpha_{\mathsf{f}}.\ u'[\overline{\gamma}_{\mathsf{f}}(a)/\overline{x}], \top\big) \\
\quad \text{else if } t = \forall \overline{x}.\ u \text{ then} \\
\quad\quad \text{let } (u', \chi) = \mathcal{A}_0(u, p) \text{ in } \big(\forall \overline{x}.\ u', \forall \overline{x}.\ \chi\big) \\
\quad \text{else} \\
\quad\quad \big(t, \bigwedge\{\exists a : \alpha_{\mathsf{f}}.\ \overline{\gamma}_{\mathsf{f}}(a) \approx \overline{s} \mid \mathsf{f}(\overline{s}) \in \mathcal{T}(\mathsf{t}), \mathsf{f} \in \Sigma^{\mathrm{dfn}}\}\big) \\
\\
\mathcal{A}(\varphi) = \text{let } (\varphi', \chi) = \mathcal{A}_0(\varphi, \mathsf{pos}) \text{ in } \varphi'
\end{array}
$$

Fig. 1. Definition of translation $\mathcal{A}$

We construct the signature $\mathcal{E}(\Sigma)$ so that, for each $\mathsf{f} : \tau_1 \times \cdots \times \tau_n \to \tau \in \Sigma^{\mathrm{dfn}}$, it contains an uninterpreted *abstract type* α_{f}, abstracting the Cartesian product $\tau_1 \times \cdots \times \tau_n$, and n uninterpreted *concretization functions* $\gamma_{\mathsf{f},1} : \alpha_{\mathsf{f}} \to \tau_1, \ldots, \gamma_{\mathsf{f},n} : \alpha_{\mathsf{f}} \to \tau_n$.

The translation $\mathcal{A}$ defined in Fig. 1 converts the Σ-formula φ into the $\mathcal{E}(\Sigma)$-formula φ'. It relies on the auxiliary function $\mathcal{A}_0$, which takes two arguments: the term t to translate and a polarity p for t, which can be pos, neg, or none. $\mathcal{A}_0$ returns a pair (t', χ), where t' is a term of the same type as t and χ is an $\mathcal{E}(\Sigma)$-formula.

The translation alters the formula φ in two ways. First, it restricts the quantification on function definitions for f to the corresponding uninterpreted type α_{f}, inserting applications of the concretization functions $\gamma_{\mathsf{f},i}$ as needed. Second, it augments φ with additional constraints of the form $\exists a : \alpha_{\mathsf{f}}.\ \overline{\gamma}_{\mathsf{f}}(a) \approx \overline{s}$, where $\overline{\gamma}_{\mathsf{f}}(a) \approx \overline{s}$ abbreviates the formula $\bigwedge_{i=1}^{n} \gamma_{\mathsf{f},i}(a) \approx s_i$ with $\overline{s} = (s_1, \ldots, s_n)$. These existential constraints ensure that the restricted definition for f covers all relevant tuples of terms, namely those occurring in applications of f that are relevant to the satisfiability of φ. The constraints are generated as deep in the formula as possible, based on the polarities of Boolean connectives, to allow models where the sets denoted by the α_{f} types are as small as possible.

In the call $\mathcal{A}_0(t, p)$, if t's top symbol is a predicate symbol b, including the operators $\neg$, $\wedge$, $\vee$, $\approx$, and ite, $\mathcal{A}_0$ calls itself recursively on the arguments t_i of b and polarity $\mathsf{pol}(\mathsf{b}, i, p)$ with pol defined as

$$
\mathsf{pol}(\mathsf{b}, i, p) = \begin{cases} p & \text{if } \mathsf{b} \in \{\wedge, \vee\}, \text{ or } \mathsf{b} = \mathsf{ite} \text{ and } i \in \{2, 3\} \\ -p & \text{if } \mathsf{b} = \neg \\ \mathsf{none} & \text{otherwise} \end{cases}
$$

where $-p$ is neg if p is pos, pos if p is neg, and none otherwise. The term t is then reconstructed as $\mathsf{b}(t'_1, \ldots, t'_n)$ where each t'_i is the result of the recursive call with argument t_i. If the polarity p of t is pos, $\mathcal{A}_0$ conjunctively adds to $\mathsf{b}(t'_1, \ldots, t'_n)$ the constraint χ derived from the subterms, and returns $\top$ as the constraint. Dually, if p is neg, it adds a disjunction with the negated constraint to produce the same net effect (since $\neg(\phi \vee \neg\chi) \Leftrightarrow \neg\phi \wedge \chi$). It p is none, it returns the constraint χ.

If t is a function definition $\forall_{\mathsf{f}}\, \bar{x}.\ u$, then $\mathcal{A}_0$ recursively constructs a formula u' from u, replaces all occurrences of $\bar{x}$ in u' with $\overline{\gamma}_{\mathsf{f}}(a)$ where a is single variable of type α_{f}, and then quantifies a. (Since function definitions are top-level conjuncts, χ must be $\top$ and can be ignored.) If t is an unannotated quantified formula $\forall \bar{x}.\ u$, then $\mathcal{A}_0$ calls itself on u with the same polarity p and returns the quantification over $\bar{x}$ of the formula u' and of the constraint χ returned by the recursive call. Finally, if t is an application of an uninterpreted predicate symbol or a term of a type other than Bool, $\mathcal{A}_0$ returns t together with a conjunction of constraints of the form $\exists a : \alpha_{\mathsf{f}}.\ \overline{\gamma}_{\mathsf{f}}(a) \approx \bar{s}$ for each subterm $\mathsf{f}(\bar{s})$ of t such that $\mathsf{f} \in \Sigma^{\mathrm{dfn}}$. Such constraints, when asserted positively, ensure that some element in the abstract domain α_{f} is the preimage of the argument tuple $\bar{s}$.

Example 5. Let T be linear integer arithmetic with the uninterpreted symbols $\{\mathsf{c} : \mathsf{Int},\ \mathsf{s} : \mathsf{Int} \to \mathsf{Int}\}$. Let φ be the Σ-formula

$$\forall_{\mathsf{s}}\, x : \mathsf{Int}.\ \mathsf{ite}\big(x \le 0,\ \mathsf{s}(x) \approx 0,\ \mathsf{s}(x) \approx x + \mathsf{s}(x-1)\big) \wedge \mathsf{s}(\mathsf{c}) > 100 \qquad (1)$$

The definition of s specifies that it returns the sum of all positive integers up to x. The formula φ, which is in definitional form with respect to Σ^{dfn}, states that the sum of all positive numbers up to some constant c is greater than 100. It is satisfied in a model of T that interprets c as 14 or more. Due to the universal quantifier, SMT solvers cannot find a model for φ. The signature $\mathcal{E}(\Sigma)$ extends Σ with the type α_{s} and the function symbol $\gamma_{\mathsf{s}} : \alpha_{\mathsf{s}} \to \mathsf{Int}$. The result of $\mathcal{A}(\varphi)$, after simplification, is the $\mathcal{E}(\Sigma)$-formula

$$\begin{aligned}&\big(\forall a : \alpha_{\mathsf{s}}.\ \mathsf{ite}\big(\gamma_{\mathsf{s}}(a) \le 0,\ \mathsf{s}(\gamma_{\mathsf{s}}(a)) \approx 0,\\ &\qquad \mathsf{s}(\gamma_{\mathsf{s}}(a)) \approx \gamma_{\mathsf{s}}(a) + \mathsf{s}(\gamma_{\mathsf{s}}(a) - 1) \wedge \exists b : \alpha_{\mathsf{s}}.\ \gamma_{\mathsf{s}}(b) \approx \gamma_{\mathsf{s}}(a) - 1\big)\big) \qquad (2)\\ &\wedge \mathsf{s}(\mathsf{c}) > 100 \wedge \exists a : \alpha_{\mathsf{s}}.\ \gamma_{\mathsf{s}}(a) \approx \mathsf{c}\end{aligned}$$

The universal quantifier in Formula (2) ranges over an uninterpreted type α_{s}, making it amenable to the finite model finding techniques by Reynolds et al. [32,33], implemented in CVC4, which search for a finite interpretation for α_{s}. Furthermore, since all occurrences of the quantified variable a are beneath applications of the uninterpreted function γ_{s}, the formula is in the essentially uninterpreted fragment, for which Ge and de Moura [19] provide an instantiation procedure, implemented in Z3. Both CVC4 and Z3 run indefinitely on Formula (1), as expected. However, they both produce a model for (2) within 100 ms. ■

Note that the translation $\mathcal{A}$ results in formulas whose models (i.e., satisfying interpretations) are generally different from those of φ. One model $\mathcal{I}$ for Formula (2) in the above example interprets α_{s} as a finite set $\{u_0, \ldots, u_{14}\}$, γ_{s} as a finite map $u_i \mapsto i$ for $i = 0, \ldots, 14$, c as 14, and s as the almost constant function

$\lambda x : \mathsf{Int}.\ \mathsf{ite}(x \approx 0,\ 0,\ \mathsf{ite}(x \approx 1, 1,\ \mathsf{ite}(x \approx 2,\ 3,\ \mathsf{ite}(\ldots,\ \mathsf{ite}(x \approx 13,\ 91,\ 105)\ldots))))$

In other words, $\mathscr{I}$ interprets s as a function mapping x to the sum of all positive integers up to x when $0 \leq x \leq 13$, and 105 otherwise. The Σ-reduct of $\mathscr{I}$ is not a model of the original Formula (1), since $\mathscr{I}$ interprets $\mathsf{s}(n)$ as 105 when $n < 0$ or $n > 14$.

However, under the assumption that the function definitions in Σ^{dfn} are admissible, $\mathcal{A}(\varphi)$ is equisatisfiable with φ for any φ. Moreover, the models of $\mathcal{A}(\varphi)$ contain pertinent information about the models of φ. For example, the model $\mathscr{I}$ for Formula (2) given above interprets c as 14 and $\mathsf{s}(n)$ as $\sum_{i=1}^{n} i$ for $0 \leq n \leq 14$, and there exists a model of Formula (1) that also interprets c and $\mathsf{s}(n)$ in the same way (for $0 \leq n \leq 14$). In general, for every model of $\mathcal{A}(\varphi)$, there exists a model of φ that coincides with it on its interpretation of all function symbols in $\Sigma^{\mathrm{f}} \setminus \Sigma^{\mathrm{dfn}}$. Furthermore, the model of $\mathcal{A}(\varphi)$ will also give correct information for the defined functions at all points belonging to the domains of the corresponding abstract types α_{f}. This can sometimes help users debug their function definitions.

We sketch the correctness of translation $\mathcal{A}$. For a set of ground literals L, we write $\mathrm{X}(L)$ to denote the set of constraints that force the concretization functions to have enough elements in their range to determine the satisfiability of L with respect to the function definitions in the translation. Formally,

$$\mathrm{X}(L) = \{\exists a : \alpha_{\mathsf{f}}.\ \overline{\gamma}_{\mathsf{f}}(a) \approx \bar{t} \mid \mathsf{f}(\bar{t}) \in \mathcal{T}(L),\ \mathsf{f} \in \Sigma^{\mathrm{dfn}}\} \tag{3}$$

The following lemma states the central invariant behind the translation $\mathcal{A}$.

Lemma 6. *Let ψ be a Σ-formula not containing function definitions, and let $\mathscr{I}$ be an $\mathcal{E}(\Sigma)$-model of T. Then, $\mathscr{I}$ satisfies $\mathcal{A}(\psi)$ if and only if it satisfies $L \cup \mathrm{X}(L)$ for some set L of ground Σ-literals such that $L \models_T \psi$.*

Proof Sketch. By definition of $\mathcal{A}$ and case analysis on the return values of $\mathcal{A}_0$. □

Lemma 7. *If ψ is a Σ-formula not containing function definitions, then $\mathcal{A}(\psi) \models_T \psi$.*

Theorem 8. *If φ is a Σ-formula in definitional form with respect to Σ^{dfn}, the set of function definitions Δ corresponding to Σ^{dfn} is admissible, and the goal formula φ_0 of φ is ground, then φ and $\mathcal{A}(\varphi)$ are equisatisfiable in T.*

Proof Sketch. First, we show that if φ is satisfied by a Σ-model $\mathscr{I}$ of T, then $\mathcal{A}(\varphi)$ is satisfied by an $\mathcal{E}(\Sigma)$-model $\mathscr{I}'$. Given such a model $\mathscr{I}$, let $\mathscr{I}'$ be the $\mathcal{E}(\Sigma)$-interpretation that interprets all types $\tau \in \Sigma^{\mathrm{ty}}$ as $\tau^{\mathscr{I}}$, all function symbols $\mathsf{f} \in \Sigma^{\mathrm{f}}$ as $\mathsf{f}^{\mathscr{I}}$, and for each $\mathsf{f} : \tau_1 \times \cdots \times \tau_n \to \tau$ in Σ^{dfn}, interprets α_{f} as $\tau_1^{\mathscr{I}} \times \cdots \times \tau_n^{\mathscr{I}}$ and each $\gamma_{\mathsf{f},i}$ as the ith projection on such tuples for $i = 1, \ldots, n$. Since $\mathscr{I}'$ satisfies φ and T is Herbrand, $\mathscr{I}'$ satisfies a set of ground literals L that entail φ. Furthermore, $\mathscr{I}'$ satisfies every constraint of the form $\exists a : \alpha_{\mathsf{f}}.\ \overline{\gamma}_{\mathsf{f}}(a) \approx \bar{t}$, since by its construction there is a value $v \in {\alpha_{\mathsf{f}}}^{\mathscr{I}'}$ such that $v = \bar{t}^{\mathscr{I}'}$. Thus, $\mathscr{I}'$ satisfies $L \cup \mathrm{X}(L)$, and by Lemma 6 we conclude $\mathscr{I}'$ satisfies $\mathcal{A}(\varphi)$.

Second, we show that if $\mathcal{A}(\varphi)$ is satisfied by a $\mathcal{E}(\Sigma)$-interpretation $\mathcal{I}'$, then φ is satisfied by a Σ-interpretation $\mathcal{I}$. Since φ is in definitional form with respect to the functions defined by Δ, it must be of the form $\Delta \wedge \varphi_0$. First, we define a sequence of Σ-literals sets $L_0 \subseteq L_1 \subseteq \cdots$ such that $\mathcal{I}'$ satisfies $L_i \cup \mathrm{X}(L_i)$ for $i \geq 0$. Since $\mathcal{I}'$ satisfies $\mathcal{A}(\varphi_0)$, by Lemma 6, $\mathcal{I}'$ satisfies a set of literals $L \cup \mathrm{X}(L)$ where L is a set of Σ-literals that entail φ_0. Let $L_0 = L$. For each $i \geq 0$, let ψ_i be the formula $\bigwedge\{\mathcal{A}(\varphi_\mathsf{f}[\bar{t}/\bar{x}]) \mid \mathsf{f}(\bar{t}) \in \mathcal{T}(L_i),\ \mathsf{f} \in \Sigma^{\mathrm{dfn}}\}$, where $\forall_\mathsf{f}\,\bar{x}.\ \varphi_\mathsf{f} \in \Delta$. Since $\mathcal{I}'$ satisfies $\mathcal{A}(\forall_\mathsf{f}\,\bar{x}.\ \varphi_\mathsf{f})$ and $\mathrm{X}(L_i)$, we know that $\mathcal{I}'$ also satisfies ψ_i. Thus by Lemma 6, $\mathcal{I}'$ satisfies a set of literals $L \cup \mathrm{X}(L)$ where L is a set of Σ-literals that entail ψ_i. Let $L_{i+1} = L_0 \cup L$. Let L_∞ be the limit of this sequence (i.e., $\ell \in L_\infty$ if and only if $\ell \in L_i$ for some i), and let ψ be the Σ-formula $\bigwedge L_\infty$. To show that ψ is closed under function expansion with respect to Δ, we first note that by construction ψ entails ψ_∞. For any function symbol f and terms $\bar{t}$, since $\varphi_\mathsf{f}[\bar{t}/\bar{x}]$ does not contain function definitions, by Lemma 7, $\mathcal{A}(\varphi_\mathsf{f}[\bar{t}/\bar{x}])$ entails $\varphi_\mathsf{f}[\bar{t}/\bar{x}]$. Thus, ψ entails $\{\varphi_\mathsf{f}[\bar{t}/\bar{x}] \mid \mathsf{f}(\bar{t}) \in \mathcal{T}(\psi), \mathsf{f} \in \Sigma^{\mathrm{dfn}}\}$, meaning that ψ is closed under function expansion with respect to Δ. Furthermore, ψ entails φ_0 since $L_0 \subseteq L_\infty$. Since ψ is a T-satisfiable formula that is closed under function expansion and Δ is admissible, by definition there exists a Σ-interpretation $\mathcal{I}$ satisfying $\psi \wedge \Delta$, which entails $\Delta \wedge \varphi_0$, i.e., φ. □

The intuition of the above proof is as follows. First, $\mathcal{A}(\varphi)$ cannot be unsatisfiable when φ is satisfiable since any Σ-interpretation that satisfies φ can be extended in a straightforward way to an $\mathcal{E}(\Sigma)$-interpretation that satisfies $\mathcal{A}(\varphi)$, by interpreting the abstract types in the same way as the Cartesian products they abstract, thereby satisfying all existential constraints introduced by $\mathcal{A}$. Conversely, if a model is found for $\mathcal{A}(\varphi)$, existential constraints introduced by $\mathcal{A}$ ensure that this model also satisfies a Σ-formula that is closed under function expansion and that entails the goal of φ. This implies the existence of a model for φ provided that Δ is admissible.

We give an intuition of Theorem 8 in the context of an example.

Example 9. Let us revisit the formulas in Example 5. If the original Formula (1) is T-satisfiable, the translated Formula (2) is clearly also T-satisfiable since α_s can be interpreted as the integers and γ_s as the identity function. Conversely, we claim that (2) is T-satisfiable only if (1) is T-satisfiable, noting that the set $\{\forall_\mathsf{s}\, x.\ \ \varphi_\mathsf{s}\}$ is admissible, where φ_s is the formula $\mathsf{ite}\big(x \leq 0,\ \mathsf{s}(x) \approx 0,\ \mathsf{s}(x) \approx x + \mathsf{s}(x-1)\big)$. Clearly, any interpretation $\mathcal{I}$ satisfying Formula (2) satisfies $L_0 \cup \mathrm{X}(L_0)$, where $L_0 = \{\mathsf{s}(\mathsf{c}) > 100\}$ and $\mathrm{X}(L_0)$, defined by Eq. (3), consists of the single constraint $\exists a : \alpha_\mathsf{s}.\ \gamma_\mathsf{s}(a) \approx \mathsf{c}$. Since $\mathcal{I}$ also satisfies both the translated function definition for s (the first conjunct of (2)) and $\mathrm{X}(L_0)$, it must also satisfy

$$\mathsf{ite}\big(\mathsf{c} \leq 0,\ \mathsf{s}(\mathsf{c}) \approx 0,\ \mathsf{s}(\mathsf{c}) \approx \mathsf{c} + \mathsf{s}(\mathsf{c}-1) \wedge \exists b : \alpha_\mathsf{s}.\ \gamma_\mathsf{s}(b) \approx \mathsf{c} - 1\big)$$

The existential constraint in the above formula ensures that whenever $\mathcal{I}$ satisfies the set $L_1 = L_0 \cup \{\neg\, \mathsf{c} \leq 0,\ \mathsf{s}(\mathsf{c}) \approx \mathsf{c} + \mathsf{s}(\mathsf{c}-1)\}$, $\mathcal{I}$ satisfies $\mathrm{X}(L_1)$ as well. Hence, by repeated application of this reasoning, it follows that a model of Formula (2) that interprets c as n must also satisfy ψ:

$$\mathsf{s}(\mathsf{c}) > 100 \wedge \bigwedge_{i=0}^{n-1} (\neg\,(\mathsf{c} - i \leq 0) \wedge \mathsf{s}(\mathsf{c} - i) \approx \mathsf{c} - i + \mathsf{s}(\mathsf{c} - i - 1)) \\ \wedge\, \mathsf{c} - n \leq 0 \wedge \mathsf{s}(\mathsf{c} - n) \approx 0$$

This formula is closed under function expansion since it entails $\varphi_\mathsf{s}[(\mathsf{c} - i)/x]$ for $i = 0, \ldots, n$ and contains only s applications corresponding to $\mathsf{s}(\mathsf{c} - i)$ for $i = 0, \ldots, n$. Since $\{\forall_\mathsf{s}\, x.\ \varphi_\mathsf{s}\}$ is admissible, there exists a Σ-interpretation satisfying $\psi \wedge \forall_\mathsf{s}\, x.\ \varphi_\mathsf{s}$, which entails Formula (1). ∎

4 Implementations

We have implemented the translation $\mathcal{A}$ in two separate systems, as a preprocessor in CVC4 (version 1.5 prerelease) and in the CVC4-based higher-order model finder Nunchaku. This section describes how the translation is implemented in each system, as well as optimizations used by CVC4 to find models of translated problems.

4.1 CVC4

In CVC4, function definitions $\forall_\mathsf{f}\, \overline{x}.\ \varphi$ can be written using the define-fun-rec command from SMT-LIB 2.5 [3]. Formula (1) from Example 5 can be specified as

```
(define-fun-rec s ((x Int)) Int (ite (<= x 0) 0 (+ x (s (- x 1)))))
(declare-fun c () Int)
(assert (> (s c) 100))
```

When asked to check the satisfiability of the assertions above, CVC4 adds the formula $\forall_\mathsf{s}\, x.\ \mathsf{s}(x) \approx \mathsf{ite}(x \leq 0,\ 0,\ \mathsf{s}(x-1))$ to its list of assertions, which after rewriting becomes $\forall_\mathsf{s}\, x.\ \mathsf{ite}(x \leq 0,\ \mathsf{s}(x) \approx 0,\ \mathsf{s}(x) \approx \mathsf{s}(x-1))$. By specifying the command-line option `--fmf-fun`, users can enable CVC4's finite model finding mode for recursive functions. In this mode, CVC4 will rewrite the asserted formulas according to the $\mathcal{A}$ translation before checking for satisfiability. Accordingly, it will output the approximation of the interpretation it used for recursive function definitions. For the example above, CVC4 outputs a model of s where only the values of $\mathsf{s}(x)$ for $x = 0, \ldots, 14$ are correctly given:

```
(model
  (define-fun s (($x1 Int)) Int
    (ite (= $x1 14) 105 (ite (= $x1 13) 91 (ite (= $x1 12) 78
      (ite (= $x1 11) 66 (ite (= $x1 10) 55 (ite (= $x1 4) 10
        (ite (= $x1 9) 45 (ite (= $x1 8) 36 (ite (= $x1 7) 28
          (ite (= $x1 6) 21 (ite (= $x1 3) 6 (ite (= $x1 5) 15
            (ite (= $x1 2) 3 (ite (= $x1 1) 1 0))))))))))))))
  (define-fun c () Int 14))
```

With the `--fmf-fun` option enabled, CVC4 assumes that functions introduced using `define-fun-rec` are admissible. Admissibility must be proved externally by the user—e.g., manually, using a syntactic criterion, or with the help of a termination prover. If some function definitions are not admissible, CVC4 may answer *sat* for an unsatisfiable problem. For example, if we add the inconsistent definition

```
(define-fun-rec h ((x Int)) Int (+ (h x) x))
```

to the above problem and run CVC4 with `--fmf-fun`, it wrongly answers *sat*.

CVC4 implements a few optimizations designed to help finding finite models of $\mathcal{A}(\varphi)$. As in other systems, the finite model finding capability of CVC4 incrementally fixes bounds on the cardinalities of uninterpreted types and increases these bounds until it encounters a model. When multiple types are present, it uses a fairness scheme that bounds the sum of cardinalities of all uninterpreted types [34]. For example, if a signature has two uninterpreted types τ_1 and τ_2, it will first search for models where $|\tau_1| + |\tau_2|$ is at most 2, then 3, 4, and so on. To accelerate the search for models, we implemented an optimization based on statically inferring *monotonic* types. Intuitively, a type of a theory T is monotonic if every model of T can be extended with additional elements of that type and remain a model of T [9,13]. Types α_{f} introduced by our translation $\mathcal{A}$ are monotonic, because $\approx$ is never used directly on such types [13]. CVC4 takes advantage of this by fixing the bounds for all monotonic types simultaneously. That is, if τ_1 and τ_2 are inferred to be monotonic (regardless of whether they are present in the original problem or introduced by our translation), the solver fixes the bound for both types to be 1, then 2, and so on. This scheme allows the solver greater flexibility compared with the default scheme, and comes with no loss of generality with respect to models, since monotonic types can always be extended to have equal cardinalities.

By default, CVC4 uses techniques to minimize the number of literals it considers when constructing propositional satisfying assignments for formulas [16]. However, we have found that such techniques degrade performance for finite model finding on problems with recursive functions defined by cases. For this reason, we disable the techniques for problems produced from our translation.

4.2 Nunchaku

Nunchaku is a new higher-order model finder designed to be integrated with several proof assistants. The first version was released in January 2016 with support for (co)algebraic datatypes, (co)recursive functions, and (co)inductive predicates. Support for higher-order functions is in the works. We have developed an Isabelle frontend and are planning further frontends for Coq, the TLA$^+$ Proof System, and other proof assistants.

Nunchaku is a spiritual successor to Nitpick [10] for Isabelle/HOL, but is developed as a standalone OCaml program, with its own input language. Whereas Nitpick generates a succession of problems where the cardinalities of finite types grow at each step, Nunchaku translates its input to one first-order logic program that targets the finite model finding fragment of CVC4, including (co)algebraic datatypes [29]. Using CVC4 also allows Nunchaku to provide efficient arithmetic reasoning and to detect unsatisfiability in addition to satisfiability. We plan to integrate other tools as backends, to exploit the strengths of competing approaches.

The input syntax was inspired by that of other systems based on higher-order logic (e.g., Isabelle/HOL) and by functional programming languages (e.g., OCaml). The following simple problem gives a taste of it:

```
data nat := 0 | Suc nat.

pred even : nat -> prop :=
  even 0;
  forall n. odd n => even (Suc n)
and odd : nat -> prop :=
  forall n. even n => odd (Suc n).

val m : nat.
goal even m && ~ (m = 0).
```

The problem defines a datatype (`nat`) and two mutually recursive inductive predicates (`even` and `odd`), declares a constant `m`, and specifies a goal to satisfy ("`m` is even and nonzero"). Nunchaku quickly finds the following partial model:

```
val m := Suc (Suc 0).
val odd := fun x. if x = Suc 0 then true else ?__.
val even := fun x. if x = Suc (Suc 0) || x = 0 then true else ?__.
```

The partial model gives sufficient information to the user to evaluate the goal: "2 is even if 1 is odd, 1 is odd if 0 is even, and 0 is even." Our experience with Nitpick is that users are mostly interested in the values assigned to uninterpreted constants (e.g., `m`). Occasionally, the models of underspecified recursive functions are instructive. A typical example is the `head` function that returns the first element of a nonempty list:

```
data list A := Nil | Cons A (list A).

rec head : pi A. list A -> A :=
  forall y ys. head (Cons y ys) = y.

goal ~(head Nil = 0).
```

Nunchaku transforms the definition of `head` into

```
head xs = match xs with Nil -> head xs | Cons y ys -> y end
```

where the unspecified `Nil` case is expressed via nonterminating recursion (`head xs = head xs`). The tool exhibits a model in which `head Nil` is interpreted as a nonzero value.

Internally, Nunchaku parses and types the input problem before applying a sequence of translations, each reducing the distance to the target fragment. In our example, the predicates `even` and `odd` are *polarized* (specialized into a pair of predicates such that one is used in positive positions and the other in negative positions), then translated into admissible recursive functions, before another pass applies the encoding described in this paper. If a model is found, it is translated back to the input language, with `?__` placeholders indicating unknown values.

Conceptually, the sequence of transformation is a bidirectional pipeline built by composing pairs (*Encode*, *Decode*) of transformations. For each such pair, *Encode* translates a Σ-problem in a logic $\mathscr{L}$ to a Σ'-problem in a logic $\mathscr{L}'$, and *Decode* translates a model in $\mathscr{L}'$ over Σ' into a model in $\mathscr{L}$ over Σ, in the spirit of institution theory [20]. The pipeline includes the following phases:

Type inference infers types and checks definitions;
Monomorphization specializes polymorphic definitions on their type arguments and removes unused definitions;
Elimination of equations translates multiple-equation definitions of recursive functions into a single nested pattern matching;
Specialization creates instances of functions with static arguments (i.e., an argument that is passed unchanged to all recursive calls);
Polarization specializes predicates into a version used in positive positions and a version used in negative positions;
Unrolling adds a decreasing argument to possibly ill-founded predicates;
Skolemization introduces Skolem symbols for term variables;
Elimination of (co)inductive predicates recasts a multiple-clause (co)inductive predicate definition into a recursive equation;
Elimination of higher-order constructs eliminates λ-abstractions and substitutes arrays for higher-order functions;
Elimination of recursion performs the encoding from Sect. 3;
Elimination of pattern matching rewrites pattern-matching expressions using datatype discriminators and selectors;
CVC4 invocation runs CVC4 to obtain a model.

5 Evaluation

In this section, we evaluate both the overall impact of the translation introduced in Sect. 3 and the performance of individual SMT techniques. We gathered 602 benchmarks from three sources, which we will refer to as IsaPlanner, Leon, and Nunchaku-Mut:

- The IsaPlanner set consists of the 79 benchmarks from the IsaPlanner suite [22] that do not contain higher-order functions. These benchmarks have been used recently as challenge problems for a variety of inductive theorem provers. They heavily involve recursive functions and are limited to a theory of algebraic datatypes with a signature that contains uninterpreted function symbols over these datatypes.

- The Leon set consists of 166 benchmarks from the Leon repository,[1] which were constructed from verification conditions on simple Scala programs. These benchmarks also heavily involve recursively defined functions over algebraic datatypes, but cover a wide variety of additional theories, including bit vectors, arrays, and both linear and nonlinear arithmetic.
- The Nunchaku-Mut set consists of 357 benchmarks originating from Isabelle/HOL. They involve (co)recursively defined functions over (co)algebraic datatypes and uninterpreted functions but no other theories. They were obtained by mutation of negated Isabelle theorems, as was done for evaluating Nitpick [10]. Benchmarks created by mutation have a high likelihood of having small, easy-to-find models.

The IsaPlanner and Leon benchmarks are expressed in SMT-LIB 2.5 and are in definitional form with respect to a set of well-founded functions. The Leon tool was used to generate SMT-LIB files. A majority of these benchmarks are unsatisfiable. For each of the 245 benchmarks, we considered up to three randomly selected mutated forms of its goal ψ. In particular, we considered unique formulas that are obtained as a result of exchanging a subterm of ψ at one position with another of the same type at another position. In total, we considered 213 mutated forms of theorems from IsaPlanner and 427 mutated forms of theorems from Leon. We will call these sets IsaPlanner-Mut and Leon-Mut, respectively. Each of these benchmarks exists in two versions: with and without the $\mathcal{A}$ translation. Problems with $\mathcal{A}$ were produced by running CVC4's preprocessor.

For Nunchaku-Mut, the Isabelle Nunchaku frontend was used to generate thousands of Nunchaku problems from Isabelle/HOL theory files involving lists, trees, and other functional data structures. Nunchaku was then used to generate SMT-LIB files, again in two versions: with and without the $\mathcal{A}$ translation. Problems requiring higher-order logic were discarded, since Nunchaku does not yet support them, leaving 357 problems.

Among SMT solvers, we considered Z3 and CVC4. Z3 runs heuristic methods for quantifier instantiation [15] as well as methods for finding models for quantified formulas [19]. For CVC4, we considered four configurations, referred to as CVC4h, CVC4f, CVC4fh, and CVC4fm here. Configuration CVC4h runs heuristic and conflict-based techniques for quantifier instantiation [31], but does not include techniques for finding models. The other configurations run the finite model finding procedure due to Reynolds et al. [32,33]. Configuration CVC4fh additionally incorporates heuristic quantifier instantiation as described in Sect. 2.3 of [33], and CVC4fm incorporates the fairness scheme for monotonic types as described in Sect. 4.1.

The results are summarized in Figs. 2 and 3. The bold font indicates the maximum value of a row. All the benchmarks and more detailed results are available online. The figures are divided into benchmarks triggering *unsat* and *sat* responses and further into benchmarks before and after the translation $\mathcal{A}$.

[1] https://github.com/epfl-lara/leon/.

	Z3		CVC4h		CVC4f		CVC4fh		CVC4fm	
	φ	$\mathcal{A}(\varphi)$	φ	$\mathcal{A}(\varphi)$	φ	$\mathcal{A}(\varphi)$	φ	$\mathcal{A}(\varphi)$	φ	$\mathcal{A}(\varphi)$
IsaPlanner	0	0	0	0	0	0	0	0	0	0
IsaPlanner-Mut	0	41	0	0	0	**153**	0	**153**	0	**153**
Leon	0	2	0	0	0	9	0	9	0	**10**
Leon-Mut	11	78	6	6	6	**189**	6	**189**	6	**189**
Nunchaku-Mut	3	27	0	0	3	199	2	**200**	2	199
Total	14	148	6	6	8	550	8	**551**	8	**551**

Fig. 2. Number of *sat* responses on benchmarks without and with $\mathcal{A}$ translation

	Z3		CVC4h		CVC4f		CVC4fh		CVC4fm	
	φ	$\mathcal{A}(\varphi)$	φ	$\mathcal{A}(\varphi)$	φ	$\mathcal{A}(\varphi)$	φ	$\mathcal{A}(\varphi)$	φ	$\mathcal{A}(\varphi)$
IsaPlanner	14	**15**	**15**	**15**	1	**15**	**15**	**15**	1	**15**
IsaPlanner-Mut	**18**	**18**	**18**	**18**	4	**18**	**18**	**18**	4	**18**
Leon	74	79	**80**	**80**	17	78	**80**	77	17	78
Leon-Mut	84	98	**104**	98	24	100	**104**	98	24	100
Nunchaku-Mut	**61**	59	46	53	45	59	44	59	45	59
Total	251	269	263	264	91	**270**	261	267	91	**270**

Fig. 3. Number of *unsat* responses on benchmarks without and with $\mathcal{A}$ translation

The raw evaluation data reveals no cases in which a solver answered *unsat* on a benchmark φ and *sat* on its corresponding benchmark $\mathcal{A}(\varphi)$, or vice versa. This is consistent with our expectations and Theorem 8, since these benchmarks contain only well-founded function definitions.

Figure 2 shows that for untranslated benchmarks (the "φ" columns), the number of *sat* responses is very low across all configurations. This confirms the shortcomings of existing SMT techniques for finding models for benchmarks containing recursively defined functions. The translation $\mathcal{A}$ (the "$\mathcal{A}(\varphi)$" columns) has a major impact. CVC4f finds 550 of the 1242 benchmarks to be satisfiable, including 9 benchmarks in the nonmutated Leon benchmark set. The two optimizations for finite model finding in CVC4 (configurations CVC4fh and CVC4fm) lead to a net gain of one satisfiable benchmark each with respect to CVC4f. The performance of Z3 for countermodels also improves dramatically, as Z3 finds 134 more benchmarks to be satisfiable, including 5 that are not solved by CVC4f. We conclude that the translation $\mathcal{A}$ enables SMT solvers to find countermodels for conjectures involving recursively defined functions.

Interestingly, the translation $\mathcal{A}$ helps all configurations for *unsat* responses as well. Z3 solves a total of 269 with the translation, whereas it solves only 251 without it. Surprisingly, the configuration CVC4f, which is not tailored for handling unsatisfiable benchmarks, solves 270 *unsat* benchmarks overall, which is more than either CVC4h or Z3. These results suggest that the translation does not degrade the performance of SMT solvers for unsatisfiable problems involving recursive functions, and in fact it often improves it. They also suggest that it might be interesting to use this translation in Sledgehammer [8] and to try Nunchaku also as a proof tool.

6 Related Work

We have already described the most closely related work, by Ge and de Moura [19] and by Reynolds et al. [32,33], earlier in this paper. The finite model finding support in the instantiation-based iProver [23] is also close, given the similarities with SMT.

Some finite model finders are based on a reduction to a decidable logic, typically propositional logic. The translation is parameterized by upper or exact finite bounds on the cardinalities of the atomic types. This procedure was pioneered by McCune in the earlier versions of Mace (originally styled MACE) [28]. Other conceptually similar finders are Paradox [14] and FM-Darwin [5] for first-order logic with equality; the Alloy Analyzer and its backend Kodkod [37] for first-order relational logic; and Refute [39] and Nitpick [10] for higher-order logic. An alternative is to perform an exhaustive model search directly on the original problem. Given fixed cardinalities, the search space is represented as multidimensional tables. The procedure tries different values in the function and predicate tables, checking each time if the problem is satisfied. This approach was pioneered by FINDER [36] and SEM [40] and serves as the basis of the Alloy Analyzer's precursor [21] and later versions of Mace [27].

Most of the above tools cannot cope with infinite types. Kuncak and Jackson [25] presented an idiom for encoding algebraic datatypes and recursive functions in Alloy, by approximating datatypes by finite subterm-closed substructures. The approach finds sound (fragments of) models for formulas in the existential–bounded-universal fragment (i.e., formulas whose prenex normal forms contain no unbounded universal quantifiers ranging over datatypes). This idiom was refined by Dunets et al. [18], who presented a translation scheme for primitive recursion. Their definedness guards play a similar role to the existential constraints generated by our translation $\mathcal{A}$.

The higher-order model finder Nitpick [10] for the Isabelle/HOL proof assistant relies on another variant of Kuncak and Jackson's approach inside a Kleene-style three-valued logic, inspired by abstract interpretation. It was also the first tool of its kind to support corecursion and coalgebraic datatypes [7]. The three-valued logic approach extends each approximated type with an unknown value, which is propagated by function application. This scheme works reasonably well in Nitpick, but experiments with CVC4 suggest that it is more efficient to avoid unknowns by adding existential constraints.

The Leon system [6] implements a procedure that can produce both proofs and counterexamples for properties of terminating functions written in a subset of Scala. Leon is based on an SMT solver. It avoids quantifiers altogether by unfolding recursive definitions up to a certain depth, which is increased on a per-need basis. Our translation $\mathcal{A}$ works in an analogous manner, but the SMT solver is invoked only once and quantifier instantiation is used in lieu of function unfolding. It would be worth investigating how existing approaches for function unfolding can inform approaches for dedicated quantifier instantiation techniques for function definitions, and vice versa.

Model finding is concerned with satisfying arbitrary logical constraints. Some tools are tailored for problems that correspond to total functional programs. QuickCheck [12] for Haskell is an early example, based on random testing. Bounded exhaustive testing [35] and narrowing [26] are other successful strategies. These tools are often much faster than model finders, but they typically cannot cope with unspecified or underspecified functions (e.g., the `head` function from Sect. 4.2). Another approach, which also fails in the face of underspecification, is to take the conjecture as an axiom and to attempt to derive a contradiction using an automatic theorem prover [4]. If the other axioms are consistent (which can be checked syntactically in some cases), a contradiction imples the existence of countermodels. Compared with these approaches, the main advantage of our approach is that it can cope with underspecification and that it exploits the SMT solver (and its SAT solver) to enumerate candidate models efficiently.

7 Conclusion

We presented a translation scheme that extends the scope of finite model finding techniques in SMT, allowing one to use them to find models of quantified formulas over infinite types, such as integers and algebraic datatypes. In future work, it would be interesting to evaluate the approach against other counterexample generators, notably Leon, Nitpick, and Quickcheck, and enrich the benchmark suite with more problems exercising CVC4's support for coalgebraic datatypes [29]. We are also working on an encoding of higher-order functions in SMT-LIB, as a generalization to the current translation scheme, for Nunchaku. Further work would also include identifying additional sufficient conditions for admissibility, thereby enlarging the applicability of the translation scheme presented here.

Acknowledgments. Viktor Kuncak and Stephan Merz have made this work possible. We would also like to thank Damien Busato-Gaston and Emmanouil Koukoutos for providing the set of Leon benchmarks used in the evaluation, and Mark Summerfield for suggesting several textual improvements. Cruanes is supported by the Inria technological development action "Contre-exemples utilisables par Isabelle et Coq" (CUIC). Reynolds and Tinelli are partially supported by grant 1228765 from the National Science Foundation.

References

1. Atkey, R., McBride, C.: Productive coprogramming with guarded recursion. In: Morrisett, G., Uustalu, T. (eds.) ICFP 2013, pp. 197–208. ACM (2013)
2. Barrett, C., Conway, C.L., Deters, M., Hadarean, L., Jovanović, D., King, T., Reynolds, A., Tinelli, C.: CVC4. In: Gopalakrishnan, G., Qadeer, S. (eds.) CAV 2011. LNCS, vol. 6806, pp. 171–177. Springer, Heidelberg (2011)
3. Barrett, C., Fontaine, P., Tinelli, C.: The SMT-LIB standard–Version 2.5. Technical report, The University of Iowa (2015). http://smt-lib.org/
4. Baumgartner, P., Bax, J.: Proving infinite satisfiability. In: McMillan, K., Middeldorp, A., Voronkov, A. (eds.) LPAR-19 2013. LNCS, vol. 8312, pp. 86–95. Springer, Heidelberg (2013)
5. Baumgartner, P., Fuchs, A., de Nivelle, H., Tinelli, C.: Computing finite models by reduction to function-free clause logic. J. Appl. Log. **7**(1), 58–74 (2009)
6. Blanc, R., Kuncak, V., Kneuss, E., Suter, P.: An overview of the Leon verification system–Verification by translation to recursive functions. In: Scala 2013. ACM (2013)
7. Blanchette, J.C.: Relational analysis of (co)inductive predicates, (co)inductive datatypes, and (co)recursive functions. Softw. Qual. J. **21**(1), 101–126 (2013)
8. Blanchette, J.C., Böhme, S., Paulson, L.C.: Extending sledgehammer with SMT solvers. J. Autom. Reasoning **51**(1), 109–128 (2013)
9. Blanchette, J.C., Krauss, A.: Monotonicity inference for higher-order formulas. J. Autom. Reasoning **47**(4), 369–398 (2011)
10. Blanchette, J.C., Nipkow, T.: Nitpick: a counterexample generator for higher-order logic based on a relational model finder. In: Kaufmann, M., Paulson, L.C. (eds.) ITP 2010. LNCS, vol. 6172, pp. 131–146. Springer, Heidelberg (2010)
11. Blanchette, J.C., Popescu, A., Traytel, D.: Foundational extensible corecursion: a proof assistant perspective. In: Reppy, J. (ed.) ICFP 2015. ACM (2015)
12. Claessen, K., Hughes, J.: QuickCheck: a lightweight tool for random testing of Haskell programs. In: ICFP 2000, pp. 268–279. ACM (2000)
13. Claessen, K., Lillieström, A., Smallbone, N.: Sort it out with monotonicity. In: Bjørner, N., Sofronie-Stokkermans, V. (eds.) CADE 2011. LNCS, vol. 6803, pp. 207–221. Springer, Heidelberg (2011)
14. Claessen, K., Sörensson, N.: New techniques that improve MACE-style model finding. In: MODEL (2003)
15. de Moura, L., Bjørner, N.S.: Efficient E-Matching for SMT solvers. In: Pfenning, F. (ed.) CADE 2007. LNCS (LNAI), vol. 4603, pp. 183–198. Springer, Heidelberg (2007)
16. de Moura, L., Bjørner, N.: Relevancy propagation. Technical report, Microsoft Research, October 2007
17. de Moura, L., Bjørner, N.S.: Z3: an efficient SMT solver. In: Ramakrishnan, C.R., Rehof, J. (eds.) TACAS 2008. LNCS, vol. 4963, pp. 337–340. Springer, Heidelberg (2008)
18. Dunets, A., Schellhorn, G., Reif, W.: Automated flaw detection in algebraic specifications. J. Autom. Reasoning **45**(4), 359–395 (2010)
19. Ge, Y., de Moura, L.: Complete instantiation for quantified formulas in satisfiabiliby modulo theories. In: Bouajjani, A., Maler, O. (eds.) CAV 2009. LNCS, vol. 5643, pp. 306–320. Springer, Heidelberg (2009)
20. Goguen, J.A., Burstall, R.M.: Institutions: abstract model theory for specification and programming. J. ACM **39**(1), 95–146 (1992)

21. Jackson, D.: Nitpick: a checkable specification language. In: FMSP 1996, pp. 60–69 (1996)
22. Johansson, M., Dixon, L., Bundy, A.: Case-analysis for rippling and inductive proof. In: Kaufmann, M., Paulson, L.C. (eds.) ITP 2010. LNCS, vol. 6172, pp. 291–306. Springer, Heidelberg (2010)
23. Korovin, K.: Non-cyclic sorts for first-order satisfiability. In: Fontaine, P., Ringeissen, C., Schmidt, R.A. (eds.) FroCoS 2013. LNCS, vol. 8152, pp. 214–228. Springer, Heidelberg (2013)
24. Krauss, A.: Automating recursive definitions and termination proofs in higher-order logic. Ph.D. thesis, Technische Universität München (2009)
25. Kuncak, V., Jackson, D.: Relational analysis of algebraic datatypes. In: Wermelinger, M., Gall, H. (eds.) ESEC/FSE 2005. ACM (2005)
26. Lindblad, F.: Property directed generation of first-order test data. In: Morazán, M. (ed.) TFP 2007, pp. 105–123. Intellect (2008)
27. McCune, W.: Prover9 and Mace4. http://www.cs.unm.edu/mccune/prover9/
28. McCune, W.: A Davis-Putnam program and its application to finite first-order model search: quasigroup existence problems. Technical report, Argonne National Laboratory (1994)
29. Reynolds, A., Blanchette, J.C.: A decision procedure for (co)datatypes in SMT solvers. In: Felty, A., Middeldorp, A. (eds.) CADE-25. LNCS, vol. 9195, pp. 197–213. Springer, Heidelberg (2015)
30. Reynolds, A., Blanchette, J.C., Tinelli, C.: Model finding for recursive functions in SMT. In: Ganesh, V., Jovanović, D. (eds.) SMT 2015 (2015)
31. Reynolds, A., Tinelli, C., de Moura, L.: Finding conflicting instances of quantified formulas in SMT. In: FMCAD 2014, pp. 195–202. IEEE (2014)
32. Reynolds, A., Tinelli, C., Goel, A., Krstić, S.: Finite model finding in SMT. In: Sharygina, N., Veith, H. (eds.) CAV 2013. LNCS, vol. 8044, pp. 640–655. Springer, Heidelberg (2013)
33. Reynolds, A., Tinelli, C., Goel, A., Krstić, S., Deters, M., Barrett, C.: Quantifier instantiation techniques for finite model finding in SMT. In: Bonacina, M.P. (ed.) CADE 2013. LNCS, vol. 7898, pp. 377–391. Springer, Heidelberg (2013)
34. Reynolds, A.J.: Finite model finding in satisfiability modulo theories. Ph.D. thesis, The University of Iowa (2013)
35. Runciman, C., Naylor, M., Lindblad, F.: Smallcheck and lazy smallcheck: automatic exhaustive testing for small values. In: Gill, A. (ed.) Haskell 2008, pp. 37–48. ACM (2008)
36. Slaney, J.K.: FINDER: finite domain enumerator system description. In: Bundy, A. (ed.) CADE 1994. LNCS, vol. 814, pp. 798–801. Springer, Heidelberg (1994)
37. Torlak, E., Jackson, D.: Kodkod: a relational model finder. In: Grumberg, O., Huth, M. (eds.) TACAS 2007. LNCS, vol. 4424, pp. 632–647. Springer, Heidelberg (2007)
38. Turner, D.A.: Elementary strong functional programming. In: Hartel, P.H., Plasmeijer, R. (eds.) FPLE 1995. LNCS, vol. 1022, pp. 1–13. Springer, Heidelberg (1995)
39. Weber, T.: SAT-based finite model generation for higher-order logic. Ph.D. thesis, Technische Universität München (2008)
40. Zhang, J., Zhang, H.: SEM: a system for enumerating models. In: Mellish, C.S. (ed.) IJCAI 1995, vol. 1, pp. 298–303. Morgan Kaufmann (1995)

Colors Make Theories Hard

Roberto Sebastiani(✉)

DISI, University of Trento, Trento, Italy
roberto.sebastiani@unitn.it

Abstract. The satisfiability problem for conjunctions of quantifier-free literals in first-order theories $\mathcal{T}$ of interest–"*$\mathcal{T}$-solving*" for short–has been deeply investigated for more than three decades from both theoretical and practical perspectives, and it is currently a core issue of state-of-the-art SMT solving. Given some theory $\mathcal{T}$ of interest, a key theoretical problem is to establish the computational *(in)tractability* of $\mathcal{T}$-solving, or to identify intractable fragments of $\mathcal{T}$.

In this paper we investigate this problem from a general perspective, and we present a simple and general criterion for establishing the NP-hardness of $\mathcal{T}$-solving, which is based on the novel concept of "*colorer*" for a theory $\mathcal{T}$.

As a proof of concept, we show the effectiveness and simplicity of this novel criterion by easily producing very simple proofs of the NP-hardness for many theories of interest for SMT, or of some of their fragments.

1 Introduction

Since the pioneering works of the late 70's and early 80's by Nelson, Oppen, Shostak and others [16,17,19–21,25,26], the satisfiability problem for conjunctions of quantifier-free literals in first-order theories $\mathcal{T}$ of interest–hereafter "*$\mathcal{T}$-solving*" for short–has been deeply investigated from both theoretical and practical perspectives, and it is currently a core issue of state-of-the-art SMT solving.

Given some theory $\mathcal{T}$ of interest, or some fragment thereof, a key theoretical problem is that of establishing the computational *(in)tractability* of $\mathcal{T}$-solving, or to identify (in)tractable fragments of $\mathcal{T}$. Although in the pool of theories of interest $\mathcal{T}$-solving presents many levels of intractability, the main divide is between polynomiality and NP-hardness. Despite a wide literature studying the complexity of single theories or of families of theories (e.g. [5,7,8,10,11,13–15, 17,19–21]) and some more general work on complexity of $\mathcal{T}$-solving [3,20,21], we are not aware of any previous work explicitly addressing NP-hardness of $\mathcal{T}$-solving for a generic theory $\mathcal{T}$.

In this paper we try to fill this gap, and we present a simple and general criterion for establishing the NP-hardness of $\mathcal{T}$-solving for theories with equality–and in some cases also for theories without equality–which is based on the novel concept of "*colorer*" for a theory $\mathcal{T}$, inducing the notion of "*colorable*" theory.

This work is supported by SRC under GRC Research Project 2012-TJ-2266 WOLF. I thank Silvio Ghilardi, Alberto Griggio and Stefano Tonetta for fruitful discussions.

N. Olivetti and A. Tiwari (Eds.): IJCAR 2016, LNAI 9706, pp. 152–170, 2016.
DOI: 10.1007/978-3-319-40229-1_11

Our work started from the heuristic observation that the *graph k-colorability problem*, which is NP-complete for $k \geq 3$, fits very naturally as a candidate problem to be polynomially encoded into $\mathcal{T}$-solving for theories with equality. (We believe, more naturally than the very frequently-used 3-SAT problem.) In fact, we notice that the set of the arcs in a graph and the coloring of the vertexes can be encoded respectively into a conjunction of disequalities between "vertex" variables and into a conjunction of equalities between "vertex" and "color" variables, *both of which are theory-independent.* Therefore, in designing a reduction from k-colorability to $\mathcal{T}$-solving, the only facts one needs formalizing by $\mathcal{T}$-specific literals is a coherent definition of k distinct "colors" and the fact that a generic vertex can be "colored" with and only with k colors.

Following this line of thought, in this paper we present a general framework for producing reductions from graph k-colorability with $k \geq 3$ to $\mathcal{T}$-solving for generic theories $\mathcal{T}$ with equality. This framework decouples the $\mathcal{T}$-specific part of a reduction from its $\mathcal{T}$-independent part: the former is formalized into the definition of a $\mathcal{T}$-specific object, called "*k-colorer*", the latter is formalized and proven once forall in this paper. Thus, the task of proving the NP-hardness of a theory $\mathcal{T}$ via reduction from k-colorability reduces to that of finding a k-colorer for $\mathcal{T}$.

To this extent, we also provide some general criteria for producing k-colorers, with hints and tips to achieve this simplified task. As a proof of concept, we show the effectiveness and simplicity of this novel approach by easily producing k-colorers with $k \geq 3$ for many theories of interest for SMT, or for some of their fragments.

We notice that this technique can be used not only to investigate the intractability of major theories, but also to investigate that of *fragments* of such theories, so that to pinpoint the subsets of constructs (i.e. functions and predicates in the signature) which cause a theory to be intractable. We stress the fact that the problem of identifying such intractable fragments is not only of theoretical interest, but also of practical importance in the development of SMT solvers, in order to drive the activation of ad-hoc techniques–including e.g. *weakened early pruning*, *layering*, *splitting-on-demand* [1,4]–which partition the search load among distinct specialized $\mathcal{T}$-solvers and between the $\mathcal{T}$-solvers and the underlining SAT solver [2,23].

Note. An extended version of this paper with more details is publicly available [24].

Content. The rest of the paper is organized as follows: Sect. 2 provides the necessary background knowledge and terminology for logic and graph coloring; Sect. 3 introduces our main definitions of k-colorer and k-colorability and presents our main results; Sect. 4 explains how to produce k-colorers for given theories, providing a list of examples; Sect. 5 provides some discussion about k-colorability vs. non-convexity; Sect. 6 extends the framework to theories without equality; Sect. 7 discusses ongoing and future developments.

2 Background and Terminology

Logic. We assume the reader is familiar with the standard syntax and semantics of first-order logic. (We report a full description in [24].) We add some terminology.

Given a signature Σ, we call *Σ-theory $\mathcal{T}$* a class of Σ-models. Given a theory $\mathcal{T}$, we call *$\mathcal{T}$-interpretation* an extension of some Σ-model $\mathcal{M}$ in $\mathcal{T}$ which maps free variables into elements of the domain of $\mathcal{M}$. (The map is denoted by $\langle . \rangle^{\mathcal{I}}$.) A Σ-formula φ–possibly with free variables–is *$\mathcal{T}$-satisfiable* if $\mathcal{I} \models \varphi$ for some $\mathcal{T}$-interpretation $\mathcal{I}$. (Hereafter we will use the symbol "$\models_{\mathcal{T}}$" to denote the $\mathcal{T}$-satisfiability relation; we will also drop the prefix "Σ-" when the signature is implicit by context.) We say that a set/conjunction of formulas Ψ *$\mathcal{T}$-entails* another formula φ, written $\Psi \models_{\mathcal{T}} \varphi$, if every $\mathcal{T}$-interpretation $\mathcal{T}$-satisfying Ψ also $\mathcal{T}$-satisfies φ. We say that φ is *$\mathcal{T}$ -valid*, written $\models_{\mathcal{T}} \varphi$, if $\emptyset \models_{\mathcal{T}} \varphi$. We call a *cube* any finite quantifier-free conjunction of literals. For short, we call "*$\mathcal{T}$-solving*" the problem of deciding the $\mathcal{T}$-satisfiability of a cube.

Finally, a theory $\mathcal{T}$ is *convex* if for all cubes μ and all sets E of equalities between variables, $\mu \models_{\mathcal{T}} \bigvee_{e \in E} e$ iff $\mu \models_{\mathcal{T}} e$ for some $e \in E$.

Remark 1. In SMT and other contexts it is often convenient to use formulas with *uninterpreted symbols* (see e.g. [2]). Notice, however, that the presence of uninterpreted function or predicate symbols of arity >0 may cause the complexity of $\mathcal{T}$-solving scale up (see e.g. the example in [21]). Thus, when not explicitly specified otherwise, we implicitly assume that a theory $\mathcal{T}$ does *not* admit such symbols. ◇

We are often interested in fragments of a theory obtained by restricting its signature. Let Σ, Σ' be two signatures s.t. $\Sigma' \subseteq \Sigma$; we say that a Σ'-model $\mathcal{M}'$ is a *restriction to Σ'* of a Σ-model $\mathcal{M}$ iff $\mathcal{M}'$ and $\mathcal{M}$ agree on all the symbols in Σ', and that a Σ'-theory $\mathcal{T}'$ is the *signature-restriction fragment* of a Σ-theory $\mathcal{T}$ wrt. Σ' iff $\mathcal{T}'$ is the set of the restrictions to Σ' of the Σ-models in $\mathcal{T}$.

Graph Coloring. We recall a few notions from [9].

Definition 1 (k-Colorability of a graph (see [9])). *Let $\mathcal{G} \stackrel{\text{def}}{=} \langle \mathcal{V}, \mathcal{E} \rangle$ be an un-directed graph, where $\mathcal{V} \stackrel{\text{def}}{=} \{V_1, ..., V_n\}$ is the set of vertexes and $\mathcal{E} \stackrel{\text{def}}{=} \{E_1, ..., E_m\}$ is the set of edges in the form $\langle V_i, V_{i'} \rangle$ for some i, i'. Let $\mathcal{C} \stackrel{\text{def}}{=} \{C_1, ..., C_k\}$ be a set of distinct values, namely "colors", for k>0. Then $\mathcal{G}$ is k-***Colorable** if and only if there exists a total map color : $\mathcal{V} \longmapsto \mathcal{C}$ s.t. color$(V_i) \neq$ color$(V_{i'})$ for every $\langle V_i, V_{i'} \rangle \in \mathcal{E}$. The problem of deciding if $\mathcal{G}$ is k-colorable is called the k-*colorability problem for $\mathcal{G}$.

Lemma 1 (see [9]). *The k-colorability problem for un-directed graphs is NP-complete for $k \geq 3$, it is in P for k<3.*

Figure 1 (top) shows two small graph 3-colorability problems.

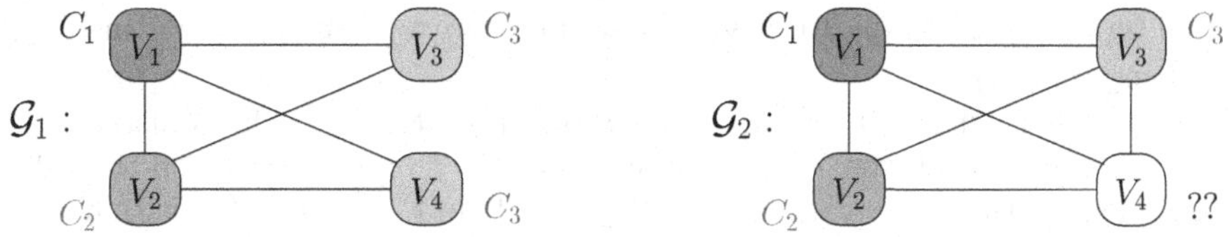

$$\mathsf{Enc}_{[\mathcal{G}_1 \Rightarrow \mathcal{LA}(\mathbb{Z})]} \stackrel{\text{def}}{=} \left(\begin{array}{l} (c_1 = 1) \wedge (c_2 = 2) \wedge (c_3 = 3) \wedge \bigwedge_{i=1}^{4}((v_i \geq 1) \wedge (v_i \leq 3)) \wedge \\ \neg(v_1 = v_2) \wedge \neg(v_1 = v_3) \wedge \neg(v_1 = v_4) \wedge \neg(v_2 = v_3) \wedge \neg(v_2 = v_4) \end{array} \right)$$

$$\mathsf{Enc}_{[\mathcal{G}_2 \Rightarrow \mathcal{LA}(\mathbb{Z})]} \stackrel{\text{def}}{=} \mathsf{Enc}_{[\mathcal{G}_1 \Rightarrow \mathcal{LA}(\mathbb{Z})]} \wedge \neg(v_3 = v_4)$$

Fig. 1. Top Left: a small 3-colorable graph ($\mathcal{G}_1$), with $C_1 = blue$, $C_2 = red$, $C_3 = green$. Top Right: the same graph augmented with the vertex $\langle V_3, V_4 \rangle$ ($\mathcal{G}_2$) is no more 3-colorable. Bottom: example of encodings of the 3-colorability of $\mathcal{G}_1$ and $\mathcal{G}_2$ into $\mathcal{LA}(\mathbb{Z})$ -solving. (Color figure online)

3 *k*-Colorers and *k*-Colorable Theories with Equality

Hereafter we focus w.l.o.g. on theories $\mathcal{T}$ of domain size ≥ 2, i.e., s.t. $\neg(v_1 = v_2)$ is $\mathcal{T}$-consistent. In fact, if not so, then it is easy to see that $\mathcal{T}$-solving is in P (see [24]).

Definition 2 (*k*-Colorer, *k*-Colored Theory). *Let $\mathcal{T}$ be some theory with equality and k be some integer value s.t. $k \geq 2$. Let v_i be a variable, called* **vertex variable**, *(implicitly) denoting the i-th vertex in an un-directed graph; let $\underline{\mathbf{c}} \stackrel{def}{=} \{c_1, .., c_k\}$ be a set of variables, called* **color variables**, *denoting the set of colors; let $\underline{\mathbf{y}_i} \stackrel{def}{=} \{y_{i1}, ..., y_{il}\}$ denote a possibly-empty set of variables, which is indexed with the same index i of the vertex variable v_i. Let $\mathsf{AllDifferent}_k(\underline{\mathbf{c}}) \stackrel{def}{=} \bigwedge_{j=1}^{k} \bigwedge_{j'=j+1}^{k} \neg(c_j = c_{j'})$.*

We call **k-colorer** *for $\mathcal{T}$, namely $\mathsf{Colorer}_k(v_i, \underline{\mathbf{c}}|\underline{\mathbf{y}_i})$, a finite quantifier-free conjunction of $\mathcal{T}$-literals (cube) over v_i, $\underline{\mathbf{c}}$ and $\underline{\mathbf{y}_i}$ which verify the following properties:*

$$\mathsf{Colorer}_k(v_i, \underline{\mathbf{c}}|\underline{\mathbf{y}_i}) \models_{\mathcal{T}} \mathsf{AllDifferent}_k(\underline{\mathbf{c}}), \quad (1)$$

$$\mathsf{Colorer}_k(v_i, \underline{\mathbf{c}}|\underline{\mathbf{y}_i}) \models_{\mathcal{T}} \bigvee_{j=1}^{k} (v_i = c_j), \quad (2)$$

there exist k $\mathcal{T}$-interpretations $\{\mathcal{I}_{i,1}, ..., \mathcal{I}_{i,k}\}$ s.t. (3)

for every $j \in [1..k]$, $\langle c_j \rangle^{\mathcal{I}_{i,1}} = \langle c_j \rangle^{\mathcal{I}_{i,2}} = ... = \langle c_j \rangle^{\mathcal{I}_{i,k}}$, and
for every $j \in [1..k]$, $\mathcal{I}_{i,j} \models_{\mathcal{T}} \mathsf{Colorer}_k(v_i, \underline{\mathbf{c}}|\underline{\mathbf{y}_i}) \wedge (v_i = c_j)$.

We say that $\mathcal{T}$ is **k-colorable** *if and only if it has a k-colorer.*

$\underline{\mathbf{y}_i}$ is a (possibly-empty) set of auxiliary variables, one distinct set for each vertex variable v_i, which sometimes may be needed to express (1), (2) and (3) (see Examples 7 and 9), or to make $\mathsf{Colorer}_k(v_i, \underline{\mathbf{c}}|\underline{\mathbf{y}_i})$ more readable by renaming internal terms (see Example 9). If $\underline{\mathbf{y}_i} = \emptyset$, we may write "$\mathsf{Colorer}_k(v_i, \underline{\mathbf{c}})$" instead of "$\mathsf{Colorer}_k(v_i, \underline{\mathbf{c}}|\emptyset)$".[1]

$\{\mathcal{I}_{i,1}, ..., \mathcal{I}_{i,k}\}$ denotes a set of $\mathcal{T}$-interpretations each satisfying $\mathsf{Colorer}_k(v_i, \underline{\mathbf{c}}|\underline{\mathbf{y}_i})$ s.t. all the $\mathcal{T}$-interpretations in $\{\mathcal{I}_{i,1}, ..., \mathcal{I}_{i,k}\}$ agree on the values assigned to the color variables in $\{c_1, ..., c_k\}$ and s.t. each $\mathcal{I}_{i,j}$ assigns to the vertex variable v_i the same value assigned to the jth color variable c_j. The condition $\langle c_j \rangle^{\mathcal{I}_{i,1}} = ... = \langle c_j \rangle^{\mathcal{I}_{i,k}}$ of (3) expresses the fact that, when passing from the scenario $\mathcal{I}_{i,j}$ in which v_i is assigned the color c_j–expressed by the equality $(v_i = c_j)$ in (3)–to the scenario $\mathcal{I}_{i,j'}$ in which v_i is assigned the color $c_{j'}$–expressed by the equality $(v_i = c_{j'})$– it is the value of the vertex variable v_i who must change, not those of the color variables $c_1, ..., c_k$.

Intuitively, $\mathsf{Colorer}_k(v_i, \underline{\mathbf{c}}|\underline{\mathbf{y}_i})$ expresses the following facts: (1) that $c_1, ..., c_k$ represent the names of distinct "color" values, (2) that each vertex represented by the variable v_i can be tagged ("colored") only with one of such color names c_j, (3) that the values associated to the color names are not affected by the choice of the color name c_j tagged to v_i–represented by the index j in $\mathcal{I}_{i,j}$–and that each tagging choice is admissible.

There may be many distinct k-colorers for a theory $\mathcal{T}$, as shown in Example 1.

Example 1 ($\mathcal{LA}(\mathbb{Z})$). We consider the theory of linear arithmetic over the integers ($\mathcal{LA}(\mathbb{Z})$), assuming the standard model of integers, so that the symbols $+, -, \leq, \geq$ and the interpreted constants $0, 1, ...$ are interpreted in the standard way by all $\mathcal{LA}(\mathbb{Z})$-interpretations. $\mathcal{LA}(\mathbb{Z})$ is 3-colorable, since we can define, e.g., $k \stackrel{\text{def}}{=} 3$, $\underline{\mathbf{y}_i} \stackrel{\text{def}}{=} \emptyset$, and

$$\mathsf{Colorer}_3(v_i, c_1, c_2, c_3) \stackrel{\text{def}}{=} (c_1 = 1) \wedge (c_2 = 2) \wedge (c_3 = 3) \wedge (v \geq 1) \wedge (v \leq 3). \quad (4)$$

It is straightforward to see that $\mathsf{Colorer}_3(v_i, c_1, c_2, c_3)$ verifies (1), (2) and (3), with $\mathcal{I}_{i,j} \stackrel{\text{def}}{=} \{c_1 \to 1, c_2 \to 2, c_3 \to 3, v_i \to j\}$ for every $j \in [1..3]$. Notice that in this case $\underline{\mathbf{y}_i} = \emptyset$, i.e. $\mathsf{Colorer}_k(v_i, \underline{\mathbf{c}}|\underline{\mathbf{y}_i})$ requires no auxiliary variables. Notice also that $\mathsf{AllDifferent}_k(\underline{\mathbf{c}})$ is implied by the usage of the interpreted constants $1, 2, 3$.

An alternative 3-colorer which does not explicitly assign fixed values to the c_j's is:

$$\mathsf{Colorer}_3(v_i, c_1, c_2, c_3) \stackrel{\text{def}}{=} \begin{pmatrix} \mathsf{AllDifferent}_3(\underline{\mathbf{c}}) \wedge \bigwedge_{j=1}^{3}((c_j \geq 1) \wedge (c_j \leq 3)) \wedge \\ (v \geq 1) \wedge (v \leq 3) \end{pmatrix}, \quad (5)$$

which verifies (1), (2) and (3), e.g., with the same $\mathcal{I}_{i,j}$'s as above. Consider instead:

$$\mathsf{Colorer}_3(v_i, c_1, c_2, c_3) \stackrel{\text{def}}{=} \begin{pmatrix} \mathsf{AllDifferent}_3(\underline{\mathbf{c}}) \wedge \bigwedge_{j=1}^{3}((c_j \geq 1) \wedge (c_j \leq 3)) \wedge \\ (v_i = 1) \end{pmatrix}. \quad (6)$$

This is not a 3-colorer, because it does not verify (3): there is no pair of $\mathcal{LA}(\mathbb{Z})$-interpretations $\mathcal{I}_{i,1}$ and $\mathcal{I}_{i,2}$ s.t. $\mathcal{I}_{i,1} \models_{\mathcal{LA}(\mathbb{Z})} \mathsf{Colorer}_3(v_i, c_1, c_2, c_3) \wedge (v_i = c_1)$ and $\mathcal{I}_{i,2} \models_{\mathcal{LA}(\mathbb{Z})} \mathsf{Colorer}_3(v_i, c_1, c_2, c_3) \wedge (v_i = c_2)$ which agree on the values of c_1, c_2, c_3. ◇

[1] The symbol "|" is used to separate color and node variables from auxiliary ones.

Remark 2. The choice of using *variables* $c_1, ..., c_k$ to represent colors is due to the fact that some theories do not provide k distinct interpreted constant symbols in their signature (see Example 9). If this is not the case, then $\mathsf{Colorer}_k(v_i, \underline{\mathbf{c}}|\underline{\mathbf{y}_i})$ can be built to force $c_1, ..., c_k$ to assume fixed values expressed by interpreted constant symbols, like $1, 2, 3$ in (4), so that the condition $\langle c_j\rangle^{\mathcal{I}_{i,1}} = ... = \langle c_j\rangle^{\mathcal{I}_{i,k}}$ of (3) is verified a priori.

The following properties of k-colorable theories follow straightforwardly.

Property 1. Let $\mathcal{T}$ be a k-colorable theory for some $k \geq 2$. Then we have that:

(*a*) $\exists\underline{\mathbf{c}}.\mathsf{AllDifferent}_k(\underline{\mathbf{c}})$ is $\mathcal{T}$-valid;
(*b*) $\mathcal{T}$ is non-convex.

Proof. Consider the definition of $\mathsf{Colorer}_k(v_i, \underline{\mathbf{c}}|\underline{\mathbf{y}_i})$ in Definition 2.

(*a*) By (3) $\mathsf{Colorer}_k(v_i, \underline{\mathbf{c}}|\underline{\mathbf{y}_i})$ is $\mathcal{T}$-satisfiable; thus by (1) $\mathsf{AllDifferent}_k(\underline{\mathbf{c}})$ is $\mathcal{T}$-satisfiable, so that $\models_{\mathcal{T}} \exists\underline{\mathbf{c}}.\mathsf{AllDifferent}_k(\underline{\mathbf{c}})$;
(*b*) By (2), $\mathsf{Colorer}_k(v_i, \underline{\mathbf{c}}|\underline{\mathbf{y}_i}) \models_{\mathcal{T}} \bigvee_{j=1}^{k}(v_i = c_j)$. By (3), for every $j_1 \in [1..k]$ there exists an interpretation $\mathcal{I}_{i,j_1}$ s.t. $\mathcal{I}_{i,j_1} \models_{\mathcal{T}} \mathsf{Colorer}_k(v_i, \underline{\mathbf{c}}|\underline{\mathbf{y}_i}) \wedge (v_i = c_{j_1})$. Then, by (1), for every $j_2 \in [1..k]$ s.t. $j_2 \neq j_1$ we have that $\mathcal{I}_{i,j_1} \models_{\mathcal{T}} \mathsf{Colorer}_k(v_i, \underline{\mathbf{c}}|\underline{\mathbf{y}_i}) \wedge \neg(v_i = c_{j_2})$. Thus for every $j \in [1..k]$ $\mathsf{Colorer}_k(v_i, \underline{\mathbf{c}}|\underline{\mathbf{y}_i}) \not\models (v_i = c_j)$. Therefore $\mathcal{T}$ is non-convex. □

Property 2. If $\mathcal{T}'$ is a k-colorable theory with equality for some $k \geq 2$, and $\mathcal{T}'$ is a signature-restriction fragment of another theory $\mathcal{T}$, then $\mathcal{T}$ is k-colorable.

Proof. If $\mathsf{Colorer}_k(v_i, \underline{\mathbf{c}}|\underline{\mathbf{y}_i})$ is a k-colorer for $\mathcal{T}'$, then by definition of signature-restriction fragment it is also a k-colorer for $\mathcal{T}$. □

Lemma 2. *Let k be an integer value s.t. $k \geq 3$. Let $\mathcal{G}$ and $\mathcal{C}$ be respectively an un-directed graph with n vertexes $V_1, ..., V_n$ and a set of k distinct colors $C_1, ..., C_k$, like in Definition 1. Let $\mathcal{T}$ be a k-colorable theory with equality. We consider the following conjunctions of $\mathcal{T}$-literals:*

$$\mathsf{Colorable}(v_1, ..., v_n, \underline{\mathbf{c}}|\underline{\mathbf{y}_1}, ..., \underline{\mathbf{y}_n}) \stackrel{def}{=} \bigwedge_{V_i \in \mathcal{V}} \mathsf{Colorer}_k(v_i, \underline{\mathbf{c}}|\underline{\mathbf{y}_i}) \tag{7}$$

$$\mathsf{Graph}_{[\mathcal{G}]}(v_1, ..., v_n) \stackrel{def}{=} \bigwedge_{\langle V_{i_1}, V_{i_2}\rangle \in \mathcal{E}} \neg(v_{i_1} = v_{i_2}) \tag{8}$$

$$\begin{aligned}\mathsf{Enc}_{[\mathcal{G}\Rightarrow\mathcal{T}]}(v_1, ..., v_n, \underline{\mathbf{c}}|\underline{\mathbf{y}_1}, ..., \underline{\mathbf{y}_n}) \stackrel{def}{=} &\ \mathsf{Colorable}(v_1, ..., v_n, \underline{\mathbf{c}}|\underline{\mathbf{y}_1}, ..., \underline{\mathbf{y}_n}) \wedge \\ &\ \mathsf{Graph}_{[\mathcal{G}]}(v_1, ..., v_n),\end{aligned} \tag{9}$$

where $v_1, ..., v_n$, $c_1, ..., c_k$ and $y_{11}, ..., y_{1l}, ...y_{i1}, ..., y_{il}, ..., y_{n1}, ..., y_{nl}$ are free variables,[2] *and all the k-colorers $\mathsf{Colorer}_k(v_i, \underline{\mathbf{c}}|\underline{\mathbf{y}_i})$ in (7) are identical modulo the renaming of the variables v_i and $\underline{\mathbf{y}_i}$, but not of the color variables $\underline{\mathbf{c}}$.*

Then $\mathcal{G}$ is k-colorable iff $\mathsf{Enc}_{[\mathcal{G}\Rightarrow\mathcal{T}]}(v_1, ..., v_n, \underline{\mathbf{c}}|\underline{\mathbf{y}_1}, ..., \underline{\mathbf{y}_n})$ is $\mathcal{T}$-satisfiable.

[2] Notice that each c_j is implicitly associated with the color $C_j \in \mathcal{C}$ for every $j \in [1..k]$ and each v_i and $\underline{\mathbf{y}_i}$ is implicitly associated to the vertex $V_i \in \mathcal{V}$ for every $i \in [1..n]$.

Proof.

If: Suppose $\mathsf{Enc}_{[\mathcal{G}\Rightarrow\mathcal{T}]}(v_1,...,v_n,\underline{\mathbf{c}}|\underline{\mathbf{y}_1},...,\underline{\mathbf{y}_n})$ is $\mathcal{T}$-satisfiable, that is, there exist an interpretation $\mathcal{I}$ in $\mathcal{T}$ s.t. $\mathcal{I} \models_{\mathcal{T}} \mathsf{Colorable}(v_1,...,v_n,\underline{\mathbf{c}}|\underline{\mathbf{y}_1},...,\underline{\mathbf{y}_n})$ and $\mathcal{I} \models_{\mathcal{T}} \mathsf{Graph}_{[\mathcal{G}]}(v_1,...,v_n)$. Thus:

(i) By (7) and (1), $\langle c_{j_1}\rangle^{\mathcal{I}} \neq \langle c_{j_2}\rangle^{\mathcal{I}}$ for every $j_1, j_2 \in [1,...,k]$ s.t. $j_1 \neq j_2$.

(ii) By (7), (2) and (1), for every $i \in [1...n]$ there exists some $j \in [1...k]$ s.t. $\langle v_i\rangle^{\mathcal{I}} = \langle c_j\rangle^{\mathcal{I}}$ and s.t. $\langle v_i\rangle^{\mathcal{I}} \neq \langle c_{j'}\rangle^{\mathcal{I}}$ for every $j' \neq j$.

(iii) By (8), $\langle v_{i_1}\rangle^{\mathcal{I}} \neq \langle v_{i_2}\rangle^{\mathcal{I}}$ for every $\langle V_{i_1}, V_{i_2}\rangle \in \mathcal{E}$.

Then by (i) and (ii) we can build a map $color : \mathcal{V} \longmapsto \mathcal{C}$ s.t., for every $V_i \in \mathcal{V}$, $color(V_i) = C_j$ iff $\langle v_i\rangle^{\mathcal{I}} = \langle c_j\rangle^{\mathcal{I}}$. By (iii) we have that $color(V_{i_1}) \neq color(V_{i_2})$ for every $\langle V_{i_1}, V_{i_2}\rangle \in \mathcal{E}$. Thus $\mathcal{G}$ is k-colorable.

Only if: Suppose $\mathcal{G}$ is k-colorable, that is, there exist a map $color : \mathcal{V} \longmapsto \mathcal{C}$ s.t. $color(V_{i_1}) \neq color(V_{i_2})$ for every $\langle V_{i_1}, V_{i_2}\rangle \in \mathcal{E}$.

Consider $i = 1$, and let $\{\mathcal{I}_{1,1},...,\mathcal{I}_{1,k}\}$ be the set of $\mathcal{T}$-interpretations for $\mathsf{Colorer}_k(v_1,\underline{\mathbf{c}}|\underline{\mathbf{y}_1})$ as in (3), so that:

(a) for every $j \in [1..k]$, $\mathcal{I}_{1,j} \models_{\mathcal{T}} \mathsf{Colorer}_k(v_1,\underline{\mathbf{c}}|\underline{\mathbf{y}_1}) \wedge (v_1 = c_j)$,

(b) for every $j \in [1..k]$, $\langle c_j\rangle^{\mathcal{I}_{1,1}} = ... = \langle c_j\rangle^{\mathcal{I}_{1,k}}$.

For every $i \in [1..n]$ we consider $\mathsf{Colorer}_k(v_i,\underline{\mathbf{c}}|\underline{\mathbf{y}_i})$ and we build a replica $\{\mathcal{I}_{i,1},...,\mathcal{I}_{i,k}\}$ of the set of $\mathcal{T}$-interpretations $\{\mathcal{I}_{1,1},...,\mathcal{I}_{1,k}\}$ in such a way that:

(i) $\langle v_i\rangle^{\mathcal{I}_{i,j}} \stackrel{\text{def}}{=} \langle v_1\rangle^{\mathcal{I}_{1,j}} = \langle c_j\rangle^{\mathcal{I}_{1,j}}$ (each $\mathcal{I}_{i,j}$ maps its vertex variable v_i into the same color as $\mathcal{I}_{1,j}$ maps its vertex variable v_1);

(ii) $\langle c_j\rangle^{\mathcal{I}_{i,1}} \stackrel{\text{def}}{=} \langle c_j\rangle^{\mathcal{I}_{1,1}}, ..., \langle c_j\rangle^{\mathcal{I}_{i,k}} \stackrel{\text{def}}{=} \langle c_j\rangle^{\mathcal{I}_{1,k}}$, so that, by (a), $\langle c_j\rangle^{\mathcal{I}_{i,1}} = ... = \langle c_j\rangle^{\mathcal{I}_{i,k}} = \langle c_j\rangle^{\mathcal{I}_{1,1}} = ... = \langle c_j\rangle^{\mathcal{I}_{1,k}}$ (all $\mathcal{I}_{i,j}$ agree on the values of the color variables, for every $i \in [1..n]$ and $j \in [1..k]$);

(iii) $\langle y_{i1}\rangle^{\mathcal{I}_{i,j}} \stackrel{\text{def}}{=} \langle y_{11}\rangle^{\mathcal{I}_{1,j}}, ..., \langle y_{il}\rangle^{\mathcal{I}_{i,j}} \stackrel{\text{def}}{=} \langle y_{1l}\rangle^{\mathcal{I}_{1,j}}$ (each $\mathcal{I}_{i,j}$ maps its auxiliary variables $\mathbf{y}_i$ into the same domain values as $\mathcal{I}_{1,j}$ maps $\mathbf{y}_1$).

Consequently, by (3), for every $v_i \in \{v_1,...,v_n\}$, $\{\mathcal{I}_{i,1},...,\mathcal{I}_{i,k}\}$ are s.t.

(a) for every $j \in [1..k]$, $\mathcal{I}_{i,j} \models_{\mathcal{T}} \mathsf{Colorer}_k(v_i,\underline{\mathbf{c}}|\underline{\mathbf{y}_i}) \wedge (v_i = c_j)$,

(b) for every $j \in [1..k]$, $\langle c_j\rangle^{\mathcal{I}_{i,1}} = ... = \langle c_j\rangle^{\mathcal{I}_{i,k}}$.

For every $i \in [1...n]$, let $j_i \in [1..k]$ be the index s.t. $C_{j_i} = color(V_i)$, and we pick the $\mathcal{T}$-interpretation $\mathcal{I}_{i,j_i}$. Thus, since all the $\mathcal{I}_{i,j_i}$s agree on the common variables $\underline{\mathbf{c}}$, we can merge them and create a global $\mathcal{T}$-interpretation $\mathcal{I}$ as follows:

(i) $\langle v_i\rangle^{\mathcal{I}} \stackrel{\text{def}}{=} \langle v_i\rangle^{\mathcal{I}_{i,j_i}} = \langle c_{j_i}\rangle^{\mathcal{I}_{i,j_i}} = \langle c_{j_i}\rangle^{\mathcal{I}}$, for every $i \in [1..n]$;

(ii) $\langle c_j\rangle^{\mathcal{I}} \stackrel{\text{def}}{=} \langle c_j\rangle^{\mathcal{I}_{i,j_i}}$, for every $j \in [1..k]$;

(iii) $\langle y_{ir}\rangle^{\mathcal{I}} \stackrel{\text{def}}{=} \langle y_{ir}\rangle^{\mathcal{I}_{i,j_i}}$, for every $i \in [1..n]$ and for every $r \in [1..l]$.

By construction, for every $i \in 1..n$, $\mathcal{I}$ agrees with $\mathcal{I}_{i,j_i}$ on $\underline{\mathbf{c}}$, v_i, and $\underline{\mathbf{y}_i}$, so that, by point (a), $\mathcal{I} \models_{\mathcal{T}} (\mathsf{Colorer}_k(v_i,\underline{\mathbf{c}}|\underline{\mathbf{y}_i}) \wedge (v_i = c_{j_i}))$.

Thus $\mathcal{I} \models_{\mathcal{T}} \mathsf{Colorable}(v_1,...,v_n,\underline{\mathbf{c}}|\underline{\mathbf{y}_1},...,\underline{\mathbf{y}_n})$.

Since the values $\langle c_1\rangle^{\mathcal{I}},...,\langle c_k\rangle^{\mathcal{I}}$ are all distinct, we can build a bijection linking each domain value $\langle c_j\rangle^{\mathcal{I}}$ to the color C_j, for every $j \in [1..k]$. Hence $\langle c_j\rangle^{\mathcal{I}} = \langle c_{j'}\rangle^{\mathcal{I}}$ iff $C_j = C_{j'}$. For every $\langle V_i, V_{i'}\rangle \in \mathcal{E}$, $color(V_i) \neq color(V_{i'})$, that is, $C_{j_i} \neq C_{j_{i'}}$. Therefore $\langle c_{j_i}\rangle^{\mathcal{I}} \neq \langle c_{j_{i'}}\rangle^{\mathcal{I}}$, and $\langle v_i\rangle^{\mathcal{I}} = \langle c_{j_i}\rangle^{\mathcal{I}} \neq \langle c_{j_{i'}}\rangle^{\mathcal{I}} = \langle v_{i'}\rangle^{\mathcal{I}}$.

Consequently $\mathcal{I} \models_{\mathcal{T}} \mathsf{Graph}_{[\mathcal{G}]}(v_1,...,v_n)$.

Thus $\mathsf{Enc}_{[\mathcal{G}\Rightarrow\mathcal{T}]}(v_1,...,v_n,\underline{\mathbf{c}}|\underline{\mathbf{y}_1},...,\underline{\mathbf{y}_n})$ is $\mathcal{T}$-satisfiable. □

Example 2. Figure 1 shows a simple example of encoding a graph 3-colorability problem into $\mathcal{LA}(\mathbb{Z})$-solving, using the k-colorer (4) of Example 1. (Notice that the literals which do not contain v_i and $\underline{\mathbf{y}_i}$ can be moved out of the conjunction $\bigwedge_{V_i \in \mathcal{V}} \ldots$ in (7).) The first formula is $\mathcal{LA}(\mathbb{Z})$-satisfied, e.g., by an interpretation $\mathcal{I}$ s.t. $\langle c_j \rangle^{\mathcal{I}} \stackrel{\text{def}}{=} j$ for every $j \in [1..3]$, $\langle v_1 \rangle^{\mathcal{I}} \stackrel{\text{def}}{=} 1$, $\langle v_2 \rangle^{\mathcal{I}} \stackrel{\text{def}}{=} 2$, $\langle v_3 \rangle^{\mathcal{I}} \stackrel{\text{def}}{=} 3$ and $\langle v_4 \rangle^{\mathcal{I}} \stackrel{\text{def}}{=} 3$, which mimics the coloring in Fig. 1 (left). The second formula is $\mathcal{LA}(\mathbb{Z})$-unsatisfiable, as expected. ◇

Lemma 3. *Let k, n, $\mathcal{G}$, $\mathcal{C}$, $\mathcal{T}$ and $\mathsf{Enc}_{[\mathcal{G} \Rightarrow \mathcal{T}]}(v_1, \ldots, v_n, \underline{\mathbf{c}} | \underline{\mathbf{y}_1}, \ldots, \underline{\mathbf{y}_n})$ be as in Lemma 2. Then $||\mathsf{Enc}_{[\mathcal{G} \Rightarrow \mathcal{T}]}(v_1, \ldots, v_n, \underline{\mathbf{c}} | \underline{\mathbf{y}_1}, \ldots, \underline{\mathbf{y}_n})||$ is polynomial in $||\mathcal{G}|| \stackrel{\text{def}}{=} ||\mathcal{V}|| + ||\mathcal{E}||$.*[3]

Proof. By Definition 2 we have that $||\mathsf{Colorer}_k(v_i, \underline{\mathbf{c}} | \underline{\mathbf{y}_i})||$ is constant wrt. $||\mathcal{V}||$ or $||\mathcal{E}||$. From (7), (8) and (9), $||\mathsf{Enc}_{[\mathcal{G} \Rightarrow \mathcal{T}]}(v_1, \ldots, v_n, \underline{\mathbf{c}} | \underline{\mathbf{y}_1}, \ldots, \underline{\mathbf{y}_n})||$ is $O(||\mathcal{V}|| + ||\mathcal{E}||)$. □

Combining Lemmas 1, 2 and 3 we have directly the following main result.

Theorem 1. *If a theory with equality $\mathcal{T}$ is k-colorable for some $k \geq 3$, then the problem of deciding the $\mathcal{T}$-satisfiability of a quantifier-free conjunction of $\mathcal{T}$-literals is NP-hard.*

Notice that the key source of hardness is condition (2) in Definition 2: intuitively, a k-colorable theory is expressive enough to represent with a quantifier-free conjunction of $\mathcal{T}$-literals–without disjunctions!–the fact that one variable must assume a value among a choice of $k \geq 3$ possible candidates–in addition to the fact that a list of pairs of variables cannot pairwise assume the same value. This source of non-deterministic choices has a high computational cost, as stated in Theorem 1.

4 Proving k-Colorabilty

Theorem 1 suggests a general technique for proving the NP-hardness of a theory $\mathcal{T}$: pick some $k \geq 3$ and then try to build a k-colorer $\mathsf{Colorer}_k(v_i, \underline{\mathbf{c}} | \underline{\mathbf{y}_i})$. Also, when $\mathcal{T}$ is known to be NP-hard, one may want to identify smaller –and possibly minimal– signature-restriction fragments $\mathcal{T}'$ which are k-colorable for some k, by identifying increasingly-smaller subsets of the signature of $\mathcal{T}$ which are needed to define a k-colorer.

We introduce some sufficient criteria for a theory to be k-colorable with some $k \geq 3$. As a proof of concept, we use these criteria to prove the k-colorability with some $k \geq 3$, and hence the NP-hardness, of some theories $\mathcal{T}$ of practical interest, and of some of their signature-restriction fragments.

We remark that the ultimate goal here is not to provide fully-detailed proofs of NP-hardness–all the main theories presented here are already well-known to

[3] Notice that k is fixed a priori and as such it is a *constant value* for the input graph k-colorability problem: e.g., depending on $\mathcal{T}$, we are speaking of reducing graph 3-colorability–or 4-colorability, or even 2^{64}-colorability–to $\mathcal{T}$-solving.

be NP-hard, although to the best of our knowledge the complexity of not all of their fragments has been investigated explicitly–rather to present proof of concept of the convenience and effectiveness of our proposed colorability-based technique, using various theories/fragments as examples. To this extent, for the sake of simplicity and space needs, and when this does not affect comprehension, sometimes we skip some formal details of the syntax and semantics of the theories under analysis, referring the reader to the proper literature. Rather, we dedicate a few lines to give some hints and tips on how to apply our colorability-based technique in potentially-typical scenarios.

4.1 Exploiting Interpreted Constants, Closed Terms and Provably-Distinct Terms

Proposition 1. *Let* $\mathcal{T}$ *be a theory which admits at least* $k \geq 3$ *terms* $t_1(\underline{\mathbf{x}}_i), ...,$ $t_k(\underline{\mathbf{x}}_i)$*, where* $\underline{\mathbf{x}}_i$ *are the set of variables which are free in* t_j *(if any), let* $\underline{\mathbf{y}_i}$ *being a possibly-empty set of auxiliary variables, and let*

$$\mathsf{Colorer}_k(v_i, \underline{\mathbf{c}}|\underline{\mathbf{x}}_i, \underline{\mathbf{y}_i}) \stackrel{def}{=} \bigwedge_{j=1}^{k}(c_j = t_j(\underline{\mathbf{x}}_i)) \wedge \Psi(v_i|\underline{\mathbf{x}}_i, \underline{\mathbf{y}_i}) \quad (10)$$

be a quantifier-free conjunction of literals s.t.

$$\models_{\mathcal{T}} \forall \underline{\mathbf{x}}_i.\ \mathsf{AllDifferent}_k(\{t_1(\underline{\mathbf{x}}_i), ..., t_k(\underline{\mathbf{x}}_i)\}) \quad (11)$$

$$\Psi(v_i|\underline{\mathbf{x}}_i, \underline{\mathbf{y}_i}) \models_{\mathcal{T}} \bigvee_{j=1}^{k}(v_i = t_j(\underline{\mathbf{x}}_i)) \quad (12)$$

there exist k $\mathcal{T}$*-interpretations* $\{\mathcal{I}_{i,1}, ..., \mathcal{I}_{i,k}\}$ *s.t.* (13)

for every $j \in [1..k]$, $\langle c_j \rangle^{\mathcal{I}_{i,1}} = \langle c_j \rangle^{\mathcal{I}_{i,2}} = ... = \langle c_j \rangle^{\mathcal{I}_{i,k}}$, *and*
for every $j \in [1..k]$, $\mathcal{I}_{i,j} \models_{\mathcal{T}} \mathsf{Colorer}_k(v_i, \underline{\mathbf{c}}|\underline{\mathbf{x}}_i, \underline{\mathbf{y}_i}) \wedge (v_i = t_j(\underline{\mathbf{x}}_i))$.

Importantly, if $t_1, .., t_k$ *are closed terms, then* (13) *reduces to he following:*

there exist k $\mathcal{T}$*-interpretations* $\{\mathcal{I}_{i,1}, ..., \mathcal{I}_{i,k}\}$ *s.t.* (14)
for every $j \in [1..k]$, $\mathcal{I}_{i,j} \models_{\mathcal{T}} \mathsf{Colorer}_k(v_i, \underline{\mathbf{c}}|\underline{\mathbf{y}_i}) \wedge (v_i = t_j)$.

Then $\mathsf{Colorer}_k(v_i, \underline{\mathbf{c}}|\underline{\mathbf{x}}_i, \underline{\mathbf{y}_i})$ *is a* k*-colorer for* $\mathcal{T}$*.*

Proof. By (11), $\bigwedge_{j=1}^{k}(c_j = t_j(\underline{\mathbf{x}}_i)) \models_{\mathcal{T}} \mathsf{AllDifferent}_k(\underline{\mathbf{c}})$, s.t. (1) holds. By (10) and (12), $\mathsf{Colorer}_k(v_i, \underline{\mathbf{c}}|\underline{\mathbf{x}}_i, \underline{\mathbf{y}_i})$ verifies (2). By (10) and (13) we have that (3) holds. □

Theories of Arithmetic. We use Proposition 1–where $t_1, ..., t_k$ are numerical constants–to prove the k-colorability of (various signature-restriction fragments of) the theories of arithmetic.

Example 3 ($\mathcal{A}^{\{\geq,=\}}(\mathbb{Z})$, $\mathcal{LA}(\mathbb{Z})$, $\mathcal{NLA}(\mathbb{Z})$). Let $\mathcal{A}^{\{\geq,=\}}(\mathbb{Z})$ be the basic theory of integers under successor [20,21], that is, whose atoms are in the form $(s_1 \odot s_2)$, where $\odot \in \{\geq, =\}$ and s_1, s_2 are variables or positive numerical constants.

Then $\mathcal{A}^{\{\geq,=\}}(\mathbb{Z})$ is 3-colorable, because we can define a 3-colorer like that of (4) in Example 1. (Notice that this is an instance of Proposition 1.) $\mathcal{A}^{\{\geq,=\}}(\mathbb{Z})$ is a signature-restriction fragment of $\mathcal{LA}(\mathbb{Z})$ and $\mathcal{NLA}(\mathbb{Z})$ (see e.g. [24]), which are then 3-colorable by Proposition 2. Therefore, $\mathcal{T}$-solving for all these theories is NP-hard by Theorem 1.[4] ◇

Notice that conjunctions of only *positive* equalities and inequalities in the form $(s_1 \odot s_2)$, without negated literals, are instead well-known to be solvable in polynomial time (see e.g. [2,18]). Notice also that, on the rational domain, the corresponding theories $\mathcal{A}^{\{\geq,=\}}(\mathbb{Q})$ and $\mathcal{LA}(\mathbb{Q})$ are convex and hence they are not colorable by Property 1. In fact, $\mathcal{T}$-solving for such theories is notoriously in P [10].

Example 4 ($\mathcal{NLA}(\mathbb{R})^{\setminus\{\geq,>\}}$,$\mathcal{NLA}(\mathbb{R})$). We consider $\mathcal{NLA}(\mathbb{R})^{\setminus\{\geq,>\}}$, the signature-restriction fragment of the non-linear arithmetic over the reals ($\mathcal{NLA}(\mathbb{R})$) without inequality symbols $\{\geq,\leq\}$. As an instance of Proposition 1, we show that $\mathcal{NLA}(\mathbb{R})^{\setminus\{\geq,>\}}$ is 3-colorable, because we can define, e.g., $k \stackrel{\text{def}}{=} 3$, $\underline{\mathbf{y}} \stackrel{\text{def}}{=} \emptyset$, and

$$\mathsf{Colorer}_3(v_i, c_1, c_2, c_3) \stackrel{\text{def}}{=} \begin{pmatrix} (c_1 = -1) \wedge (c_2 = 0) \wedge (c_3 = 1) \wedge \\ (v_i \cdot (v_i - 1) \cdot (v_i + 1) = 0) \end{pmatrix}.$$

By Proposition 1, it is straightforward to see that $\mathsf{Colorer}_3(v_i, c_1, c_2, c_3)$ verifies (1), (2) and (3), with $\langle c_1\rangle^{\mathcal{I}_{i,j}} \stackrel{\text{def}}{=} -1, \langle c_2\rangle^{\mathcal{I}_{i,j}} \stackrel{\text{def}}{=} 0, \langle c_3\rangle^{\mathcal{I}_{i,j}} \stackrel{\text{def}}{=} 1$, and $\langle v_i\rangle^{\mathcal{I}_{i,j}} \stackrel{\text{def}}{=} \langle c_j\rangle^{\mathcal{I}_{i,j}}$ s.t. $j \in [1..3]$. Then by Proposition 2 the full $\mathcal{NLA}(\mathbb{R})$ is 3-colorable, so that $\mathcal{T}$-solving for both theories is NP-hard by Theorem 1. ◇

4.2 Exploiting Finite Domains of Fixed Size

Proposition 2. *Let $\mathcal{T}$ be some theory with finite domain of fixed size $k \geq 3$. Then $\mathsf{Colorer}_k(v_i, \underline{\mathbf{c}}) \stackrel{\text{def}}{=} \mathsf{AllDifferent}_k(\underline{\mathbf{c}})$ is a k-colorer for $\mathcal{T}$.*

Proof. Let $\underline{\mathbf{c}} \stackrel{\text{def}}{=} \{c_1, ..., c_k\}$. Since the domain of $\mathcal{T}$ has fixed size $k \geq 3$, we have:

$$\mathsf{AllDifferent}_k(\underline{\mathbf{c}}) \not\models_{\mathcal{T}} \bot \quad (15)$$

$$\mathsf{AllDifferent}_{k+1}(\underline{\mathbf{c}} \cup \{v_i\}) \models_{\mathcal{T}} \bot. \quad (16)$$

$\mathsf{AllDifferent}_k(\underline{\mathbf{c}})$ entails itself, so that (1) holds. $\mathsf{AllDifferent}_k(\underline{\mathbf{c}}) \wedge \bigwedge_{j=1}^{k} \neg(v_i = c_j)$ is the same as $\mathsf{AllDifferent}_{k+1}(\underline{\mathbf{c}} \cup \{v_i\})$ which is $\mathcal{T}$-unsatisfiable by (16), so that $\mathsf{AllDifferent}_k(\underline{\mathbf{c}}) \models_{\mathcal{T}} \bigvee_{j=1}^{k}(v_i = c_j)$. Hence (2) holds. By (15) there exists some $\mathcal{T}$-interpretation $\mathcal{I}$ s.t. $\mathcal{I} \models_{\mathcal{T}} \mathsf{AllDifferent}_k(\underline{\mathbf{c}})$. For every $j \in [1..k]$ we build an extension $\mathcal{I}_{i,j}$ of $\mathcal{I}$ with the same domain s.t. $\langle c_1\rangle^{\mathcal{I}_{i,j}} \stackrel{\text{def}}{=} \langle c_1\rangle^{\mathcal{I}}, ..., \langle c_k\rangle^{\mathcal{I}_{i,j}} \stackrel{\text{def}}{=} \langle c_k\rangle^{\mathcal{I}}$, and $\langle v_i\rangle^{\mathcal{I}_{i,j}} \stackrel{\text{def}}{=} \langle c_j\rangle^{\mathcal{I}}$. Hence (3) holds. □

[4] Notice that $\mathcal{NLA}(\mathbb{Z})$-solving is undecidable.

Theories of Fixed-Width Bit-Vectors and Floating-Point Arithmetic. We prove the k-colorability of (the signature-restriction fragments of) the theories of Fixed-width Bit-vectors and Floating-point Arithmetic by instantiating Proposition 2.

Example 5 ($\mathcal{BV}_w$, $w{>}1$). Let $\mathcal{BV}_w^{\{=\}}$ be the simplest possible signature-restriction fragment of the fixed-width bit-vectors theory with equality $=$ and width $w{>}1$, with no interpreted constant, function or predicate symbol in its signature. Then by Proposition 2, $\mathcal{BV}_w^{\{=\}}$ is k-colorable, where $k = 2^w$. Hence, by Property 2 all theories $\mathcal{BV}_w^*$ obtained by augmenting the signature of $\mathcal{BV}_w^{\{=\}}$ with various combinations of interpreted constants (e.g. $\mathsf{bv_w\_0...00}$, $\mathsf{bv_w\_0...01}$,...), functions (e.g. $\mathsf{bv_w\_and}$, $\mathsf{bv_w\_or}$,...) and predicates (e.g. $\mathsf{bv_w\_}\geq$,...)–are k-colorable with $k = 2^w$. Hence, when $w{>}1$, by Theorem 1, $\mathcal{T}$-solving is NP-hard for all such theories. ◇

[7] shows that the $\mathcal{T}$-satisfiability of quantifier-free conjunctions of atoms for the fragment of $\mathcal{BV}$ involving only concatenation and partition of words is in P. Notice however that neither Example 5 contradicts the results in [7], nor Example 5 plus [7] build a proof of $P = NP$, because the polynomial procedure in [7] does not admit *negative equalities* $\neg(v_i = v_i')$ in the conjunction.

Example 6 ($\mathcal{FPA}_{e,s}$). Let $\mathcal{FPA}_{e,s}$ be the theory of floating-point arithmetic s.t. $e \geq 1$ and $s \geq 1$ are the number of available bits for the exponent and the significant respectively [22]. (E.g., $\mathcal{FPA}_{11,53}$ represents the binary64 format of IEEE 754-2008 [22].) As with Example 5, let $\mathcal{FPA}_{e,s}^{=}$ be the simplest possible signature-restriction fragment of $\mathcal{FPA}_{e,s}^{=}$ with equality $=$,[5] with no interpreted constant, function and predicate symbol in its signature. Then by Proposition 2, $\mathcal{FPA}_{e,s}^{=}$ is k-colorable, where $k = 2^{e+s}$. Hence, by Property 2, all theories $\mathcal{FPA}_{e,s}^{*}$ obtained by augmenting the signature of $\mathcal{FPA}_{e,s}^{=}$ with various combinations of interpreted constants, functions or predicates are k-colorable with $k \geq 4$, so that $\mathcal{T}$-solving is NP-hard. ◇

4.3 Dealing with Collection Datatypes

A class of theories of big interest in SMT-based formal verification are these describing collection datatypes (see e.g. [6,12])–e.g., lists, arrays, sets, etc. In general these are "families" of theories, each being a combination of a "basic" theory (e.g., the basic theory of lists) with one or more theories describing the elements or the indexes of the datatype. In what follows we consider the basic theories, where elements are represented by generic variables representing values in some infinite domain.

One potential problem if finding k-colorers for most of these "basic" theories is that neither we have interpreted constants in the domain of the elements, so that we cannot apply Proposition 1 as we did with arithmetical theories, nor

[5] Here "=" is the equality symbol and it is not the $\mathcal{FPA}_{e,s}$-specific symbol "==", see [22].

we have any information on the size of the domain of the elements, so that we cannot apply Proposition 2.

We analyze different potential scenarios. One first scenario is where we have at least one "structural" interpreted constant–e.g., that representing the empty collection–plus some function symbols, which we can use to build $k \geq 3$ closed terms $t_1, ..., t_k$ and then use the schema of Proposition 1 to build a k-colorer.

Theories of Lists. The above scenario is illustrated in the next example.

Example 7 ($\mathcal{L}^+$). Let $\mathcal{L}$ be the simplest theory of lists of generic elements, with the signature $\Sigma \stackrel{\text{def}}{=} \{\mathsf{nil}, \mathsf{car}(\cdot), \mathsf{cdr}(\cdot), \mathsf{cons}(\cdot,\cdot)\}$ and described by the axioms:

$$\forall xy.(\mathsf{car}(\mathsf{cons}(x,y) = x)),\ \forall xy.(\mathsf{cdr}(\mathsf{cons}(x,y) = y)), \tag{17}$$
$$\forall xy.(\neg(\mathsf{cons}(x,y) = \mathsf{nil})),\ \forall x.(\neg(x = \mathsf{nil}) \rightarrow (\mathsf{cons}(\mathsf{car}(x), \mathsf{cdr}(x)) = x)), \tag{18}$$

and let $\mathcal{L}^+$ be $\mathcal{L}$ enriched by the axioms

$$(\mathsf{car}(\mathsf{nil}) = \mathsf{nil}),\quad (\mathsf{cdr}(\mathsf{nil}) = \mathsf{nil}). \tag{19}$$

$\mathcal{L}^+$-solving is NP-complete whilst $\mathcal{L}$-solving is in P [17]. A more general theory of lists, which has $\mathcal{L}^+$ as a signature-restriction fragment, is described in [6,12]. Following Proposition 1, we prove that $\mathcal{L}^+$ is 4-colorable, by setting $k \stackrel{\text{def}}{=} 4$, $\underline{\mathbf{y}} \stackrel{\text{def}}{=} \{x_1, x_2, y_1, y_2\}$,

$$\mathsf{Colorer}_4(v_i, c_{11}, c_{21}, c_{12}, c_{22} | x_1, x_2, y_1, y_2) \stackrel{\text{def}}{=} \tag{20}$$
$$\left(\begin{array}{l} (c_{11} = \mathsf{cons}(\mathsf{nil},\mathsf{nil})) \wedge (c_{21} = \mathsf{cons}(\mathsf{cons}(\mathsf{nil},\mathsf{nil}),\mathsf{nil})) \wedge \\ (c_{12} = \mathsf{cons}(\mathsf{nil},\mathsf{cons}(\mathsf{nil},\mathsf{nil}))) \wedge (c_{22} = \mathsf{cons}(\mathsf{cons}(\mathsf{nil},\mathsf{nil}),\mathsf{cons}(\mathsf{nil},\mathsf{nil}))) \wedge \\ \bigwedge_{i=1}^{2} ((\mathsf{car}(x_i) = \mathsf{car}(y_i)) \wedge (\mathsf{cdr}(x_i) = \mathsf{cdr}(y_i)) \wedge \neg(x_i = y_i)) \wedge \\ (v_i = \mathsf{cons}(x_1, x_2)). \end{array}\right)$$

To prove (11) we notice that we can deduce $\neg(\mathsf{cons}(\mathsf{nil},\mathsf{nil}) = \mathsf{nil})$ from (18), so that, by construction, all the c_i's are pairwise different. Let $\Psi(v_i \underline{\mathbf{y}_i})$ be the formula given by the last two rows in (20), so that (20) matches the definition in Proposition 1. Then we derive (12) from the following observation [17], with $i \in \{1,2\}$:

$$((\mathsf{car}(x_i) = \mathsf{car}(y_i)) \wedge (\mathsf{cdr}(x_i) = \mathsf{cdr}(y_i)) \wedge \neg(x_i = y_i)) \tag{21}$$
$$\models_{\mathcal{L}^+} (x_i = \mathsf{nil}) \vee (x_i = \mathsf{cons}(\mathsf{nil},\mathsf{nil})),$$

which derives from the fact that (18) and (19) imply that either $(x_i = \mathsf{nil})$ or $(y_i = \mathsf{nil})$ must hold. Therefore $v_i \stackrel{\text{def}}{=} \mathsf{cons}(x_1, x_2)$ can consistently assume one and only one of the values $c_{11}, ..., c_{22}$ in the first two rows in (20).

To prove (14), since the c_is are closed, we deterministically define each $\mathcal{I}_{i,j}$'s using the standard interpretation of nil, cons, car, and cdr: $\langle c_{11} \rangle^{\mathcal{I}_{i,j}} \stackrel{\text{def}}{=} (\mathsf{NIL}.\mathsf{NIL})$, $\langle c_{21} \rangle^{\mathcal{I}_{i,j}} \stackrel{\text{def}}{=} ((\mathsf{NIL}.\mathsf{NIL}).\mathsf{NIL}), \ldots \langle v_i \rangle^{\mathcal{I}_{i,j}} \stackrel{\text{def}}{=} \langle c_j \rangle^{\mathcal{I}_{i,j}}$, checking that, for every $j \in [1..k]$,

$$\mathcal{I}_{i,j} \models_{\mathcal{L}^+} \mathsf{Colorer}_4(v_i, c_{11}, c_{21}, c_{12}, c_{22} | x_1, x_2, y_1, y_2) \wedge (v_i = c_j).$$

Thus $\mathcal{L}^+$-solving is NP-hard by Theorem 1, so that also the more general theory described in [6,12] is NP-hard. ◇

Remark 3. The k-colorer (20) was produced along the following heuristic process.

1. Look for an entailment in the form: $\mu_1(x_1, \underline{\mathbf{y}_1}) \models_{\mathcal{T}} (x_1 = t_1) \vee (x_1 = t_2)$, s.t. t_1, t_2 are closed terms representing distinct values in the domain (21).
2. Define $(v_i = \mathsf{cons}(x_1, x_2))$ and $(c_{r_1 r_2} = \mathsf{cons}(t_{r_1}, t_{r_2}))$, s.t. $r_1, r_2 \in \{1, 2\}$
3. Define the k-colorer as

$$\bigwedge_{i \in \{1,2\}} \mu_i(x_i, \underline{\mathbf{y}_i}) \wedge \bigwedge_{r_1, r_2 \in \{1,2\}} (c_{r_1 r_2} = \mathsf{cons}(t_{r_1}, t_{r_2})) \wedge (v_i = \mathsf{cons}(x_1, x_2)).$$

4. Check (11), (12), (14).

Notice that the only non-obvious step is 1, the other come out nearly deterministically.

Theories of Finite Sets. Another scenario is where we cannot use interpreted constants to build closed terms, but we can build k non-closed terms $t_1(\underline{\mathbf{x}_i}), ..., t_k(\underline{\mathbf{x}_i})$ which match the requirements of Proposition 1 anyway, which allows to build a k-colorer. This scenario is illustrated in the next example.

Example 8. Let $\mathcal{S}$ be the theory of finite sets as defined, e.g., in [6,12].[6] Let $\mathcal{S}^{\{\subseteq, \{\}\}}$ be the signature-restriction fragment of the $\mathcal{S}$ which considers only the subset and the enumerator operators $\{\subseteq, \{\}\}$. We show that $\mathcal{S}^{\{\subseteq, \{\}\}}$ is 4-colorable by Proposition 1.

In fact, consider the following set of literals:

$$\mathsf{Colorer}_4(v_i, \underline{\mathbf{c}} | y_1, y_2) \stackrel{\text{def}}{=} \begin{pmatrix} (c_1 = \{y_1, y_2\}) & \wedge (c_2 = \{y_1\}) & \wedge \\ (c_3 = \{y_2\}) & \wedge (c_4 = \{\}) & \wedge \\ \neg (y_1 = y_2) & \wedge (v_i \subseteq c_1) & \end{pmatrix}. \tag{22}$$

(22) is a 4-colorer. It is easy to see from the semantics of $\{\subseteq, \{\}\}$ that (11) and (12) hold. Let Y_1, Y_2 s.t. $Y_1 \neq Y_2$ be two domain elements so that we can set $\langle y_r \rangle^{\mathcal{I}_{i,j}} \stackrel{\text{def}}{=} Y_r$ for every $r \in [1..2]$ and $j \in [1..k]$. Then, for every $j \in [1..k]$, we define $\mathcal{I}_{i,j}$ s.t. $\langle c_1 \rangle^{\mathcal{I}_{i,j}} \stackrel{\text{def}}{=} \{Y_1, Y_2\}$, $\langle c_2 \rangle^{\mathcal{I}_{i,j}} \stackrel{\text{def}}{=} \{Y_1\}$, $\langle c_3 \rangle^{\mathcal{I}_{i,j}} \stackrel{\text{def}}{=} \{Y_2\}$, $\langle c_4 \rangle^{\mathcal{I}_{i,j}} \stackrel{\text{def}}{=} \{\}$, $\langle v_i \rangle^{\mathcal{I}_{i,j}} \stackrel{\text{def}}{=} \langle c_j \rangle^{\mathcal{I}_{i,j}}$. Then $\mathcal{I}_{i,1}, ..., \mathcal{I}_{i,k}$ verify (13). ◇

In this case the k-colorer (22) was really immediate to build, upon the observation that the operator $\subseteq$ can produce 4 distinct subsets of a 2-element set.

[6] $\mathcal{S}$ includes the operators $\{\{...\}), (\cdot \subseteq \cdot), (\cdot \cup \cdot), (\cdot \cap \cdot), (\cdot \setminus \cdot), (\cdot \mathcal{P} \cdot), |\cdot|, (\cdot \in \cdot)\}$, following their standard semantics. We refer the reader to [6,12] for a precise description of the theory.

Theories of Arrays. In the following case we cannot apply Proposition 1, so that we apply Definition 2 directly.

Example 9. ($\mathcal{AR}$). Let $\mathcal{AR}$ be the theory of arrays of generic elements and indexes, with the signature $\Sigma \stackrel{\text{def}}{=} \{\cdot[\cdot], \cdot\langle\cdot \leftarrow \cdot\rangle\}$ [7] and described by the axioms:

$$\forall Aijv.\ ((i = j) \rightarrow (A\langle i \leftarrow v\rangle[j] = v), \tag{23}$$
$$\forall Aijv.\ (\neg(i = j) \rightarrow (A\langle i \leftarrow v\rangle[j] = A[j]), \tag{24}$$
$$\forall AB.\ ((\forall i.\ A[i] = B[i]) \rightarrow (A = B)). \tag{25}$$

$\mathcal{AR}$ is 3-colorable, because we can define, e.g., $k \stackrel{\text{def}}{=} 3, \underline{\mathbf{y}} \stackrel{\text{def}}{=} \{A_1, ..., A_4, i_1, ..., i_3\}$ and

$$\mathsf{Colorer}_3(v_i, c_1, c_2, c_3 | A_1, ..., A_4, i_1, ..., i_3) \stackrel{\text{def}}{=} \begin{pmatrix} \mathsf{AllDifferent}_3(\underline{\mathbf{c}}) & \wedge \\ \neg(i_2 = i_3) & \wedge \\ (A_2 = A_1\langle i_1 \leftarrow c_1\rangle) \wedge \\ (A_3 = A_2\langle i_2 \leftarrow c_2\rangle) \wedge \\ (A_4 = A_3\langle i_3 \leftarrow c_3\rangle) \wedge \\ (v_i = A_4[i_1]) \end{pmatrix} \tag{26}$$

so that obviously (1) holds, and also (2) holds, because $\mathsf{Colorer}_3(v_i, \underline{\mathbf{c}}|\underline{\mathbf{y}})$ entails $(v_i = c_1)$ when $\langle i_1\rangle^{\mathcal{I}} \neq \langle i_3\rangle^{\mathcal{I}}$ and $\langle i_1\rangle^{\mathcal{I}} \neq \langle i_2\rangle^{\mathcal{I}}$, entails $(v_i = c_2)$ when $\langle i_1\rangle^{\mathcal{I}} = \langle i_2\rangle^{\mathcal{I}}$, and entails $(v_i = c_3)$ when $\langle i_1\rangle^{\mathcal{I}} = \langle i_3\rangle^{\mathcal{I}}$. Also (3) holds: given three distinct domain values C_1, C_2, C_3, the $\mathcal{T}$-interpretations $\mathcal{I}_{i,j}$ can be built straightforwardly as follows:

	c_1	c_2	c_3	v_i	i_1	i_2	i_3	A_4	...
$\mathcal{I}_{i,1}$	C_1	C_2	C_3	C_1	1	2	3	$[C_1, C_2, C_3, ...]$	
$\mathcal{I}_{i,2}$	C_1	C_2	C_3	C_2	2	2	3	$[**, C_2, C_3, ...]$	
$\mathcal{I}_{i,3}$	C_1	C_2	C_3	C_3	3	2	3	$[**, C_2, C_3, ...]$	

◇

Notice that in Example 9, $\mathsf{Colorer}_k(v_i, \underline{\mathbf{c}}|\underline{\mathbf{y}}_i)$ uses the auxiliary variables $A_1, ..., A_4$ representing arrays and $i_1, ..., i_3$ representing indexes. The A_2, A_3, A_4, however, are not strictly necessary and can be eliminated by inlining. Notice also that $\mathsf{Colorer}_k(v_i, \underline{\mathbf{c}}|\underline{\mathbf{y}}_i)$ includes explicitly $\mathsf{AllDifferent}_3(\underline{\mathbf{c}})$ because no interpreted constants come into play.

The k-colorer (26) was produced straightforwardly by noticing that the combination of (23) and (24) produces a case-split in the form "*if* $i = j$ *then* $(A\langle i \leftarrow v\rangle[j] = v)$ *else* $(A\langle i \leftarrow v\rangle[j] = A[j])$", which could be reiterated so that to produce a 3-branch decision tree, producing 3 different expressions for the term $A[i_1]$. This could be rewritten into k-colorer by means of some term renaming.

---

[7] We use the following notation: "$A[i]$" (aka "$\mathsf{read}(A, i)$") is the value returned by reading the i-th element of the array A, whilst "$A\langle i \leftarrow v_i\rangle$" (aka "$\mathsf{write}(A, i, v)$") is the array resulting from assigning the value v to the i-th element of array A.

5 k-Colorability vs. Non-Convexity

Although related by Property 1, k-colorability and non-convexity are distinct properties. First, we recall that the non-convexity of a theory $\mathcal{T}$ does not imply the NP-hardness of $\mathcal{T}$-solving. (In [24] we report a simple example.) Second, by Property 1, having domain size ≥ 3 is a strict requirement for proving NP-hardness via colorability, whereas there exist non-convex theories with domain size 2 whose $\mathcal{T}$-solving is NP-Hard. (E.g., the theory $\mathcal{BV}_1$ of bit vectors with fixed width 1, see [24].)

In what follows we introduce a theory with domain size ≥ 3 whose $\mathcal{T}$-solving is NP-hard, which is non-convex and which is not k-colorable for any $k \geq 3$. This shows that not every theory with domain size ≥ 3 can be proven NP-hard by k-colorability. The same example shows also that k-colorability is strictly stronger than non-convexity, even when the theory has domain size ≥ 3.

Example 10. Consider the theory $\mathcal{T}$ with equality whose signature consists in the interpreted constant symbols $\{0, 1, 2, ...\}$ with the standard meaning plus the function symbols $\{\mathsf{and}(\cdot,\cdot), \mathsf{not}(\cdot)\}$ which are interpreted as follows:

$$\langle \mathsf{and}(x,y)\rangle^{\mathcal{I}} \stackrel{\text{def}}{=} \begin{cases} 1 \text{ if } \langle x\rangle^{\mathcal{I}}{>}0 \text{ and } \langle y\rangle^{\mathcal{I}}{>}0 \\ 0 \text{ otherwise} \end{cases}, \quad \langle \mathsf{not}(x)\rangle^{\mathcal{I}} \stackrel{\text{def}}{=} \begin{cases} 0 \text{ if } \langle x\rangle^{\mathcal{I}}{>}0 \\ 1 \text{ otherwise.} \end{cases} \tag{27}$$

(Importantly, the $\geq, >, \leq, <$ predicates are not part of the signature.) $\mathcal{T}$-satisfiability is NP-complete since you can polynomially reduce SAT to it and you can always have a polynomial-size witness for every $\mathcal{T}$-satisfied formula.

Also, as with $\mathcal{BV}_1$, $\mathcal{T}$ is non-convex, because we have:

$$(x_0 = 0) \wedge (\mathsf{and}(x_1, x_2) = 0) \models_{\mathcal{T}} (((x_0 = x_1) \vee (x_0 = x_2)) \tag{28}$$

$$(x_0 = 0) \wedge (\mathsf{and}(x_1, x_2) = 0) \not\models_{\mathcal{T}} (x_0 = x_i) \quad i \in \{1, 2\}. \tag{29}$$

We show that $\mathcal{T}$ is not k-colorable for any $k \geq 3$. We notice that every literal l including v_i must be in one of the following forms (modulo the symmetry of $=$ and and): $(v_i = t)$, $(v_i = \mathsf{not}(t))$, $(v_i = \mathsf{and}(t_1, t_2))$, $(t = t^*(v_i, ...))$, and their negations, where t, t_1, t_2 are generic terms in $\mathcal{T}$ and $t^*(v_i, ...)$ is any term in $\mathcal{T}$ containing v_i. Looking at the above literal forms, we notice that the presence of the subterms $\mathsf{not}(v_i)$ and $\mathsf{and}(v_i, t_2)$ in a term entails either $\langle v_i\rangle^{\mathcal{I}}{>}\langle 0\rangle^{\mathcal{I}}$, or $\langle v_i\rangle^{\mathcal{I}} = \langle 0\rangle^{\mathcal{I}}$ or $\langle v_i\rangle^{\mathcal{I}} \geq \langle 0\rangle^{\mathcal{I}}$, so that one single literal l can express only the following facts about one variable v_i:[8]

(i) for every $\mathcal{T}$-interpretation $\mathcal{I}$ s.t. $\mathcal{I} \models_{\mathcal{T}} l$, $\langle v_i\rangle^{\mathcal{I}} = \langle n\rangle^{\mathcal{I}}$ for some $n \in \{0, 1, 2, 3, ...\}$;
(ii) for every $\mathcal{T}$-interpretation $\mathcal{I}$ s.t. $\mathcal{I} \models_{\mathcal{T}} l$, $\langle v_i\rangle^{\mathcal{I}} \neq \langle n\rangle^{\mathcal{I}}$ for some $n \in \{0, 1, 2, 3, ...\}$;

[8] Whereas (i) and (ii) can be also written as $l \models_{\mathcal{T}} (v_i = n)$ and $l \models_{\mathcal{T}} (v_i \neq n)$, (iii) and (iv) cannot be rewritten as $l \models_{\mathcal{T}} (v_i \geq 0)$ and $l \models_{\mathcal{T}} (v_i{>}0)$ because $\geq$ and $>$ are not part of the signature.

(iii) for every $\mathcal{T}$-interpretation $\mathcal{I}$ s.t. $\mathcal{I} \models_{\mathcal{T}} l$, $\langle v_i \rangle^{\mathcal{I}} \geq \langle 0 \rangle^{\mathcal{I}}$ (equivalent to true);
(iv) for every $\mathcal{T}$-interpretation $\mathcal{I}$ s.t. $\mathcal{I} \models_{\mathcal{T}} l$, $\langle v_i \rangle^{\mathcal{I}} > \langle 0 \rangle^{\mathcal{I}}$ (equivalent to $\langle v_i \rangle^{\mathcal{I}} \neq 0$);
(v) for every $\mathcal{T}$-interpretation $\mathcal{I}$ s.t. $\mathcal{I} \models_{\mathcal{T}} l$, $\langle v_i \rangle^{\mathcal{I}} = \langle v_i \rangle^{\mathcal{I}}$ (equivalent to true);
(vi) for every $\mathcal{T}$-interpretation $\mathcal{I}$ s.t. $\mathcal{I} \models_{\mathcal{T}} l$, $\langle v_i \rangle^{\mathcal{I}} \neq \langle v_i \rangle^{\mathcal{I}}$ (equivalent to false).

Thus, for $k \geq 3$, no finite conjunction of $\mathcal{T}$-literals $\mathsf{Colorer}_k(v_i, \underline{\mathbf{c}}|\underline{\mathbf{y}_i})$ complying with (1) and (3) can also comply with (2). ◇

6 Colorable Theories without Equality

In previous sections we have restricted our interest to theories *with equality*. In this section we extend the technique by dropping this restriction. The following definition extends Definition 2 to the case of general theories.

Definition 3 (k-Colorer, k-Colored Theory). *Let $\mathcal{T}$ be some theory and k be some integer value s.t. $k \geq 2$. Let v_i be a variable, called* **vertex variable**, *(implicitly) denoting the i-th vertex in an un-directed graph; let $\underline{\mathbf{c}} \stackrel{\text{def}}{=} \{c_1, .., c_k\}$ be a set of variables, called* **color variables**, *denoting the set of colors; let $\underline{\mathbf{y}_i} \stackrel{\text{def}}{=} \{y_{i1}, ..., y_{il}\}$ denote a possibly-empty set of variables, which is indexed with the same index i of the vertex variable v_i. We call* **k-colorer** *for $\mathcal{T}$, namely $\mathsf{Colorer}_k(v_i, \underline{\mathbf{c}}|\underline{\mathbf{y}_i})$, a finite quantifier-free conjunction of $\mathcal{T}$-literals (cube) over v_i, $\underline{\mathbf{c}}$ and $\underline{\mathbf{y}_i}$ which verify the following properties:*

- *For every $\mathcal{T}$-intepretation $\mathcal{I}$, if $\mathcal{I} \models_{\mathcal{T}} \mathsf{Colorer}_k(v_i, \underline{\mathbf{c}}|\underline{\mathbf{y}_i})$, then:*

$$\text{for every } j, j' \in [1..k] \text{ s.t. } j \neq j', \quad \langle c_j \rangle^{\mathcal{I}} \neq \langle c_{j'} \rangle^{\mathcal{I}}, \tag{30}$$

$$\text{for some } j \in [1..k], \quad \langle v \rangle^{\mathcal{I}} = \langle c_j \rangle^{\mathcal{I}}, \tag{31}$$

- *There exist k $\mathcal{T}$-interpretations $\{\mathcal{I}_{i,1}, ..., \mathcal{I}_{i,k}\}$ s.t.*

$$\text{for every } j \in [1..k], \quad \langle c_j \rangle^{\mathcal{I}_{i,1}} = \langle c_j \rangle^{\mathcal{I}_{i,2}} = ... = \langle c_j \rangle^{\mathcal{I}_{i,k}}, \text{ and} \tag{32}$$

$$\text{for every } j \in [1..k], \quad \begin{cases} \langle v \rangle^{\mathcal{I}_{i,j}} = \langle c_j \rangle^{\mathcal{I}_{i,j}} \text{ and} \\ \mathcal{I}_{i,j} \models_{\mathcal{T}} \mathsf{Colorer}_k(v_i, \underline{\mathbf{c}}|\underline{\mathbf{y}_i}). \end{cases}$$

We say that $\mathcal{T}$ is **k-colorable** *iff it has a k-colorer.*

Notice that if $\mathcal{T}$ is a theory with equality, then Definitions 2 and 3 are equivalent.

Definition 4. *We say that a theory $\mathcal{T}$* **emulates equality** *[resp.* **disequality***] if and only if there exists a finite quantifier-free conjunction of $\mathcal{T}$-literals $\mathsf{Eq}(x_1, x_2)$ [resp. $\mathsf{Neq}(x_1, x_2)$] such that, for every $\mathcal{T}$-interpretation $\mathcal{I}$, $\mathcal{I} \models_{\mathcal{T}} \mathsf{Eq}(x_1, x_2)$ [resp. $\mathcal{I} \models_{\mathcal{T}} \mathsf{Neq}(x_1, x_2)$] if and only if $\langle x_1 \rangle^{\mathcal{I}} = \langle x_2 \rangle^{\mathcal{I}}$ [resp. $\langle x_1 \rangle^{\mathcal{I}} \neq \langle x_2 \rangle^{\mathcal{I}}$].*

Obviously every theory $\mathcal{T}$ with equality emulates both equality and disequality, with $\mathsf{Eq}(x_1, x_2) \stackrel{\text{def}}{=} (x_1 = x_2)$ and $\mathsf{Neq}(x_1, x_2) \stackrel{\text{def}}{=} \neg(x_1 = x_2)$.

Theorem 2. *If a theory $\mathcal{T}$ is k-colorable for some $k \geq 3$ and $\mathcal{T}$ emulates equality and inequality, then the problem of deciding the $\mathcal{T}$-satisfiability of a finite conjunction of quantifier-free $\mathcal{T}$-literals is $\mathcal{T}$-satisfiable is NP-hard.*

Proof. Identical to that of Theorem 1, referring to Definition 3 instead of Definition 2 and substituting every positive equality in the form $(x_1 = x_2)$ with $\mathsf{Eq}(x_1, x_2)$ and every negative equality in the form $\neg(x_1 = x_2)$ with $\mathsf{Neq}(x_1, x_2)$. □

Example 11. Let $\mathcal{NLA}(\mathbb{R})^{\setminus\{=\}}$ be the signature-restriction fragment of $\mathcal{NLA}(\mathbb{R})$ without equality. We notice that $\mathcal{NLA}(\mathbb{R})^{\setminus\{=\}}$ emulates both equality and inequality:

$$\mathsf{Eq}(x_1, x_2) \stackrel{\text{def}}{=} (x_1 \geq x_2) \wedge (x_2 \geq x_1) \quad (33)$$

$$\mathsf{Neq}(x_1, x_2) \stackrel{\text{def}}{=} ((x_1 - x_2) * (x_1 - x_2) > 0). \quad (34)$$

$\mathcal{T}$ is 3-colorable because, like in Example 4, we can define, e.g., $k \stackrel{\text{def}}{=} 3$, $\underline{\mathbf{y}} \stackrel{\text{def}}{=} \emptyset$, and

$$\mathsf{Colorer}_3(v_i, c_1, c_2, c_3) \stackrel{\text{def}}{=} \mathsf{Eq}(c_1, -1) \wedge \mathsf{Eq}(c_2, 0) \wedge \mathsf{Eq}(c_3, 1) \wedge \mathsf{Eq}(v_1 * (v_2 - 1) * (v_1 + 1), 0).$$

Like in Example 4, it is straighforward to see that $\mathsf{Colorer}_3(v, c_1, c_2, c_3)$ verifies (30), (31) and (32), with $\langle c_1 \rangle^{\mathcal{I}_{i,j}} \stackrel{\text{def}}{=} -1$, $\langle c_2 \rangle^{\mathcal{I}_{i,j}} \stackrel{\text{def}}{=} 0$, $\langle c_3 \rangle^{\mathcal{I}_{i,j}} \stackrel{\text{def}}{=} 1$, and $\langle v_i \rangle^{\mathcal{I}_{i,j}} \stackrel{\text{def}}{=} \langle c_j \rangle^{\mathcal{I}_{i,j}}$ for every $j \in [1..3]$. Thus $\mathcal{NLA}(\mathbb{R})^{\setminus\{=\}}$-solving is NP-hard by Theorem 2. ◇

7 Open Issues, Ongoing and Future Work

We believe that our framework can be generalized along the following directions, which we are currently working on: (i) adopt some more general notion of fragment, so that to extend the range of applicability of Property 2; (ii) extend the applicability of our technique for the case of theories without equality by providing a more general definition of $\mathsf{Eq}(.,.)$ and $\mathsf{Neq}(.,.)$ enriched with auxiliary variables –or uninterpreted function/predicate symbols– adapting Theorem 2 accordingly; (iii) extend $\mathsf{Colorer}_k(v_i, \underline{\mathbf{c}}|\underline{\mathbf{y}}_i)$ so that to use also uninterpreted function/predicate symbols as auxiliary symbols $\underline{\mathbf{y}}_i$; (iv) to overcome the restriction of domain size ≥ 3, extend $\mathsf{Colorer}_k(v_i, \underline{\mathbf{c}}|\underline{\mathbf{y}}_i)$ to use pairs of variables $\underline{\mathbf{v}}_i$ $\underline{\mathbf{c}}_1, .., \underline{\mathbf{c}}_k$ instead of single variables to encode vertexes and colors, including ad hoc $\mathsf{Neq}(.,.)$ functions.

The above work should be run in parallel and interleaved with an extensive exploration of the pool of available NP-hard theories, proving the k-colorability of as many theories/fragments as possible. To this extent, we would like to investigate the boundary of k-colorability, looking for theories of domain size ≥ 3 which are not k-colorable.

References

1. Barrett, C.W., Nieuwenhuis, R., Oliveras, A., Tinelli, C.: Splitting on demand in SAT modulo theories. In: Hermann, M., Voronkov, A. (eds.) LPAR 2006. LNCS (LNAI), vol. 4246, pp. 512–526. Springer, Heidelberg (2006)
2. Barrett, C., Sebastiani, R., Seshia, S.A., Tinelli, C.: Satisfiability Modulo Theories, chapter 26, pp. 825–885. IOS Press, Amsterdam (2009)

3. Berman, L.: The complexity of logical theories. Theor. Comput. Sci. **11**, 71 (1980)
4. Bozzano, M., Bruttomesso, R., Cimatti, A., Junttila, T.A., van Rossum, P., Schulz, S., Sebastiani, R.: Mathsat: tight integration of sat and mathematical decision procedures. J. Autom. Reasoning **35**(1–3), 265–293 (2005)
5. Bradley, A., Manna, Z.: The Calculus of Computation: Decision Procedures with Applications to Verification. Springer, Heidelberg (2010)
6. Bruun, H., Damm, F., Dawes, J., Hansen, B., Larsen, P., Parkin, G., Plat, N., Toetenel, H.: A formal definition of vdm-sl. Technical report, University of Leicester (1998)
7. Cyrluk, D., Möller, M.O., Rueß, H.: An efficient decision procedure for the theory of fixed-sized bit-vectors. In: Grumberg, O. (ed.) Computer Aided Verification. LNCS, vol. 1254, pp. 60–71. Springer, Heidelberg (1997)
8. Fröhlich, A., Kovásznai, G., Biere, A.: More on the complexity of quantifier-free fixed-size bit-vector logics with binary encoding. In: Bulatov, A.A., Shur, A.M. (eds.) CSR 2013. LNCS, vol. 7913, pp. 378–390. Springer, Heidelberg (2013)
9. Garey, M.R., Johnson, D.S.: Computers and Intractability. Freeman and Company, New York (1979)
10. Karmakar, N.: A new polynomial-time algorithm for linear programming. Combinatorica **4**(4), 373 (1984)
11. Kovásznai, G., Fröhlich, A., Biere, A.: On the complexity of fixed-size bit-vector logics with binary encoded bit-width. In: Proceedings of the SMT (2012)
12. Kroening, D., Rümmer, P., Weissenbacher, G.: A proposal for a theory of finite sets, lists, and maps for the SMT-LIB standard. In: Proceedings of the SMT Workshop (2007)
13. Kroening, D., Strichman, O.: Decision Procedures. Springer, Heidelberg (2008)
14. Kuncak, V., Rinard, M.: Towards efficient satisfiability checking for Boolean algebra with Presburger arithmetic. In: Pfenning, F. (ed.) CADE 2007. LNCS (LNAI), vol. 4603, pp. 215–230. Springer, Heidelberg (2007)
15. Lahiri, S.K., Musuvathi, M.: An efficient decision procedure for UTVPI constraints. In: Gramlich, B. (ed.) FroCos 2005. LNCS (LNAI), vol. 3717, pp. 168–183. Springer, Heidelberg (2005)
16. Nelson, G., Oppen, D.: Simplification by cooperating decision procedures. ACM Trans. Program. Lang. Syst. **1**(2), 245–257 (1979)
17. Nelson, G., Oppen, D.: Fast decision procedures based on congruence closure. J. ACM **27**(2), 356–364 (1980)
18. Nieuwenhuis, R., Oliveras, A.: DPLL(T) with exhaustive theory propagation and its application to difference logic. In: Etessami, K., Rajamani, S.K. (eds.) CAV 2005. LNCS, vol. 3576, pp. 321–334. Springer, Heidelberg (2005)
19. Oppen, D.: Reasoning about recursively defined data structures. J. ACM **27**(3), 403–411 (1980)
20. Oppen, D.C.: Complexity, convexity and combinations of theories. Theor. Comput. Sci. **12**, 291–302 (1980)
21. Pratt, V.R.: Two Easy Theories Whose Combination is Hard. Technical report, M.I.T. (1977)
22. Rümmer, P., Wahl, T.: An SMT-LIB theory of binary floating-point arithmetic. In: Proceedings of the SMT (2010)
23. Sebastiani, R.: Lazy satisfiability modulo theories. J. Satisfiability Boolean Model. Comput. (JSAT) **3**(3–4), 141–224 (2007)

24. Sebastiani, R.: Colors Make Theories hard. Technical Report DISI-16-001, DISI, University of Trento, Extended version. http://disi.unitn.it/rseba/ijcar16extended.pdf
25. Shostak, R.: An algorithm for reasoning about equality. In: Proceedings of the 7th International Joint Conference on Artificial Intelligence, pp. 526–527 (1977)
26. Shostak, R.: A pratical decision procedure for arithmetic with function symbols. J. ACM **26**(2), 351–360 (1979)

Rewriting

Nominal Confluence Tool

Takahito Aoto[1(✉)] and Kentaro Kikuchi[2]

[1] Faculty of Engineering, Niigata University, Niigata, Japan
aoto@ie.niigata-u.ac.jp
[2] RIEC, Tohoku University, Sendai, Japan
kentaro@nue.riec.tohoku.ac.jp

Abstract. Nominal rewriting is a framework of higher-order rewriting introduced in (Fernández, Gabbay & Mackie, 2004; Fernández & Gabbay, 2007). Recently, (Suzuki et al., 2015) revisited confluence of nominal rewriting in the light of feasibility. We report on an implementation of a confluence tool for (non-closed) nominal rewriting, based on (Suzuki et al., 2015) and succeeding studies.

Keywords: Confluence · Nominal rewriting · Automation · Variable binding · Higher-order rewriting

1 Introduction

Rewriting captures various computational aspects in equational reasoning [4]. *Higher-order rewriting* deals with rewriting of expressions with higher-order functions and variable binding. Various formalisms for higher-order rewriting have been considered e.g. [12,14]. *Nominal rewriting* [6,7] is a formalism of higher-order rewriting, based on the nominal approach for terms and unification [9,16,21].

Confluence is a central property in rewriting [4]. Confluence tools for various rewriting formalisms have been developed [2,10,17,22], and a yearly competition for confluence tools has emerged from 2012 [1]. Some basic confluence results for nominal rewriting have been mentioned in [6]. Recently, these results have been revisited and extended by the authors [11,19,20] in the light of feasibility and more-in-depth analysis. In this paper, we report on a confluence tool for nominal rewriting based on those confluence studies.

2 Preliminaries

In this section, we recall basic notions and fix notations on nominal terms and rewriting. We refer to [6,7,19] for omitted definitions and intuitive explanations.

A *nominal signature* Σ is a set of *function symbols* ranged over by $f, g, \ldots$. We fix a countably infinite set $\mathcal{X}$ of *term variables* ranged over by $X, Y, \ldots$, and

This work is partially supported by JSPS KAKENHI (Nos. 15K00003, 16K00091).

N. Olivetti and A. Tiwari (Eds.): IJCAR 2016, LNAI 9706, pp. 173–182, 2016.
DOI: 10.1007/978-3-319-40229-1_12

a countably infinite set $\mathcal{A} = \{\mathsf{a}, \mathsf{b}, \mathsf{c}, \dots\}$ of *atoms* ranged over by $a, b, c \dots$ (i.e. $\mathsf{a}, \mathsf{b}, \mathsf{c}, \dots$ stand for objects and $a, b, c, \dots$ stand for meta-variables). A *swapping* is a pair $(a\ b)$ of atoms. *Permutations* π are bijections on $\mathcal{A}$ with finite $support(\pi) = \{a \in \mathcal{A} \mid a \neq \pi(a)\}$; permutations are represented by compositions of swappings. $\mathcal{P}$ stands for the set of permutations. We put $ds(\pi, \pi') = \{a \in \mathcal{A} \mid \pi{\cdot}a \neq \pi'{\cdot}a\}$ for any $\pi, \pi' \in \mathcal{P}$. *Terms* are generated by the grammar

$$s, t \in \mathcal{T} ::= a \mid \pi{\cdot}X \mid [a]t \mid f\ t \mid \langle t_1, \dots, t_n \rangle$$

A term of form $\pi{\cdot}X$ is called a *suspension*. A suspension $Id{\cdot}X$ is abbreviated as X, where Id denotes the identity. We write $\mathcal{A}(t)$ and $\mathcal{X}(t)$ for the sets of atoms and term variables occurring in a term t (or any expression t, in general) where the former includes the atoms in abstractions $[a]$ and in $support(\pi)$ of suspensions $\pi{\cdot}X$. The *subterm* of t at a position p is written as $t|_p$. The term obtained from a term s by replacing the subterm at position p by a term t is written as $s[t]_p$. *Action* $\pi{\cdot}t$ and *meta-action* t^π are defined as follows:

$$\begin{array}{rlrl}
\pi{\cdot}a &= \pi(a) & a^\pi &= \pi(a) \\
\pi{\cdot}(\pi'{\cdot}X) &= (\pi \circ \pi'){\cdot}X & (\pi'{\cdot}X)^\pi &= (\pi \circ \pi' \circ \pi^{-1}){\cdot}X \\
\pi{\cdot}([a]t) &= [\pi{\cdot}a](\pi{\cdot}t) & ([a]t)^\pi &= [a^\pi]t^\pi \\
\pi{\cdot}(f\ t) &= f\ \pi{\cdot}t & (f\ t)^\pi &= f\ t^\pi \\
\pi{\cdot}\langle t_1, \dots, t_n \rangle &= \langle \pi{\cdot}t_1, \dots, \pi{\cdot}t_n \rangle & \langle t_1, \dots, t_n \rangle^\pi &= \langle t_1^\pi, \dots, t_n^\pi \rangle
\end{array}$$

A *substitution* is a map $\sigma : \mathcal{X} \to \mathcal{T}$ with finite $dom(\sigma) = \{X \in \mathcal{X} \mid \sigma(X) \neq X\}$. The application of a substitution σ on a term t is written as $t\sigma$.

A finite set of pairs $a\#X$ of $a \in \mathcal{A}$ and $X \in \mathcal{X}$ is called a *freshness context*. For a freshness context ∇, $a \in \mathcal{A}$ and $s, t \in \mathcal{T}$, the relations $\nabla \vdash a\#t$ and $\nabla \vdash s \approx_\alpha t$ are defined as follows:

$$\begin{array}{ccc}
\dfrac{}{\nabla \vdash a\#b}\ a \neq b & \dfrac{\nabla \vdash a\#t}{\nabla \vdash a\#f\ t} & \dfrac{\nabla \vdash a\#t_1 \quad \cdots \quad \nabla \vdash a\#t_n}{\nabla \vdash a\#\langle t_1, \dots, t_n \rangle} \\[2ex]
\dfrac{}{\nabla \vdash a\#[a]t} & \dfrac{\nabla \vdash a\#t}{\nabla \vdash a\#[b]t}\ a \neq b & \dfrac{\pi^{-1}{\cdot}a\#X \in \nabla}{\nabla \vdash a\#\pi{\cdot}X}
\end{array}$$

$$\begin{array}{cc}
\dfrac{}{\nabla \vdash a \approx_\alpha a} & \dfrac{\nabla \vdash t_1 \approx_\alpha s_1 \quad \cdots \quad \nabla \vdash t_n \approx_\alpha s_n}{\nabla \vdash \langle t_1, \dots, t_n \rangle \approx_\alpha \langle s_1, \dots, s_n \rangle} \\[2ex]
\dfrac{\nabla \vdash t \approx_\alpha s}{\nabla \vdash f\ t \approx_\alpha f\ s} & \dfrac{\nabla \vdash t \approx_\alpha (a\ b){\cdot}s \quad \nabla \vdash a\#s}{\nabla \vdash [a]t \approx_\alpha [b]s}\ a \neq b \\[2ex]
\dfrac{\nabla \vdash t \approx_\alpha s}{\nabla \vdash [a]t \approx_\alpha [a]s} & \dfrac{\forall a \in ds(\pi, \pi').\ a\#X \in \nabla}{\nabla \vdash \pi{\cdot}X \approx_\alpha \pi'{\cdot}X}
\end{array}$$

Here $a\#t$ is called a *freshness constraint*, and $s \approx_\alpha t$ an ***α**-equivalence constraint.* For (freshness or **α**-equivalence) constraints $\gamma_1, \dots, \gamma_n$, we write $\Delta \vdash \gamma_1, \dots, \gamma_n$ if $\Delta \vdash \gamma_i$ for all $1 \leq i \leq n$. We put $(a\#t)\sigma = a\#t\sigma$ and $(s \approx_\alpha t)\sigma = s\sigma \approx_\alpha t\sigma$. *Nominal unification* finds a pair $\langle \Delta, \sigma \rangle$ of a freshness context Δ and a substitution σ such that $\Delta \vdash \gamma_1\sigma, \dots, \gamma_n\sigma$ from $\mathcal{C} = \{\gamma_1, \dots, \gamma_n\}$; a most general such pair is an *mgu* of $\mathcal{C}$ [21].

A triple $\nabla \vdash l \to r$ of a freshness context ∇ and $l, r \in \mathcal{T}$ such that l is not a suspension and $\mathcal{X}(\nabla) \cup \mathcal{X}(r) \subseteq \mathcal{X}(l)$ is called a *nominal rewrite rule*, or simply *rewrite rule*. Rewrite rules are identified modulo renaming of term variables. A *nominal rewriting system* (*NRS* for short) is a finite set of rewrite rules. Let $R = \nabla \vdash l \to r$ be a rewrite rule. For a freshness context Δ and $s, t \in \mathcal{T}$, the *rewrite relation* is defined by

$$\Delta \vdash s \to_{\langle R,\pi,p,\sigma\rangle} t \overset{\text{def}}{\iff} \Delta \vdash \nabla^{\pi}\sigma,\ \Delta \vdash s|_p \approx_{\alpha} l^{\pi}\sigma,\ t = s[r^{\pi}\sigma]_p$$

where $\mathcal{X}(l) \cap (\mathcal{X}(\Delta) \cup \mathcal{X}(s)) = \emptyset$. Here, $\nabla^{\pi} = \{\pi(a)\#X \mid a\#X \in \nabla\}$. For an NRS $\mathcal{R}$, we write $\Delta \vdash s \to_{\mathcal{R}} t$ if there exist $R \in \mathcal{R}$, π, p and σ such that $\Delta \vdash s \to_{\langle R,\pi,p,\sigma\rangle} t$. We define $\Delta \vdash s_1 \bowtie_1 s_2 \bowtie_2 \cdots (\bowtie_{n-1} s_n)$ $(\bowtie_i \in \{\to_{\mathcal{R}}, \approx_{\alpha}, \ldots\})$ in the obvious way. $\Delta \vdash s \to_{\mathcal{R}}^* t$ stands for $\Delta \vdash s \to_{\mathcal{R}} \cdots \to_{\mathcal{R}} t$, and $\Delta \vdash s \downarrow_{\approx_{\alpha}} t$ stands for $\Delta \vdash s \to_{\mathcal{R}}^* \circ \approx_{\alpha} \circ \leftarrow_{\mathcal{R}}^* t$. An NRS $\mathcal{R}$ is *Church-Rosser modulo* $\approx_{\alpha}$ if $\Delta \vdash s\ (\leftarrow_{\mathcal{R}} \cup \to_{\mathcal{R}} \cup \approx_{\alpha})^*\ t$ implies $\Delta \vdash s \downarrow_{\approx_{\alpha}} t$. An NRS $\mathcal{R}$ is *terminating* if there is no infinite rewrite sequence $\Delta \vdash s_1 \to_{\mathcal{R}} s_2 \to_{\mathcal{R}} \cdots$.

3 Computing Rewrite Steps and Basic Critical Pairs

A most fundamental ingredient in automation of confluence checking is the computation of rewrite steps, that is, to compute a term t such that $\Delta \vdash s \to_{\mathcal{R}} t$ or even (representatives of) all t such that $\Delta \vdash s \to_{\mathcal{R}} t$, from a given NRS $\mathcal{R}$, a freshness context Δ and a term s. The main challenge here is to find suitable π and σ such that $\Delta \vdash \nabla^{\pi}\sigma$ and $\Delta \vdash s|_p \approx_{\alpha} l^{\pi}\sigma$, when fixing $\nabla \vdash l \to r \in \mathcal{R}$ and a position p in s. Another key ingredient is the computation of basic critical pairs:

Definition 3.1 (Basic critical pair [20]**).** *Let $R_i = \nabla_i \vdash l_i \to r_i$ $(i = 1, 2)$ be rewrite rules. We assume w.l.o.g. $\mathcal{X}(l_1) \cap \mathcal{X}(l_2) = \emptyset$. Let $\nabla_1 \cup \nabla_2^{\pi} \cup \{l_1 \approx l_2^{\pi}|_p\}$ be unifiable for some permutation π and a non-variable position p and let $\langle \Gamma, \sigma\rangle$ be an mgu. Then, $\Gamma \vdash \langle l_2^{\pi}\sigma[r_1\sigma]_p, r_2^{\pi}\sigma\rangle$ is called a* basic critical pair *(*BCP *for short) of R_1 and R_2. The set of BCP of rules in $\mathcal{R}$ is denoted by* BCP($\mathcal{R}$).

Again, the main challenge for the computation of (representatives of) all BCPs is to find suitable π and σ when fixing $R_1, R_2 \in \mathcal{R}$ and a position p.

Since π is not fixed here, these problems are not computed by nominal unification but by *equivariant nominal unification* [5]. In what follows, we present our formalization of equivariant nominal unification and then explain how BCPs are computed. (The computation of rewrite steps is done by replacing equivariant unification by *equivariant matching*, obtained by adding constraints on instantiation.)

3.1 Equivariant Unification

We extend our language by countably infinite sets $\mathcal{X}_A$ and $\mathcal{X}_P$ of *atom variables* ranged over by $A, B, \ldots$ and *permutation variables* ranged over by $P, Q \ldots$. Elements of $\mathcal{A} \cup \mathcal{X}_A$ are ranged over by $\alpha, \beta, \ldots$ and called *atom expressions. Permutation/atomic/term expressions* ($\mathcal{E}_P/\mathcal{E}_A/\mathcal{E}_T$) are generated by the grammar:

$$\begin{array}{ll} \Pi, \Psi \in \mathcal{E}_P & := P \mid \mathsf{Id} \mid (v\ w) \mid \Pi \circ \Psi \mid \Pi^{-1} \\ v, w \in \mathcal{E}_A & := \Pi{\cdot}\alpha \\ S, T \in \mathcal{E}_T & := v \mid \Pi{\cdot}X \mid [v]T \mid f\ T \mid \langle T_1, \ldots, T_n \rangle \end{array}$$

Note here that "Id" etc. are not meta-operations but new constructs. For example, we have $(((P \circ Q)^{-1} \cdot A)\ B) \in \mathcal{E}_P$, $(((P \circ Q)^{-1} \cdot A)\ B) \cdot \mathsf{c} \in \mathcal{E}_A$ and $[(((P \circ Q)^{-1} \cdot A)\ B) \cdot \mathsf{c}](\mathtt{f}\ \langle P^{-1} \cdot X, Q^{-1} \cdot \mathsf{c} \rangle) \in \mathcal{E}_T$.

An instantiation is a pair $\theta = \langle \theta_A, \theta_P \rangle$ of mappings $\theta_A : \mathcal{X}_A \to \mathcal{A}$ and $\theta_P : \mathcal{X}_P \to \mathcal{P}$. For each $\Pi \in \mathcal{E}_P$, $v \in \mathcal{E}_A$, $S \in \mathcal{E}_T$, their interpretations $[\![\Pi]\!]_\theta \in \mathcal{P}$, $[\![v]\!]_\theta \in \mathcal{A}$, $[\![S]\!]_\theta \in \mathcal{T}$ by an instantiation θ are defined by the following:

$$\begin{array}{rclrclrcl} [\![P]\!]_\theta &=& \theta_P(P) & [\![\Pi{\cdot}\alpha]\!]_\theta &=& [\![\Pi]\!]_\theta{\cdot}[\![\alpha]\!]_\theta & [\![a]\!]_\theta &=& a \\ [\![\mathsf{Id}]\!]_\theta &=& Id & [\![\Pi{\cdot}X]\!]_\theta &=& [\![\Pi]\!]_\theta{\cdot}X & [\![A]\!]_\theta &=& \theta_A(A) \\ [\![(v\ w)]\!]_\theta &=& ([\![v]\!]_\theta\ [\![w]\!]_\theta) & [\![[v]T]\!]_\theta &=& [[\![v]\!]_\theta][\![T]\!]_\theta && & \\ [\![\Pi \circ \Psi]\!]_\theta &=& [\![\Pi]\!]_\theta \circ [\![\Psi]\!]_\theta & [\![f\ T]\!]_\theta &=& f\ [\![T]\!]_\theta && & \\ [\![\Pi^{-1}]\!]_\theta &=& [\![\Pi]\!]_\theta^{-1} & [\![\langle T_1, \ldots, T_n \rangle]\!]_\theta &=& \langle [\![T_1]\!]_\theta, \ldots, [\![T_n]\!]_\theta \rangle && & \end{array}$$

Note here that "*Id*" etc. in the rhs's of the definitions are not constructs but meta-operations. For example, if we take $\theta_P(P) = (\mathsf{a}\ \mathsf{b})$, $\theta_P(Q) = (\mathsf{b}\ \mathsf{c})$ and $\theta_A(A) = \mathsf{a}$, $\theta_A(B) = \mathsf{b}$ then we have $[\![(((P \circ Q)^{-1} \cdot A)\ B)]\!]_\theta = (\mathsf{c}\ \mathsf{b}) \in \mathcal{P}$, $[\![(((P \circ Q)^{-1} \cdot A)\ B) \cdot \mathsf{c}]\!]_\theta = \mathsf{b} \in \mathcal{A}$ and $[\![[(((P \circ Q)^{-1} \cdot A)\ B) \cdot \mathsf{c}](\mathtt{f}\ \langle P^{-1} \cdot X, Q^{-1} \cdot \mathsf{c} \rangle)]\!]_\theta = [\mathsf{b}](\mathtt{f}\ \langle (\mathsf{a}\ \mathsf{b}) \cdot X, \mathsf{b} \rangle) \in \mathcal{T}$. For a permutation expression $\Pi \in \mathcal{E}_P$ and a term expression $T \in \mathcal{E}_T$, we define action $\Pi{\cdot}T \in \mathcal{E}_T$ and meta-action $T^\Pi \in \mathcal{E}_T$ as follows:

$$\begin{array}{rclrcl} \Pi{\cdot}(\Pi'{\cdot}\alpha) &=& (\Pi \circ \Pi'){\cdot}\alpha & (\Pi'{\cdot}\alpha)^\Pi &=& (\Pi \circ \Pi'){\cdot}\alpha \\ \Pi{\cdot}(\Pi'{\cdot}X) &=& (\Pi \circ \Pi'){\cdot}X & (\Pi'{\cdot}X)^\Pi &=& (\Pi \circ \Pi' \circ \Pi^{-1}){\cdot}X \\ \Pi{\cdot}([v]T) &=& [\Pi{\cdot}v](\Pi{\cdot}T) & ([v]T)^\Pi &=& [v^\Pi]T^\Pi \\ \Pi{\cdot}(f\ T) &=& f\ \Pi{\cdot}T & (f\ T)^\Pi &=& f\ T^\Pi \\ \Pi{\cdot}\langle T_1, \ldots, T_n \rangle &=& \langle \Pi{\cdot}T_1, \ldots, \Pi{\cdot}T_n \rangle & \langle T_1, \ldots, T_n \rangle^\Pi &=& \langle T_1^\Pi, \ldots, T_n^\Pi \rangle \end{array}$$

A *freshness constraint expression* is a pair $v\#T$ of $v \in \mathcal{E}_A$ and $T \in \mathcal{E}_T$ and an *α-equivalence constraint expression* is a pair $S \approx T$ of $S, T \in \mathcal{E}_T$. An *equivariant unification problem (EUP)* is a finite set of (freshness or α-equivalence) constraint expressions. We put $[\![v\#T]\!]_\theta = [\![v]\!]_\theta \# [\![T]\!]_\theta$ and $[\![S \approx T]\!]_\theta = [\![S]\!]_\theta \approx_\alpha [\![T]\!]_\theta$. A *model* of an EUP $\mathcal{C} = \{\gamma_1, \ldots, \gamma_n\}$ is a triple $\langle \theta, \sigma, \Delta \rangle$ of an instantiation θ, a substitution σ and a freshness context Δ such that $\Delta \vdash [\![\gamma_i]\!]_\theta \sigma$ for all $1 \leq i \leq n$. We write $\langle \theta, \sigma, \Delta \rangle \models \mathcal{C}$ if $\langle \theta, \sigma, \Delta \rangle$ is a model of $\mathcal{C}$.

An *answer constraint* is a finite set of expressions of the following forms:

$$A \mapsto v \mid P : \alpha \mapsto \beta \mid \alpha \not\approx \beta \mid X \mapsto T \mid \alpha\#X \mid \#(X, \Pi, \Pi')$$

A triple $\langle \theta, \sigma, \Delta \rangle$ is a model of an answer constraint $\mathcal{S}$, written as $\langle \theta, \sigma, \Delta \rangle \models \mathcal{S}$, if $\theta_A(A) = [\![v]\!]_\theta$ for any $A \mapsto v \in \mathcal{S}$, $\theta_P(P)([\![\alpha]\!]_\theta) = [\![\beta]\!]_\theta$ for any $P : \alpha \mapsto \beta \in \mathcal{S}$, $[\![\alpha]\!]_\theta \neq [\![\beta]\!]_\theta$ for any $\alpha \not\approx \beta \in \mathcal{S}$, $\sigma(X) = [\![T]\!]_\theta$ for all $X \mapsto T \in \mathcal{S}$,

$\Delta \vdash [\![\alpha]\!]_\theta \# X\sigma$ for all $\alpha \# X \in \mathcal{S}$, and $\Delta \vdash a \# X\sigma$ for any $a \in ds([\![\Pi]\!]_\theta, [\![\Pi']\!]_\theta)$ and $\#(X, \Pi, \Pi') \in \mathcal{S}$. For a given EUP $\mathcal{C}$, *equivariant unification* [5] computes a finite set $\mathcal{M} = \mathsf{Sol}(\mathcal{C})$ of answer constraints such that, for any triple $\langle \theta, \sigma, \Delta \rangle$, $\langle \theta, \sigma, \Delta \rangle \models \mathcal{C}$ iff $\exists \mathcal{S} \in \mathcal{M}.\ \langle \theta, \sigma, \Delta \rangle \models \mathcal{S}$.

3.2 Computing Basic Critical Pairs

We now proceed to explain how the representative set of BCPs are computed using equivariant unification, from two given rewrite rules $R_i = \nabla_i \vdash l_i \to r_i$ $(i = 1, 2)$ and a position p. The procedure consists of the following two steps.

1. *Equivariant Unification.* We solve the following EUP:

$$\mathcal{C} = \nabla_1 \cup \nabla_2^P \cup \{l_1 \approx l_2^P|_p\} \cup \{P \cdot a_i \approx A_i \mid a_i \in \mathcal{A}(l_2[\,]_p) \cup \mathcal{A}(r_1) \cup \mathcal{A}(r_2)\}$$

   where $P \in \mathcal{X}_P$, and each A_i is a fresh atom variable. The last component of the union is added to specify $P(a)$ for all a required to construct $l_2^\pi \sigma[r_1\sigma]_p$ and $r_2^\pi \sigma$. If $\mathsf{Sol}(\mathcal{C}) = \emptyset$ then we return the empty set of BCPs.
2. *Instantiation.* For each $\mathcal{S} \in \mathsf{Sol}(\mathcal{C})$, we compute all (representative of) BCPs obtained by models of answer constraints $\mathcal{S} \in \mathsf{Sol}(\mathcal{C})$, more formally, a finite set $T_\mathcal{S}$ representing $\{\Gamma \vdash \langle l_2^{\theta_P(P)} \sigma[r_1\sigma]_p, r_2^{\theta_P(P)} \sigma \rangle \mid \langle \theta, \sigma, \Gamma \rangle \models \mathcal{S}\}$. We obtain a set $\mathrm{BCP}_\mathcal{S}$ of BCPs from $\mathcal{S}$, l_2, r_1 and r_2 by successively instantiating each atom variable and atomic expression $P{\cdot}\alpha$ in $\mathcal{S}$ by all atoms already used and one new fresh atom (as the representative of all other non-used atoms), where any instantiation must satisfy $\Gamma \vdash \gamma$ for all freshness constraints γ obtained from $\alpha \# X \in \mathcal{S}$ and $\#(X, \Pi, \Psi) \in \mathcal{S}$. Note also that due to the form of the input, all occurrences of P in $\#(X, \Pi, \Psi) \in \mathcal{S}$ have the form $P{\cdot}\alpha$. Therefore, any $\#(X, \Pi, \Psi)$ can be replaced with $\{a \# X \mid a \in ds(\Pi, \Psi)\}$ when instantiations are completed. (This is not always possible for general equivariant unification problems e.g. consider $\#(X, (\mathsf{a}\ \mathsf{b}), P)$.) Finally, we put $\mathrm{BCP}_\mathcal{C} = \bigcup_{\mathcal{S} \in \mathsf{Sol}(\mathcal{C})} \mathrm{BCP}_\mathcal{S}$.

Example 3.2. Let $\texttt{forall} \in \Sigma$ and consider the following NRS:

$$\mathcal{R}_{\mathsf{com}\forall} = \{\ \vdash \texttt{forall}\ [\mathsf{a}]\texttt{forall}\ [\mathsf{b}]X \to \texttt{forall}\ [\mathsf{b}]\texttt{forall}\ [\mathsf{a}]X\}$$

Consider the overlap at position 11. In the first step, we solve an EUP:

$$\mathcal{C} = \{\texttt{forall}\ [\mathsf{a}]\texttt{forall}\ [\mathsf{b}]X \approx (\texttt{forall}\ [P{\cdot}\mathsf{b}]Y)\} \cup \{P \cdot \mathsf{a} \approx A\}$$

Then we obtain

$$\mathsf{Sol}(\mathcal{C}) = \left\{ \begin{array}{l} \{Y \mapsto (\texttt{forall}\ [\mathsf{b}]X), P : \mathsf{a} \mapsto A, P : \mathsf{b} \mapsto \mathsf{a}\}, \\ \{Y \mapsto (\texttt{forall}\ [\mathsf{a}][(\mathsf{a}\ \mathsf{b})]X), P : \mathsf{a} \mapsto A, P : \mathsf{b} \mapsto \mathsf{b}\}, \\ \{Y \mapsto (\texttt{forall}\ [(\mathsf{a}\ C){\cdot}\mathsf{b}][(\mathsf{a}\ C)]X), C \# X, C \not\approx \mathsf{a}, C \not\approx \mathsf{b}, P : \mathsf{a} \mapsto A, P : \mathsf{b} \mapsto C\} \end{array} \right\}$$

By instantiating A (by a, b and c) and C (by a, b, c and d) successively, we obtain the following seven BCPs from this overlap:

$$\mathrm{BCP}_C = \left\{ \begin{array}{l} \vdash \langle \mathtt{forall[b]forall[b]forall[a]}X,\ \mathtt{forall[a]forall[b]forall[b]}X \rangle \\ \vdash \langle \mathtt{forall[c]forall[b]forall[a]}X,\ \mathtt{forall[a]forall[c]forall[b]}X \rangle \\ \vdash \langle \mathtt{forall[a]forall[b]forall[a]}X,\ \mathtt{forall[b]forall[a]forall[a][(a\ b)]}X \rangle \\ \vdash \langle \mathtt{forall[c]forall[b]forall[a]}X,\ \mathtt{forall[b]forall[c]forall[a][(a\ b)]}X \rangle \\ \mathtt{c}\#X \vdash \langle \mathtt{forall[b]forall[b]forall[a]}X,\ \mathtt{forall[c]forall[b]forall[b](a\ c)}\cdot X \rangle \\ \mathtt{c}\#X \vdash \langle \mathtt{forall[a]forall[b]forall[a]}X,\ \mathtt{forall[c]forall[a]forall[b](a\ c)}\cdot X \rangle \\ \mathtt{d}\#X \vdash \langle \mathtt{forall[c]forall[b]forall[a]}X,\ \mathtt{forall[d]forall[c]forall[b](a\ d)}\cdot X \rangle \end{array} \right\}$$

4 Proving Confluence Automatically

4.1 Confluence Criteria

We prove (non-)confluence based on the following confluence criteria.

Proposition 4.1 [19]. *Let $\mathcal{R}$ be an orthogonal NRS that is abstract skeleton preserving (ASP). Then, $\mathcal{R}$ is Church-Rosser modulo $\approx_\alpha$.*

Proposition 4.2 [20]. *Let $\mathcal{R}$ be a linear uniform NRS. Then $\mathcal{R}$ is Church-Rosser modulo $\approx_\alpha$ if $\Gamma \vdash u \to^= \circ \approx_\alpha \circ \leftarrow^* v$ and $\Gamma \vdash u \to^* \circ \approx_\alpha \circ \leftarrow^= v$ for any $\Gamma \vdash \langle u, v\rangle \in \mathrm{BCP}(\mathcal{R})$.*

Proposition 4.3 [20]. *Let $\mathcal{R}$ be a terminating uniform NRS. Then $\mathcal{R}$ is Church-Rosser modulo $\approx_\alpha$ if and only if $\Gamma \vdash u \downarrow_{\approx_\alpha} v$ for any $\Gamma \vdash \langle u, v\rangle \in$ $\mathrm{BCP}(\mathcal{R})$.*

Proposition 4.4 [11]. *Let $\mathcal{R}$ be a left-linear uniform NRS. Then $\mathcal{R}$ is Church-Rosser modulo $\approx_\alpha$ if $\Gamma \vdash u \mathrel{+\!\!\!\!\!\!\to\!\!\!\!\to} \circ \approx_\alpha v$ for any $\Gamma \vdash \langle u, v\rangle \in \mathrm{BCP}_{in}(\mathcal{R})$ and $\Gamma \vdash u \mathrel{+\!\!\!\!\!\!\to\!\!\!\!\to} \circ \approx_\alpha \circ \leftarrow^* v$ for any $\Gamma \vdash \langle u, v\rangle \in \mathrm{BCP}_{out}(\mathcal{R})$.*

Here, an NRS is *orthogonal* if it is left-linear and has no proper BCPs [19]; $\Gamma \vdash s \to^= t$ stands for $\Gamma \vdash s \to t$ or $s = t$; $\Gamma \vdash s \mathrel{+\!\!\!\!\!\!\to\!\!\!\!\to} t$ stands for the parallel rewrite relation [19]; and $\mathrm{BCP}_{in}(\mathcal{R})$ and $\mathrm{BCP}_{out}(\mathcal{R})$ denote the sets of inner and outer BCPs [11], respectively.

The ASP condition and uniformness of NRSs are decidable [6,19]. To check the joinability conditions in Propositions 4.2 and 4.4, sets $\{w \mid \Gamma \vdash u \to^= w\}$ and $\{w \mid \Gamma \vdash u \mathrel{+\!\!\!\!\!\!\to\!\!\!\!\to} w\}$ are computed using the procedure for computing rewrite steps. For checking confluence criteria of Proposition 4.3, termination checking is required, which we explain in the next subsection.

4.2 Proving Termination

In this subsection, we present a simple technique to show termination of NRSs.

Definition 4.5. *Let Σ be a nominal signature, and $\mathcal{F}$ a arity-fixed first-order signature given by $\mathcal{F} = \{f \mid f \in \Sigma\} \cup \{\diamond, \lambda\} \cup \{\mathsf{pair}_n \mid n \geq 0\}$, where $\diamond$ is of arity 0, λ and all $f \in \Sigma$ are of arity 1, and pair_n is of arity n for each n. We define a translation Φ from nominal terms over Σ to first-order terms over $\mathcal{F}$ (with the set $\mathcal{X}$ of variables) as follows:*

$$\Phi(a) = \diamond \qquad \Phi(\pi{\cdot}X) = X \qquad \Phi([a]t) = \lambda(\Phi(t))$$
$$\Phi(f\ t) = f(\Phi(t)) \qquad \Phi(\langle t_1, \dots, t_n\rangle) = \mathsf{pair}_n(\Phi(t_1), \dots, \Phi(t_n))$$

For an NRS $\mathcal{R}$, we define a first-order term rewriting system $\Phi(\mathcal{R})$ by: $\Phi(\mathcal{R}) = \{\Phi(l) \to \Phi(r) \mid \nabla \vdash l \to r \in \mathcal{R}\}$.

Theorem 4.6. *If $\Phi(\mathcal{R})$ is terminating then $\mathcal{R}$ is terminating modulo $\approx_\alpha$.*

Proof. The claim follows from the fact that for any Δ, s, t, (i) $\Delta \vdash s \approx_\alpha t$ implies $\Phi(s) = \Phi(t)$ and (ii) $\Delta \vdash s \to_{\mathcal{R}} t$ implies $\Phi(s) \to_{\Phi(\mathcal{R})} \Phi(t)$. □

Remark 4.7. In [8], nominal terms are given by the following grammar:

$$t, s ::= a \mid \pi{\cdot}X \mid [a]t \mid f(t_1, \dots, t_n)$$

It is easy to modify the translation Φ to adapt to this definition. In [8, Definition 6], recursive path order on nominal terms for proving termination of "closed rewriting" has been given. It is easy to see that the order can be obtained by combining the translation Φ and recursive path order on first-order terms.

5 Implementation and Experiments

Our tool nrbox (**n**ominal **r**ewriting tool**box**) is implemented in Standard ML of New Jersey[1]. It reads an NRS $\mathcal{R}$ from the input and tries to prove whether it is Church-Rosser modulo $\approx_\alpha$ or not—it prints out "YES" ("NO") if it successfully proves that $\mathcal{R}$ is (resp. is not) Church-Rosser modulo $\approx_\alpha$ and "MAYBE" if it fails to prove or disprove that $\mathcal{R}$ is Church-Rosser modulo $\approx_\alpha$.

The source code of the tool is obtained from http://www.nue.ie.niigata-u.ac.jp/tools/nrbox/. It consists of about 4500 lines of code, and roughly one third of the code is devoted to equivariant unification. The format of input NRSs follows a specification bundled in the distribution. To prove the termination of NRSs by the method described in Sect. 4.2, the tool requires an external termination prover for first-order term rewriting systems.

We have tested our confluence prover with 30 NRSs, collected from the literature [3,6,8] and constructed during our studies [11,18–20]. All tests have been performed in a PC with one 2.50 GHz CPU and 4G memory. We have used TTT [13] with 20 s timeout as the external termination prover for first-order term rewriting systems.

Summary of experiments is shown in Table 1. The column below "NRS" shows descriptions of the input NRSs. The columns below "Orth.", "Strong",

[1] http://www.smlnj.org/

Table 1. Summary of experiments

	NRS	Orth.	Strong	K.-B.	Parallel
1	α-reduction rule ([6] Intro.)	MAYBE	YES	MAYBE	YES
2	Eta: η-reduction rule ([6] Intro.)	YES	YES	YES	YES
3	η-expansion rule ([6] Intro.)	MAYBE	MAYBE	MAYBE	MAYBE
4	$\mathcal{R}^*_\sigma$: subst. for λ with σ_ε (Ex. 43 [6])	MAYBE	MAYBE	YES	YES
5	β-reduction $\{\mathsf{Beta}\} \cup \mathcal{R}^*_\sigma$ (Ex. 43 [6])	MAYBE	MAYBE	MAYBE	MAYBE
6	a fragment of ML (Ex. 43 [6])	MAYBE	MAYBE	MAYBE	MAYBE
7	PNF of FOF (Ex. 44 [6])	MAYBE	MAYBE	NO	MAYBE
8	PNF of FOF with addition (Ex. 44 [6])	MAYBE	MAYBE	NO	MAYBE
9	non-joinable trivial CP (Lem. 56 [6])	MAYBE	MAYBE	MAYBE	MAYBE
10	$\{a\#X \vdash X \to [a]X\}$ (Lem. 56 [6])	MAYBE	MAYBE	MAYBE	MAYBE
11	$\{\mathsf{Eta}, \bot\}$ (Ex. 5 [8])	MAYBE	MAYBE	NO	MAYBE
12	$\{\mathsf{Eta}, \bot\}$ with CP (Ex. 5 [8])	MAYBE	YES	YES	YES
13	summation (Ex. 6 [8])	MAYBE	MAYBE	NO	MAYBE
14	summation with CP (Ex. 6 [8])	MAYBE	MAYBE	YES	MAYBE
15	$\{\vdash f(X) \to [a]X\}$ (Ex. 1.2 [18])	MAYBE	MAYBE	NO	MAYBE
16	$\{a\#X \vdash f(X) \to [a]X\}$ (Ex. 4.7 [18])	YES	YES	YES	YES
17	$\mathcal{R}_\sigma$: subst. for λ with $\sigma_{\mathtt{var}\varepsilon}$ (Ex. 8 [19])	YES	MAYBE	YES	YES
18	β-reduction $\{\mathsf{Beta}\} \cup \mathcal{R}_\sigma$	MAYBE	MAYBE	MAYBE	MAYBE
19	$\beta\eta$-reduction $\{\mathsf{Beta}\} \cup \{\mathsf{Eta}\} \cup \mathcal{R}_\sigma$	MAYBE	MAYBE	MAYBE	MAYBE
20	$\mathcal{R}_{\mathsf{uc}\text{-}\eta}$ (Ex. 17 [19])	MAYBE	MAYBE	MAYBE	MAYBE
21	$\mathcal{R}_{\mathsf{uc}\text{-}\eta-exp}$ (Ex. 19 [19])	MAYBE	MAYBE	NO	MAYBE
22	μ-substitution for $\lambda\mu$-term ([15])	YES	MAYBE	YES	YES
23	$\{\vdash f(X) \to f([a]X)\}$ (Ex. 4.3 [3])	MAYBE	MAYBE	MAYBE	MAYBE
24	NNF of $\{\neg, \forall, \wedge\}$-form. with swap (Ex. 5.5 [3])	MAYBE	YES	YES	YES
25	$\mathsf{Com}_\forall$: com. rule for $\forall$ (Ex. 5 [20])	MAYBE	YES	MAYBE	MAYBE
26	PNF of $\{\forall, \wedge\}$-form. (Ex. 7 [20])	MAYBE	MAYBE	NO	MAYBE
27	PNF of $\{\forall, \wedge\}$-form. + $\mathsf{Com}_\forall$ (Ex. 12 [20])	MAYBE	MAYBE	MAYBE	MAYBE
28	NNF of $\{\neg, \forall, \exists\}$-form. (Ex. 29 [20])	MAYBE	MAYBE	YES	MAYBE
29	NNF of FOF	MAYBE	MAYBE	YES	MAYBE
30	NNF of FOF without DNE	YES	YES	YES	YES
	(♯YES, ♯NO)	(5,0)	(7,0)	(11,7)	(9,0)
	$\sum$ time (msec.)	611	1367	4377	2217

"K.-B." and "Parallel" show the results of applying the confluence proving methods from Propositions 4.1, 4.2 (with an approximation of $\rightarrow^*$ by $\rightarrow^=$), 4.3 and 4.4 (with an approximation of $\rightarrow^*$ by $\twoheadrightarrow\!\!\!\!\!\!\!+\ $), respectively—`YES` denotes for the success for proving, `NO` denotes for the success of disproving, and `MAYBE` denotes failure. For each method, the last two lines of the table show the number of successes for proving/disproving confluence and the total time for checking all of the examples.

Using the combination of all the methods, our prover succeeded in proving confluence of 13 examples and non-confluence of 7 examples. All details of the experiments are available on the webpage http://www.nue.ie.niigata-u.ac.jp/tools/nrbox/experiments/ijcar16/.

References

1. Confluence competition. http://coco.nue.riec.tohoku.ac.jp/
2. Aoto, T., Yoshida, J., Toyama, Y.: Proving confluence of term rewriting systems automatically. In: Treinen, R. (ed.) RTA 2009. LNCS, vol. 5595, pp. 93–102. Springer, Heidelberg (2009)
3. Ayala-Rincón, M., Fernández, M., Gabbay, M.J., Rocha-Oliveira, A.C.: Checking overlaps of nominal rewrite rules. In: Pre-proceedings of the 10th LSFA, pp. 199–214 (2015)
4. Baader, F., Nipkow, T.: Term Rewriting and All That. Cambridge University Press, Cambridge (1998)
5. Cheney, J.: Equivariant unification. J. Autom. Reasoning **45**, 267–300 (2010)
6. Fernández, M., Gabbay, M.J.: Nominal rewriting. Inform. Comput. **205**, 917–965 (2007)
7. Fernández, M., Gabbay, M.J., Mackie, I.: Nominal rewriting systems. In: Proceedings of the 6th PPDP, pp. 108–119. ACM Press (2004)
8. Fernández, M., Rubio, A.: Nominal completion for rewrite systems with binders. In: Czumaj, A., Mehlhorn, K., Pitts, A., Wattenhofer, R. (eds.) ICALP 2012, Part II. LNCS, vol. 7392, pp. 201–213. Springer, Heidelberg (2012)
9. Gabbay, M.J., Pitts, A.M.: A new approach to abstract syntax with variable binding. Formal Aspects Comput. **13**, 341–363 (2002)
10. Hirokawa, N., Klein, D.: Saigawa: a confluence tool. In: Proceedings of the 1st IWC, p. 49 (2012)
11. Kikuchi, K., Aoto, T., Toyama, Y.: Parallel closure theorem for left-linear nominal rewriting systems. http://www.nue.riec.tohoku.ac.jp/user/kentaro/cr-nominal/pct.pdf
12. Klop, J.W., van Oostrom, V., van Raamsdonk, F.: Combinatory reduction systems: introduction and survey. Theoret. Comput. Sci. **121**, 279–308 (1993)
13. Korp, M., Sternagel, C., Zankl, H., Middeldorp, A.: Tyrolean termination tool 2. In: Treinen, R. (ed.) RTA 2009. LNCS, vol. 5595, pp. 295–304. Springer, Heidelberg (2009)
14. Mayr, R., Nipkow, T.: Higher-order rewrite systems and their confluence. Theoret. Comput. Sci. **192**, 3–29 (1998)
15. Parigot, M.: $\lambda\mu$-calculus: an algorithmic interpretation of classical natural deduction. In: Voronkov, A. (ed.) Logic Programming and Automated Reasoning. LNCS, vol. 624, pp. 190–201. Springer, Heidelberg (1992)

16. Pitts, A.M.: Nominal logic, a first order theory of names and binding. Inform. Comput. **186**, 165–193 (2003)
17. Sternagel, T., Middeldorp, A.: Conditional confluence (system description). In: Dowek, G. (ed.) RTA-TLCA 2014. LNCS, vol. 8560, pp. 456–465. Springer, Heidelberg (2014)
18. Suzuki, T., Kikuchi, K., Aoto, T., Toyama, Y.: On confluence of nominal rewriting systems. In: Proceedings of the 16th PPL, in Japanese (2014)
19. Suzuki, T., Kikuchi, K., Aoto, T., Toyama, Y.: Confluence of orthogonal nominal rewriting systems revisited. In: Proceedings of the 26th RTA. LIPIcs, vol. 36, pp. 301–317 (2015)
20. Suzuki, T., Kikuchi, K., Aoto, T., Toyama, Y.: Critical pair analysis in nominal rewriting. In: Proceedings of the 7th SCSS. EPiC, vol. 39, pp. 156–168. EasyChair (2016)
21. Urban, C., Pitts, A.M., Gabbay, M.J.: Nominal unification. Theoret. Comput. Sci. **323**, 473–497 (2004)
22. Zankl, H., Felgenhauer, B., Middeldorp, A.: CSI – a confluence tool. In: Bjørner, N., Sofronie-Stokkermans, V. (eds.) CADE 2011. LNCS, vol. 6803, pp. 499–505. Springer, Heidelberg (2011)

Built-in Variant Generation and Unification, and Their Applications in Maude 2.7

Francisco Durán[1], Steven Eker[2], Santiago Escobar[3(✉)], Narciso Martí-Oliet[4], José Meseguer[5], and Carolyn Talcott[2]

[1] Universidad de Málaga, Málaga, Spain
duran@lcc.uma.es
[2] SRI International, Menlo Park, CA, USA
eker@csl.sri.com, clt@cs.stanford.edu
[3] Universitat Politècnica de València, Valencia, Spain
sescobar@dsic.upv.es
[4] Universidad Complutense de Madrid, Madrid, Spain
narciso@ucm.es
[5] University of Illinois at Urbana-Champaign, Champaign, USA
meseguer@illinois.edu

Abstract. This paper introduces some novel features of Maude 2.7. We have added support for: (i) built-in order-sorted unification modulo associativity, commutativity, and identity, (ii) built-in variant generation, (iii) built-in order-sorted unification modulo a finite variant theory, and (iv) symbolic reachability modulo a finite variant theory.

1 Introduction

Maude[1] is a language and a system based on rewriting logic [5]. Maude provides a precise mathematical model thanks to its logical basis and its initial model semantics, allowing its formal tool environment to be used in three, mutually reinforcing ways: as a declarative programming language, as an executable formal specification language, and as a formal verification system.

Order-sorted unification and narrowing modulo axioms were first available in 2009 as part of the Maude 2.4 release [4]. Unification was available as a built-in feature in Maude while narrowing was available in Full Maude, an extension of Maude written in Maude itself. Unification worked for any combination of

F. Durán was partially supported by Spanish MINECO under grant TIN 2014-52034-R and Universidad de Málaga (Campus de Excelencia Internacional Andalucía Tech). S. Eker and C. Talcott were partially supported by NSF grant CNS-1318848. S. Escobar was partially supported by Spanish MINECO under grants TIN 2015-69175-C4-1-R and TIN 2013-45732-C4-1-P, and by Generalitat Valenciana under grant PROMETEOII/2015/013. N. Martí-Oliet was partially supported by Spanish MINECO under grant StrongSoft (TIN 2012-39391-C04-04) and Comunidad de Madrid program N-GREENS Software (S2013/ICE-2731). J. Meseguer was partially supported by NSF grant CNS-1319109.

[1] Maude is publicly available at http://maude.cs.illinois.edu.

N. Olivetti and A. Tiwari (Eds.): IJCAR 2016, LNAI 9706, pp. 183–192, 2016.
DOI: 10.1007/978-3-319-40229-1_13

symbols being either free or associative-commutative (AC). Narrowing worked for modules having only rules and axioms and relied on the built-in unification algorithm. It supported the concept of symbolic reachability analysis of terms with logical variables, computing suitable substitutions for the variables in both the origin and the destination terms [11].

Unification and narrowing were updated in 2011 as part of the Maude 2.6 release [7]. First, the built-in unification was extended to allow any combination of symbols being either free, commutative (C), associative-commutative (AC), or associative-commutative with an identity symbol (ACU). The performance was dramatically improved, allowing further development of other techniques in Maude. Second, the concept of *variant* [6] was added to Maude. The introduction of variants led to a significant improvement in the reasoning capabilities in Maude: variant generation, variant-based unification, and symbolic reachability based on variant-based unification were all available for the first time. However, all the variant-based features and the narrowing-based reachability were only available in Full Maude, and for a restricted class of theories called *strongly right irreducible.*

In this paper, we present the new unification and narrowing features available in the most recent Maude 2.7 version. First, the built-in unification algorithm allows any combination of symbols being free, C, AC, ACU, CU (commutativity and identity), U (identity), Ul (left identity), and Ur (right identity). Second, variant generation and variant-based unification are implemented as built-in features in Maude. This built-in implementation works for any convergent theory modulo the axioms described above, both allowing very general equational theories (beyond the strongly right irreducible) and boosting the performance not only of these features but of their applications, described in Sect. 6. Third, narrowing-based reachability is still only available in Full Maude but uses the built-in variant-based unification.

2 Built-in Order-Sorted Unification Modulo Axioms

Maude currently provides an order-sorted Ax-unification algorithm for all order-sorted theories (Σ, Ax) such that the order-sorted signature Σ is *preregular* modulo Ax (see [9, Footnote 2]) and the axioms Ax associated to function symbols can have any combination (even empty) of the following equational attributes: the `comm` attribute (C), the `assoc comm` attributes (AC), the `assoc comm id` attributes (ACU), the `comm id` attributes (CU), the `id` attribute (U), the `left id` attribute (Ul), and the `right id` attribute (Ur). The reason for excluding the `assoc` attribute without `comm` is the fact that associative unification is not finitary. Maude 2.7 provides an Ax-unification command of the form

```
unify [n] in ⟨ModId⟩ :
  ⟨Term-1⟩ =? ⟨Term'-1⟩ /\ ... /\ ⟨Term-k⟩ =? ⟨Term'-k⟩ .
```

where $k \geq 1$, n is an optional argument providing a bound on the number of unifiers requested, and `ModId` is the module where the command takes place.

Let us show some examples of unification with an identity attribute, which is the new feature available in Maude 2.7. Let us consider first a module using the `left id` attribute.

```
mod LEFTID-UNIFICATION-EX is
    sorts Magma Elem . subsorts Elem < Magma .
    op __ : Magma Magma -> Magma [left id: e] .
    ops a b c d e : -> Elem .
endm
```

Then the following two unification problems have a different meaning, where we have swapped the position of the variables. First, when we unify two terms where variables of sort `Magma` are at the left of the terms, we have both a syntactic unifier and a unifier modulo identity; note that unification may require the introduction of new variables in the modulo case and they are indicated in Maude using the notation `#n:Sort`, where new variables start with number 1.

```
Maude> unify in LEFTID-UNIFICATION-EX : X:Magma a =? (Y:Magma a) a .
Solution 1                Solution 2
X:Magma --> a             X:Magma --> #1:Magma a
Y:Magma --> e             Y:Magma --> #1:Magma
```

When the variables are instead at the right side of the terms of sort `Magma` , there is clearly no unifier.

```
Maude> unify in LEFTID-UNIFICATION-EX : a X:Magma =? (a a) Y:Magma .
No unifier.
```

Symmetric results could be obtained for a module with right identity (`right id: e`) instead of left identity. And similar results could be obtained for a module with an identity symbol (`id: e`) instead of left or right identity. A different result is obtained when we add commutativity.

```
mod COMM-ID-UNIFICATION-EX is
    sorts Magma Elem . subsorts Elem < Magma .
    op __ : Magma Magma -> Magma [comm id: e] .
    ops a b c d e : -> Elem .
endm
```

When we unify two terms where variables of sort `Magma` are at the left of the terms, we have both a syntactic unifier (Solution 2) and a unifier modulo identity and commutativity (Solution 1), but the latter is duplicated (Solution 3) because most general unifiers may not always be returned.

```
Maude> unify in COMM-ID-UNIFICATION-EX : X:Magma a =? (Y:Magma a) a .
Solution 1         Solution 2                Solution 3
X:Magma --> a      X:Magma --> a #1:Magma    X:Magma --> a
Y:Magma --> e      Y:Magma --> #1:Magma      Y:Magma --> e
```

3 Built-in Variant Generation

Given an equational theory $(\Sigma, E \cup Ax)$ where $\overrightarrow{E}$ is a set of convergent oriented equations modulo the axioms Ax, the (E, Ax)-*variants* [6,12] of a term t are the set of all pairs consisting, each one, of a substitution σ and the (E, Ax)-canonical form of $t\sigma$. A preorder relation of generalization that holds between such pairs provides a notion of most general variants and also of completeness of a set of variants. An equational theory has the *finite variant property* (or it is called a *finite variant theory*) iff there is a finite and complete set of most general variants for each term. Whether an equational theory has the finite variant property is undecidable [2] but a technique based on the dependency pair framework has been developed in [12] and a semi-decision procedure that works well in practice was introduced in [3].

At a practical level, variants are generated using a narrowing strategy. Narrowing with oriented equations E (with or without modulo Ax) enjoys well-known completeness results. But narrowing can be quite inefficient, generating a huge search space, and different narrowing strategies have been devised to reduce the search space while remaining complete. The *folding variant narrowing strategy* is proved in [12] to be complete for variants and it is able to terminate for all inputs if the theory has the finite variant property.

The equational theories that are admissible for variant generation are as follows. Let `fmod` $(\Sigma, E \cup Ax)$ **`endfm`** be an order-sorted functional module where E is a set of equations specified with the `eq` keyword and the attribute `variant`, and Ax is a set of axioms such that the axioms satisfy the restrictions explained in Sect. 2. Furthermore, the equations E must be unconditional, not using the `owise` attribute, and confluent, terminating, sort-decreasing, and coherent modulo Ax (we then call the equational theory *convergent*).

Any system module `mod` $(\Sigma, G \cup E \cup Ax, R)$ **`endm`** where G is an additional set of equations and R is a set of rules, is also considered admissible for variant generation if the equational part $(\Sigma, E \cup Ax)$ satisfies the conditions described above. Note that Maude requires that the equations E used for variant generation (and variant-based unification) should be clearly distinguished from the standard equations G in Maude by using the attribute `variant` (both E and G are used for term simplification but R not).

Maude provides a variant generation command of the form:

```
get variants [ n ] in ⟨ModId⟩ : ⟨Term⟩ .
```

where n is an optional argument providing a bound on the number of variants requested, so that if the cardinality of the set of variants is greater than the specified bound, the variants beyond that bound are omitted; and `ModId` is the module where the command takes place.

For example, consider the following equational theory for exclusive or.

```
fmod EXCLUSIVE-OR is
  sorts Nat NatSet .  subsort Nat < NatSet .
  op 0 : -> Nat .
  op s : Nat -> Nat .
  op mt : -> NatSet .
  op _*_ : NatSet NatSet -> NatSet [assoc comm] .
  vars X Z : [NatSet] .
  eq [idem] :      X * X = mt    [variant] .
  eq [idem-Coh] : X * X * Z = Z [variant] .
  eq [id] :        X * mt = X    [variant] .
endfm
```

We can check that the `EXCLUSIVE-OR` module above has the finite variant property by simply generating the variants for the exclusive-or symbol $*$.

```
Maude> get variants in EXCLUSIVE-OR : X * Y .
Variant 1                                          Variant 7
[NatSet]: #1:[NatSet] * #2:[NatSet]   .........   [NatSet]: %1:[NatSet]
X --> #1:[NatSet]                                  X -->  %1:[NatSet]
Y --> #2:[NatSet]                                  Y --> mt
```

The above output illustrates a difference between unifiers returned by the built-in unification modulo axioms and substitutions (or unifiers) returned by variant generation or variant-based unification: two forms of fresh variables, the former *#n:Sort* and the new *%n:Sort*. Note that the two forms have different counters.

We can consider a more complex equational theory such as the one of Abelian groups specified in the following module; this theory could not be handled by Maude 2.6 because it is not strongly right irreducible.

```
fmod ABELIAN-GROUP is
  sorts Elem .
  op _+_ : Elem Elem -> Elem [comm assoc] .
  op -_ : Elem -> Elem .
  op 0 : ->  Elem .
  vars X Y Z : Elem  .
  eq X + 0 = X [variant] .
  eq X + (- X) = 0 [variant] .
  eq X + (- X) + Y = Y [variant] .
  eq - (- X) = X [variant] .
  eq - 0 = 0 [variant] .
  eq (- X) + (- Y) = -(X + Y) [variant] .
  eq -(X + Y) + Y = - X [variant] .
  eq -(- X + Y) = X + (- Y) [variant] .
  eq (- X) + (- Y) + Z  = -(X + Y) + Z [variant] .
  eq -(X + Y) + Y + Z = (- X) + Z [variant] .
endfm
```

The generation of the variants for the addition symbol provides 47 variants:

```
Maude> get variants in ABELIAN-GROUP : X + Y .
Variant 1                                        Variant 47
Elem: #1:Elem + #2:Elem    .................    Elem: - (%2:Elem + %3:Elem)
X --> #1:Elem                                    X --> %4:Elem + - (%1:Elem + %2:Elem)
Y --> #2:Elem                                    Y --> %1:Elem + - (%3:Elem + %4:Elem)
```

And the minus sign symbol has four variants:

```
Maude> get variants in ABELIAN-GROUP : - X .
Variant 1             Variant 2             Variant 3     Variant 4
Elem: - #1:Elem       Elem: %1:Elem         Elem: 0       Elem: %1:Elem + - %2:Elem
X --> #1:Elem         X --> - %1:Elem       X --> 0       X --> %2:Elem + - %1:Elem
```

Another interesting feature is that variant generation is *incremental*. In this way we are able to support general convergent equational theories modulo axioms that need not have the finite variant property. Let us consider the following functional module for addition `NAT-VARIANT` that does not have the finite variant property.

```
fmod NAT-VARIANT is
  sort Nat .
  op 0 : -> Nat .
  op s : Nat -> Nat .
  op _+_ : Nat Nat -> Nat .
```

```
  vars X Y : Nat .
  eq [base] : 0 + Y = Y [variant] .
  eq [ind] :  s(X) + Y = s(X + Y) [variant] .
endfm
```

On the one hand, it is possible to have a term with a finite number of most general variants although the theory does not have the finite variant property. For instance, the term `s(0) + X` has the single variant `s(X)`.

```
Maude> get variants in NAT-VARIANT : s(0) + X .
Variant 1
Nat: s(#1:Nat)
X --> #1:Nat
```

On the other hand, we can incrementally generate the variants of a term that we suspect does not have a finite number of most general variants. For instance, the term `X + s(0)` has an infinite number of most general variants. In such a case, Maude can either output all the variants to the screen (and the user can stop the process whenever she wants), or generate the first n variants by including a bound n in the command.

```
Maude> get variants [10] in NAT-VARIANT : X + s(0) .
Variant 1                                                  Variant 10
Nat: #1:Nat + s(0)   ..................................   Nat: s(s(s(s(s(0)))))
X --> #1:Nat                                               X --> s(s(s(s(0))))
```

Note that a third approach is to incrementally increase the bound and, if we obtain a number of variants smaller than the bound, then we know for sure that it had a finite number of most general variants.

4 Built-in Variant-Based Unification

The most natural application of variant generation is unification in an equational theory $(\Sigma, E \cup Ax)$ where the equations E can be oriented into convergent rules $\overrightarrow{E}$ modulo Ax. Intuitively, when we extend such an equational theory $(\Sigma, E \cup Ax)$ with a new equation `eq(x,x) = true`, two terms t and t' unify with substitution α modulo the equational theory if and only if $(\texttt{true}, \alpha)$ is a variant of the term $\texttt{eq}(t, t')$. The key distinction is one between *dedicated* unification algorithms for a limited set of axioms Ax (as in Sect. 2) and *generic* unification algorithms which can be applied to a much wider range of user-definable theories (namely convergent theories modulo axioms) and can even deal with incremental generation of infinite sets of unifiers.

Given a module `ModId` satisfying the requirements of Sect. 3 and being a finite variant theory, Maude provides a command for equational unification:

```
variant unify [ n ] in ⟨ModId⟩ :
    ⟨Term-1⟩ =? ⟨Term'-1⟩ /\ ... /\ ⟨Term-k⟩ =? ⟨Term'-k⟩ .
```

where $k \geq 1$ and n is an optional argument providing a bound on the number of unifiers requested, so that if the cardinality of the set of unifiers is greater than the specified bound, the unifiers beyond that bound are omitted.

Similarly to the incremental generation of variants, one can obtain an incremental number of unifiers for a given unification problem. Let us consider again the NAT-VARIANT module that does not have the finite variant property. On the one hand, it is possible to have a finite number of most general unifiers for a unification problem although the theory does not have the finite variant property.

```
Maude> variant unify in NAT-VARIANT : s(0) + X =? s(s(s(0)))  .
Unifier #1
X --> s(s(0))
```

On the other hand, we can approximate the number of unifiers of a unification problem that we suspect does not have a finite number of most general unifiers. For instance, the unification problem between terms `X + s(0)` and `s(s(s(0)))` has only one solution `X` $\mapsto$ `s(s(0))` and we can obtain that solution by including a bound in the command, as it is also done for variant generation.

```
Maude> variant unify [1] in NAT-VARIANT : X + s(0) =? s(s(s(0)))  .
Unifier #1
X --> s(s(0))
```

However, if we tried to obtain a second unifier, Maude would not stop because it would keep trying to generate a second unifier for a unification problem that has only one unifier, without knowing that it could stop.

5 Narrowing-Based Symbolic Reachability Analysis

The modern application of narrowing, when the rules R are understood as *transition rules*, is that of *symbolic reachability analysis* [15]. Specifically, we consider transition systems specified by order-sorted rewrite theories of the form mod $(\Sigma, E \cup Ax, R)$ endm where: (i) $E \cup Ax$ satisfies the requirements of Sect. 3, and (ii) the transition rules R are $E \cup Ax$-coherent and *topmost* (so that rewriting is always done at the top of the term). Then, narrowing modulo $E \cup Ax$ is a *complete* deductive method [15] for symbolic reachability analysis, i.e., for solving existential queries of the form $\exists \overline{x} : t \rightarrow^* t'$ where $\overline{x}$ are all the variables appearing in t and t', in the sense that the formula holds for $(\Sigma, E \cup Ax, R)$ iff there is a narrowing sequence $t \rightsquigarrow^*_{R,E \cup Ax} u$ such that u and t' have an $(E \cup Ax)$-unifier. Narrowing-based reachability was already introduced in Maude 2.4 [4] and Maude 2.6 [7] but now can be performed modulo theories with the finite variant property.

This symbolic reachability is supported by Full Maude's search command:

```
(search [ n,m ] in ⟨ModId⟩ : ⟨Term-1⟩ ⟨SearchArrow⟩ ⟨Term-2⟩ .)
```

where: n and m are optional arguments providing, respectively, a bound on the number of solutions and the maximum depth of the search; *ModId* is the module where the search takes place; *Term-1* is the starting term, which cannot be a variable but may contain variables; *Term-2* is the term specifying the pattern that has to be reached (some variables possibly shared with the starting term);

and *SearchArrow* is an arrow indicating the form of the narrowing proof, where `~>1` indicates a narrowing proof consisting of exactly one step; `~>+` indicates a proof of one or more steps; `~>*` indicates a proof of none, one, or more steps; and `~>!` indicates that the reached term cannot be further narrowed.

Consider again the typical example in Maude of a vending machine (e.g. in [7]) but now extended with the theory for Abelian groups shown above; the rules are coherent modulo the Abelian group theory by using a generic variable `Money`.

```
(mod AG-VENDING is
  sorts Item Items State Coin Money .
  subsort Item < Items . subsort Coin < Money .
  op __ : Items Items -> Items [assoc comm id: mt] .   op mt : -> Items .
  op <_|_> : Money Items -> State .
  ops a c : -> Item .   ops q $ : -> Coin .

  rl < M:Money | I:Items > => < M:Money + - $             | I:Items c > .
  rl < M:Money | I:Items > => < M:Money + - q + - q + - q | I:Items a > .

  eq $ = q + q + q + q [variant] . --- Property of the original vending
  machine example

  op _+_ : Money Money -> Money [comm assoc] .
  op -_ : Money -> Money .
  op 0 : -> Money .
  vars X Y Z : Money  .
  ... (here come the variant equations shown before for Abelian Group)
endm)
```

We can use the narrowing search command to answer the question: *Is there any combination of one or more coins that returns exactly an apple and a cake?* This can be done by searching for states that are reachable from a term `< M:Money | mt >` and match the pattern `< 0 | a c >` at the end.

```
Maude> (search [1] in AG-VENDING : < M:Money | mt > ~>* < 0 | a c > .)
Solution 1
M:Money --> q + q + q + q + q + q + q
```

Note that we must restrict the search to just one solution, because narrowing does not terminate for this reachability problem.

6 Applications

Unification and narrowing in Maude have opened up many applications. First, variant-based unification itself as described in Sect. 4. Several formal reasoning tools that either rely on unification capabilities, such as termination proofs [8] and proofs of local confluence and coherence [9], or rely on narrowing capabilities such as narrowing-based theorem proving [17] or testing [16]. Also, narrowing-based reachability analysis has evolved into *logical model checking* [1,11], where standard model checking cannot handle either infinite sets of initial states or infinite sets of reachable states but performing model checking from initial states with logical variables can handle these broader possibilities symbolically. The area of cryptographic protocol analysis has also benefited: the Maude-NPA tool [10] is the most successful example of combining narrowing and unification features in Maude. The Tamarin tool [13] also uses a variant-generation algorithm, although only for the Diffie-Hellman theory. Finally, several decision procedures for formula satisfiability modulo equational theories have been provided based on narrowing [18] or by variant generation in finite variant theories [14].

References

1. Bae, K., Escobar, S., Meseguer, J.: Abstract logical model checking of infinite-state systems using narrowing. In: van Raamsdonk, F., (ed.) 24th International Conference on Rewriting Techniques and Applications, RTA 2013, June 24–26, 2013, Eindhoven, The Netherlands. LIPIcs vol. 21, pp. 81–96. Schloss Dagstuhl - Leibniz-Zentrum fuer Informatik (2013)
2. Bouchard, C., Gero, K.A., Lynch, C., Narendran, P.: On forward closure and the finite variant property. In: Fontaine, P., Ringeissen, C., Schmidt, R.A. (eds.) FroCoS 2013. LNCS, vol. 8152, pp. 327–342. Springer, Heidelberg (2013)
3. Cholewa, A., Meseguer, J., Escobar, S.: Variants of variants and the finite variant property. Technical report, University of Illinois at Urbana-Champaign (2014). http://hdl.handle.net/2142/47117
4. Clavel, M., Durán, F., Eker, S., Escobar, S., Lincoln, P., Martí-Oliet, N., Meseguer, J., Talcott, C.: Unification and narrowing in maude 2.4. In: Treinen, R. (ed.) RTA 2009. LNCS, vol. 5595, pp. 380–390. Springer, Heidelberg (2009)
5. Clavel, M., Durán, F., Eker, S., Lincoln, P., Martí-Oliet, N., Meseguer, J., Talcott, C.: All About Maude - A High-Performance Logical Framework. LNCS, vol. 4350. Springer, Heidelberg (2007)
6. Comon-Lundh, H., Delaune, S.: The finite variant property: how to get rid of some algebraic properties. In: Giesl, J. (ed.) RTA 2005. LNCS, vol. 3467, pp. 294–307. Springer, Heidelberg (2005)
7. Durán, F., Eker, S., Escobar, S., Meseguer, J., Talcott, C.L.: Variants, unification, narrowing, and symbolic reachability inMaude 2.6. In: Schmidt-Schauß, M., (ed.) Proceedings of the 22ndInternational Conference on Rewriting Techniques and Applications, RTA2011, May 30 - June 1, 2011, Novi Sad, Serbia. LIPIcs, vol. 10, pp. 31–40. Schloss Dagstuhl - Leibniz-Zentrum fuer Informatik (2011)
8. Durán, F., Lucas, S., Meseguer, J.: Termination modulo combinations of equational theories. In: Ghilardi, S., Sebastiani, R. (eds.) FroCoS 2009. LNCS, vol. 5749, pp. 246–262. Springer, Heidelberg (2009)
9. Durán, F., Meseguer, J.: On the Church-Rosser and coherence properties of conditional order-sorted rewrite theories. J. Logic Algebraic Program. **81**(7–8), 816–850 (2012)
10. Escobar, S., Meadows, C., Meseguer, J.: Maude-NPA: Cryptographic protocol analysis modulo equationalproperties. In: Aldini, A., Barthe, G., Gorrieri, R. (eds.) FOSAD 2007. LNCS, vol. 5705, pp. 1–50. Springer, Heidelberg (2007)
11. Escobar, S., Meseguer, J.: Symbolic model checking of infinite-state systems using narrowing. In: Baader, F. (ed.) RTA 2007. LNCS, vol. 4533, pp. 153–168. Springer, Heidelberg (2007)
12. Escobar, S., Sasse, R., Meseguer, J.: Folding variant narrowing and optimal variant termination. J. Logic Algebraic Program. **81**(7–8), 898–928 (2012)
13. Meier, S., Schmidt, B., Cremers, C., Basin, D.: The TAMARIN prover for the symbolic analysis of security protocols. In: Sharygina, N., Veith, H. (eds.) CAV 2013. LNCS, vol. 8044, pp. 696–701. Springer, Heidelberg (2013)

14. Meseguer, J.: Variant-based satisfiability in initial algebras. Technical report, University of Illinois at Urbana-Champaign (2015). http://hdl.handle.net/2142/88408
15. Meseguer, J., Thati, P.: Symbolic reachability analysis using narrowing and its application to verification of cryptographic protocols. High. Order Symbolic Comput. **20**(1–2), 123–160 (2007)
16. Riesco, A.: Using big-step and small-step semantics in maude to perform declarative debugging. In: Codish, M., Sumii, E. (eds.) FLOPS 2014. LNCS, vol. 8475, pp. 52–68. Springer, Heidelberg (2014)
17. Rusu, V.: Combining theorem proving and narrowing for rewriting-logic specifications. In: Fraser, G., Gargantini, A. (eds.) TAP 2010. LNCS, vol. 6143, pp. 135–150. Springer, Heidelberg (2010)
18. Tushkanova, E., Giorgetti, A., Ringeissen, C., Kouchnarenko, O.: A rule-based system for automatic decidability and combinability. Sci. Comput. Program. **99**, 3–23 (2015)

Arithmetic Reasoning and Mechanizing Mathematics

Interpolant Synthesis for Quadratic Polynomial Inequalities and Combination with *EUF*

Ting Gan[1,4], Liyun Dai[1], Bican Xia[1], Naijun Zhan[2(✉)], Deepak Kapur[3], and Mingshuai Chen[2]

[1] LMAM & School of Mathematical Sciences, Peking University, Beijing, China
[2] State Key Laboratory of Computer Science, Institute of Software, Chinese Academy of Sciences, Beijing, China
znj@ios.ac.cn
[3] Department of Computer Science, University of New Mexico, Albuquerque, USA
[4] State Key Laboratory of Software Engineering, Wuhan University, Wuhan, China

Abstract. An algorithm for generating interpolants for formulas which are conjunctions of quadratic polynomial inequalities (both strict and nonstrict) is proposed. The algorithm is based on a key observation that quadratic polynomial inequalities can be linearized if they are concave. A generalization of Motzkin's transposition theorem is proved, which is used to generate an interpolant between two mutually contradictory conjunctions of polynomial inequalities, using semi-definite programming in time complexity $\mathcal{O}(n^3 + nm)$, where n is the number of variables and m is the number of inequalities (This complexity analysis assumes that despite the numerical nature of approximate SDP algorithms, they are able to generate correct answers in a fixed number of calls.). Using the framework proposed in [22] for combining interpolants for a combination of quantifier-free theories which have their own interpolation algorithms, a combination algorithm is given for the combined theory of concave quadratic polynomial inequalities and the equality theory over uninterpreted functions (*EUF*).

Keywords: Program verification · Interpolant · Concave quadratic polynomial · Motzkin's theorem · SOS · Semi-definite programming

1 Introduction

It is well known that the bottleneck of existing verification techniques including theorem proving, model-checking, abstraction and so on is the scalability. Interpolation-based technique provide a powerful mechanism for local and modular reasoning, which provides an effective solution to this challenge. The study of interpolation was pioneered by Krajíček [14] and Pudlák [19] in connection with theorem proving, by McMillan in connection with model-checking [16], by Graf and Saïdi [9], McMillan [17] and Henzinger et al. [10] in connection with abstraction like CEGAR, by Wang et al. [11] in connection with machine-learning

N. Olivetti and A. Tiwari (Eds.): IJCAR 2016, LNAI 9706, pp. 195–212, 2016.
DOI: 10.1007/978-3-319-40229-1_14

based invariant generation. Since then, developing efficient algorithms for generating interpolants for various theories and their use in verification have become an active research area [3,10,12,13,17,18,20,26,26]. In addition, D'Silva et al. [6] investigated strengths of various interpolants.

Methods have been developed for generating interpolants for Presburger arithmetic, decidable fragments of first-order logic, theory of equality over uninterpreted functions as well as their combination. However, in the literature, there is little work on how to synthesize non-linear interpolants, although nonlinear polynomials inequalities have been found useful to express invariants for software involving number theoretic functions as well as hybrid systems [27,28]. In [5], Dai et al. had a first try and gave an algorithm for generating interpolants for conjunctions of mutually contradictory nonlinear polynomial inequalities based on the existence of a witness guaranteed by Stengle's Positivstellensatz [23] that can be computed using semi-definite programming (SDP). Their algorithm is incomplete in general but if every variable ranges over a bounded interval (called Archimedean condition), then their algorithm is complete. A major limitation of their work is that two mutually contradictory formulas α, β must have the same set of variables.

We propose an algorithm to generate interpolants for quadratic polynomial inequalities (including strict inequalities). Based on the insight that for analyzing the solution space of concave quadratic polynomial inequalities, it suffices to linearize them. A generalization of Motzkin's transposition theorem is proved to be applicable for concave quadratic polynomial inequalities (both strict and nonstrict). Using this, we prove the existence of an interpolant for two mutually contradictory conjunctions α, β of concave quadratic polynomial inequalities. The proposed algorithm is recursive with the basis step of the algorithm relying on an additional condition (called the **NSC** condition). In this case, an interpolant output by the algorithm is a strict or a nonstrict inequality similar to the linear case. If **NSC** is not satisfied, then linear equalities on variables are derived resulting in simpler interpolation problems over fewer variables; the algorithm is recursively invoked on these smaller problem. The output of this recursive algorithm is in general an interpolant that is a disjunction of conjunction of polynomial inequalities. **NSC** can be checked in polynomial time by SDP algorithms; even though such algorithms are not exact and produce numerical errors, they often generate acceptable results in a few calls. It is proved that the interpolation algorithm is of polynomial time complexity in the number of variables and polynomial inequalities given that the time complexity of SDP algorithms is polynomial in the size of their input; this assumes that an SDP tool returns an approximate answer sufficient to generate a correct interpolant in a fixed number of calls.

Later, we develop a combination algorithm for generating interpolants for the combination of quantifier-free theory of concave quadratic polynomial inequalities and equality theory over uninterpreted function symbols (EUF). We use the hierarchical calculus framework proposed in [22] and used in [20] for combining linear inequalities with EUF. We show that concavity condition on quadratic polynomials inequalities disallows derivation of nonlinear equalities of degree ≥ 2; further, under **NSC** on concave quadratic polynomial inequalities, only linear inequalities can be used to derive possible linear equalities. As a result, the algorithm for deducing equalities from linear inequalities in [20] as well as the SEP algorithm for separating terms expressed in common symbols in α, β can be used for interpolation generation for the combined theory of quadratic polynomial inequalities and *EUF*.

A prototypical implementation indicates the scalability and efficiency of the proposed approach.

The paper is organized as follows. After introducing some preliminaries in the next section, Sect. 3 discusses the linearization of concave quadratic polynomial. Section 4 presents an approach for computing an interpolant for two mutually contradictory conjunctions α, β of concave quadratic polynomial inequalities using SDP. Section 5 extends this algorithm to the combined theory of concave quadratic inequalities and *EUF*. Section 6 presents a preliminary implementation of the proposed algorithms and gives some comparison with related work. We draw a conclusion in Sect. 7. Because of space limit, we omit all proofs, please refer to the full version [8] for the details.

2 Preliminaries

Let $\mathbb{Q}$ and $\mathbb{R}$ be the set of rational and real numbers, respectively. Let $\mathbb{R}[\mathbf{x}]$ be the polynomial ring over $\mathbb{R}$ with variables $\mathbf{x} = (\mathbf{x}_1, \cdots, \mathbf{x}_n)$. An atomic polynomial formula φ is of the form $p(\mathbf{x}) \diamond 0$, where $p(\mathbf{x}) \in \mathbb{R}[\mathbf{x}]$, and $\diamond$ can be any of $>, \geq$. Let $\mathbf{PT}(\mathbb{R})$ be a first-order theory of polynomials with real coefficients. In this paper, we are focusing on quantifier-free fragment of $\mathbf{PT}(\mathbb{R})$. Later we discuss quantifier-free theory of equality of terms over uninterpreted function symbols and its combination with the quantifier-free fragment of $\mathbf{PT}(\mathbb{R})$. Let Σ be a set of (new) function symbols and $\mathbf{PT}(\mathbb{R})^{\Sigma}$ be the extension of the quantifier-free theory with uninterpreted function symbols in Σ.

For convenience, we use $\bot$ to stand for *false* and $\top$ for *true* in what follows.

Craig showed that given two formulas ϕ and ψ in a first-order logic $\mathcal{T}$ s.t. $\phi \models \psi$, there always exists an *interpolant* I over the common symbols of ϕ and ψ s.t. $\phi \models I, I \models \psi$. In the verification literature, this terminology has been abused following [17], where a *reverse interpolant* (coined by Kovács and Voronkov in [13]) I over the common symbols of ϕ and ψ is defined by

Definition 1. *Given ϕ and ψ in a theory $\mathcal{T}$ s.t. $\phi \wedge \psi \models_{\mathcal{T}} \bot$, a formula I is a* (reverse) interpolant *of ϕ and ψ if* (i) *$\phi \models_{\mathcal{T}} I$;* (ii) *$I \wedge \psi \models_{\mathcal{T}} \bot$; and* (iii) *$I$ only contains common symbols and free variables shared by ϕ and ψ.*

Clearly, $\phi \models_{\mathcal{T}} \psi$ iff $\phi \wedge \neg\psi \models_{\mathcal{T}} \bot$. Thus, I is an interpolant of ϕ and ψ iff I is a reverse interpolant of ϕ and $\neg\psi$. We abuse the terminology by calling reverse interpolants as interpolants.

2.1 Motzkin's Transposition Theorem

Motzkin's transposition theorem [21] is one of the fundamental results about linear inequalities; it also served as a basis of the interpolant generation algorithm for the quantifier-free theory of linear inequalities in [20].

Theorem 1 (Motzkin's transposition theorem [21]). *Let A and B be matrices and let $\boldsymbol{\alpha}$ and $\boldsymbol{\beta}$ be column vectors. Then there exists a vector $\mathbf{x}$ with $A\mathbf{x} \geq \boldsymbol{\alpha}$ and $B\mathbf{x} > \boldsymbol{\beta}$, iff for all row vectors $\mathbf{y}, \mathbf{z} \geq 0$:*

$$(i) \text{ if } \mathbf{y}A + \mathbf{z}B = 0 \text{ then } \mathbf{y}\boldsymbol{\alpha} + \mathbf{z}\boldsymbol{\beta} \leq 0;$$
$$(ii) \text{ if } \mathbf{y}A + \mathbf{z}B = 0 \text{ and } \mathbf{z} \neq 0 \text{ then } \mathbf{y}\boldsymbol{\alpha} + \mathbf{z}\boldsymbol{\beta} < 0.$$

The following variant of Theorem 1 is used later.

Corollary 1. *Let $A \in \mathbb{R}^{r \times n}$ and $B \in \mathbb{R}^{s \times n}$ be matrices and $\boldsymbol{\alpha} \in \mathbb{R}^r$ and $\boldsymbol{\beta} \in \mathbb{R}^s$ be column vectors, where $A_i, i = 1, \ldots, r$ is the ith row of A and $B_j, j = 1, \ldots, s$ is the jth row of B. There does not exist a vector $\mathbf{x}$ with $A\mathbf{x} \geq \boldsymbol{\alpha}$ and $B\mathbf{x} > \boldsymbol{\beta}$, iff there exist real numbers $\lambda_1, \ldots, \lambda_r \geq 0$ and $\eta_0, \eta_1, \ldots, \eta_s \geq 0$ s.t.*

$$\sum_{i=1}^{r} \lambda_i (A_i \mathbf{x} - \alpha_i) + \sum_{j=1}^{s} \eta_j (B_j \mathbf{x} - \beta_j) + \eta_0 \equiv 0 \quad \textit{with} \quad \sum_{j=0}^{s} \eta_j > 0. \tag{1}$$

3 Concave Quadratic Polynomials and their Linearization

Given $n \times n$-matrix A, we say A is *negative semi-definite*, written as $A \preceq 0$, if for every vector $\mathbf{x}$, $\mathbf{x}^{\mathrm{T}} A \mathbf{x} \leq 0$, and *positive semi-definite*, written as $A \succeq 0$, if for every vector $\mathbf{x}$, $\mathbf{x}^{\mathrm{T}} A \mathbf{x} \geq 0$. Let $A = (a_{ij})$ and $B = (b_{ij})$ be two matrices in $\mathbb{R}^{m \times n}$, the *inner product* of A and B, denoted by $\langle A, B \rangle$, is defined as $\langle A, B \rangle = \sum_{i=1}^{m} \sum_{j=1}^{n} a_{ij} \times b_{ij}$.

Definition 2 (Concave Quadratic). *A polynomial $f \in \mathbb{R}[\mathbf{x}]$ is called* concave quadratic (CQ) *if the following two conditions hold:*

(i) *f has total degree at most 2, i.e., it has the form $f = \mathbf{x}^{\mathrm{T}} A \mathbf{x} + 2\boldsymbol{\alpha}^{\mathrm{T}} \mathbf{x} + a$, where A is a real symmetric matrix, $\boldsymbol{\alpha}$ is a column vector and $a \in \mathbb{R}$;*
(ii) the matrix A is negative semi-definite.

It is easy to see that if $f \in \mathbb{R}[\mathbf{x}]$ is linear, then f is CQ because its total degree is 1 and the corresponding A is 0 which is of course negative semi-definite.

A quadratic polynomial $f(\mathbf{x}) = \mathbf{x}^{\mathrm{T}} A \mathbf{x} + 2\boldsymbol{\alpha}^{\mathrm{T}} \mathbf{x} + a$ can also be represented as an inner product of matrices, *i.e.*, $\left\langle P, \begin{pmatrix} 1 & \mathbf{x}^{\mathrm{T}} \\ \mathbf{x} & \mathbf{x}\mathbf{x}^{\mathrm{T}} \end{pmatrix} \right\rangle$, *with* P as $\begin{pmatrix} a & \alpha^{\mathrm{T}} \\ \alpha & A \end{pmatrix}$.

3.1 Linearization

Given a quadratic polynomial $f(\mathbf{x}) = \left\langle P, \begin{pmatrix} 1 & \mathbf{x}^{\mathrm{T}} \\ \mathbf{x} & \mathbf{x}\mathbf{x}^{\mathrm{T}} \end{pmatrix} \right\rangle$, its *linearization* is defined as $f(\mathbf{x}) = \left\langle P, \begin{pmatrix} 1 & \mathbf{x}^{\mathrm{T}} \\ \mathbf{x} & \boldsymbol{X} \end{pmatrix} \right\rangle$, where $\boldsymbol{X}$ is a symmetric matrix and $\begin{pmatrix} 1 & \mathbf{x}^{\mathrm{T}} \\ \mathbf{x} & \boldsymbol{X} \end{pmatrix} \succeq 0$.

Let $\overline{\boldsymbol{X}} = (\boldsymbol{X}_{(1,1)}, \boldsymbol{X}_{(2,1)}, \boldsymbol{X}_{(2,2)}, \ldots, \boldsymbol{X}_{(k,1)}, \ldots, \boldsymbol{X}_{(k,k)}, \ldots, \boldsymbol{X}_{(n,1)}, \ldots, \boldsymbol{X}_{(n,n)})$ be the vector variable with dimension $\frac{n(n+1)}{2}$ corresponding to the matrix $\boldsymbol{X}$. Since $\boldsymbol{X}$ is a symmetric matrix, $\left\langle P, \begin{pmatrix} 1 & \mathbf{x}^{\mathrm{T}} \\ \mathbf{x} & \boldsymbol{X} \end{pmatrix} \right\rangle$ is a linear expression in $\mathbf{x}, \overline{\boldsymbol{X}}$.

Consider quadratic polynomials f_i and g_j $(i = 1, \ldots, r,\ j = 1, \ldots, s)$,

$$f_i = \mathbf{x}^{\mathrm{T}} A_i \mathbf{x} + 2\boldsymbol{\alpha}_i^{\mathrm{T}} \mathbf{x} + a_i, \quad g_j = \mathbf{x}^{\mathrm{T}} B_j \mathbf{x} + 2\boldsymbol{\beta}_j^{\mathrm{T}} \mathbf{x} + b_j,$$

where A_i, B_j are symmetric $n \times n$ matrices, $\boldsymbol{\alpha}_i, \boldsymbol{\beta}_j \in \mathbb{R}^n$, and $a_i, b_j \in \mathbb{R}$. Then

$$f_i(\mathbf{x}) = \left\langle P_i, \begin{pmatrix} 1 & \mathbf{x}^{\mathrm{T}} \\ \mathbf{x} & \mathbf{x}\mathbf{x}^{\mathrm{T}} \end{pmatrix} \right\rangle, \ g_j(\mathbf{x}) = \left\langle Q_j, \begin{pmatrix} 1 & \mathbf{x}^{\mathrm{T}} \\ \mathbf{x} & \mathbf{x}\mathbf{x}^{\mathrm{T}} \end{pmatrix} \right\rangle,$$

where $P_i = \begin{pmatrix} a_i & \boldsymbol{\alpha}_i^{\mathrm{T}} \\ \boldsymbol{\alpha}_i & A_i \end{pmatrix}$, $Q_j = \begin{pmatrix} b_j & \boldsymbol{\beta}_j^{\mathrm{T}} \\ \boldsymbol{\beta}_j & B_j \end{pmatrix}$ are $(n+1) \times (n+1)$ matrices.

For CQ polynomials f_is and g_js, let

$$K \hat{=} \{\mathbf{x} \in \mathbb{R}^n \mid f_1(\mathbf{x}) \geq 0, \ldots, f_r(\mathbf{x}) \geq 0, g_1(\mathbf{x}) > 0, \ldots, g_s(\mathbf{x}) > 0\}, \tag{2}$$

$$K_1 \hat{=} \{\mathbf{x} \mid \exists \boldsymbol{X}. \begin{pmatrix} 1 & \mathbf{x}^{\mathrm{T}} \\ \mathbf{x} & \boldsymbol{X} \end{pmatrix} \succeq 0 \wedge \bigwedge_{i=1}^{r} \left\langle P_i, \begin{pmatrix} 1 & \mathbf{x}^{\mathrm{T}} \\ \mathbf{x} & \boldsymbol{X} \end{pmatrix} \right\rangle \geq 0 \wedge \bigwedge_{j=1}^{s} \left\langle Q_j, \begin{pmatrix} 1 & \mathbf{x}^{\mathrm{T}} \\ \mathbf{x} & \boldsymbol{X} \end{pmatrix} \right\rangle > 0\}. \tag{3}$$

In [7,15], when K and K_1 are defined only with f_is without g_js, *i.e.*, only with nonstrict inequalities, it is proved that $K = K_1$. By Theorem 2 below, we show that $K = K_1$ also holds even in the presence of strict inequalities when f_i and g_j are CQ. So, when f_is and g_js are CQ, the CQ polynomial inequalities can be transformed equivalently to a set of linear inequality constraints and a positive semi-definite constraint.

Theorem 2. *Let $f_1, \ldots, f_r$, $g_1, \ldots, g_s \in \mathbb{R}[\mathbf{x}]$ be CQ, K and K_1 as above, then $K = K_1$.*

3.2 Motzkin's Theorem in Matrix Form

If $\left\langle P, \begin{pmatrix} 1 & \mathbf{x}^{\mathrm{T}} \\ \mathbf{x} & \boldsymbol{X} \end{pmatrix} \right\rangle$ is seen as a linear expression in $\mathbf{x}, \overline{\boldsymbol{X}}$, then Corollary 1 can be reformulated as:

Corollary 2. *Let $\mathbf{x}$ be a column vector variable with dimension n and $\boldsymbol{X}$ be an $n \times n$ symmetric matrix variable. Suppose $P_1, \ldots, P_r$ and $Q_1, \ldots, Q_s$ are $(n+1) \times (n+1)$ symmetric matrices. Let*

$$V \hat{=} \{(\mathbf{x}, \boldsymbol{X}) \mid \bigwedge_{i=1}^{r} \left\langle P_i, \begin{pmatrix} 1 & \mathbf{x}^{\mathrm{T}} \\ \mathbf{x} & \boldsymbol{X} \end{pmatrix} \right\rangle \geq 0, \bigwedge_{i=1}^{s} \left\langle Q_j, \begin{pmatrix} 1 & \mathbf{x}^{\mathrm{T}} \\ \mathbf{x} & \boldsymbol{X} \end{pmatrix} \right\rangle > 0\},$$

then $V = \emptyset$ *iff there exist* $\lambda_1, \ldots, \lambda_r \geq 0$ *and* $\eta_0, \eta_1, \ldots, \eta_s \geq 0$ *such that*

$$\sum_{i=1}^{r} \lambda_i \left\langle P_i, \begin{pmatrix} 1 & \mathbf{x}^{\mathrm{T}} \\ \mathbf{x} & \boldsymbol{X} \end{pmatrix} \right\rangle + \sum_{j=1}^{s} \eta_j \left\langle Q_j, \begin{pmatrix} 1 & \mathbf{x}^{\mathrm{T}} \\ \mathbf{x} & \boldsymbol{X} \end{pmatrix} \right\rangle + \eta_0 \equiv 0, \quad \eta_0 + \eta_1 + \ldots + \eta_s > 0.$$

4 Interpolants for Concave Quadratic Polynomial Inequalities

Problem 1: Given two formulas ϕ and ψ on n variables with $\phi \wedge \psi \models \bot$, where

$$\phi = f_1 \geq 0 \wedge \ldots \wedge f_{r_1} \geq 0 \wedge g_1 > 0 \wedge \ldots \wedge g_{s_1} > 0,$$
$$\psi = f_{r_1+1} \geq 0 \wedge \ldots \wedge f_r \geq 0 \wedge g_{s_1+1} > 0 \wedge \ldots \wedge g_s > 0,$$

in which $f_1, \ldots, f_r, g_1, \ldots, g_s$ are all CQ. Our goal is to develop an algorithm to generate a (reverse) Craig interpolant I for ϕ and ψ, on the common variables of ϕ and ψ, s.t. $\phi \models I$ and $I \wedge \psi \models \bot$. We use $\mathbf{x} = (x_1, \ldots, x_d)$ to stand for the common variables appearing in both ϕ and ψ, $\mathbf{y} = (y_1, \ldots, y_u)$ for the variables appearing only in ϕ and $\mathbf{z} = (z_1, \ldots, z_v)$ for the variables appearing only in ψ, where $d+u+v = n$. We call the conjunctive theory of CQ polynomial inequalities as *CQI*.

The proposed Algorithm IG-CQI in Sect. 4.5 is recursive: the base case is when no sum of squares (SOS) polynomial can be generated by a nonpositive constant combination of the polynomials in nonstrict inequalities in $\phi \wedge \psi$. When this condition is not satisfied, then identify variables which can be eliminated by replacing them by linear expressions in terms of other variables and generate equisatisfiable problem with fewer variables on which the algorithm can be recursively invoked.

4.1 NSC Condition and Generalization of Motzkin's Theorem

Definition 3. *Formulas* ϕ *and* ψ *in Problem 1 satisfy the non-existence of an SOS polynomial condition (***NSC***) iff there do not exist* $\delta_1 \geq 0, \ldots, \delta_r \geq 0$*, s.t.* $-(\delta_1 f_1 + \ldots + \delta_r f_r)$ *is a non-zero SOS.*

Note that nonnegative quadratic polynomials are all SOS. So, the above condition implies that there is no nonnegative constant combination of nonstrict inequalities which is always *nonpositive*. If quadratic polynomials appearing in ϕ and ψ are linearized, then the above condition is equivalent to requiring that every nonnegative linear combination of the linearization of nonstrict inequalities in ϕ and ψ is *negative*.

The following theorem is a generalization of Motzkin's theorem to CQI and gives a method when **NSC** is satisfied, for generating an interpolant by considering linearization of ϕ, ψ in Problem 1 and using Corollary 2.

Theorem 3. *Let $f_1, \ldots, f_r, g_1, \ldots, g_s$ be CQ polynomials in Problem 1. If* **NSC** *holds, then there exist $\lambda_i \geq 0$ $(i = 1, \cdots, r)$, $\eta_j \geq 0$ $(j = 0, 1, \cdots, s)$ and a quadratic SOS polynomial $h \in \mathbb{R}[\mathbf{x}, \mathbf{y}, \mathbf{z}]$ such that*

$$\sum_{i=1}^{r} \lambda_i f_i + \sum_{j=1}^{s} \eta_j g_j + \eta_0 + h \equiv 0, \quad \eta_0 + \eta_1 + \ldots + \eta_s = 1. \tag{4}$$

4.2 Base Case: Generating Interpolant when NSC is Satisfied

Using Theorem 3, an interpolant for ϕ and ψ is generated from the SOS polynomial h by splitting it into two SOS polynomials as shown below.

Theorem 4. *Let ϕ and ψ be as in Problem 1 with $\phi \wedge \varphi \models \perp$, which satisfy* **NSC**. *Then there exist $\lambda_i \geq 0$ $(i = 1, \cdots, r)$, $\eta_j \geq 0$ $(j = 0, 1, \cdots, s)$ and two quadratic SOS polynomial $h_1 \in \mathbb{R}[\mathbf{x}, \mathbf{y}]$ and $h_2 \in \mathbb{R}[\mathbf{x}, \mathbf{z}]$ such that*

$$\sum_{i=1}^{r} \lambda_i f_i + \sum_{j=1}^{s} \eta_j g_j + \eta_0 + h_1 + h_2 \equiv 0, \quad \eta_0 + \eta_1 + \ldots + \eta_s = 1. \tag{5}$$

Let $I = \sum_{i=1}^{r_1} \lambda_i f_i + \sum_{j=1}^{s_1} \eta_j g_j + \eta_0 + h_1$. Then $I \in \mathbb{R}[\mathbf{x}]$, and if $\sum_{j=0}^{s_1} \eta_j > 0$, then $I > 0$ is an interpolant otherwise $I \geq 0$ is an interpolant.

Further, we can prove that h, h_1, h_2 have the following form:

$$\begin{aligned}
(\mathrm{H}): h(\mathbf{x}, \mathbf{y}, \mathbf{z}) &= a_1(y_1 - l_1(\mathbf{x}, y_2, \ldots, y_u))^2 + \cdots + a_u(y_u - l_u(\mathbf{x}))^2 + \\
& a_{u+1}(z_1 - l_{u+1}(\mathbf{x}, z_2, \ldots, z_v))^2 + \cdots + a_{u+v}(z_v - l_{u+v}(\mathbf{x}))^2 + a_{u+v+1}(x_1 - \\
& l_{u+v+1}(x_2, \ldots, x_d))^2 + \cdots + a_{u+v+d}(x_d - l_{u+v+d})^2 + a_{u+v+d+1}, \\
(\mathrm{H1}): h_1(\mathbf{x}, \mathbf{y}) &= a_1(y_1 - l_1(\mathbf{x}, y_2, \ldots, y_u))^2 + \cdots + a_u(y_u - l_u(\mathbf{x}))^2 + \tfrac{a_{u+v+1}}{2}(x_1 - \\
& l_{u+v+1}(x_2, \ldots, x_d))^2 + \cdots + \tfrac{a_{u+v+d}}{2}(x_d - l_{u+v+d})^2 + \tfrac{a_{u+v+d+1}}{2}, \\
(\mathrm{H2}): h_2(\mathbf{x}, \mathbf{z}) &= a_{u+1}(z_1 - l_{u+1}(\mathbf{x}, z_2, \ldots, z_v))^2 + \cdots + a_{u+v}(z_v - l_{u+v}(\mathbf{x}))^2 + \\
& \tfrac{a_{u+v+1}}{2}(x_1 - l_{u+v+1}(x_2, \ldots, x_d))^2 + \cdots + \tfrac{a_{u+v+d}}{2}(x_d - l_{u+v+d})^2 + \tfrac{a_{u+v+d+1}}{2},
\end{aligned}$$

where $a_i \geq 0$ and l_j's are linear expressions. These forms of h_1, h_2 are used to generate equalities among variables later in the algorithm when **NSC** is not satisfied.

4.3 Computing Interpolant Using Semi-definite Programming

Let $W = \begin{pmatrix} 1 & \mathbf{x}^{\mathrm{T}} & \mathbf{y}^{\mathrm{T}} & \mathbf{z}^{\mathrm{T}} \\ \mathbf{x} & \mathbf{x}\mathbf{x}^{\mathrm{T}} & \mathbf{x}\mathbf{y}^{\mathrm{T}} & \mathbf{x}\mathbf{z}^{\mathrm{T}} \\ \mathbf{y} & \mathbf{y}\mathbf{x}^{\mathrm{T}} & \mathbf{y}\mathbf{y}^{\mathrm{T}} & \mathbf{y}\mathbf{z}^{\mathrm{T}} \\ \mathbf{z} & \mathbf{z}\mathbf{x}^{\mathrm{T}} & \mathbf{z}\mathbf{y}^{\mathrm{T}} & \mathbf{z}\mathbf{z}^{\mathrm{T}} \end{pmatrix}$, $f_i = \langle P_i, W \rangle$, $g_j = \langle Q_j, W \rangle$, where P_i and Q_j are $(n+1) \times (n+1)$ matrices, and $h_1 = \langle M, W \rangle$, $h_2 = \langle \hat{M}, W \rangle$, $M = (M_{ij})_{4\times 4}, \hat{M} = (\hat{M}_{ij})_{4\times 4}$ with appropriate dimensions, e.g., $M_{12} \in \mathbb{R}^{1\times d}$ and $\hat{M}_{34} \in \mathbb{R}^{u\times v}$. Then, with **NSC**, by Theorem 4, computing the interpolant is reduced to the following **SDP** feasibility problem.

Find: $\lambda_1, \ldots, \lambda_r, \eta_1, \ldots, \eta_s \in \mathbb{R}$ and symmetric matrices $M, \hat{M} \in \mathbb{R}^{(n+1)\times(n+1)}$ s.t.

$$\begin{cases} \sum_{i=1}^{r} \lambda_i P_i + \sum_{j=1}^{s} \eta_j Q_j + \eta_0 E_{(1,1)} + M + \hat{M} = 0, \sum_{j=1}^{s} \eta_j = 1, \\ M_{41} = (M_{14})^{\mathrm{T}} = 0, M_{42} = (M_{24})^{\mathrm{T}} = 0, M_{43} = (M_{34})^{\mathrm{T}} = 0, M_{44} = 0, \\ \hat{M}_{31} = (\hat{M}_{13})^{\mathrm{T}} = 0, \hat{M}_{32} = (\hat{M}_{23})^{\mathrm{T}} = 0, \hat{M}_{33} = 0, \hat{M}_{34} = (\hat{M}_{43})^{\mathrm{T}} = 0, \\ M \succeq 0, \hat{M} \succeq 0, \lambda_i \geq 0, \eta_j \geq 0, \text{ for } i = 1, \ldots, r, j = 1, \ldots, s, \end{cases}$$

where $E_{(1,1)}$ is a $(n+1)\times(n+1)$ matrix, whose all entries are 0 except for $(1,1) = 1$.

This standard **SDP** feasibility problem can be efficiently solved by **SDP** solvers such as CSDP [1], SDPT3 [24], etc. A major weakness of these algorithms is their incompleteness, however.

Approximate Nature of SDP Algorithms. Even though known SDP algorithms are of polynomial complexity, they are numerical and are not guaranteed to produce exact answers; they are however able to generate results within a very small threshold in a fixed number of iterations. Such techniques are thus considerably more attractive than solving the Problem 1 using exact symbolic methods of high complexity. This is especially critical for scaling our approach. To guarantee the soundness of our approach, we check results produced by approximate numerical algorithms by symbolic checking [4] and numeric-symbolic method [25] to verify whether an interpolant so computed does indeed satisfy the conditions in Definition 1. If not, we can tone down the threshold of the **SDP** and repeat the above procedure.

4.4 General Case

The case of $Var(\phi) \subset Var(\psi)$ is easy: ϕ itself serves as an interpolant of ϕ and ψ. We thus assume that $Var(\phi) \not\subseteq Var(\psi)$. If ϕ and ψ do not satisfy **NSC**, then an SOS polynomial $h(\mathbf{x}, \mathbf{y}, \mathbf{z}) = -(\sum_{i=1}^{r} \lambda_i f_i)$ can be computed which can be split into two SOS polynomials $h_1(\mathbf{x}, \mathbf{y})$ and $h_2(\mathbf{x}, \mathbf{z})$ as discussed in Subsect. 4.2. Then an SOS polynomial $f(\mathbf{x})$ s.t. $\phi \models f(\mathbf{x}) \geq 0$ and $\psi \models -f(\mathbf{x}) \geq 0$ can be constructed by setting $f(\mathbf{x}) = (\sum_{i=1}^{r_1} \delta_i f_i) + h_1 = -(\sum_{i=r_1+1}^{r} \delta_i f_i) - h_2, \delta_i \geq 0$. We show below how "simpler" interpolation subproblems ϕ', ψ' are constructed from ϕ and ψ using f.

Lemma 1. *If* **NSC** *is not satisfied, then there exists $f \in \mathbb{R}[\mathbf{x}]$ s.t. $\phi \Leftrightarrow \phi_1 \vee \phi_2$ and $\psi \Leftrightarrow \psi_1 \vee \psi_2$, where,*

$$\phi_1 = (f > 0 \wedge \phi),\ \phi_2 = (f = 0 \wedge \phi),\ \psi_1 = (-f > 0 \wedge \psi),\ \psi_2 = (f = 0 \wedge \psi). \quad (6)$$

It easily follows that an interpolant I for ϕ and ψ can be constructed from an interpolant $I_{2,2}$ for ϕ_2 and ψ_2.

Theorem 5. *Let ϕ, ψ, $\phi_1, \phi_2, \psi_1, \psi_2$ be same as in Lemma 1, $I_{2,2}$ be an interpolant for ϕ_2 and ψ_2, then $I := (f > 0) \vee (f \geq 0 \wedge I_{2,2})$ is an interpolant for ϕ and ψ.*

If h and hence h_1, h_2 have a positive constant $a_{u+v+d+1} > 0$, then f cannot be 0, implying that ϕ_2 and ψ_2 are $\bot$. We thus have

Theorem 6. *With ϕ, ψ, $\phi_1, \phi_2, \psi_1, \psi_2$ as in Lemma 1 and h has $a_{u+v+d+1} > 0$, then $f > 0$ is an interpolant for ϕ and ψ.*

In case h does not have a constant, *i.e.*, $a_{u+v+d+1} = 0$, from the fact that h_1 is an SOS and has the form (H1), each nonzero square term in h_1 is identically 0. This implies that some of the variables in $\mathbf{x}, \mathbf{y}$ can be linearly expressed in term of other variables; the same argument applies to h_2 as well. In particular, at least one variable is eliminated in both ϕ_2 and ψ_2, reducing the number of variables appearing in ϕ and ψ, which ensures the termination of the algorithm.

Theorem 7. *If h above does not have a constant, i.e., if $a_{u+v+d+1} = 0$, by eliminating (at least one) variables in ϕ and ψ in terms of other variables as derived from $h_1 = 0$, $h_2 = 0$, mutually contradictory formulas ϕ', ψ' with fewer variables are derived by*

$$\phi' = \bigwedge_{i=1}^{r_1} \hat{f}_i \geq 0 \wedge \bigwedge_{j=1}^{s_1} \hat{g}_j > 0, \quad \psi' = \bigwedge_{i=r_1+1}^{r} \hat{f}_i \geq 0 \wedge \bigwedge_{j=s_1+1}^{s} \hat{g}_j > 0,$$

where $\hat{f}_i$s and $\hat{g}_j$s are derived from the respective f_i and g_i by replacing the eliminated variable(s) with the corresponding resulting expression(s).

The following simple example illustrates how the above construction works.

Example 1. Let $f_1 = x_1, f_2 = x_2, f_3 = -x_1^2 - x_2^2 - 2x_2 - z^2, g_1 = -x_1^2 + 2x_1 - x_2^2 + 2x_2 - y^2$. Two formulas $\phi := (f_1 \geq 0) \wedge (f_2 \geq 0) \wedge (g_1 > 0)$, $\psi := (f_3 \geq 0)$. $\phi \wedge \psi \models \bot$. **NSC** does not hold, since $h = -(0f_1 + 2f_2 + f_3) = x_1^2 + x_2^2 + z^2$ is an SOS; h is split into $h_1 = \frac{1}{2}x_1^2 + \frac{1}{2}x_2^2$, $h_2 = \frac{1}{2}x_1^2 + \frac{1}{2}x_2^2 + z^2$. Thus $f = 0f_1 + 2f_2 + h_1 = \frac{1}{2}x_1^2 + \frac{1}{2}x_2^2 + 2x_2$.

For the recursive call, we construct ϕ' from ϕ by adding $x_1 = 0, x_2 = 0$ derived from $h_1 = 0$; similarly ψ' is constructed from ψ by adding $x_1 = x_2 = 0, z = 0$ derived from $h_2 = 0$. That is, $\phi' = 0 \geq 0 \wedge 0 \geq 0 \wedge -y^2 > 0 = \bot$, $\psi' = 0 \geq 0 = \top$. Thus, $I(\phi', \psi') := (0 > 0) = \bot$ is an interpolant for (ϕ', ψ').

An interpolant for ϕ and ψ is thus $(f(x) > 0) \vee (f(x) = 0 \wedge I(\phi', \psi'))$, which is $\frac{1}{2}x_1^2 + \frac{1}{2}x_2^2 + 2x_2 > 0$.

4.5 Algorithms

The above recursive approach is formally described as Algorithm 2. For the base case when ϕ, ψ satisfy **NSC**, it invokes Algorithm 1 using known SDP algorithms. For a predefined threshold, an **SDP** problem can be solved in polynomial time, say $g(k)$, where k is the input size [7]. Further its solution can be checked to determine whether a formula thus generated is indeed an interpolant; in case of failure, the process is repeated typically leading to convergence in a few iterations.

Algorithm 1. `Interpolation Generation for NSC Case (IG-NSC)`

input : ϕ and ψ satisfying **NSC**, and $\phi \wedge \psi \models \bot$, where
$\phi = f_1 \geq 0 \wedge \ldots \wedge f_{r_1} \geq 0 \wedge g_1 > 0 \wedge \ldots \wedge g_{s_1} > 0$,
$\psi = f_{r_1+1} \geq 0 \wedge \ldots \wedge f_r \geq 0 \wedge g_{s_1+1} > 0 \wedge \ldots \wedge g_s > 0$,
$f_1, \ldots, f_r, g_1, \ldots, g_s$ are all CQ polynomials,
$f_1, \ldots, f_{r_1}, g_1, \ldots, g_{s_1} \in \mathbb{R}[\mathbf{x}, \mathbf{y}]$, $f_{r_1+1}, \ldots, f_r, g_{s_1+1}, \ldots, g_s \in \mathbb{R}[\mathbf{x}, \mathbf{z}]$

output: A formula I to be an interpolant for ϕ and ψ

Find $\lambda_1, \ldots, \lambda_r \geq 0, , \eta_0, \eta_1, \ldots, \eta_s \geq 0, h_1 \in \mathbb{R}[\mathbf{x}, \mathbf{y}], h_2 \in \mathbb{R}[\mathbf{x}, \mathbf{z}]$ by SDP s.t.

$$\sum_{i=1}^{r} \lambda_i f_i + \sum_{j=1}^{s} \eta_j g_j + \eta_0 + h_1 + h_2 \equiv 0, \quad \eta_0 + \eta_1 + \ldots + \eta_s = 1,$$

where h_1, h_2 are SOS polynomials;

$f := \sum_{i=1}^{r_1} \lambda_i f_i + \sum_{j=1}^{s_1} \eta_j g_j + \eta_0 + h_1$;
if $\sum_{j=0}^{s_1} \eta_j > 0$ **then** $I := (f > 0)$; **else** $I := (f \geq 0)$;
return I

Algorithm 2. `Interpolation Generation for CQ Formulas (IG-CQI)`

input : ϕ and ψ with $\phi \wedge \psi \models \bot$, where
$\phi = f_1 \geq 0 \wedge \ldots \wedge f_{r_1} \geq 0 \wedge g_1 > 0 \wedge \ldots \wedge g_{s_1} > 0$,
$\psi = f_{r_1+1} \geq 0 \wedge \ldots \wedge f_r \geq 0 \wedge g_{s_1+1} > 0 \wedge \ldots \wedge g_s > 0$,
$f_1, \ldots, f_r, g_1, \ldots, g_s$ are all CQ polynomials,
$f_1, \ldots, f_{r_1}, g_1, \ldots, g_{s_1} \in \mathbb{R}[\mathbf{x}, \mathbf{y}]$, and
$f_{r_1+1}, \ldots, f_r, g_{s_1+1}, \ldots, g_s \in \mathbb{R}[\mathbf{x}, \mathbf{z}]$

output: A formula I to be an interpolant for ϕ and ψ

if $Var(\phi) \subseteq Var(\psi)$ **then** $I := \phi$; **return** I;
Find $\delta_1, \ldots, \delta_r \geq 0, h \in \mathbb{R}[\mathbf{x}, \mathbf{y}, \mathbf{z}]$ by SDP s.t. $\sum_{i=1}^{r} \delta_i f_i + h \equiv 0$ and h is SOS;
`/* Check the condition` **NSC** `*/`
if *no solution* **then** $I :=$ IG-NSC(ϕ, ψ); **return** I;
`/*` **NSC** `holds */`
Construct $h_1 \in \mathbb{R}[\mathbf{x}, \mathbf{y}]$ and $h_2 \in \mathbb{R}[\mathbf{x}, \mathbf{z}]$ with the forms (H1) and (H2);
$f := \sum_{i=1}^{r_1} \delta_i f_i + h_1 = -\sum_{i=r_1+1}^{r} \delta_i f_i - h_2$;
Construct ϕ' and ψ' using Theorems 6 & 7 by eliminating variables due to $h_1 = h_2 = 0$;
$I' :=$ IG-CQI(ϕ', ψ') ;
$I := (f > 0) \vee (f \geq 0 \wedge I')$;
return I

Theorem 8 (Soundness and Completeness). *Algorithm 2 computes an interpolant I if it exists for any given ϕ and ψ with $\phi \wedge \psi \models \bot$.*

Theorem 9. *The complexity of* IG-NSC *and* IG-CQI *are* $\mathcal{O}(g(r+s+n^2))^1$, *and* $\mathcal{O}(ng(r+s+n^2))$, *respectively, where* r *is the number of nonstrict inequalities,* s *is the number of strict inequalities, and* n *is the number of variables.*

5 Combination: *CQI* with *EUF*

This section combines the conjunctive theory of concave quadratic polynomial inequalities (*CQI*) with the theory of equality over uninterpreted function symbols (*EUF*). Algorithm 4 for generating interpolants for the combined theories is patterned after the algorithm $\text{INTER}_{LI(Q)^{\Sigma}}$ in Fig. 3 in [20] following the hierarchical reasoning and interpolation generation framework in [22] with the following key differences[2]:

1. To generate interpolants for CQI, Algorithm 2 is called.
2. If **NSC** is satisfied by nonstrict polynomial inequalities, linear equalities are deduced only from the linear inequalities; it is thus possible to use $\text{INTER}_{LI(Q)^{\Sigma}}$ in Fig. 3 in [20] for deducing equalities; separating terms for mixed equalities are computed in the same way as in the algorithm SEP in [20]. Further, it can be proved that a nonlinear polynomial equality of degree ≥ 2 cannot be generated from CQI.
3. If **NSC** is not satisfied, as in Algorithm 2, a polynomial $f(\mathbf{x})$ s.t. $\phi \models f(\mathbf{x}) \geq 0$ and $\psi \models -f(\mathbf{x}) \geq 0$ can be constructed by letting $f(\mathbf{x}) = (\sum_{i=1}^{r_1} \delta_i f_i) + h_1 = -(\sum_{i=r_1+1}^{r} \delta_i f_i) - h_2, \delta_i \geq 0$, as discussed in Sect. 4.4. Using Lemma 1, reduce the interpolation problem for ϕ and ψ to a simpler interpolation problem for ϕ' and ψ' with fewer variables.

5.1 Problem Formulation

Let $\Omega = \Omega_1 \cup \Omega_2 \cup \Omega_3$ be a finite set of uninterpreted function symbols in *EUF*; further, denote $\Omega_1 \cup \Omega_2$ by Ω_{12} and $\Omega_1 \cup \Omega_3$ by Ω_{13}. Let $\mathbb{R}[\mathbf{x},\mathbf{y},\mathbf{z}]^{\Omega}$ be the extension of $\mathbb{R}[\mathbf{x},\mathbf{y},\mathbf{z}]$ in which polynomials can have terms built using function symbols in Ω and variables in $\mathbf{x},\mathbf{y},\mathbf{z}$.

Problem 2: Suppose two formulas ϕ and ψ with $\phi \wedge \psi \models \bot$, where $\phi = f_1 \geq 0 \wedge \ldots \wedge f_{r_1} \geq 0 \wedge g_1 > 0 \wedge \ldots \wedge g_{s_1} > 0$, $\psi = f_{r_1+1} \geq 0 \wedge \ldots \wedge f_r \geq 0 \wedge g_{s_1+1} > 0 \wedge \ldots \wedge g_s > 0$, in which $f_1, \ldots, f_r, g_1, \ldots, g_s$ are all CQ polynomials, $f_1, \ldots, f_{r_1}, g_1, \ldots, g_{s_1} \in \mathbb{R}[\mathbf{x},\mathbf{y}]^{\Omega_{12}}$, $f_{r_1+1}, \ldots, f_r, g_{s_1+1}, \ldots, g_s \in \mathbb{R}[\mathbf{x},\mathbf{z}]^{\Omega_{13}}$, the goal is to generate an interpolant I for ϕ and ψ, over the common symbols $\mathbf{x}$, Ω_1, *i.e.*, I contains only polynomials in $\mathbb{R}[\mathbf{x}]^{\Omega_1}$.

---

1 Under the assumption that SDP tool returns an approximate but correct answer in a fixed number of calls.

2 The proposed algorithm and its way of handling of combined theories do not crucially depend upon using algorithms in [20]; however, adopting their approach makes proofs and presentation totally on *CQI*.

Flatten and Purify: Flatten and purify ϕ and ψ by introducing fresh variables for each term starting with uninterpreted symbols as well as for the terms containing uninterpreted symbols. Keep track of new variables introduced exclusively for ϕ and ψ as well as new common variables.

Let $\overline{\phi} \wedge \overline{\psi} \wedge \bigwedge D$ be obtained from $\phi \wedge \psi$ by flattening and purification where D consists of unit clauses of the form $\omega(c_1, \ldots, c_n) = c$, where $c_1, \ldots, c_n$ are variables and $\omega \in \Omega$. Following [20,22], using the axiom of an uninterpreted function symbol, a set N of Horn clauses are generated as follows, $N = \{\bigwedge_{k=1}^{n} c_k = b_k \rightarrow c = b \mid \omega(c_1, \ldots, c_n) = c \in D, \omega(b_1, \ldots, b_n) = b \in D\}$. The set N is partitioned into $N_\phi, N_\psi, N_{\text{mix}}$ with all symbols in N_ϕ, N_ψ appearing in $\overline{\phi}$, $\overline{\psi}$, respectively, and N_{mix} consisting of symbols from both $\overline{\phi}, \overline{\psi}$. It is easy to see that for every Horn clause in N_{mix}, each of equalities in the hypothesis as well as the conclusion is also mixed.

$$\phi \wedge \psi \models \bot \ \text{ iff } \ \overline{\phi} \wedge \overline{\psi} \wedge D \models \bot \ \text{ iff } \ (\overline{\phi} \wedge N_\phi) \wedge (\overline{\psi} \wedge N_\psi) \wedge N_{\text{mix}} \models \bot. \tag{7}$$

Notice that $(\overline{\phi} \wedge N_\phi) \wedge (\overline{\psi} \wedge N_\psi) \wedge N_{\text{mix}} \models \bot$ has no uninterpreted function symbols. If N_{mix} can be replaced by N_{sep}^{ϕ} and N_{sep}^{ψ} as in [20] using separating terms, then IG-CQI can be applied. An interpolant generated for this problem[3] can be used to generate an interpolant for ϕ, ψ after uniformly replacing all new symbols by their corresponding expressions from D.

5.2 Combination Algorithm

If N_{mix} is empty, Algorithm 4 invokes Algorithm 2 (IG-CQI) on a finite set of subproblems generated from a disjunction of conjunction of polynomial inequalities by expanding Horn clauses in N_ϕ and N_ψ, and applying De Morgan's rules. The resulting interpolant is a disjunction of conjunction of the interpolants generated for each subproblem.

The case when N_{mix} is nonempty has the same structure as the algorithm $\text{INTER}_{LI(Q)^\Sigma}$ in [20]. The following lemma proves that if a conjunction of polynomial inequalities satisfies **NSC** and an equality on variables can be deduced from it, then it suffices to consider only linear inequalities in the conjunction. This property enables us to use Algorithm $\text{INTER}_{LI(Q)^\Sigma}$ in Fig. 3 in [20] for deducing equalities; separating terms for the constants appearing in mixed equalities are computed in the same way as in Algorithm SEP in [20] (Lines 2 and 3 in Algorithm 3 where INTERp, a modified version of $\text{INTER}_{LI(Q)^\Sigma}$, is used solely to deduce equalities and separating terms and not interpolants, thus generating $N_{\text{sep}}^{\phi}, N_{\text{sep}}^{\psi}$). Then Algorithm 4 is called.

[3] After properly handling N_{mix} since Horn clauses have symbols both from $\overline{\phi}$ and $\overline{\psi}$.

Algorithm 3. `IG-CQI-EUF`

input : $\overline{\phi}$ and $\overline{\psi}$ constructed respective from ϕ and ψ by flattening and purification,
D : definitions of fresh variables introduced during flattening and purifying ϕ, ψ,
N : instances of functionality axioms for functions in D,
$\overline{\phi} = f_1 \geq 0 \wedge \ldots \wedge f_{r_1} \geq 0 \wedge g_1 > 0 \wedge \ldots \wedge g_{s_1} > 0$,
$\overline{\psi} = f_{r_1+1} \geq 0 \wedge \ldots \wedge f_r \geq 0 \wedge g_{s_1+1} > 0 \wedge \ldots \wedge g_s > 0$,
where $\phi \wedge \psi \models \bot$, $f_1, \ldots, f_r, g_1, \ldots, g_s$ are all CQ polynomials,
$f_1, \ldots, f_{r_1}, g_1, \ldots, g_{s_1} \in \mathbb{R}[\mathbf{x}, \mathbf{y}]$, and $f_{r_1+1}, \ldots, f_r, g_{s_1+1}, \ldots, g_s \in \mathbb{R}[\mathbf{x}, \mathbf{z}]$

output: A formula I to be a Craig interpolant for ϕ and ψ

if NSC *holds* **then**
  $L_1 := \text{LP}(\overline{\phi})$; $L_2 := \text{LP}(\overline{\psi})$;
  $\text{INTERp}(L_1, L_2, N, \emptyset, \emptyset, D, \emptyset)$;
  /* INTERp is a modified version of the $\text{INTER}^{\Sigma}_{\text{LI}(\mathbb{Q})}$ algorithm given in Figure 3 in [20] which is used here to separate every mixed Horn clause in N of the form $\wedge_{i=1}^{n} c_i = d_i \Rightarrow c = d$ into $\wedge_{i=1}^{n} c_i = t_i^+ \Rightarrow c = f(t_1^+, \cdots, t_n^+)$, $\wedge_{i=1}^{n} d_i = t_i^+ \Rightarrow d = f(t_1^+, \cdots, t_n^+)$. It does not call $\text{INTER}_{\text{LI}(Q)}$ to generate an interpolant (line 29 of $\text{INTER}^{\Sigma}_{\text{LI}(\mathbb{Q})}$). When INTERp terminates N_{mix} with initial value N is separated into N_ϕ and N_ψ with entailed equalities in Δ. Because of space limitations, we are not reproducing lines 1-28 of the code in $\text{INTER}^{\Sigma}_{\text{LI}(\mathbb{Q})}$. */
  $\overline{I} := \text{IG-NMIX}(\overline{\phi}, \overline{\psi}, N_\phi, N_\psi)$;
else
  Find $\delta_1, \ldots, \delta_r \geq 0$ and an SOS polynomial h by SDP s.t. $\sum_{i=1}^{r} \delta_i f_i + h \equiv 0$;
  Construct $h_1 \in \mathbb{R}[\mathbf{x}, \mathbf{y}]$ and $h_2 \in \mathbb{R}[\mathbf{x}, \mathbf{z}]$ with form (H1) and (H2);
  $f := \sum_{i=1}^{r_1} \delta_i f_i + h_1 = -\sum_{i=r_1+1}^{r} \delta_i f_i - h_2$;
  Construct $\overline{\phi'}$ and $\overline{\psi'}$ by Theorem 7 by eliminating variables from $h_1 = h_2 = 0$;
  $I' := \text{IG-CQI-EUF}(\overline{\phi'}, \overline{\psi'}, D, N_0)$; $\overline{I} := (f > 0) \vee (f \geq 0 \wedge I')$;
end
Obtain I from $\overline{I}$; **return** I

Algorithm 4. `Invariant Generation without` N_{mix} `(IG-NMIX)`

input : $\overline{\phi}$ and $\overline{\psi}$, constructed respectively from ϕ and ψ by flattening and purification,
N_ϕ : instances of functionality axioms for functions in D_ϕ,
N_ψ : instances of functionality axioms for functions in D_ψ,
where $\overline{\phi} \wedge \overline{\psi} \wedge N_\phi \wedge N_\psi \models \bot$

output: A formula I to be a Craig interpolant for ϕ and ψ

Transform $\overline{\phi} \wedge N_\phi$ to a DNF $\vee_i \phi_i$;
Transform $\overline{\psi} \wedge N_\psi$ to a DNF $\vee_j \psi_j$;
return $I := \vee_i \wedge_j \text{IG-CQI}(\phi_i, \psi_j)$

Lemma 2. *Let $\overline{\phi}$ and $\overline{\psi}$ be obtained as above satisfying* **NSC**. *If $\overline{\phi} \wedge \overline{\psi}$ is satisfiable, $\overline{\phi} \wedge \overline{\psi} \models c_k = b_k$, then $LP(\overline{\phi}) \wedge LP(\overline{\psi}) \models c_k = b_k$, where $LP(\theta)$ is the conjunction of the linear constraints in θ.*

If **NSC** is not satisfied, then linear equalities from SOS polynomials h, h_1, h_2 and f as explained above and discussed in Sect. 4.4 (Lines 6–8 in Algorithm 3) are used to generate simpler subproblems $\overline{\phi'}$ and $\overline{\psi'}$ from $\overline{\phi}$ and $\overline{\psi}$, and Algorithm 3 is recursively called (Lines 9–10 in Algorithm 3).

Theorem 10 (Soundness and Completeness). IG-CQI-EUF *computes an interpolant I of mutually contradictory ϕ, ψ with CQ polynomial inequalities and EUF if it exists.*

Example 2. Let $\phi := (f_1 = -(y_1 - x_1 + 1)^2 - x_1 + x_2 \geq 0) \wedge (y_2 = \alpha(y_1) + 1) \wedge (g_1 = -x_1^2 - x_2^2 - y_2^2 + 1 > 0)$, $\psi := (f_2 = -(z_1 - x_2 + 1)^2 + x_1 - x_2 \geq 0) \wedge (z_2 = \alpha(z_1) - 1) \wedge (g_2 = -x_1^2 - x_2^2 - z_2^2 + 1 > 0)$. Flattening and purification gives $\overline{\phi} := (f_1 \geq 0 \wedge y_2 = y + 1 \wedge g_1 > 0)$, $\overline{\psi} := (f_2 \geq 0 \wedge z_2 = z - 1 \wedge g_2 > 0)$, where $D = \{y = \alpha(y_1), z = \alpha(z_1)\}$, $N = (y_1 = z_1 \rightarrow y = z)$.

NSC is not satisfied, since $h = -f_1 - f_2 = (y_1 - x_1 + 1)^2 + (z_1 - x_2 + 1)^2$ is an SOS. We follow the steps given in Sect. 4.4 (Lines 6–8 of IG-CQI-EUF) and obtain $h_1 = (y_1 - x_1 + 1)^2$, $h_2 = (z_1 - x_2 + 1)^2$. This gives $f := f_1 + h_1 = -f_2 - h_2 = -x_1 + x_2$.

By Lemma 1, an interpolant for ϕ, ψ is an interpolant of $((\phi \wedge f > 0) \vee (\phi \wedge f = 0))$ and $((\psi \wedge -f > 0) \vee (\phi \wedge f = 0))$, that is $(f > 0) \vee (f \geq 0 \wedge I_2)$, where I_2 is an interpolant for $\phi \wedge f = 0$ and $\psi \wedge f = 0$. It is easy to see that $\phi \wedge f = 0 \models y_1 = x_1 - 1$, $\psi \wedge f = 0 \models z_1 = x_2 - 1$. Thus, it follows $\overline{\phi'} : -x_1 + x_2 \geq 0 \wedge y_2 = y + 1 \wedge g_1 > 0 \wedge y_1 = x_1 - 1$, and $\overline{\psi'} : x_1 - x_2 \geq 0 \wedge z_2 = z - 1 \wedge g_2 > 0 \wedge z_1 = x_2 - 1$.

At Line 10, recursively call IG-CQI-EUF. Now **NSC** holds (Line 1); from linear inequalities in $\overline{\phi'}$ and $\overline{\psi'}$, $y_1 = z_1$ is deduced. Separating terms for y_1, z_1 are constructed by: $\overline{\phi'} \models x_1 - 1 \leq y_1 \leq x_2 - 1, \overline{\psi'} \models x_2 - 1 \leq z_1 \leq x_1 - 1$. Let $t = \alpha(x_2 - 1)$, then $y_1 = z_1 \rightarrow y = z$ is separated into two parts, *i.e.*, $y_1 = t^+ \rightarrow y = t$ and $t^+ = z_1 \rightarrow t = z$. Add them to $\overline{\phi'}$ and $\overline{\psi'}$ respectively, we have $\overline{\phi'}_1 = -x_1 + x_2 \geq 0 \wedge y_2 = y + 1 \wedge g_1 > 0 \wedge y_1 = x_1 - 1 \wedge y_1 = x_2 - 1 \rightarrow y = t$, $\overline{\psi'}_1 = x_1 - x_2 \geq 0 \wedge z_2 = z - 1 \wedge g_2 > 0 \wedge z_1 = x_2 - 1 \wedge x_2 - 1 = z_1 \rightarrow t = z$. Then $\overline{\phi'}_1 = -x_1 + x_2 \geq 0 \wedge y_2 = y + 1 \wedge g_1 > 0 \wedge y_1 = x_1 - 1 \wedge (x_2 - 1 > y_1 \vee y_1 > x_2 - 1 \vee y = t)$, $\overline{\psi'}_1 = x_1 - x_2 \geq 0 \wedge z_2 = z - 1 \wedge g_2 > 0 \wedge z_1 = x_2 - 1 \wedge t = z$. Thus, $\overline{\phi'}_1 = \overline{\phi'}_2 \vee \overline{\phi'}_3 \vee \overline{\phi'}_4$, where $\overline{\phi'}_2 = -x_1 + x_2 \geq 0 \wedge y_2 = y + 1 \wedge g_1 > 0 \wedge y_1 = x_1 - 1 \wedge x_2 - 1 > y_1$, $\overline{\phi'}_3 = -x_1 + x_2 \geq 0 \wedge y_2 = y + 1 \wedge g_1 > 0 \wedge y_1 = x_1 - 1 \wedge y_1 > x_2 - 1$, $\overline{\phi'}_4 = -x_1 + x_2 \geq 0 \wedge y_2 = y + 1 \wedge g_1 > 0 \wedge y_1 = x_1 - 1 \wedge y = t$. Since $\overline{\phi'}_3 = \bot$, it follows $\overline{\phi'}_1 = \overline{\phi'}_2 \vee \overline{\phi'}_4$. Find interpolants $I(\overline{\phi'}_2, \overline{\psi'}_1)$ and $I(\overline{\phi'}_4, \overline{\psi'}_1)$, then $I(\overline{\phi'}_2, \overline{\psi'}_1) \vee I(\overline{\phi'}_4, \overline{\psi'}_1)$ is an interpolant.

6 Implementation and Experimental Results

We are currently developing a state of the art implementation of the above algorithms using C. In the meantime, for experimentation purposes, we have developed a prototype for putting together existing tools in *Mathematica*. An optimization library *AiSat* [5] built on *CSDP* [1] is used for solving SOS and SDP problems. We give some performance data about this prototype on some

Table 1. The output formulas in the last column have been verified using the approach given in [4] to be the true interpolants w.r.t. their corresponding problems in the third column.

Ex	Type	Problem	Synthesized interpolant
5	NLA	$\phi: -y_1 + x_1 - 2 \geq 0 \wedge 2x_2 - x_1 - 1 > 0$ $\wedge - y_1^2 - x_1^2 + 2x_1y_1 - 2y_1 + 2x_1 \geq 0$ $\wedge - y_2^2 - y_1^2 - x_2^2 - 4y_1 + 2x_2 - 4 \geq 0$ $\psi: -z_1 + 2x_2 + 1 \geq 0 \wedge 2x_1 - x_2 - 1 > 0$ $\wedge - z_1^2 - 4x_2^2 + 4x_2z_1 + 3z_1 - 6x_2 - 2 \geq 0$ $\wedge - z_2^2 - x_1^2 - x_2^2 + 2x_1 + z_1 - 2x_2 - 1 \geq 0$	$-x_1 + x_2 > 0$
6	NLA	$\phi: 4 - x^2 - y^2 \geq 0 \wedge y \geq 0 \wedge x + y - 1 > 0$ $\psi: x \geq 0 \wedge 1 - x^2 - (y+1)^2 \geq 0$	$\frac{1}{2}(x^2 + y^2 + 4y) > 0$
7	LA	$\phi: z - x \geq 0 \wedge x - y \geq 0 \wedge -z > 0$ $\psi: x + y \geq 0 \wedge -y \geq 0$	$-0.8x - 0.2y > 0$
8	LA+EUF	$\phi: f(x) \geq 0 \wedge x - y \geq 0 \wedge y - x \geq 0$ $\psi: -f(y) > 0$	$f(y) \geq 0$
9	Ellipsoid	$\phi: -{x_1}^2 + 4x_1 + x_2 - 4 \geq 0$ $\wedge - x_1 - x_2 + 3 - y^2 > 0$ $\psi: -3{x_1}^2 - {x_2}^2 + 1 \geq 0 \wedge x_2 - z^2 \geq 0$	$-3 + 2x_1 + {x_1}^2 + \frac{1}{2}{x_2}^2 > 0$
10	Ellipsoid	$\phi: 4 - (x-1)^2 - 4y^2 \geq 0 \wedge y - \frac{1}{2} \geq 0$ $\psi: 4 - (x+1)^2 - 4y^2 \geq 0 \wedge x + 2y \geq 0$	$-15.93 + 19.30x - 9.65x^2$ $+91.76y - 38.60y^2 > 0$
11	Octagon	$\phi: -3 \leq x \leq 1 \wedge -2 \leq y \leq 2 \wedge -4 \leq x - y \leq 2$ $\wedge - 4 \leq x + y \leq 2 \wedge x + 2y + 1 \leq 0$ $\psi: -1 \leq x \leq 3 \wedge -2 \leq y \leq 2 \wedge -2 \leq x - y \leq 4$ $\wedge - 2 \leq x + y \leq 4 \wedge 2x - 5y + 6 \leq 0$	$-88.08 - 649.94x$ $-1432.44y > 0$
12	Octagon	$\phi: 2 \leq x \leq 7 \wedge 0 \leq y \leq 3 \wedge 0 \leq x - y \leq 6$ $\wedge 3 \leq x + y \leq 9 \wedge 23 - 3x - 8y \leq 0$ $\psi: 0 \leq x \leq 5 \wedge 2 \leq y \leq 5 \wedge -4 \leq x - y \leq 2$ $\wedge 3 \leq x + y \leq 9 \wedge y - 3x - 2 \leq 0$	$562.10 + 1244.11x$ $-869.83y > 0$

examples (see Table 1), which have been evaluated on a 64-bit Linux computer with a 2.93 GHz Intel Core-i7 processor and 4 GB of RAM.

The performance of the prototype is compared on the same platform to those of three publicly available interpolation procedures for linear-arithmetic cases, *i.e.* Rybalchenko's tool CLP-Prover in [20], McMillan's procedure Foci in [17], and Beyer's tool CSIsat in [2]. As Table 2 shows, our approach can successfully solve all these examples rather efficiently. It is especially the completeness and generality that makes the approach competitive for synthesizing interpolants. In particular, the prototype performs, in linear cases, with the same complexity as CSIsat and even better than CLP-Prover and Foci. Whilst in nonlinear cases, the method developed in [5] is limited and incomplete even though it works for nonlinear polynomials (using SDP) since it requires bounds on variables as well as uncommon variables are not allowed.

Table 2. Evaluation results of the presented examples

Example	Type	Time (sec)			
		CLP-prover	Foci	CSIsat	Our approach
Example 1	NLA	–	–	–	0.003
Example 2	NLA+*EUF*	–	–	–	0.036
Example 5	NLA	–	–	–	0.014
Example 6	NLA	–	–	–	0.003
Example 7	LA	0.023	×	0.003	0.003
Example 8	LA+*EUF*	0.025	0.006	0.007	0.003
Example 9	Ellipsoid	–	–	–	0.002
Example 10	Ellipsoid	–	–	–	0.002
Example 11	Octagon	0.059	×	0.004	0.004
Example 12	Octagon	0.065	×	0.004	0.004

– means interpolant generation fails, and × specifies particularly wrong answers (satisfiable).

7 Conclusion

The paper proposes a polynomial time algorithm for generating interpolants from mutually contradictory conjunctions of concave quadratic polynomial inequalities over the reals. Under a technical condition that if no nonpositive constant combination of nonstrict inequalities is a sum of squares polynomials, then such an interpolant can be generated essentially using the linearization of concave quadratic polynomials. Otherwise, if this condition is not satisfied, then the algorithm is recursively called on smaller problems after deducing linear equalities relating variables. The resulting interpolant is a disjunction of conjunction of polynomial inequalities.

Using the hierarchical calculus framework proposed in [22], we give an interpolation algorithm for the combined quantifier-free theory of concave quadratic polynomial inequalities and equality over uninterpreted function symbols. The combination algorithm is patterned after a combination algorithm for the combined theory of linear inequalities and equality over uninterpreted function symbols.

A prototype has been built, and experimental results indicate our approach is applicable to all existing abstract interpretation domains widely used in verification for programs and hybrid systems like *octagon*, *polyhedra*, *ellipsoid* and so on, which is encouraging for using this approach in the state of the art of verification techniques based on interpolation[4].

[4] The tool and all case studies can be found at http://lcs.ios.ac.cn/~chenms/tools/InterCQI_v1.1.tar.bz2.

Acknowledgement. The first three authors are supported partly by NSFC under grants 11290141, 11271034 and 61532019; the fourth and sixth authors are supported partly by "973 Program" under grant No. 2014CB340701, by NSFC under grant 91418204, by CDZ project CAP (GZ 1023), and by the CAS/SAFEA International Partnership Program for Creative Research Teams; the fifth author is supported partly by NSF under grant DMS-1217054 and by the CAS/SAFEA International Partnership Program for Creative Research Teams.

References

1. *CSDP*. http://projects.coin-or.org/Csdp/
2. Beyer, D., Zufferey, D., Majumdar, R.: CSIsat: interpolation for LA+EUF. In: Gupta, A., Malik, S. (eds.) CAV 2008. LNCS, vol. 5123, pp. 304–308. Springer, Heidelberg (2008)
3. Cimatti, A., Griggio, A., Sebastiani, R.: Efficient interpolant generation in satisfiability modulo theories. In: Ramakrishnan, C.R., Rehof, J. (eds.) TACAS 2008. LNCS, vol. 4963, pp. 397–412. Springer, Heidelberg (2008)
4. Dai, L., Gan, T., Xia, B., Zhan, N.: Barrier certificate revisited. J. Symbolic Comput. (2016, to appear)
5. Dai, L., Xia, B., Zhan, N.: Generating non-linear interpolants by semidefinite programming. In: Sharygina, N., Veith, H. (eds.) CAV 2013. LNCS, vol. 8044, pp. 364–380. Springer, Heidelberg (2013)
6. D'Silva, V., Kroening, D., Purandare, M., Weissenbacher, G.: Interpolant strength. In: Barthe, G., Hermenegildo, M. (eds.) VMCAI 2010. LNCS, vol. 5944, pp. 129–145. Springer, Heidelberg (2010)
7. Fujie, T., Kojima, M.: Semidefinite programming relaxation for nonconvex quadratic programs. J. Global Optim. **10**(4), 367–380 (1997)
8. Gan, T., Dai, L., Xia, B., Zhan, N., Kapur, D., Chen, M.: Interpolation synthesis for quadratic polynomial inequalities and combination with EUF. CoRR, abs/1601.04802 (2016)
9. Graf, S., Saïdi, H.: Construction of abstract state graphs with PVS. In: Grumberg, O. (ed.) CAV 1997. LNCS, vol. 1254, pp. 72–83. Springer, Heidelberg (1997)
10. Henzinger, T., Jhala, R., Majumdar, R., McMillan, K.: Abstractions from proofs. In: POPL 2004, pp. 232–244 (2004)
11. Jung, Y., Lee, W., Wang, B.-Y., Yi, K.: Predicate generation for learning-based quantifier-free loop invariant inference. In: Abdulla, P.A., Leino, K.R.M. (eds.) TACAS 2011. LNCS, vol. 6605, pp. 205–219. Springer, Heidelberg (2011)
12. Kapur, D., Majumdar, R., Zarba, C.: Interpolation for data structures. In: FSE 2006, pp. 105–116 (2006)
13. Kovács, L., Voronkov, A.: Interpolation and symbol elimination. In: Schmidt, R.A. (ed.) CADE-22. LNCS, vol. 5663, pp. 199–213. Springer, Heidelberg (2009)
14. Krajíček, J.: Interpolation theorems, lower bounds for proof systems, and independence results for bounded arithmetic. J. Symbolic Logic **62**(2), 457–486 (1997)
15. Laurent, M.: Sums of squares, moment matrices and optimization over polynomials. In: Putinar, M., Sullivant, S. (eds.) Emerging Applications of Algebraic Geometry. The IMA Volumes in Mathematics and its Applications, vol. 149, pp. 157–270. Springer, New York (2009)
16. McMillan, K.L.: Interpolation and SAT-based model checking. In: Hunt Jr., W.A., Somenzi, F. (eds.) CAV 2003. LNCS, vol. 2725, pp. 1–13. Springer, Heidelberg (2003)

17. McMillan, K.: An interpolating theorem prover. Theor. Comput. Sci. **345**(1), 101–121 (2005)
18. McMillan, K.L.: Quantified invariant generation using an interpolating saturation prover. In: Ramakrishnan, C.R., Rehof, J. (eds.) TACAS 2008. LNCS, vol. 4963, pp. 413–427. Springer, Heidelberg (2008)
19. Pudlák, P.: Lower bounds for resolution and cutting plane proofs and monotone computations. J. Symbolic Logic **62**(3), 981–998 (1997)
20. Rybalchenko, A., Sofronie-Stokkermans, V.: Constraint solving for interpolation. J. Symb. Comput. **45**(11), 1212–1233 (2010)
21. Schrijver, A.: Theory of Linear and Integer Programming. Wiley, Chichester (1998)
22. Sofronie-Stokkermans, V.: Interpolation in local theory extensions. Logical Methods Comput. Sci. **4**(4), 1–31 (2008)
23. Stengle, G.: A nullstellensatz and a positivstellensatz in semialgebraic geometry. Ann. Math. **207**, 87–97 (1974)
24. Tütüncü, R.H., Toh, K.C., Todd, M.J.: Solving semidefinite-quadratic-linear programs using SDPT3. J. Math. Program. **95**(2), 189–217 (2003)
25. Yang, Z., Lin, W., Wu, M.: Exact safety verification of hybrid systems based on bilinear SOS representation. ACM Trans. Embed. Comput. Syst. **14**(1), 16:1–16:19 (2015)
26. Yorsh, G., Musuvathi, M.: A combination method for generating interpolants. In: Nieuwenhuis, R. (ed.) CADE 2005. LNCS (LNAI), vol. 3632, pp. 353–368. Springer, Heidelberg (2005)
27. Zhao, H., Zhan, N., Kapur, D.: Synthesizing switching controllers for hybrid systems by generating invariants. In: Liu, Z., Woodcock, J., Zhu, H. (eds.) Theories of Programming and Formal Methods. LNCS, vol. 8051, pp. 354–373. Springer, Heidelberg (2013)
28. Zhao, H., Zhan, N., Kapur, D., Larsen, K.G.: A "Hybrid" approach for synthesizing optimal controllers of hybrid systems: a case study of the oil pump industrial example. In: Giannakopoulou, D., Méry, D. (eds.) FM 2012. LNCS, vol. 7436, pp. 471–485. Springer, Heidelberg (2012)

Race Against the Teens – Benchmarking Mechanized Math on Pre-university Problems

Takuya Matsuzaki[1,2(✉)], Hidenao Iwane[2,3], Munehiro Kobayashi[4], Yiyang Zhan[5], Ryoya Fukasaku[6], Jumma Kudo[6], Hirokazu Anai[3,7], and Noriko H. Arai[2]

[1] Nagoya University, Nagoya, Japan
matuzaki@nuee.nagoya-u.ac.jp
[2] National Institute of Informatics, Chiyoda, Japan
[3] Fujitsu Laboratories, Ltd., Kawasaki, Japan
[4] University of Tsukuba, Tsukuba, Japan
[5] Université Paris Diderot, Paris, France
[6] Tokyo University of Science, Shinjuku, Japan
[7] Kyushu University, Fukuoka, Japan

Abstract. This paper introduces a benchmark problem library for mechanized math technologies including computer algebra and automated theorem proving. The library consists of pre-university math problems taken from exercise problem books, university entrance exams, and the International Mathematical Olympiads. It thus includes problems in various areas of pre-university math and with a variety of difficulty. Unlike other existing benchmark libraries, this one contains problems that are formalized so that they are obtainable as the result of mechanical translation of the original problems expressed in natural language. In other words, the library is designed to support the integration of the technologies of mechanized math and natural language processing towards the goal of end-to-end automatic math problem solving. The paper also presents preliminary experimental results of our prototype reasoning component of an end-to-end system on the library. The library is publicly available through the Internet.

1 Introduction

One of the ultimate goals of automated theorem proving is to produce computer programs that allow a machine to conduct mathematical reasoning like human beings. It seems that a tacit understanding exists on how we should interpret this goal. First, the input of the programs is assumed to be expressed in some formal language, but not in a natural language. Second, the term "human beings" is used to mean gifted mathematicians rather than ordinary people. In this paper, we propose a different interpretation of the goal by providing a new problem library for benchmarking automated math reasoners, and showing experimental results on the problem set.

Though traditionally ignored in the framework of automated theorem proving and computer algebra, interpreting given problems is as important as solving

N. Olivetti and A. Tiwari (Eds.): IJCAR 2016, LNAI 9706, pp. 213–227, 2016.
DOI: 10.1007/978-3-319-40229-1_15

```
(Find (x)
  (exists (A B C D E)
    (& (exists (F)
         (& (exists (U T S R Q P)
              (& (= (polygon (list-of A B C D E F))
                    (polygon (list-of P Q R S T U)))
                 (is-regular-polygon (polygon (list-of P Q R S T U)))))
            (is-diagonal-of (seg A C) (polygon (list-of A B C D E F)))
            (exists (U T S R Q P)
              (& (= (polygon (list-of A B C D E F))
                    (polygon (list-of P Q R S T U)))
                 (is-regular-polygon (polygon (list-of P Q R S T U)))))
            (is-diagonal-of (seg C E) (polygon (list-of A B C D E F)))))
       (exists (M N)
         (& (exists (r)
              (& (= (/ (length-of (seg A M)) (length-of (seg A C)))
                    (/ (length-of (seg C N)) (length-of (seg C E))))
                 (= (/ (length-of (seg C N)) (length-of (seg C E))) r)
                 (on M (seg A C))
                 (= (/ (length-of (seg A M)) (length-of (seg A C)))
                    (/ (length-of (seg C N)) (length-of (seg C E))))
                 (= (/ (length-of (seg C N)) (length-of (seg C E))) r)
                 (on N (seg C E))
                 (= x r)))
            (~ (= M A)) (~ (= M C))
            (~ (= N C)) (~ (= N E))
            (points-colinear (list-of B M N)))))))
```

Fig. 1. Mechanical translation result of IMO 1982, Problem 5

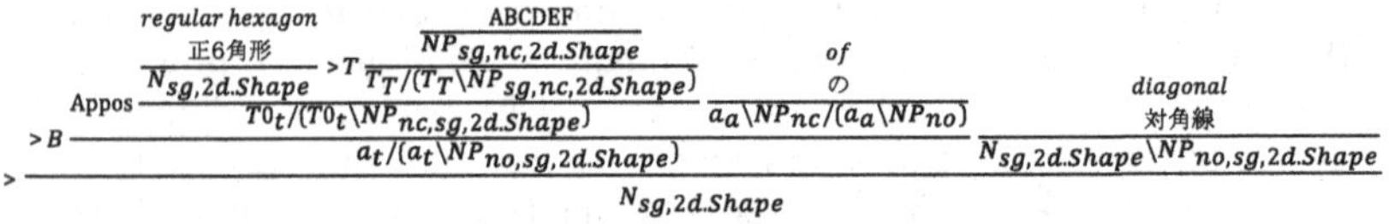

Fig. 2. Syntactic/Semantic analysis of problem (CCG derivation tree)

them in mathematical reasoning; it took almost a century to determine the language and axioms required to express the Jordan curve theorem, and this is exactly how long it took to solve the problem. This is also the case in curriculum math. Hence, it is fair to assume the problems are expressed in a natural language but not in a formal language if we want to seriously argue about whether or not a machine is as intelligent as high school graduates in math problem solving.

Given this situation, we developed a new problem library of real pre-university math problems. It is designed to cover various sub-areas of curriculum math and a diverse range of difficulty. The initial release of the data set includes more than 700 problems taken from three sources: popular high school math exercise book series, entrance examinations of seven top universities in Japan, and the past problems from International Mathematical Olympiads (IMO). Our choice of the three problem sources is motivated by the desire to measure the performance of mechanized math systems with high school students of different skill and intellectual levels as reference points.

Although problems in the library are formalized in a formal language so that the automatic reasoning (AR) and computer algebra system (CAS) communities

Table 1. Number of problems and directives

		Chart	Univ	IMO	Total
#Problems		288	245	212	745
#Directives	`Find`	473	438	110	1021
	`Draw`	28	16	0	44
	`Show`	78	73	134	285

Table 2. Subject areas (IMO)

Algebra	57
Number Theory	38
Analysis	1
Geometry	105
Combinatorics	11

will find it appealing to challenge the problems, the formalization is designed so that the problems can be obtained as the result of mechanical translation of their originals. This might have sounded unrealistic in the previous century but is now within the range of contemporary research, thanks to the recent progress that has been made in deep linguistic processing (e.g., [2,3,9]). Figure 1 presents an output from the translation module under development and Fig. 2 depicts a part of the process that derives the logical translation of a Japanese phrase "6/regular-hexagon $ABCDEF$ /of /diagonal", that corresponds to "diagonal(s) of the regular hexagon $ABCDEF$." Problems in the library were formalized manually according to the design of the aforementioned translation module. That is, they were translated manually into the formal language by word-by-word and sentence-by-sentence basis without any inference and paraphrasing.

The formalized problem set and the accompanying axioms are publicly available at http://github.com/torobomath/benchmark. The problems and the axioms are formulated in a higher-order language that is mostly compatible with the TPTP's typed higher-order format (THF) [13]. The data and the axioms are distributed both in the TPTP's THF syntax and an S-expression format. For readability, we use the S-expression format for presenting the data. Several basic elements such as logical connectives and quantifiers are renamed following the TPTP's convention.

The rest of the paper is structured as follows. We first describe how we collected and formalized curriculum math problems in Sect. 2. Several problems are shown in Sect. 3 to exemplify what aspects of mechanized math are necessary. Besides proof problems, the benchmark includes many "Find X"-type problems. Technical issues in formalizing such problems are discussed in Sect. 4. Section 5 provides an overview of a prototype solver system built on an integration of a simple logical inference system with computer algebra systems. Experimental results on the initial release of the data by our solver system are presented in Sect. 6. Finally, we conclude the paper and discuss future directions in Sect. 7.

Table 3. Subject areas (Chart & Univ)

	Chart	Univ
Algebra	51	10
Linear Algebra	28	62
Geometry	136	65
Pre-Calculus	15	74
Calculus	42	33
Combinatorics	16	1

Table 4. Distribution of theory labels

	Chart	Univ	IMO
PA	84	0	42
RCF	174	245	115
ZF	30	0	55
RCF+PA	1	0	8
Transc	23	0	6
PA+Transc	5	0	1
Comb	0	0	10
other	1	0	30

Table 5. Statistics on the syntactic properties (min/avg/max/**median**)

	Todai robot project math benchmark				TPTP-THF
	Chart	Univ	IMO	All	
# of formulae	1/ 2/ 7/ **2**	1/ 2/ 8/ **2**	1/ 1/ 5/ **1**	1/ 2/ 8/ **1**	1/103/ 5639/**10**
# of atoms	14/80/485/**65**	14/125/652/**95**	11/85/2658/**65**	11/97/2658/**72**	1/819/64867/**88**
Avg atoms/formula	9/42/161/**38**	14/ 58/232/**54**	10/77/2658/**56**	9/57/2658/**48**	1/ 22/ 811/ **6**
# of symbols	3/16/ 31/**16**	6/ 19/ 34/**19**	4/19/1332/**12**	3/18/1332/**15**	1/ 45/ 1442/ **9**
# of variables	1/12/ 55/ **9**	1/ 17/ 72/**13**	0/ 9/ 35/ **8**	0/13/ 72/ **9**	0/154/11290/**19**
λ	0/ 4/ 22/ **3**	0/ 4/ 23/ **3**	0/ 1/ 9/ **1**	0/ 3/ 23/ **2**	0/ 22/ 385/ **2**
$\forall$	0/ 2/ 49/ **0**	0/ 2/ 24/ **0**	0/ 5/ 22/ **4**	0/ 3/ 49/ **0**	0/123/10753/ **9**
$\exists$	0/ 5/ 38/ **4**	0/ 9/ 50/ **6**	0/ 2/ 20/ **1**	0/ 6/ 50/ **4**	0/ 8/ 496/ **2**
# of connectives	11/67/416/**55**	13/105/476/**78**	11/77/2655/**58**	11/82/2655/**61**	0/574/51044/**52**
Max formula depth	8/20/ 50/**19**	12/ 25/ 59/**23**	9/28/1327/**20**	8/24/1327/**21**	2/ 36/ 359/**11**
Avg formula depth	0/ 5/ 9/ **4**	0/ 5/ 9/ **4**	0/ 5/ 9/ **5**	0/ 5/ 9/ **5**	0/ 5/ 9/ **6**

2 Pre-university Math Problems as a Benchmark for Mechanized Math Systems

In this section, we first describe the sources and the types of the benchmark problems. We then explain how we encoded the problems other than proof problems. Finally, the representation language is described.

2.1 The Problem Library

The initial release of the data set consists of 745 problem files containing 1,353 directives, and 1,897 axioms defining 1,040 symbols (functions, predicates, and constants). The problems were taken from three sources: "Chart-shiki" (**Chart**), Japanese university entrance exams (**Univ**), and International Mathematical Olympiads (**IMO**).

"Chart-shiki" is a popular problem book series containing more than ten thousand problems in total. In the first release of the data set, the **Chart** division consists of arithmetic problems and various types of geometry problems (including those involving calculus and linear algebra) (Table 3). Every problem in "Chart-shiki" is marked with one to five stars by the editors of the book series according to its difficulty. We sampled the problems so that their levels of difficulty would be uniformly distributed.

The **Univ** division of the data consists of the past entrance exams of seven top Japanese national universities. Unlike in most countries, in Japan each national university prepares its entrance exam by itself. As a result, several hundreds of brand-new problems are produced every year for the entrance exams. In the first release, the **Univ** division includes the problems that were manually classified as 'most likely expressible' in the first-order theory of real-closed fields (RCF) (Table 3). Two hundred more **Univ** problems involving transcendental functions and integers arithmetic (often as a mixture with reals) are currently under preparation for the second release of the data set.

The **IMO** division consists of about 2/3 of the past IMO problems. The initial release includes all of the geometry and real algebra problems, and some of the problems in number theory, function equations, and combinatorics.

Each problem is labeled by its subject domain name such as geometry or calculus (Tables 2 and 3), and also by its formal theory name. The problems that are naturally expressible (by humans) in the theories of RCF or Peano Arithmetic (PA) are labeled so, and the rest of the problems are tentatively labeled ZF, standing for Zermelo-Fraenkel Set Theory. We scrutinized the problems labeled ZF and classified them into several groups such as 'RCF+PA' (mixture of integer and real arithmetics) and 'Transc' (problems involving transcendental functions that cannot be reformulated in RCF), though they are not formal theories

IMO 1982, Problem 5

The diagonals AC and CE of the regular hexagon $ABCDEF$ are divided by the inner points M and N, respectively, so that $\frac{AM}{AC} = \frac{CN}{CE} = r$. Determine r if B, M, and N are collinear.

```
;;-------------------------------------------------------------
(def-directive problem_IMO_1982_2
  (Find (r)
    (exists (A B C D E F M N)
        (& (is-regular-polygon (polygon (list-of A B C D E F)))
           (on M (seg A C))
           (on N (seg C E))
           (~ (= M A)) (~ (= M C))
           (~ (= N C)) (~ (= N E))
           (= (/ (length-of (seg A M)) (length-of (seg A C)))
              (/ (length-of (seg C N)) (length-of (seg C E))))
           (= (/ (length-of (seg A M)) (length-of (seg A C)))
              r)
           (colinear B M N)))))

(def-answer problem_IMO_1982_2
   (lambda r (= r (/ 1 3))))
;;-------------------------------------------------------------
```

Fig. 3. Problem file example (IMO 1982, problem 5)

Table 6. Logical translations in ZF set theory and Peano arithmetic

(a) There are infinitely many prime numbers greater than 4
ZF: $\|\{n \in \mathbb{N} \mid \text{prime}(n) \wedge n > 4\}\| = \omega$
PA: $\forall N \exists n (n > N \wedge \text{prime}(n) \wedge n > 4)$
(b) There is an even number of prime numbers less than 4
ZF: $\exists k \in \mathbb{N}(\text{even}(k) \wedge \|\{n \mid \text{prime}(n) \wedge n < 4\}\| = k)$ PA: $\exists k(\text{even}(k) \wedge \text{num\_of}(\text{prime\_less\_than}(4)) = k)$
(c) There are two prime numbers less than 4
ZF: $\|\{n \in \mathbb{N} \mid \text{prime}(n) \wedge n < 4\}\| = 2$
PA$_1$: $\exists n_1 \exists n_2 \left(\begin{array}{l} \text{prime}(n_1) \wedge \text{prime}(n_2) \wedge n_1 < 4 \wedge n_2 < 4 \wedge n_1 \neq n_2 \\ \hline \wedge \forall m((\text{prime}(m) \wedge m < 4) \rightarrow (m = n_1 \vee m = n_2)) \end{array} \right)$
PA$_2$: $\text{num\_of}(\text{prime\_less\_than}(4)) = 2$

(Table 4). Table 5 lists statistics for the formalized problems. For reference, it also lists those for the typed-higher order format (THF) problems in TPTP version 6.1.0.

2.2 A Formalization of Curriculum Math Problems

We formalize a problem as a pair of a *directive* and its *answer*. By surveying the problems, we identified three major types of directives:

- `Show` $[\phi]$ is a proof problem to prove ϕ.
- `Find`(v) $[\phi(v)]$ is a problem to find all values for v that satisfy condition $\phi(v)$.
- `Draw`(v) $[\phi(v)]$ requests a geometric object v defined by $\phi(v)$ be drawn.

`Show` directives must be familiar to the reader, though the set of problems requiring proofs is a minority in curriculum math. Table 1 shows that students are asked to find some values more frequently than to prove propositions. The answer to a `Find` problem, `Find`$(v)[\phi(v)]$, is expected to be a characteristic function $f(v)$ that returns `true` if v satisfies $\phi(v)$. The answer to a `Draw` problem `Draw`$(v)[\phi(v)]$ should be the geometric object v expressed as a characteristic function on R^2. Figure 3 is an example of a `Find` problem taken from IMO 1982.

To use a problem set in the above format as benchmark data, we need a *rule* to judge whether a system's output is acceptable or not. It is clear for the `Show` directives: true or false. We regard a `Draw` directive as a variant of `Find` problems for which the system is supposed to find a formula that defines the geometric object. Then, what is "to solve a `Find` problem?" Roughly speaking, a solver is supposed to give a *correct* solution in its *simplest* form. We will discuss the properties an answer formula for a `Find` problem has to satisfy in Sect. 4.

2.3 Representation Language

We formalized all the problems in a single theory on the basis of ZF regardless of their context. In formality, it is a typed lambda calculus with parametric

```
;; tangent(S1, S2, P) <-> geometric objects
;; S1 and S2 are tangent at point P
(def-pred
  tangent :: Shape -> Shape -> Point => Bool)

(axiom
  def_tangent_line_and_circle
  (p q c r P)
  (<-> (tangent (line p q) (circle c r) P)
       (& (on P (line p q))
          (perpendicular (line c P) (line p q))
          (= (distance^2 P c) (^ r 2)))))
```

```
;; maximum(S, m) <->
;; m is the maximum element of set S
(def-pred
  maximum :: (SetOf R) -> R => Bool)

(axiom
  def_maximum
  (set max)
  (<-> (maximum set max)
       (& (elem max set)
          (forall (v)
              (-> (elem v set)
                  (<= v max))))))
```

Fig. 4. Type definitions and axioms

polymorphism. This is again due to the fully automatic, end-to-end task setting. Table 6 demonstrates why a higher-order language is appropriate as the target language. Mechanical translation assumes a systematic correspondence between the syntactic structures of the input and output languages; the results of the mechanical translations of (a), (b), and (c) into ZF in Table 6 are expected to have the same or at least a similar structure thanks to the set builder notation such as $\{n \in \mathbb{N} \mid \text{prime}(n) \wedge n > 4\}$, which is expressed using λ-abstraction in the dataset. However, the expressions of the three sentences in PA must be different since the concept of finiteness cannot be expressed in first-order logic. Meanwhile, the expressibility of ZF allows almost word-by-word translations for all sentences. Parametric polymorphism is utilized to have polymorphic lists and sets in the language and define various operations on them while keeping the axioms and the lexicon (i.e., the mapping table from words to their semantic representations) concise.

We believe the vast majority of our benchmark problems can be eventually expressed in first-order logic. To mechanically fill the gap between the heavy-duty language and the relatively simple content is however a mandatory step to connect natural language processing and automated reasoning together for end-to-end automatic problem solving.

Since our mechanical translator is still under development, the problems were formalized manually at the current stage. Operators, all majored in computer science and/or mathematics, were trained to translate the problems as faithfully as possible to the original natural language statements following the NLP design. The sets of new symbols and their defining axioms were introduced in parallel with the problem formalization, to match the problem formula as close as possible to the problem text.

In the language, we currently have 31 types including `Bool`(ean), `Z` (integers), `Q` (rational numbers), `R`(eals), `C`(omplex numbers), `ListOf`(α) (polymorphic lists), `SetOf`(α) (polymorphic sets), `Point` (in 2D and 3D spaces), `Shape` (sets of `Point`s), `Equation` (in real domain), and so on. The types are somewhat redundant in that we can represent, e.g., `Equation` simply by a function of type $\mathtt{R} \to \mathtt{R}$ by regarding $f : \mathtt{R} \to \mathtt{R}$ as representing $f(x) = 0$. The abundance of types, however, helped a lot in organizing the axioms and debugging the formalized problems. Figure 4 presents an excerpt from an axiom file that includes two type definitions (two `def-pred`s) and two axioms.

All in all, the language shall be understood as a conservative extension of ZF set theory. It thus has some overlap with previous efforts toward formalizing a large part math, such as Mizar's math library [5]. However, some essential parts of the system (e.g., the definition of the real numbers and arithmetic) are left undefined although nothing prevents the users of the problem library from doing so. Instead of writing all the inference rules explicitly, we delegated computer algebra systems to take care of it. Although it is not within our current research focus, full formalization of the system (maybe by embedding it into an existing formalized math library) is an interesting future direction.

2.4 Related Work

Development of a well-designed benchmark is doubtlessly a crucial part of AR. The most notable example is the "Thousands of Problems for Theorem Provers" (TPTP, [12]), which covers various domains and several problem formats including CNF, first-order formula with quantifiers, and typed higher-order logic. Previous efforts have also accumulated benchmarks for various branches of AR, such as SAT [6], satisfiability modulo theory [1], inductive theorem proving [4], and geometry problems [11]. However, the current study is the first attempt to offer a large collection of curriculum math problems including not only proof problems but also `Find` and `Draw` problems with a wide range of difficulties as a benchmark for AR technologies.

3 Problem Samplers

We provide several sample problems taken from the first release of the library.

Hokkaido University, 2011, Science Course, Problem 3 (2)

Let ℓ be the trajectory of $(t+2, t+2, t)$ for t ranging over the real numbers. $\mathrm{O}(0,0,0)$, $\mathrm{A}(2,1,0)$, and $\mathrm{B}(1,2,0)$ are on a sphere S, centered at $\mathrm{C}(a,b,c)$. Determine the condition on a, b, c for which S intersects with ℓ.

In the data set, the above problem is formalized as shown in Fig. 5.

It is not difficult to obtain an equivalent formula in the language of first-order RCF by rewriting the predicates and functions using their defining axioms. However, it results in a formula including 22 variables and 22 atoms, that is way above the ability of existing RCF-QE solvers to deal with. It is not very surprising seeing that the time complexity of RCF-QE, a key step in the solution process, is doubly exponential in the number of variables in a given formula. We enhanced existing RCF-QE algorithms to overcome the difficulty. Fortunately, our prototype system successfully solved this problem. We will explain the enhancement in detail in Sect. 5.

Chart-shiki, Math 3+C, Problem 09CBCE011

Consider $0 < \frac{|ax-y|}{\sqrt{1+a^2}} < \frac{2\sqrt{2}}{x+y}$ for $x > 0$ and $y > 0$. Prove that there are only finitely many pairs of positive integers (x, y) that satisfy the above inequalities when a is a rational number.

```
;; FILE: Univ-Hokkaido-2011-Ri-3.lsp
(def-directive
  hokudai_2011_Ri_3_2
   (Find (abc)
     (exists (a b c O A B C l S)
       (& (= abc (list-of a b c))
          (line-type l)
          (= l (shape-of-cpfun (lambda p (exists (t) (= p (point (+ t 2) (+ t 2) t))))))
          (sphere-type S)
          (= O (point 0 0 0)) (= A (point 2 1 0)) (= B (point 1 2 0))
          (on O S) (on A S) (on B S)
          (= C (point a b c))
          (= C (center-of S))
          (intersect l S)))))))
```

Fig. 5. Hokkaido university, 2011, science course, problem 3 (2)

In the data set, the above problem is formalized as follows:

$$\forall a \in \mathbb{Q} \exists n \in \mathbb{Z}(n > 0 \wedge n = |\{(x, y) \in \mathbb{Z}^2 \mid P(\texttt{int2real}(x), \texttt{int2real}(y))\}|)$$

where $P = \lambda(x, y) \in \mathbb{R}^2(x > 0 \wedge y > 0 \wedge 0 < \frac{|ax-y|}{\sqrt{1+a^2}} < \frac{2\sqrt{2}}{x+y})$. Several $\in$'s preceding to domain names in the formula signify their types. Despite the seeming mixture of reals, integers, and rational numbers, we can easily find an equivalent formula in the language of PA. The mechanization of processes such as this is one of our ongoing research topics.

IMO 2012, Problem 2

Let $n \geq 3$ be an integer, and let $a_2, a_3, \ldots, a_n$ be positive real numbers such that $a_2 a_3 \cdots a_n = 1$. Prove that $(1 + a_2)^2(1 + a_3)^3 \ldots (1 + a_n)^n > n^n$.

In the data set, this problem is formalized using a higher-order function $\texttt{prod\_from\_to} :: (\texttt{Z} \to \texttt{R}) \to \texttt{Z} \to \texttt{Z} \to \texttt{R}$, which corresponds to $\Pi_{\text{from}}^{\text{to}}$ in the common notation. This problem apparently requires some kind of inductive reasoning but the domain includes both real numbers and integers. Problems of this type are abundant in curriculum math. We believe they will prove to be new and interesting and challenging problems for automated inductive reasoning, both theoretically (e.g., formalizing them in a suitable local theory other than ZF) and practically.

IMO 2003, Problem 1

> S is the set $\{1, 2, 3, \ldots, 1000000\}$. Show that for any subset A of S with 101 elements we can find 100 distinct elements x_i of S, such that the sets $\{a + x_i \mid a \in A\}$ are all pairwise disjoint.

It is straightforward to translate the above-mentioned problem in ZF:

$$\forall A \begin{pmatrix} A \subset S \land |A| = 101 \\ \quad \to \exists X \, (X \subset S \land |X| = 100 \land \text{pairwise\_disjoint}(\{\{a + x \mid a \in A\} \mid x \in X\})) \end{pmatrix}$$

where $S = \{n \in \mathbb{N} \mid 1 \leq n \leq 1000000\}$. Moreover, it can be expressed in PA, too. However, the effort to reformulate it in PA does little help in solving it.

Table 7. Preference hierarchy on answer form

Directive type	Syntactic condition on the answer formula
$\texttt{Find}(v : \texttt{R})[\phi(v, \boldsymbol{p})]$	1. $\bigvee_i \left(\bigwedge_j (v \; \rho_{ij} \; \alpha_{ij}) \land \psi_i(\boldsymbol{p})\right)$
	2. $\bigvee_i \left(\bigwedge_j f_{ij}(v) \; \rho_{ij} \; 0 \land \psi_i(\boldsymbol{p})\right)$
	3. $\bigvee_i \left(\exists (n : \texttt{Z}).\left(\bigwedge_j (v \; \rho_{ij} \; \alpha_{ij}(n)) \land (n \; \rho_i \; \gamma_i)\right) \land \psi_i(\boldsymbol{p})\right)$
	4. $\bigvee_i \left(\exists (r : \texttt{R}).\left(\bigwedge_j (v \; \rho_{ij} \; \alpha_{ij}(r)) \land (r \; \rho_i \; \gamma_i)\right) \land \psi_i(\boldsymbol{p})\right)$
$\texttt{Find}(v : \texttt{Z})[\phi(v, \boldsymbol{p})]$	1. $\bigvee_i \left(\bigwedge_j (v \; \rho_{ij} \; \alpha_{ij}) \land \psi_i(\boldsymbol{p})\right)$
	2. $\bigvee_i (\exists (n : \texttt{Z}).\, (v = \alpha_i(n) \land (n \; \rho_i \; \gamma_i)) \land \psi_i(\boldsymbol{p}))$
$\texttt{Find}(v : \texttt{SetOf(Point)})[\phi(v, \boldsymbol{p})]$	$\bigvee_i (v = \{(x, y) \mid \xi_i(x, y)\} \land \psi_i(\boldsymbol{p}))$

($\rho_* \in \{=, <, \leq, \geq, >\}$; α_*, $\alpha_*(\cdot)$, γ_*, $\xi_*(\cdot, \cdot)$: first-order terms not including v, x, y; $f_{ij}(v)$: first-order term; $\psi_i(\boldsymbol{p})$: quantifier-free first-order formula)

4 What Constitutes an Answer to a Find Problem?

In Subsect. 2.2, the properties an answer formula for a `Find` problem has to satisfy for it to be regarded as acceptable (correctness and simplicity) were briefly discussed. Now we will discuss these in detail. In [14], Sutcliffe et al. proposed the conditions which answers of answer-extraction problems have to satisfy. Our definition of 'answer' encompasses theirs in spirit and covers more complicated cases beyond the extraction of a finite number of answers.

The definition of the *correctness* of an answer is straightforward. Given a problem $\texttt{Find}(x)[\psi(x, \boldsymbol{p})]$, where $\boldsymbol{p}$ stands for zero or more free parameters, an answer formula $\phi(x, \boldsymbol{p})$ must satisfy:

$$\forall x \forall \boldsymbol{p} (\psi(x, \boldsymbol{p}) \leftrightarrow \phi(x, \boldsymbol{p})). \tag{1}$$

An example of a correct answer formula $\phi'(x, \boldsymbol{p})$ is provided for each `Find` problem in the library. If $\phi'(x, \boldsymbol{p})$ is used instead of $\psi(x, \boldsymbol{p})$, the proof task for (1) should generally be easy.

The *simplicity* of an answer is harder to define. Suppose that you are given a problem, $\texttt{Find}(v : \mathbb{R})[v^2 = a]$, in a math test. Then, $\lambda v.(v^2 = a)$ is of course not an acceptable answer. However, test-takers are expected to answer, for example,

$$\lambda v.\left((a \geq 0 \wedge v = a^{1/2}) \vee (a \geq 0 \wedge v = -a^{1/2})\right).$$

An answer to a problem asking to find all real numbers v satisfying a formula $\phi(v)$ in the first-order language of RCF is called *simple* when it is in the form $\lambda v.\psi(v)$ satisfying the following conditions.

- $\psi(v)$ is a quantifier-free formula in disjunctive normal form, and
- each dual clause in $\psi(v)$ consists of atoms of the form of $v\ \rho\ \alpha$ or $\beta\ \rho\ 0$, where $\rho \in \{=, <, >, \leq, \geq\}$, and α and β are first-order terms not including v and comprises numbers, variables (i.e., parameters) and functions in $\{+,\ -,\ \cdot,\ /,\ \hat{}\,(\text{power})\}$.

The aforementioned syntactic conditions for a problem classified in RCF should be acceptable because RCF allows quantifier elimination [15]. Furthermore, the statistics tell us that almost all pre-university math problems have explicit solutions (i.e., in the form of $x = \alpha, \beta > x > \gamma$, etc.)

For problems other than those expressible in RCF, we tried our best to capture a loose, common understanding in the form of acceptable answers by examining the model answers (for humans) to the benchmark problems. Our tentative definition of 'simple answers' is as follows:

- Simplicity of the sub-language: an answer formula should be in a language consisting of Boolean connectives, equality and inequalities, numbers, variables, and the four arithmetic operations and power calculations, sin, cos, tan, exp, log, 'type coercion functions' such as `int_to_real`, and *a minimal use of lambda abstractions and quantifications.*
- Explicitness: whenever possible within the above restriction imposed on the language, the answer to a problem of the form $\texttt{Find}(x)[\phi(x)]$ should be given using atoms such as $x = \alpha$ and $x > \alpha$, where α does not include x.

Note that we need quantification in general unless the problem is expressible in a theory that allows quantifier elimination. For instance, in the sub-language defined above, there is no way to express the answer to "Determine all positive numbers v that are divisible by three and also by two," other than, e.g., $\exists k(v = 6k \wedge k > 0)$. As for "*minimal use of* $\lambda, \forall, \exists$", we define the preference of answer form tentatively (Table 7). The answer-check routine compares a solver's answer and the model answer in the data set, and checks whether the solver's answer ranks equal (or higher) in the hierarchy.

5 Prototype Solver

While developing the benchmark data set, we also developed a prototype math problem solver system (overviewed in Fig. 6). Given a formalized problem, the

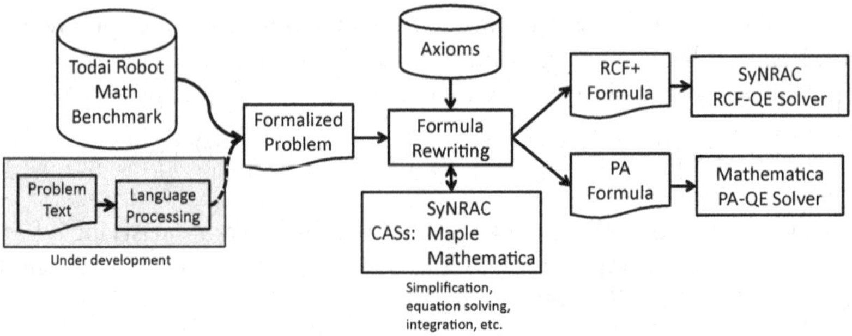

Fig. 6. System overview

Table 8. Overall results

		Succeeded		Failed		
		Success %	Time (sec) Min/Med/Avg/Max	Timeout	Wrong	Other
Chart	RCF	63.8 % (111/174)	13/18.0/ 37.4/ 343	10.9 %	1.7 %	23.6 %
	PA	57.1 % (48/ 84)	12/17.0/ 20.3/ 172	0.0 %	0.0 %	42.9 %
	Other	10.0 % (3/ 30)	13/14.0/ 17.7/ 26	0.0 %	0.0 %	90.0 %
	All	56.3 % (162/288)	12/17.0/ 32.0/ 343	6.6 %	1.0 %	36.1 %
Univ	All (RCF only)	58.0 % (142/245)	12/26.5/ 85.5/1417	15.5 %	2.9 %	23.7 %
IMO	RCF	16.5 % (19/115)	14/25.0/ 51.8/ 197	29.6 %	0.9 %	53.0 %
	PA	4.8 % (2/ 42)	25/29.5/ 29.5/ 34	16.7 %	0.0 %	78.6 %
	Other	3.6 % (2/ 55)	17/24.5/ 24.5/ 32	12.7 %	0.0 %	83.6 %
	All	10.8 % (23/212)	14/25.0/ 47.5/ 197	22.6 %	0.5 %	66.0 %

system first rewrites it iteratively using the axioms and several equivalence-preserving transformation rules such as beta-reduction, extensional equality between functions ($\lambda x.M = \lambda x.N \Leftrightarrow \forall x(M = N)$), variable elimination by substitution ($\forall x(x = f \rightarrow \phi(x)) \Leftrightarrow \phi(f)$, and $(\exists x(x = f \wedge \phi(x)) \Leftrightarrow \phi(f)$ where x does not occur free in f). In the course of the rewriting process, several types of terms, such as multiplication and division of polynomials and integration, are evaluated (simplified) by CASs such as Mathematica 9.0 and Maple 18. Once the input is rewritten to a formula in the language of RCF, quantifier-elimination (QE) algorithms are invoked; we utilized the RCF-QE algorithm implemented in SyNRAC [8]. When QE is proceeded successfully, the remaining tasks, solving equations and inequalities in many cases, will be taken care by the CASs. When the input is rewritten in the language of PA, we apply the `Reduce` command of Mathematica.

As mentioned in Sect. 3, the first-order formulas generated by mechanical translation are much larger than expected [7,10]. We enhanced the RCF-QE algorithms by numerous techniques to handle them: choice of the computation order of sub-formulas, specialized QE algorithms for restricted input formulas, simplification of the intermediate formulas by utilizing the interim results, and

Table 9. Breakdown of results on **Chart** RCF problem by number of stars

# of Stars	Succeeded		Failed		
	Success %	Time (sec) Min/Med/Avg/Max	Timeout	Wrong	Other
1	82.4 % (28/34)	13/17.0/20.4/ 65	2.9 %	0.0 %	14.7 %
2	79.4 % (27/34)	16/18.0/28.1/230	2.9 %	2.9 %	14.7 %
3	57.6 % (19/33)	15/17.0/36.1/341	6.1 %	0.0 %	36.4 %
4	47.4 % (18/38)	15/19.0/62.1/343	23.7 %	2.6 %	26.3 %
5	54.3 % (19/35)	16/28.0/53.6/279	17.1 %	2.9 %	25.7 %

Table 10. Breakdown of results on **Univ** RCF problems by university

University	# of All Problems	RCF Problems %	Overall Success %	Success % on RCF Problems
Hokkaido	72	44.4 % (32/ 72)	25.0 % (18/ 72)	56.3 % (18/32)
Tohoku	80	52.5 % (42/ 80)	30.0 % (24/ 80)	57.1 % (24/42)
Tokyo	160	38.8 % (62/160)	18.8 % (30/160)	48.4 % (30/62)
Nagoya	72	41.7 % (30/ 72)	20.8 % (15/ 72)	50.0 % (15/30)
Osaka	64	37.5 % (24/ 64)	32.8 % (21/ 64)	87.5 % (21/24)
Kyoto	88	43.2 % (38/ 88)	33.0 % (29/ 88)	76.3 % (29/38)
Kyushu	96	36.5 % (35/ 96)	18.8 % (18/ 96)	51.4 % (18/35)

Table 11. Results for **IMO** problems by decade

Years	Human efficency	Machine efficiency	Succeeded	Failed		
				Timeout	Wrong	Other
1959-69	58.23 %	21.11 %	26.3 % (15/57)	22.8 %	1.8 %	49.1 %
1970-79	46.57 %	7.00 %	13.3 % (4/30)	26.7 %	0.0 %	60.0 %
1980-89	44.35 %	1.85 %	3.1 % (1/32)	31.2 %	0.0 %	65.6 %
1990-99	38.27 %	3.33 %	5.7 % (2/35)	11.4 %	0.0 %	82.9 %
2000-13	34.31 %	1.19 %	1.9 % (1/54)	22.2 %	0.0 %	75.9 %

so on. Additionally, we developed an algorithm for computing *the area enclosed by a set of curves* and an extended RCF-QE command to reduce some of the problems involving trigonometric functions to RCF-QE problems.

6 Experiments

The prototype system was run on the benchmark problems with a time limit of 3600 s per problem (including the time spent on checking the correctness of the answers). Table 8 shows the number of successfully solved problems, minimum, median, average, and maximum (wallclock) time spent on solved problems, number of failures due to timeout, wrong answers (disproofs for `Show` or wrong answers for `Find` or `Draw` directives), and those not solved due to various reasons (the column headed 'Other'). Approximately one-third of the 'Other' cases were due to a failure in the problem reformulation phase; i.e., for those problems, the system could not find an equivalent formula expressible in either RCF or PA. Explicitly wrong answers were due to bugs in our formula rewriting system module and/or malfunctions of Maple's equation/inequality solving command.

Overall, the performances for the **Chart**, **Univ**, and **IMO** divisions seem to well reflect the inherent differences in their difficulty levels. Tables 9, 10, and 11 show further analysis of the results obtained for the three divisions. Table 9 lists the performance figures for the RCF problem subsets in the **Chart** division that are rated level 1 to 5 in the exercise books. We see a clear difference between those rated level 1 or 2, and 4 or 5, especially in the percentages of the problems that had a timeout. Table 10 lists the performance figures for each university from which the exam problems were taken. Although average scores etc. of the entrance exams are not published, a statistical analysis undertaken by major prep schools tells us that the average score of successful applicants to the top universities is around 30–60% depending on schools and departments. Hence, it is very plausible that a machine will come to have the ability to pass the entrance math exams of top universities if it is able to cover areas other than RCF.

Finally, Table 11 lists the results on **IMO** problems taken from different time periods. Human and Machine Efficiency in the table shows the ratio between the attained points (by all contestants in a year and by our system, respectively) and all possible points[1]. It seems that the IMO problems are getting harder year by year not only for human participants but more so for our system.

We believe that these experimental results support our decision on the library organization, and encourage us to further proceed toward the goal of end-to-end math problem solving with the monolithic logical language based on ZF.

7 Conclusion and Prospects

In this paper, we introduced a benchmark problem library for mechanized math technologies. The library consists of curriculum math problems taken from exercise problem books, university entrance exams, and International Mathematical Olympiads. Unlike other existing benchmark libraries, this one contains problems that are formalized so that they are obtainable as the result of mechanical translation of the original problems expressed in natural language. Preliminary experimental results we obtained for our prototype system on the benchmark show that its performance is comparable to that of candidates for admission to top universities, at least for problems in real-closed fields.

[1] The statistics were taken from the official IMO website: https://www.imo-official.org/results_year.aspx.

Our future plan includes the expansion of the library with more problems on integer arithmetic, transcendental functions, combinatorics, and a mixture of real and integer arithmetics as well as development of the natural language processing module for an end-to-end system.

References

1. Barrett, C., Stump, A., Tinelli, C.: The Satisfiability Modulo Theories Library (SMT-LIB) (2010). www.SMT-LIB.org
2. Bos, J.: Wide-coverage semantic analysis with boxer. In: Bos, J., Delmonte, R. (eds.) Semantics in Text Processing, STEP 2008 Conference Proceedings, pp. 277–286. Research in Computational Semantics, College Publications (2008)
3. Clark, S., Curran, J.R.: Wide-coverage efficient statistical parsing with CCG and log-linear models. Comput. Linguist. **33**, 493–552 (2007)
4. Dennis, L.A., Gow, J., Schürmann, C.: Challenge problems for inductive theorem provers v1.0. Technical report ULCS-07-004, University of Liverpool, Department of Computer Science (2007)
5. Grabowski, A., Korni lowicz, A., Naumowicz, A.: Mizar in a nutshell. J. Formalized Reasoning **3**(2), 153–245 (2010)
6. Hoos, H.H., Stützle, T.: SATLIB: An Online Resource for Research on SAT. In: Sat2000: Highlights of Satisfiability Research in the Year 2000, pp. 283–292. IOS Press, Amsterdam (2000)
7. Iwane, H., Matsuzaki, T., Arai, N., Anai, H.: Automated natural language geometry math problem solving by real quantier elimination. In: Proceedings of the 10th International Workshop on Automated Deduction (ADG2014), pp. 75–84 (2014)
8. Iwane, H., Yanami, H., Anai, H., Yokoyama, K.: An effective implementation of symbolic-numeric cylindrical algebraic decomposition for quantifier elimination. Theor. Comput. Sci. **479**, 43–69 (2013)
9. Kwiatkowksi, T., Zettlemoyer, L., Goldwater, S., Steedman, M.: Inducing probabilistic CCG grammars from logical form with higher-order unification. In: Proceedings of the 2010 Conference on Empirical Methods in Natural Language Processing, pp. 1223–1233. Association for Computational Linguistics (2010)
10. Matsuzaki, T., Iwane, H., Anai, H., Arai, N.H.: The most uncreative examinee: A first step toward wide coverage natural language math problem solving. In: Proceedings of the Twenty-Eighth AAAI Conference on Artificial Intelligence, pp. 1098–1104 (2014)
11. Quaresma, P.: Thousands of geometric problems for geometric theorem provers (TGTP). In: Schreck, P., Narboux, J., Richter-Gebert, J. (eds.) ADG 2010. LNCS, vol. 6877, pp. 169–181. Springer, Heidelberg (2011)
12. Sutcliffe, G.: The TPTP problem library and associated infrastructure: the FOF and CNF Parts, v3.5.0. J. Autom. Reasoning **43**(4), 337–362 (2009)
13. Sutcliffe, G., Benzmüller, C.: Automated reasoning in higher-order logic using the TPTP THF infrastructure. J. Formalized Reasoning **3**(1), 1–27 (2010)
14. Sutcliffe, G., Stickel, M., Schulz, S., Urban, J.: Answer extraction for TPTP. http://www.cs.miami.edu/~tptp/TPTP/Proposals/AnswerExtraction.html
15. Tarski, A.: A Decision Method for Elementary Algebra and Geometry. University of California Press, Berkeley (1951)

raSAT: An SMT Solver for Polynomial Constraints

Vu Xuan Tung[1(✉)], To Van Khanh[2], and Mizuhito Ogawa[1]

[1] Japan Advanced Institute of Science and Technology, Nomi, Japan
{tungvx,mizuhito}@jaist.ac.jp
[2] University of Engineering and Technology, Vietnam National University, Hanoi, Vietnam
khanhtv@vnu.edu.vn

Abstract. This paper presents the **raSAT** SMT solver for polynomial constraints, which aims to handle them over both reals and integers with simple unified methodologies: (1) **raSAT** *loop* for inequalities, which extends the *interval constraint propagation* with testing to accelerate SAT detection, and (2) a non-constructive reasoning for equations over reals, based on the generalized intermediate value theorem.

1 Introduction

Polynomial constraint solving is to find an instance that satisfies a given system of polynomial inequalities/equations. Various techniques for solving such a constraint are implemented in SMT solvers, e.g., **Cylindrical algebraic decomposition** (RAHD [18,19], Z3 4.3 [13]), **Virtual substitution** (SMT-RAT [5], Z3 3.1), **Interval constraint propagation** [2] (iSAT3 [7], dReal [9,10], RSolver [20], RealPaver [11]), and **CORDIC** (CORD [8]). For integers, **Bit-blasting** (MiniSmt [23]) and **Linearization** (Barcelogic [3]) can be used.

This paper presents the **raSAT** SMT solver[1] for polynomial constraints over reals. For inequalities, it applies a simple iterative approximation refinement, **raSAT** *loop*, which extends the interval constraint propagation (ICP) with testing to boost SAT detection (Sect. 3). For equations, a non-constructive reasoning based on the generalized intermediate value theorem [17] is applied (Sect. 4). Implementation with soundness guarantee and optimizing strategies is evaluated by experiments (Sect. 5).

Although **raSAT** has been developed for constraints over reals, constraints over integers are easily adopted, e.g., by stopping interval decompositions when the width becomes smaller than 1, and generating integer-valued test instances.

raSAT has participated SMT Competition 2015, in two categories of main tracks, *QF_NRA* and *QF_NIA*. The results, in which **Z3 4.4** is a reference, are,

- 3rd in *QF_NRA*, **raSAT** solved 7952 over 10184 (where **Z3 4.4**, **Yices-NL** and **SMT-RAT** solved 10000, 9854 and 8759, respectively.)

[1] Available at http://www.jaist.ac.jp/~s1310007/raSAT/index.html.

N. Olivetti and A. Tiwari (Eds.): IJCAR 2016, LNAI 9706, pp. 228–237, 2016.
DOI: 10.1007/978-3-319-40229-1_16

- 2nd in *QF_NIA*, **raSAT** solved 7917 over 8475 (where **Z3 4.4** and **AProVE** solved 8459 and 8270, respectively).

A preliminary version of **raSAT** was orally presented at *SMT workshop 2014* [22].

2 SMT Solver for Polynomial Constraints

Definition 1. *A polynomial constraint ψ is defined as follow*

$$\psi ::= g(x_1, ..., x_n) \diamond 0 \mid \psi \wedge \psi \mid \psi \vee \psi \mid \neg\psi \tag{1}$$

where ($\diamond \in \{>, \geq, <, \leq, =, \neq\}$) and $g(x_1, \cdots, x_n)$ is a polynomial with integer coefficients over variables $x_1, \cdots, x_n$. We call $g(x_1, \cdots, x_n) \diamond 0$ an atomic polynomial constraint (APC). When $x_1, \cdots, x_n$ are clear from the context, we denote g for $g(x_1, \cdots, x_n)$, and $var(g)$ for the set of variables appearing in g.

An SMT solver decides whether ψ is satisfiable (SAT), i.e., whether there exists an assignment of reals (resp. integers) to variables that makes ψ *true*. We organize the **raSAT** SMT solver in a very lazy approach for an arithmetic theory T over reals (resp. integers). As a preprocessing, **raSAT** converts a polynomial constraint into conjunctive normal form (CNF) by Tseitin conversion [21]. In addition, the APCs are preprocessed so that the constraint becomes a CNF containing only $>$ and $=$. Then, first, each APC is assigned a Boolean value (*true* or *false*) by an SAT solver such that ψ is evaluated to *true*. Second, the boolean assignment is checked for consistency against the theory T.

raSAT is one of the interval constraint propagation (ICP) based SMT solvers, as well as **iSAT** [7] and **dReal** [10]. In ICP [2], *interval arithmetic* (IA) [16] plays a central role. **raSAT** implements Classical Interval (CI) [16] and four kinds of Affine Intervals (AI) [4,14]. We fix their notations. Let $\mathbb{R}$ be the set of real numbers and $\mathbb{R}^\infty = \mathbb{R} \cup \{-\infty, \infty\}$. We naturally extend the standard arithmetic operations on $\mathbb{R}$ to those on $\mathbb{R}^\infty$ as in [16]. The set of all intervals is denoted by $\mathbb{I} = \{[l, h] \mid l \leq h \in \mathbb{R}^\infty\}$. A *box* for a sequence of variables $x_1, \cdots, x_n$ is $B = I_1 \times \cdots \times I_n$ for $I_1, \cdots, I_n \in \mathbb{I}$.

A conjunction φ of APCs is *IA-valid* (resp. *IA-UNSAT*) in a box B if φ is evaluated to *true* (resp. *false*) by IA over B. In this case, B is called a *IA-valid* (resp. *IA-UNSAT*) box with respect to φ. Since IA is an over approximation of arithmetical results, IA-valid (resp. IA-UNSAT) in B implies valid (resp. UNSAT) in B. If neither of them holds, we call *IA-SAT* (as shown below), which cannot decide the satisfiability at the moment. Note that if φ is IA-valid in B, φ is SAT.

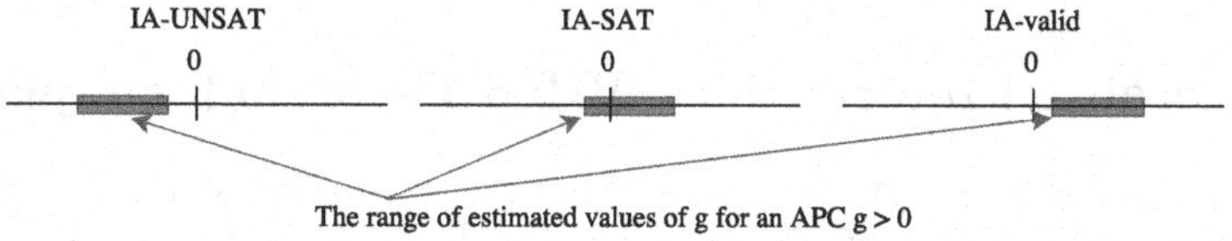

3 ICP and raSAT Loop for Inequality

Since ICP is based on IA, which is an over-approximation, it can be applied to decide SAT/UNSAT of inequalities and UNSAT of equalities, but not for SAT of equalities. We first explain ICP for (a conjunction of) inequalities and then extend it as a **raSAT** loop for SAT detection acceleration. Handling the presence of equations will be shown in Sect. 4.

Starting with a box B ($(-\infty, \infty)^n$ by default), ICP [2] tries to detect SAT of φ in B by iteratively contracting boxes (by backward propagation of interval constraints) and decomposing boxes (when neither IA-valid nor IA-UNSAT detected) until either an IA-valid box is found or no boxes remain to explore.

The **raSAT** loop [14] intends to accelerate ICP for SAT detection by testing. Figure below illustrates the **raSAT** loop, in which "*Test-SAT*" in B means that a satisfiable instance is found by testing in B, and "*Test-UNSAT*", otherwise.

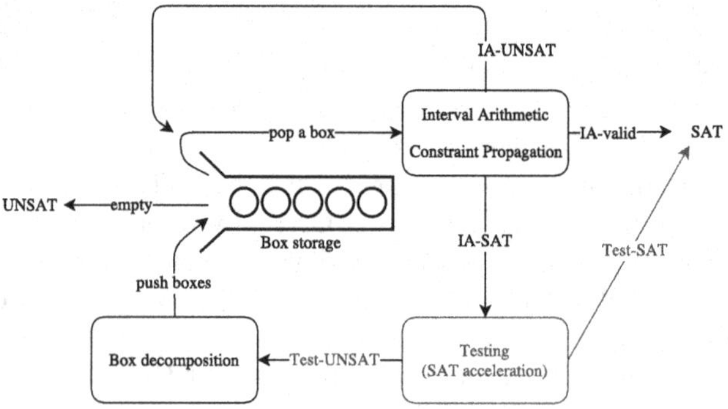

Limitation of ICP and raSAT Loop for Inequality. ICP concludes SAT when it identifies a valid box by IA. Although the number of boxes may be exponential, if $I_1, \cdots, I_n$ are bounded, ICP always detects SAT of the inequalities ψ as Fig. (a) and detects UNSAT of ψ if not touching as illustrated in Fig. (b,c). If $I_1, \cdots, I_n$ are not bounded, adding to touching cases, a typical case of failure in UNSAT detection is a converging case as Fig. (d).

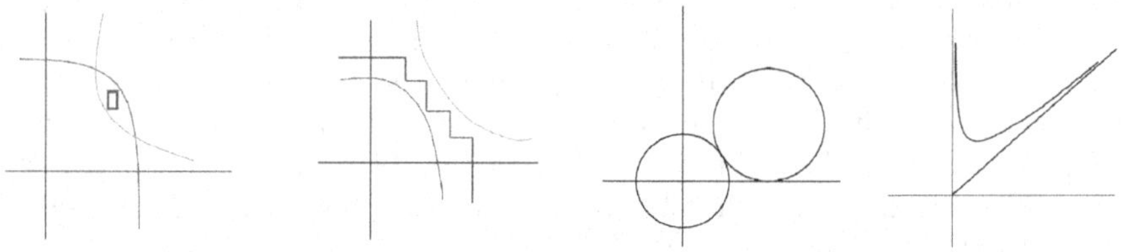

(a) SAT detection (b) UNSAT detection (c) Touching case (d) Convergent case

4 Generalized Intermediate Value Theorem for Equations

Handle equations in **raSAT** is illustrated by the *intermediate value theorem* (IVT) for a single equation $g(x) = 0$. If we find t_1, t_2 with $g(t_1) > 0$ and $g(t_2) < 0$,

$g = 0$ holds in between. For multi-variant equations, we apply a custom version (Theorem 1) of the generalized IVT [17, Theorem 5.3.7].

4.1 Generalized Intermediate Value Theorem

Let $B = [l_1, h_1] \times \cdots [l_n, h_n]$ be a box over $V = \{x_1, \cdots, x_n\}$, and let $V' = \{x_{i_1}, \cdots, x_{i_k}\}$ be a subset of V. We denote $B{\downarrow_{V'}} = \{(r_1, \cdots, r_n) \in B \mid r_i = l_i \text{ for } i = i_1, ..., i_k\}$ and $B{\uparrow_{V'}} = \{(r_1, \cdots, r_n) \in B \mid r_i = h_i \text{ for } i = i_1, ..., i_k\}$. Given an assignment $\theta : V' \mapsto \mathbb{R}$, which assigns a real value to each variable in V', $B|_\theta = \{(r_1, \cdots, r_n) \in B \mid r_i = \theta(x_i) \text{ if } x_i \in V'\}$.

Definition 2. *Let* $\bigwedge_{j=1}^{m} g_j = 0$ *be a conjunction of equations over* V. *A sequence* $(V_1, \cdots, V_m)$ *is a* check basis *of* $(g_1, \cdots, g_m)$ *in* B, *if, for each* $j, j' \leq m$,

1. $\emptyset \neq V_j \subseteq var(g_j)$,
2. $V_j \cap V_{j'} = \emptyset$ *if* $j \neq j'$, *and*
3. *either* $g_j < 0$ *on* $B{\uparrow_{V_j}}$ *and* $g_j > 0$ *on* $B{\downarrow_{V_j}}$, *or* $g_j < 0$ *on* $B{\uparrow_{V_j}}$ *and* $g_j > 0$ *on* $B{\downarrow_{V_j}}$.

Theorem 1. *For a conjunction of polynomial inequalities/equations*

$$\varphi = \bigwedge_{j=1}^{m} g_j > 0 \wedge \bigwedge_{j=m+1}^{m'} g_j = 0$$

and $B = [l_1, h_1] \times \cdots [l_n, h_n]$, *assume that the followings hold.*

1. *For* $\varphi_1 \wedge \varphi_2 = \bigwedge_{j=1}^{m} g_j > 0$, φ_1 *is IA-valid in* B *and* φ_2 *is Test-SAT in* B *with an assignment* $\theta_{\varphi_2} : V_{\varphi_2} \mapsto \mathbb{R}$ *such that* $\theta_{\varphi_2}(x_i) \in [l_i, h_i]$ *for each* $x_i \in V_{\varphi_2}$, *where* V_{φ_2} *is the set of variables in* φ_2.
2. *A* check basis $(V_{m+1}, \cdots, V_{m'})$ *over* $V \backslash V_{\varphi_2}$ *of* $(g_{m+1}, \cdots, g_{m'})$ *in* $B|_{\theta_{\varphi_2}}$ *exists.*

Then, φ *has a SAT instance in* B.

Example 1 illustrates Theorem 1 for $V = \{x, y\}$ with $m = 0$ and $m' = n = 2$.

Example 1. Given two equations $g_1(x, y) = 0$ and $g_2(x, y) = 0$. Assuming that there exists a box $B = [c_1, d_1] \times [c_2, d_2]$ such that

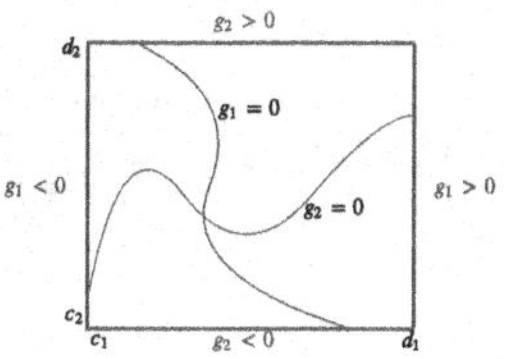

- $g_1(c_1, y) < 0$ for $y \in [c_2, d_2]$, $g_1(d_1, y) > 0$ for $y \in [c_2, d_2]$, and
- $g_2(x, c2) < 0$ for $x \in [c_1, d_1]$, $g_2(x, d2) > 0$ for $x \in [c_1, d_1]$.

Thus, $g_1(x, y) = 0$ and $g_2(x, y) = 0$ share a root in B.

Limitation of the Generalized IVT for Equality. There are two limitations on applying Theorem 1.

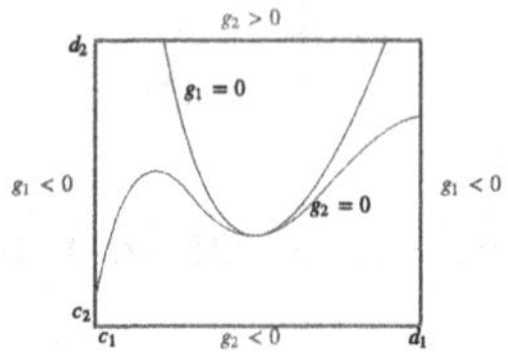

- The number of variables (dimensions) must be no less than the number of equations.
- Trajectories of equations must be crossing. For instance, it may fail to show SAT if two equation $g_1 = 0$, $g_2 = 0$ are touching, as in the right figure.

4.2 raSAT Loop with Generalized IVT

Theorem 1 is added into the **raSAT** loop as in Figure below. We borrow notations φ, φ_1, and φ_2 from Theorem 1. The label "$>$: IA-valid" means that the conjunction of inequalities appearing in the input is IA-valid. Similar for "=: IA-SAT" and "$>$: Test-SAT". The label "Test-SAT over $V_{\varphi_2} \subseteq V$" means that a test instance to conclude Test-SAT of φ_2 is generated on V_{φ_2} and the generalized IVT is applied over $V \setminus V_{\varphi_2}$ in the box $B|_{\theta_{\varphi_2}}$ (described by "IVT over $V \setminus V_{\varphi_2}$").

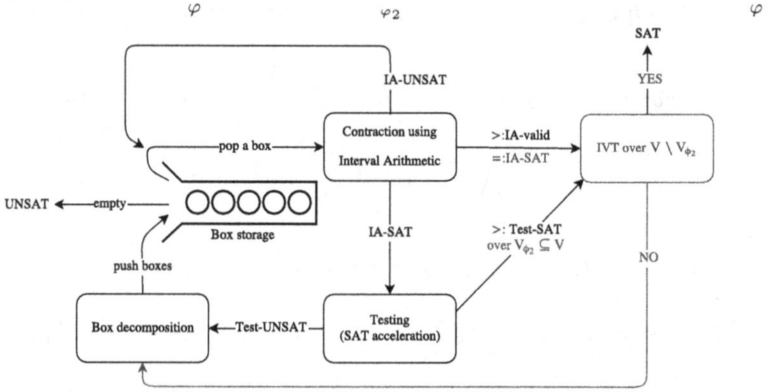

Example 2. Suppose φ is $g_1 > 0 \wedge g_2 = 0 \wedge g_3 = 0$ where $g_1 = cd - d$, $g_2 = a - c - 2$, and $g_3 = bc - ad - 2$. The initial box storage contains only $B = [-2, 3.5] \times [-5, 0] \times [0, 1.5] \times [-5, -0.5]$ as the initial range of (a, b, c, d).

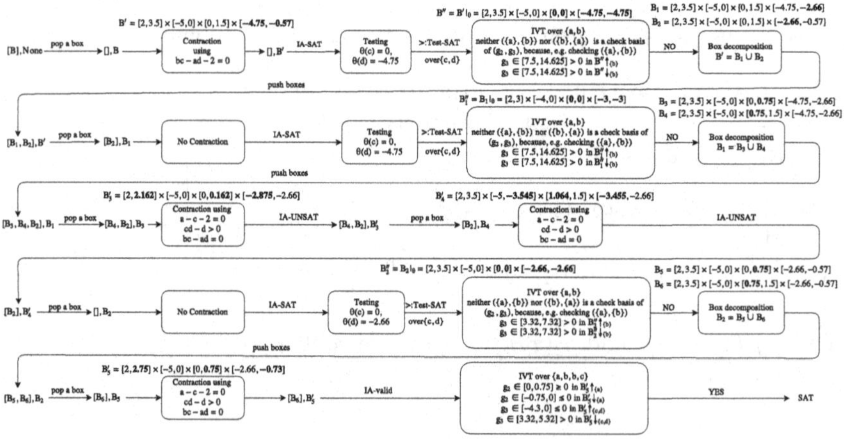

Figure above shows the flow of the **raSAT** loop with IVT, where a label [...], B is for a pair of a box storage and a currently exploring box B, and θ for a test instance. The backward interval constraint propagation reduces B, B_1, and B_3 to B', B'_1, and B'_3, respectively.

5 Implementation and Experiments

5.1 Implementation of raSAT

In **raSAT** implementation, the SAT solver miniSAT [6] manages the Boolean part of the DPLL procedure. There are several notable features of **raSAT**.

Soundness. raSAT uses the floating point arithmetic, and round-off errors may violate the soundness. To get rid of such pitfalls, **raSAT** integrates an IA library [1] which applies outward rounding [12] of intervals. For the soundness of Test-SAT, **iRRAM**[2], which guarantees the round-off error bounds, confirms that a SAT instance found by the floating point arithmetic is indeed SAT.

Affine Interval. Various IAs, including Classical Interval (CI) [16] and 4 variations, AF_1, AF_2, EAI, CAI, of Affine Intervals (AI) [4,14,15], are implemented as a part of **raSAT**. At the moment, AF2 and CI are used by default, and the choice option will be prepared in the future releases.

AI introduces noise symbols ϵ's, which are interpreted as values in $[-1, 1]$. Variations of AIs come from how to (over) approximate the multiplication of noise symbols in a linear formula. Although the precision is incomparable, AI partially preserve the dependency among values, which is lost in CI. For instance, let $x \in [2, 4] = 3 + \epsilon$. Then, $x - x$ is evaluated to $[-2, 2]$ by CI, but $[0, 0]$ by AI. The example below shows the value dependency. Let $h(x, y) = x^3 - 2xy$ for $x = [0, 2] = 1 + \epsilon_1$ and $y = [1, 3] = 2 + \epsilon_2$. CI estimates $h(x, y)$ as $[-12, 8]$, and AF_2 does as $-3 - \epsilon_1 - 2\epsilon_2 + 3\epsilon_+ + 3\epsilon_\pm$ (evaluated to $[-9, 6]$). Such information is used to design SAT-directed heuristics for choosing a variable at a box decomposition.

SAT-Directed Heuristics. The variable selection strategy is, (1) select the least likely satisfiable *APC* with respect to *SAT-likelihood*, and (2) choose the most likely influential variable in the APC with respect to the *sensitivity*.

Suppose AI estimates the range $range(g, B)$ of a polynomial g in a box B as $[c_1, d_1]\epsilon_1 + \cdots + [c_n, d_n]\epsilon_n$, which is evaluated by instantiating $[-1, 1]$ to ϵ_i.

- The *SAT-likelihood* of an APC $g > 0$ is $|range(g, B) \cap (0, \infty)|/|range(g, B)|$.
- The *sensitivity* of a variable x_i in $g > 0$ is $max(|c_i|, |d_i|)$.

For instance, the SAT-likelihood of $h(x, y)$ above is $0.4 = \frac{6}{9-(-6)}$ by AF_2 and the sensitivity of x and y are 1 and 2 by AF_2, respectively.

When selecting a box, **raSAT** adopts the largest *SAT-likelihood*, where the *SAT-likelihood* of a box is the least *SAT-likelihood* among APCs on it. Thus, the box storage in the **raSAT** loop with IVT is implemented as a priority queue.

[2] Available at http://irram.uni-trier.de.

The effect of the heuristics is examined with 18 combinations of the least, largest (with respect to measures), and random variable/box choices. Among them, only the combination above shows visible differences from the random choices, especially on SAT detection for quite large problems, such that it detects 11 SAT (including 5 problems marked "*unknown*") in Zankl/Matrix2∼5, whereas others detect at most 5 SAT (with at most 1 problem marked "*unknown*").

5.2 Experiments

Comparison with Other SMT Solvers. Our comparison has two views, (1) ICP-based solvers, e.g., **iSAT3** and **dReal**, and (2) other SMT-solvers, which are superior than **raSAT** at the SMT competition 2015, e.g., **Z3 4.4** and **SMT-RAT 2.0**[3]. After the competition, **raSAT** has been improved on the backward interval constraint propagation [2]. They are compared on SMT-LIB benchmarks 2015-06-01[4] with timeout of 2500 s on an Intel Xeon E7-8837 2.66 GHz and 8 GB RAM. Note that

- **iSAT3** requires bounded intervals, and its bound of variables is set to $[-1000, 1000]$. For other tools (including **raSAT**), it is kept $(-\infty, \infty)$.
- **dReal** decides δ-SAT, instead of SAT, which allows δ-deviation on the evaluation of polynomials for some $\delta > 0$. Note that δ-SAT does not imply SAT. δ for **dReal** is set to its default value (0.001).

Table 1 shows the numbers of solved problems in each benchmark of the QF_NRA category in SMT-LIB. The "Time" row shows the cumulative running time of successful cases. In the "Benchmark" column, the numbers of SAT/UNSAT problems are associated if already known. "*" means δ-SAT.

Unknown Problems in SMT-LIB. In SMT-LIB benchmark, many problems are marked "*unknown*". Among such unknown inequality problems, **raSAT** solves 15 (5 SAT, 10 UNSAT), **Z3 4.4** solves 36 (13 SAT, 23 UNSAT), and **SMT-RAT 2.0** solves 15 (3 SAT, 12 UNSAT). For problems with equations, **raSAT** and **SMT-RAT 2.0** solve 3 UNSAT problems, and **Z3 4.4** solves 492 (276 SAT, 216 UNSAT). For large problems, UNSAT can be detected by finding a small UNSAT core among APCs, whereas SAT detection requires to check all APCs.

For unknown problems, SAT results are easy to check. Although **Z3 4.4** outperforms others, it is worth mentioning that **raSAT** also detects SAT on several quite large problems in Zankl/Matrix-2∼5, which often have more than 50 variables (Meta-Tarski and Matrix-1 have mostly less than 10 and 30 variables, respectively). For instance, **Z3 4.4** solely solves Matrix-3-7, 4-12, and 5-6 (which have 75, 200, and 258 variables), and **raSAT** solely solves Matrix-2-3, 2-8, 3-5, 4-3, and 4-9 (which have 57, 17, 81, 139, and 193 variables). **SMT-RAT 2.0** shows no new SAT detection in Zankl/Matrix-2∼5.

[3] https://github.com/smtrat/smtrat/releases/download/v2.0/rat1_linux64.zip.
[4] http://smtlib.cs.uiowa.edu/benchmarks.shtml.

Table 1. Comparison among SMT solvers on SMT-LIB benchmark ($* = \delta$-SAT)

Benchmark (inequality only)	raSAT	iSAT3	dReal	Z3 4.4	SMT-RAT
zankl (SAT)	28	16	103*	**54**	15
zankl (UNSAT)	10	12	0	**23**	13
meti-Tarski (SAT)(3220)	2940	2774	3534*	**3220**	3055
meti-Tarski (UNSAT)(1526)	1138	1242	1172	**1523**	1298
hong (UNSAT)(20)	**20**	**20**	**20**	8	3
Total	4136	4064	1192	**4828**	4384
Time(s)	12363.34	1823.83	11145.23	64634.91	124823.17
Benchmark (with equations)	raSAT	iSAT3	dReal	Z3 4.4	SMT-RAT
zankl (SAT)(11)	**11**	0	11*	**11**	**11**
zankl (UNSAT)(4)	**4**	**4**	**4**	**4**	**4**
meti-Tarski (SAT)(1805)	1313	1	1994*	**1805**	1767
meti-Tarski (UNSAT)(1162)	1011	1075	965	**1162**	1114
kissing (SAT)(42)	6	0	18*	**36**	7
kissing (UNSAT)(3)	0	0	**1**	0	0
hycomp (SAT)	0	0	317*	**254**	33
hycomp (UNSAT)	1931	**2279**	2130	2200	1410
LassoRanker (SAT)	0	16	0*	**120**	0
LassoRanker (UNSAT)	0	27	0	**118**	0
Total	4276	3750	3100	**5710**	4346
Time(s)	5978.58	4522.84	32376.47	124960.95	102940.90

6 Conclusion

This paper presented an SMT solver **raSAT** for polynomial constraints over reals using simple techniques, i.e., *interval arithmetic* and *the generalized intermediate value theorem*. Among ICP based SMT solvers, **iSAT3** requires bounded intervals for inputs and SAT detection of equations is limited (e.g., a SAT instance in integers). **dReal** handles only δ-SAT. **raSAT** pursues the theoretical limitation of SAT/UNSAT detection based on ICP.

ICP-based techniques have essential limitations on completeness. These limitations often appear with multiple roots and/or 0-dimensional ideals, and our next step is to combine computer algebraic techniques as a last resort. For instance, we observe during experiments that **raSAT** fails the touching cases with generally a rapid convergence until a box cannot be decomposed further (e.g., a box becomes smaller than the roundoff error limit). When such a box is detected, we plan to apply an existing package of Gröbner basis.

Acknowledgements. The authors would like to thank Pascal Fontain and Nobert Müller for valuable comments and kind instructions on **veriT** and iRRAM. They also thank to anonymous referees for fruitful suggestions. This work is supported by JSPS KAKENHI Grant-in-Aid for Scientific Research(B) (23300005,15H02684).

References

1. Alliot, J.M., Gotteland, J.B., Vanaret, C., Durand, N., Gianazza, D.: Implementing an interval computation library for OCaml on x86/amd64 architectures. In: ICFP. ACM (2012)
2. Benhamou, F., Granvilliers, L.: Continuous and interval constraints. In: van Beek, P., Rossi, F., Walsh, T. (eds.) Handbook of Constraint Programming, pp. 571–604. Elsevier, Amsterdam (2006)
3. Bofill, M., Nieuwenhuis, R., Oliveras, A., Rodríguez-Carbonell, E., Rubio, A.: The barcelogic SMT solver. In: Gupta, A., Malik, S. (eds.) CAV 2008. LNCS, vol. 5123, pp. 294–298. Springer, Heidelberg (2008)
4. Comba, J.L.D., Stolfi, J.: Affine arithmetic and its applications to computer graphics. In: SIBGRAPI 1993, pp. 9–18 (1993)
5. Corzilius, F., Loup, U., Junges, S., Ábrahám, E.: SMT-RAT: an SMT-compliant nonlinear real arithmetic toolbox. In: Cimatti, A., Sebastiani, R. (eds.) SAT 2012. LNCS, vol. 7317, pp. 442–448. Springer, Heidelberg (2012)
6. Eén, N., Sörensson, N.: An extensible SAT-solver. In: Giunchiglia, E., Tacchella, A. (eds.) SAT 2003. LNCS, vol. 2919, pp. 502–518. Springer, Heidelberg (2004)
7. Fränzle, M., Herde, C., Teige, T., Ratschan, S., Schubert, T.: Efficient solving of large non-linear arithmetic constraint systems with complex boolean structure. JSAT **1**, 209–236 (2007)
8. Ganai, M., Ivancic, F.: Efficient decision procedure for non-linear arithmetic constraints using cordic. In: FMCAD 2009, pp. 61–68, November 2009
9. Gao, S., Kong, S., Clarke, E.M.: Satisfiability modulo odes. In: FMCAD 2013, pp. 105–112, October 2013
10. Gao, S., Kong, S., Clarke, E.M.: dReal: an SMT solver for nonlinear theories over the reals. In: Bonacina, M.P. (ed.) CADE 2013. LNCS, vol. 7898, pp. 208–214. Springer, Heidelberg (2013)
11. Granvilliers, L., Benhamou, F.: Realpaver: an interval solver using constraint satisfaction techniques. ACM Trans. Math. Softw. **32**, 138–156 (2006)
12. Hickey, T., Ju, Q., Van Emden, M.H.: Interval arithmetic: from principles to implementation. J. ACM **48**(5), 1038–1068 (2001)
13. Jovanović, D., de Moura, L.: Solving non-linear arithmetic. In: Gramlich, B., Miller, D., Sattler, U. (eds.) IJCAR 2012. LNCS, vol. 7364, pp. 339–354. Springer, Heidelberg (2012)
14. Khanh, T.V., Ogawa, M.: SMT for polynomial constraints on real numbers. In: TAPAS 2012. ENTCS, vol. 289, pp. 27–40 (2012)
15. Messine, F.: Extentions of affine arithmetic: application to unconstrained global optimization. J. UCS **8**(11), 992–1015 (2002)
16. Moore, R.: Interval Analysis. Prentice-Hall Series in Automatic Computation. Prentice-Hall, Upper Saddle River (1966)
17. Neumaier, A.: Interval Methods for Systems of Equations. Cambridge Middle East Library. Cambridge University Press, Cambridge (1990)
18. Passmore, G.O.: Combined decision procedures for nonlinear arithmetics, real and complex. Dissertation, School of Informatics, University of Edinburgh (2011)

19. Passmore, G.O., Jackson, P.B.: Combined decision techniques for the existential theory of the reals. In: Carette, J., Dixon, L., Coen, C.S., Watt, S.M. (eds.) MKM 2009, Held as Part of CICM 2009. LNCS, vol. 5625, pp. 122–137. Springer, Heidelberg (2009)
20. Ratschan, S.: Efficient solving of quantified inequality constraints over the real numbers. ACM Trans. Comput. Logic **7**(4), 723–748 (2006)
21. Tseitin, G.: On the complexity of derivation in propositional calculus. In: Siekmann, J.H., Wrightson, G. (eds.) Automation of Reasoning. Symbolic Computation, pp. 466–483. Springer, Heidelberg (1983)
22. Tung, V.X., Khanh, T.V., Ogawa, M.: raSAT: SMT for polynomial inequality. In: SMT Workshop 2014, p. 67 (2014)
23. Zankl, H., Middeldorp, A.: Satisfiability of non-linear irrational arithmetic. In: Clarke, E.M., Voronkov, A. (eds.) LPAR-16 2010. LNCS, vol. 6355, pp. 481–500. Springer, Heidelberg (2010)

First-Order Logic and Proof Theory

Schematic Cut Elimination and the Ordered Pigeonhole Principle

David M. Cerna[1(✉)] and Alexander Leitsch[2]

[1] Research Institute for Symbolic Computation (RISC), Johannes Kepler University, Linz, Austria
dcerna@risc.uni-linz.ac.at
[2] Logic and Theory Group, Technical University of Vienna, Vienna, Austria
leitsch@logic.at

Abstract. Schematic cut-elimination is a method of cut-elimination which can handle certain types of inductive proofs. In previous work, an attempt was made to apply the *schematic CERES method* to a formal proof with an arbitrary number of Π_2 cuts (a recursive proof encapsulating the infinitary pigeonhole principle). However the derived schematic refutation for the *characteristic clause set* of the proof could not be expressed in the schematic resolution calculus developed so far. Without this formalization a *Herbrand system* cannot be algorithmically extracted. In this work, we provide a restriction of infinitary pigeonhole principle, the *ECA-schema (Eventually Constant Assertion), or ordered infinitary pigeonhole principle*, whose analysis can be completely carried out in the existing framework of schematic CERES. This is the first time the framework is used for proof analysis. From the refutation of the clause set and a substitution schema we construct a *Herbrand system.*

1 Introduction

For his famous *Hauptsatz* [14], Gerhard Gentzen developed the sequent calculus **LK**. Gentzen went on to show that the *cut* inference rule is redundant and in doing so, was able to show several results on consistency and decidability. The method he developed for eliminating cuts from **LK**-derivations works by inductively reducing the cuts in a given **LK**-derivation to cuts which either have a reduced *formula complexity* and/or reduced *rank* [17]. This method of cut elimination is sometimes referred to as *reductive cut elimination.* A useful consequence of cut elimination for the **LK**-calculus is that cut-free **LK**-derivations have the *subformula property*, i.e. every formula occurring in the derivation is a subformula of some formula in the end sequent. This property admits the construction of *Herbrand sequents* and other objects which are essential in proof analysis.

By using the technique of cut-elimination, it is also possible to gain mathematical knowledge concerning the connection between different proofs of the same theorem. For example, Jean-Yves Girard's application of cut elimination to the Fürstenberg-Weiss' proof of van der Waerden's theorem [15] resulted in the *analytic* proof of van der Waerden's theorem as found by van der Waerden

N. Olivetti and A. Tiwari (Eds.): IJCAR 2016, LNAI 9706, pp. 241–256, 2016.
DOI: 10.1007/978-3-319-40229-1_17

himself. From the work of Girard, it is apparent that interesting results can be derived by eliminating cuts in "mathematical" proofs.

A more recently developed method of cut elimination, the CERES method [3], provides the theoretic framework to directly study the cut structure of **LK**-derivations, and in the process reduces the computational complexity of deriving a cut-free proof. The cut structure is transformed into a clause set allowing for a clausal analysis of the resulting clause form. Methods of reducing clause set complexity, such as *subsumption* and *tautology elimination* can be applied to the characteristic clause set to increase the efficiency. It was shown by Baaz & Leitsch in "Methods of cut Elimination" [4] that this method of cut elimination has a *non-elementary speed up* over reductive cut elimination.

In the same spirit as Girard's work, the CERES method was applied to a formalization of Fürstenberg's proof of the infinitude of primes [1]. Instead of formalizing the proof as a single proof (in second-order arithmetic) it was represented as a sequence of first-order proofs enumerated by a single numeric parameter indexing the number of primes assumed to exist (leading to a contradiction). The resulting schema of clause sets was refuted for the first few instances by the system CERES. The general refutation schema, resulting in Euclid's method of prime construction (constructing a larger prime from the primes already constructed), was specified on the mathematical meta-level. At that time no object-level construction of the resolution refutation schema existed.

A straightforward mathematical formalization of Fürstenberg's proof requires induction. In higher-order logic, induction is easily formalized via the comprehension principle. However in first-order logic, an induction rule or induction axioms have to be added to the **LK**-calculus. As was shown in [13], ordinary reductive cut elimination does not work in the presence of an induction rule in the **LK**-calculus. There are, however, other systems [16] which provide cut-elimination in the presence of an induction rule; but these systems do not produce proofs with the subformula property, which is necessary for Herbrand system extraction. Also, there have been other investigations focusing on similar forms of proof representation [5–7,11], though cut-elimination, and especially CERES cut-elimination was not the primary focus of these works.

In "Cut-Elimination and Proof Schemata" [13], a version of the **LK**-calculus was introduced (**LKS**-calculus) allowing for the formalization of sequences of proofs as a single object level construction, i.e. *proof schema*, as well as a framework for performing a CERES-type cut elimination on proof schemata. Cut elimination performed within the framework of [13] results in cut-free proof schemata with the subformula property.

In previous work, we applied the schematic CERES method of [13] to a proof formalized in the **LKS**-calculus [8,10]. We referred to this formal proof as the *Non-injectivity Assertion* (NiA) schema. A well known variation of the NiA-schema, of which has been heavily studied in literature, is the *infinitary Pigeonhole Principle* (PHP). Though a resolution refutation schema was found and mathematically specified [8], it was not possible to express this refutation schema within the language of [13]. The main problem was the specification of a unification and refutation schema. This issue points to a fundamental property of CERES-based schematic cut-elimination, namely that the language for specifying the refutation schema is more complex than that specifying the proof schema.

In this work we construct a formal proof for a weaker variant of the NiA-schema which we call the *Eventually Constant Assertion schema* (ECA-schema). The ECA-schema is an encapsulation of the infinitary pigeonhole principle where the holes are ordered. For the ECA-schema a specification of the resolution refutation schema within the formalism of [13] turned out to be successful. In particular, we are able to extract a Herbrand system and complete the proof analysis of the ECA-schema.

The paper is structured as follows: In Sect. 2, we introduce the **LKS**-calculus and the essential concepts from [13] concerning the schematic clause set analysis. In Sect. 3, we mathematically prove the ECA-schema. The formal proof written in the **LKS**-calculus can be found in [9]. In Sect. 4, we extract the characteristic clause set from the ECA-schema and perform *normalization* and tautology elimination. In Sect. 5, we provide a refutation of the extracted characteristic clause set. In Sect. 6, we extract a Herbrand system for the refutation of Sect. 5. In Sect. 7, we conclude the paper and discuss our conjecture concerning sufficient conditions allowing for the application of the schematic CERES method to a given proof schema.

2 The LKS-Calculus and Clause Set Schema

In this section we introduce the **LKS**-calculus, which will be used to formalize the ECA-schema, and the schematic CERES method.

2.1 Schematic Language, Proofs, and the LKS-Calculus

The **LKS**-calculus is based on the **LK**-calculus constructed by Gentzen [14]. When one grounds the *parameter* indexing an **LKS**-derivation, the result is an **LK**-derivation [13]. The term language used is extended to accommodate the schematic constructs of **LKS**-derivations. We work in a two-sorted setting containing a *schematic sort* ω and an *individual sort* ι. The schematic sort contains numerals constructed from the constant $0 : \omega$, a monadic function $s(\cdot) : \omega \rightarrow \omega$ as well as ω-variables $\mathcal{N}_v$, of which one variable, the *free parameter*, will be used to index **LKS**-derivations. When it is not clear from context, we will represent numerals as $\overline{m}$. The free parameter will be represented by n unless otherwise noted.

The individual sort is constructed in a similar fashion to the standard first order language [17] with the addition of schematic functions. Thus, ι contains countably many constant symbols, countably many *constant function symbols*, and *defined function symbols*. The constant function symbols are part of the standard first order language and the defined function symbols are used for schematic terms. However, defined function symbols can also unroll to numerals and thus can be of type $\omega^n \rightarrow \omega$. The ι sort also has *free* and *bound* variables and an additional concept, *extra variables* [13]. These are variables introduced during the unrolling of defined function (*predicate*) symbols. We do not use extra variables in the formalization of the ECA-schema. Also important are the *schematic variable symbols* which are variables of type $\omega \rightarrow \iota$. Essentially second order variables, though, when evaluated with a *ground term* from the ω sort we treat them as first order variables. Our terms are built inductively using constants and variables as a base.

Formulae are constructed inductively using countably many *predicate constants*, logical operators $\vee,\wedge,\rightarrow,\neg,\forall$, and $\exists$, as well as *defined predicate symbols* which are used to construct schematic formulae. In this work *iterated* $\bigvee$ is the only defined predicate symbol used. Its formal specification is:

$$\varepsilon_{\vee} = \bigvee_{i=0}^{s(y)} P(i) \equiv \begin{cases} \bigvee_{i=0}^{s(y)} P(i) \Rightarrow \bigvee_{i=0}^{y} P(i) \vee P(s(y)) \\ \bigvee_{i=0}^{0} P(i) \Rightarrow P(0) \end{cases} \tag{1}$$

As proof theoretical framework we use the sequent calculus **LK** [14,17] (note that, as we using CERES for proof analysis the specific form of structural rules does not matter). To obtain more flexibility in formalizing mathematical theorems we extend **LK** to a calculus **LKE**, essentially the **LK**-calculus [17] under an equational theory ε (in our case $\varepsilon_{\vee}$ Eq. 1). This equational theory, concerning our particular usage, is a primitive recursive term algebra describing the structure of the defined function (predicate) symbols. The **LKE**-calculus is the base calculus for the **LKS**-calculus which also includes *proof links*.

Definition 1 (ε-inference rule)

$$\frac{S[t]}{S[t']}\ (\varepsilon)$$

In the ε inference rule, the term t in the sequent S is replaced by a term t' such that, given the equational theory ε, $\varepsilon \models t = t'$.

To extend the **LKE**-calculus with proof links we need a countably infinite set of *proof symbols* denoted by $\varphi, \psi, \varphi_i, \psi_j \ldots$. Let $S(\bar{x})$ by a sequent with a vector of schematic variables $\bar{x}$, by $S(\bar{t})$ we denote the sequent $S(\bar{x})$ where each of the variables in $\bar{x}$ is replaced by the terms in the vector $\bar{t}$ respectively, assuming that they have the appropriate type. Let φ be a proof symbol and $S(\bar{x})$ a sequent, then the expression $\frac{(\varphi(\bar{t}))}{S(\bar{t})}$ (dashed line) is called a *proof link*. For a variable $n : \omega$, proof links such that the only ω-variable is n are called *n-proof links*.

Definition 2 (LKE-calculus [13]). *The sequent calculus* **LKS** *consists of the rules of* **LKE**, *where proof links may appear at the leaves of a proof.*

Definition 3 (Proof schemata [13]**).** *Let ψ be a proof symbol and $S(n,\bar{x})$ be a sequent such that $n:\omega$. Then a* proof schema pair *for ψ is a pair of* **LKS**-*proofs $(\pi,\nu(k))$ with end-sequents $S(0,\bar{x})$ and $S(k+1,\bar{x})$ respectively such that π may not contain proof links and $\nu(k)$ may contain only proof links of the form $\dfrac{(\psi(k,\bar{a}))}{S(k,\bar{a})}$ and we say that it is a proof link to ψ. We call $S(n,\bar{x})$ the end sequent of ψ and assume an identification between the formula occurrences in the end sequents of π and $\nu(k)$ so that we can speak of occurrences in the end sequent of ψ. Finally a proof schema Ψ is a tuple of proof schema pairs for $\psi_1,\cdots\psi_\alpha$ written as $\langle\psi_1,\cdots\psi_\alpha\rangle$, such that the* **LKS**-*proofs for ψ_β may also contain n-proof links to ψ_γ for $1\leq\beta<\gamma\leq\alpha$. We also say that the end sequent of ψ_1 is the end sequent of Ψ.*

We will not delve further into the structure of proof schemata and instead refer the reader to [13]. We now introduce the *characteristic clause set schema.*

2.2 Characteristic Clause Set Schema

The construction of the characteristic clause set as described for the CERES method [3] required inductively following the formula occurrences of *cut ancestors* up the proof tree to the leaves. The cut ancestors are sub-formulas of any cut in the given proof. However, in the case of proof schemata, the concept of ancestors and formula occurrence is more complex. A formula occurrence might be an ancestor of a cut formula in one recursive call and in another it might not. Additional machinery is necessary to extract the characteristic clause term from proof schemata. A set Ω of formula occurrences from the end-sequent of an **LKS**-proof π is called *a configuration for* π. A configuration Ω for π is called relevant w.r.t. a proof schema Ψ if π is a proof in Ψ and there is a $\gamma\in\mathbb{N}$ such that π induces a subproof $\pi\downarrow\gamma$ of $\Psi\downarrow\gamma$ such that the occurrences in Ω correspond to cut-ancestors below $\pi\downarrow\gamma$ [12]. By $\pi\downarrow\gamma$, we mean substitute the free parameter of π with $\gamma\in\mathbb{N}$ and unroll the proof schema to an **LKE**-proof. We note that the set of relevant cut-configurations can be computed given a proof schema Ψ. To represent a proof symbol φ and configuration Ω pairing in a clause set we assign them a *clause set symbol* $cl^{\varphi,\Omega}(a,\bar{x})$, where a is a term of the ω sort.

Definition 4 (Characteristic clause term [13]**).** *Let π be an* **LKS**-*proof and Ω a configuration. In the following, by Γ_Ω, Δ_Ω and Γ_C, Δ_C we will denote multisets of formulas of Ω- and cut-ancestors respectively. Let r be an inference in π. We define the clause-set term $\Theta_r^{\pi,\Omega}$ inductively:*

- *if r is an axiom of the form $\Gamma_\Omega,\Gamma_C,\Gamma\vdash\Delta_\Omega,\Delta_C,\Delta$, then $\Theta_r^{\pi,\Omega}=\{\Gamma_\Omega,\Gamma_C\vdash\Delta_\Omega,\Delta_C\}$*
- *if r is a proof link of the form $\dfrac{\psi(a,\bar{u})}{\Gamma_\Omega,\Gamma_C,\Gamma\vdash\Delta_\Omega,\Delta_C,\Delta}$ then define Ω' as the set of formula occurrences from $\Gamma_\Omega,\Gamma_C\vdash\Delta_\Omega,\Delta_C$ and $\Theta_r^{\pi,\Omega}=cl^{\psi,\Omega}(a,\bar{u})$*
- *if r is a unary rule with immediate predecessor r' , then $\Theta_r^{\pi,\Omega}=\Theta_{r'}^{\pi,\Omega}$*
- *if r is a binary rule with immediate predecessors r_1, r_2, then*

- *if the auxiliary formulas of r are Ω- or cut-ancestors, then* $\Theta^{\pi,\Omega}_{r} = \Theta^{\pi,\Omega}_{r_1} \oplus \Theta^{\pi,\Omega}_{r_2}$
- *otherwise,* $\Theta^{\pi,\Omega}_{r} = \Theta^{\pi,\Omega}_{r_1} \otimes \Theta^{\pi,\Omega}_{r_2}$

Finally, define $\Theta^{\pi,\Omega} = \Theta^{\pi,\Omega}_{r_0}$ *where* r_0 *is the last inference in* π *and* $\Theta^{\pi} = \Theta^{\pi,\emptyset}$. *We call* Θ^{π} *the characteristic term of* π.

Clause terms evaluate to sets of clauses by $|\Theta| = \Theta$ for clause sets Θ, $|\Theta_1 \oplus \Theta_2| = |\Theta_1| \cup |\Theta_2|$, $|\Theta_1 \otimes \Theta_2| = \{C \circ D \mid C \in |\Theta_1|, D \in |\Theta_2|\}$.

The characteristic clause term is extracted for each proof symbol in a given proof schema Ψ, and together they make the *characteristic clause set schema* for Ψ, $CL(\Psi)$.

Definition 5 (Characteristic Term Schema [13]**).** *Let* $\Psi = \langle \psi_1, \cdots, \psi_\alpha \rangle$ *be a proof schema. We define the rewrite rules for clause-set symbols for all proof symbols* ψ_β *and configurations* Ω *as* $cl^{\psi_\beta,\Omega}(0,\overline{u}) \rightarrow \Theta^{\pi_\beta,\Omega}$ *and* $cl^{\psi_\beta,\Omega}(k+1,\overline{u}) \rightarrow \Theta^{\nu_\beta,\Omega}$ *where* $1 \leq \beta \leq \alpha$. *Next, let* $\gamma \in \mathbb{N}$ *and* $cl^{\psi_\beta,\Omega}\downarrow_\gamma$ *be the normal form of* $cl^{\psi_\beta,\Omega}(\gamma,\overline{u})$ *under the rewrite system just given extended by rewrite rules for defined function and predicate symbols. Then define* $\Theta^{\psi_\beta,\Omega} = cl^{\psi_\beta,\Omega}$ *and* $\Theta^{\Psi,\Omega} = cl^{\psi_1,\Omega}$ *and finally the characteristic term schema* $\Theta^{\Psi} = \Theta^{\Psi,\emptyset}$.

2.3 Resolution Proof Schemata

From the characteristic clause set we can construct *clause schemata* which are an essential part of the definition of *resolution terms* and *resolution proof schema* [13]. Clause schemata serve as the base for the resolution terms used to construct a resolution proof schema. One additional notion needed for defining resolution proof schema is that of *clause variables.* The idea behind clause variables is that parts of the clauses at the leaves can be passed down a refutation to be used later on. The definition of resolution proof schemata uses clause variables as a way to handle this passage of clauses. Substitutions on clause variables are defined in the usual way.

Definition 6 (Clause Schema [13]**).** *Let* b *be a numeric term,* $\overline{u}$ *a vector of schematic variables and* $\overline{X}$ *a vector of clause variables. Then* $c(b,\overline{u},\overline{X})$ *is a clause schema w.r.t. the rewrite system* R:

$$c(0,\overline{u},\overline{X}) \rightarrow C \circ X \text{ and } c(k+1,\overline{u},\overline{X}) \rightarrow c(k,\overline{u},\overline{X}) \circ D$$

where C *is a clause with* $V(C) \subseteq \{\overline{u}\}$ *and* D *is a clause with* $V(D) \subseteq \{k,\overline{u}\}$. *Clauses and clause variables are clause schemata w.r.t. the empty rewrite system.*

Definition 7 (Resolution Term [13]**).** *Clause schemata are resolution terms; if* ρ_1 *and* ρ_2 *are resolution terms, then* $r(\rho_1;\rho_2;P)$ *is a resolution term, where* P *is an atom formula schema.*

The idea behind the resolution terms is that in the term $r(\rho_1;\rho_2;P)$, P is the resolved atom of the resolvents ρ_1, ρ_2. The notion of most general unifier has not yet been introduced since we introduce the concept as a separate schema from the resolution proof schema.

Definition 8 (Resolution Proof Schema [13]**).** *A resolution proof schema* $\mathcal{R}(n)$ *is a structure* $(\varrho_1, \cdots, \varrho_\alpha)$ *together with a set of rewrite rules* $\mathcal{R} = \mathcal{R}_1 \cup \cdots \cup \mathcal{R}_\alpha$ *, where the* $\mathcal{R}_i$ *(for* $1 \leq i \leq \alpha$*) are pairs of rewrite rules*

$$\varrho_i(0, \overline{w}, \overline{u}, \overline{X}) \rightarrow \eta_i$$
$$and$$
$$\varrho_i(k+1, \overline{w}, \overline{u}, \overline{X}) \rightarrow \eta_i'$$

where, $\overline{w}, \overline{u}$*, and* $\overline{X}$ *are vectors of* ω*, schematic, and clause variables respectively,* η_i *is a resolution term over terms of the form* $\varrho_j(a_j, \overline{m}, \overline{t}, \overline{C})$ *for* $i < j \leq \alpha$*, and* η_i' *is a resolution term over terms of the form* $\varrho_j(a_j, \overline{m}, \overline{t}, \overline{C})$ *and* $\varrho_i(k, \overline{m}, \overline{t}, \overline{C})$ *for* $i < j \leq \alpha$*; by* a_j*, we denote a term of the* ω *sort.*

The idea behind the definition of resolution proof schema is that the definition simulates a recursive construction of a resolution derivation tree and can be unfolded into a tree once the free parameter is instantiated. The expected properties of resolution and resolution derivations hold for resolution proof schema, more detail can be found in [13].

Definition 9 (Substitution Schema [13]**).** *Let* $u_1, \cdots, u_\alpha$ *be schematic variable symbols of type* $\omega \rightarrow \iota$ *and* $t_1, \cdots, t_\alpha$ *be term schemata containing no other* ω*-variables than* k*. Then a substitution schema is an expression of the form* $[u_1/\lambda k.t_1, \cdots, u_\alpha/\lambda k.t_\alpha]$.

Semantically, the meaning of the substitution schema is for all $\gamma \in \mathbb{N}$ we have a substitution of the form $[u_1(\gamma)/\lambda k.t_1 \downarrow_\gamma, \cdots, u_\alpha(\gamma)/\lambda k.t_\alpha \downarrow_\gamma]$. For the resolution proof schema the semantic meaning is as follows, Let $R(n) = (\varrho_1, \cdots, \varrho_\alpha)$ be a resolution proof schema, θ be a clause substitution, ν an ω-variable substitution, ϑ be a substitution schema, and $\gamma \in \mathbb{N}$, then $R(\gamma) \downarrow$ denotes a resolution term which has a normal form of $\varrho_1(n, \overline{w}, \overline{u}, \overline{X})\theta\nu\vartheta[n/\gamma]$ w.r.t. R extended by rewrite rules for defined function and predicate symbols.

2.4 Herbrand Systems

From the resolution proof schema and the substitution schema we can exact a so-called *Herbrand system.* The idea is to generalize the mid sequent theorem of Gentzen to proof schemata [4,17]. This theorem states that a proof (cut-free or with quantifier-free cuts) of a prenex end-sequent can be transformed in a way that there is a midsequent separating quantifier inferences from propositional ones. The mid-sequent is propositionally valid (w.r.t. the axioms) and contains (in general several) instances of the matrices of the prenex formulae; it is also called a *Herbrand sequent.* The aim of this paper is to extract schematic Herbrand sequents from schematic cut-elimination via CERES. We restrict the sequents further to skolemized ones. In the schematization of these sequents we allow only the matrices of the formulae to contain schematic variables (the number of formulae in the sequents and the quantifier prefixes are fixed).

Definition 10 (skolemized prenex sequent schema). *Let*

$$S(n) = \Delta_n, \varphi_1(n), \cdots, \varphi_k(n) \vdash \psi_1(n), \cdots, \psi_l(n), \Pi_n, \text{ for } k, l \in \mathbb{N} \text{ where}$$

$$\varphi_i(n) = \forall x_1^i \cdots \forall x_{\alpha_i}^i F_i(n, x_1^i, \cdots, x_{\alpha_i}^i), \quad \psi_j(n) = \exists x_1^j \cdots \exists y_{\beta_j}^j E_j(n, y_1^j, \cdots, y_{\beta_j}^j),$$

for $\alpha_i, \beta_j \in \mathbb{N}$, F_i and E_j are quantifier-free schematic formulae and Δ_n, Π_n are multisets of quantifier-free formulae of fixed size; moreover, the only free variable in any of the formulae is $n : \omega$. Then $S(n)$ is called a skolemized prenex sequent schema (sps-schema).

Definition 11 (Herbrand System). *Let $S(n)$ be a sps-schema as in Definition 10. Then a Herbrand system for $S(n)$ is a rewrite system $\mathcal{R}$ (containing the list constructors and unary function symbols w_i^x, for $x \in \{\varphi, \psi\}$), such that for each $\gamma \in \mathbb{N}$, the normal form of $w_i^x(\gamma)$ w.r.t $\mathcal{R}$ is a list of list of terms $t_{i,x,\gamma}$ (of length $m(i,x)$) such that the sequent*

$$\Delta_\gamma, \Phi_1(\gamma), \ldots, \Phi_k(\gamma) \vdash \Psi_1(\gamma), \ldots, \Psi_l(\gamma)$$

for

$$\Phi_j(\gamma) = \bigwedge_{p=1}^{m(j,\varphi)} E_j(\gamma, t_{j,\varphi,\gamma}(p,1), \ldots, t_{j,\varphi,\gamma}(p,\alpha_j)) \ (j = 1, \ldots, k),$$

$$\Psi_j(\gamma) = \bigvee_{p=1}^{m(j,\psi)} F_j(\gamma, t_{j,\psi,\gamma}(p,1), \ldots, t_{j,\psi,\gamma}(p,\beta_j)) \ (j = 1, \ldots, l),$$

*is **LKE**-provable.*

Though our definition of a Herbrand system differs from the definition introduced in [13] (where only purely existential schemata are treated), it is only a minor syntactic generalization. All results proven in [13] carry over to this more general form above.

3 "Mathematical" Proof of the ECA Statement and Discussion of Formal Proof

For lack of space, we will not provide a formal proof of the ECA-schema in the **LKS**-calculus (see [9]), but rather a mathematical argument proving the statement, of which closely follows the intended formal proof. The ECA-schema can be stated as follows:

Theorem 1 (Eventually Constant Assertion). *Given a total monotonically decreasing function $f : \mathbb{N} \to \{0, \cdots, n\}$, for $n \in \mathbb{N}$, there exists an $x \in \mathbb{N}$ such that for all $y \in \mathbb{N}$, where $x \leq y$, it is the case that $f(x) = f(y)$.*

Proof. If the range only contains 0 then the theorem trivially holds. Let us assume it holds for a codomain with n elements and show that it holds for a codomain with $n+1$ elements. If for all positions x, $f(x) = n$ then the theorem holds, else if at some y, $f(y) \neq n$ then from that point on f cannot map to n because the function is monotonically decreasing, thus, f will only have n elements in its codomain and the theorem holds in this case by the induction hypothesis.

The cut consists of the case distinction made in the stepcase. When written in the **LKS**-calculus, it is as follows:

$$\exists x \forall y\left(((x \leq y) \rightarrow n+1 = f(y)) \vee f(y) < n+1\right)$$

Notice that if we are to formalize the statement in the **LKS**-calculus the consequent has a $\exists\forall$ quantifier prefix:

$$\forall x(\bigvee_{i=0}^{n+1} i = f(x)), \forall x \forall y\Big(x \leq y \rightarrow f(y) \leq f(x)\Big) \vdash \exists x \forall y(x \leq y \rightarrow f(x) = f(y))$$

The CERES method (as well as the schematic CERES method) was designed for proofs without *strong quantification* in the end sequent. To get around this problem the proofs have to be *skolemized* [2]. We will not go into details of proof skolemization in this work, but to note, in the formal proof $g(\cdot)$, is the introduced skolem symbol.

4 Extraction of the Characteristic Term Schema

Each of the proof schema pairs of the formal proof (see [9]) have one cut configuration. In the case of ψ it is the empty configuration, and in the case of $\varphi(n)$ it is

$$\Omega(n) \equiv \exists x \forall y\left(((x \leq y) \rightarrow n+1 = f(y)) \vee f(y) < n+1\right).$$

This holds for the base cases as well as the step cases. Thus, we have the following clause set terms:

$$CL_{ECA}(0) \equiv \Theta^{\psi,\emptyset}(0) \equiv cl^{\varphi,\Omega(0)}(0) \oplus (\{\vdash f(\alpha) < 0\} \otimes \{\vdash 0 = f(\alpha)\} \otimes \{0 \leq \beta \vdash\}) \tag{2a}$$

$$cl^{\varphi,\Omega(0)}(0) \equiv \Theta^{\varphi,\Omega(0)}(0) \equiv \{f(\alpha) < 0 \vdash\} \oplus \{f(g(\alpha)) < 0 \vdash\} \oplus \{\vdash \alpha \leq \alpha\} \oplus \{\vdash \alpha \leq g(\alpha)\} \oplus \{0 = f(\alpha), 0 = f(g(\alpha)) \vdash\} \tag{2b}$$

$$CL_{ECA}(n+1) \equiv \Theta^{\psi,\emptyset}(n+1) \equiv cl^{\varphi,\Omega(n+1)}(n+1) \oplus (\{\vdash f(\alpha) < n+1\} \otimes \{\vdash n+1 = f(\alpha)\} \otimes \{0 \leq \beta \vdash\}) \tag{2c}$$

$$cl^{\varphi,\Omega(n+1)}(n+1) \equiv \Theta^{\varphi,\Omega(n+1)}(n+1) \equiv cl^{\varphi,\Omega(n)}(n) \oplus \{n+1=f(\alpha), n+1=f(g(\alpha)) \vdash\} \oplus \{\vdash \alpha \leq \alpha\} \oplus \{\alpha \leq g(\alpha)\} \oplus \{n+1=f(\beta) \vdash n+1=f(\beta)\} \oplus \{\alpha \leq \beta \vdash \alpha \leq \beta\} \oplus \{f(\beta)<n+1 \vdash f(\beta)<n+1\} \oplus \{f(\alpha)<n+1, \alpha \leq \beta \vdash n=f(\beta), f(\beta)<n\} \tag{2d}$$

In the characteristic clause set schema $CL_{ECA}(n{+}1)$ presented in Eq. 2 tautology and subsumption elimination have not been applied. Applying both types of elimination to $CL_{ECA}(n)$ and normalizing the clause set yields the following clause set $C(n)$:

$$\begin{array}{rcl}
C1(x,k) & \equiv & \vdash x(k) \leq x(k) \\
C2(x,k) & \equiv & \vdash x(k) \leq g(x(k)) \\
C3(x,i,k) & \equiv & i = f(x(k)), i = f(g(x(k))) \vdash \\
C4(x,y,i,k) & \equiv & y(k) \leq x(k), f(y(k)) < i+1 \vdash \\
 & & f(x(k)) < i, i = f(x(k)) \\
C4'(x,y,i,k) & \equiv & y(k) \leq x(k+1), f(y(k)) < i+1 \vdash \\
 & & f(x(k+1)) < i, i = f(x(k+1)) \\
C5(x,k) & \equiv & f(x(k)) < 0 \vdash \\
C6(x,k) & \equiv & f(g(x(k))) < 0 \vdash \\
C7(x,k) & \equiv & 0 \leq x(k) \vdash f(x(k)) < n, f(x(k)) = n
\end{array}$$

We have introduced clause names, schematic variables, and an additional ω-variable which will be used in the refutation of Sect. 5.

5 Refutation of the Characteristic Clause Set of the ECA-Schema

We discovered the resolution refutation schema which we present here with the help of the SPASS theorem prover [18] in default mode, and with the flags for standard resolution and ordered resolution set. Various other modes of the theorem prover were tested, however, given that we needed to translate the resulting proof into the simple resolution language of [13], the chosen modes provided the easiest proofs for translation. After running the theorem prover on five instances of the clause set, we were able to extract an invariant for the resolution refutation schema. Essentially, the refutation differentiates between the symbols occurring in the codomain of f and not occurring. This is denoted using the function g. The excerpt from the SPASS output in Table 1 indicates the invariant. However, even though SPASS was able to provide a refutation for each instance, we could not use these refutations directly in the resolution refutation schema being that the SPASS output ignores the structural importance of the ω sort. Unlike the ordering problem of the NiA-schema [8,10], this choice made by SPASS was not necessary to the refutation of the ECA-schema and we were able find a suitable refutation.

Our resolution refutation schema of the ECA-schema is $\mathcal{R} = (\varrho_1, \cdots, \varrho_{10})$, where we use one clause variable Y, two schematic variables, and one ω-variable. Our substitution schema is as follows:

$$\vartheta = \{x(k) \leftarrow \lambda k.(h(k)), y(k) \leftarrow \lambda k.(h(k))\}$$

where $h(\cdot)$ is defined as $h(0) \rightarrow 0,\ h(s(k)) \rightarrow g(h(k))$. The components are as follows:

Table 1. Excerpt from SPASS output for the clause set instance $C(5)$ indicating the invariant.

310[0:MRR:309.0,306.1]	$\vdash f(\alpha) < 3$
311[0:MRR:10.1,310.0]	$\alpha \leq \beta \vdash 2 = f(\beta) \quad f(\beta) < 2$
312[0:Res:2.0,311.0]	$\vdash 2 = f(\alpha) \quad f(\alpha) < 2$
314[0:Res:312.0,6.1]	$2 = f(\alpha) \vdash f(g(\beta)) < 2$
315[0:Res:314.1,11.1]	$2 = f(\alpha) \quad g(\alpha) \leq \beta \vdash 1 = f(\beta) \quad f(\beta) < 1$
316[0:Res:312.0,315.0]	$g(\alpha) \leq \beta \vdash f(\alpha) < 2 \quad 1 = f(\beta) \quad f(\beta) < 1$
317[0:Res:2.0,316.0]	$\vdash f(\alpha) < 2 \quad 1 = f(g(\alpha)) \quad f(g(\alpha)) < 1$
318[0:Res:3.0,316.0]	$\vdash f(\alpha) < 2 \quad 1 = f(g(g(\alpha))) \quad f(g(g(\alpha))) < 1$
321[0:Res:318.1,7.1]	$1 = f(g(\alpha)) \vdash f(\alpha) < 2 \quad f(g(g(\alpha))) < 1$
322[0:Res:321.2,14.1]	$1 = f(g(\alpha)) \quad g(g(\alpha)) \leq \beta \vdash f(\alpha) < 2 \quad 0 = f(\beta)$
325[0:Res:317.1,322.0]	$g(g(\alpha)) \leq \beta \vdash f(\alpha) < 2 \quad f(g(\alpha)) < 1 \quad f(\alpha) < 2 \quad 0 = f(\beta)$
327[0:Obv:325.1]	$g(g(\alpha)) \leq \beta \vdash f(g(\alpha)) < 1 \quad f(\alpha) < 2 \quad 0 = f(\beta)$
328[0:Res:2.0,327.0]	$\vdash f(g(\alpha)) < 1 \quad f(\alpha) < 2 \quad 0 = f(g(g(\alpha)))$
329[0:Res:3.0,327.0]	$\vdash f(g(\alpha)) < 1 \quad f(\alpha) < 2 \quad 0 = f(g(g(g(\alpha))))$
335[0:Res:329.2,8.1]	$0 = f(g(g(\alpha))) \vdash f(g(\alpha)) < 1 \quad f(\alpha) < 2$
336[0:MRR:335.0,328.2]	$\vdash f(g(\alpha)) < 1 \quad f(\alpha) < 2$
337[0:Res:336.0,14.1]	$g(\alpha) \leq \beta \vdash f(\alpha) < 2 \quad 0 = f(\beta)$
338[0:Res:2.0,337.0]	$\vdash f(\alpha) < 2 \quad 0 = f(g(\alpha))$
339[0:Res:3.0,337.0]	$\vdash f(\alpha) < 2 \quad 0 = f(g(g(\alpha)))$
344[0:Res:339.1,8.1]	$0 = f(g(\alpha)) \vdash f(\alpha) < 2$
345[0:MRR:344.0,338.1]	$\vdash f(\alpha) < 2$

$$\varrho_1(n+1,k,x,y,Y) \Rightarrow r(\varrho_2(n+1,k,x,y,Y); \varrho_5(n,k,x,y,Y \circ (f(x(k)) < n+1 \vdash)); f(x(k)) < n+1)$$

$$\varrho_1(0,k,x,y,Y) \Rightarrow \quad r(\varrho_2(0,k,x,y,Y); C5(x,k); f(x(k)) < 0)$$

$$\varrho_2(n+1,k,x,y,Y) \Rightarrow r(\varrho_3(n+1,k,x,y,Y); r(C1(x,k); C7(x,k); x(k) \leq x(k)); n+1 = f(x(k)))$$

$$\varrho_2(0,k,x,y,Y) \Rightarrow \quad r(\varrho_3(0,k,x,y,Y); r(C1(x,k); C7(x,k); x(k) \leq x(k)); n+1 = f(x(k)))$$

$$\varrho_3(n+1,k,x,y,Y) \Rightarrow r(\varrho_4(n+1,k,x,y,Y); C3(x,n+1,k); \\ n+1 = f(g(x(k))))$$

$$\varrho_3(0,k,x,y,Y) \Rightarrow \quad r(\varrho_4(0,k,x,y,Y); C3(x,0,k); 0 = f(g(x(k))))$$

$$\varrho_4(n+1,k,x,y,Y) \Rightarrow r(\varrho_5(n,k+1,x,y,Y \circ f(x(k+1)) < n+1 \vdash); \\ r(C2(x,k); C7(x,k+1); f(x(k+1)) < n+1)$$

$$\varrho_4(0,k,x,y,Y) \Rightarrow \quad r(C6(x,k); r(C2(x,k); C7(x,k+1); f(g(x(k))) < 0)$$

$$\varrho_5(n+1,k,x,y,Y) \Rightarrow r(\varrho_6(n+1,k,x,y,Y); \varrho_5(n,k,x,y,Y \circ \\ (f(x(k)) < n+1 \vdash)); f(x(k)) < n+1)$$

$$\varrho_5(0,k,x,y,Y) \Rightarrow \quad r(\varrho_6(0,k,x,y,Y); C5(x,k); f(x(k)) < 0)$$

$$\varrho_6(n+1,k,x,y,Y) \Rightarrow r(\varrho_7(n+1,k,x,y,Y); \varrho_8(n+1,k,x,y,Y); \\ n+1 = f(x(k)))$$

$$\varrho_6(0,k,x,y,Y) \Rightarrow \quad r(\varrho_7(0,k,x,y,Y); \varrho_8(0,k,x,y,Y); 0 = f(x(k))$$

$$\varrho_7(n+1,k,x,y,Y) \Rightarrow r(\varrho_9(n+1,k,x,y,Y); C3(x,n+1,k); \\ n+1 = f(g(x(k))))$$

$$\varrho_7(0,k,x,y,Y) \Rightarrow \quad r(\varrho_9(0,k,x,y,Y); C3(x,0,k); 0 = f(g(x(k))))$$

$$\varrho_8(n+1,k,x,Y) \Rightarrow \quad r(C1(x,k); Y \circ C4(x,y,n,k); x(k) \leq x(k))$$

$$\varrho_8(0,k,x,y,Y) \Rightarrow \quad r(C1(x,k); Y \circ C4(x,y,0,k); x(k) \leq x(k))$$

$$\varrho_9(n+1,k,x,y,Y) \Rightarrow \quad r(\varrho_5(n,k+1,x,y,Y' \circ f(g(x(k))) < n+1 \vdash); \\ \varrho_{10}(n+1,x,y,Y); f(g(x(k))) < n+1)$$

$$\varrho_9(0,k,x,y,Y) \Rightarrow \quad r(C6(x,k); \varrho_{10}(0,k,x,y,Y); f(g(x(k))) < 0)$$

$$\varrho_{10}(n+1,k,x,y,Y) \Rightarrow r(C2(x,k); Y \circ C4'(x,y,n,k); x(k) \leq g(x(k)))$$

$$\varrho_{10}(0,k,x,y,Y) \Rightarrow \quad r(C2(x,k); Y \circ C4'(x,y,0,k); x(k) \leq g(x(k)))$$

One can find a graphical representation of the refutation in Fig. 1. The clause substitution is $\theta = \{Y \leftarrow \vdash\}$, the ω-variable substitution is $\nu = \{k \leftarrow \overline{\mu}\}$ for any $\overline{\mu} \in \mathbb{N}$. The normal form of the refutation for $\gamma \in \mathbb{N}$ is

$$\varrho_1(n,k,x,y,Y)\theta\nu\vartheta\,[n \leftarrow \gamma] = \varrho_1(\gamma, \overline{\mu}, \lambda_k.(i_s(k)), \lambda_k.(h(k)), \vdash),$$

where $i_s(0) = 0$, $i_s(s(k)) = s(i_s(k))$. Substitution of the empty clause into Y suffices for every instance, i.e. $\{Y \leftarrow \vdash\}$. This property makes extraction of the Herbrand system much easier.

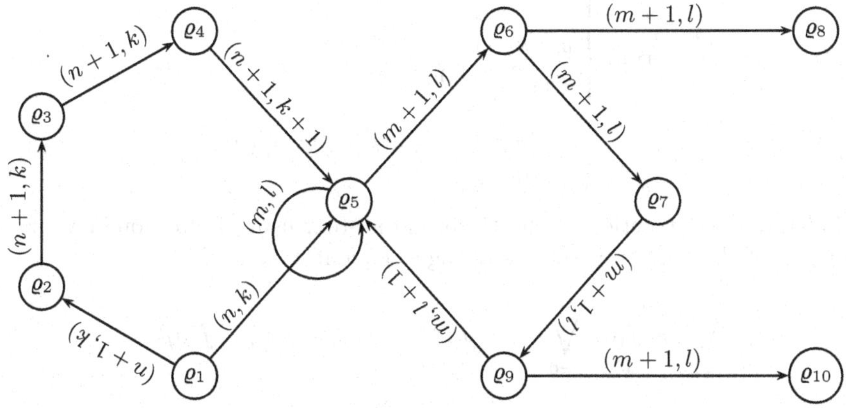

Fig. 1. A graph representation of the resolution refutation. The variable n is the free parameter, k is the ω-variable used in the refutation and the variables m and l are dependent on the position in the computation.

6 The Herbrand System for the ECA-Schema

Now we move on to the construction of a Herbrand system for the sequent

$$S(n) \equiv (\forall x \bigvee_{i=0}^{n} i = f(x), \forall y\,(0 \leq y \rightarrow f(y) \leq f(0))) \vdash \\ \exists x\,(x \leq g(x) \rightarrow f(x) = f(g(x)))$$

based on our proof analysis. The sequent $S(n)$ is an sps-schema of the form $\varphi_1(n), \varphi_2(n) \vdash \psi_1(n)$. Note that we dropped one of the quantifiers from the antecedent since it is obvious from the proof itself what the substitution would be, see [9]. Each formula in $S(n)$ is derived along with a set of clauses in the proof schemata $\Psi = \langle(\psi(n+1), \psi(0)), (\varphi(n+1), \varphi(0))\rangle$. By observing the construction of the formulae in the formal proof [9], one can see that $\varphi_1(n)$, $\varphi_2(n)$, and $C7(x,k)$ as constructed together, while $\psi_1(n)$, $C2(x,k)$, and $C3(x,i,k)$ are constructed together. We will only consider the case when the ω-variable substitution is $\nu = \{k \rightarrow 0\}$ to simplify the derivation.

Notice that $C7(x,k)$ is used at the top of the refutation and only twice. Once as $C7(x,0)$ and once as $C7(x,1)$. On the other hand, $C2(x,k)$ is used in ϱ_{10} and $C3(x,i,k)$ is use in ϱ_7. For every pair (i,l) in the ranges $0 \leq i \leq n+1, 0 \leq l < n+1$, the clauses $C2(x,l)$ and $C3(x,i,l)$, and $C2(x,l+1)$ and $C3(x,i,l+1)$ are used in the refutation. This implies, by the substitution schema that $\psi_1(n)$ will have its quantifier replaced by the term derived from $h(i)$, for all $i \in [0,n]$, in the Herbrand system. This information can be used to construct the required rewrite system:

$$\mathcal{R} = \begin{cases} w_1^{\varphi}(k+1) \Rightarrow [[0];[g(0)]] \\ w_1^{\varphi}(0) \Rightarrow [[0];[g(0)]] \\ \\ w_2^{\varphi}(k+1) \Rightarrow [[0];[g(0)]] \\ w_2^{\varphi}(0) \Rightarrow [[0];[g(0)]] \\ \\ w_1^{\psi}(k+1) \Rightarrow [[h(k+1)];w_1^{\psi}(k)] \\ w_1^{\psi}(0) \Rightarrow [0] \end{cases}$$

To finish our construction of the Herbrand system using Definition 11 we need to put all of the parts together as a single sequent as follows

$$\bigvee_{i=0}^{n} i = f(0), \bigvee_{i=0}^{n} i = f(g(0)), (0 \leq 0 \rightarrow f(0) \leq f(0)),$$

$$(0 \leq g(0) \rightarrow f(g(0)) \leq f(0)) \vdash \bigvee_{i=0}^{n} (h(i) \leq g(h(i)) \rightarrow f(h(i)) = f(g(h(i)))).$$

At first this does not seem to be **LKE** provable, However, one has to remember that for the construction of our cut formula we made an assumption that f is monotonically decreasing and has a codomain consisting of elements in the interval $[0,n]$. These assumptions are represented by the following axiom found in [9]:

$$AX \equiv f(\alpha) < n+1, \alpha \leq \beta \vdash n = f(\beta), f(\beta) < n$$

It is not used in the construction of the end sequent but is used for the construction of the cut formulae. We just need to find a set of axioms which correspond to these semantic assumptions, the following set suffices:

$$\begin{array}{l} A1(i): \bigvee_{i=0}^{j-1} i = f(\alpha), j = f(g(\alpha)), f(g(\alpha)) < f(\alpha) \vdash \\ A2(i): i = f(\alpha), \bigvee_{j=0}^{i-1} j = f(g(\alpha)), \alpha \leq g(\alpha) \vdash \\ A3(i): i = f(\alpha), i = f(g(\alpha)) \vdash f(\alpha) = f(g(\alpha)) \\ A4(i): f(g(\alpha)) = f(\alpha) \vdash f(\alpha) = f(g(\alpha)) \\ A5(i): \vdash \alpha \leq \alpha \\ A6(i): f(\alpha) < f(\alpha) \vdash \end{array}$$

The first pair of axioms enforce the required properties of f and g, the next pair provide the needed properties of equality, and the last pair provide the needed properties of linear orderings. Interesting enough, using these axioms, we are able to prove the derived Herbrand sequent using only a single nesting of g, thus making the majority of the consequent redundant. This is a result of our usage of the clause $C7(x,k)$. Thus, it turns out that a minimal Herbrand sequent is the following:

$$\bigvee_{i=0}^{n} i = f(0), \bigvee_{i=0}^{n} i = f(g(0)), (0 \leq g(0) \rightarrow f(g(0)) \leq f(0)),$$

$$(0 \leq 0 \rightarrow f(0) \leq f(0)) \vdash 0 \leq g^1(0) \rightarrow f(0) = f(g^1(0)).$$

The Herbrand sequent can be derived for deeper nestings of g by changing the ω-variable substitution used.

7 Conclusion

Weakening the NiA-schema of [8] by reducing the complexity of the cuts allowed for extraction of the Herbrand system using the concepts of [13]. As a case study of the schematic CERES method, to the best of our knowledge this is the first one. From the analysis of the ECA-schema there are two issues which seem to influence the applicability of the schematic CERES method. The first issue, as we pointed out earlier, is the ordering of the terms in the ω sort. However, a second issue arising in this work is the complexity of the terms, specifically what is the highest arity function symbol allowed. In the case of the NiA-schema, terms were constructed from both an arity two and an arity one function symbol, but in the case of the ECA-schema only arity one function symbols were used. When only arity one function symbols are used nesting of the function symbols does not require the addition of *extra variables* in a given term, of which were used in the NiA-schema [8,10]. This seems to allow for the creation of more complex orderings of the ω sort. We conjecture a sufficient condition that proof schema containing only arity one function symbols can be analysed using the schematic CERES method. Also, an open problem we plan to address in future work is a generalization of the resolution refutation calculus of [13] which can handle more complex ordering structures [10]. It seems necessary to handle more complex ordering structure if one wants to formalize and analyse more complex mathematical arguments such as Fürstenberg's proof of the infinitude of primes.

References

1. Baaz, M., Hetzl, S., Leitsch, A., Richter, C., Spohr, H.: Ceres: An analysis of Fürstenberg's proof of the infinity of primes. Theoret. Comput. Sci. **403**(2–3), 160–175 (2008)
2. Baaz, M., Leitsch, A.: On skolemization and proof complexity. Fundamenta Informaticae **20**(4), 353–379 (1994)
3. Baaz, M., Leitsch, A.: Cut-elimination and redundancy-elimination by resolution. J. Symbolic Comput. **29**, 149–176 (2000)
4. Baaz, M., Leitsch, A.: Methods of Cut-Elimination. Springer Publishing Company, Incorporated (2013)
5. Brotherston, J.: Cyclic proofs for first-order logic with inductive definitions. In: Beckert, B. (ed.) TABLEAUX 2005. LNCS (LNAI), vol. 3702, pp. 78–92. Springer, Heidelberg (2005)

6. Brotherston, J., Simpson, A.: Sequent calculi for induction and infinite descent. J. Logic Comput. **22**, 1177–1216 (2010)
7. Bundy, A.: The automation of proof by mathematical induction. In: Robinson, J.A., Voronkov, A. (eds.) Handbook of Automated Reasoning, pp. 845–911. Elsevier, Amsterdam (2001)
8. Cerna, D., Leitsch, A.: Analysis of clause set schema aided by automated theorem proving: Acase study (2015). arXiv:1503.08551v1 [cs.LO]
9. Cerna, D., Leitsch, A.: Schematic cut elimination and the ordered pigeonhole principle [extended version] (2016). arXiv:1601.06548 [math.LO]
10. Cerna, D.M.: Advances in schematic cut elimination. Ph.D. thesis, Technical University of Vienna (2015). http://media.obvsg.at/p-AC12246421-2001
11. Comon, H.: Inductionless induction. In: Robinson, A., Voronkov, A. (eds.) Handbook of Automated Reasoning, pp. 913–962. Elsevier, Amsterdam (2001)
12. Dunchev, C.: Automation of cut-elimination in proof schemata. Ph.D. thesis, Technical University of Vienna (2012)
13. Dunchev, C., Leitsch, A., Rukhaia, M., Weller, D.: Cut-elimination and proof schemata. In: Aher, M., Hole, D., Jeřábek, E., Kupke, C. (eds.) TbiLLC 2013. LNCS, vol. 8984, pp. 117–136. Springer, Heidelberg (2015)
14. Gentzen, G.: Untersuchungen über das logische Schließen I. Math. Z. **39**(1), 176–210 (1935)
15. Girard, J.-Y.: Proof Theory and Logical Complexity. Studies in Proof Theory, vol. 1. Bibliopolis, Naples (1987)
16. Mcdowell, R., Miller, D.: Cut-elimination for a logic with definitions and induction. Theoret. Comput. Sci. **232**, 2000 (1997)
17. Takeuti, G.: Proof Theory. Studies in Logic and the Foundations of Mathematics, vol. 81. American Elsevier Pub., New York (1975)
18. Weidenbach, C., Dimova, D., Fietzke, A., Kumar, R., Suda, M., Wischnewski, P.: SPASS version 3.5. In: Schmidt, R.A. (ed.) CADE-22. LNCS, vol. 5663, pp. 140–145. Springer, Heidelberg (2009)

Subsumption Algorithms for Three-Valued Geometric Resolution

Hans de Nivelle(✉)

Instytut Informatyki Uniwersytetu Wrocławskiego, Wrocław, Poland
nivelle@ii.uni.wroc.pl

Abstract. In an implementation of geometric resolution, the most costly operation is subsumption (or matching): One has to decide for a three-valued, geometric formula, whether this formula is false in a given interpretation. The formula contains only atoms with variables, equality, and existential quantifiers. The interpretation contains only atoms with constants.

Because the atoms have no term structure, matching for geometric resolution is a hard problem. We translate the matching problem into a generalized constraint satisfaction problem, and give an algorithm that solves it efficiently. The algorithm uses learning techniques, similar to clause learning in propositional logic. Secondly, we adapt the algorithm in such a way that it finds solutions that use a minimal subset of the interpretation.

The techniques presented in this paper may have applications in constraint solving.

1 Introduction

Geometric logic as a theorem proving strategy was introduced in [1]. (The infinitary variant is called *coherent logic.*) Bezem and Coquand were motivated mostly by the desire to obtain a theorem proving strategy with a simple normal form transformation, which makes that many natural problems need no transformation at all, others have a much simpler transformation, and which makes that in all cases Skolemization can be avoided. This results in more readable proofs, and proofs that can be backtranslated more easily.

Our motivation for using geometric resolution is different, more engineering-oriented. We hope that three-valued, geometric resolution can be used as a generic reasoning core, into which different kinds of two- or three-valued reasoning or decision problems (e.g. problems representing type correctness, two-valued decision problems, simply typed classical problems) can be solved. Because we want the geometric reasoning core to be generic, we are willing to accept transformations that do not preserve much of the structure of the original formula. Subformulas are freely renamed, and functional expressions are flattened and replaced by relations.

We start by giving a definition of three-valued, geometric formulas. The definition that we give here is too general, but it is easier to understand than the

N. Olivetti and A. Tiwari (Eds.): IJCAR 2016, LNAI 9706, pp. 257–272, 2016.
DOI: 10.1007/978-3-319-40229-1_18

correct definition in [3], which contains some additional, technical restrictions that are required by other parts of the search algorithm.

Definition 1. *A* geometric literal *has one of the following four forms:*

1. *A simple atom of form* $p_\lambda(x_1, \ldots, x_n)$, *where* $x_1, \ldots, x_n$ *are variables (with repetitions allowed) and* $\lambda \in \{\mathbf{f},\ \mathbf{e},\ \mathbf{t}\}$. *(denoting* false, error *and* true.*)*
2. *An equality atom of form* $x_1 \approx x_2$, *with* x_1, x_2 *distinct variables.*
3. *A domain atom* $\#_{\mathbf{f}}\, x$, *with* x *a variable.*
4. *An* existential atom *of form* $\exists y\ p_\lambda(x_1, \ldots, x_n, y)$ *with* $\lambda \in \{\mathbf{f},\ \mathbf{e},\ \mathbf{t}\}$, *and such that* y *occurs at least once in the atom, not necessarily on the last place.*

A geometric formula *has form* $A_1, \ldots, A_p \mid B_1, \ldots, B_q$, *where the* A_i *are simple or domain atoms, and the* B_j *are atoms of arbitrary type.*

We require that geometric formulas are range restricted, *which means that every variable that occurs free in a* B_j *must occur in an* A_i *as well.*

The intuitive meaning of $A_1, \ldots, A_p \mid B_1, \ldots, B_q$ *is* $\forall \overline{x}\ A_1 \vee \cdots \vee A_p \vee B_1 \vee \cdots \vee B_q$, *where* $\overline{x}$ *are all the free variables. The vertical bar* ($\mid$) *has no logical meaning. Its only purpose is to separate the two types of atoms.*

A geometric formula that is not range restricted, can always be made range restricted by inserting suitable $\#_{\mathbf{f}}$ atoms into the left hand side. This is the only purpose of the #-predicate. Interpretations contain predicates of form $\#_{\mathbf{t}}\ c$, for every domain element c. Atoms in geometric formulas are variable-only, and are labeled with truth-values, as in [11]. It is shown in [3,4] that formulas in classical logic with partial functions ([2]) can be translated into sets of geometric formulas.

Definition 2. *We define an* interpretation I *as a finite set of atoms of forms* $\#\, c$ *with* c *a constant, or form* $p_\lambda(c_1, \ldots, c_n)$, *where* $c_1, \ldots, c_n$ *are constants (repetitions allowed). Interpretations must be* range restricted *as well. This means that every constant* c *occurring in the interpretation must occur in an atom of form* $\#_{\mathbf{t}}\, c$.

Matching searches for false formulas. These are formulas whose premises $A_1, \ldots, A_p$ clash with I, while none of the B_j is true in I.

Definition 3. *Let* I *be an interpretation. Let* A *be a geometric literal. Let* Θ *be a substitution that assigns constants to variables, and that is defined on the variables in* A. *We say that* $A\Theta$ conflicts*(or* is in conflict with*)* I *if* **(1)** A *has form* $p_\lambda(x_1, \ldots, x_n)$, *and there is an atom of form* $p_\mu(x_1\Theta, \ldots, x_n\Theta) \in I$ *with* $\lambda \neq \mu$, **(2)** A *has form* $x_1 \approx x_2$ *and* $x_1\Theta \neq x_2\Theta$, *or* **(3)** A *has form* $\#_{\mathbf{f}}\, x$ *and* $(\#_{\mathbf{t}}\, x\Theta) \in I$.
We say that $A\Theta$ is true *in* I *if*
(1) A *has form* $p_\lambda(x_1, \ldots, x_n)$ *and* $p_\lambda(x_1\Theta, \ldots, x_n\Theta) \in I$, **(2)** A *has form* $x_1 \approx x_2$ *and* $x_1\Theta = x_2\Theta$, **(3)** A *has form* $\#_{\mathbf{t}}\, x$ *and* $(\#_{\mathbf{t}}\, x\Theta) \in I$, *or* **(4)** A *has form* $\exists y\ B_\lambda(x_1, \ldots, x_n, y)$ *and there exists a constant* c, *s.t.* $B_\lambda(x_1\Theta, \ldots, x_n\Theta, c) \in I$.

In the definitions of truth and conflict, # is treated as a usual predicate.

Definition 4. *Let I be an interpretation. Let B be a geometric atom. Let Θ be a substitution that instantiates all free variables of B, and for which $B\Theta$ is not true in I. We define the* extension set *$E(B,\Theta)$ as follows:*

- *If B has form $p_\lambda(x_1,\ldots,x_n)$ or $\#_{\mathbf{t}}\ x$, then $E(B,\Theta) = \{B\Theta\}$.*
- *If B has form $x_1 \approx x_2$, then $E(B,\Theta) = \emptyset$.*
- *If B has form $\exists y\ B_\lambda(x_1,\ldots,x_n,y)$, then*

$$E(B,\Theta) = \{\ B\Theta\{y := c\} \mid c \in I\ \} \cup \{\ B\Theta\{y := \hat{c}\}\ \}.$$

By $c \in I$ we mean: c is a constant occurring in an atom of I. We assume that $\hat{c}$ is the first constant for which $\hat{c} \notin I$.

Intuitively, if for a geometric formula $\phi = A_1,\ldots,A_p \mid B_1,\ldots,B_q$ and a substitution Θ, the $A_i\Theta$ are in conflict with I, while none of the $B_j\Theta$ is true in I, then $\phi\Theta$ is false in I. If there exist a B_j and an atom $C \in E(B_j,\Theta)$ that is not in conflict with I, then $\phi\Theta$ can be made true by adding C. If no such C exists, a conflict was found. If more than one C exists, the search algorithm has to backtrack through all possibilities. The search algorithm tries to extend an initial interpretation I into an interpretation $I' \supset I$ that makes all formulas true. At each stage of the search, it looks for a formula and a substitution that make the formula false. If no formula and substitution can be found, the current interpretation is a model. Otherwise, search continues either by extending I, or by backtracking. Details of the procedure are described in [5] for the two-valued case, and in [3] for the three-valued case. Experiments with the current three-valued version, and the previous two-valued version ([6]) show that the search for false formulas consumes nearly all of the resources of the prover.

Definition 5. *An instance of* the matching problem *consists of an interpretation I and a geometric formula $A_1,\ldots,A_p \mid B_1,\ldots,B_q$.*

Determine if there exists a substitution Θ that brings all A_i in conflict with I, and makes none of the B_j true in Θ. If yes, then return such substitution.

Example 1. Consider an interpretation I consisting of atoms

$$P_{\mathbf{t}}(c_0,c_0),\ P_{\mathbf{e}}(c_0,c_1),\ P_{\mathbf{t}}(c_1,c_1),\ P_{\mathbf{e}}(c_1,c_2),\ Q_{\mathbf{t}}(c_2,c_0).$$

The formula $\phi_1 = P_{\mathbf{f}}(X,Y),\ P_{\mathbf{f}}(Y,Z) \mid Q_{\mathbf{t}}(Z,X)$ can be matched in five ways:

$$\begin{aligned}
\Theta_1 &= \{\ X := c_0,\ Y := c_0,\ Z := c_0\ \}\\
\Theta_2 &= \{\ X := c_0,\ Y := c_0,\ Z := c_1\ \}\\
\Theta_3 &= \{\ X := c_0,\ Y := c_1,\ Z := c_1\ \}\\
\Theta_4 &= \{\ X := c_1,\ Y := c_1,\ Z := c_1\ \}\\
\Theta_5 &= \{\ X := c_1,\ Y := c_1,\ Z := c_2\ \}
\end{aligned}$$

The substitution $\Theta_6 = \{\ X := c_0,\ Y := c_1,\ Z := c_2\ \}$ would make the conclusion $Q_{\mathbf{t}}(Z,X)$ true. Next consider the formula $\phi_2 = P_{\mathbf{f}}(X,Y),\ P_{\mathbf{t}}(Y,Z) \mid X \approx Y$. The substitution $\Theta = \{\ X := c_0,\ Y := c_1,\ Z := c_2\ \}$ is the only matching of ϕ_2 into I. Finally, the formula $\phi_3 = P_{\mathbf{t}}(X,Y) \mid \exists Z\ Q_{\mathbf{t}}(Y,Z)$ can be matched with $\Theta = \{\ X := c_0,\ Y := c_1\ \}$, and in no other way.

The first formula ϕ_1 in Example 1 has five matchings. In case there exists more than one matching, it matters for the geometric prover which matching is returned. This is because the prover analyses which ground atoms in the interpretation I contributed to the matching, and will consider only those in backtracking. In general, the set of conflicting atoms in I should be as small as possible, and should depend on as few as possible decisions. (Decisions in the sense of propositional reasoning, see [10].) The simplest solution for finding the best matching would be to enumerate all matchings, and use some preference relation $\preceq$ to keep the best one. Unfortunately, this approach is not practical because the number of matchings can be extremely high. We will address this problem in Sect. 5.

Even if one is interested in the decision problem only, matching is still intractable because the decision problem is already NP-complete. This can be shown by a simple reduction from SAT.

In this paper, we introduce an algorithm for efficiently solving the matching problem. Earlier versions of the algorithm have been implemented in the two-valued version of **Geo** ([6]). The three-valued version of **Geo** that took part in CASC 25 (see [13]) used a very naive implementation of matching.

In Sect. 2, we will translate the matching problem into a structure called *generalized constraint satisfaction problem* (GCSP). The generalization consists of the fact that it contains additional constraints, that a solution must not make true. These constraints correspond to the conclusions of the geometric formula that one is trying to match.

After that, we give in Sect. 3 a backtracking algorithm for solving GCSP, which is based on local consistency checking and backtracking. It makes use of a data structure that we will call *choice stack*. Choice stacks can be used for controlling the backtracking process, but also for keeping track of changes that occur during local consistency checking, and which may induce further changes.

In Sect. 4, we add lemma learning to the algorithm of Sect. 3. In Sect. 5, we show how the algorithm of Sect. 3, or any other algorithm for solving GCSP, can be modified for efficiently finding optimal matchings.

2 Translation into Generalized Constraint Satisfaction Problem

We introduce the generalized constraint satisfaction problem, and show how instances of the matching problem can be translated. It is 'generalized' because there are additional, negative constraints (called *blockings*), which a solution is not allowed to satisfy. The blockings originate from translations of the $B_1, \ldots, B_q$.

Definition 6. *A* substlet *s is a (small) substitution. We usually write s in the form $\overline{v}/\overline{c}$, where $\overline{v}$ is a sequence of variables without repetitions, and $\overline{c}$ is a sequence of constants of same length as $\overline{v}$.*

We say that two substlets $\overline{v}_1/\overline{c}_1$ and $\overline{v}_2/\overline{c}_2$ are in conflict *if there exist i, j s.t. $v_{1,i} = v_{2,j}$ and $c_{1,i} \neq c_{2,j}$.*

If $\overline{v}_1/\overline{c}_1, \ldots, \overline{v}_n/\overline{c}_n$ *is a sequence of substlets not containing a conflicting pair, then one can merge them into a substitution as follows:* $\bigcup\{\overline{v}_1/\overline{c}_1, \ldots, \overline{v}_n/\overline{c}_n\} = \{v_{i,j} := c_{i,j} \mid 1 \leq i \leq n,\ 1 \leq j \leq \|\overline{v}_i\|\}$.

If Θ *is a substitution and* $s = \overline{v}/\overline{c}$ *is a substlet, we say that* Θ *makes* s true *if every* $v_i := c_i$ *is present in* Θ.

We say that Θ *and* s *are in conflict if there is a* v_i/c_i *with* $1 \leq i \leq \|v\|$, *s.t.* $v_i\Theta$ *is defined and distinct from* c_i.

A clause C *is a finite set of substlets. We say that a substitution* Θ makes C true *(notation* $\Theta \models C$*) if* Θ *makes a substlet* $(\overline{v}/\overline{c}) \in C$ *true. We say that* Θ makes C false *(notation* $\Theta \models \neg C$*) if every substlet* $(\overline{v}/\overline{c}) \in C$ *is in conflict with* Θ*. In the remaining case, we call* C undecided by Θ.

Definition 7. *A* generalized constraint satisfaction problem *(GCSP) is a pair of form* (Σ^+, Σ^-) *in which* Σ^+ *is a finite set of clauses, and* Σ^- *is a finite set of substlets.*

A substitution Θ *is a* solution *of* (Σ^+, Σ^-)*, if every clause in* Σ^+ *is true in* Θ*, and there is no* $\sigma \in \Sigma^-$*, s.t.* Θ *makes* σ *true.*

Definition 8. *Let* (Σ^+, Σ^-) *a GCSP. We call* (Σ^+, Σ^-) range restricted *if for every variable* v *that occurs in a substlet* $\sigma \in \Sigma^-$*, there exists a clause* $c \in \Sigma^+$ *s.t. every substlet* $s \in c$ *has* v *in its domain.*

We now explain how a matching instance is translated into a generalized constraint satisfaction problem.

Definition 9. *Assume that* I *and* $\phi = A_1, \ldots, A_p \mid B_1, \ldots, B_q$ *together form an instance of the matching problem. The* translation (Σ^+, Σ^-) *of* (I, ϕ) into *GCSP is obtained as follows:*

- *For every* A_i*, let* $\overline{v}_i$ *denote the variables of* A_i*. Then* Σ^+ *contains the clause*

$$\{\ \overline{v}_i/\overline{v}_i\Theta \mid A_i\Theta \text{ is in conflict with } I\ \}.$$

- *For every* B_j*, let* $\overline{w}_j$ *denote the variables of* B_j*. For every* Θ *that makes* $B_j\Theta$ *true in* I*,* Σ^- *contains the substlet* $\overline{w}_j/(\overline{w}_j\Theta)$*.*

Theorem 1. *A matching instance* (I, ϕ) *has a matching iff its corresponding GCSP has a solution.*

In theory, the set of blockings Σ^- can be removed, because a blocking σ can always be replaced by a clause as follows: Let σ be a blocking, let $\overline{v}$ be its variables. Define $\sigma_1 = \sigma$, and let $\sigma_2, \ldots, \sigma_n \in \Sigma^-$ be the blockings whose domain is also $\overline{v}$. One can replace $\sigma_1, \ldots, \sigma_n$ by the clause $\{\ \overline{v}/\overline{c} \mid \overline{v}/\overline{c}$ conflicts all σ_i $(1 \leq i \leq n)\ \}$.

We prefer to keep Σ^-, because in the worst case, the resulting clause has size $m^{\|\overline{v}\|}$, where m is the size of the domain. For example, if $\sigma_1, \ldots, \sigma_n$ result from an equality $X \approx Y$, then σ_i has form $(X, Y)/(c_i, c_i)$. The resulting clause $C = \{(X, Y)/(c_i, c_j) \mid i \neq j\}$ has size $n(n-1) \approx n^2$.

Clauses resulting from a matching problem have the following trivial, but essential property:

Lemma 1. *Let (Σ^+, Σ^-) be obtained by the translation in Definition 9. Let $s_1, s_2 \in C \in \Sigma^+$. Then either $s_1 = s_2$, or s_1 and s_2 are in conflict with each other.*

Lemma 1 holds because s_1 and s_2 have the same domain.

Example 2. In Example 1, the matching problem (I, ϕ_1) can be translated into the GCSP below. The clauses are above the horizontal line, and the blockings are below it. Because substlets in the same clause always have the same variables, we write the variables of a clause only once.

$$\begin{array}{l}
(X, Y) \ / \ (c_0, c_0) \mid (c_0, c_1) \mid (c_1, c_1) \mid (c_1, c_2) \\
(Y, Z) \ / \ (c_0, c_0) \mid (c_0, c_1) \mid (c_1, c_1) \mid (c_1, c_2) \\
\hline
(X, Z) \ / \ (c_0, c_2)
\end{array}$$

Translating (I, ϕ_2) results in:

$$\begin{array}{l}
(X, Y) \ / \ (c_0, c_0) \mid (c_0, c_1) \mid (c_1, c_1) \mid (c_1, c_2) \\
(Y, Z) \ / \ (c_0, c_1) \mid (c_1, c_2) \\
\hline
(X, Y) \ / \ (c_0, c_0) \\
(X, Y) \ / \ (c_1, c_1) \\
(X, Y) \ / \ (c_2, c_2)
\end{array}$$

Translation of (I, ϕ_3) results in:

$$\begin{array}{l}
(X, Y) \ / \ (c_0, c_1) \mid (c_1, c_2) \\
\hline
(Y) \ / \ (c_2)
\end{array}$$

Before one runs any algorithms on a GCSP, it is useful to do some simplifications. If the GCSP contains a propositional clause (not containing any variables), this clause is either the empty clause, or a tautology. In the first case, the problem is trivially unsolvable. In the second case, the clause can be removed.

Similarly, if Σ^- contains a propositional blocking, then (Σ^+, Σ^-) is trivially unsolvable. Such blockings originate from a B_j that is purely propositional, or that has form $\exists y \ P_\lambda(y)$.

A third important preprocessing step is *removal of unit blockings.* Let $\sigma \in \Sigma^-$ be a blocking whose domain is included in the domain of some clause $C \in \Sigma^+$. In that case, one can remove every substlet $\overline{v}/\overline{c}$ from C, that has $\bigcup\{\overline{v}/\overline{c}\} \models \sigma$. If this results in C being empty, then (Σ^+, Σ^-) trivially has no solution. If no $\overline{v}/\overline{c}$ in any clause $C \in \Sigma^+$ implies σ, then σ can be removed from Σ^-, because of Lemma 1.

Applying removal of unit blockings to the translation of (I, ϕ_2) above results in

$$\begin{array}{l}
(X, Y) \ / \ (c_0, c_1) \mid (c_1, c_2) \\
(Y, Z) \ / \ (c_0, c_1) \mid (c_1, c_2) \\
\hline
\end{array}$$

It is worth noting that removal of propositional blockings can be viewed as a special case of removal of unit blockings.

A GCSP can be solved by backtracking, similar to SAT solving. A backtracking algorithm for GCSP can be either variable or clause based. A variable based algorithm maintains a substitution Θ, which it tries to extend into a solution. It backtracks by picking a variable v and trying to assign it in all possible ways. It backtracks when Θ makes a clause $C \in \Sigma^+$ false, or a blocking $\sigma \in \Sigma^-$ true.

A clause based algorithm maintains a consistent set S of substlets (whose union defines a substitution). It backtracks by picking an undecided clause $C \in \Sigma^+$, and consecutively inserting all substlets that are consistent with S into S. It backtracks when there is a clause C all of whose atoms are in conflict with S, or when $\bigcup S$ makes a blocking true.

Our experiments suggest that there is no significant difference in performance, nor in programming effort, between the two variants. We will stick with clause based algorithms, because it seems that they can be more easily combined with local consistency checking.

Local consistency checking (see [7,9,12]) is a pre-check that comes in many variations. They all have in common that one enumerates small subsets Σ' of Σ^+, for each subset Σ' generates all solutions, and then checks for each substlet occurring in a $C \in \Sigma'$ whether it occurs in a solution for Σ'. If some substlet $s \in C$ does not occur in a solution for Σ', then it certainly does not occur in a solution for Σ, so that it can be removed from C.

Experiments with **geo** show that a simple filtering using all subsets Σ' of size 2 whose clauses share a variable, a priori rejects a large fraction of matching instances. On those that are not immediately rejected, the next stage of search based on backtracking is much more efficient.

In [7] (Chap. 3), local consistency checking is done with subsets of variables (instead of clauses). Using subsets of two variables is called *arc consistency checking*, while considering subsets of three variables is called *path consistency checking*. In general, using bigger subsets is a more effective precheck, but also more costly because it gets closer to the original problem. It still may be worth it to try subsets of bigger size than size 2.

3 Matching Using Choice Stacks

We first present a matching algorithm without learning, and introduce learning in the next section. The non-learning algorithm is based on a combination of local consistency checking and backtracking. At each level, it first tries to reduce the set of clauses by local consistency checking. During local consistency checking, it checks all subsets of size two for variable conflicts, and subsets of arbitrary size, of which all but one are unit, for conflicts based on blockings. After local checking has been exhaustively applied, it picks a clause, splits it into two parts, and backtracks using the two resulting clauses.

Local consistency checking is a change-driven process. Whenever some clause C gets replaced by a $C' \subset C$, one has to check for all clauses D that share a variable with C', whether D now contains substlets that are in conflict with every substlet in C'. If yes, then D can be replaced by a $D' \subset D$, which in turn may result in further replacements. This means that the search algorithm needs to maintain a set of changed clauses, and using this set of changed clauses, check which more clauses it can change. When the set of changed clauses is empty, it either has a solution, or it needs to backtrack, which will introduce a new changed clause. Instead of maintaining a set of changed clauses, we use a data structure that we call choice stack.

Definition 10. *A* choice stack $\overline{C}$ *is a data structure consisting of a sequence of clauses* $C_1, \ldots, C_n$.

A choice stack supports refinement *of clauses: Refining* C_i *into* C' *means replacing* $\overline{C} = C_1, \ldots, C_n$ *by* $C_1, \ldots, [C_i], \ldots, C_n, C'$, *where* $C' \subset C_i$, *and the square brackets indicate that* C_i *has been refined. We write* $\overline{C}[i/C']$ *for the resulting choice stack.*

Restoring a choice stack to size n *means removing all clauses at positions* $> n$, *and removing all square brackets that resulted from adding the clauses at positions* $> n$. *We use a predicate* A_i *for querying whether the* i*-th clause is* actual *(is not in brackets). We use the notation* R_i *for the original, initial clause from which* C_i *is obtained by successive refinements.*

The size *of a choice stack is the total number of clauses in it (with or without brackets).*

Given a choice stack $\overline{C}$ *and a substitution* Θ, *we say that* Θ agrees with $\overline{C}$ *if* Θ *does not make any* C_i *false.*

Choice stacks can be efficiently implemented without need to copy clauses. They support all features needed by a search algorithm based on backtracking and local consistency checking. Change driven inspection is implemented by the fact that changed clauses are blocked and reinserted at a higher position. If one starts at the position of the latest change, and iteratively proceeds towards the end of the choice stack, checking all clauses on positions i that have A_i true, one will exhaustively inspect all changes.

Definition 11. *If* $\overline{v}/\overline{c}$ *is a substlet, and* C *a clause, we call* $\overline{v}/\overline{c}$ *in conflict with* C *if* $\overline{v}/\overline{c}$ *is in conflict with every* $s \in C$.

Definition 12. *A call to* **findmatch**$(\overline{C}, k, \Sigma^-)$ *either returns* $\bot$, *or constructs a solution of the GCSP* $(\overline{C}, \Sigma^-)$. *For simplicity, we assume that* **findmatch** *stops when it finds a solution, instead of returning it.* **findmatch**$(\overline{C}, k, \Sigma^-)$ *is recursively defined as follows:*

LOCAL: *As long as* $k \leq \|\overline{C}\|$, *do the following: If* A_k *holds, then*

LOC-CONFL: *For every* C_i *for which* A_i *holds, and which shares a variable with* C_k, *partition* C_i *into two parts as follows:*

$$C_i^- = \{s \in C_i \mid s \textit{ is in conflict with } C_k\},$$
$$C_i^+ = \{s \in C_i \mid s \textit{ is not in conflict with } C_k\}.$$

If $C_i^+ = \emptyset$, *then return* $\bot$. *Otherwise, refine* $\overline{C}$ *into* $\overline{C}[i/C_i^+]$.

LOC-BLOCKING: *If* $\|C_k\| = 1$, *then define* $\Theta = \bigcup\{s_j \mid 1 \leq j \leq \|\overline{C}\|$ *and* C_j *has form* $\{s_j\}$ $\}$. *(Note that the unique element of* C_k *also contributes to* Θ*.) For every* $\sigma \in \Sigma^-$ *that shares a variable with* C_k, *for every* C_i, *for which* A_i *holds and which shares a variable with* σ, *partition* C_i *into two parts as follows:*

$$C_i^- = \{s' \in C_i \mid \Theta \cup \bigcup\{s'\} \models \sigma\},$$
$$C_i^+ = \{s' \in C_i \mid \Theta \cup \bigcup\{s'\} \not\models \sigma\}.$$

If $C_i^+ = \emptyset$, *then return* $\bot$. *Otherwise, refine* $\overline{C}$ *into* $\overline{C}[i/C_i^+]$.

Assign $k = k + 1$.

SOLUTION: *If all* i *for which* A_i *holds, have* $\|C_i\| = 1$, *then* $\Theta = \bigcup\{s_j \mid 1 \leq j \leq \|\overline{C}\|$ *and* C_j *has form* $\{s_j\}\}$ *is a solution.*

BACKTRACK: *Otherwise, pick an* i *for which* A_i *holds, and which has* $\|C_i\| > 1$. *Partition* C_i *into two parts* C_1, C_2, *s.t. neither* C_1 *nor* C_2 *is empty. Recursively call* **findmatch**$(\overline{C}[i/C_1], \|\overline{C}\|, \Sigma^-)$. *If this call does not result in a solution, then also call* **findmatch**$(\overline{C}[i/C_2], \|\overline{C}\|, \Sigma^-)$.

Before **findmatch** *can be called on* (Σ^+, Σ^-), *one must first check for propositional clauses, propositional blockings and apply unit blocking removal. Let* (Σ'^+, Σ'^-) *be the result. Call* **findmatch**$(\Sigma'^+, 1, \Sigma'^-)$.

Algorithm **findmatch** is similar to DPLL in that it tries to postpone backtracking as long as possible by giving preference to deterministic reasoning. It differs from DPLL in the fact that it does not rely on interpretations. (which would be substitutions Θ in our case.) Instead of gradually building an interpretation by adding assignments, algoritm **findmatch** gradually refines a set of clauses into an interpretation by removing substlets from the clauses, until every clause is unit.

Deterministic reasoning is done by deleting substlets from clauses that are in conflict with one of the other clauses. When deterministic reasoning fails to solve the problem, we pick a non-unit clause C_i, and remove some part C_1 from it, and apply deterministic reasoning again. If this fails to give a solution, we replace C_1 by $C \backslash C_1$.

4 Matching with Conflict Learning

The matching algorithm in the two-valued version of **Geo** ([6]) was already equipped with a weak form of conflict learning. Before releasing it, we had experimented with naive matching, the algorithm in [8], and a lot of ad hoc methods. Matching with conflict learning is the only approach that gives acceptable performance. Despite this, matching is still a critical operation in the two-valued version of **Geo**, which will need significant improvement in the three-valued version. The algorithm in the current paper tries to obtain this in several ways: Firstly, algorithm **findmatch** mixes local consistency checking with backtracking. The algorithm in the two-valued version never attempted any deterministic reasoning. It always backtracked on a randomly picked variable. Secondly, in the two-valued version of **Geo**, lemmas have form $v_1/c_1, \ldots, v_n/c_n \rightarrow \bot$, i.e. they consist of a single, negated substlet. The lemmas that we will introduce shortly, are more expressive because they are positive. A negative substlet refutes only the substlets that it is included in, while a positive substlet rejects all substlets that it conflicts with. We hence expect that a single lemma will reject more matching attempts, and that, as a consequence, less lemmas will be generated. Thirdly, the algorithm in the two-valued version of **Geo** is unable to find optimal solutions. It always stops on the first solution, which frequently causes unpredictable behaviour. In Sect. 5, we give an algorithm that will turn every algoritm that can find some solution, into an algorithm that can find an optimal solution.

Definition 13. *A* lemma *is defined in the same way as a clause. For a substitution Θ, the notions of truth, falsehood, and undecidedness are defined in the same way as for clauses.*

We distinguish between clauses and lemmas because they serve different functions in the algorithm, which makes it useful to implement them as different classes. There is also a technical distinction, namely that all substlets in a clause always have the same domain, while in a lemma they can be different.

Definition 14. *Let (Σ^+, Σ^-) be a GCSP. Let λ be a lemma. We say that λ is* valid in *(Σ^+, Σ^-) if every solution Θ of (Σ^+, Σ^-) makes λ true.*

Let $\overline{C}$ be a choice stack, let λ be a lemma. We say that $\overline{C}$ makes λ false if λ is false in every substitution Θ that makes all clauses in $\overline{C}$ true.

If $\overline{C}$ is choice stack, and there exists a valid lemma that is false in $\overline{C}$, then there exists no solution Θ of (Σ^+, Σ^-) that agrees with $\overline{C}$, and one can backtrack.

Our extended matching algorithm will be similar to DPLL with conflict learning. It derives valid lemmas that are used to guide the search process. For the derivation of lemmas, we use two variants of lemma resolution that we will introduce shortly.

There is a technical complication arising from the fact that algorithm **find-match** is not based on substitutions, but on choice stacks. A substitution can play the same role as an interpretation in DPLL, and checking whether a lemma is false in a substitution is easy. A choice stack lazily refines a set of clauses into a substitution. This has the advantage that commitment can be postponed, but it has a major disadvantage that checking falsehood of a lemma becomes harder. In fact, if the choice stack $\overline{C}$ has no solutions, then every lemma is false in $\overline{C}$, which renders checking falsehood NP-complete. It follows that one needs a stronger notion than falsehood.

Definition 15. *Let $\overline{C} = (C_1, \ldots, C_n)$ be a choice stack. Let λ be a lemma. We call λ* a conflict lemma of *$\overline{C}$ if for every substlet $s \in \lambda$, there is a C_i in $\overline{C}$, such that s is in conflict with C_i.*

Checking whether a given λ is a conflict lemma is a cheap operation. In addition, one can use a watching scheme based on the changes in $\overline{C}$, and recheck only the lemmas whose watched substlets might be affected by a refinement. (See [10]). Conflict lemmas are created by resolution rules:

Definition 16. *Let* λ_1 *and* λ_2 *be lemmas. Let* $\mu_1 \subseteq \lambda_1$. *We define* $\text{RES}(\lambda_1, \mu_1, \lambda_2)$, *the* conflict resolvent of λ_1 *and* λ_2 based on μ_1. *First define* μ_2 *and* μ_1' *as follows:*

$$\begin{cases} \mu_2 = \{\ (\overline{v}_2/\overline{c}_2) \in \lambda_2 \mid (\overline{v}_2/\overline{c}_2)\ conflicts\ every\ (\overline{v}_1/\overline{c}_1) \in \mu_1\ \}, \\ \mu_1' = \{\ (\overline{v}_1/\overline{c}_1) \in \lambda_1 \mid (\overline{v}_1/\overline{c}_1)\ conflicts\ every\ (\overline{v}_2/\overline{c}_2) \in \mu_2\ \}. \end{cases}$$

Then $\text{RES}(\lambda_1, \mu_1, \lambda_2) = (\lambda_1 \backslash \mu_1') \cup (\lambda_2 \backslash \mu_2)$.

Suppose that one wants to resolve $\lambda_1 = \{\ (x,y)/(1,2),\ (x,y)/(2,1),\ (x,y)/(3,3)\ \}$ with $\lambda_2 = \{\ (y,z)/(1,2),\ (y,z)/(2,1)\ \}$ based on $\mu_1 = \{\ (x,y)/(1,2)\ \}$. In that case, $\mu_2 = \{\ (y,z)/(1,2)\ \}$. It turns out that $(x,y)/(3,3)$ can also resolve with μ_2, so we have $\mu_1' = \{\ (x,y)/(1,2),\ (x,y)/(3,3)\ \}$. The resolvent is $\{\ (x,y)/(2,1),\ (y,z)/(1,2)\ \}$.

Definition 17. *Let* $\sigma \in \Sigma^-$ *be a blocking. Let* $C_1, \ldots, C_n$ *be a sequence of clauses. Let* $\mu_1 \subseteq C_1, \ldots, \mu_n \subseteq C_n$ *be subsets of* $C_1, \ldots, C_n$, *s.t. for every sequence of substlets* $s_1 \in \mu_1, \ldots, s_n \in \mu_n$, *we have* $\bigcup\{s_1, \ldots, s_n\} \models \sigma$. *Then* $(C_1 \backslash \mu_1) \cup \cdots \cup (C_n \backslash \mu_n)$ *is a* σ-resolvent *of* $C_1, \ldots, C_n$.
We will write $\text{RES}(C_1, \ldots, C_n, \mu_1, \ldots, \mu_n, \sigma)$ *for the result.*

It is easy to see that both conflict resolution and σ-resolution are valid reasoning rules.

In order to extend algorithm **findmatch**, so that it will generate conflict lemmas, we assume a set of conflict lemmas Λ, which is initially empty. Whenever the modified algorithm **findmatch** backtracks, it extends Λ with a conflict lemma λ that conflicts the current choice stack $\overline{C}$. The following extensions are be made to **findmatch**:

- If at any stage, there is a lemma $\lambda \in \Lambda$, that is in conflict with $\overline{C}$, the algorithm backtracks.
- Assume that the algorithm passes through **LOC-CONFL**. If $C^+ = \emptyset$, the algorithm can insert the original version R_i of C_i into Λ. This is a conflict lemma of C, because it can be written as $C_i \cup (R_i \backslash C_i)$. Choice stack $\overline{C}$ contains the clause C_i, with which every $s \in (R_i \backslash C_i)$ is in conflict, by Lemma 1. Every $s \in C_i$ is in conflict with C_k, which follows from the fact that $C_i = C_i^-$.

  If $C^+ \neq \emptyset$, algorithm **findmatch** continues, and by induction, one can assume that it inserts a conflict lemma of $\overline{C}[i/C_i^+]$ into to Λ.

  If λ already is a closing lemma of $\overline{C}$, nothing needs to be done. Otherwise, let $\mu \subseteq \lambda$ be the substlets in λ that are in conflict with C_i^+. Let $\lambda' = \lambda \backslash \mu$. Every substlet in λ' is in conflict with $\overline{C}$, because λ conflicts $\overline{C}[i/C_i^+]$. We can insert $\text{RES}(\lambda, \mu, C_i)$ into Λ. The resolvent has form $(\lambda \backslash \mu') \cup (C_i \backslash \mu_2)$, where $\mu \subseteq \mu'$ and $\mu_2 \subseteq C_i^+$. It follows that $\lambda \backslash \mu' \subseteq \lambda'$ and $(C_i \backslash \mu_2) \subseteq C_i^-$, which is by its construction in conflict with C_k.

- Assume that clause C_i was partitioned into two parts (C_i^+, C_i^-) at **LOC-BLOCKING**. We first apply σ-resolution: Let

$$S \subseteq \{s_j \mid 1 \leq j \leq \|\overline{C}\| \text{ and } C_j \text{ has form } \{s_j\}\}$$

be the set of singletons that contribute to truth of the blocking σ, i.e. for every $s' \in C_i^-$, $\bigcup S \cup \{s'\} \models \sigma$, and there is no $S' \subset S$ for which this is still the case.

Let $j_1, \ldots, j_m$ be an enumeration of the clauses from which the elements of S originate. We can construct the σ-resolvent

$$\lambda_\sigma = (R_{j_1} \backslash \{s_{j_1}\}) \cup \cdots \cup (R_{j_m} \backslash \{s_{j_m}\}) \cup (R_i \backslash C_i^-).$$

It follows from the construction of S, that λ_σ is indeed a σ-resolvent. By Lemma 1, the elements in each $R_{j_z} \backslash \{s_{j_z}\}$ are in conflict with $\{s_{j_z}\}$. By the same lemma, the substlets in $R_i \backslash C_i^-$ that are not in conflict with C_i, are contained in C_i^+.

If C_i^+ is empty, we are done. Otherwise, algorithm **findmatch** continues, and by induction it inserts a conflict lemma λ of $\overline{C}[i/C^+]$ into Λ. If λ is a closing lemma of $\overline{C}$, then nothing needs to be done. (and we constructed λ_σ without reason.)

If λ is not a closing lemma of $\overline{C}$, we can define $\mu \subseteq \lambda$ as the substlets in λ that are in conflict with C_i^+. Construct $\lambda' = \text{RES}(\lambda, \mu, \lambda_\sigma)$ and insert λ' into Λ. The argument that λ' is a conflict lemma, is similar to the case for LOC-CONFLICT.
- At **BACKTRACK**, one can assume by induction that the first recursive call inserts a closing lemma λ_1 of $\overline{C}[i/C_1]$ into Λ. If λ_1 is a closing lemma of C, nothing more needs to be done. Otherwise, the second recursive call will insert a closing lemma λ_2 of $\overline{C}[i/C_2]$ into Λ. Again, if λ_2 is a closing lemma of C, we are done. Otherwise, we can insert $\lambda' = \text{RES}(\text{RES}(R_i, C_1, \lambda_1), C_2, \lambda_2)$ into Λ. The proof that λ' is indeed a closing lemma is analogous to the proof for LOC-CONFLICT.

Example 3. Consider the choice stack $\overline{C}$, consisting of $C_1 = \{ (X, Y) \;/\; (c_0, c_1) \mid (c_1, c_2)\}$ and $C_2 = \{ (Y, Z) \;/\; (c_0, c_1) \mid (c_1, c_2)\}$. At $k = 1$, C_2 will be refined into $C_3 = \{ (Y, Z) \;/\; (c_1, c_2)\}$. At $k = 2$, we have A_2 is false, so it is skipped. At $k = 3$, the clause C_1 is refined into $C_4 = \{ (X, Y) \;/\; (c_0, c_1)\}$. At $k = 4$, nothing is done, and after that we reach SOLUTION.

Example 4. Consider $C_1 = \{ (X, Y) \;/\; (c_0, c_1) \mid (c_1, c_2)\}$, $C_2 = \{ (Y, Z, T) \;/\; (c_1, c_2, c_3) \mid (c_1, c_2, c_4)\}$ with a blocking $\sigma = (X, Z) \;/\; (c_0, c_2)$.

At $k = 1$, nothing is changed. At $k = 2$, clause C_1 is refined into $C_3 = \{ (X, Y) \;/\; (c_0, c_1) \}$ by LOC-CONFLICT. At $k = 3$, clause C_2 can be refined into the empty clause by LOC-BLOCKING.

In order to obtain a conflict lemma, we σ-resolve $R_3 = C_1$ with $R_2 = C_2$. The result is $\lambda_1 = \{ (X, Y) \;/\; (c_1, c_2) \}$ which conflicts (C_2, C_3). Back at level 2, the returned lemma λ_1 is not a conflict lemma of (C_1, C_2). It can resolve with $R_2 = C_2$ and the result is $\{ \}$.

5 Finding Optimal Matchings

In this section we address the problem of finding optimal matchings. For the effectiveness of geometric resolution, it is important that a minimal matching is returned, in case more than one exists. A minimal matching is a matching that uses the smallest possible set of assumptions. In terminology of DPLL, assumptions represent decision levels. The assumptions contributing to a conflict represent choice options, which will be replaced by other options during backtracking. In addition to being as few as possible, assumptions at a lower decision level should always be preferred over assumptions at a higher decision level. The reason for this is the fact that in other branches of the search tree, there is a risk that more assumptions will be used, and when assumptions are at a lower level, there is less room for this.

Definition 18. *Let I be an interpretation. A weight function α is a function that assigns finite subsets of natural numbers to the atoms of I.*

Let A be a geometric literal. Let Θ be a substitution such that $A\Theta$ is in conflict with I. Referring to Definition 3, we define $\alpha(p_\lambda(x_1,\ldots,x_n)\Theta, I) = \alpha(p_\mu(x_1\Theta,\ldots,x_n\Theta))$, $\alpha((x_1 \approx x_2)\Theta, I) = \{\}$, and $\alpha((\#_{\mathbf{f}}x)\Theta, I) = \alpha((\#_{\mathbf{t}}x\Theta))$.

Definition 19. *Let I and $\phi = A_1,\ldots,A_p \mid B_1,\ldots,B_q$ together form an instance of the matching problem (Definition 5). Assume that Θ is a solution. The* weight of Θ, *for which we write $\alpha(I,\phi,\Theta)$, is defined as*

$$\bigcup \begin{cases} \{\ \alpha(A_i\Theta, I) \mid 1 \leq i \leq p\} \\ \{\ \alpha(C, I) \mid 1 \leq j \leq q,\ \ C \in E(B_j, \Theta),\ \textit{and}\ C\ \textit{conflicts}\ I\ \} \end{cases}$$

Solving optimal matching means: First establish if (I,ϕ) has a solution. If it has, then find a solution Θ for which $\alpha(I,\phi,\Theta)$ is multiset minimal.

One could try to impose further selection criteria that are harder to explain and whose advantage is less evident.

Solving the minimal matching problem is non-trivial, because the number of possible solutions can be very large. The straightforward solution is to use some efficient algorithm (e.g. the one in this paper) that enumerates all solutions, and keeps the best solution. Unfortunately, this approach is completely impractical because some instances have a very high number of solutions. One frequently encounters instances with $> 10^9$ solutions.

In order to find a minimal solution without enumerating all solutions, one can use any algorithm that stops on the first solution in the following way: The first call is used to find out whether a solution exists. If not, then we are done. Otherwise, the algorithm is called again with its input restricted in such a way that it has to find a better solution than the previous. One can continue doing this, until all possibilities to improve the solution have been exhausted. It can be shown that the number of calls needed to obtain an optimal solution is linear in the size of the assumption set of solution. In this way, it can be avoided that all solutions have to be enumerated.

Definition 20. *Let I be an interpretation that is equipped with a weight function α. Let $\phi = A_1, \ldots, A_p \mid B_1, \ldots, B_q$ be a geometric formula. Let α be a fixed set of natural numbers. We define the α-restricted translation (Σ^+, Σ^-) of (I, ϕ) as follows:*

- *For every A_i, let $\overline{v}_i$ be the variables of A_i. Then Σ^+ contains the clause*

$$\{\overline{v}_i/\overline{v}_i\Theta \mid A_i\Theta \textit{ is in conflict with } I \textit{ and } \alpha(A_i\Theta, I) \subseteq \alpha\}.$$

- *For each B_j, let $\overline{w}_j$ denote the variables of B_j. For every Θ that makes $B_j\Theta$ true in I, Σ^- contains the substlet $\overline{w}_j/(\overline{w}_j\Theta)$. In addition, if there exists a $C \in E(B_j, \Theta)$ that is in conflict with I and for which $\alpha(C, \Theta) \nsubseteq \alpha$, then Σ^- contains the substlet $\overline{w}_j/(\overline{w}_j\Theta)$.*

The α-restricted translation ensures that only conflicts involving atoms C with $\alpha(C) \subseteq \alpha$ are considered, and (independently of α), that no B_j is made true. The translation of Definition 9 can be viewed as a special case of α-restricted translation with $\alpha = \mathcal{N}$.

Theorem 2. *Let (Σ^+, Σ^-) be obtained by α-restricted translation of (I, ϕ). For every substitution Θ, Θ is a solution of (Σ^+, Σ^-) iff Θ is a solution of (I, ϕ), and it has $\alpha(I, \phi, \Theta) \subseteq \alpha$.*

Using α-restricted translation, we can define the **optimal** matching algorithm:

Definition 21. *Let* **solve**(Σ^+, Σ^-) *be a function that returns some solution of (Σ^+, Σ^-) if it has a solution, and $\perp$ otherwise.*

We define the algorithm **optimal**(I, ϕ) *that returns an optimal solution of (I, ϕ) if one exists and $\perp$ otherwise.*

- *Let (Σ^+, Σ^-) be the GCSP obtained by the translation of Definition 9. If Σ^+ contains an empty clause, then return $\perp$. If Σ^- contains a propositional blocking, then return $\perp$. Otherwise, remove unit blockings from (Σ^+, Σ^-). If this results in Σ^+ containing an empty clause, then return $\perp$.*
- *Let $\Theta =$* **solve**(Σ^+, Σ^-). *If $\Theta = \perp$, then* **return** $\perp$.
- *Let $\alpha = \alpha(I, \phi, \Theta)$, and let $k := \sup(\alpha)$.*
- *As long as $k \neq 0$, do the following:*
  - *Set $k = k - 1$. If $k \in \alpha$, then do*
    - *Let $\alpha' = (\alpha\backslash\{k\}) \cup \{0, 1, 2, \ldots, k-1\}$.*
    - *Let (Σ^+, Σ^-) be the α'-restricted translation of (I, ϕ).*
    - *If Σ^+ contains an empty clause or Σ^- contains a propositional blocking, then skip the rest of the loop. Otherwise, remove the unit blockings from (Σ^+, Σ^-). If this results in Σ^+ containing the empty clause, then skip the rest of the loop.*
    - *Let $\Theta' =$* **solve**(Σ^+, Σ^-). *If $\Theta' \neq \perp$, then set $\Theta = \Theta'$ and $\alpha = \alpha(I, \phi, \Theta)$.*
- *Now Θ is an optimal solution, so we can* **return** Θ.

Algorithm **optimal** first solves (I, ϕ) without restriction. If this results in a solution Θ, it checks for each $k \in \alpha(I, \phi, \Theta)$ if k can be removed. The invariant of the main loop is: There exists no $k' \geq k$ that occurs in $\alpha(I, \phi, \Theta)$, and no Θ' that is a solution of (I, ϕ) with $k' \notin \alpha(I, \phi, \Theta')$. In addition, the invariant $\alpha = \alpha(I, \phi, \Theta)$ is maintained.

Example 5. Assume that in Example 1, the atoms have weights as follows:

$$\alpha(\ P_{\mathbf{t}}(c_0, c_0)\ ) = \{1\},\ \ \alpha(\ P_{\mathbf{e}}(c_0, c_1)\ ) = \{2\},\ \ \alpha(\ P_{\mathbf{t}}(c_1, c_1)\ ) = \{3\},$$
$$\alpha(\ P_{\mathbf{e}}(c_1, c_2)\ ) = \{4\},\ \ \alpha(\ Q_{\mathbf{t}}(c_2, c_0)\ ) = \{5\}.$$

We have $\alpha(I, \phi_1, \Theta_1) = \{1\}$, $\alpha(I, \phi_1, \Theta_2) = \{1, 2\}$, and $\alpha(I, \phi_1, \Theta_3) = \{2, 3\}$. If Θ_3 is the first solution generated, **solve** will construct the $\{1, 2\}$-restricted translation of (I, ϕ_1), which equals

$$\frac{\begin{array}{l}(X, Y)\ /\ (c_0, c_0) \mid (c_0, c_1)\\ (Y, Z)\ /\ (c_0, c_0) \mid (c_0, c_1)\end{array}}{(X, Z)\ /\ (c_0, c_2)}$$

If the next solution found is Θ_2, then **solve** will construct the $\{1\}$-restricted translation

$$\frac{\begin{array}{l}(X, Y)\ /\ (c_0, c_0)\\ (Y, Z)\ /\ (c_0, c_0)\end{array}}{(X, Z)\ /\ (c_0, c_2)}$$

whose only solution is Θ_1.

6 Conclusions

The problem of matching a geometric formula into an interpretation is currently the most time consuming part of our implementations of geometric resolution. We gave a method for solving the matching problem by translating it into GCSP, and by providing efficient algorithms for GCSP. The algorithm in this paper will be implemented in the nearest future.

One might argue that a calculus that uses an NP-complete problem as its basic operation is not viable, but there is room for interpretation: The complexity of the matching problem is caused by the fact that as result of flattening, geometric formulas and interpretations have DAG-structure instead of tree-structure. This increased expressiveness means that a geometric formula possibly represents exponentially many formulas with tree-structure. This may very well result in shorter proofs. Only experiments can determine which of the two effects will be stronger.

Acknowledgment. We gratefully acknowledge that this work was supported by the Polish National Science Center (Narodowe Centrum Nauki) under grant number DEC-2011/03/B/ST6/00346.

References

1. Bezem, M., Coquand, T.: Automating coherent logic. In: Sutcliffe, G., Voronkov, A. (eds.) LPAR 2005. LNCS (LNAI), vol. 3835, pp. 246–260. Springer, Heidelberg (2005)
2. de Nivelle, H.: Classical logic with partial functions. J. Autom. Reasoning **47**(4), 399–425 (2011)
3. de Nivelle, H.: Theorem proving for classical logic with partial functions by reduction to Kleene logic. J. Logic Comput. (2014). (http://logcom.oxfordjournals.org/. April 2014)
4. de Nivelle, H.: Theorem proving for logic with partial functions by reduction to Kleene logic. In: Benzmüller, C., Otten, J. (eds.) VSL Workshop Proceedings of Automated Reasoning in Quantified Non-Classical Logics (ARQNL), pp. 71–85 (2014)
5. de Nivelle, H., Meng, J.: Geometric resolution: a proof procedure based on finite model search. In: Furbach, U., Shankar, N. (eds.) IJCAR 2006. LNCS (LNAI), vol. 4130, pp. 303–317. Springer, Heidelberg (2006)
6. de Nivelle, H., Meng, J.: Theorem prover Geo 2007f, September 2006. http://www.ii.uni.wroc.pl/~nivelle/
7. Dechter, R.: Constraint Processing. Morgan Kaufmann Publishers, San Francisco (2003)
8. Gottlob, G., Leitsch, A.: On the efficiency of subsumption algorithms. J. ACM **32**(2), 280–295 (1985)
9. Maloberti, J., Sebag, M.: Fast theta-subsumption with constraint satisfaction algorithms. Mach. Learn. **55**, 137–174 (2004)
10. Marques-Silva, J., Lynce, I., Malik, S.: Conflict-driven clause learning SAT solvers. In: Biere, A., Heule, M., van Maaren, H., Walsh, T. (eds.) Handbook of Satisfiability, Chap. 4, pp. 131–153. IOS Press (2009)
11. Murray, N., Rosenthal, E.: Signed formulas: a liftable meta-logic for multiple-valued logics. In: Komorowski, J., Raś, Z.W. (eds.) ISMIS 1993. LNCS, vol. 689, pp. 275–284. Springer, Heidelberg (1993)
12. Scheffer, T., Herbrich, R., Wysotzki, F.: Efficient Θ-subsumption based on graph algorithms. In: Muggleton, S. (ed.) ILP 1996. LNCS(LNAI), vol. 1314, pp. 212–228. Springer, Heidelberg (1996)
13. Sutcliffe, G.: The CADE ATP system competition, August 2015. http://www.cs.miami.edu/~tptp/CASC/25/

On Interpolation and Symbol Elimination in Theory Extensions

Viorica Sofronie-Stokkermans(✉)

University Koblenz-Landau, Koblenz, Germany
sofronie@uni-koblenz.de

Abstract. In this paper we study possibilities of interpolation and symbol elimination in extensions of a theory $\mathcal{T}_0$ with additional function symbols whose properties are axiomatised using a set of clauses. We analyze situations in which we can perform such tasks in a hierarchical way, relying on existing mechanisms for symbol elimination in $\mathcal{T}_0$. This is for instance possible if the base theory allows quantifier elimination. We analyze possibilities of extending such methods to situations in which the base theory does not allow quantifier elimination but has a model completion which does. We illustrate the method on various examples.

1 Introduction

Many problems in computer science (e.g. in program verification) can be reduced to checking satisfiability of ground formulae w.r.t. a theory which can be a standard theory (e.g. linear arithmetic) or a complex theory (typically the extension of a base theory $\mathcal{T}_0$ with additional function symbols axiomatized by a set $\mathcal{K}$ of formulae, or a combination of theories). SMT solvers are tuned for efficiently checking satisfiability of ground formulae in increasingly complex theories; the output can be "satisfiable", "unsatisfiable", or "unknown" (if incomplete methods are used, or termination cannot be guaranteed). More interesting is to go beyond yes/no answers, i.e. to consider parametric systems and infer constraints on parameters (which can be values or functions) which guarantee that certain properties are met (e.g. guarantee the unsatisfiability of ground clauses in suitable theory extensions). In [22,23] – in a context specially tailored for the parametric verification of safety properties in increasingly more complex systems – we showed that such constraints could be generated in extensions of a theory allowing quantifier elimination. In this paper, we propose a symbol elimination method in theory extensions and analyze its properties. We also discuss possibilities of applying such methods to extensions of theories which do not allow quantifier elimination provided that they have a model completion which does.

Another problem we analyze is interpolation (widely used in program verification [11,15–17]). Intuitively, interpolants can be used for describing separations between the sets of "good" and "bad" states; they can help to discover relevant predicates in predicate abstraction with refinement and for over-approximation in model checking. It often is desirable to obtain "ground" interpolants of ground

N. Olivetti and A. Tiwari (Eds.): IJCAR 2016, LNAI 9706, pp. 273–289, 2016.
DOI: 10.1007/978-3-319-40229-1_19

formulae. The first algorithms for interpolant generation in program verification required explicit constructions of proofs [12,16] (in general a relatively difficult task). In [11] the existence of ground interpolants for *arbitrary formulae* is studied – which is proved to be equivalent to the theory having quantifier elimination. This limits the applicability of the results in [11] to situations in which the involved theories allow quantifier elimination. Symbol elimination (e.g. using resolution and/or superposition) has been used for interpolant generation in e.g. [7]. In [21] we identify classes of local theory extensions in which interpolants can be computed hierarchically, using a method of computing interpolants in the base theory. [18] proposes an algorithm for the generation of interpolants for linear arithmetic with uninterpreted function symbols which reduces the problem to constraint solving in linear arithmetic. In both cases, when considering theory extensions $\mathcal{T}_0 \subseteq \mathcal{T}_0 \cup \mathcal{K}$ we devise ways of "separating" the instances of axioms in $\mathcal{K}$ and of the congruence axioms. There also exist results which relate ground interpolation to amalgamation (cf. e.g. [1,2]). We use such results for obtaining criteria which allow us to recognize theories with ground interpolation. However, in general just knowing that ground interpolants exist is not sufficient: we want to construct the interpolants fast (in a hierarchical or modular way) and characterize situations in which we know which (extension) terms these interpolants contain. For this, [25] introduces the notion of W-separability and studies its links to a form of hierarchical interpolation. We here make the results in [25] more precise, and extend them.

The main results of this paper can be summarized as follows:

- We link the existence (and computation) of ground interpolants in a theory $\mathcal{T}$ to their existence (and computation) in a model completion $\mathcal{T}^*$ of $\mathcal{T}$.
- We study possibilities of effective symbol elimination in theory extensions (based on quantifier elimination in the base theory or in a model completion thereof) and analyze the properties of the formulae obtained this way.
- We analyze possibilities of hierarchical interpolation in local theory extensions. Our analysis extends both results in [21] and results in [25] by avoiding the restriction to convex base theories. We explicitly point out all conditions needed for hierarchical interpolation and show how to check them.

The paper is structured as follows. In Sect. 2 we present the main results on model theory needed in the paper. In Sect. 3 we first present existing results linking (sub-)amalgamation, quantifier elimination and the existence of ground interpolants, which we then combine to obtain efficient ways of proving ground interpolation and computing ground interpolants. Section 4 contains the main definitions and results on local theory extensions; these are used in Sect. 5 for symbol elimination and in Sect. 6 for ground interpolation in theory extensions.

2 Preliminaries

We assume known standard definitions from first-order logic such as Π-structures, models, homomorphisms, logical entailment, satisfiability, unsatisfiability. We consider signatures of the form $\Pi = (\Sigma, \mathsf{Pred})$, where Σ is a family of function symbols and Pred a family of predicate symbols. In this paper, (logical) theories are simply sets of sentences. We denote "falsum" with $\perp$. If F and G are formulae we write $F \models G$ (resp. $F \models_{\mathcal{T}} G$ – also written as $\mathcal{T} \cup F \models G$) to express the fact that every model of F (resp. every model of F which is also a model of $\mathcal{T}$) is a model of G. $F \models\perp$ means that F is unsatisfiable; $F \models_{\mathcal{T}}\perp$ means that there is no model of $\mathcal{T}$ in which F is true. If there is a model of $\mathcal{T}$ which is also a model of F we say that F is $\mathcal{T}$-consistent.

If $\mathcal{T}$ is a theory over a signature $\Pi = (\Sigma, \mathsf{Pred})$ we denote by $\mathcal{T}_\forall$ (the universal theory of $\mathcal{T}$) the set of all universal sentences which are logical consequences of $\mathcal{T}$. For Π-structures $\mathcal{A}$ and $\mathcal{B}$, $\varphi : \mathcal{A} \to \mathcal{B}$ is an embedding if and only if it is an injective homomorphism and has the property that for every $P \in \mathsf{Pred}$ with arity n and all $(a_1, \dots, a_n) \in A^n$, $(a_1, \dots, a_n) \in P_{\mathcal{A}}$ iff $(\varphi(a_1), \dots, \varphi(a_n)) \in P_{\mathcal{B}}$. In particular, an embedding preserves the truth of all literals. An elementary embedding between two Π-structures is an embedding that preserves the truth of all first-order formulae over Π. Two Π-structures are elementarily equivalent if they satisfy the same first-order formulae over Π.

Let $\mathcal{A} = (A, \{f_{\mathcal{A}}\}_{f\in\Sigma}, \{P_{\mathcal{A}}\}_{P\in\mathsf{Pred}})$ be a Π-structure. In what follows, we will sometimes denote the universe A of the structure $\mathcal{A}$ by $|\mathcal{A}|$. The *diagram* $\Delta(\mathcal{A})$ of $\mathcal{A}$ is the set of all literals true in the extension $\mathcal{A}^A$ of $\mathcal{A}$ where we have an additional constant for each element of A (which we here denote with the same symbol) with the natural expanded interpretation mapping the constant a to the element a of $|\mathcal{A}|$ (this is a set of sentences over the signature Π^A obtained by expanding Π with a fresh constant a for every element a from $|\mathcal{A}|$). Note that if $\mathcal{A}$ is a Π-structure and $\mathcal{T}$ a theory and $\Delta(\mathcal{A})$ is $\mathcal{T}$-consistent then there exists a Π-structure $\mathcal{B}$ which is a model of $\mathcal{T}$ and in which $\mathcal{A}$ embeds.

A theory $\mathcal{T}$ over signature Π *allows quantifier elimination* if for every formula ϕ over Π there exists a quantifier-free formula ϕ^* over Π which is equivalent to ϕ modulo $\mathcal{T}$. Quantifier elimination can, in particular, be used for eliminating certain constants from ground formulae:

Theorem 1. *Let $\mathcal{T}$ be a theory with signature Π and $A(c_1, \dots, c_n, d_1, \dots, d_m)$ a ground formula over an extension Π^C of Π with additional constants $c_1, \dots, c_n$, $d_1, \dots, d_m$. If $\mathcal{T}$ has quantifier elimination then there exists a ground formula $\Gamma(c_1, \dots, c_n)$ containing only constants $c_1, \dots, c_n$, which is satisfiable w.r.t. $\mathcal{T}$ iff $A(c_1, \dots, c_n, d_1, \dots, d_m)$ is satisfiable w.r.t. $\mathcal{T}$.*

Proof Idea: $\Gamma(c_1, \dots, c_n)$ can be obtained from $\exists d_1, \dots, d_m A(c_1, \dots, c_n, d_1, \dots, d_m)$ by quantifier elimination (where $d_1, \dots, d_n$ are now regarded as variables). □

A *model complete* theory has the property that all embeddings between its models are elementary. Every theory which allows quantifier elimination (QE) is model complete (cf. [8], Theorem 7.3.1).

Example 2. Presburger arithmetic with congruence mod. n, rational linear arithmetic, the theories of real closed fields and of algebraically closed fields, the theory of finite fields and the theory of acyclic lists in the signature $\{\text{car}, \text{cdr}, \text{cons}\}$ ([6, 13]) allow QE, hence are model complete.

A theory $\mathcal{T}^*$ is called a *model companion* of $\mathcal{T}$ if (i) $\mathcal{T}$ and $\mathcal{T}^*$ are co-theories (i.e. every model of $\mathcal{T}$ can be extended to a model of $\mathcal{T}^*$ and vice versa), (ii) $\mathcal{T}^*$ is model complete. $\mathcal{T}^*$ is a *model completion* of $\mathcal{T}$ if it is a model companion of $\mathcal{T}$ with the additional property (iii) for every model $\mathcal{A}$ of $\mathcal{T}$, $\mathcal{T}^* \cup \Delta(\mathcal{A})$ is a complete theory (where $\Delta(\mathcal{A})$ is the diagram of $\mathcal{A}$). Several examples of model completions are mentioned later, in Example 12.

Lemma 3. *Let $\mathcal{T}_1$ and $\mathcal{T}_2$ be two co-theories with signature Π, and $A(c_1, \dots, c_n)$ be a ground formula over an extension Π^C of Π with new constants $c_1, \dots, c_n$. Then A is satisfiable w.r.t. $\mathcal{T}_1$ if and only if it is satisfiable w.r.t. $\mathcal{T}_2$.*

Proof: Consequence of the fact that if $\mathcal{T}$ and $\mathcal{T}'$ are co-theories then $\mathcal{T}_\forall = \mathcal{T}'_\forall$. □

3 Ground Interpolation and Quantifier Elimination

A Π-theory $\mathcal{T}$ has interpolation if, for all Π-formulae ϕ and ψ if $\phi \models_\mathcal{T} \psi$ then there exists a formula I containing only symbols common to ϕ and ψ such that $\phi \models_\mathcal{T} I$ and $I \models_\mathcal{T} \psi$. First order logic has interpolation [4]. It is often important to identify theories for which ground formulae have ground interpolants.

Definition 4. *A theory $\mathcal{T}$ has the* ground interpolation property *(for short: $\mathcal{T}$ has* ground interpolation*) if for every pair of ground formulae $A(\overline{c}, \overline{a})$ (containing constants $\overline{c}, \overline{a}$) and $B(\overline{c}, \overline{b})$ (containing constants $\overline{c}, \overline{b}$), if $A(\overline{c}, \overline{a}) \wedge B(\overline{c}, \overline{b}) \models_\mathcal{T} \perp$ then there exists a ground formula $I(\overline{c})$, containing only the constants $\overline{c}$ occurring both in A and B, such that $A(\overline{c}, \overline{a}) \models_\mathcal{T} I(\overline{c})$ and $B(\overline{c}, \overline{b}) \wedge I(\overline{c}) \models_\mathcal{T} \perp$.*

Let $\mathcal{T}$ be a theory in a signature Σ and Σ' a signature disjoint from Σ. We denote by $\mathcal{T} \cup \mathsf{UIF}_{\Sigma'}$ the extension of $\mathcal{T}$ with uninterpreted symbols in Σ'.

Definition 5 ([2])**.** *We say that a theory $\mathcal{T}$ in a signature Σ has the* general ground interpolation property *(or, shorter, that $\mathcal{T}$ has* general ground interpolation*) if for every signature Σ' disjoint from Σ and every pair of ground $\Sigma \cup \Sigma'$-formulae A and B, if $A \wedge B \models_{\mathcal{T} \cup \mathsf{UIF}_{\Sigma'}} \perp$ then there exists a ground formula I, such that (i) all predicate, constants and function symbols from Σ' occurring in I also occur in A and B; (ii) $A \models_{\mathcal{T} \cup \mathsf{UIF}_{\Sigma'}} I$ and (iii) $B \wedge I \models_{\mathcal{T} \cup \mathsf{UIF}_{\Sigma'}} \perp$.*

There exist results which relate ground interpolation to amalgamation (cf. e.g. [1,2]) and thus allow us to recognize many theories with ground interpolation.

Definition 6 ([2])**.** *A theory $\mathcal{T}$ has the* sub-amalgamation property *iff whenever we are given models M_1 and M_2 of $\mathcal{T}$ with a common substructure A, there exists a further model M of $\mathcal{T}$ endowed with embeddings $\mu_i : M_i \rightarrow M$, $i = 1, 2$ whose restrictions to A coincide. $\mathcal{T}$ has the* strong sub-amalgamation property *if the preceding embeddings μ_1, μ_2 and the preceding model M can be chosen so as to satisfy the following additional condition: if for some m_1, m_2 we have $\mu_1(m_1) = \mu_2(m_2)$, then there exists an element $a \in A$ such that $m_1 = m_2 = a$.*

Definition 7 ([2]). *A theory $\mathcal{T}$ is equality interpolating iff it has the ground interpolation property and has the property that for all tuples $x = x_1, \dots, x_n$, $y^1 = y^1_1, \dots, y^1_{n_1}$, $z^1 = z^1_1, \dots, z^1_{m_1}$, $y^2 = y^2_1, \dots, y^2_{n_2}$, $z^2 = z^2_1, \dots, z^2_{m_2}$ of constants, and for every pair of ground formulae $A(x, z^1, y^1)$ and $B(x, z^2, y^2)$ such that $A(x, z^1, y^1) \wedge B(x, z^2, y^2) \models_{\mathcal{T}} \bigvee_{i=1}^{n_1} \bigvee_{j=1}^{n_2} y^1_i = y^2_j$ there exists a tuple of terms containing only the constants in x, $v(x) = v_1, \dots, v_k$ such that*

$$A(x, z^1, y^1) \wedge B(x, z^2, y^2) \models_{\mathcal{T}} \bigvee_{i=1}^{n_1} \bigvee_{u=1}^{k} y^1_i = v_k \vee \bigvee_{j=1}^{n_2} \bigvee_{u=1}^{k} v_k = y^2_j$$

Theorem 8 ([1–3]). *The following hold for any theory $\mathcal{T}$:*

(1) If $\mathcal{T}$ is universal and has the amalgamation property then $\mathcal{T}$ has ground interpolation [1].
(2) $\mathcal{T}$ has the sub-amalgamation property iff it has ground interpolation [2].
(3) $\mathcal{T}$ is strongly sub-amalgamable iff it has general ground interpolation [2].
(4) If $\mathcal{T}$ has ground interpolation, then $\mathcal{T}$ is strongly sub-amalgamable iff it is equality interpolating [2].
(5) If $\mathcal{T}$ is universal and allows QE, $\mathcal{T}$ is equality interpolating [2].
(6) If $\mathcal{T}^*$ is a model companion of $\mathcal{T}$ then (i) $\mathcal{T}^*$ is a model completion of $\mathcal{T}$ iff (ii) $\mathcal{T}$ has the amalgamation property. If, additionally, $\mathcal{T}$ has universal axiomatization, either of the equivalent conditions (i), (ii) above is equivalent to the fact that $\mathcal{T}^*$ allows quantifier elimination [3].

Clearly, if a theory $\mathcal{T}$ allows quantifier elimination then it has ground interpolation (Assume $A \wedge B \models_{\mathcal{T}} \perp$. We can simply use quantifier elimination to eliminate the non-shared constants from A w.r.t. $\mathcal{T}$ and obtain an interpolant). The converse is not true (the theory of uninterpreted function symbols over a signature Σ has ground interpolation but does not allow quantifier elimination).

Theorem 9. *If $\mathcal{T}$ is a universal theory which allows quantifier elimination then $\mathcal{T}$ has* general ground interpolation.

Proof: If $\mathcal{T}$ allows QE, it has ground interpolation. By Theorem 8(5), $\mathcal{T}$ is equality interpolating. But by Theorem 8(4), a theory that has ground interpolation and is equality interpolating has the strong sub-amalgamation property, hence by Theorem 8(3) it has general ground interpolation. □

Example 10.

(1) All theories in Example 2 allow QE, so have ground interpolation.
(2) The theory of pure equality has the strong (sub-)amalgamation property, hence by Theorem 8 it allows general ground interpolation.
(3) The theory of absolutely-free data structures [13] is universal and has quantifier elimination, hence by Theorem 9 it has general ground interpolation.

Model Companions and Ground Interpolation. In what follows we establish links between ground interpolation resp. quantifier elimination in a theory and in its model companions (if they exist).

Theorem 11. *Let $\mathcal{T}$ be a theory and $\mathcal{T}^*$ a model companion of $\mathcal{T}$.*

(1) If $\mathcal{T}$ is universal and has ground interpolation, then $\mathcal{T}^*$ allows QE.
(2) If $\mathcal{T}^*$ has ground interpolation then so does $\mathcal{T}$; the ground interpolants computed w.r.t. $\mathcal{T}^*$ are also interpolants w.r.t. $\mathcal{T}$.
(3) A universal theory $\mathcal{T}$ has ground interpolation iff $\mathcal{T}^*$ has ground interpolation.
(4) If $\mathcal{T}^*$ allows quantifier elimination then $\mathcal{T}$ has ground interpolation.

Proof: (1) By Theorem 8(1), $\mathcal{T}$ has the amalgamation property so Theorem 8(6) can be used. (2) Let A, B be ground formulae s.t. $\mathcal{T} \cup A \cup B \models \perp$. As $\mathcal{T}_\forall = \mathcal{T}^*_\forall$, by Lemma 3, $\mathcal{T}^* \cup A \cup B \models \perp$, so an interpolant I (in theory $\mathcal{T}^*$) exists. By Lemma 3, I is an interpolant w.r.t. $\mathcal{T}$. (3) The direct implication follows from (1), the converse from (2). (4) follows from Theorem 9 and (2). □

Example 12. The following theories have ground interpolation:

(1) The pure theory of equality (its model completion is the theory of an infinite set, which allows QE).
(2) The theory of total orderings (its model completion is the theory of dense total orders without endpoints, which allows QE [6]).
(3) The theory of Boolean algebras (its model completion is the theory of atomless Boolean algebras, which allows QE [3]).
(4) The theory of fields (its model completion is the theory of algebraically closed fields, which allows QE [3,8]).

Until now, we discussed possibilities for symbol elimination and ground interpolation in arbitrary theories. However, often the theories we consider are extensions of a "base" theory with additional function symbols satisfying certain properties axiomatized using clauses; we now analyze such theories. In Sect. 4 we recall the main definitions and results related to (local) theory extensions. We use these results in Sect. 5 to study possibilities of symbol elimination in theory extensions and in Sect. 6 to identify theory extensions with ground interpolation.

4 Local Theory Extensions

Let $\Pi_0{=}(\Sigma_0, \mathsf{Pred})$ be a signature, and $\mathcal{T}_0$ be a "base" theory with signature Π_0. We consider extensions $\mathcal{T} := \mathcal{T}_0 \cup \mathcal{K}$ of $\mathcal{T}_0$ with new function symbols Σ (*extension functions*) whose properties are axiomatized using a set $\mathcal{K}$ of (universally closed) clauses in the extended signature $\Pi = (\Sigma_0 \cup \Sigma, \mathsf{Pred})$, which contain function symbols in Σ. Let C be a fixed countable set of fresh constants. We will denote by Π^C the extension of Π with constants in C. If G is a finite set of ground Π^C-clauses and $\mathcal{K}$ a set of Π-clauses, we will denote by $\mathsf{st}(\mathcal{K}, G)$ (resp. $\mathsf{est}(\mathcal{K}, G)$) the set of all ground terms (resp. extension ground terms, i.e. terms starting with a function in Σ) which occur in G or $\mathcal{K}$. In this paper we regard every finite set G of ground clauses as the ground formula $\bigwedge_{C \in G} C$. If T is a set of ground terms in the signature Π^C, we denote by $\mathcal{K}[T]$ the set of all instances of $\mathcal{K}$ in which the terms starting with a function symbol in Σ are in T. Let Ψ be a map associating with every finite set T of ground terms a finite set $\Psi(T)$ of ground terms. For any set G of ground Π^C-clauses we write $\mathcal{K}[\Psi_\mathcal{K}(G)]$ for $\mathcal{K}[\Psi(\mathsf{est}(\mathcal{K}, G))]$. We define:

(Loc_f^Ψ) For every finite set G of ground clauses in Π^C it holds that $\mathcal{T}_0 \cup \mathcal{K} \cup G \models \bot$ if and only if $\mathcal{T}_0 \cup \mathcal{K}[\Psi_\mathcal{K}(G)] \cup G$ is unsatisfiable.

Extensions satisfying condition (Loc_f^Ψ) are called *Ψ-local*. If Ψ is the identity (in which case $\mathcal{K}[\Psi_\mathcal{K}(G)] = \mathcal{K}[G] = \mathcal{K}[\mathsf{est}(\mathcal{K}, G)]$) we obtain the notion of *local theory extensions* [19,20], which generalizes the notion of *local theories* [5,14].

Hierarchical Reasoning. Consider a Ψ-local theory extension $\mathcal{T}_0 \subseteq \mathcal{T}_0 \cup \mathcal{K}$. Condition ($\mathsf{Loc}_f^\Psi$) requires that for every finite set G of ground Π^C clauses, $\mathcal{T}_0 \cup \mathcal{K} \cup G \models \bot$ iff $\mathcal{T}_0 \cup \mathcal{K}[\Psi_\mathcal{K}(G)] \cup G \models \bot$. In all clauses in $\mathcal{K}[\Psi_\mathcal{K}(G)] \cup G$ the function symbols in Σ only have ground terms as arguments, so $\mathcal{K}[\Psi_\mathcal{K}(G)] \cup G$ can be flattened and purified by introducing, in a bottom-up manner, new constants $c_t \in C$ for subterms $t = f(c_1, \dots, c_n)$ where $f \in \Sigma$ and c_i are constants, together with definitions $c_t = f(c_1, \dots, c_n)$. We thus obtain a set of clauses $\mathcal{K}_0 \cup G_0 \cup \mathsf{Def}$, where $\mathcal{K}_0$ and G_0 do not contain Σ-function symbols and Def contains clauses of the form $c = f(c_1, \dots, c_n)$, where $f \in \Sigma$, $c, c_1, \dots, c_n$ are constants.

Theorem 13 ([9,19,20]). *Let $\mathcal{K}$ be a set of clauses. Assume that $\mathcal{T}_0 \subseteq \mathcal{T}_1 = \mathcal{T}_0 \cup \mathcal{K}$ is a Ψ-local theory extension. For any finite set G of ground clauses, let $\mathcal{K}_0 \cup G_0 \cup \mathsf{Def}$ be obtained from $\mathcal{K}[\Psi_\mathcal{K}(G)] \cup G$ by flattening and purification, as explained above. Then the following are equivalent to $\mathcal{T}_1 \cup G \models \bot$:*

(1) $\mathcal{T}_0 \cup \mathcal{K}[\Psi_\mathcal{K}(G)] \cup G \models \bot$.

(2) $\mathcal{T}_0 \cup \mathcal{K}_0 \cup G_0 \cup \mathsf{Con}_0 \models \bot$, *where* $\mathsf{Con}_0 = \{ \bigwedge_{i=1}^{n} c_i \approx d_i \rightarrow c \approx d \mid {}^{f(c_1, \dots, c_n) \approx c \in \mathsf{Def}}_{f(d_1, \dots, d_n) \approx d \in \mathsf{Def}} \}$.

Since local extensions can be recognized by showing that certain partial models embed into total ones, we introduce the main definitions here.

Partial Structures. Let $\Pi = (\Sigma, \mathsf{Pred})$ be a first-order signature with set of function symbols Σ and set of predicate symbols Pred. A *partial Π-structure* is a structure $\mathcal{A} = (A, \{f_\mathcal{A}\}_{f \in \Sigma}, \{P_\mathcal{A}\}_{P \in \mathsf{Pred}})$, where A is a non-empty set, for every n-ary $f \in \Sigma$, $f_\mathcal{A}$ is a partial function from A^n to A, and for every n-ary $P \in \mathsf{Pred}$, $P_\mathcal{A} \subseteq A^n$. We consider constants (0-ary functions) to be always defined. $\mathcal{A}$ is called a *total structure* if the functions $f_\mathcal{A}$ are all total. Given a (total or partial) Π-structure $\mathcal{A}$ and $\Pi_0 \subseteq \Pi$ we denote the reduct of $\mathcal{A}$ to Π_0 by $\mathcal{A}|_{\Pi_0}$.

The notion of evaluating a term t with variables X w.r.t. an assignment $\beta : X \rightarrow A$ for its variables in a partial structure $\mathcal{A}$ is the same as for total algebras, except that the evaluation is undefined if $t = f(t_1, \dots, t_n)$ and at least one of $\beta(t_i)$ is undefined, or else $(\beta(t_1), \dots, \beta(t_n))$ is not in the domain of $f_\mathcal{A}$.

A *weak Π-embedding* between partial Π-structures $\mathcal{A} = (A, \{f_{\mathcal{A}}\}_{f\in\Sigma}, \{P_{\mathcal{A}}\}_{P\in\mathsf{Pred}})$ and $\mathcal{B} = (B, \{f_{\mathcal{B}}\}_{f\in\Sigma}, \{P_{\mathcal{B}}\}_{P\in\mathsf{Pred}})$ is a total map $\varphi : A \to B$ such that (i) φ is an embedding w.r.t. $\mathsf{Pred} \cup \{=\}$ and (ii) whenever $f_{\mathcal{A}}(a_1, \dots, a_n)$ is defined (in $\mathcal{A}$), then $f_{\mathcal{B}}(\varphi(a_1), \dots, \varphi(a_n))$ is defined (in $\mathcal{B}$) and $\varphi(f_{\mathcal{A}}(a_1, \dots, a_n)) = f_{\mathcal{B}}(\varphi(a_1), \dots, \varphi(a_n))$, for all $f \in \Sigma$.

Let $\mathcal{A}$ be a partial Π-structure and $\beta : X \to A$ be a variable assignment. $(\mathcal{A}, \beta)$ *weakly satisfies a clause* C (notation: $(\mathcal{A}, \beta) \models_w C$) if either some of the literals in $\beta(C)$ are not defined or otherwise all literals are defined and for at least one literal L in C, L is true in $\mathcal{A}$ w.r.t. β. $\mathcal{A}$ is a *weak partial model* of a set of clauses $\mathcal{K}$ if $(\mathcal{A}, \beta) \models_w C$ for every variable assignment β and every clause C in $\mathcal{K}$.

Recognizing Ψ-local theory extensions. In [19] we proved that if all weak partial models of an extension $\mathcal{T}_0 \cup \mathcal{K}$ of a base theory $\mathcal{T}_0$ with total base functions can be embedded into a total model of the extension, then the extension is local. In [9] we lifted these results to Ψ-locality.

Let $\mathcal{A} = (A, \{f_{\mathcal{A}}\}_{f\in\Sigma_0\cup\Sigma\cup C}, \{P_{\mathcal{A}}\}_{P\in\mathsf{Pred}})$ be a partial Π^C-structure with total Σ_0-functions. Let Π^A be the extension of the signature Π with constants from A. We denote by $T(\mathcal{A})$ the following set of ground Π^A-terms:

$$T(\mathcal{A}) := \{f(a_1, ..., a_n) \mid f \in \Sigma, a_i \in A, i = 1, \dots, n, f_{\mathcal{A}}(a_1, ..., a_n) \text{ is defined}\}.$$

Let $\mathsf{PMod}^{\Psi}_{w,f}(\Sigma, \mathcal{T})$ be the class of all weak partial models $\mathcal{A}$ of $\mathcal{T}_0 \cup \mathcal{K}$, such that $\mathcal{A}_{|\Pi^0}$ is a total model of $\mathcal{T}_0$, the Σ-functions are partial, $T(\mathcal{A})$ is finite and all terms in $\Psi(\mathsf{est}(\mathcal{K}, T(\mathcal{A})))$ are defined (in the extension $\mathcal{A}^A$ with constants from A). We consider the following embeddability property of partial structures:

$(\mathsf{Emb}^{\Psi}_{w,f})$ Every $\mathcal{A} \in \mathsf{PMod}^{\Psi}_{w,f}(\Sigma, \mathcal{T})$ weakly embeds into a total model of $\mathcal{T}$.

We also consider the properties $(\mathsf{EEmb}^{\Psi}_{w,f})$, which additionally requires the embedding to be *elementary* and (Comp_f) which requires that every structure $\mathcal{A} \in \mathsf{PMod}^{\Psi}_{w,f}(\Sigma, \mathcal{T})$ embeds into a total model of $\mathcal{T}$ *with the same support.*

When establishing links between locality and embeddability we require that the clauses in $\mathcal{K}$ are *flat* and *linear* w.r.t. Σ-functions. We distinguish between ground and non-ground clauses. An *extension clause D is flat* when all symbols below a Σ-function symbol in D are variables. D is *linear* if whenever a variable occurs in two terms of D starting with Σ-functions, the terms are equal, and no term contains two occurrences of a variable. A *ground clause D is flat* if all symbols below a Σ-function in D are constants. A *ground clause D is linear* if whenever a constant occurs in two terms in D whose root symbol is in Σ, the two terms are identical, and if no term which starts with a Σ-function contains two occurrences of the same constant.

Theorem 14 ([9,10])**.** *Let $\mathcal{T}_0$ be a first-order theory and $\mathcal{K}$ a set of universally closed flat clauses in the signature Π. The following hold:*

(1) If all clauses in $\mathcal{K}$ are linear and Ψ is a term closure operator (for definition cf. [10]) with the property that for every flat set of ground terms T, $\Psi(T)$ is flat then either of the conditions $(\mathsf{Emb}^{\Psi}_{w,f})$ *and* $(\mathsf{EEmb}^{\Psi}_{w,f})$ *implies* (Loc^{Ψ}_f).
(2) If the extension $\mathcal{T}_0 \subseteq \mathcal{T}{=}\mathcal{T}_0 \cup \mathcal{K}$ satisfies (Loc^{Ψ}_f) *then* $(\mathsf{Emb}^{\Psi}_{w,f})$ *holds.*

Property $(\mathsf{EEmb}^{\Psi}_{\mathsf{w,f}})$ is preserved if we enrich $\mathcal{T}_0$ (cf. [10]).

5 Symbol Elimination in Theory Extensions

We show that in theory extensions $\mathcal{T}_0 \subseteq \mathcal{T} = \mathcal{T}_0 \cup \mathcal{K}$ for which $\mathcal{T}_0$ (resp. its model completion $\mathcal{T}_0^*$) allows quantifier elimination, for every ground formula G containing function symbols considered to be "parameters" we can generate a (universal) constraint Γ on the parameters of G such that $\mathcal{T} \cup \Gamma \cup G \models \perp$.

Let $\Pi_0 = (\Sigma_0, \mathsf{Pred})$. Let $\mathcal{T}_0$ be a Π_0-theory and Σ_P be a set of parameters (function and constant symbols). Let Σ be a signature such that $\Sigma \cap (\Sigma_0 \cup \Sigma_P) = \emptyset$, containing functions not in $(\Sigma_0 \cup \Sigma_P)$. Let $\mathcal{K}$ be a set of clauses in the signature $\Pi_0 \cup \Sigma_P \cup \Sigma$ in which all variables occur also below functions in $\Sigma_1 = \Sigma_P \cup \Sigma$. Let G be a finite set of ground clauses, and T a finite set of ground terms over the signature $\Pi_0 \cup \Sigma_P \cup \Sigma \cup C$, where C is a set of additional constants.

We construct a universal formula $\forall y_1 \dots y_n \Gamma_T(y_1, \dots, y_n)$ over the signature $\Pi_0 \cup \Sigma_P$ by following the Steps 1–5 below:

Step 1: Let $\mathcal{K}_0 \cup G_0 \cup \mathsf{Con}_0$ be the set of Π_0^C clauses obtained from $\mathcal{K}[T] \cup G$ after the purification step in Theorem 13 (with set of extension symbols Σ_1).

Step 2: Let $G_1 = \mathcal{K}_0 \cup G_0 \cup \mathsf{Con}_0$. Among the constants in G_1, we identify (i) the constants $\overline{c}_f$, $f \in \Sigma_P$, where either $c_f = f \in \Sigma_P$ is a constant parameter, or is introduced by a definition $c_f := f(c_1, \dots, c_k)$ in the hierarchical reasoning method, and (ii) all constants $\overline{c}_p$ occurring as arguments of functions in Σ_P in such definitions. Let $\overline{c}$ be the remaining variables. We replace the constants in $\overline{c}$ with existentially quantified variables $\overline{x}$ in G_1, i.e. replace $G_1(\overline{c}_p, \overline{c}_f, \overline{c})$ with $G_1(\overline{c}_p, \overline{c}_f, \overline{x})$, and consider the formula $\exists \overline{x} G_1(\overline{c}_p, \overline{c}_f, \overline{x})$.

Step 3: Using the method for quantifier elimination in $\mathcal{T}_0$ we can construct a formula $\Gamma_1(\overline{c}_p, \overline{c}_f)$ equivalent to $\exists \overline{x} G_1(\overline{c}_p, \overline{c}_f, \overline{x})$ w.r.t. $\mathcal{T}_0$.

Step 4: Let $\Gamma_2(\overline{c}_p)$ be the formula obtained by replacing back in $\Gamma_1(\overline{c}_p, \overline{c}_f)$ the constants c_f introduced by definitions $c_f := f(c_1, \dots, c_k)$ with the terms $f(c_1, \dots, c_k)$. We replace $\overline{c}_p$ with existentially quantified variables $\overline{y}$ and obtain the formula $\exists \overline{y} \Gamma_2(\overline{y})$.

Step 5: Let $\forall \overline{y} \Gamma_T(\overline{y})$ be $\neg(\exists \overline{y} \Gamma_2(\overline{y}))$, i.e. $\forall \overline{y} \neg \Gamma_2(\overline{y})$.

A similar approach is used in [22] for generating constraints on parameters which guarantee safety of parametric systems. We show that $\forall \overline{y} \Gamma_T(\overline{y})$ guarantees unsatisfiability of G and further study the properties of these formulae. At the end of Sect. 6 we briefly indicate how this can be used for interpolant generation.

We first analyze the case in which $\mathcal{T}_0$ allows quantifier elimination.

Theorem 15. *Assume that $\mathcal{T}_0$ allows quantifier elimination. For every finite set of ground clauses G, and every finite set T of terms over the signature $\Pi_0 \cup \Sigma \cup \Sigma_P \cup C$ with* $\mathsf{est}(G) \subseteq T$, *Steps 1–5 yield a universally quantified $\Pi_0 \cup \Sigma_P$-formula $\forall \overline{x} \Gamma_T(\overline{x})$ with the following properties:*

(1) For every structure $\mathcal{A}$ with signature $\Pi_0 \cup \Sigma \cup \Sigma_P \cup C$ which is a model of $\mathcal{T}_0 \cup \mathcal{K}$, if $\mathcal{A} \models \forall \overline{y} \Gamma_T(\overline{y})$ then $\mathcal{A} \models \neg G$.

(2) $\mathcal{T}_0 \cup \forall \overline{y} \Gamma_T(\overline{y}) \cup \mathcal{K} \cup G$ is unsatisfiable.

Proof: The proof is given in [24]. □

From the proof of Theorem 15 we can see that (with the notation used in Steps 1–5) the formulae $\exists \overline{x} G_1(\overline{c}_p, \overline{c}_f, \overline{x}) \wedge \mathsf{Def}$ and $\Gamma_2(\overline{c}_p)$ are equivalent w.r.t. $\mathcal{T}_0 \cup \mathsf{UIF}_{\Sigma_P}$.

Theorem 16. *If $T_1 \subseteq T_2$ then $\forall \overline{y} \Gamma_{T_1}(\overline{y})$ entails $\forall \overline{y} \Gamma_{T_2}(\overline{y})$ (modulo $\mathcal{T}_0$).*

Proof: The proof is given in [24]. □

In what follows, $\forall \overline{y} \Gamma_G(\overline{y})$ denotes the formula obtained when $T = \mathsf{est}(\mathcal{K}, G)$.

Theorem 17. *If the extension $\mathcal{T}_0 \subseteq \mathcal{T}_0 \cup \mathcal{K}$ satisfies condition (Comp_f) and $\mathcal{K}$ is flat and linear then $\forall y \Gamma_G(y)$ is entailed by every conjunction Γ of clauses such that $\mathcal{T}_0 \cup \Gamma \cup \mathcal{K} \cup G$ is unsatisfiable (i.e. it is the weakest such constraint).*

Proof: The proof is given in [24]. □

Example 18. Let $\mathcal{T}_0$ be the theory of dense total orderings without endpoints. Consider the extension of $\mathcal{T}_0$ with functions $\Sigma_1 = \{f, g, h, c\}$ whose properties are axiomatized by $\mathcal{K} := \{\forall x(x \leq c \rightarrow g(x) = f(x)), \forall x(c < x \rightarrow g(x) = h(x))\}$. Assume $\Sigma_P = \{f, h, c\}$ and $\Sigma = \{g\}$. We are interested in generating a set of constraints on the parameters f, h and c which ensure that g is monotone, e.g. satisfies $\mathsf{Mon}(g) : \; \forall x, y(x \leq y \rightarrow g(x) \leq g(y))$, i.e. a set Γ of $\Sigma_0 \cup \Sigma_P$-constraints such that $\mathcal{T}_0 \cup \Gamma \cup \mathcal{K} \cup \{c_1 \leq c_2, g(c_1) > g(c_2)\}$ is unsatisfiable, where $G = \{c_1 \leq c_2, g(c_1) > g(c_2)\}$ is the negation of $\mathsf{Mon}(g)$.

Step 1: We compute $\mathcal{T}_0 \cup \mathcal{K}[G] \cup G$, then purify it. We obtain $\mathsf{Def} = \{g_1 = g(c_1), g_2 = g(c_2), f_1 = f(c_1), f_2 = f(c_2), h_1 = h(c_1), h_2 = h(c_2)\}$ and:

$$\begin{aligned} \mathcal{K}_0 \cup \mathsf{Con}_0 \cup G_0 := \{ & c_1 \leq c \rightarrow g_1 = f_1, c_2 \leq c \rightarrow g_2 = f_2, c < c_1 \rightarrow g_1 = h_1, \\ & c < c_2 \rightarrow g_2 = h_2, c_1 = c_2 \rightarrow g_1 = g_2, c_1 = c_2 \rightarrow f_1 = f_2, \\ & c_1 = c_2 \rightarrow h_1 = h_2, c_1 \leq c_2, g_1 > g_2\} \end{aligned}$$

Step 2: $\Sigma_P = \{f, h, c\}$. To eliminate g we replace g_1, g_2 with existentially quantified variables z_1, z_2 and obtain $\exists z_1, z_2 G_1(c_1, c_2, c, f_1, f_2, h_1, h_2, z_1, z_2) =$

$$\begin{aligned} \exists z_1, z_2 (& c_1 \leq c \rightarrow z_1{=}f_1 \wedge c_2 \leq c \rightarrow z_2{=}f_2 \wedge c_1{=}c_2 \rightarrow f_1{=}f_2 \wedge c_1 = c_2 \rightarrow h_1{=}h_2 \\ & c < c_1 \rightarrow z_1{=}h_1 \wedge c < c_2 \rightarrow z_2{=}h_2 \wedge c_1{=}c_2 \rightarrow z_1{=}z_2 \wedge c_1 \leq c_2 \wedge z_1 > z_2) \end{aligned}$$

Step 3: After simplification followed by the method for QE for dense total orderings without endpoints we obtain the formula $\Gamma_1(c_1, c_2, c, f_1, f_2, h_1, h_2)$

$$\begin{aligned} (& (c_1 \leq c \;\wedge\; c_2 \leq c \;\wedge\; c_1 \leq c_2 \;\wedge\; f_1 > f_2 \;\wedge\; c_1 \neq c_2) \;\vee \\ & (c_1 \leq c \;\wedge\; c < c_2 \;\wedge\; c_1 \leq c_2 \;\wedge\; f_1 > h_2 \;\wedge\; c_1 \neq c_2) \;\vee \\ & (c < c_1 \;\wedge\; c_2 \leq c \;\wedge\; c_1 \leq c_2 \;\wedge\; h_1 > f_2 \;\wedge\; c_1 \neq c_2) \;\vee \\ & (c < c_1 \;\wedge\; c < c_2 \;\wedge\; c_1 \leq c_2 \;\wedge\; h_1 > h_2 \;\wedge\; c_1 \neq c_2)) \end{aligned}$$

Step 4: We construct the formula $\Gamma_2(c_1, c_2, c)$ from Γ_1 by replacing f_i by $f(c_i)$ and h_i by $h(c_i)$, $i = 1, 2$. We obtain (after further minor simplification and rearrangement for facilitating reading):

$$((c_1{<}c_2 \leq c \wedge f(c_1){>}f(c_2)) \vee (c_1{\leq}c < c_2 \wedge f(c_1){>}h(c_2)) \vee (c{<}c_1{<}c_2 \wedge h(c_1){>}h(c_2)))$$

Step 5: Then $\forall z_1, z_2 \Gamma_T(z_1, z_2)$ is $\forall z_1, z_2[(z_1 < z_2 \leq c \rightarrow f(z_1) \leq f(z_2)) \wedge (z_1 \leq c < z_2 \rightarrow f(z_1) \leq h(z_2)) \wedge (c < z_1 < z_2 \rightarrow h(z_1) \leq h(z_2))]$.

We now analyze the case in which $\mathcal{T}_0$ does not necessarily allow quantifier elimination, but has a model completion which allows quantifier elimination.

Theorem 19. *Let $\mathcal{T}_0$ be a theory having a model completion $\mathcal{T}_0^*$ with $\mathcal{T}_0 \subseteq \mathcal{T}_0^*$. Let $\mathcal{T} = \mathcal{T}_0 \cup \mathcal{K}$ be an extension of $\mathcal{T}_0$ with new function symbols $\Sigma_1 = \Sigma_P \cup \Sigma$ whose properties are axiomatized by a set of clauses $\mathcal{K}$ in which all variables occur also below extension functions in Σ_1. Assume that (i) every model of $\mathcal{T}_0 \cup \mathcal{K}$ embeds into a model of $\mathcal{T}_0^* \cup \mathcal{K}$, and (ii) $\mathcal{T}_0^*$ allows quantifier elimination.*

Then, for every finite set of ground clauses G, and every finite set T of terms over the signature $\Pi^C = \Pi_0 \cup \Sigma \cup \Sigma_P \cup C$ with $\mathsf{est}(G) \subseteq T$ we can construct a universally quantified $\Pi_0 \cup \Sigma_P$-formula $\forall \overline{x} \Gamma_T(\overline{x})$ such that:

(1) For every structure $\mathcal{A}$ with signature $\Pi_0 \cup \Sigma \cup \Sigma_P \cup C$ which is a model of $\mathcal{T}_0 \cup \mathcal{K}$, if $\mathcal{A} \models \forall \overline{x} \Gamma_T(\overline{x})$ then $\mathcal{A} \models \neg G$.
(2) $\mathcal{T}_0 \cup \forall y \Gamma_T(y) \cup \mathcal{K} \cup G$ is unsatisfiable.

Proof: This follows from Theorem 15 (applied to $\mathcal{T}_0^*$) and the properties of $\mathcal{T}_0^*$, details are given in [24]. □

Example 20. Consider the problem in Example 18 when the base theory $\mathcal{T}_0$ is the theory of total orderings. It can be checked that (i) and (ii) hold. By Theorem 19, the formula $\forall z_1, z_2 \Gamma_T(z_1, z_2)$ constructed in Example 18 ensures that g is monotone also in this case.

Unfortunately, under the assumptions of Theorem 19 we cannot guarantee that $\forall y \Gamma_G(y)$ is the weakest set of constraints in the set of all Γ with $\mathcal{T}_0 \cup \Gamma \cup \mathcal{K} \cup G \models \perp$.

Example 21. Let $\mathcal{T}_0$ be the theory of total orderings and $G := \{a < g(a), g(a) < h(a)\}$. Using Steps 1–5 for $\mathcal{T}_0^*$ we obtain the following formula $\forall y \Gamma_G(y) = \forall x (h(x) \leq x)$. Let $\Gamma := \forall x, y, z (x < y \rightarrow y \geq z)$. Then $\Gamma \wedge G$ is unsatisfiable, but there exists a structure with two elements $a_1 < a_2$ such that $h(a_1) = a_2$ which satisfies Γ but not Γ_G. (Note that this situation cannot occur when $\mathcal{T}_0$ has quantifier elimination: Then the formula $\exists \overline{x} G_1(\overline{x})$ is either true or false in $\mathcal{T}_0$. If it is true then to achieve unsatisfiability we have to add $\Gamma = \perp$, which entails any other constraint. If it is false then we do not need to add any constraints to achieve unsatisfiability. so $\Gamma = \top$, which is entailed by any other constraint).

6 Ground Interpolation in Theory Extensions

In this section we present criteria for recognizing whether a theory extension $\mathcal{T} = \mathcal{T}_0 \cup \mathcal{K}$ has ground interpolation provided that $\mathcal{T}_0$ has (general) ground interpolation. In [21] we identified classes of local extensions in which ground interpolants can be computed hierarchically (for this, we had to find ways of separating the instances of axioms in $\mathcal{K}$ and of the congruence axioms). Criteria linking hierarchical ground interpolation to an amalgamability property for partial algebras were given in [25]. We here extend the results in [21, 25].

Definition 22 ([25]). *An amalgamation closure for a theory extension $\mathcal{T} = \mathcal{T}_0 \cup \mathcal{K}$ is a function W associating with finite sets of ground terms T_A and T_B, a finite set $W(T_A, T_B)$ of ground terms such that*

(1) all ground subterms in $\mathcal{K}$ and T_A are in $W(T_A, T_B)$;
(2) W is monotone, i.e., for all $T_A \subseteq T'_A$, $T_B \subseteq T'_B$, $W(T_A, T_B) \subseteq W(T'_A, T'_B)$;
(3) W is a closure, i.e., $W(W(T_A, T_B), W(T_B, T_A)) \subseteq W(T_A, T_B)$;
(4) W is compatible with any map h between constants satisfying $h(c_1) \neq h(c_2)$, for all constants $c_1 \in \mathsf{st}(T_A), c_2 \in \mathsf{st}(T_B)$ that are not shared between T_A and T_B, i.e., for any such h we require $W(h(T_A), h(T_B)) = h(W(T_A, T_B))$; and
(5) $W(T_A, T_B)$ only contains T_A-pure terms.

For sets of ground clauses A, B we write $W(A, B)$ for $W(\mathsf{st}(A), \mathsf{st}(B))$. In this paper when we use W we always refer to an amalgamation closure.

Definition 23 ([25]). *A theory extension $\mathcal{T} = \mathcal{T}_0 \cup \mathcal{K}$ is W-separable if for all sets of ground clauses A and B, $\mathcal{T}_0 \cup \mathcal{K} \cup A \cup B \models \perp$ iff $\mathcal{T}_0 \cup \mathcal{K}[W(A, B)] \cup A \cup \mathcal{K}[W(B, A)] \cup B \models \perp$.*

Example 24. Let $\mathcal{T}_0$ be the theory TOrd of total orderings. We consider the extension of $\mathcal{T}_0$ with $\mathcal{K} = \{\mathsf{SGC}(f, g), \mathsf{Mon}(f, g)\}$ (cf. also [21]), where:

- $\mathsf{SGC}(f, g) : \forall x, y (x \leq g(y) \rightarrow f(x) \leq y)$;
- $\mathsf{Mon}(f, g) : \forall x, y (x \leq y \rightarrow f(x) \leq f(y)) \wedge \forall x, y (x \leq y \rightarrow g(x) \leq g(y))$.

TOrd is $\leq$-interpolating [21]: If A_0 and B_0 are sets of ground clauses in the signature of TOrd and $A_0 \wedge B_0 \models_{\mathsf{TOrd}} a \leq b$, where a is a constant in A_0 and b a constant in B_0 then there exists a constant d (common to A_0 and B_0) such that $A_0 \wedge B_0 \models_{\mathsf{TOrd}} a \leq d \wedge d \leq b$. Let A and B sets of ground clauses in the signature of $\mathcal{T}_0 \cup \mathcal{K}$. If C_I are the terms corresponding to the common constants used for $\leq$-interpolating the premises of the mixed instances of $\mathsf{SGC}(f, g)[A \cup B] \wedge \mathsf{Mon}(f, g)[A \cup B]$ after the hierarchical reduction, the results in [21] show that $\mathcal{T}_0 \cup \mathcal{K}$ is W-separable where $W(A, B) = \mathsf{st}(A) \cup \{f(c) \mid c \in C_I\}$ and $W(B, A) = \mathsf{st}(B) \cup \{f(c) \mid c \in C_I\}$.

W-Separability and Partial Amalgamation. In [25] it is shown that if $\mathcal{T} = \mathcal{T}_0 \cup \mathcal{K}$ is W-separable, and $\mathcal{K}$ flat and linear, then the extension $\mathcal{T}_0 \subseteq \mathcal{T}_0 \cup \mathcal{K}$ is Ψ-local where $\Psi(T) = W(T, T)$ for all sets of ground terms T. Also, a notion of partial amalgamability is defined; it is shown that if $\mathcal{T}_0 \subseteq \mathcal{T}_1 = \mathcal{T}_0 \cup \mathcal{K}$ is a theory extension with $\mathcal{K}$ flat and linear and $\mathcal{T}_1$ has the partial amalgamation property w.r.t. W, then $\mathcal{T}_1$ is W-separable. We make the last result more precise.

Definition 25. *A theory extension $\mathcal{T} = \mathcal{T}_0 \cup \mathcal{K}$ has the partial amalgamation property for models with the same Π_0-reduct if whenever $M_A, M_B, M_C \in \mathsf{PMod}_{w,f}(\Sigma, \mathcal{T})$ are such that:*

(1) M_A, M_B and M_C have the same reduct to Π_0;
(2) M_C is a (partial) substructure of M_A, M_B in which $f_{M_C}(m_1, \dots, m_n)$ is defined and is equal to m iff both $f_{M_A}(m_1, \dots, m_n)$ and $f_{M_B}(m_1, \dots, m_n)$ are defined and equal to m;

(3) *The sets* $T_{M_A} = \{f(a_1, \ldots, a_n) \mid a_1, \ldots, a_n \in M_A, f_{M_A}(a_1, \ldots, a_n) \text{ defined}\}$ *and* $T_{M_B} = \{f(a_1, \ldots, a_n) \mid a_1, \ldots, a_n \in M_B, f_{M_B}(a_1, \ldots, a_n) \text{ defined}\}$ *of terms which are defined in* M_A *resp.* M_B *are closed under the operator* W*, i.e.* $W(T_{M_A}, T_{M_B}) \subseteq T_{M_A}$ *and* $W(T_{M_B}, T_{M_A}) \subseteq T_{M_B}$*;*

there exists a model M_D *of* $\mathcal{T}_0 \cup \mathcal{K}$ *and weak embeddings* $h_A : M_A \to M_D, h_B : M_B \to M_D$ *such that* $h_A|_{M_C} = h_B|_{M_C}$*.*

Theorem 26. *Assume that* $\mathcal{T}_0$ *is a first-order theory and let* $\mathcal{K}$ *be a set of clauses over* $\Pi_0 \cup \Sigma$*. If* $\mathcal{T}_0 \cup \mathcal{K}$ *has the partial amalgamation property for models with the same* Π_0*-reduct then* $\mathcal{T}_0 \cup \mathcal{K}$ *is* W*-separable.*

Proof: Proof similar to the one in [25] (but we dropped some hypotheses). □

Example 27. In [25] it was proved that the theory of arrays with difference function and the theory of linked lists with reachability have partial amalgamation, hence are W-separable (for suitable versions of W, described in [25]).

Theorem 28. *Assume that* $\mathcal{T}_0$ *is a first-order theory which allows general ground interpolation and has the property that for every set* Σ *of additional function symbols and ground* $\Sigma_0 \cup \Sigma$*-formulae* A, B*, the interpolant* I *contains only* Σ*-terms in* $W(A, B) \cap W(B, A)$*. Let* $\mathcal{K}$ *be a set of clauses over* $\Pi_0 \cup \Sigma$*, such that all variables occur below an extension symbol.*

If $\mathcal{T} = \mathcal{T}_0 \cup \mathcal{K}$ *is* W*-separable then it has the partial amalgamation property for models with the same* Π_0*-reduct.*

Proof: The proof is given in [24]. □

Separability, Locality and Interpolant Computation. We now show that if the extension $\mathcal{T}_0 \subseteq \mathcal{T}_0 \cup \mathcal{K}$ is W-separable and $\mathcal{T}_0$ has ground interpolation, then we can hierarchically compute interpolants in $\mathcal{T}_0 \subseteq \mathcal{T}_0 \cup \mathcal{K}$.

Theorem 29. *Assume that the theory* $\mathcal{T}_0$ *has general ground interpolation, and there is a method for effectively computing general ground interpolants w.r.t.* $\mathcal{T}_0$*. Let* $\mathcal{T}_0 \cup \mathcal{K}$ *be a* W*-separable extension of* $\mathcal{T}_0$ *with a set of clauses* $\mathcal{K}$ *in which every variable occurs below an extension function. Let* A *and* B *be two ground* $\Sigma_0 \cup \Sigma$*-formulae. Assume that* $A \wedge B \models_{\mathcal{T}_0 \cup \mathcal{K}} \perp$*. Then we can effectively compute a ground interpolant for* A *and* B*, by computing an interpolant of* $\mathcal{K}[W(A, B)] \cup A$ *and* $\mathcal{K}[W(B, A)] \cup B$*.*

Proof: By W-separability $A \wedge B \models_{\mathcal{T}_0 \cup \mathcal{K}} \perp$ iff $\mathcal{K}[W(A, B)] \cup A \cup \mathcal{K}[W(B, A)] \cup B \models_{\mathcal{T}_0} \perp$, a ground interpolation problem; it can be shown that the interpolant I_0 w.r.t. $\mathcal{T}_0$ is an interpolant for A and B w.r.t. $\mathcal{T}_0 \cup \mathcal{K}$. □

Corollary 30. *Let* $\mathcal{T}_0 \cup \mathcal{K}$ *be a* W*-separable extension of* $\mathcal{T}_0$ *with a set of clauses* $\mathcal{K}$ *in which every variable occurs below an extension function. Then* $\mathcal{T}_0 \cup \mathcal{K}$ *has ground interpolation in each of the following cases:*

(1) $\mathcal{T}_0$ *has ground interpolation and is equality interpolating.*
(2) $\mathcal{T}_0$ *allows quantifier elimination and is equality interpolating.*

(3) $\mathcal{T}_0$ is universal and allows quantifier elimination.

Ground Interpolation and Model Completions. It is sometimes difficult to check directly whether the theory $\mathcal{T}_0$ has ground interpolation. If $\mathcal{T}_0$ has a model completion with good properties, this becomes easier to check (we can then also use quantifier elimination in the model completion to compute the interpolant).

Theorem 31. *Let $\mathcal{T}_0 \cup \mathcal{K}$ be a W-separable extension of $\mathcal{T}_0$ with a set of clauses $\mathcal{K}$ in which every variable occurs below an extension function. Assume that $\mathcal{T}_0$ has a model companion $\mathcal{T}_0^*$ with the following properties: (i) $\mathcal{T}_0 \subseteq \mathcal{T}_0^*$; (ii) every model of $\mathcal{T}_0 \cup \mathsf{UIF}_\Sigma$ embeds into a model of $\mathcal{T}_0^* \cup \mathsf{UIF}_\Sigma$; and (iii) $\mathcal{T}_0^*$ has general ground interpolation. (This can happen for instance when $\mathcal{T}_0^*$ allows quantifier elimination and is equality interpolating.)*

Then $\mathcal{T}_0 \cup \mathcal{K}$ has ground interpolation.

Proof: The proof is given in [24]. □

Example 32. Consider the theory in Example 24. Let $A : d \leq g(a) \wedge a \leq c$ and $B : b \leq d \wedge c < f(b)$. It is easy to see that $A \wedge B \models_{\mathcal{T}_0 \cup \mathcal{K}} \perp$. To show this, as $\mathcal{T}_0 \cup \mathcal{K}$ is a local extension of $\mathcal{T}_0$, after instantiation and purification we obtain:

Extension $D_A \wedge D_B$	Base $A_0 \wedge B_0 \wedge \mathsf{SGc}_0 \wedge \mathsf{Mon}_0 \wedge \mathsf{Con}_0$	
$a_1 \approx g(a)$	$A_0 = d \leq a_1 \wedge a \leq c$	$\mathsf{SGc}_0 = b \leq a_1 \rightarrow b_1 \leq a$
$b_1 \approx f(b)$	$B_0 = b \leq d \wedge c < b_1$	$\mathsf{Con}_A \wedge \mathsf{Mon}_A = a \lhd a \rightarrow a_1 \lhd a_1, \lhd \in \{\approx, \leq\}$
		$\mathsf{Con}_B \wedge \mathsf{Mon}_B = b \lhd b \rightarrow b_1 \lhd b_1,\ \lhd \in \{\approx, \leq\}$

Then $A_0 \wedge B_0 \models b \leq a_1$, and we can show that $A_0 \wedge B_0 \models b \leq d \wedge d \leq a_1$, where d is shared. $W(A,B) = \{a, c, d, g(a), f(d)\}$ and $W(B,A) = \{b, c, d, f(b), f(d)\}$. After W-separation and purification (using d_1 for $f(d)$) we obtain:

(1) $\overline{A}_0 = d \leq a_1 \wedge a \leq c \wedge (d \leq a_1 \rightarrow d_1 \leq a)$ equiv. to $(d \leq a_1 \wedge a \leq c \wedge d_1 \leq a)$
(2) $\overline{B}_0 = b \leq d \wedge c < b_1 \wedge (b \leq d \rightarrow b_1 \leq d_1)$ equiv. to $(b \leq d \wedge c < b_1 \wedge b_1 \leq d_1)$

We can use a method for ground interpolation in TOrd to obtain an interpolant I_0. However, it might be more efficient to use QE in the model completion of TOrd (the theory of dense total orderings without endpoints) for eliminating the constants a, a_1 from $\overline{A}_0$. We obtain the interpolant $I_0 = d_1 \leq c$. Since d_1 is an abbreviation for $f(d)$, we replace it back and obtain the interpolant $I = f(d) \leq c$.

Symbol Elimination and Interpolation. For W-separable theories we can use the method for symbol elimination in Sect. 5 for computing interpolants. If $\mathcal{T}_0 \cup \mathcal{K}[W(A,B)] \cup A \cup \mathcal{K}[W(B,A)] \cup B \models \perp$, the formula Γ_2 obtained using Steps 1–4 in Sect. 5 for $\mathcal{T}_0 \cup \mathcal{K}[W(A,B)] \cup A$ (with Σ_P consisting of the common constants) is an interpolant. The following Theorem is proved in [24].

Theorem 33. *If $\mathcal{T}_0 \cup \mathcal{K}[W(A,B)] \cup A \cup \mathcal{K}[W(B,A)] \cup B \models \perp$, the formula Γ_2 obtained using Steps 1–4 in Sect. 5 for $\mathcal{T}_0 \cup \mathcal{K}[W(A,B)] \cup A$ (with Σ_P consisting of the common constants) is an interpolant of A and B w.r.t. $\mathcal{T}_0 \cup \mathcal{K}$.*

Example 34. Consider the theory $\mathcal{T}_0 \cup \mathcal{K}$ in Example 18. Let $A := \{c_1 \leq c_2, g(c_1) = a_1, g(c_2) = a_2, a_1 > a_2\}$ and $B := \{c_1 \leq c \leq c_2, f(c_1) = b_1, f(c_2) = b_2, b_1 \leq b_2\}$. It is easy to check that $\mathcal{T}_0 \cup \mathcal{K} \cup A \cup B \models \perp$. We can compute an interpolant as follows. Let Γ_2 be the formula computed in Step 4 in Example 18, namely: $((c_1{<}c_2 \leq c \wedge f(c_1){>}f(c_2)) \vee (c_1{\leq}c < c_2 \wedge f(c_1){>}h(c_2)) \vee (c{<}c_1{<}c_2 \wedge h(c_1){>}h(c_2)))$. By Theorem 33, this formula is an interpolant of A and B.

7 Conclusions

In this paper we studied several problems related to symbol elimination and ground interpolation in theories and theory extensions. It is well-known that if a theory has quantifier elimination then this can be used for symbol elimination and also for computing ground interpolants of ground formulae. However, the great majority of logical theories do not have quantifier elimination. We showed that if a theory $\mathcal{T}$ has a model completion $\mathcal{T}^*$, then ground interpolants computed w.r.t. $\mathcal{T}^*$ are also interpolants w.r.t. $\mathcal{T}$. As there are many examples of model completions of theories $\mathcal{T}$ which allow quantifier elimination, this can be used for computing interpolants w.r.t. $\mathcal{T}$. We analyzed how this approach can be lifted to *extensions* of a theory $\mathcal{T}$, by identifying situations in which we can use existing methods for symbol elimination in $\mathcal{T}$ for symbol elimination or for ground interpolation in the extension. If $\mathcal{T}$ has a model completion $\mathcal{T}^*$, we analyzed under which conditions we can use possibilities of symbol elimination in $\mathcal{T}^*$ for such tasks. In the study of ground interpolation in extensions $\mathcal{T} \cup \mathcal{K}$ of a theory $\mathcal{T}$ with a set of clauses $\mathcal{K}$ we followed an approach proposed in [25], in which the terms needed to separate the instances of $\mathcal{K}$ are described using a closure operator. Our analysis extends both the results in [21] and those in [25] mainly by avoiding the restriction to convex base theories. (In addition, when formulating our theorems we explicitly pointed out all conditions needed for hierarchical interpolation which were missing or only implicit in [25]). In future work we would like to also apply these ideas in the study of *uniform interpolation* in logical theories and theory extensions.

Acknowledgments. Many thanks to the reviewers for their helpful comments. This work was partly supported by the German Research Council (DFG) as part of the Transregional Collaborative Research Center "Automatic Verification and Analysis of Complex Systems" (SFB/TR 14 AVACS) www.avacs.org.

References

1. Bacsich, P.: Amalgamation properties and interpolation theorem for equational theories. Algebra Universalis **5**, 45–55 (1975)
2. Bruttomesso, R., Ghilardi, S., Ranise, S.: Quantifier-free interpolation in combinations of equality interpolating theories. ACM Trans. Comput. Log. **15**(1), 1–34 (2014)
3. Chang, C., Keisler, J.: Model Theory. North-Holland, Amsterdam (1990)

4. Craig, W.: Linear reasoning. A new form of the Herbrand-Gentzen theorem. J. Symb. Log. **22**(3), 250–268 (1957)
5. Ganzinger, H.: Relating semantic and proof-theoretic concepts for polynomial time decidability of uniform word problems. In: Logic in Computer Science, LICS 2001, pp. 81–92. IEEE Computer Society Press (2001)
6. Ghilardi, S.: Model-theoretic methods in combined constraint satisfiability. J. Autom. Reasoning **33**(3–4), 221–249 (2004)
7. Hoder, K., Kovács, L., Voronkov, A.: Interpolation and symbol elimination in Vampire. In: Giesl, J., Hähnle, R. (eds.) IJCAR 2010. LNCS, vol. 6173, pp. 188–195. Springer, Heidelberg (2010)
8. Hodges, W.: A Shorter Model Theory. Cambridge University Press, Cambridge (1997)
9. Ihlemann, C., Jacobs, S., Sofronie-Stokkermans, V.: On local reasoning in verification. In: Ramakrishnan, C.R., Rehof, J. (eds.) TACAS 2008. LNCS, vol. 4963, pp. 265–281. Springer, Heidelberg (2008)
10. Ihlemann, C., Sofronie-Stokkermans, V.: On hierarchical reasoning in combinations of theories. In: Giesl, J., Hähnle, R. (eds.) IJCAR 2010. LNCS, vol. 6173, pp. 30–45. Springer, Heidelberg (2010)
11. Kapur, D., Majumdar, R., Zarba, C.G.: Interpolation for data structures. In: 14th ACM SIGSOFT International Symposium Foundations of Software Engineering, pp. 105–116. ACM (2006)
12. Krajícek, J.: Interpolation theorems, lower bounds for proof systems, and independence results for bounded arithmetic. J. Symb. Log. **62**(2), 457–486 (1997)
13. Mal'cev, A.: Axiomatizable classes of locally free algebras of various types. In: The Metamathematics of Algebraic Systems. Collected Papers: 1936–1967. Studies in Logic and the Foundation of Mathematics, vol. 66, chap. 23. North-Holland, Amsterdam (1971)
14. McAllester, D.: Automatic recognition of tractability in inference relations. J. ACM **40**(2), 284–303 (1993)
15. McMillan, K.L.: Interpolation and SAT-based model checking. In: Hunt Jr., W.A., Somenzi, F. (eds.) CAV 2003. LNCS, vol. 2725, pp. 1–13. Springer, Heidelberg (2003)
16. McMillan, K.L.: An interpolating theorem prover. In: Jensen, K., Podelski, A. (eds.) TACAS 2004. LNCS, vol. 2988, pp. 16–30. Springer, Heidelberg (2004)
17. McMillan, K.L.: Applications of Craig interpolants in model checking. In: Halbwachs, N., Zuck, L.D. (eds.) TACAS 2005. LNCS, vol. 3440, pp. 1–12. Springer, Heidelberg (2005)
18. Rybalchenko, A., Sofronie-Stokkermans, V.: Constraint solving for interpolation. J. Symb. Comput. **45**(11), 1212–1233 (2010)
19. Sofronie-Stokkermans, V.: Hierarchic reasoning in local theory extensions. In: Nieuwenhuis, R. (ed.) CADE 2005. LNCS (LNAI), vol. 3632, pp. 219–234. Springer, Heidelberg (2005)
20. Sofronie-Stokkermans, V.: Hierarchical and modular reasoning in complex theories: the case of local theory extensions. In: Konev, B., Wolter, F. (eds.) FroCos 2007. LNCS (LNAI), vol. 4720, pp. 47–71. Springer, Heidelberg (2007)
21. Sofronie-Stokkermans, V.: Interpolation in local theory extensions. Log. Methods Comput. Sci. **4**(4), 1–31 (2008)
22. Sofronie-Stokkermans, V.: Hierarchical reasoning for the verification of parametric systems. In: Giesl, J., Hähnle, R. (eds.) IJCAR 2010. LNCS, vol. 6173, pp. 171–187. Springer, Heidelberg (2010)

23. Sofronie-Stokkermans, V.: Hierarchical reasoning and model generation for the verification of parametric hybrid systems. In: Bonacina, M.P. (ed.) CADE 2013. LNCS, vol. 7898, pp. 360–376. Springer, Heidelberg (2013)
24. Sofronie-Stokkermans, V.: On interpolation and symbol elimination in theory extensions. AVACS Technical Report 102, SFB/TR 14 AVACS (2016)
25. Totla, N., Wies, T.: Complete instantiation-based interpolation. In: Giacobazzi, R., Cousot, R. (eds.) POPL 2013. ACM (2013)

First-Order Theorem Proving

System Description: GAPT 2.0

Gabriel Ebner[1(✉)], Stefan Hetzl[1], Giselle Reis[2], Martin Riener[3], Simon Wolfsteiner[1], and Sebastian Zivota[1]

[1] Vienna University of Technology, Vienna, Austria
gebner@gebner.org, {stefan.hetzl,simon.wolfsteiner}@tuwien.ac.at, sebastian.zivota@mailbox.org

[2] Inria & LIX/École Polytechnique, Palaiseau, France
giselle.reis@inria.fr

[3] Inria Nancy & MSR-Inria Joint Centre, Palaiseau, France
riener@logic.at

Abstract. GAPT (General Architecture for Proof Theory) is a proof theory framework containing data structures, algorithms, parsers and other components common in proof theory and automated deduction. In contrast to automated and interactive theorem provers whose focus is the construction of proofs, GAPT concentrates on the transformation and further processing of proofs. In this paper, we describe the current 2.0 release of GAPT.

1 Introduction

This paper describes the system GAPT (General Architecture for Proof Theory). GAPT is a versatile proof theory framework containing data structures, algorithms, parsers and other components common in proof theory and automated deduction. In contrast to automated and interactive theorem provers whose focus is the construction of proofs, GAPT concentrates on the transformation and further processing of proofs.

We are convinced that such a system is of importance to computational proof theory and automated deduction because of the growing interest in the *output* of provers. It is no longer enough for a prover to answer with yes or no as often a proof object (or a countermodel) is sought for further processing. For example, the use of SAT-solvers for solving various problems in NP needs the solver to return a propositional interpretation or a refutation as certificate of unsatisfiability. The use of interpolation in software verification needs proofs (or interpolants) as output. The use of automated reasoning systems in proof assistants—e.g., Sledgehammer in Isabelle—needs to provide proofs to incorporate in the proof script. This change in role of automated theorem provers is also reflected in the growing interest in proof certificates, e.g., in research projects like ProofCert [16], Dedukti [4], in common formats for proofs shared

Supported by the Vienna Science Fund (WWTF) project VRG12-04, the Austrian Science Fund (FWF) projects P25160 and W1255-N23, and the ERC Advanced Grant *ProofCert*.

N. Olivetti and A. Tiwari (Eds.): IJCAR 2016, LNAI 9706, pp. 293–301, 2016.
DOI: 10.1007/978-3-319-40229-1_20

between provers like TPTP derivations [21] and OpenTheory [13] and in conference series like Certified Programs and Proofs (CPP). It is also reflected in CASC (the CADE ATP System Competition) evaluating theorem provers considering the number of problems solved presenting a solution, i.e. a proof [20].

GAPT provides a rich reservoir of functionality for the transformation and further processing of formal proofs in a uniform framework. GAPT contains interfaces to a variety of automated reasoning systems including first-order provers, SAT-solvers and SMT-solvers. Thus it provides a platform which is well-suited not only for computational proof theory but also for the cooperation of automated provers.

GAPT has been used as an environment to experiment with the implementation of several specific algorithms and tools: cut elimination by resolution [2,3,10], post-processing of resolution proofs [12,14], cut introduction [7–9,11], and inductive theorem proving based on tree grammars [6]. In addition to these applications, the graphical user interface has been described in detail in [5,14] and GAPT's use of expansion trees for proof import in [17]. Using a single system for these applications had a synergistic effect since all these algorithms share a common basis. This basis has been developed and extended into the GAPT system which has now reached a level of maturity to be of interest for its own sake. We mark this occasion by the release of version 2.0[1] and the first system description of GAPT as a whole. GAPT is implemented in Scala and licensed under the GNU General Public License. It is available at https://logic.at/gapt.

2 Features

Formulas. Terms and formulas are uniformly represented as expressions in a simply typed lambda calculus with multiple base sorts. This representation allows considerable code reuse: for example, substitutions are only defined once for terms, atoms, formulas, etc. While these are all represented as lambda expressions, they are each instance of a more specific Scala type as well: `FOLAtom` is a subtype of `HOLFormula`, which is in turn a subtype of `LambdaExpression`. These Scala types are determined at run-time using smart constructors. In this way, we support type-safe programming with defined subsets of `LambdaExpression`. GAPT allows arbitrary Unicode strings as names for constants, variables, predicate symbols, etc.

Proofs. GAPT contains an implementation of a standard sequent calculus LK for classical higher-order logic as well as a version of the sequent calculus using Skolem terms instead of eigenvariables (LKsk, see [10] for details). In addition, it contains resolution calculi: $\mathcal{R}_{al}$ (see [10]) which is a labelled variant of Andrew's $\mathcal{R}$ [1] and a standard first-order resolution calculus. GAPT

[1] For a list of changes and new features in the 2.0 release specifically, please refer to the release notes: https://github.com/gapt/gapt/blob/master/RELEASE-NOTES.md.

also contains expansion proofs [15], a generalisation of the notion of Herbrand-disjunction to arbitrary formulas in higher-order logic. The proof objects in these calculi are automatically validated during the construction of each inference, preventing ill-formed proofs. This eager validation has been highly valuable for the early detection of bugs. Our main focus is on tree-like proofs in first and higher-order logic, these usually have less than 1000 inferences. In resolution, GAPT can work with dag-like proofs of about 10000 inferences.

Algorithms. GAPT contains a number of basic algorithms like transformations between the above-mentioned proof calculi, Skolemisation and regularisation of sequent calculus proofs, naive and structural first-order clause normal form transformations, proof pruning, etc. More advanced algorithms include: Gentzen-style cut elimination in the sequent calculus, interpolation in first-order proofs and a built-in tableaux prover for (classical) propositional logic as a quick way to generate propositional sequent calculus proofs.

First-order Theorem Proving. GAPT interfaces with several first-order theorem provers: it can invoke and import proofs from Vampire, the E prover, and Prover9. There is specific proof import code for Prover9, which successfully imports more than 99 % of the Prover9 solutions in the TSTP [19] as GAPT resolution proofs. In addition, there is a general purpose import for TPTP-proofs based on proof replay, which currently imports 34 % of the FOF and CNF solutions in the TSTP from a total of 12 different provers. We are currently also developing leanCoP-specific import code [17] to have reliable import for non-resolution first-order provers.

SAT- and SMT-solving. GAPT is able to export formulas as SMT-LIB benchmarks and can check their satisfiability modulo `QF_UF` with an arbitrary SMT-LIB compliant SMT-solver. This interface natively supports many-sorted logic, and works with at least Z3, CVC4, and veriT. Proof import is implemented for the `QF_UF` logic for veriT, see [17]. For propositional formulas, GAPT writes DIMACS files and can use any DIMACS-compliant SAT-solver to check their satisfiability and import satisfying assignments. We support Glucose, Sat4j, and miniSAT out of the box—adding support for other solvers usually only requires specifying the executable path. In addition, GAPT also provides an interface to solvers for the MaxSAT optimization problem, such as OpenWBO and the MaxSAT solver in Sat4j.

User Interfaces. GAPT comes with two user interfaces: the system's full functionality is available via a customised Scala shell, thus providing a flexible and scriptable command-line interface. In addition, GAPT provides a graphical user interface, prooftool, to conveniently display large proofs and other objects. For example, prooftool also includes a viewer for expansion trees with a point-and-click interface to selectively expand quantifiers, see [12]. Large proofs in LK can be visualized using a so-called Sunburst viewer [14]. Sunburst visualisations are radial, space-filling representations of hierarchical information [18]: instead of a tree, the inferences in a proof are displayed as concentric rings.

These logics, proof systems and interfaces with other provers are not intended to be a final fixed set of features in GAPT. The system's architecture allows the implementation of extensions, so new logics and proof systems can be added as they become necessary while having a versatile collection of tools readily available for tests and analysis.

3 Example

Figure 1 shows a first-order prover utilising the GAPT API. This example is not meant to implement a practically relevant, efficient, or short prover, but to illustrate the features provided by GAPT. The prover continuously generates new instances of clauses in the input clause set by unifying literals of opposite polarity. The `done` set contains the clauses where the pairwise unifiers have already been computed, in each iteration we pick a clause from the `todo` queue and unify it with all clauses in `done`. When the set of instances becomes propositionally unsatisfiable (which we check using the Sat4j SAT solver[2]), we minimize the number of instances using `minimalExpansionSequent` and convert the instances to a proof in LK using `ExpansionProofToLK`. The resulting proof is then displayed in a GUI window using `prooftool`.

Utilising the functionality already provided by GAPT, we can concentrate on the actual algorithm, while the interface and "glue code" is already implemented for us, such as:

- formula parsing
- robust structural clausification (including Skolemisation)
- unification, matching and substitution
- SAT solver interface
- proof construction (and validation)
- proof simplification
- graphical visualisation of the resulting proof

This example prover can be immediately executed from the binary distribution of GAPT[3], without installing any other extra dependencies. It will read the problem from standard input (Fig. 2), refute it, and then open the resulting proof in the graphical user interface (Fig. 3):

```
./gapt.sh instprover.scala <example.in
```

This usage of GAPT scripts is convenient for early prototyping. But should our prototype develop into a larger project, we are not stuck with developing

[2] We use Sat4j as it is bundled with GAPT. To use another solver, it is enough to replace `Sat4j` with `Glucose` or `MiniSAT` in the source code.

[3] This example is included in the `examples/scriptability` directory in the binary distribution of GAPT.

```
import scala.collection.mutable

// Parse input
val sequent = Stream.continually(Console.in.readLine()).
  takeWhile(_ != null).map(_.trim).filter(_.nonEmpty).
  map(parseFormula).map(univclosure(_)) ++: Sequent()

// Transform into clause normal form
val (cnf, justifications, definitions) = structuralCNF(sequent,
    generateJustifications = true, propositional = false)

// Main loop
val done = mutable.Set[FOLClause]()
val todo = mutable.Queue[FOLClause](cnf.toSeq: _*)
while (Sat4j solve (done ++ todo) isDefined) {
  val next = todo.dequeue()
  if (!done.contains(next)) for {
    clause2 <- done
    clause1 = FOLSubstitution(
      rename(freeVariables(next), freeVariables(clause2)))(next)
    (atom1,index1) <- clause1.zipWithIndex.elements
    (atom2,index2) <- clause2.zipWithIndex.elements
    if !index2.sameSideAs(index1)
    mgu <- syntacticMGU(atom1, atom2)
  } todo ++= Seq(mgu(clause1), mgu(clause2))
  done += next
}
// Postprocessing
val instances = for (clause <- cnf) yield clause ->
  (for { inst <- done ++ todo
         subst <- syntacticMatching(
           clause.toFormula, inst.toFormula)
       } yield subst).toSet
val expansion = expansionProofFromInstances(
  instances.toMap, sequent, justifications, definitions)
val Some(minimized) = minimalExpansionSequent(expansion, Sat4j)
val lkProof = ExpansionProofToLK(minimized)

// Visualisation
prooftool(lkProof)
```

Fig. 1. instprover.scala: Instantiation-based first-order prover with graphical proof output

```
p(0,y) & (p(x,f(y)) -> p(s(x),y))
(p(x,c) -> q(x,g(x))) & (q(x,y) -> r(x)) & -r(s(s(s(s(0)))))
```

Fig. 2. example.in: Example input for the prover from Fig. 1

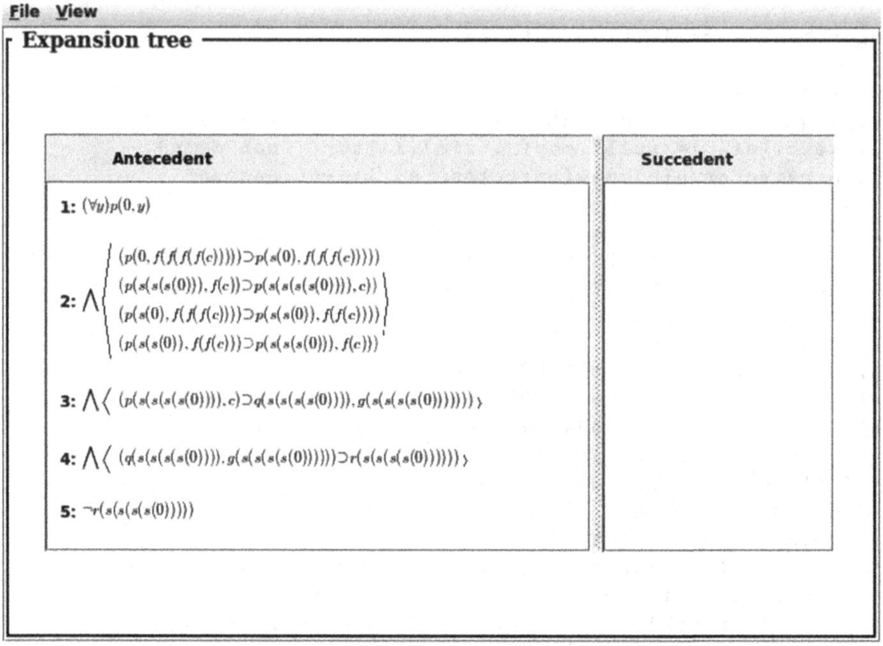

Fig. 3. Graphical visualisation of the resulting proof using the expansion tree viewer.

it as a single file. Since GAPT is available as a Scala library from the JCenter repository, it can be added as a dependency for another project by adding a single line to its `sbt` build script. This way, we can seamlessly move from a small prototype to a full-fledged separate project.

4 Applications

We have used GAPT primarily as a basis for prototype implementations of newly developed algorithms. We briefly review these applications here, highlighting the aspects of the GAPT-system which are of particular relevance.

Cut Elimination by Resolution (CERES). This is a method for cut elimination which is based on using a resolution theorem prover to generate a skeleton structure for a cut-free proof. This method has been applied to show that Fürstenberg's topological proof of the infinity of primes can be transformed into Euclid's original proof by cut elimination [2]. The CERES method depends heavily on several non-trivial proof transformation like the Skolemisation of proofs with cut or the combination of a resolution refutation of a clause set $\mathcal{C}$ with cut-free sequent calculus proofs ψ_C of $\Gamma \vdash \Delta \circ C$ for $C \in \mathcal{C}$ to a sequent calculus proof with only atomic cuts. To analyse the results produced by CERES, expansion tree extraction and visualisation is used.

Cut Introduction. GAPT has been used as basis for the implementation of a method for cut introduction (i.e., lemma generation) [8,9,11]. This method is based on a structural analysis of expansion trees using tree grammars. It relies heavily on the flexible use of expansion trees and on the interface to MaxSAT-solvers which are used to compute minimal tree grammars. Hence GAPT also contains an implementation of some tree grammars. As database for the testing and evaluation this algorithm, the TSTP and the reliable Prover9 import have proved indispensable.

Inductive Theorem Proving Based on Tree Grammars. Currently, GAPT is used for a prototype implementation of an inductive theorem prover based on the method described in [6]. This being a generalisation of the method for cut introduction to induction, it also benefits from the availability of proof transformations and the flexible handling of expansion trees and tree grammars as described above. In addition, for this application, the use of resolution provers and SMT-solvers for generating instance proofs is necessary.

Teaching. GAPT is used, along with several automated theorem provers, in a graduate course on automated deduction taught at the Vienna University of Technology. The students are asked to perform various computational experiments relating run-time, size of output, and other parameters of various algorithms. For example: naive clause form transformation by distributivity vs. Tseitin transformation, a SAT-solver on sequences of propositional tautologies of varying proof complexity, a first-order resolution prover vs. a SAT-solver on a propositional clause set and on ground instances of a first-order clause set. Such comparisons crucially rely on having a uniform framework with interfaces to different automated reasoning systems.

5 Future Work and Conclusion

We are currently implementing a tactics language for the more convenient input of formal proofs which will make it into the next release. A built-in superposition-based theorem prover will be included in the next release as well, enabling more efficient proof replay without external dependencies. As further future work, we plan to implement support for a wider variety of different proof calculi (e.g., natural deduction) and logics (e.g., intuitionistic logic). In addition, we are looking to extend the existing support for multiple uninterpreted base sorts to interpreted sorts such as integers and arrays, and interface them with the built-in theories of SMT solvers. We will continue to use the system for the applications described in Sect. 4.

The power of GAPT comes from the integration of a wide variety of different systems (e.g. SAT-solvers, SMT-solvers, resolution and connection provers) and the flexibility of combining them using a large number of standard algorithms and transformations, all within one uniform framework. GAPT is developed in Scala which, on the one hand, permits elegant functional code close to mathematical definitions, but on the other hand also provides access to the whole Java library, including, e.g., Swing, on which prooftool is based. GAPT has already proved very useful for the development of and experiments with new algorithms in computational proof theory and automated deduction and we are convinced that it will continue to do so.

While GAPT already interfaces with a large number of external provers, we always try to expand our support to other provers. As next steps we plan to add support for the DRUP format used by SAT solvers, and to add proof import for first-order provers that employ inferences rules that go beyond the standard resolution calculus, such as the splitting rule in SPASS. Adding support for a new prover takes a considerable amount of work, ranging from minute details such as recognizing different headers in the output files to supporting new proof systems. We hope that GAPT will benefit from further efforts in the standardisation of proof output.

Acknowledgements. The authors would like to thank the following students, researchers, and software developers for their contributions to the development of GAPT (in alphabetic order): Alexander Birch, Cvetan Dunchev, Alexander Leitsch, Tomer Libal, Bernhard Mallinger, Olivier Roland, Mikheil Rukhaia, Christoph Spörk, Janos Tapolczai, Daniel Weller, and Bruno Woltzenlogel Paleo.

References

1. Andrews, P.B.: Resolution in type theory. J. Symbolic Log. **36**(3), 414–432 (1971). doi:10.2307/2269949
2. Baaz, M., Hetzl, S., Leitsch, A., Richter, C., Spohr, H.: CERES: an analysis of fürstenberg's proof of the infinity of primes. Theoret. Comput. Sci. **403**(2–3), 160–175 (2008)
3. Baaz, M., Leitsch, A.: Cut-elimination and redundancy-elimination by resolution. J. Symbolic Comput. **29**(2), 149–176 (2000)
4. Boespflug, M., Carbonneaux, Q., Hermant, O.: The $\lambda\Pi$-calculus modulo as a universal proof language. In: Pichardie, D., Weber, T. (eds.) Proceedings of PxTP2012: Proof Exchange for Theorem Proving, pp. 28–43 (2012)
5. Dunchev, C., Leitsch, A., Libal, T., Riener, M., Rukhaia, M., Weller, D., Paleo, B.W.: PROOFTOOL: a GUI for the GAPT framework. In: Kaliszyk, C., Lüth, C. (eds.) Proceedings 10th International Workshop on User Interfaces for Theorem Provers (UITP) 2012, EPTCS, vol. 118, pp. 1–14 (2012)
6. Eberhard, S., Hetzl, S.: Inductive theorem proving based on tree grammars. Ann. Pure Appl. Log. **166**(6), 665–700 (2015)
7. Hetzl, S.: Project presentation: algorithmic structuring and compression of proofs (ASCOP). In: Jeuring, J., Campbell, J.A., Carette, J., Dos Reis, G., Sojka, P., Wenzel, M., Sorge, V. (eds.) CICM 2012. LNCS, vol. 7362, pp. 438–442. Springer, Heidelberg (2012)
8. Hetzl, S., Leitsch, A., Reis, G., Tapolczai, J., Weller, D.: Introducing quantified cuts in logic with equality. In: Demri, S., Kapur, D., Weidenbach, C. (eds.) IJCAR 2014. LNCS, vol. 8562, pp. 240–254. Springer, Heidelberg (2014)
9. Hetzl, S., Leitsch, A., Reis, G., Weller, D.: Algorithmic introduction of quantified cuts. Theoret. Comput. Sci. **549**, 1–16 (2014)

10. Hetzl, S., Leitsch, A., Weller, D.: CERES in higher-order logic. Ann. Pure Appl. Log. **162**(12), 1001–1034 (2011)
11. Hetzl, S., Leitsch, A., Weller, D.: Towards algorithmic cut-introduction. In: Bjørner, N., Voronkov, A. (eds.) LPAR-18 2012. LNCS, vol. 7180, pp. 228–242. Springer, Heidelberg (2012)
12. Hetzl, S., Libal, T., Riener, M., Rukhaia, M.: Understanding resolution proofs through Herbrand's theorem. In: Galmiche, D., Larchey-Wendling, D. (eds.) TABLEAUX 2013. LNCS, vol. 8123, pp. 157–171. Springer, Heidelberg (2013)
13. Hurd, J.: The OpenTheory standard theory library. In: Bobaru, M., Havelund, K., Holzmann, G.J., Joshi, R. (eds.) NFM 2011. LNCS, vol. 6617, pp. 177–191. Springer, Heidelberg (2011)
14. Libal, T., Riener, M., Rukhaia, M.: Advanced Proof Viewing in ProofTool. In: Benzmüller, C., Paleo, B.W. (eds.) Proceedings of the 11th Workshop on User Interfaces for Theorem Provers (UITP) 2014, EPTCS, vol. 167, pp. 35–47 (2014)
15. Miller, D.: A compact representation of proofs. Stud. Logica **46**(4), 347–370 (1987)
16. Miller, D.: ProofCert: broad spectrum proof certificates. An ERC Advanced Grant funded for the five years 2012–2016. http://www.lix.polytechnique.fr/Labo/Dale.Miller/ProofCert.pdf
17. Reis, G.: Importing SMT and connection proofs as expansion trees. In: Kaliszyk, C., Paskevich, A. (eds.) Proceedings Fourth Workshop on Proof eXchange for Theorem Proving (PxTP), EPTCS, vol. 186, pp. 3–10 (2015)
18. Stasko, J., Zhang, E.: Focus+context display and navigation techniques for enhancing radial, space-filling hierarchy visualizations. In: IEEE Symposium on Information Visualization, 2000, InfoVis 2000, pp. 57–65 (2000)
19. Sutcliffe, G.: The TPTP world – infrastructure for automated reasoning. In: Clarke, E.M., Voronkov, A. (eds.) LPAR-16 2010. LNCS, vol. 6355, pp. 1–12. Springer, Heidelberg (2010)
20. Sutcliffe, G., Suttner, C.: The State of CASC. AI Commun. **19**(1), 35–48 (2006)
21. Sutcliffe, G., Schulz, S., Claessen, K., Van Gelder, A.: Using the TPTP language for writing derivations and finite interpretations. In: Furbach, U., Shankar, N. (eds.) IJCAR 2006. LNCS (LNAI), vol. 4130, pp. 67–81. Springer, Heidelberg (2006)

nanoCoP: A Non-clausal Connection Prover

Jens Otten[1,2](✉)

[1] Institut für Informatik, University of Potsdam, August-Bebel-Str. 89, 14482 Potsdam-Babelsberg, Germany
jeotten@leancop.de
[2] Department of Informatics, University of Oslo, PO Box 1080 Blindern, 0316 Oslo, Norway

Abstract. Most of the popular efficient proof search calculi work on formulae that are in clausal form, i.e. in disjunctive or conjunctive normal form. Hence, most state-of-the-art fully automated theorem provers require a translation of the input formula into clausal form in a preprocessing step. Translating a proof in clausal form back into a more readable non-clausal proof of the original formula is not straightforward. This paper presents a non-clausal theorem prover for classical first-order logic. It is based on a non-clausal connection calculus and implemented with a few lines of Prolog code. By working entirely on the original structure of the input formula, the resulting non-clausal proofs are not only shorter, but can also be more easily translated into, e.g., sequent proofs. Furthermore, a non-clausal proof search is more suitable for some non-classical logics.

1 Introduction

Automated theorem proving in classical first-order logic is a core research area in the field of Automated Reasoning. Most efficient fully automated theorem provers implement proof search calculi that require the input formula to be in a *clausal form*, i.e. disjunctive or conjunctive normal form. In the core first-order category "FOF" at the most recent ATP competition, CASC-25, only the Muscadet prover implements a proof search that works on the original formula structure. First-order formulae that are not in this *clausal* form are translated into clausal form in a preprocessing step. While the use of a clausal form technically simplifies the proof search and the required data structures, it also has some disadvantages. The standard translation into clausal form as well as the definitional translation [19], which introduces definitions for subformulae, introduce a significant overhead for the proof search [14]. Furthermore, a translation into clausal form modifies the structure of the original formula and the translation of the clausal proof back into one of the original formula is not straightforward [20]. On the other hand, fully automated theorem provers that use non-clausal calculi, such as standard tableau or sequent calculi, are usually not suitable for an efficient proof search.

N. Olivetti and A. Tiwari (Eds.): IJCAR 2016, LNAI 9706, pp. 302–312, 2016.
DOI: 10.1007/978-3-319-40229-1_21

The present paper describes the non-clausal connection prover nanoCoP for classical first-order logic. By performing the proof search on the original structure of the input formula, it combines the advantages of more *natural* non-clausal provers with a more efficient *goal-oriented* connection-based proof search. The prover is based on a non-clausal connection calculus for classical first-order logic [15] (Sect. 2) that generalizes the clausal connection (tableau) calculus [4,5]. This non-clausal calculus is implemented in a very compact way (Sect. 3) following the *lean* methodology. An experimental evaluation (Sect. 4) shows a solid performance of nanoCoP.

2 The Non-clausal Connection Calculus

The standard notation for first-order formulae is used. Terms (denoted by t) are built up from functions (f, g, h, i), constants (a, b, c), and variables (x, y, z). An atomic formula (denoted by A) is built up from predicate symbols (P, Q, R, S) and terms. A *(first-order) formula* (denoted by F, G, H) is built up from atomic formulae, the connectives $\neg$, $\wedge$, $\vee$, $\Rightarrow$, and the standard first-order quantifiers $\forall$ and $\exists$. A *literal* L has the form A or $\neg A$. Its *complement* $\overline{L}$ is A if L is of the form $\neg A$; otherwise $\overline{L}$ is $\neg L$.

A *connection* is a set $\{A, \neg A\}$ of literals with the same predicate symbol but different polarity. A *term substitution* σ assigns terms to variables. A formula in *clausal form* has the form $\exists x_1 \ldots \exists x_n (C_1 \vee \ldots \vee C_n)$, where each clause C_i is a conjunction of literals $L_1, \ldots, L_{m_i}$. It is usually represented as a set of clauses $\{C_1, \ldots, C_n\}$, which is called a (clausal) *matrix*. The *polarity* 0 or 1 is used to represent negation in a matrix, i.e. literals of the form A and $\neg A$ are represented by A^0 and A^1, respectively,

The non-clausal connection calculus uses non-clausal matrices. In a non-clausal matrix a clause consists of literals *and* (sub)matrices. Let F be a formula and *pol* be a polarity. The *non-clausal matrix* $M(F^{pol})$ of a formula F^{pol} is a set of clauses, in which a clause is a set of literals and (sub-)matrices, and is defined inductively according to Table 1. In Table 1, x^* is a new variable, t^* is the skolem term $f^*(x_1, \ldots, x_n)$ in which f^* is a new function symbol and $x_1, \ldots, x_n$ are the free variables in $\forall x G$ or $\exists x G$. The *non-clausal matrix* $M(F)$ of a formula F is the matrix $M(F^0)$. In the *graphical representation* its clauses are arranged horizontally, while the literals and (sub-)matrices of each clause are arranged vertically.

For example, the formula $F_\#$

$$P(a) \wedge (\neg((Q(f(f(c))) \wedge \forall x(Q(f(x)) \Rightarrow Q(x))) \Rightarrow Q(c)) \vee \forall y(P(y) \Rightarrow P(g(y)))) \Rightarrow \exists z\, P(g(g(z)))$$

has the simplified (i.e. redundant brackets are removed) non-clausal matrix $M_\# = M(F_\#)$:

$$\{\{P(a)^1\}, \{\{\{Q(f(f(c)))^1\}, \{Q(f(x))^0, Q(x)^1\}, \{Q(c)^0\}\}, \{\{P(y)^0, P(g(y))^1\}\}\}, \{P(g(g(z)))^0\}\}.$$

The graphical representation of the matrix $M_\#$ is depicted in Fig. 1. It already contains two clause copies using the fresh variables x' and y' and represents

Table 1. The definition of the non-clausal matrix

Type	F^{pol}	$M(F^{pol})$
Atomic	A^0	$\{\{A^0\}\}$
	A^1	$\{\{A^1\}\}$
α	$(\neg G)^0$	$M(G^1)$
	$(\neg G)^1$	$M(G^0)$
	$(G \wedge H)^1$	$\{\{M(G^1)\},\{M(H^1)\}\}$
	$(G \vee H)^0$	$\{\{M(G^0)\},\{M(H^0)\}\}$
	$(G \Rightarrow H)^0$	$\{\{M(G^1)\},\{M(H^0)\}\}$
β	$(G \wedge H)^0$	$\{\{M(G^0), M(H^0)\}\}$
	$(G \vee H)^1$	$\{\{M(G^1), M(H^1)\}\}$
	$(G \Rightarrow H)^1$	$\{\{M(G^0), M(H^1)\}\}$
γ	$(\forall x G)^1$	$M(G[x\backslash x^*]^1)$
	$(\exists x G)^0$	$M(G[x\backslash x^*]^0)$
δ	$(\forall x G)^0$	$M(G[x\backslash t^*]^0)$
	$(\exists x G)^1$	$M(G[x\backslash t^*]^1)$

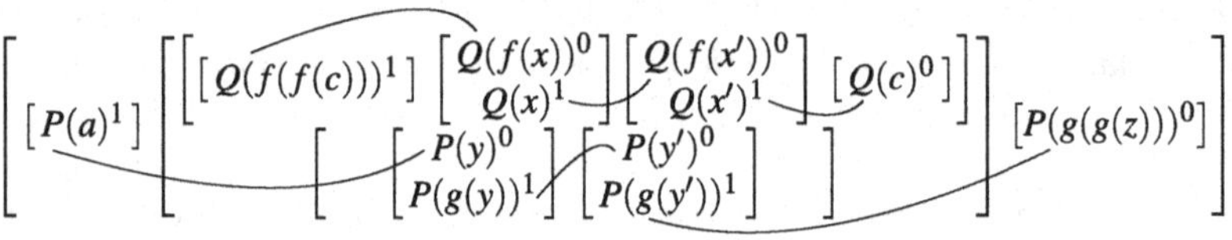

Fig. 1. Graphical representation of a non-clausal matrix and its non-clausal connection proof

a non-clausal connection proof, in which the literals of each connection are connected with a line, using the substitution σ with $\sigma(x) = f(c)$, $\sigma(x') = c$, $\sigma(y) = \sigma(z) = a$, $\sigma(y') = g(a)$.

The axiom and the rules of the *non-clausal connection calculus* [15] are given in Fig. 2. It works on tuples "$C, M, Path$", where M is a non-clausal matrix, C is a (subgoal) clause or ε and (the active) $Path$ is a set of literals or ε; σ is a term substitution. A *non-clausal connection proof* of M is a non-clausal connection proof of $\varepsilon, M, \varepsilon$.

The non-clausal connection calculus for classical logic is *sound* and *complete* [15]. It has the same axiom, start rule, and reduction rule as the formal *clausal* connection calculus [18]. The extension rule is slightly modified and a decomposition rule is added. A few additional concepts are required as follows in order to specify which clauses C_1 can be used within the non-clausal extension rule. See [15] for details and examples.

A clause C *contains* a literal L if and only if *(iff)* $L \in C$ or C' contains L for a matrix $M' \in C$ with $C' \in M$. A clause C is *α-related* to a literal L iff

Axiom (A)	$\frac{}{\{\},M,Path}$	*Start (S)*	$\frac{C_2,M,\{\}}{\varepsilon,M,\varepsilon}$ and C_2 is copy of $C_1 \in M$
Reduction (R)	$\frac{C,M,Path \cup \{L_2\}}{C \cup \{L_1\},M,Path \cup \{L_2\}}$	and $\sigma(L_1)=\sigma(\overline{L_2})$	
Extension (E)	$\frac{C_3,M[C_1 \backslash C_2],Path \cup \{L_1\} \quad C,M,Path}{C \cup \{L_1\},M,Path}$		and C_3:=β-$clause_{L_2}(C_2)$, C_2 is copy of C_1, C_1 is e-clause of M wrt. $Path \cup \{L_1\}$, C_2 contains L_2 with $\sigma(L_1)=\sigma(\overline{L_2})$
Decomposition (D)	$\frac{C \cup C_1,M,Path}{C \cup \{M_1\},M,Path}$	and $C_1 \in M_1$	

Fig. 2. The non-clausal connection calculus

$\{C', C''\} \subseteq M'$ for a clause C' and matrix M' such that C' contains L and C'' contains C. A *copy of the clause* C in the matrix M is made by renaming all *free variables* in C. $M[C_1 \backslash C_2]$ denotes the matrix M, in which the clause C_1 is replaced by the clause C_2. C' is a *parent clause* of C iff $M' \in C'$ and $C \in M'$ for some matrix M'. C is an *extension clause (e-clause) of the matrix M with respect to* a set of literals $Path$, only if either (a) C contains a literal of $Path$, or (b) C is α-related to all literals of $Path$ occurring in M and if C has a parent clause, it contains a literal of $Path$. In the *β-clause* of C_2 with respect to L_2, denoted by β-$clause_{L_2}(C_2)$, L_2 and all clauses that are α-related to L_2 are deleted from C_2, as these clauses do not need to be considered in the subgoal clause C_3 in the premise of the extension rule.

The analytic, i.e. bottom-up *proof search* in the non-clausal calculus is carried out in the same way as in the clausal calculus. Additional *backtracking* might be required when choosing C_1 in the decomposition rule; no backtracking is required when choosing M_1. The *rigid* term substitution σ is calculated whenever a connection is identified in an application of the reduction or extension rule. On formulae in clausal form, the non-clausal connection calculus coincides with the clausal connection calculus. *Optimization techniques*, such as positive start clauses, regularity, lemmata and restricted backtracking, can be employed in a way similar to the clausal connection calculus [14].

3 The Implementation

The implementation of the non-clausal connection calculus of Fig. 2 follows the *lean methodology* [3], which is already used for the clausal connection prover leanCoP [18]. It uses very compact Prolog code to implement the basic calculus and adds a few essential optimization techniques in order to prune the search space. The resulting *na*tural *no*nclausal *co*nnection *p*rover nanoCoP is available under the GNU General Public License and can be downloaded at http://www.leancop.de/nanocop/.

Non-clausal Matrices. In a first step the input formula F is translated into a non-clausal (indexed) matrix $M(F)$ according to Table 1; redundant brackets of the form "$\{\{\ldots\}\}$" are removed [15]. Additionally, every (sub-)clause $(I,V):C$ and (sub-)matrix $J:M$ is marked with a unique *index* I and J; clause C is also marked with a set of variables V that are newly introduced in C but not in any subclause of C. Atomic formulae are represented by Prolog atoms, term variables by Prolog variables and the polarity 1 by "-". Sets, e.g. clauses and matrices, are represented by Prolog lists (representing multisets). For example, the matrix $M_{\#}$ from Sect. 2 is represented by the Prolog term

```
[(1^K)^[]:[-(p(a))],
 (2^K)^[]:[3^K:[(4^K)^[]:[-(q(f(f(c))))],(5^K)^[X]:[q(f(X)), -(q(X))],
                (6^K)^[]:[q(c)]], 7^K:[(8^K)^[Y]:[p(Y), -(p(g(Y)))]]],
 (9^K)^[Z]:[p(g(g(Z)))]]
```

in which the Prolog variable `K` is used to enumerate clause copies. In the second step the matrix $M = M(F)$ is written into Prolog's database. For every literal `Lit` in M the fact

```
lit(Lit,ClaB,ClaC,Grnd)
```

is asserted into the database where `ClaC` $\in M$ is the (largest) clause in which `Lit` occurs and `ClaB` is the β-clause of `ClaC` with respect to `Lit`. `Grnd` is set to `g` if the smallest clause in which `Lit` occurs is ground, i.e. does not contain any variables; otherwise `Grnd` is set to `n`. Storing literals of M in the database in this way is called *lean Prolog technology* [14] and integrates the advantages of the Prolog technology approach [23] into the lean theorem proving framework. No other modifications or simplifications of the original formula (structure) are done during these two preprocessing steps.

Non-clausal Proof Search. The nanoCoP source code is shown in Fig. 3. It uses only the standard Prolog predicates `member`, `append`, `length`, `assert`, `retract`, `copy_term`, `unify_with_occurs_check`, and the additional predicate `positiveC(Cla,Cla1)`, which returns the clause `Cla1` in which all clauses that are not positive in `Cla` are deleted. A clause is positive if all of its elements (matrices and literals) are positive; a matrix is positive if it contains at least one positive clause; a literal is positive if its polarity is 0.

The predicate `prove(Mat,PathLim,Set,Proof)` implements the start rule (lines *1–8*). `Mat` is the matrix generated in the preprocessing step. `PathLim` is the maximum size of the active path used for iterative deepening, `Set` is a list of options used to control the proof search, and `Proof` contains the returned connection proof. Start clauses are restricted to positive clauses (line *2*) before the actual proof search is invoked (line *3*). If no proof is found with the current active path limit `PathLim` and this limit was reached, then `PathLim` is increased and the proof search starts over again (lines *4–8*).

The predicate `prove(Cla,Mat,Path,PathI,PathLim,Lem,Set,Proof)` implements the axiom (line 9), the decomposition rule (lines 10–14), the reduction rule (lines 15–18, 21–22, 31), and the extension rule (lines 15–18, 24–42) of the non-clausal connection calculus in Fig. 2. `Cla`, `Mat`, and `Path` represent the subgoal clause C, the (indexed) matrix M and the (active) *Path*. The *indexed path* `PathI` contains the indices of all clauses and matrices that contain literals of `Path`; it is used for calculating extension clauses. The list `Lem` is used for the lemmata rule and contains all literals that have been "solved" already [14]. `Set` is a list of options and may contain the elements "`cut`" and "`comp(`I`)`" for $I \in I\!N$, which are used to control the restricted backtracking technique [14]. This `prove` predicate succeeds iff there is a connection proof for the tuple (`Cla`, `Mat`, `Path`) with $|$`Path`$|<$ `PathLim`. In this case `Proof` contains a compact connection proof. The input matrix `Mat` has to be stored in Prolog's database (as explained above).

When the decomposition rule is applied, a clause `Cla1` of the first matrix of the subgoal clause `[J:Mat|Cla]` is selected (line 11). The search continues with clause `Cla1` (line 12) using the extended indexed path `[I,J|PI]`, and the remaining elements of `Cla` (line 13). For the reduction and extension rules the complement `NegLit` of the first literal `Lit` of the subgoal clause is calculated (line 18) and used for the following reduction and extension step. When the reduction rule is applied, it is checked whether the active `Path` contains a literal `NegL` that unifies with `NegLit` (line 21). In this case the proof search continues with the clause `Cla` for the premise of the reduction rule (line 31). When the extension rule is applied, the predicate `lit(NegLit,ClaB,Cla,Grnd1)` is called to find a clause in Prolog's database that contains the complement `NegLit` of the literal `Lit` (line 24). For this operation sound unification has to be switched on (in, e.g., ECLiPSe Prolog this is done by calling "`set_flag(occur_check,on)`" before the proof search starts). The predicate `prove_ec` calculates an appropriate extension clause and returns its β-clause `ClaB1` with respect to `NegLit` (line 27). The proof search continues with `ClaB1` as new subgoal clause for the left premise of the extension rule with the literal `Lit` added to the active `Path` (line 28), and with the remaining subgoal clause `Cla` for the right premise (line 31). The substitution σ is stored implicitly by Prolog.

The predicate `prove_ec(ClaB,Cla1,Mat,ClaB1,Mat1)` is used to calculate extension clauses (lines 32–42). Starting with the (largest possible) extension clause `Cla1`, its β-clause `ClaB`, and the current (indexed) matrix `Mat`, this predicate returns an appropriate extension clause `Cla`, copies it into `Mat` and returns its β-clause `ClaB1` and the new matrix `Mat1`. The extension clause has to fulfil the conditions described in Sect. 2: it has to be (a) large enough to contain a literal of `Path` or (b) small enough to be α-related to all literals of `Path` occurring in `Mat` and again large enough that in case it has a parent clause, this contains a literal of `Path`; in both cases the extension clause has to be large enough such that the literal `Lit` unifies with the literal `NegLit` in the current matrix. As an optimization only extension clauses that introduce new variables are considered.

Prolog depth-first search results in an incomplete proof search. In order to regain completeness nanoCoP performs an *iterative deepening* on the size of the active path. When the extension rule is applied and the extension clause is not ground, it is checked whether the size `K` of the active `Path` exceeds the current path limit `PathLim` (line 25). In this case the predicate `pathlim` is written into

```
         % start rule
(1)      prove(Mat,PathLim,Set,[(I^0)^V:Cla1|Proof]) :-
(2)          member((I^0)^V:Cla,Mat), positiveC(Cla,Cla1), Cla1\=!,
(3)          prove(Cla1,Mat,[],[I^0],PathLim,[],Set,Proof).
(4)      prove(Mat,PathLim,Set,Proof) :-
(5)          retract(pathlim) ->
(6)          ( member(comp(PathLim),Set) -> prove(Mat,1,[],Proof) ;
(7)            PathLim1 is PathLim+1, prove(Mat,PathLim1,Set,Proof) ) ;
(8)          member(comp(_),Set) -> prove(Mat,1,[],Proof).

         % axiom
(9)      prove([],_,_,_,_,_,_,[]).

         % decomposition rule
(10)     prove([J:Mat1|Cla],MI,Path,PI,PathLim,Lem,Set,Proof) :- !,
(11)         member(I^_:Cla1,Mat1),
(12)         prove(Cla1,MI,Path,[I,J|PI],PathLim,Lem,Set,Proof1),
(13)         prove(Cla,MI,Path,PI,PathLim,Lem,Set,Proof2),
(14)         append(Proof1,Proof2,Proof).

         % reduction and extension rules
(15)     prove([Lit|Cla],MI,Path,PI,PathLim,Lem,Set,Proof) :-
(16)         Proof=[[I^V:[NegLit|ClaB1]|Proof1]|Proof2], copy_term(Lit,LitV),
(17)         \+ (member(LitC,[Lit|Cla]), member(LitP,Path), LitC==LitP),
(18)         (-NegLit=Lit;-Lit=NegLit) ->
(19)            ( member(LitL,Lem), Lit==LitL, ClaB1=[], Proof1=[]
(20)              ;
(21)              member(NegL,Path), unify_with_occurs_check(NegL,NegLit),
(22)              ClaB1=[], Proof1=[]
(23)              ;
(24)              lit(NegLit,ClaB,Cla1,Grnd1),
(25)              ( Grnd1=g -> true ; length(Path,K), K<PathLim -> true ;
(26)                \+ pathlim -> assert(pathlim), fail ),
(27)              prove_ec(ClaB,Cla1,MI,PI,I^V:ClaB1,MI1),
(28)              prove(ClaB1,MI1,[Lit|Path],[I|PI],PathLim,Lem,Set,Proof1)
(29)            ),
(30)            ( (member(cut,Set);Lit==LitV) -> ! ; true ),
(31)            prove(Cla,MI,Path,PI,PathLim,[Lit|Lem],Set,Proof2).

         % extension clause (e-clause)
(32)     prove_ec((I^K)^V:ClaB,IV:Cla,MI,PI,ClaB1,MI1) :-
(33)         append(MIA,[(I^K1)^V1:Cla1|MIB],MI), length(PI,K),
(34)         ( ClaB=[J^K:[ClaB2]|_], member(J^K1,PI),
(35)           unify_with_occurs_check(V,V1), Cla=[_:[Cla2|_]|_],
(36)           append(ClaD,[J^K1:MI2|ClaE],Cla1),
(37)           prove_ec(ClaB2,Cla2,MI2,PI,ClaB1,MI3),
(38)           append(ClaD,[J^K1:MI3|ClaE],Cla3),
(39)           append(MIA,[(I^K1)^V1:Cla3|MIB],MI1)
(40)           ;
(41)           (\+member(I^K1,PI);V\==V1;V\=[]) ->
(42)           ClaB1=(I^K)^V:ClaB, append(MIA,[IV:Cla|MIB],MI1) ).
```

Fig. 3. Source code of the nanoCoP prover

Prolog's database (line 26) indicating the need to increase the path limit if the proof search fails for the current path limit.

nanoCoP uses additional optimization techniques that are already used in the classical (clausal) connection prover leanCoP [14]: *regularity* (line 17), *lemmata* (line 19), and *restricted backtracking* (line 30). *Regularity* ensures that no literal occurs more than once in the active path. The idea of *lemmata* (or factorization) is to reuse subproofs during the proof search. *Restricted backtracking* is a very effective technique for pruning the search space in connection calculi [14]. It is switched on if the list `Set` contains the element "`cut`". If it also contains "`comp(`I`)`" for $I \in I\!N$, then the proof search restarts again without restricted backtracking if the path limit `PathLim` exceeds I.

4 Experimental Evaluation

These evaluations were conducted on a 3.4 GHz Xeon system with 4 GB of RAM running Linux 3.13.0 and ECLiPSe Prolog 5.10, and a CPU time limit of 100 s.

The following formula F_n is a slightly extended version of the formula $F_{\#}$ in Sect. 2, where f^n, g^n, h^n, and i^n are abbreviations for n nested applications of these functions:

$$
\begin{aligned}
F_n \equiv\ & P(a) \ \wedge\ (\neg((Q(f^n(c)) \wedge \forall x(Q(f(x)) \Rightarrow Q(x))) \Rightarrow Q(c)) \\
& \vee \neg((R(h^n(c)) \wedge \forall x(R(h(x)) \Rightarrow R(x))) \Rightarrow R(c)) \vee \neg((S(i^n(c)) \wedge \forall x(S(i(x)) \Rightarrow S(x))) \Rightarrow S(c)) \\
& \vee\ \ \forall y(P(y) \Rightarrow P(g(y))))\ \ \Rightarrow\ \ \exists z\, P(g^n(z)) .
\end{aligned}
$$

Table 2 shows the results on this (valid) formula class for $n = 10$, $n = 30$, and $n = 90$ for the following provers: the lean (non-clausal) tableau prover leanTAP [3], the resolution prover Prover9 [12], the superposition prover E [22] (using options "`--auto --tptp3-format`"), leanCoP [14,18], and nanoCoP. The leanCoP *core prover* with the standard ("[nodef]") and the definitional ("[def]") translation into clausal form were tested. For nanoCoP restricted backtracking was switched off (`Set=[]`). Times are given in seconds; "size" is the number of nodes in the returned proof tree.

Table 2. Results on formula class F_n

$n=$	leanTAP	Prover9	E	leanCoP 2.2		nanoCoP
	2.3	2009-02A	1.9	[nodef]	[def]	[]
10 time (size)	0.17 (128)	–	1.22 (2916)	–	–	**0.09 (45)**
30 time (size)	–	–	84.57 (57628)	–	–	**0.12 (125)**
90 time (size)	–	–	–	–	–	**0.42 (365)**

Table 3. Results on TPTP library v3.7.0

	leanTAP	Prover9	E	leanCoP 2.2			nanoCoP 1.0	
	2.3	2009-02A	1.9	[nodef]	[def]	"full"	[]	[cut,comp(6)]
Proved	404	1611	2782	1134	1065	1710	**1232**	**1485**
0 to 1 sec.	379	1285	2104	938	865	1215	**1001**	**1172**
1 to 10 sec.	13	200	338	113	125	216	**139**	**157**
10 to 100 sec.	4	126	340	83	75	279	**92**	**156**

Table 3 shows the test results on all 5051 first-order (FOF) problems in the TPTP library v3.7.0 [24]. For leanTAP, leanCoP, and nanoCoP, the required equality axioms were added in a preprocessing step (which is included in the timings). The full leanCoP prover ("full") additionally uses strategy scheduling [14]. For nanoCoP, a restricted backtracking strategy, i.e. `Set=[cut,comp(6)]`, was tested as well. The nanoCoP core prover perform significantly better than both clausal form translations of the leanCoP core prover. 40 %/51 % of the proofs found by nanoCoP (without restricted backtracking) are shorter than those of leanCoP [nodef]/[def], respectively; as many of these problems are (mostly) in clausal form, 56 %/47 % of the proofs have the same size. The nanoCoP proofs are up to 96 %/74 % shorter than those of leanCoP [nodef]/[def], respectively. The classical version of the non-clausal connection prover JProver [21] has a lower performance than leanTAP (also reflected in its intuitionistic performance [13]).

5 Conclusion

This paper presents nanoCoP, a non-clausal connection prover for classical first-order logic. Using non-clausal matrices the proof search works directly on the original structure of the input formula. No translation steps to any clausal or other normal form are required. This combines the advantages of more *natural* non-clausal tableau or sequent provers with the goal-oriented *efficiency* of connection provers.

Even though the non-clausal inferences introduce a slight overhead, nanoCoP outperforms both clausal form translations of the leanCoP core prover on a large set of TPTP problems. It is expected that the integration of strategy scheduling will also outperform the "full" leanCoP prover. About half of the returned non-clausal proofs are up to 96 % shorter than their clausal counterparts. By using the standard translation, i.e. applying the distributive laws, the size of the resulting formula might grow exponentially with respect to the size of the original formula, which is not feasible for some formulae. The definitional translation [19] introduces definitions for subformulae, which results in a significant overhead for the proof search as well [14]. Other clausal form translations that work well for resolution provers, e.g. the ones used in E or Flotter, have a significant lower performance when used in with a connection prover [14].

Both clausal form translations modify the structure of the original formula, which makes it difficult to translate the (clausal) proof back into a proof of the original formula [20]. nanoCoP returns a compact non-clausal connection proof, which directly represents a free-variable tableau proof. A connection in the nanoCoP proof corresponds to a closed branch in the tableau calculus [7] or an axiom in the sequent calculus [6]. The translation into, e.g., a sequent proof is straightforward, when skolemization is seen as a way to encode the eigenvariable condition of the sequent calculus. This close relationship to the sequent calculus makes nanoCoP an ideal tool to be used within interactive proof systems, such as Coq, Isabelle, HOL or NuPRL. The compact size of nanoCoP makes it also a suitable tool for the development of verifiably correct software [17], as its correctness can be proven much more easily than that of a large proof system consisting of several thousand lines of source code.

Only few research work investigates non-clausal connection calculi and their implementations. Other non-clausal calculi [1,5,8,11] work only on ground formulae. For first-order formulae, copies of subformulae have to be added iteratively, which introduces a huge redundancy into the proof search, as already observed with JProver [21]. For an efficient proof search, clauses have to be added dynamically *during* the proof search. Some older non-clausal implementations [9] are not available anymore.

Another important application of nanoCoP is its usage within non-classical logics, such as intuitionistic or modal *first-order* logic, for which the use of a clausal form is either not desirable or not possible. Hence, future work includes the combination of the non-clausal approach with the prefix (unification) technique for some non-classical logics, as already done for leanCoP [13,16]. In order to improve performance, further optimization techniques need to be integrated into nanoCoP, such as strategy scheduling [14], learning [10] or variable splitting [2].

Acknowledgements. The author would like to thank Wolfgang Bibel for his helpful comments on a preliminary version of this paper.

References

1. Andrews, P.B.: Theorem proving via general matings. J. ACM **28**, 193–214 (1981)
2. Antonsen, R., Waaler, A.: Liberalized variable splitting. J. Autom. Reasoning **38**, 3–30 (2007)
3. Beckert, B., Posegga, J.: leanTAP: lean, tableau-based deduction. J. Autom. Reasoning **15**(3), 339–358 (1995)
4. Bibel, W.: Matings in matrices. Communications of the ACM, pp. 844–852 (1983)
5. Bibel, W.: Automated Theorem Proving, 2nd edn. Vieweg, Wiesbaden (1987)
6. Gentzen, G.: Untersuchungen über das logische Schließen. Math. Z. **39**, 176–210 (1935)
7. Hähnle, R.: Tableaux and related methods. In: Handbook of Automated Reasoning, pp. 100–178. Elsevier, Amsterdam (2001)
8. Hähnle, R., Murray, N., Rosenthal, E.: Linearity and regularity with negation normal form. Theoret. Comput. Sci. **328**, 325–354 (2004)
9. Issar, S.: Path-focused duplication: a search procedure for general matings. In: AAAI 1990 Proceedings, pp. 221–226 (1990)
10. Kaliszyk, C., Urban, J.: FEMaLeCoP: fairly efficient machine learning connection prover. In: Davis, M., et al. (eds.) LPAR-20 2015. LNCS, vol. 9450, pp. 88–96. Springer, Heidelberg (2015)
11. Kreitz, C., Otten, J.: Connection-based theorem proving in classical and non-classical logics. J. Univ. Comput. Sci. **5**, 88–112 (1999)
12. McCune, W.: Release of prover9. Mile high conference on quasigroups, loops and nonassociative systems (2005)

13. Otten, J.: leanCoP2.0 and ileanCoP1.2: high performance lean theorem proving in classical and intuitionistic logic (system descriptions). In: Armando, A., Baumgartner, P., Dowek, G. (eds.) IJCAR 2008. LNCS (LNAI), vol. 5195, pp. 283–291. Springer, Heidelberg (2008)
14. Otten, J.: Restricting backtracking in connection calculi. AI Commun. **23**, 159–182 (2010)
15. Otten, J.: A non-clausal connection calculus. In: Brünnler, K., Metcalfe, G. (eds.) TABLEAUX 2011. LNCS, vol. 6793, pp. 226–241. Springer, Heidelberg (2011)
16. Otten, J.: MleanCoP: a connection prover for first-order modal logic. In: Demri, S., Kapur, D., Weidenbach, C. (eds.) IJCAR 2014. LNCS, vol. 8562, pp. 269–276. Springer, Heidelberg (2014)
17. Otten, J., Bibel, W.: Advances in connection-based automated theorem proving. In: Bowen, J., Hinchey, M., Olderog, E.-R. (eds.) Provably Correct Systems (to appear)
18. Otten, J., Bibel, W.: leanCoP: lean connection-based theorem proving. J. Symbolic Comput. **36**, 139–161 (2003)
19. Plaisted, D., Greenbaum, S.: A structure-preserving clause form translation. J. Symbolic Comput. **2**, 293–304 (1986)
20. Reis, G.: Importing SMT and connection proofs as expansion trees. In: Kaliszyk, C., Paskevich, A. (eds.) 4th Workshop on Proof eXchange for Theorem Proving (PxTP15), EPTCS 186, pp. 3–10 (2015)
21. Schmitt, S., Lorigo, L., Kreitz, C., Nogin, A.: *JProver*: integrating connection-based theorem proving into interactive proof assistants. In: Goré, R.P., Leitsch, A., Nipkow, T. (eds.) IJCAR 2001. LNCS (LNAI), vol. 2083, pp. 421–426. Springer, Heidelberg (2001)
22. Schulz, S.: E - a Brainiac theorem prover. AI Commun. **15**(2), 111–126 (2002)
23. Stickel, M.: A prolog technology theorem prover: implementation by an extended prolog compiler. J. Autom. Reasoning **4**, 353–380 (1988)
24. Sutcliffe, G.: The TPTP problem library and associated infrastructure: the FOF and CNF parts, v3.5.0. J. Autom. Reasoning **43**(4), 337–362 (2009)

Selecting the Selection

Kryštof Hoder, Giles Reger[1(✉)], Martin Suda[1], and Andrei Voronkov[1,2,3]

[1] University of Manchester, Manchester, UK
giles.reger@manchester.ac.uk
[2] Chalmers University of Technology, Gothenburg, Sweden
[3] EasyChair

Abstract. Modern saturation-based Automated Theorem Provers typically implement the *superposition calculus* for reasoning about first-order logic with or without equality. Practical implementations of this calculus use a variety of *literal selections* and *term orderings* to tame the growth of the search space and help steer proof search. This paper introduces the notion of *lookahead* selection that estimates (*looks ahead*) the effect of selecting a particular literal on the number of immediate children of the given clause and selects to minimize this value. There is also a case made for the use of *incomplete* selection strategies that attempt to restrict the search space instead of satisfying some completeness criteria. Experimental evaluation in the VAMPIRE theorem prover shows that both lookahead selection and incomplete selection significantly contribute to solving hard problems unsolvable by other methods.

1 Introduction

This paper considers the usage of literal selection strategies in practical implementations of the superposition calculus (and its extensions). The role of literal selection in arguments for completeness have been known for a long time [1], but there has been little written about their role in proof search. This paper is concerned with the properties of literal selections that lead to the *quick* proofs i.e. those that restrict proof search in a way that can make finding a proof quickly more likely. In fact, our disregard for completeness is strong enough to suggest *incomplete* literal selections as a fruitful route to such fast proofs. Our approach is based on the (experimental) observation that it is generally most helpful to perform inferences that lead to as few new clauses as possible. The main conclusion of this is a notion of *lookahead selection* that selects exactly the literal that is estimated to take part in as few inferences as possible.

The setting of this work is saturation-based first-order theorem provers based on the superposition calculus. These are predominant in the area of first-order theorem proving (see the latest iteration of the CASC competition [16]). Provers such as E [11], SPASS [18], and VAMPIRE [7,10] work by saturating a clause search space with respect

Kryštof Hoder's contribution was carried out while at the University of Manchester. This work was supported by EPSRC Grant EP/K032674/1. Martin Suda and Andrei Voronkov were partially supported by ERC Starting Grant 2014 SYMCAR 639270. Andrei Voronkov was also partially supported by the Wallenberg Academy Fellowship 2014 - TheProSE.

N. Olivetti and A. Tiwari (Eds.): IJCAR 2016, LNAI 9706, pp. 313–329, 2016.
DOI: 10.1007/978-3-319-40229-1_22

to an inference system (the superposition calculus) with the aim of deriving the empty clause (witnessing unsatisfiability of the initial clause set). Various techniques are vital to avoid explosion of the search space. Predominant among these is *redundancy elimination* (such as subsumption) used to remove clauses. One can also consider methods to restrict the number of generated clauses, this is where we will consider the role of *literal selection*. The idea is that inferences are only performed on selected literals and literals are selected in a way to restrict the growth of the search space. Another effect of literal selection is to avoid obtaining the same clauses by permutations of inferences.

For the resolution calculus there is a famous result about completeness with respect to selection and term orderings [1] that supposes properties of the selection strategy to construct a model given a saturated set of clauses. This result carries over to superposition. As a consequence, particular selections and orderings can be used to show decidability of certain fragments of first-order logic, see e.g. [3,5]. However, the requirements placed on selection by this completeness result are some times at odds with the aim of taming proof search. This paper presents different selection strategies (including the aforementioned lookahead selection) that aim to effectively control proof search and argues that dropping the completeness requirements can further this goal.

The main contributions of this paper can be summarised as follows: (a) we formulate a new version of the superposition calculus which captures the notion of incomplete selections while being general enough to subsume the standard presentation (Sect. 3); (b) we introduce *quality selections*, an easy to implement compositional mechanism for defining literal selections based on a notion of quality (Sect. 4), and (c) we introduce *lookahead selection* and describe how it can be efficiently implemented (Sect. 5). These ideas have been realised within VAMPIRE and complemented by several selections adapted from other theorem provers (Sect. 6). Our experimental evaluation (Sect. 7) shows that these new selections (incomplete and lookahead) are good at both solving the most problems overall and solving problems uniquely.

2 Preliminaries

We consider the standard first-order predicate logic with equality. Terms are of the form $f(t_1, \ldots, t_n)$, c or x where f is a *function symbol* of arity $n \geq 1$, $t_1, \ldots, t_n$ are terms, c is a zero arity function symbol (i.e. a constant) and x is a variable. The *weight* of a term t is defined as $w(t) = 1$ if t is a variable or a constant and as $w(t) = 1 + \sum_{i=1,\ldots,n} w(t_i)$ if t is of the form $f(t_1, \ldots, t_n)$. In other words, the weight or a term is the number of symbols in it. Atoms are of the form $p(t_1, \ldots, t_n)$, q or $t_1 \simeq t_2$ where p is a *predicate symbol* of arity n, $t_1, \ldots, t_n$ are terms, q is a zero arity predicate symbol and $\simeq$ is the *equality symbol*. The weight function naturally extends to atoms: $w(p(t_1, \ldots, t_n)) = 1 + \sum_{i=1,\ldots,n} w(t_i)$, $w(t_1 \simeq t_2) = w(t_1) + w(t_2)$, and $w(q) = 1$. A literal is either an atom A, in which case we call it *positive*, or a negation $\neg A$, in which case we call it *negative*. We write negated equalities as $t_1 \not\simeq t_2$. The weight of a literal is the weight of the corresponding atom. We write $t[s]_p$ and $L[s]_p$ to denote that a term s occurs in a term t (in a literal L) at a position p.

A *clause* is a disjunction of literals $L_1 \vee \ldots \vee L_n$ for $n \geq 0$. We disregard the order of literals and treat a clause as a multiset. When $n = 0$ we speak of the *empty clause*, which is always false. When $n = 1$ a clause is called a unit clause. Variables in clauses are considered to be universally quantified. Standard methods exist to transform an arbitrary first-order formula into clausal form.

A *substitution* is any expression θ of the form $\{x_1 \mapsto t_1, \ldots, x_n \mapsto t_n\}$, where $n \geq 0$, and $E\theta$ is the expression obtained from E by the simultaneous replacement of each x_i by t_i. By an expression here we mean a term, an atom, a literal, or a clause. An expression is *ground* if it contains no variables.

A *unifier* of two expressions E_1 and E_2 different from clauses is a substitution θ such that $E_1\theta = E_2\theta$. It is known that if two expressions have a unifier, then they have a so-called most general unifier. Let mgu be a function returning a most general unifier of two expressions if it exists.

A *simplification ordering* (see, e.g. [4]) on terms is an ordering that is *well-founded, monotonic, stable under substitutions* and has the *subterm property*. Such an ordering captures a notion of *simplicity* i.e. $t_1 \prec t_2$ implies that t_1 is in some way simpler than t_2. VAMPIRE uses the Knuth-Bendix ordering [6]. Such term orderings are usually total on ground terms and partial on non-ground ones. There is a simple extension of the term ordering to literals, the details of which are not relevant here.

3 The Superposition Calculus and Literal Selection

The superposition calculus as implemented in modern theorem provers usually derives from the work of Bachmair and Ganzinger [1] (see also [2,8]). There, the inference rules of the calculus come equipped with a list of side conditions which restrict the applicability of each rule. The rules are sound already in their pure form, but the additional side conditions are essential in practice as they prevent the clause search space from growing too fast. At the same time, it is guaranteed that the calculus remains refutationally complete, i.e. able to derive the empty clause from every unsatisfiable input clause set.

Here we are particularly interested in side conditions concerning individual literals within a clause on which an inference should be performed. The formulation by Bachmair and Ganzinger derives these conditions from a simplification ordering $\prec$ on terms and its extension to literals, and from a so called *selection function* S which assigns to each clause C a possibly empty multiset $S(C)$ of negative literals in C, which are called selected. The ordering and the selection function should be understood as parameters of the calculus.

The calculus is designed in such a way that an inference on a positive literal L within a clause C must only be performed when L is a maximal literal in C (i.e. there is no literal L' in C such that $L \prec L'$) and there is no selected literal in C. Complementarily, an inference on a negative literal L within a clause C must only be performed when L is a maximal literal in C and there is no selected literal in C or L is selected in C. Such conditions are shown to be compatible with completeness.

In this paper, we take a different perspective on literal selection. We propose the notion of a *literal selection strategy*, or *literal selection* for short, which is a procedure that assigns to a non-empty clause C a non-empty multiset of its literals. We avoid the use of the word "function" on purpose, since it is not guaranteed that we select the same

Resolution **Factoring**

$$\frac{\underline{A} \vee C_1 \quad \underline{\neg A'} \vee C_2}{(C_1 \vee C_2)\theta}, \qquad \frac{\underline{A} \vee A' \vee C}{(A \vee C)\theta},$$

where, for both inferences, $\theta = \mathsf{mgu}(A, A')$ *and* A *is not an equality literal*

Superposition

$$\frac{\underline{l \simeq r} \vee C_1 \quad \underline{L[s]_p} \vee C_2}{(L[r]_p \vee C_1 \vee C_2)\theta} \quad or \quad \frac{\underline{l \simeq r} \vee C_1 \quad \underline{t[s]_p \otimes t'} \vee C_2}{(t[r]_p \otimes t' \vee C_1 \vee C_2)\theta},$$

where $\theta = \mathsf{mgu}(l, s)$ *and* $r\theta \not\succeq l\theta$ *and, for the left rule* $L[s]$ *is not an equality literal, and for the right rule* $\otimes$ *stands either for* $\simeq$ *or* $\not\simeq$ *and* $t'\theta \not\succeq t[s]\theta$

EqualityResolution **EqualityFactoring**

$$\frac{\underline{s \not\simeq t} \vee C}{C\theta}, \qquad \frac{\underline{s \simeq t} \vee s' \simeq t' \vee C}{(t \not\simeq t' \vee s' \simeq t' \vee C)\theta},$$

where $\theta = \mathsf{mgu}(s, t)$ *where* $\theta = \mathsf{mgu}(s, s')$, $t\theta \not\succeq s\theta$, *and* $t'\theta \not\succeq s'\theta$

Fig. 1. The rules of the superposition and resolution calculus.

multiset even if the same clause occurs in a search space again after being deleted. In addition, we do not want the selection to depend just on the clause itself, but potentially also on a broader context including the current state of the search space.

We formulate the inference rules of superposition such that an inference on a literal within a clause is only performed when that literal is selected. This is evidently a simpler concept, which primarily decouples literal selection from completeness considerations as it also allows incomplete literal selection. At the same time, however, it is general enough so that completeness can be easily taken into account when a particular selection strategy is designed.

The Calculus. Our formulation of the superposition and resolution calculus with literal selection is presented in Fig. 1. It consists of the resolution and factoring rules for dealing with non-equational literals and the superposition, equality resolution and equality factoring rules for equality reasoning. Although resolution and factoring can be simulated by the remaining rules provided non-equational atoms are encoded in a suitable way, we prefer to present them separately, because they also have separate implementations in VAMPIRE for efficiency reasons.

The calculus in Fig. 1 is parametrised by a simplification ordering $\prec$ and a literal selection strategy, which we indicate here (and also in the rest of the paper) by underlining. In more detail, literals underlined in a clause must be selected by the strategy. Literals without underlying may be selected as well. Generally, inferences are only performed between selected literals with the exception of the two factoring rules.

There only one atom needs to be selected and factorings are performed with other unifiable atoms.

We remark that further restrictions on the calculus can be added on top of those mentioned in Fig. 1. In particular, if literal selection captures the maximality condition of a specific literal in a premise, this maximality may be required to also hold for the instance of the premise obtained by applying the $\mathsf{mgu}\ \theta$. We observed that these additional restrictions did not affect the practical performance of our prover in a significant way and for simplicity kept them disabled during our experiments.

We also note that the calculi based on the standard notion of selection function can be captured by our calculi – all we have to do is to select all maximal literals in clauses with no literals selected by the function.

Selection and Completeness. We now reformulate the previously mentioned side conditions on literals which are required by the completeness proof of Bachmair and Ganzinger [1] in terms of literal selection strategies. In the rest of the paper we refer to strategies satisfying the following *completeness condition* as *complete selections*:

$$\text{Select either a negative literal or all maximal literals with respect to } \prec. \qquad (1)$$

Although selections which violate Condition (1) cannot be used for showing satisfiability of a clause set by saturation, our experimental results will demonstrate that incomplete selections are invaluable ingredients for solving many problems.

As an example of what can happen if Condition (1) is violated, consider the following unsatisfiable set of clauses where *all* selected literals are underlined.

$$p \vee \underline{q} \qquad \underline{p} \vee \neg q \qquad \underline{\neg p} \vee q \qquad \neg p \vee \underline{\neg q}$$

Note that this set is clearly unsatisfiable as one can easily derive p and $\neg p$ and then the empty clause. However, using the given selection it is only possible to derive tautologies. The selection strategy does not fulfill the above requirements as either $p \succ q$ and p must be selected in $p \vee q$, or $q \succ p$ and $\neg q$ must be selected in $p \vee \neg q$.

4 Quality Selections

Vampire implements various literal selections in a uniform way, using preorders on literals, which try to reflect certain notions of quality. We convert such a preorder to a linear order by breaking ties in an arbitrary but fixed way. This order on literals (a *quality order*) induces two selections, one incomplete and one complete. Essentially, the incomplete one simply selects the literal greatest in this order and the complete one modifies the incomplete literal selection where the latter violates the sufficient conditions for completeness. We call the resulting class of selections *quality selections*. We believe that this is a new way of defining literal selections that has not been reported in the literature or observed in other systems before.

The preorders we use capture various notions of quality the literals we want to select should have. Let us now discuss what it is that we want to achieve from selection. The perfect selection strategy contains an oracle that knows the exact inferences necessary to derive the empty clause in the shortest possible time. Without such an oracle we can employ heuristics to suggest those inferences that are more desirable.

There is a general insight that a slowly growing search space is superior to a faster growing one, provided completeness is not compromised too much. It should be evident that a search space that grows too quickly will soon become unmanageable, reducing the likelihood that a proof is found. This has been repeatedly observed in practice. This insight holds despite the fact that the shortest proofs for some formulas may theoretically become much longer in the restricted (slowly growing) setting. *Therefore, the aim of a selection strategy in our setting is to generate the fewest new clauses.*

4.1 Quality Orderings

Let us consider several preorders $\rhd$ on literals that capture notions of preference for selection i.e. $l_1 \rhd l_2$ means we should prefer selecting l_1 to l_2. If they are equally preferable, that is $l_1 \rhd l_2$ and $l_2 \rhd l_1$, we will write $l_1 \equiv l_2$. We are interested in preorders that prefer literals having as few children as possible, this means decreasing the likelihood that we can apply the inferences in Fig. 1.

Unifiability. Firstly we note that all inferences require the selected literal (or one of its subterms) to unify with something in another clause. Therefore, we prefer literals that are potentially unifiable with fewer literals in the search space.

To this end, we first note that a heavy literal is likely to have a complex structure containing multiple function symbols. It is therefore unlikely that two heavy literals will be unifiable. This observation is slightly superficial because, for example, a literal $p(x_1, \ldots, x_n)$ for large n has a large weight but unifies with all negative literals containing p. Let $l_1 \rhd_{weight} l_2$ if the weight of l_1 is greater than the weight of l_2.

Next, we note that the fewer variables a literal contains the less chance it has to unify with other literals e.g. $p(f(x), y)$ will unify with every literal that $p(f(a), y)$ will unify with, and potentially many more. Let $l_1 \rhd_{vars} l_2$ if l_1 has fewer variables than l_2.

However, we can observe that not all variables are equal, the literal $p(x)$ will unify with more than $p(f(f(x)))$. As a simple measure of this we can consider only variables that occur at the top-level i.e. immediately below a predicate symbol. Let $l_1 \rhd_{top} l_2$ if l_1 has fewer top-level variables than l_2. Similarly, $p(f(x), f(y))$ will unify with more than $p(f(x), f(x))$ as the repetition of x constrains the unifier. To capture this effect we can prefer literals with fewer *distinct* variables. Let $l_1 \rhd_{dvar} l_2$ if l_1 has fewer distinct variables than l_2.

Equality and Polarity. We can observe from the inference rules in Fig. 1 that positive equality is required for superposition, which can be a prolific inference as it can rewrite inside a clause many times. Therefore, we should prefer not to select positive equality where possible. Let $L \rhd_{nposeq} s \simeq t$, where L is a non-equality literal, and $s \not\simeq t \rhd_{nposeq} s' \simeq t'$.

In a similar spirit, we observe that negative equality otherwise only appears in Equality Resolution which is in general a non-problematic inference as it is performed on a single clause and decreases the number of its literals. Therefore, in certain cases we should prefer negative equalities. Let $s \not\simeq t \rhd_{neq} L$ where L is a non-equality literal.

Finally, for non-equality literals it is best to default to selecting a single polarity as literals with the same polarity cannot resolve. Furthermore, selecting negative literals seems to be preferable as it keeps the corresponding selection strategy from compromising the completeness condition. We let $\neg A \rhd_{neg} A'$.

4.2 Quality-Based Selections

We want to compose different notions of quality so that we can break ties when the first notion is too coarse to distinguish literals. We define the composition of two preorders $\rhd_a$ and $\rhd_b$, denoted by $\rhd_a \circ \rhd_b$, by $l_1\ (\rhd_a \circ \rhd_b)\ l_2$ if and only $l_1 \rhd_a l_2$, or $l_1 \equiv_a l_2$ and $l_1 \rhd_b l_2$. Evidently, a composition of two preorders is also a preorder.

Given a preorder $\rhd$ we define a selection strategy $\pi_{\rhd}$ that selects the greatest (highest quality) literal with respect to $\rhd$ breaking ties arbitrarily, but in a deterministic fashion. We call such strategies *quality selections.*

4.3 Completing the Selection

Quality selections are not necessarily complete i.e. they do not satisfy the completeness Condition (1) introduced in Sect. 3. It is our hypothesis that these incomplete selection strategies are practically useful. However, there are cases where complete selection is desirable. One obvious example is where we are attempting to establish satisfiability.[1]

Given a quality selection $\pi_{\rhd}$, it is possible to also define a *complete* selection strategy using the following steps. Let N initially be the set of all literals in a clause and M be the subset of N consisting of all its literals maximal in the simplification ordering.

1. **If** $\pi_{\rhd}(N)$ is negative **then** select $\pi_{\rhd}(N)$
2. **If** $\pi_{\rhd}(N) \in M$ and all literals in M are positive **then** select M
3. **If** M contains a negative literal **then** set N to be the set of all negative literals in M and **goto** 1
4. Remove $\pi_{\rhd}(N)$ from N and **goto** 1

This attempts to, where possible, select a single negative literal that is maximal with respect to the quality ordering. The hypothesis being that it is always preferable to select a single negative literal rather than several maximal ones.

5 Lookahead Selection

In this section we introduce a general notion of *lookahead selection* and describe an efficient implementation of the idea. Our discussion in the previous section suggested that we try to find preorders that potentially minimize the number of children of a selected literal. Essentially, lookahead selection tries to select literals that result in the smallest number of children. Note that this idea requires a considerable change in the design and implementation, because the number of children depends on the current state of the search space rather than on measures using only the clause we are dealing with.

[1] It should be noted that VAMPIRE always knows when it is incomplete and therefore returns Unknown when obtaining a saturated set with the help of an incomplete strategy.

5.1 Given-Clause Algorithms and Term Indexing

Before we can describe lookahead selection we give some context about how VAMPIRE and other modern provers implement saturation-based proof search.

VAMPIRE implements a given-clause algorithm that maintains a set of *passive* and a set of *active* clauses and executes a loop where (i) a given clause is chosen from the passive set and added to the active set, (ii) all (generating) inferences between the given clause and clauses in active are performed, and (iii) new clauses are considered for forward and backward simplifications and added to passive if they survive. The details of (iii) are not highly relevant to this discussion, but are very important for effective proof search.

Generating inferences are implemented using *term indexing* techniques (see e.g. [13]) that index a set of clauses (the active clauses in this case) and can be queried for clauses containing subexpressions that match or unify with a given expression.

We can view a term index $\mathcal{T}$ for an inference rule as a map that takes a clause $\underline{l} \vee D$ with a selected literal l and returns a list of *candidate clauses*, which is a set containing all clauses that can have this inference against $\underline{l} \vee D$. VAMPIRE maintains two term indexes for superposition and a separate one for binary resolution. Term indexes are not required for factoring or equality resolution as these are performed on a single clause.

5.2 General Idea Behind Lookahead Selection

The idea of lookahead selection is that we directly estimate for each literal l in C how many children the clause C would have when selecting l and applying inferences on l against active clauses.

Ideally we would have access to a function $\mathsf{children}(C, l)$ that would return the number of children of clause C resulting from inferences with active clauses, given that the literal l was selected in C. We discuss how we practically estimate such a value below.

Given this value we can define a preorder[2] that minimises the number of children:

$$l_1 \rhd_{lmin} l_2 \ \textit{iff} \ \mathsf{children}(C, l_1) < \mathsf{children}(C, l_2)$$

This is based on our previous assertion that we want to produce as few children as possible. But now we have an effective way of steering this property we can also consider the opposite i.e. introduce a quality ordering that maximises the number of children:

$$l_1 \rhd_{lmax} l_2 \ \textit{iff} \ \mathsf{children}(C, l_1) > \mathsf{children}(C, l_2)$$

Our hypothesis is that a selection strategy based on this second ordering will perform poorly, as the search space would grow too quickly.

---

[2] Note that this is not a preorder in the same sense as before as it requires the context of a clause and active clause set. In other words, this preorder is a relation that changes during the proof search process.

5.3 Completing the Selection... Differently

In Sect. 4.3 selection strategies were made complete by searching for the best negative literal where possible. The same approach is taken for selection strategies based on lookahead selection but because it is now relatively much more expensive to compare literals it is best to decide on the literals to compare beforehand.

Firstly, if there are no negative literals all maximal literals must be selected and no lookahead selection is performed. Otherwise, selection is performed on all negative literals and a single maximal positive literal (if there is only one). This ignores the complex case where the combination of all maximal literals would lead to fewer children than the best negative literal.

5.4 Efficiently Estimating Children

To efficiently estimate the number of children that would arise from selecting a particular literal in a clause we make use of the term indexing structures.

Let $\mathcal{T}_1, \ldots, \mathcal{T}_n$ be a set of term indexes capturing the current active clause set. An estimate for $\mathsf{children}(C, l)$ can then be given by:

$$\mathsf{estimate}(l) = \Sigma_{i=1}^{n} |\mathcal{T}_i[l]|.$$

This is an overestimate as the term indexes do not check side-conditions related to orderings after substitution. For example, if we apply a superposition from $l \simeq r$ with $\theta = mgu(l, r)$ and we have $r \not\succeq l$, $r\theta \succeq l\theta$, the index will select $l \simeq r \vee C_1$ as a candidate clause but the rule does not apply. In addition, the number of children is not the same as the number of children that survive retention tests (those neither deleted nor simplified away). However, applying all rules and simplifying children for every literal can be very time-consuming, so we use an easier-to-compute approximation instead.

It is possible to extend the estimate to include inferences that do not rely on indexes. We have done this for equality resolution but not factoring, due to the comparative effort required. In general, our initial hypothesis was that selection should be a cheap operation and so it is best to perform as few additional checks as possible.

In VAMPIRE term indexes return iterators over clauses. This allows us to compute `estimate` in a fail-fast fashion where we search all literals at once and terminate as soon as the estimate for a single literal is finished. This assumes we are minimising (i.e. computing maximal literals with respect to $\rhd_{lmin}$), otherwise we must exhaust the iterators of all but one of the literals.

Of course, as selecting literals in this way now depends on the active clauses it is desirable to do selection as late as possible to maximise accuracy of the estimate. Therefore, VAMPIRE performs literal selection at the point when it chooses a clause from the passive set for activation.

Note that the technique described here can be extended to any setting that uses indexes for generating inferences.

6 Concrete Literal Selection Strategies

In this section we briefly describe concrete literal selection strategies. To have a more general view of selections, we also implemented some selections found in other systems. Of course, when considering selections adapted from other systems we cannot draw conclusions about their utility in the original system as the general implementation is different. But it is useful to compare the general ideas. Strategies have been given numbers to identify them that is based on an original numbering in VAMPIRE, these numbers are used in the next section.

6.1 Vampire

We give a brief overview of the selection strategies currently implemented in VAMPIRE.

Total Selection. The most trivial literal selection strategy is to select everything. This corresponds to the calculus without a notion of selection and is obviously complete. This is referred to by number 0.

Maximal Selection. VAMPIRE's version of maximal selection either selects one maximal negative literal, if one of the maximal literals is negative, or all maximal literals, in which case they will all be positive. This is referred to by number 1.

Quality Selections. VAMPIRE uses four quality selections obtained by combining preorders defined in the previous section as follows:

$$\begin{aligned}
\rhd_2 &= \rhd_{weight} \\
\rhd_3 &= \rhd_{noposeq} \circ \rhd_{top} \circ \rhd_{dvar} \\
\rhd_4 &= \rhd_{noposeq} \circ \rhd_{top} \circ \rhd_{var} \circ \rhd_{weight} \\
\rhd_{10} &= \rhd_{neq} \circ \rhd_{weight} \circ \rhd_{neg}
\end{aligned}$$

VAMPIRE uses both the incomplete versions of the selection strategies, which it numbers 1002, 1003, 1004 and 1010, and the complete versions, which it numbers 2, 3, 4 and 10. We note that not all combinations of the preorders discussed in Sect. 4 are used. As may be suggested by the numbering, previous experimentation introduced and removed various combinations thereof, leaving the current four.

Lookahead Selection. VAMPIRE uses two lookahead selections based on preorders defined as follows

$$\begin{aligned}
\rhd_{11} &= \rhd_{lmin} \circ \rhd_3 \\
\rhd_{12} &= \rhd_{lmax} \circ \rhd_3
\end{aligned}$$

The incomplete versions of the associated strategies are numbered 1011 and 1012 whilst the complete versions are numbered 11 and 12.

6.2 SPASS Inspired

We consider three literal selection strategies adapted from SPASS (as found in the prover's source code)[3]:

[3] SPASS also has "select from list", which requires the user to specify predicates that will be preferred for selection. We did not implement this for the obvious reason.

- **Selection off (20)** selects all the maximal literals. From the perspective of the original Bachmair and Ganzinger theory nothing is selected, but in our setting this effectively amounts to selecting all the maximal literals.
- **Selection always (22)** selects a negative literal with maximal weight, if there is one. Otherwise it selects all the maximal ones.
- **If several maximal (21)** selects a unique maximal, if there is one. Otherwise it selects a negative literal with maximal weight, if there is one. And otherwise it selects all the maximal ones.

6.3 E Prover Inspired

We consider the following five literal selection strategies adapted from E (as mentioned in the prover's manual [12]):

- **SelectNegativeLiterals (30)** selects all negative literals, if there are any. Otherwise it selects all the maximal ones.
- **SelectPureVarNegLiterals (31)** selects a negative equality between variables, if there is one. Otherwise it selects all the maximal literals.
- **SelectSmallestNegLit (32)** selects a negative literal with minimal weight, if there is one. Otherwise it selects all the maximal literals.
- **SelectDiffNegLit (33)** selects a negative literal which maximises the difference between the weight of the left-hand side and the right-hand side,[4] if there is a negative literal at all. Otherwise it selects all the maximal literals.
- **SelectGroundNegLit (34)** selects a negative ground literal for which the weight difference between the left-hand side and the right-hand side terms is maximal, if there is a negative literal at all. Otherwise it selects all the maximal literals.
- **"SelectOptimalLit" (35)** selects as (34) if there is a ground negative literal and as (33) otherwise.

It should be noted that our adaptations of E's selections are only approximate, because E uses a different notion of term weight than VAMPIRE, defining constants and function symbols to have basic weight 2 and variables to have weight 1. Also we do not consider E's **NoSelection** strategy separately as it is the same as SPASS's **Selection off** and E's **SelectLargestNegLit** strategy as it is the same as SPASS's **Selection always** (modulo the notion of term weight).

7 Experimental Evaluation

Here we report on our experiments with selection strategies using the theorem prover VAMPIRE. Our aim is to look for strategies which help to solve many problems, but also for strategies which solve problems other strategies cannot solve. This is because we are ultimately interested in constructing a portfolio combining several strategies which solve as many problems as possible within a reasonably short amount of time.

[4] In E, all literals are represented as equalities. A non-equational atom $p(t)$ is represented as $p(t) = \top$, where $\top$ is a special constant *true*. Thus it makes sense to talk about left-hand and right-hand side of a literal even in the non-equational case.

Experimental Setup. For our experiments we took all the problems from the TPTP [15] library version 6.3.0 which are in the FOF or CNF format, excluding only unit equality problems (for which literal selection does not play any role) and problems of rating 0.0 (which are trivial to solve). This resulted in a collection of 11 107 problems.[5]

We ran VAMPIRE on these problems with saturation algorithm set to *discount* and *age-weight ratio* to 1 : 5 (cf. [7,10]), otherwise keeping the default settings and varying the choice of literal selection. By default, VAMPIRE employs the AVATAR architecture to perform clause splitting [9,17]. AVATAR was also enabled in our experiments.

The time limit was set to 10 seconds for a strategy-problem pair. This should be sufficient for obtaining a realistic picture of relative usefulness of each selection strategy, given the empirical observation pertaining to first-order theorem proving in general, that a strategy usually solves a problem very fast if at all. The experiments were run on the StarExec cluster [14], whose nodes are equipped with Intel Xeon 2.4 GHz processors. Experiments used Vampire's default memory limit of 3GB.

Result Overview. In total, we tested 23 selection strategies i.e. those summarised in Sect. 6. With AVATAR, VAMPIRE never considers clauses with ground literals for selection, therefore selection 34 behaves the same as 20 and 35 the same as 33. Consequently, results for 34 and 35 are left out from initial discussions, but will be discussed later when we consider what happens when AVATAR is not used.

Out of our problem set, 5 908 problems were solved by at least one strategy.[6] This includes 31 problems of TPTP rating 1.0. Out of the solved problems, 5 621 are unsatisfiable and 287 satisfiable. Because we are mainly focusing on theorem proving, i.e. showing unsatisfiability, we will first restrict our attention to the unsatisfiable problems.

Ranking the Selections. Table 1 (left) shows the performance of the individual selection strategies. We report the number of problems solved by each strategy (which determines the order in the table), the percentage with respect to the above reported overall total of problems solved, the number of problems solved by only the given strategy (unique), and an indicator we named *u-score*. U-score is a more refined version of the number of uniquely solved problems. It accumulates for each problem solved by a strategy the reciprocal of the number of strategies which solve that problem. This means that each uniquely solved problem contributes 1.0, each problem solved also by one other strategy adds 0.5, etc. It also means that the sum of u-scores in the whole table equals the number of problems solved in total.

By looking at Table 1 we observe that 1011, the incomplete version of the lookahead selection, is a clear winner both with respect to the number of solved problems and the number of uniques. It solves more than 80 % of problems solvable by at least one strategy and accumulates by far the highest u-score. Other very successful selections are the incomplete 1010 and 1002, and 11, the complete version of lookahead.

Inverted lookahead in the incomplete (1012) and complete (12) version end up last in the table, which can be seen as a confirmation of our hypothesis from Sect. 5. Similarly the experimental selection 0, which selects all the literals in a clause, and the

[5] A list of the selected problems, the executable of our prover as well as the results of the experiment are available at http://www.cs.man.ac.uk/~sudam/selections.zip.

[6] And 1 952 problems were solved by every strategy.

Table 1. Left: performance of the individual selection strategies. Right: statistics collected from the runs: #child is the average number of children of an activated clause, %incomp is the average percentage of the cases when an incomplete selection violates the completeness condition. The values marked 's.o.' (solved only) are collected only from runs which solved a problem, the values marked 'all' are collected from all runs.

selection	#solved	%total	#unique	u-score	#child (s.o./all)	%incomp. (s.o./all)
1011	4718	83.9	156	563.6	4.2 / 9.9	3.3 / 4.5
1010	4461	79.3	31	384.1	9.4 / 14.6	2.1 / 2.5
11	4333	77.0	26	354.7	6.5 / 13.6	
1002	4327	76.9	62	396.1	8.7 / 15.4	9.7 / 7.6
10	4226	75.1	8	283.3	9.9 / 14.5	
21	4113	73.1	6	274.2	10.7 / 13.8	
2	4081	72.6	1	261.0	10.3 / 14.9	
1004	4009	71.3	8	276.2	6.3 / 14.1	19.5 / 7.3
4	3987	70.9	2	247.2	7.8 / 13.7	
3	3929	69.8	1	235.5	8.7 / 13.8	
1003	3907	69.5	6	258.2	6.5 / 14.7	22.6 / 8.6
33	3889	69.1	1	239.2	7.1 / 18.3	
22	3885	69.1	0	236.2	7.0 / 18.4	
1	3778	67.2	6	227.9	9.4 / 19.9	
31	3702	65.8	0	218.2	13.4 / 23.1	
20	3682	65.5	0	217.1	13.3 / 23.2	
30	3559	63.3	3	204.9	16.6 / 28.8	
32	3538	62.9	5	209.8	6.3 / 19.9	
0	3362	59.8	8	203.1	35.8 / 48.7	
12	3308	58.8	3	183.4	14.0 / 24.5	
1012	2532	45.0	5	146.1	13.9 / 30.8	7.6 / 5.8

selection 32 adapted from E, which selects the *smallest* negative literal, inverting the intuition that large (with large weight) literals should be selected, end up at the end of the table.[7] Interestingly, however, to each of these "controversial" selections we can attribute several uniquely solved problems.

Table 1 also shows that, with the exception of selection 3 (and 12), the incomplete version of a selection always solves more problems than the complete one.

Additional Statistics. Table 1 (right) displays for each selection two interesting averages obtained across the runs. The first is the average number of children of an activated clause and the results confirm that the lookahead selections (1011 and 11), in accord with their design, achieve the smallest value for this metric. This further confirms our hypothesis that preferring to generate as few children as possible leads to successful

[7] The selection strategy selecting the largest negative literal has number 22.

strategies. The second is the average number of times an incomplete strategy selects in such a way as to violate the completeness condition. We can see that there is lot of variance between the selections in this regard and that the "most complete" incomplete selection is the second best selection 1010.

Time spent on performing selection. As we might expect, lookahead selections are far more expensive to compute. On average, performing quality selection consumes roughly 0.1 % of the time spent on proof search, with other non-lookahead selections taking similar times, whereas complete and incomplete lookahead selection consumes roughly 1.74 % and 4.27 % respectively. These numbers are taken from all proof attempts, not just successful ones. Incomplete lookahed selection is more expensive than its complete counterpart as the latter is not performed when there are no negative literals. The previously observed success of lookahead selection confirms that the extra time spent on selecting is more than well spent.

Table 2. Performance of the individual selection strategies (left) and statistics collected from the runs (right) for runs with AVATAR turned off. Columns analogous to those described in Table 1.

selection	#solved	%total	#unique	u-score	#child (s.o./all)	%incomp. (s.o./all)
1010	4289	80.0	64	379.8	9.3 / 17.0	9.0 / 9.4
1011	4255	79.4	104	412.7	8.5 / 15.0	6.5 / 8.3
1002	4207	78.5	45	356.2	7.5 / 18.5	17.6 / 8.6
11	4121	76.9	25	292.9	12.1 / 25.7	
10	4116	76.8	9	251.7	13.1 / 21.2	
2	4063	75.8	0	235.7	16.5 / 23.5	
21	4055	75.7	4	244.1	16.3 / 23.6	
22	3896	72.7	0	218.0	8.8 / 30.3	
33	3895	72.7	1	218.0	9.0 / 30.1	
4	3892	72.6	3	216.5	9.6 / 19.8	
35	3858	72.0	1	211.7	9.0 / 30.2	
1004	3810	71.1	8	228.2	8.6 / 20.4	23.8 / 10.5
3	3755	70.1	3	205.4	12.0 / 20.8	
1	3744	69.9	2	207.3	13.1 / 31.5	
30	3731	69.6	11	220.0	8.9 / 33.8	
1003	3654	68.2	2	211.2	8.2 / 22.8	25.7 / 11.1
31	3517	65.6	0	184.9	22.4 / 33.2	
34	3491	65.1	1	183.0	21.5 / 31.9	
32	3482	65.0	2	188.5	7.8 / 31.9	
20	3479	64.9	0	182.2	21.7 / 33.3	
12	3313	61.8	6	173.8	25.0 / 33.9	
0	3279	61.2	24	206.4	59.2 / 83.1	
1012	2403	44.8	7	126.7	17.9 / 36.4	7.2 / 10.6

The Effect of Turning Splitting off. The previous results were obtained running VAMPIRE with splitting turned on. In order to establish how much the standing of the individual selections depends on running within the context of the AVAVAR architecture, we ran a separate experiment with the same strategies but turning AVATAR off. Arguably, these results are more relevant to implementations that do not incorporate the effective AVATAR approach.

In total, the strategies without AVATAR solved 5 563 problems (5 356 unsatisfiable, 207 satisfiable). The number of problems solved by all strategies was 1 748. Table 2 presents a view analogous to Table 1 for these strategies. Strategies 34 and 35 are now relevant (see experimental setup above).

Notably, selection 1011 has dropped to the second place in the overall ranking (after 1010). However, the incomplete lookahead still accumulated the highest u-score as a standalone strategy and we see the same general trend that incomplete versions of strategies outperform their complete counterpart (again with the exception of 3 and 12).

Focusing on Satisfiable Problems. Recall that in our experiments 287 satisfiable problems were solved by at least one strategy. Table 3 (left) shows the performance of the best 5 *complete* selections on these problems.[8] The first two places are taken by a selection from E and SPASS while the lookahead selection is third. The differences between the three places are, however, only by one problem. Moreover, in Table 3 (right) we can see that when AVATAR was turned off in a separate experiment lookahead selection came first.

Table 3. Performance of the five best complete selection strategies on satisfiable problems.

AVATAR on (total 287)					AVATAR off (total 207)				
selection	#solved	%total	#unique	u-score	selection	#solved	%total	#unique	u-score
33	248	86.4	0	24.5	11	195	94.2	0	16.7
22	247	86.0	0	24.1	4	191	92.2	0	17.1
11	246	85.7	0	23.4	3	190	91.7	0	16.9
32	241	83.9	1	23.8	32	184	88.8	0	14.7
1	238	82.9	0	21.6	35	183	88.4	0	14.6

8 Impact of Selection on Portfolio Solving

As mentioned at the beginning of Sect. 7, VAMPIRE, like most leading first-order theorem provers, will (when asked) try to use a portfolio of *strategies* to solve a problem. To make an effective portfolio we want a mix of strategies that either solve many problems and or unique problems. One measure of the usefulness of a selection strategy is its impact on the creation of a portfolio mode with respect to the second of these properties i.e. which problems can only be solved using a particular approach.

[8] The table has been shortened due to space restrictions. However, the #unique and u-score indicators still take into account all the other complete selections.

Table 4. Numbers of problems solved only by a strategy using a particular selection strategy.

Selection	Problems solved only using this selection	
	All	Problems solved only by VAMPIRE
11	151	118
1011	78	62
1	62	58
10	55	41
lookahead	278	216
non-lookahead	502	377
complete	824	691
incomplete	229	169

To find useful strategies for VAMPIRE we have a dedicated cluster using a semi-guided method to randomly search the space of strategies. At the time of writing over 786k proofs have been found of 11,354 problems out of 13,770 (unsatisfiable, non unit-equality) problems taken from TPTP 6.1.0. It took over 160 CPU-years of computation to collect these data. Table 4 gives results for the best four selection strategies and four groups of selection strategies. Numbers are given for all problems and for the subset of problems that were not solved by any other theorem prover at the time TPTP 6.1.0 was released[9]. From this we can see that both lookahead and incomplete selections are required to solve many problems unsolvable by other methods. Additionally, this shows that having a spread of different selection strategies is useful as they contribute to uniquely solving different problems.

9 Conclusion

Selection strategies can have a very large impact on proof search, often making the difference between solving and not solving a problem. Little had been written about how effective selection strategies could be designed and implemented, although most successful implementations of the superposition calculus have relied on them.

We have introduced two new ways of performing literal selection based on the observation that it is good to select those literals that lead to as few children as possible. The first approach, *quality selection*, is an easy to implement compositional mechanism for defining literal selection based on qualities of literals that lead to few children. We described different selection strategies based on concrete qualities and demonstrated their effectiveness. What may be surprising to some is how effective *incomplete* versions of such strategies can be. Experimentally establishing this phenomenon is a large contribution of this work. However, our main result is the second approach, the powerful idea of *lookahead selection* based on the observation that if we want to select literals leading to as few children as possible then the best thing to do is just that. Experimental

[9] See the ProblemAndSolutionStatistics file distributed with TPTP.

results showed that by using this approach we could solve many problems that could not otherwise be solved by any other selection strategy taken from VAMPIRE, E or SPASS.

References

1. Bachmair, L., Ganzinger, H.: Rewrite-based equational theorem proving with selection and simplification. Revised version in the J. Log. Comput. **4**(3), 217–247 (1994). Research Report MPI-I-91-208, Max-Planck-Institut für Informatik, 1991
2. Bachmair, L., Ganzinger, H.: Resolution theorem proving. In: Handbook of Automated Reasoning, vol. I, chapter 2, pp. 19–99. Elsevier Science (2001)
3. Bachmair, L., Ganzinger, H., Waldmann, U.: Superposition with simplification as a desision. In: Mundici, D., Gottlob, G., Leitsch, A. (eds.) KGC 1993. LNCS, vol. 713, pp. 83–96. Springer, Heidelberg (1993)
4. Dershowitz, N., Plaisted, D.A.: Rewriting. In: Handbook of Automated Reasoning, vol. I, chapter 9, pp. 535–610. Elsevier Science (2001)
5. Ganzinger, H., de Nivelle, H.: A superposition decision procedure for the guarded fragment with equality. In: 14th Annual IEEE Symposium on Logic in Computer Science, Trento, Italy, 2–5 July, pp. 295–303. IEEE Computer Society (1999)
6. Knuth, D., Bendix, P.: Simple word problems in universal algebra. In: Computational Problems in Abstract Algebra, pp. 263–297. Pergamon Press (1970)
7. Kovács, L., Voronkov, A.: First-order theorem proving and VAMPIRE. In: Sharygina, N., Veith, H. (eds.) CAV 2013. LNCS, vol. 8044, pp. 1–35. Springer, Heidelberg (2013)
8. Nieuwenhuis, R., Rubio, A.: Paramodulation-based theorem proving. In: Handbook of Automated Reasoning, vol. I, chapter 7, pp. 371–443. Elsevier Science (2001)
9. Reger, G., Suda, M., Voronkov, A.: Playing with AVATAR. In: Reger, G., Suda, M., Voronkov, A. (eds.) CADE-25. LNCS, vol. 9195, pp. 399–415. Springer, Switzerland (2015)
10. Reger, G., Voronkov, A.: The Vampire manual. Technical report (2016, in preperation)
11. Schulz, S.: E – a brainiac theorem prover. AI Commun. **15**(2–3), 111–126 (2002)
12. Schulz, S.: E 1.8 User Manual (2015). http://wwwlehre.dhbw-stuttgart.de/sschulz/WORK/E_DOWNLOAD/V_1.9/eprover.pdf. Accessed 22 Jan 2016
13. Sekar, R., Ramakrishnan, I., Voronkov, A.: Term indexing. In: Handbook of Automated Reasoning, vol. II, chapter 26, pp. 1853–1964. Elsevier Science (2001)
14. Stump, A., Sutcliffe, G., Tinelli, C.: StarExec, a cross community logic solving service (2012). https://www.starexec.org
15. Sutcliffe, G.: The TPTP problem library and associated infrastructure. J. Autom. Reason. **43**(4), 337–362 (2009)
16. Sutcliffe, G., Suttner, C.: The state of CASC. AI Commun. **19**(1), 35–48 (2006)
17. Voronkov, A.: AVATAR: the architecture for first-order theorem provers. In: Biere, A., Bloem, R. (eds.) CAV 2014. LNCS, vol. 8559, pp. 696–710. Springer, Heidelberg (2014)
18. Weidenbach, C.: Combining superposition, sorts and splitting. In: Handbook of Automated Reasoning, vol. II, chapter 27, pp. 1965–2013. Elsevier Science (2001)

Performance of Clause Selection Heuristics for Saturation-Based Theorem Proving

Stephan Schulz(✉) and Martin Möhrmann(✉)

DHBW Stuttgart, Stuttgart, Germany
{schulz,moehrmann}@eprover.org

Abstract. We analyze the performance of various clause selection heuristics for saturating first-order theorem provers. These heuristics include elementary first-in/first-out and symbol counting, but also interleaved heuristics and a complex heuristic with goal-directed components.

We can both confirm and dispel some parts of developer folklore. Key results include: (1) Simple symbol counting heuristics beat first-in/first-out, but by a surprisingly narrow margin. (2) Proofs are typically small, not only compared to all generated clauses, but also compared to the number of selected and processed clauses. In particular, only a small number of *given clauses* (clauses selected for processing) contribute to any given proof. However, the results are extremely diverse and there are extreme outliers. (3) Interleaving selection of the given clause according to different clause evaluation heuristics not only beats the individual elementary heuristics, but also their union - i.e. it shows a synergy not achieved by simple strategy scheduling. (4) Heuristics showing better performance typically achieve a higher ratio of *given-clause* utilization, but even a fairly small improvement leads to better outcomes. There seems to be a huge potential for further progress.

1 Introduction

Saturating theorem provers for first-order logic try to show the unsatisfiability of a clause set by systematically enumerating direct consequences and adding them to the clause set, until either no new (non-redundant) clauses can be generated, or the empty clause as an explicit witness of inconsistency is found.

At this time, the most powerful provers for first-order logic with equality are based on saturation. These provers implement saturation by variants of the *given-clause algorithm*. In this algorithm, clauses are selected for inferences one at a time. The order of selection of the *given clause* for each iteration of the main loop is a major choice point in the algorithm. While there is significant folklore about this choice point, we are not aware of a systematic evaluation of different heuristics for this choice point.

There is also little understanding of the properties of proofs and the proof induced by different strategies. Previous work was restricted to unit-equational logic and much smaller search spaces [4].

In this paper, we compare different classical and modern clause selection heuristics. In particular, we consider the following questions:

N. Olivetti and A. Tiwari (Eds.): IJCAR 2016, LNAI 9706, pp. 330–345, 2016.
DOI: 10.1007/978-3-319-40229-1_23

- How powerful are different heuristics on different classes of problems?
- How well do different heuristics perform compared to a perfect oracle that finds the same proofs? Which proportion of selected clauses is contributing to a given proof?
- How do different heuristics interact when interleaved?
- Can commonly held beliefs about clause selection be supported by data?
- What are typical properties of proofs found by a modern theorem prover?

To obtain data on these questions, we have instrumented the prover E to efficiently collect data about the ongoing proof search and to print out an analysis of both the proof object and the complete proof search graph at termination.

This paper is organized as follows. First, we introduce the concept of saturation and briefly describe the given-clause algorithm. We also discuss the basics of clause evaluation and E's flexible implementation of clause selection heuristics. In Sect. 3 we describe the design of the experiments and the particular clause selection heuristics analyzed. Section 4 contains results on the performance of different heuristics and their analysis, as well as information on properties and structures of proofs and proof search. We then conclude the paper.

2 Saturating Theorem Proving

Modern saturating theorem proving started with resolution [17]. It was also a natural framework for completion-based equational reasoning [1,6,7]. The confluence of resolution and completion, implemented e.g. in Otter [12,13], the first modern-style high-performance theorem prover, lead to the still current equality-based superposition calculus, definitively described by Bachmair and Ganzinger [2]. Today, systems based on superposition and saturation like Vampire [8,15], Prover9 [11], SPASS [23] and E [19,20] define the state of the art in theorem proving for first-order logic with equality.

Saturating calculi for first-order logic are based on a refutational paradigm, i.e. the axioms and conjecture are converted into a clause set that is unsatisfiable if and only if the conjecture is logically implied by the axioms. The calculus defines a series of inference rules which take one or more (most often two) existing clauses as premises and produce a new clause as the conclusion. This new clause is added to the original clause set and is available as a premise for future inferences. The process stops when either no new non-redundant clause can be derived (in this case, the clause set is *saturated* up to redundancy), or when the empty clause as an explicit witness of unsatisfiability is derived.

Current calculi also include simplification rules which allow the replacement of some clauses by simpler (and often syntactically smaller) clauses, or even the complete removal of redundant clauses. Examples include in particular *rewriting* (replacement of terms by smaller terms), subsumption (discarding of a clause implied by a more general clause) and tautology deletion.

In most cases, saturation can, in principle, derive an infinite number of consequences. In these cases, *completeness* of the proof search requires a certain notion of *fairness*, namely that no non-redundant inference is delayed infinitely.

The superposition calculus is the current state of the art in saturating theorem proving. It subsumes earlier calculi like resolution, paramodulation, and unfailing completion. In the superposition calculus, inferences can be restricted to maximal terms of maximal literals using a *term ordering*, and optionally to selected negative literals using a *literal selection* scheme. All systems we are currently aware of determine a fixed term ordering and literal selection scheme before saturation starts, either by user input or automatically after analyzing the problem.

2.1 Saturation Algorithms

Saturation algorithms handle the problem of organizing the search through the space of all possible derivations. The simplest and obviously fair algorithm is *level saturation*. Given a clause set C_0, level saturation computes the set of all direct consequences D_0 of clauses in C_0. The union $C_1 = C_0 \cup D_0$ then forms the basis for the next iteration of the algorithm. Level saturation does not support heuristic guidance, and we are not aware of any current or competitive system built on the basis of level saturation. To our knowledge level saturation has never been implemented with modern redundancy elimination techniques.

At the other extreme, a *single step* algorithm performs just one inference at a time, adding the consequence to the set and making it available for further

Search state: (U, P)
U contains *unprocessed* clauses, P contains *processed* clauses.
Initially, P is empty and all clauses are in U.
The *given clause* is denoted by g.

```
while U ≠ {}
  g = extract_best(U)
  g = simplify(g, P)
  if g == □
    SUCCESS, Proof found
  if g is not subsumed by any clause in P (or otherwise redundant w.r.t. P)
    P = P\{c ∈ P | c subsumed by (or otherwise redundant w.r.t.) g}
    T = {c ∈ P | c can be simplified with g}
    P = (P\T) ∪ {g}
    T = T ∪ generate(g, P)
    foreach c ∈ T
      c = cheap_simplify(c, P)
      if c is not trivial
        U = U ∪ {c}
SUCCESS, original U is satisfiable
```

Remarks: extract_best(U) finds and extracts the clause with the best heuristic evaluation from U. This is the choice point we are particularly interested in this paper.

Fig. 1. The *given-clause* algorithm as implemented in E

inferences (and potential simplification). The major disadvantage of the single-step algorithm is the necessary book-keeping. Moreover, while search heuristics can work at the finest possible granularity, the objects of heuristic evaluations are potential inferences, not concrete clauses. We are not aware of any system that uses a per-inference evaluation for search guidance, although e.g. Vampire's *limited resource strategy* [16] discards some potential inferences up-front, based on a very cursory evaluation.

The most widely used saturation algorithms are variants of the *given-clause* algorithm. They split the set of all clauses into two subsets U of *unprocessed* clauses and P of *processed* clause (initially empty). In each iteration, the algorithm selects one clause g from U and adds it to P, computing all inferences in which g is at least one premise and all other premises are from P. The algorithm adds the resulting new clauses to U, maintaining the invariant that all inferences between clauses in P have been performed.

Variants of the *given-clause* algorithm are at the heart of most of today's saturating theorem provers. The two main variants are the so-called *Otter loop* and the *DISCOUNT loop*, popularized by the eponymous theorem provers [3,13]. In the Otter loop, all clauses are used for simplification. In particular, newly generated clauses are used to back-simplify both processed and unprocessed clauses. In the DISCOUNT loop, unprocessed clauses are truly passive, i.e. only clauses that are selected for processing are used for back-simplification. As a result, the Otter loop can typically find proofs in less iterations of the main loop, but each iteration takes longer. In the DISCOUNT loop, contradictory clauses in U may not be discovered until selected for processing. However, each individual iteration of the main loop results in less work. In both variants, selection of the given clause is the main heuristic choice point. In the DISCOUNT loop this control is at a finer level of granularity, since each iteration of the main loop represents a smaller part of the proof search.

In addition to Otter, the Otter loop is implemented in Prover9, SPASS and Vampire. The DISCOUNT loop historically was implemented in systems specializing in equational reasoning, including Waldmeister [10] and E. It was also added as an alternative loop to both SPASS and Vampire. There is little evidence that one or the other variant has a systematic advantage. A comparison in Vampire [16] showed some advantage for the DISCOUNT loop over the plain Otter loop, but also some advantage of the Otter loop in combination with the *limited resource strategy* (which sacrifices completeness for efficiency by discarding some new clauses) over Vampire's DISCOUNT loop.

Figure 1 depicts the DISCOUNT loop as implemented in E. The given-clause selection is represented by the `extract_best()` function.

2.2 Clause Selection Heuristics

Once term ordering and literal selection scheme are fixed, clause selection, i.e. the order of processing of the unprocessed clauses, is the main choice point. The standard implementation assigns a heuristic weight to each clause, and processes clauses in ascending order of weight, i.e. at each iteration of the main loop the clause with the lowest weight is selected.

Most modern provers allow at least the interleaving of a best-first (lowest weight) and breadth-first (oldest clause) search, where the weight is usually based on (weighted) symbol counting. The ratio of clauses picked by size to clauses picked by age is also known as the *pick-given ratio* [12]. E generalizes this concept. It supports a large number of different parameterized clause evaluation functions and allows the user to specify an arbitrary number of priority queues and a weighted round-robin scheme that determines how many clauses are picked from each queue. This enables us to configure the prover to use nearly arbitrarily complex clause selection heuristics and makes it possible to simulate nearly every conventional clause selection heuristic.

In this study, we are, in particular, concerned with the properties of conventional clause selection schemes. Thus, we look at the following basic clause evaluation heuristics:

- *First-in/First-out* or *FIFO* clause selection always prefers the oldest unprocessed clause. In E, this is realized by giving each new clause a pseudo-evaluation based on a counter that is increased each time a new clause is generated. If one ignores simplification, a pure FIFO strategy will emulate *level saturation*, i.e. it will generate all clauses of a given level before clauses of the next level. In this case, it should find the shortest possible proof (by number of generating inferences). Integration of simplification complicates the issue, although we would still expect FIFO to find short proves. FIFO is an obviously fair heuristic.
- *Symbol counting* or *SC* clause evaluation counts the number of symbols in a clause, and prefers small clauses. Function symbols and variables can have uniform or different weights. There are several intuitive reasons why this should be a good strategy. On the most obvious level, the goal of the saturation is the derivation of the empty clause, which has zero symbols. Moreover, clauses with fewer symbols are more general, hence allowing the system to remove more redundancy via subsumption and rewriting. And finally, clauses with fewer symbols also have fewer positions, and hence likely fewer successors, keeping explosion of the search spaces lower than large clauses. As long as all symbols (or at least all function symbols with non-zero arity) have positive weight, *SC*-based strategies are fair (there is only a finite number of different clauses below any given weight).
- *Ordering-aware* evaluation functions are symbol-counting variants that are designed to prefer clauses with few maximal terms and maximal literals. In the general case, this reduces the number of inference positions (and hence potential successors), decreasing the branching factor in the search space. In the unit-equational case it will also prefer orientable equations (*rules*) to unorientable equations. Rules are much cheaper to apply for simplification. In E, the *refined weight* (*RW*) heuristic achieves the desired effect by multiplying the weight of maximal terms and maximal literals by user-selectable constant factors.

- A major feature of E is the use of *goal-directed* evaluation functions (*GD*). These give a lower weight to symbols that occur in the conjecture, and a higher weight to other symbols, thus preferring clauses which are more likely to be applicable for inferences with the conjecture.

Most of our experiments look at simple heuristics employing only one or two clause evaluation functions - see the experimental design section. However, for comparison we also include the globally best clause selection heuristic for E known to us. This scheme was created via genetic algorithm from a population of random heuristics spanning the parameter space of manually created heuristics developed over the last 15 years [18].

In addition to clause selection based on the syntactic form of the clause, the system can also select clauses based on their origin. In particular, a common recommendation is to first process all the initial clauses, before any of the derived clauses is picked.

3 Experimental Design

We added the ability to efficiently record compact internal proof objects in E 1.8. The overhead for proof recording is minimal and barely measurable [20]. We have now slightly extended the internal representation of the proof search to be able to record all processed given clauses, thus enabling the prover to provide more detailed statistics on the quality of clause selection. Other statistics were obtained by analyzing the existing proof object, and by counting operations and inferences performed during the proof search. The code is part of E version 1.9.1 (pre-release) and will be included in the next release of the prover.

3.1 Computing Environment and Test Set

We used problems from the TPTP [22] library, version v6.3.0. Since we are interested in the performance of the heuristics for proof search, and since several of our statistical measures only make sense for proofs, we restricted the problem set to full first-order (FOF) and clause normal form (CNF) problems that should be provable, i.e. CNF problems with status *Unsatisfiable* or *Unknown*[1] and FOF problems with status *Unsatisfiable*, *ContradictoryAxioms*, or *Theorem.*[2]

This selection left 13774 problems, 7082 FOF and 6692 CNF problems. FOF problems were translated to CNF by E dynamically, with (usually short) translation time included in the reported times.

We report performance results separately for unit problems (all clauses are unit), Horn problems (all clauses are Horn and at least one clause is a non-unit Horn clause) and general (there is at least one non-Horn clause), with and

[1] Status *Unknown* is assigned to problems which should be provable, but for which no machine proof is known.

[2] Two trivial syntactic test examples were excluded. They tested floating point syntax features that at the time of the experiments were incorrectly handled by E.

without equality. The classification of problems into these types refers to the clausified form and was performed by E after clausification.

The StarExec Cluster [21] was used for all benchmark runs. Each problem was executed alone and single threaded on an Intel Xeon E5-2609 processor running at 2.4 GHz base clock speed. Each node had at least 128 GB RAM. We ran the experiments with a per-problem time limit of 300 s and, given the amount of RAM available, without enforced memory limit.

3.2 Claus Selection Heuristics

We tested 40 different clause selection heuristics. From these we selected the 14 heuristics described in Table 1 as sufficiently distinct and reasonably covering the parameter space we are interested in.

The 14 selected heuristics include basic FIFO and symbol counting, ordering-aware and goal-directed heuristics, as well as combinations of symbol counting

Table 1. Clause selection heuristics used

Heuristic	Description
FIFO	First-in/First-out, i.e. oldest clause first
SC12	Symbol counting, function symbols have weight 2, variables have weight 1
SC11	Symbol counting, both function symbols and variables have weight 1
SC21	Symbol counting, function symbols have weight 2, variables have weight 1
RW212	Symbol counting, function symbols have weight 2, variables have weight 1, maximal terms receive double weight.
2SC11/FIFO	Interleaved selection: Select 2 out of every 3 clauses according to SC11, the remaining one with FIFO
5SC11/FIFO	Ditto, with a selection ration of 5:1. This is inspired by Larry Wos comment on Otter ("The optimal pick-given ratio is five")
10SC11/FIFO	Ditto, selection ratio 10:1
15SC11/FIFO	Ditto, selection ratio 15:1
GD	Individual goal-directed heuristic, extracted from *Evolved* below
5GD/FIFO	GD interleaved 5:1 with FIFO
SC11-PI	As SC11, but always process initial clauses first
10SC11/FIFO-PI	As 10SC11/FIFO, but always process initial clauses first
Evolved	Evolved heuristic, combining 2 goal-directed evaluation functions, two symbol-counting heuristics, and FIFO. See [18]

variants with FIFO. We also tested the performance of a preference for initial clauses, and include the *Evolved* heuristic as a benchmark that represents the current state of the art.

The full data for all 40 strategies and the exact parameters for the provers are archived and available at http://www.eprover.eu/E-eu/Heuristics.html.

4 Results

4.1 Global Search Performance

Table 2 summarizes the performance of the 14 different strategies on the full test set. The *Rank* column shows the ranking of strategies by total number of successes within the time limit. The third column shows the number of successes, as an absolute number and as a fraction of all 13774 problems. The next column shows how many problems were solved by the corresponding strategy only, not by any of the other strategies. Finally, the last column shows how many problems are solved within a one second search time, and the fraction of total successes by that strategy this number represents. Figure 2 visualizes the performance of a selected subset of strategies over time.

From this data, we can already draw a number of conclusions:

- All performance curves are similar in basic shape, and all strategies find the bulk of their proofs within the first few seconds. Indeed, most strategies reach around

Table 2. Global search performance

Heuristic	Rank	Successes		Successes within 1s	
		Total	Unique	Absolute	Of column 3
FIFO	14	4930 (35.8 %)	17	3941	79.9 %
SC12	13	4972 (36.1 %)	5	4155	83.6 %
SC11	9	5340 (38.8 %)	0	4285	80.2 %
SC21	10	5326 (38.7 %)	17	4194	78.7 %
RW212	11	5254 (38.1 %)	13	5764	79.8 %
2SC11/FIFO	7	7220 (52.4 %)	24	5846	79.7 %
5SC11/FIFO	5	7331 (53.2 %)	3	5781	78.3 %
10SC11/FIFO	3	7385 (53.6 %)	1	5656	77.6 %
15SC11/FIFO	6	7287 (52.9 %)	6	5006	82.5 %
GD	12	4998 (36.3 %)	12	5856	78.4 %
5GD/FIFO	4	7379 (53.6 %)	62	4213	80.2 %
SC11-PI	8	6071 (44.1 %)	13	4313	86.3 %
10SC11/FIFO-PI	2	7467 (54.2 %)	31	5934	80.4 %
Evolved	1	8423 (61.2 %)	593	6406	76.1 %

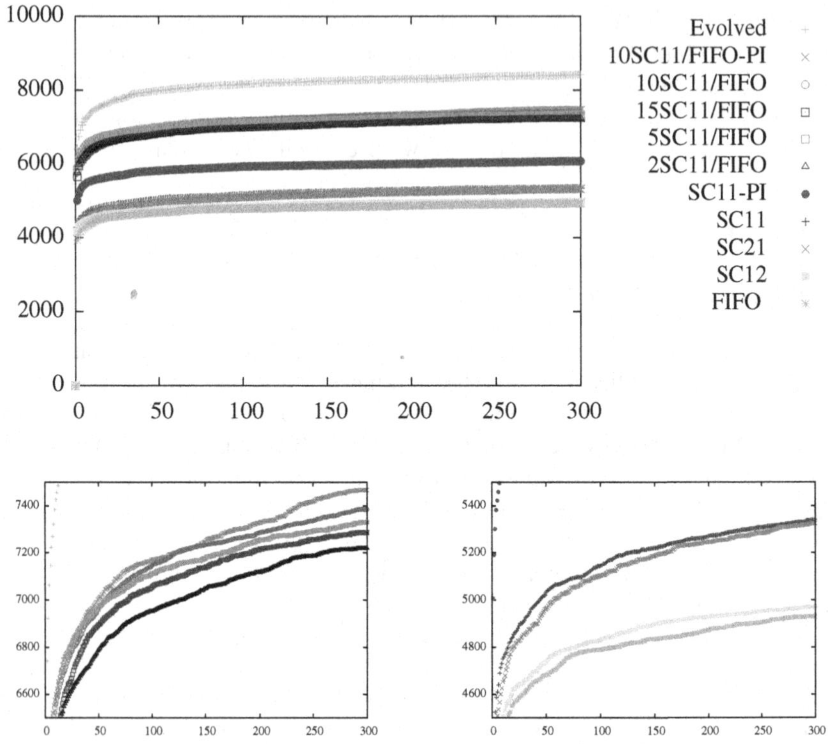

The large plot shows overall performance (vertical axis is number of proofs found up to a given time, horizontal axis is run time in seconds). The smaller plots scale interesting sections of the y-axis to differentiate strategies with similar overall performance.

Fig. 2. Solutions over time for different clause selection heuristics (Color figure online)

80 % of their successes within the first second, and even for the *Evolved* strategy, more than three quarter of the successes are achieved within one second.
- *FIFO* is the weakest of the search strategies. However, even *SC11*, the best simple symbol counting heuristic, proves less than 10 % more than *FIFO*.
- There is no evidence that using different weights for function symbols and variables increases overall performance. Indeed, using a higher weight for variables markedly decreases performance. However, it changes the part of the search space explored early, potentially adding more solutions to the performance of the ensemble of all strategies.
- The ordering-aware *RW212* has slightly lower global performance than the corresponding simple symbol-counting heuristics. This is surprising, since this and similar strategies have for a long time been major contributors to E's collection of standard heuristics.
- All four strategies interleaving simple symbol counting and FIFO perform much better than the corresponding pure symbol-counting strategy, with the

best one solving more than 2000 extra problems, an increase of nearly 40 %. On the other hand, the spread of performance over the *pick-given ratios* from 2 to 15 is very small, varying by only about 2 %. The best ratio in our tests for E is not 5 as sometimes anecdotally reported for the Otter loop, but 10.
- For the union of solutions found by SC11 and FIFO (with 300 second time limit for each), the prover finds only 6329 proofs. Thus, there is real synergy in the interleaved strategies, which beat not only the individual components but also their union. We believe this is due to two effects: Symbol counting selection builds a compact representation of the theory induced by the axioms, thus enabling the prover to traverse larger parts of search space, while FIFO ensures that no part of the search space is unduly delayed.
- The goal-directed heuristic on its own is not particularly powerful. Its performance is in line with the symbol-counting heuristics. However, it profits even more from the addition of a FIFO component than the other strategies.
- Processing initial clauses first does indeed boost performance of a strategy. However, the effect is much stronger for the pure symbol-counting heuristic than for a strategy that interleaves FIFO selection. The intuitive explanation is that FIFO selection will bring in all initial clauses relatively early anyway.
- The *Evolved* strategy significantly outperforms even the best other strategy.

4.2 Search Performance by Problem Class

Table 3 breaks down the performance of the different heuristics by problem class. Interesting observations are in particular in the unit categories. First, all strategies solved all non-equational unit problems. This is not surprising, since this category is decidable and comprises only the task of finding one pair of complementary unifiable literals. In the unit-equality category, *FIFO* is comparatively much weaker than in the other categories. Likewise, *GD* is weak, but makes a

Table 3. Number of problems solved in 300 s for different problem classes

Type	General		Horn		Unit	
Equational heuristic/size	Eq. (8626)	Non-eq. (1607)	Eq. (1011)	Non-eq. (1432)	Eq. (1037)	Non-eq. (61)
FIFO	2421 (28 %)	907 (56 %)	371 (37 %)	835 (58 %)	335 (32 %)	61 (100 %)
SC12	2160 (25 %)	842 (52 %)	432 (43 %)	828 (58 %)	649 (63 %)	61 (100 %)
SC11	2369 (27 %)	918 (57 %)	465 (46 %)	853 (60 %)	674 (65 %)	61 (100 %)
SC21	2410 (28 %)	978 (61 %)	428 (42 %)	800 (56 %)	649 (63 %)	61 (100 %)
RW212	2336 (27 %)	972 (60 %)	429 (42 %)	800 (56 %)	656 (63 %)	61 (100 %)
2SC11/FIFO	3809 (44 %)	1199 (75 %)	576 (57 %)	953 (67 %)	622 (60 %)	61 (100 %)
5SC11/FIFO	3798 (44 %)	1200 (75 %)	606 (60 %)	983 (69 %)	683 (66 %)	61 (100 %)
10SC11/FIFO	3803 (44 %)	1192 (74 %)	617 (61 %)	989 (69 %)	723 (70 %)	61 (100 %)
15SC11/FIFO	3732 (43 %)	1187 (74 %)	612 (61 %)	967 (68 %)	728 (70 %)	61 (100 %)
GD	2271 (26 %)	819 (51 %)	431 (43 %)	821 (57 %)	595 (57 %)	61 (100 %)
5GD/FIFO	3860 (45 %)	1153 (72 %)	606 (60 %)	967 (68 %)	732 (71 %)	61 (100 %)
SC11-PI	2894 (34 %)	968 (60 %)	523 (52 %)	913 (64 %)	712 (69 %)	61 (100 %)
10SC11/FIFO-PI	3929 (46 %)	1142 (71 %)	631 (62 %)	986 (69 %)	718 (69 %)	61 (100 %)
Evolved	4477 (52 %)	1201 (75 %)	712 (70 %)	1204 (84 %)	768 (74 %)	61 (100 %)

strong showing in the combination with *FIFO*. Most of the results in the non-unit problems are in line with the general performance discussed in the previous section. We do notice that general (i.e. non-Horn) problems with equality are the hardest class for the tested strategies.

4.3 Proof Size and Structure

We are interested in the properties of proofs actually found by the prover. Particular properties we are interested in are:

- Is there a substantial difference between proofs found by different heuristics?
- How many of the initial clauses are used in the proof? I.e. what is the size of the unsatisfiable core of the axioms (and negated conjecture) that the prover found? In addition to the general interest, this value also provides important information for tuning pre-search axiom pruning techniques [9] like SInE [5] and MePo [14].
- How many inferences are in a typical proof, and how many search decisions contribute to it?

In principle, we would expect *FIFO* to find shorter proofs, since the underlying search is breadth-first. However, simplification may complicate this, and

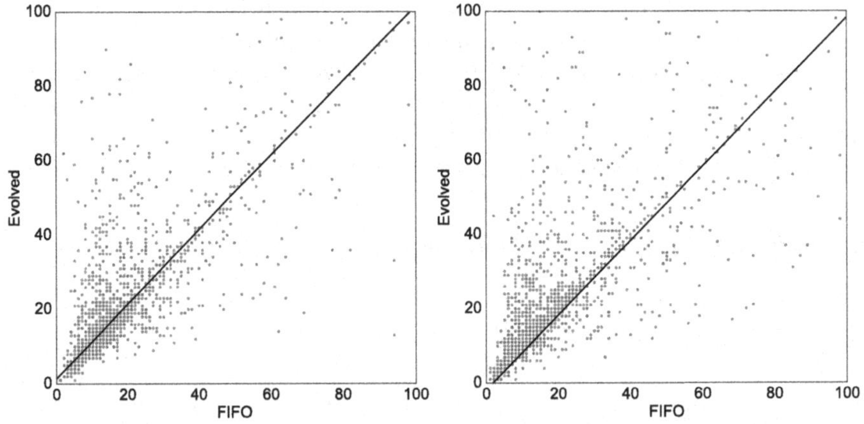

Proof size scatter plots. Each dot corresponds to one solved problem, with the size of the proof found by *Evolved* on the y-axis and the size of the *FIFO* proof on the X-axis. Proof size measure in the left is number of *given clauses* in the proof object, on the right it is total number of inferences in the proof. Both diagrams where cut off at 100 on each axis for better visibility. The left plot covers 93.8% and the right plot covers 90% of all data points. Only proofs where both strategies need at least 0.02 seconds are represented. The linear regression lines are $1.0x + 1.44$ for the left and $1.0x - 1.72$ for the right plot.

Fig. 3. Comparison of proof sizes

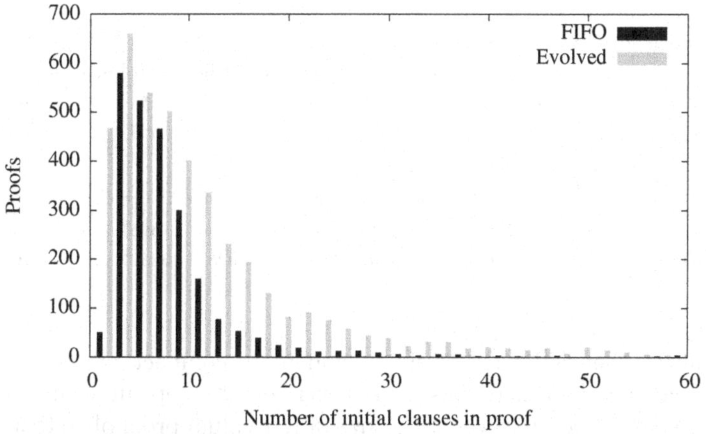

Heuristic	Proofs found	Initial clauses in proof					
		Mean	Minimum	First quartile	Median	Third quartile	Maximum
FIFO	4930	12.7	1	4	6	9	1330
Evolved	8423	18.3	1	5	9	15	1330

For *FIFO*, there are 104 proofs with more than 60 initial clauses in the proof object, i.e. the diagram covers 97.9% of all proofs. For *Evolved* there are 326 proofs with more than 60 initial clauses, i.e. the diagram covers 96.1% of all proofs.

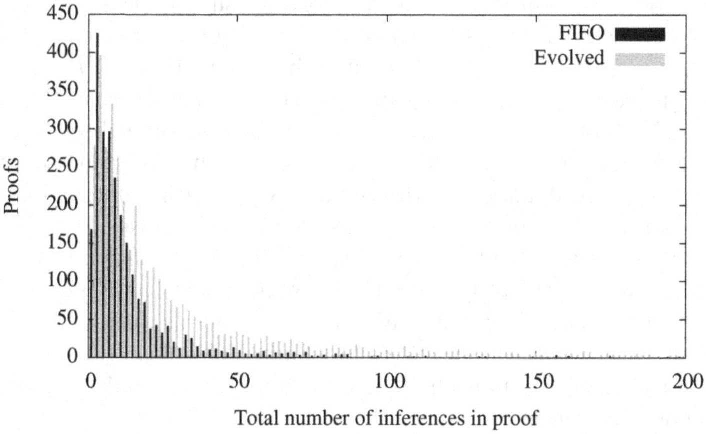

Heuristic	Proofs found	Given clauses in proof					
		Mean	Minimum	First quartile	Median	Third Quartile	Maximum
FIFO	4930	784.8	1	4	9	17	933822
Evolved	8423	587.2	1	7	17	49	933819

There are 171 *FIFO* proofs with more than 200 inferences, i.e. the diagram covers 96.5% of all proofs. There are 658 *Evolved* proofs with more than 200 inferences, i.e. the diagram covers 92.2% of all proofs.

Fig. 4. Distribution of the number of initial clauses and inferences in proofs (Color figure online)

Table 4. Number of clauses in proofs and proof searches

Heuristic	Proofs found	Given clauses in proof search					
		Mean	Minimum	First quartile	Median	Third quartile	Maximum
FIFO	4930	2302.5	1	28	157	875	209154
Evolved	8423	3598.7	1	38	188	1506	190309
Heuristic	Proofs found	Total clauses generated					
		Mean	Minimum	First quartile	Median	Third quartile	Maximum
FIFO	4930	342422.4	0	28	582	16951	21822536
Evolved	8423	356893.0	0	37	1023	38327	26187659

symbol-counting heuristics are likely to find more compact representations of the equational theory earlier, thus using fewer rewrite steps in normalization.

Figure 3 shows a comparison of the size of individual proof objects for proofs found with *FIFO* and *Evolved*, the two strategies with the widest difference in performance. While relative proof sizes are distributed over the whole diagram, there is a distinct increase in density towards the diagonal, and the computed regression is very close to the diagonal indeed. On average, *FIFO* proofs have slightly smaller number of given clauses, in line with our expectations. *Evolved* proofs have slightly fewer inferences. The difference is indeed due to the number of simplification steps. However, neither effect is very strong, and on average the proofs found by both heuristics seem to be of very similar sizes.

Figure 4 (top) shows the distribution of the number of initial clauses in proof objects. On average, there are 12.7 clauses in a non-trivial *FIFO* proof, and nearly 50 % more initial clauses in an *Evolved* proof. Note that this statistic is based on all proofs found by either strategy, not on the subset of problems solved by both strategies. The bulk of the weight of the distribution is towards small numbers of initial axioms, with the mean very much influenced by a small number of combinatorial problems that need over 1000 clauses.

A similar observation holds for the actual proof size as shown in Fig. 4 (bottom). By median, *Evolved* proofs are nearly twice as large as *FIFO* proofs, and at the third quartile, *Evolved* proofs are nearly 3 times as long as *FIFO* proofs. Thus, quite a lot of non-trivial proofs can be found. The mean proof size is again strongly influenced by a small number of combinatorial problems that require nearly a million inferences.

4.4 Proof Search Statistics and Performance

Table 4 shows the size of the search space constructed and traversed during the proof search. Comparing this with Fig. 4, we see that the number of *given clauses* actually processed to find a proof is orders of magnitude greater than the number of such clauses in the proof object. However, we also see that the number of clauses generated is again much larger, i.e. for non-trivial proofs many clauses derived by the inference engine are never processed.

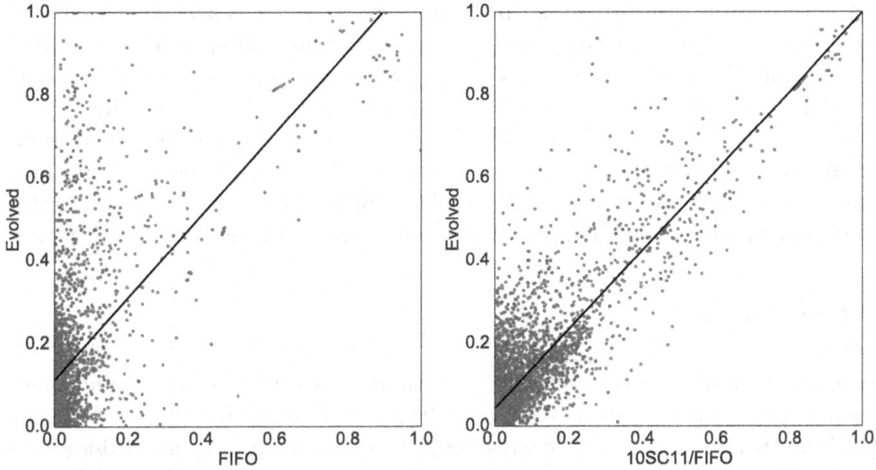

Given-clause utilization rate scatter plot. The vertical axis shows the given-clause utilization for *Evolved*, the horizontal axis for *FIFO* (left) and *10SC11/FIFO* (right). Only proofs where both strategies need at least 0.02 seconds are represented. The linear regression lines are $0.992x + 0.111$ for the left and $0.957x + 0.043$ for the right plot.

Fig. 5. Comparison of given-clause utilization ratios

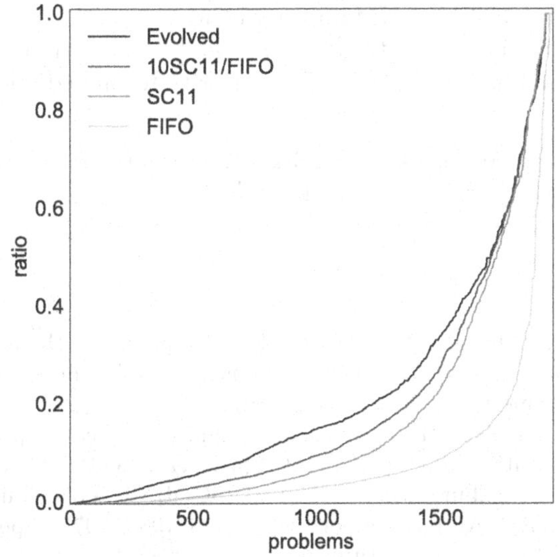

Given clause utilization rate for problems solved by four different strategies (only proofs for problems that are solved by all four strategies and where each needs at least 0.02 seconds are considered). The graph shows how many problems are solved with a given-clause utilization no better than the value on the vertical axis.

Fig. 6. Given clause utilization ratios over problem set (Color figure online)

An interesting measure is the fraction of processed *given clauses* that end up in the proof object, i.e. that represent good search decisions that contributed to the proof. We have plotted this *given-clause utilization* in Fig. 5 (comparing different heuristics pairwise) and in Fig. 6 (showing the distribution of the ratio over the set of problems solved by four representative strategies). In both diagrams it is clear that the *given-clause utilization* is, on average, quite low. Also, Fig. 6 strongly suggests that given-clause utilization is a good predictor for overall performance, with stronger strategies showing significantly better ratios.

5 Conclusion

Our analysis shows the comparative performance of several classical and modern clause selection heuristics. We can confirm that interleaving symbol-counting and FIFO selection shows significantly better performance than either does individually. We also found that preferring initial clauses is, on average, a significant advantage, and that goal-directed heuristics seem to work best in combination with other heuristics.

Proofs found by different heuristics for the same problem seem to be similar in size and complexity, however, stronger heuristics are able to find longer and more complex proofs. The average *given-clause utilization* as a measure of the quality of search decisions seems to correlate well with performance. It also shows us that even the best heuristics are far from optimal, or, to state it positively, that there still is a lot of room for improvement.

An open question is how far these results can be transferred to provers which employ the Otter loop, which places more priority to immediate simplification.

Acknowledgements. We thank the StarExec [21] team for providing the community infrastructure making these experiments possible.

References

1. Bachmair, L., Dershowitz, N., Plaisted, D.: Completion without failure. In: Ait-Kaci, H., Nivat, M. (eds.) Resolution of Equations in Algebraic Structures, vol. 2, pp. 1–30. Academic Press, New York (1989)
2. Bachmair, L., Ganzinger, H.: Rewrite-based equational theorem proving with selection and simplification. J. Log. Comput. **3**(4), 217–247 (1994)
3. Denzinger, J., Kronenburg, M., Schulz, S.: DISCOUNT: a distributed and learning equational prover. J. Autom. Reason. **18**(2), 189–198 (1997). (Special Issue on the CADE 13 ATP System Competition)
4. Denzinger, J., Schulz, S.: Recording and analysing knowledge-based distributed deduction processes. J. Symb. Comput. **21**(4/5), 523–541 (1996)
5. Hoder, K., Voronkov, A.: Sine qua non for large theory reasoning. In: Bjørner, N., Sofronie-Stokkermans, V. (eds.) CADE 2011. LNCS, vol. 6803, pp. 299–314. Springer, Heidelberg (2011)

6. Hsiang, J., Rusinowitch, M.: On word problems in equational theories. In: Ottmann, T. (ed.) Automata, Languages and Programming. LNCS, vol. 267, pp. 54–71. Springer, Heidelberg (1987)
7. Knuth, D., Bendix, P.: Simple word problems in universal algebras. In: Leech, J. (ed.) Computational Algebra, pp. 263–297. Pergamon Press, Oxford (1970)
8. Kovács, L., Voronkov, A.: First-order theorem proving and Vampire. In: Sharygina, N., Veith, H. (eds.) CAV 2013. LNCS, vol. 8044, pp. 1–35. Springer, Heidelberg (2013)
9. Kühlwein, D., van Laarhoven, T., Tsivtsivadze, E., Urban, J., Heskes, T.: Overview and evaluation of premise selection techniques for large theory mathematics. In: Gramlich, B., Miller, D., Sattler, U. (eds.) IJCAR 2012. LNCS, vol. 7364, pp. 378–392. Springer, Heidelberg (2012)
10. Löchner, B., Hillenbrand, T.: A phytography of Waldmeister. J. AI Commun. **15**(2/3), 127–133 (2002)
11. McCune, W.W.: Prover9 and Mace4 (2005–2010). http://www.cs.unm.edu/mccune/prover9/. Acccessed 29 Mar 2016
12. McCune, W.: Otter 3.0 Reference Manual and Guide. Technical report ANL-94/6, Argonne National Laboratory (1994)
13. McCune, W., Wos, L.: Otter: the CADE-13 competition incarnations. J. Autom. Reason. **18**(2), 211–220 (1997). (Special Issue on the CADE 13 ATP System Competition)
14. Meng, J., Paulson, L.C.: Lightweight relevance filtering for machine-generated resolution problems. J. Appl. Log. **7**(1), 41–57 (2009)
15. Riazanov, A., Voronkov, A.: The design and implementation of VAMPIRE. J. AI Commun. **15**(2/3), 91–110 (2002)
16. Riazanov, A., Voronkov, A.: Limited resource strategy in resolution theorem proving. J. Symb. Comput. **36**(1–2), 101–115 (2003)
17. Robinson, J.A.: A machine-oriented logic based on the resolution principle. J. ACM **12**(1), 23–41 (1965)
18. Schäfer, S., Schulz, S.: Breeding theorem proving heuristics with genetic algorithms. In: Gottlob, G., Sutcliffe, G., Voronkov, A. (eds.) Proceedings of Global Conference on Artificial Intelligence, EPiC, vol. 36, pp. 263–274. EasyChair, Tibilisi, Georgia (2015)
19. Schulz, S.: E - a brainiac theorem prover. J. AI Commun. **15**(2/3), 111–126 (2002)
20. Schulz, S.: System description: E 1.8. In: McMillan, K., Middeldorp, A., Voronkov, A. (eds.) LPAR-19 2013. LNCS, vol. 8312, pp. 735–743. Springer, Heidelberg (2013)
21. Stump, A., Sutcliffe, G., Tinelli, C.: StarExec: a cross-community infrastructure for logic solving. In: Demri, S., Kapur, D., Weidenbach, C. (eds.) IJCAR 2014. LNCS, vol. 8562, pp. 367–373. Springer, Heidelberg (2014)
22. Sutcliffe, G.: The TPTP problem library and associated infrastructure: the FOF and CNF parts, v3.5.0. J. Autom.Reason. **43**(4), 337–362 (2009)
23. Weidenbach, C., Schmidt, R.A., Hillenbrand, T., Rusev, R., Topic, D.: System description: SPASS version 3.0. In: Pfenning, F. (ed.) CADE 2007. LNCS (LNAI), vol. 4603, pp. 514–520. Springer, Heidelberg (2007)

Higher-Order Theorem Proving

Internal Guidance for Satallax

Michael Färber[1(✉)] and Chad Brown[2]

[1] Universität Innsbruck, Innsbruck, Austria
michael.faerber@uibk.ac.at
[2] Czech Technical University in Prague, Prague, Czech Republic

Abstract. We propose a new internal guidance method for automated theorem provers based on the given-clause algorithm. Our method influences the choice of unprocessed clauses using positive and negative examples from previous proofs. To this end, we present an efficient scheme for Naive Bayesian classification by generalising label occurrences to types with monoid structure. This makes it possible to extend existing fast classifiers, which consider only positive examples, with negative ones. We implement the method in the higher-order logic prover Satallax, where we modify the delay with which propositions are processed. We evaluated our method on a simply-typed higher-order logic version of the Flyspeck project, where it solves 26 % more problems than Satallax without internal guidance.

1 Introduction

Experience can be described as knowing which methods to apply in which context. It is a result of experiments, which can show a method to either fail or succeed in a certain situation. Mathematicians solve problems by experience. When solving a problem, mathematicians gain experience, which in the future can help them to solve harder problems that they would not have been able to solve without the experience gained before.

Fully automated theorem provers (ATPs) attempt to prove mathematical problems without user interaction. A thriving field of research is how to make ATPs behave more like mathematicians, by learning which decisions to take from previous proof attempts, in order to find more proofs in shorter time, and to prove problems that were previously out of reach for the ATP. Machine learning can help advance that field, for it provides techniques to model experience and to compare the quality of possible decisions. Machine learning approaches to improve ATP performance include:

- **Premise selection**: Preselecting a set of axioms for a problem can be done as a preprocessing step or inside the ATP at the beginning of proof search. Examples of this technique are the Sumo INference Engine (SInE) [HV11] and E.T. [KSUV15].
- **Internal guidance**: Unlike premise selection, internal guidance influences choices made during the proof search. The *hints* technique [Ver96] was among

N. Olivetti and A. Tiwari (Eds.): IJCAR 2016, LNAI 9706, pp. 349–361, 2016.
DOI: 10.1007/978-3-319-40229-1_24

the earliest attempts to directly influence proof search by learning from previous proofs. Other systems are E/TSM [Sch00], an extension of E [Sch13] with term space maps, and MaLeCoP [UVŠ11] respectively FEMaLeCoP [KU15], which are versions of leanCoP [Ott08] extended by Naive Bayesian learning.
- **Learning of strategies**: Finding good settings for ATPs automatically has been researched for example in the Blind Strategymaker (BliStr) project [Urb15].
- **Learning of strategy choice**: Once one has found good ATP strategies for different sets of problems, it is not directly clear which strategies to apply for which time when encountering a new problem. This problem was treated in the Machine Learning of Strategies (MaLeS) [Kü14].

In this paper, we show an internal guidance algorithm for ATPs that use (variations of) the given-clause algorithm. Specifically, we study a Naive Bayesian classification method, introduced for the connection calculus in FEMaLeCoP, and generalise it by measuring label occurrences with an arbitrary type having monoid structure, in place of a single number. This generalisation has the benefit that it can handle positive and negative occurrences. As a proof of concept, we implement the algorithm in the ATP Satallax [Bro12], using no features at all, which already solves 26 % more problems given the same amount of time, and which can solve about as many problems in 1 s than without internal guidance in 2 s.

2 Naive Bayesian Classifier with Monoids

2.1 Motivation

Many automated theorem provers have a proof state in which they make decisions, by ranking available choices (e.g. which proposition to process) and choosing the best one. This is related to the classification problem in machine learning, which takes data about previous decisions, i.e. which situation has led to which choice, and then orders choices by usefulness for the current situation.

For example, let us assume that the state of the theorem prover is modelled by the set of constants appearing in the previously processed propositions or in the conjecture. Let our conjecture be $x + y = y + x$ and let our premises include

$$\forall P.[P(0) \implies (\forall x.P(x) \implies P(s(x))) \implies \forall x.P(x)], \tag{1}$$

$$x + 0 = x. \tag{2}$$

If we first process Eq. 1, the prover state is characterised by $F = \{+, s, 0\}$. If we then continue to process Eq. 2 and it turns out that this contributes to the final proof, we register that in the situation F, Eq. 2 was useful.

In other proof searches, processing Eq. 2 in a certain prover state will not contribute towards the final proof. We call such situations negative examples.

Intuitively, we would like to apply propositions in situations that are similar to those in which the propositions were useful, and avoid processing propositions in situations similar to those where the propositions were useless. In general, examples (positive and negative) can be characterised by a prover state F and a proposition l that was processed in state F. This makes it possible to treat the choice of propositions as classification problem. In the next section, we show how to rank choices based on previous experience.

2.2 Classifiers with Positive Examples

A classifier takes pairs (F, l), relating a set of features F with a label l, and produces a function that, given a set of features, predicts a label. Classifiers can be characterised by a function $r(l, F)$, which represents the relevance of a label wrt a set of features. For internal guidance, we use r to estimate the relevance of a clause l to process in the current prover state F.

A Bayesian classifier estimates the relevance of a label by its probability to occur with a set of features, i.e. $P(l \mid F)$. By using the Naive Bayesian assumption that features are conditionally independent, the conditional probability is:

$$P(l \mid F) = \frac{P(l)P(F \mid l)}{P(F)} = \frac{P(l)\prod_{f\in F} P(f \mid l)}{P(F)} \propto P(l) \prod_{f \in F} P(f \mid l).$$

To increase numerical stability, we use sums of logarithms. Furthermore, we weight the probabilities with the inverse document frequency (IDF) of the features, and we omit the constant factor $P(F)$. The resulting classifier then is:

$$r(l, F) = \log P(l) + \sum_{f \in F} \log(\mathrm{idf}(f_i)) \log P(f \mid l).$$

In FEMaLeCoP, the simplified probability functions[1] are approximated by

$$P(l) \approx D_l, \qquad P(f \mid l) \approx \begin{cases} c & \text{if } D_{l,f} = 0 \\ \frac{D_{l,f}}{D_l} & \text{otherwise} \end{cases}$$

where $D_{l,f}$ denotes the number of times l appeared among the training examples in conjunction with f, D_l denotes how often l appeared among all training examples, and c is a constant.

2.3 Generalised Classifiers

In our experiments, we found negative training examples to be crucial for internal guidance. Therefore, we generalised the classifier to represent the type of occurrences as a *commutative monoid.*

Definition 1. *A pair $(M, +)$ is a* monoid *if there exists a neutral element* $0 \in M$ *such that for all* $x, y, z \in M$, $(x + y) + z = x + (y + z)$ *and* $x + 0 = 0 + x = x$. *If furthermore* $x + y = y + x$, *then the monoid is* commutative.

[1] We omitted several constant factors. Furthermore, FEMaLeCoP considers also features of training examples that are *not* part of the features F, albeit this is a further derivation of the theoretical model.

The generalised classifier is instantiated with a commutative monoid $(M, +)$ and reads triples (F, l, o), which in addition to features and label now store the label occurrence $o \in M$. For example, if the classifier is to support positive and negative examples, then one can use the monoid $(\mathbb{N} \times \mathbb{N}, +_2)$, where the first and second elements of the pair represent the number of positive respectively negative occurrences, the $+_2$ operation is pairwise addition, and the neutral element is $(0, 0)$. A triple learnt by this classifier could be $(F, l, (1, 2))$, meaning that l occurs with F once in a positive and twice in a negative way. Commutativity imposes that the order in which the classifier is trained does not matter.

We now formally define D_l (occurrences of label), $D_{l,f}$ (co-occurrences of label with feature) and idf (inverse document frequency):

$$\begin{aligned} D_l &= \sum \{o \mid (F, l', o) \in D, l = l'\}, \\ D_{l,f} &= \sum \{o \mid (F, l', o) \in D, l = l', f \in F\}, \\ \text{idf}(f) &= \frac{|D|}{|\{(F, l', o) \mid (F, l', o) \in D, f \in F\}|} \end{aligned}$$

With this, our classifier for positive and negative examples can be defined as follows:

$$P(l) = \frac{|p - n|}{p + n}(c_p p + c_n n), \qquad P(f_i \mid l) = \begin{cases} c & \text{if } D_{l,f} = 0 \\ c_p \frac{p_f}{p} + c_n \frac{n_f}{n} & \text{otherwise} \end{cases}$$

where $(p, n) = D_l$, $(p_f, n_f) = D_{l,f}$, and c, c_p, and c_n are constants. The term $\frac{|p-n|}{p+n}$ represents *confidence* and models our intuition that labels which appear always in the same role (say, as positive example) should have a greater influence than more ambivalent labels. For example, if a label occurs about the same number of times as positive and as negative example, confidence is approximately 0, and when a label is almost exclusively positive or negative, confidence is 1.

We call D_l, $D_{l,f}$, and idf classification data. They are precalculated to allow fast classification. Furthermore, new training examples can be added to existing classification data efficiently, similarly to [KU15].

3 Learning Scenarios

In this section, we still consider ATPs as black boxes, taking as input a problem and classification data for internal guidance, returning as output training data (empty if the ATP did not find a proof).

We propose two different scenarios to generate training data and to use it in subsequent proof searches, see Fig. 1:

- On-line learning: We run the ATP on every problem with classification data. For every problem the ATP solves, we update the classifier with the training data from the ATP proof.

– Off-line learning: We first run the ATP on all problems without classification data, saving training data for every problem solved. We then create classification data from the training data and rerun the ATP with the classifier on all problems.

While the second scenario can be parallelised, thus taking less wall-clock time, it has to treat every problem twice in the worst case (namely when every problem fails), thus taking up to double the CPU time of the first scenario.

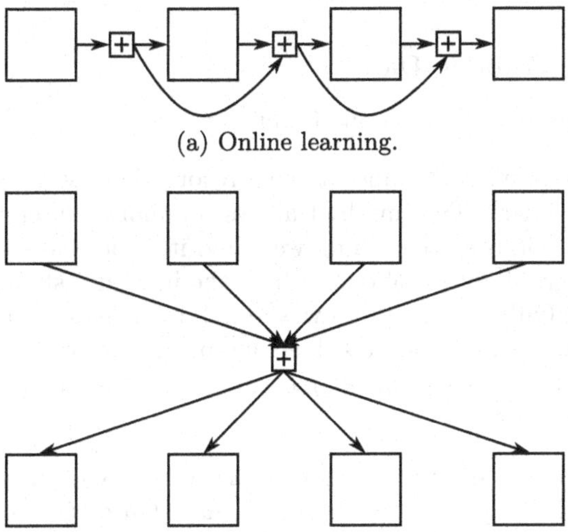

(a) Online learning.

(b) Offline learning.

Fig. 1. Comparison of online and offline learning. The large boxes symbolise an ATP proof search, which takes classifier data and returns training data (empty if no proof found). The small "+" boxes combine classifiers and training data, returning new classifier data.

4 Internal Guidance for Given-Clause Provers

Variants of the given-clause algorithm are commonly used in refutation-based ATPs, such as Vampire [KV13] or E [Sch13].[2] We introduce a simple version of the algorithm: Given an initial set of clauses to refute, the set of *unprocessed* clauses is initialised with the initial set of clauses, and the set of *processed* clauses is the empty set. At every iteration of the algorithm, a *given clause* is selected from the unprocessed clauses and moved to the processed clauses, possibly generating new clauses which are moved to the unprocessed clauses. The algorithm

[2] Technically, our reference prover Satallax does not implement a given-clause algorithm, as Satallax treats terms instead of clauses, and it interleaves the choice of unprocessed terms with other commands. However, for the sake of internal guidance, we can consider Satallax to implement a version of the given-clause algorithm. We describe the differences in more detail in Sect. 6.

terminates as soon as either the set of unprocessed clauses is empty or the empty clause was generated.

The integration of our internal guidance method into an ATP with given-clause algorithm involves two tasks: The recording of training data, and the ranking of unprocessed clauses, which influences the choice of the given clause. To reduce the amount of data an ATP has to load for internal guidance, we process training data and transform it into classification data outside of the ATP. We describe these tasks below in the order they are executed when no internal guidance data is present yet.

4.1 Recording Training Data

Recording training data can be done in different fashions:

- **In situ**: Information about clause usage is recorded every time an unprocessed clause gets processed. This method allows for more expressive prover state characterisation, on the other hand, we found it to decrease the proof success rate, as the recording of proof data makes the inference slower.
- **Post mortem**: Only when a proof was found, information about clause usage is reconstructed. As this method does not place any overhead on the proof search, we resorted to post-mortem recording, which is still sufficiently expressive for our purposes.

For every proof, we save: conjecture (if one was given), axioms A (premises given in the problem), processed clauses C, and clauses C_+ that were used in the final proof ($C_+ \subseteq C$). We call such information for a single proof a *training datum*. We ignore unprocessed clauses, as we cannot easily estimate whether they might have contributed to a proof.

4.2 Postprocessing Training Data

In our experiments, we frequently encounter clauses that are the same, differing only by containing different Skolem constants. To this end, we process the training data before creating classification data from it. We tried different techniques to handle Skolem constants, as well as other postprocessing methods:

- **Skolem filtering**: We discard clauses containing any Skolem constants.
- **Consistent Skolemisation**: We normalise Skolem constants inside all clauses, similarly to [UVŠ11]. That is, a clause $P(x, y, x)$, where x and y are Skolem constants, becomes $P(c_1, c_2, c_1)$.
- **Consistent normalisation**: Similarly to consistent Skolemisation, we normalise *all* symbols of a clause. That is, $P(x, y, x)$ as above becomes $c_1(c_2, c_3, c_2)$. This allows the ATP to discover similar groups of clauses, for example $a + b = b + a$ and $a * b = b * a$ both map to $c_1(c_2, c_3) = c_1(c_3, c_2)$, but on the other hand, this also maps possibly different clauses such as $P(x)$ and $Q(z)$ to the same clause. Still, in problem collections which do not share a common set of function constants (such as TPTP), this method is suitable.

- **Inference filtering**: An interesting experiment is to discard all clauses generated during proof search that are not part of the initial clauses.

We denote the consistent Skolemisation/normalisation of a clause c described above as $\mathcal{N}(c)$.

4.3 Transforming Training Data to Classification Data

For a given training datum with processed clauses C and proof clauses C_+, we define the corresponding classifier data to be:

$$\{(\mathcal{F}(c), c, (1,0)) \mid c \in C_+\} \cup \{(\mathcal{F}(c), c, (0,1)) \mid c \in C \setminus C_+\},$$

where $\mathcal{F}(c)$ denotes the features of a clause. We use the monoid $(\mathbb{N}\times\mathbb{N}, +_2, (0,0))$ introduced in Sect. 2, storing positive and negative examples. The classifier data of the whole training data is then the (multiset) union of the classifier data of the individual training data.

4.4 Clause Ranking

This section describes how our internal guidance method influences the choice of unprocessed clauses using a previously constructed classifier.

At the beginning of proof search, the ATP loads the classifier. Some learning ATPs, such as E/TSM [Sch00], select and prepare knowledge relevant to the current problem before the proof search. However, as we store classifier data in a hash table, filtering irrelevant knowledge to the problem at hand would require a relatively slow traversal of the whole table, whereas lookup of knowledge is fast even in the presence of a large number of irrelevant facts. For this reason we do not filter the classification data per problem.

Then, at every choice point, i.e. every time the ATP chooses a clause from the unprocessed clauses C, it picks a clause $c \in C$ that maximises the clause rank $R(c, F)$, where

$$R(c, F) = r_{\text{ATP}}(c) + r(\mathcal{N}(c), F)$$

and:

- $r_{\text{ATP}}(c)$ is an ATP function that calculates the relevance of a clause with traditional means (such as weight, age, ...),
- F is the current prover state,
- $r(c, F)$ is the Naive Bayesian ranking function as shown in Sect. 2, and
- $\mathcal{N}(c)$ is the normalisation function as introduced in Subsect. 4.2.

5 Tuning of Guidance Parameters

We employed two different methods to automatically find good parameters for internal guidance, such as c, c_p, and c_n from Sect. 2.

5.1 Off-Line Tuning

Off-line tuning analyses existing training data and attempts to find parameters that give proof-relevant clauses from the training data a high rank, while giving proof-irrelevant clauses a low rank. To do this, we evaluate for every training datum the following formula, which adds for every proof-relevant clause the number of proof-irrelevant clauses that were ranked higher:

$$\sum_{c_+ \in C_+} |\{c \mid R(c, F) > R(c_+, F_+), c \in C \setminus C_+\}|,$$

where C and C_+ come from the training datum (see Subsect. 4.1), F and F_+ are the features of the prover states when c respectively c_+ were processed (we reconstruct these from the training datum), and R is the ranking formula from Subsect. 4.4.

In the end, we sum up the results of the formula above for all training data, and take the guidance parameters which minimise that sum.

5.2 Particle Swarm Optimisation

Particle Swarm Optimisation [KE95] (PSO) is a standard optimisation algorithm that can be applied to minimise the output of a function $f(\boldsymbol{x})$, where $\boldsymbol{x}$ is a vector of continuous values. A *particle* is defined by a location $\boldsymbol{x}$ (a candidate solution for the optimisation problem) and a velocity $\boldsymbol{v}$. Initially, p particles are created with random locations and velocities. Then, at every iteration of the algorithm, a new velocity is calculated for every particle and the particle is moved by that amount. The new velocity of a particle is:

$$\boldsymbol{v}(t+1) = \omega \cdot \boldsymbol{v}(t) + \phi_p \cdot \boldsymbol{r}_p \cdot (\boldsymbol{b}_p(t) - \boldsymbol{x}(t)) + \phi_g \cdot \boldsymbol{r}_g \cdot (\boldsymbol{b}_g(t) - \boldsymbol{x}(t)),$$

where:

- $\boldsymbol{v}(t)$ is the old velocity of the particle,
- $\boldsymbol{b}_p(t)$ is the location of the best previously found solution among all particles,
- $\boldsymbol{b}_g(t)$ is the location of the best previously found solution of the particle,
- $\boldsymbol{r}_p$ and $\boldsymbol{r}_g$ are random vectors generated at every evaluation of the formula, and
- $\omega = 0.4$, $\phi_p = 0.4$, and $\phi_g = 3.6$ are constants.

We apply PSO to optimise the performance of an ATP on a problem set S. For this, we define $f(\boldsymbol{x})$ to be the number of problems in S the ATP can solve with a set of flags being set to $\boldsymbol{x}$ and with timeout t. We then run PSO and take the best global solution obtained after n iterations. We fixed $t = 1s$, $p = 300$, and $|S| = 1000$. The algorithm has worst-case execution time $t \cdot p \cdot n \cdot |S|$.

6 Implementation

We implement our internal guidance in Satallax version 2.8. Satallax is an automated theorem prover for higher-order logic, based on a tableaux calculus with extensionality and choice. It is written in OCaml by Brown [Bro12]. Satallax implements a priority queue, on which it places several kinds of proof search commands: Among the 11 different commands in Satallax 2.8, there are for example proposition processing, mating, and confrontation. Proof search works by processing the commands on the priority queue by descending priority, until a proof is found or a timeout is reached. The priorities assigned to these commands are determined by *flags*, which are the settings Satallax uses for proof search. A set of flag settings is called a *mode* (in other ATPs frequently called *strategies*) and can be chosen by the user upon the start of Satallax. Similar to other modern ATPs such as Vampire [KV13] or E [Sch13], Satallax also supports timeslicing via *strategies* (in other ATPs frequently called *schedules*), which define a set of modes together with time amounts Satallax calls each mode with. Formally, a strategy is a sequence $[(m_1, t_1), \ldots, (m_n, t_n)]$, where m_i is a mode and t_i the time to run the mode with. The total time of the strategy is the sum of times, i.e. $t_\Sigma(S) = \sum_{(m,t) \in S} t$.

As a side-effect of this work, we have extended Satallax with the capability of loading user-defined strategies, which was previously not possible as strategies were hard-coded into the program. Furthermore, we implemented modifying flags via the command line, which is useful e.g. to change a flag among all modes of a strategy without changing the flag among all files of a strategy. We used this extensively in the automatic evaluation of flag settings via PSO, as shown in Subsect. 5.2.

When running Satallax with a strategy S and a timeout t_{max}, then all the times of the strategy are multiplied by $\frac{t_{max}}{t_\Sigma(S)}$ if $t_{max} > t_\Sigma(S)$, to divide the time between modes appropriately when running Satallax for longer than what the strategy S specifies. Then, every mode m_i in the strategy is run sequentially for time t_i until a proof is found or the timeout t_{max} is hit.

An analysis of several proof searches yielded that on average, more than 90 % of commands put onto the priority queue of Satallax are proposition processing commands, which correspond to processing a clause from the set of unprocessed clauses in given-clause provers. For that reason, we decided to influence the priority of proposition processing commands, giving those propositions with a high probability of being useful a higher priority. The procedure follows the one described in Subsect. 4.4, but the ranking of a proposition is performed when the proposition processing command is put onto the priority queue, and the Naive Bayes rank is added to the priority that Satallax without internal guidance would have assigned to the command. As other types of commands are in the priority queue as well, we pay attention not to influence the priority of term processing commands too much (by choosing too large guidance parameters), as this can lead to disproportionate displacement of other commands.

To record training data, we use the terms from the proof search that contributed to the final proof. For this, Satallax uses `picomus` [Bie08] to construct a minimal unsatisfiable core.

To characterise the prover state of Satallax, we tried different kinds of features:

- Symbols of processed terms: We collect the symbols of all processed propositions at the time a proposition is inserted into the priority queue and call these symbols the features of the proposition. However, this experimentally turned out to be a bad choice, because the set of features for each proposition grows quite rapidly, as the set of processed propositions grows monotonically.
- Axioms of the problem: We associate every proposition processed in a proof search with all the axioms of the problem. In contrast to the method above, this associates the same features to all propositions processed during the proof search for a problem, and is thus more a characterisation of the problem (similar to TPTP characteristics [SB10]) than of the prover state.

In our experiments, just calculating the influence of these features without them actually influencing the priority makes Satallax prove less problems (due to the additional calculation time), and the positive impact of the features on the proof search does not compensate for the initial loss of problems. Therefore, we currently do not use features at all and associate the empty set of features to all labels, i.e. $\mathcal{F}(c) = \{\}$. However, it turns out that even without features, learning from previous proofs can be quite effective, as shown in the next section.

7 Evaluation

To evaluate the performance of our internal guidance method in Satallax, we used a THF0 [SB10] version (simply-typed higher-order logic) of the top-level theorems of the Flyspeck [HAB+15] project, as generated by Kaliszyk and Urban [KU14]. The test set consists of 14185 problems from topology, geometry, integration, and other fields. The premises of each problem are the actual premises that were used in the Flyspeck proofs, amounting to an average of 84.3 premises per problem.[3] We used an Intel Core i3-5010U CPU (2.1 GHz Dual Core, 3 MB Cache) and ran maximally one instance of Satallax at a time.

To evaluate the performance of the off-line learning scenario described in Sect. 3, we run Satallax on all Flyspeck problems, generating training data whenever Satallax finds a proof. We use the Satallax 2.5 strategy (abbreviated as "S2.5"), because the newest strategy in Satallax 2.8 can not always retrieve the terms that were used in the final proof, which is important to obtain training data.

As the off-line learning scenario involves evaluating every problem twice (once to generate training data and once to prove the problem with internal guidance), it doubles runtime in the worst case, i.e. if no problem is solved. Therefore, a user

[3] The test set, as well as our modified version of Satallax and instructions to recreate our evaluation, can be found under: http://cl-informatik.uibk.ac.at/~mfaerber/satallax.html.

might like to compare its performance to simply running the ATP with double timeout directly: When increasing the timeout from 1 s to 2 s, the number of solved problems increases from 2717 to 3394. However, this is mostly due to the fact that Satallax tries more modes, so to measure the gain in solved problems more fairly, we create a strategy "S2.5_1s" which contains only those modes that were already used during the 1 s run, and let each of them run about double the time. This strategy proves 2845 problems in 2 s.

We now compare the different postprocessing options introduced in Subsect. 4.2. For this, we create a classifier from the training data gathered during the 1 s run. We then run Satallax with internal guidance in off-line learning mode with 1 s timeout and the Satallax 2.5 strategy. We perform this procedure for each postprocessing option. We call a problem "lost" that Satallax with guidance could not solve and Satallax without guidance could. Vice versa for "gained". The results are given in Table 1. We perform best when influencing only the priority of axioms (inference filtering), solving 786 problems that could not be solved by Satallax in 1 s without internal guidance.

Table 1. Comparison of postprocessing options.

Postprocessing	Solved	Lost	Gained
Consistent normalisation	1911	920	114
Consistent Skolemisation	1939	885	107
None	2166	688	137
Skolem filtering	3395	98	776
Inference filtering	3428	75	786

To evaluate online learning, we run Satallax on all Flyspeck problems by ascending order, accumulating training data and using it for all subsequent proof searches. We filter away terms in the training data that contain Skolem variables. As result, Satallax with online learning, running 1 s per problem, solves 3374 problems (59 lost, 716 gained), which is a plus of 24 %.

In the next experiment, we evaluate the prover performance with the "S2.5_1s" strategy and a timeout of 30 s. For this, we use an 48-core server with 2.2 GHz AMD Opteron CPUs and 320 GB RAM, running 10 instances of Satallax in parallel. First, we run Satallax without internal guidance for 30 s, which solves 3097 problems. Next, we create from the training data a classifier with Skolem filtering, which takes 3 s and results in a 1.8M file. Finally, we run Satallax with internal guidance in off-line learning mode using the classifier. This proves 4028 problems in 30 s, which is a plus of 30 %. Results are shown in Fig. 2. The "jumps" in the data stem from changes of modes.

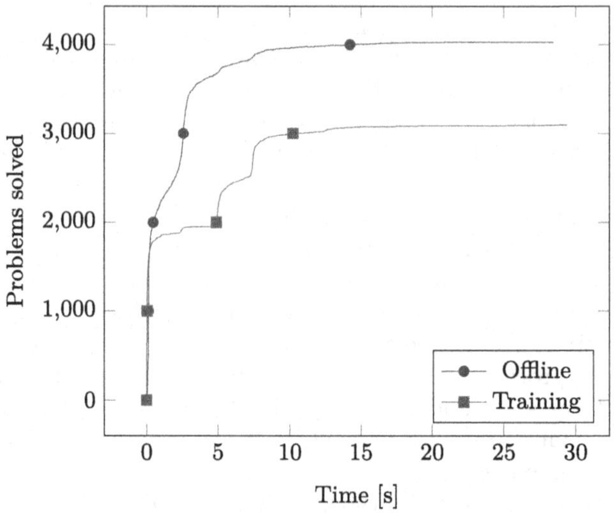

Fig. 2. Problems solved in a certain time (Color figure online).

8 Conclusion

We have shown how to integrate internal guidance into ATPs based on the given-clause algorithm, using positive as well as negative examples. We have demonstrated the usefulness of this method experimentally, showing that on a given test set, we can solve up to 26 % more problems. ATPs with internal guidance could be integrated into hammer systems such as Sledgehammer (which can already reconstruct Satallax proofs [SBP13]) or HOL(y)Hammer [KU14], continually improving their success rate with minimal overhead. It could also be interesting to learn internal guidance for ATPs from subgoals given by the user in previous proofs. Currently, we learn only from problems we could find a proof for, but in the future, we could benefit from considering also proof searches that did not yield proofs. Furthermore, it would be interesting to see the effect of negative examples on existing ATPs with internal guidance, such as FEMaLeCoP. We believe that finding good features that characterise prover state are important to further improve the learning results.

Acknowledgements. We would like to thank Sebastian Joosten and Cezary Kaliszyk for reading initial drafts of the paper, and especially Josef Urban for inspiring discussions and inviting the authors to Prague. Furthermore, we would like to thank the anonymous IJCAR referees for their valuable comments.

This work has been supported by the Austrian Science Fund (FWF) grant P26201 as well as by the European Research Council (ERC) grant AI4REASON.

References

[Bie08] Biere, A.: PicoSAT essentials. JSAT **4**(2–4), 75–97 (2008)

[Bro12] Brown, C.E.: Satallax: an automatic higher-order prover. In: Gramlich, B., Miller, D., Sattler, U. (eds.) IJCAR 2012. LNCS, vol. 7364, pp. 111–117. Springer, Heidelberg (2012)

[HAB+15] Hales, T.C., Adams, M., Bauer, G., Dang, D.T., Harrison, J., Le Hoang, T., Kaliszyk, C., Magron, V., McLaughlin, S., Nguyen, T.T., Nguyen, T.Q., Nipkow, T., Obua, S., Pleso, J., Rute, J., Solovyev, A., Ta, A.H.T., Tran, T.N., Trieu, D.T., Urban, J., Vu, K.K., Zumkeller, R.: A formal proof of the Kepler conjecture. CoRR, abs/1501.02155 (2015)

[HV11] Hoder, K., Voronkov, A.: Sine qua non for large theory reasoning. In: Bjørner, N., Sofronie-Stokkermans, V. (eds.) CADE 2011. LNCS, vol. 6803, pp. 299–314. Springer, Heidelberg (2011)

[KE95] Kennedy, J., Eberhart, R.: Particle swarm optimization. In: IEEE International Conference on Neural Networks, vol. 4, pp. 1942–1948, November 1995

[KSUV15] Kaliszyk, C., Schulz, S., Urban, J., Vyskocil, J.: System description: E.T. 0.1. In: Felty, A.P., Middeldorp, A. (eds.) CADE-25. LNCS (LNAI), vol. 9195, pp. 389–398. Springer, Heidelberg (2015)

[KU14] Kaliszyk, C., Urban, J.: Learning-assisted automated reasoning with Flyspeck. J. Autom. Reasoning **53**(2), 173–213 (2014)

[KU15] Kaliszyk, C., Urban, J.: FEMaLeCoP: fairly efficient machine learning connection prover. In: Davis, M., et al. (eds.) LPAR-20 2015. LNCS, vol. 9450, pp. 88–96. Springer, Heidelberg (2015). doi:10.1007/978-3-662-48899-7_7

[KV13] Kovács, L., Voronkov, A.: First-order theorem proving and VAMPIRE. In: Sharygina, N., Veith, H. (eds.) CAV 2013. LNCS, vol. 8044, pp. 1–35. Springer, Heidelberg (2013)

[Kü14] Daniel, A.K.: Machine learning for automated reasoning. Ph.D. thesis, Radboud Universiteit Nijmegen, April 2014

[Ott08] Otten, J.: leanCoP2.0 and ileanCoP1.2: high performance lean theorem proving in classical and intuitionistic logic (system descriptions). In: Armando, A., Baumgartner, P., Dowek, G. (eds.) IJCAR 2008. LNCS (LNAI), vol. 5195, pp. 283–291. Springer, Heidelberg (2008)

[SB10] Sutcliffe, G., Benzmüller, C.: Automated reasoning in higher-order logic using the TPTP THF infrastructure. J. Formalized Reasoning **3**(1), 1–27 (2010)

[SBP13] Sultana, N., Blanchette, J.C., Paulson, L.C.: LEO-II, Satallax on the Sledgehammer test bench. J. Appl. Logic **11**(1), 91–102 (2013)

[Sch00] Schulz, S.: Learning Search Control Knowledge for Equational Deduction. DISKI, vol. 230. Akademische Verlagsgesellschaft Aka GmbH Berlin, Berlin (2000)

[Sch13] Schulz, S.: System description: E 1.8. In: McMillan, K., Middeldorp, A., Voronkov, A. (eds.) LPAR-19 2013. LNCS, vol. 8312, pp. 735–743. Springer, Heidelberg (2013)

[Urb15] Urban, J.: BliStr: the blind Strategy maker. In: Gottlob, G., Sutcliffe, G., Voronkov, A. (eds.) GCAI 32015, Global Conference on Artificial Intelligence. EPiC Series in Computing, vol. 36, pp. 312–319. EasyChair (2015)

[UVŠ11] Urban, J., Vyskočil, J., Štěpánek, P.: MaLeCoP machine learning connection prover. In: Brünnler, K., Metcalfe, G. (eds.) TABLEAUX 2011. LNCS, vol. 6793, pp. 263–277. Springer, Heidelberg (2011)

[Ver96] Veroff, R.: Using hints to increase the effectiveness of an automated reasoning program: case studies. J. Autom. Reasoning **16**(3), 223–239 (1996)

Effective Normalization Techniques for HOL

Max Wisniewski, Alexander Steen, Kim Kern, and Christoph Benzmüller(✉)

Department of Mathematics and Computer Science, Freie Universität Berlin, Berlin, Germany
{max.wisniewski,a.steen,kim.kern,c.benzmueller}@fu-berlin.de

Abstract. Normalization procedures are an important component of most automated theorem provers. In this work we present an adaption of advanced first-order normalization techniques for higher-order theorem proving which have been bundled in a stand-alone tool. It can be used in conjunction with any higher-order theorem prover, even though the implemented techniques are primarily targeted on resolution-based provers. We evaluated the normalization procedure on selected problems of the TPTP using multiple HO ATPs. The results show a significant performance increase, in both speed and proving capabilities, for some of the tested problem instances.

1 Introduction

Problem normalization has always been an integral part of most automated theorem proving (ATP). Whereas early ATP systems relied heavily on external normalization and clausification, the normalization task has gradually been transferred to the prover themselves. The influence and success of FLOTTER [16] underlined the importance of careful employment of pre-processing techniques. Current state-of-the-art first-order ATP systems can spend a large portion of their execution time on pre-processing. Higher-order (HO) ATPs have not yet developed as sophisticated methods as their first-order counterparts and use hardly any sophisticated pre-processing techniques regarding clausification.

In this paper we present adaptations of prominent first-order techniques that improve clause normal form (CNF) calculations [12] first analyzed by Kern [8] in the context of higher-order logic (HOL; cf. [1] and the references therein). These adaptions are further augmented with HOL specific techniques and bundled in different normalization procedures. These procedures are intended as pre-processing routines for the new Leo-III theorem prover [17].

The effectiveness of these procedures is evaluated using a benchmark suite of over 500 higher-order problems. The measurements are conducted using the HO ATP systems LEO-II [5], Satallax [7] and Isabelle/HOL [10].

Furthermore, the normalization techniques are implemented in a stand-alone tool, called *Leonora*, ready to use with any TPTP-compliant HO ATP system.

This work has been supported by the DFG under grant BE 2501/11-1 (Leo-III).

N. Olivetti and A. Tiwari (Eds.): IJCAR 2016, LNAI 9706, pp. 362–370, 2016.
DOI: 10.1007/978-3-319-40229-1_25

2 Normalization Techniques

The first two techniques, *simplification* and *extensionality treatment*, are already implemented in most systems. Nonetheless, we briefly survey them in the following. They are essential to the overall normalization process since they allow, when combined with further techniques, a more thorough in-depth normalization in some cases (see e.g. Sect. 2.4).

2.1 Simplification and Extensionality Treatment

Simplification is a procedure invoked quite often during proof search. It resolves simple syntactical tautologies and antinomies, removes trivial quantifiers, and eliminates the constant symbols for truth and falsehood from a formula. In general, simplification can be used to minimize formulas and reduce the number of applicable inference rules.

Extensionality Treatment. In comparison to FOL, equalities in HOL can occur between terms of any type, especially between terms of Boolean type or functional type. To guarantee completeness, these equalities must also comply to the extensionality principle. This is often dealt with using special extensionality rules in the underlying calculus. In our context, we employ an adaption of the extensional RUE calculus rules as implemented in LEO-II [3,5]. Intuitively, the rule for equality on Booleans $\Phi = \Psi$ replaces the equality by the equivalence $\Phi \Leftrightarrow \Psi$ using the fact, that the domain of Booleans only contains truth and falsehood. For equality on functions, as in $f = g$, the rule states that two functions are equal if and only if they agree on each argument, hence we have $\forall X \,.\, f\,X = g\,X$ as result. We included both rules in the normalization framework for enabling deeply normalizing formulas: In some of the normalization procedures below, a rule can only be applied to a non-nested formula, i.e. not occurring at argument position. Using extensionality treatment, formulas can be lifted to top-level and then subsequently processed by other normalization steps.

2.2 Formula Renaming

Formula Renaming is a technique to reduce the size of the CNF [12]. Essentially, the idea is to split a clause into two separate clauses and to logically link them via a freshly introduced symbol that is added to both new clauses.

Definition 1 (Formula Renaming). *Let Φ be a formula and $\Psi_1 \circ \Psi_2$ a subterm of Φ, where $\circ$ is a binary Boolean connective. We replace Ψ_2 by $r(X_1, \ldots, X_n)$, where $\{X_1, \ldots, X_n\} = free(\Psi_2)$, r is a fresh predicate symbol (of appropriate type), and add a new clause D with*

$$D = \begin{cases} r(X_1, \ldots, X_n) \supset \Psi_2 & , \textit{if } polarity(\Psi_2) = 1 \\ \Psi_2 \supset r(X_1, \ldots, X_n) & , \textit{if } polarity(\Psi_2) = -1 \end{cases}$$

if the size of the CNF (denoted #CNF*) is decreasing, i.e.*

$$\#\text{CNF}(\Phi) > \#\text{CNF}(\Phi[\Psi_2 \setminus r(X_1, \ldots, X_n)]) + \#\text{CNF}(D).$$

The definition of $free(.)$ and $polarity(.)$ are hereby straight-forward adaptions of the their usual first-order counterparts (cf. e.g. [12]). We intentionally omitted the case of polarity(Ψ_2) = 0, since it is subsumed by a technique in Sect. 2.3.

Renaming reduces the size[1] of the CNF tremendously. In the example of a formula in disjunctive normal form, e.g. $(a \wedge b \wedge c) \vee (d \wedge e \wedge f)$, we obtain the nine multiplied cases $(a \vee d), (a \vee e), \ldots, (c \vee f)$. First renaming, however, yields the two clauses $(a \wedge b \wedge c) \vee r$ and $r \supset (d \wedge e \wedge f)$, which are normalized to six clauses $(a \vee r), (b \vee r), \ldots, (\neg r \vee e), (\neg r \vee f)$. This effect of formula renaming is mostly present in the multiplicative case of a β-rule, where the size of the CNF is reduced from a product (of the subterm sizes) to a sum. Thus, we can eliminate cases of exponential blowup in the transformation to CNF.

Reducing the search space in this manner can greatly boost the search process, as shown for first-order problems by Nonnengart et al. [11].

2.3 Argument Extraction

One major difference between higher-order and first-order logic is the shallow term-formula structure: Whereas in FOL we have the well-known distinct constructs of *formulas* and *terms*, in HOL there exist only terms (terms of Boolean type are still referred to as formulas). This allows in HOL the notion of nested formulas, that is, formulas p occurring at argument position of, e.g., an uninterpreted function symbol (in which case the polarity of p is 0). A treatment of these nested formulas is not immediately possible, since calculus rules dealing with Boolean formulas, such as clausification rules, cannot be applied to subterms. In order to apply these rules, the nested formulas have to be lifted to the top level, e.g. by decomposition rules as part of common unification procedures.

To allow immediate processing of nested formulas this lifting can be done in a pre-processing step [8]. Possible duplicated normalizations of nested formulas at a later proof search phase can thus be avoided. This *argument extraction* can be seen as a special higher-order case of *formula renaming* [12].

Definition 2 (Argument Extraction). *Let Φ be a formula with $f(p)$ occurring as subterm. We replace p if its head symbol is a logical connective. More precisely, let $\{X_1, \ldots, X_n\} = free(p)$. We introduce a new function symbol s (of appropriate type) and return $\Phi[p \setminus s(X_1, \ldots, X_n)]$ together with the definition $\forall X_1 \ldots X_n \,.\, p = s(X_1, \ldots, X_n)$.*

Consider the following theorem of HOL, which LEO-II is not able to solve:

$$\vdash \forall R \,.\, (R(\bot \Leftrightarrow (b \Leftrightarrow c)) \Rightarrow R((c \Leftrightarrow b) \Leftrightarrow \forall X \,.\, (X \wedge \neg X)))$$

[1] With *#CNF* we denote the number of clauses generated by transforming the given formula into clause normal form.

In this formula the arguments of both occurrences of the Boolean connective R can be extracted, resulting in two axioms and the remaining conjecture:

$$\begin{aligned} &s_1 \Leftrightarrow (\bot \Leftrightarrow (b \Leftrightarrow c)),\\ &s_2 \Leftrightarrow ((c \Leftrightarrow b) \Leftrightarrow \forall X \,.\, (X \wedge \neg X))\\ \vdash\, &\forall R \,.\, R(s_1) \Rightarrow R(s_2) \end{aligned}$$

It is easy to see that further normalization steps are enabled by argument extraction. In other words, some challenging HOL aspects have been eliminated from the given proof problem in a pre-processing step. In fact, the processed problem is now easily provable for LEO-II.

2.4 Extended Prenex Normal Form

A term Φ is in prenex normal form, if it is of the form $Q_1 X_1 \ldots Q_n X_n \,.\, \Psi$ where Q_i are quantifier symbols and Ψ does not contain any quantifier. Many proof calculi, especially unification-based calculi, work on clauses with implicitly bound variables. The quantifier for such an implicitly bound variable is (implicitly) always enclosing the whole formula. Hence, it is necessary to move the quantifiers outwards.

We first adapted the normalization of Nonnengart and Weidenbach [12] that first skolemizes existential quantifiers. A higher-order formula

$$\vdash (\forall X \,.\, a(X) \wedge \exists Y \,.\, Y) \wedge \forall Y \,.\, p(\forall X \,.\, b(Y) \supset a(X))$$

treated with the adopted algorithm for prenex normal form yields

$$\vdash \forall X \,.\, \forall Y \,.\, ((a(X) \wedge sk_1(X)) \wedge p(\forall X \,.\, b(Y) \supset a(X)).$$

With this simple adaption, a prenex form cannot be reached in HOL. As in Sect. 2.3 we have to cope with nested formulas. Moving the quantifiers out of the nested application is not possible. In fact, even skolemizing is impossible, since we loose track of the polarity inside the application. However, applying argument extraction will introduce a new axiom containing the nested Boolean argument. Subsequently processed with extensionality – forcing a hard polarity distinction – all quantifiers will now appear at top-level and can be treated with the standard adaption to transform the problem into a *pure* higher-order prenex form. We call this approach *extended prenex normal form* which is, up to the author's knowledge, novel in the context of HOL. Normalizing the example finally yields

$$\begin{aligned} &\forall Y \,.\, \forall X \,.\, \neg ek_1(Y) \vee (\neg b(Y) \vee a(X)),\\ &\forall Y \,.\, ek_1(Y) \vee (b(Y) \wedge \neg a(sk_2(Y)))\\ \vdash\, &\forall X \,.\, \forall Y \,.\, ((a(X) \wedge sk_1(X)) \wedge p(ek_1(Y)). \end{aligned}$$

3 Evaluation and Discussion

We have conducted several experiments to evaluate the potential benefits of the afore described normalization procedures. These experiments are designed to benchmark the number of problems that can be solved with the respective ATP as well as the time spent by the ATP on solving each individual problem.

The benchmark suite consists of overall 537 higher-order problems divided in eight domains from a broad field of application domains. The problems were taken from the TPTP library [14,15] (version 6.3.0) and coincide with the complete higher-order subsets of the corresponding TPTP problem domains[2] AGT, CSR, GEG, LCL, PHI, PUZ, QUA and SET.

For assessing the effectiveness of the proposed normalization pre-processing, we run the ATP systems on the original problems first and then on a series of differently normalized versions of the respective problems. These versions differ hereby in the number and combination of enabled normalization transformations from Sect. 2. More specifically, we investigated four different normalization procedures, denoted $N_1, \ldots, N_4$:

N_1 Enabled routines: Prenex form, argument extraction, formula renaming, simplification and extensionality processing (*i.e. full normalization*)
N_2 Enabled routines: Argument extraction, formula renaming, simplification and extensionality processing
N_3 Enabled routines: Prenex form, argument extraction, simplification and extensionality processing
N_4 Enabled routines: Argument extraction, simplification and extensionality processing

We chose to investigate different combinations of normalization techniques as pre-processing step to take into account that different ATP systems can use fundamentally different calculi and thus may benefit from different input conditions.

The measurements were primarily taken using the higher-order ATP systems LEO-II [5] and Satallax [7]. While the former system is based on higher-order resolution, the latter uses a sophisticated tableau-like approach. Selected benchmark results are additionally investigated using the automated and interactive theorem prover Isabelle/HOL [10]. In its automatic proof mode, Isabelle employs different solving tools such as the counter model finder Nitpick [6], the first-order tableau prover Blast [13], the SMT solver CVC4 [2] and several more.

As indicated before, for each original problem and each of the four normalized versions (by $N_1, \ldots, N_4$ respectively) we measure whether the system were able to solve the input problems as well as the time taken to do so. The CPU limit (timeout) for each problem and each ATP system is limited to 60 s. The measurements were taken on a 8 core (2x AMD Opteron Processor 2376 Quad Core) machine with 32 GB RAM.

Pre-processing time is not considered in the results below, since it still carries essentially little weight and does not considerably contribute to the overall CPU time.

[2] A comprehensive presentation of the different TPTP problem domains and their application domain can be found at http://www.cs.miami.edu/~tptp/cgi-bin/SeeTPTP?Category=Documents&File=OverallSynopsis.

Table 1. Measurement results for normalizations N_1 and N_2 over all benchmark domains

	AGT (23 Prob.)			**CSR** (123 Prob.)			**GEG** (18 Prob.)			**LCL** (139 Prob.)			**PHI** (10 Prob.)			**PUZ** (59 Prob.)			**QUA** (20 Prob.)			**SET** (145 Prob.)		
	Org	N_1	N_2	Org	N_1	N_2	Org	N_1	N_2	Org	N_1	N_2	Org	N_1	N_2	Org	N_1	N_2	Org	N_1	N_2	Org	N_1	N_2
LEO-II																								
Solved	23	23	23	61	69	45	12	13	11	104	93	104	6	5	5	31	32	31	0	0	0	134	134	135
-THM	23	23	23	57	65	41	12	13	11	99	88	99	6	5	5	31	32	31	0	0	0	134	134	135
Σ [s]	4.9	4.8	4.6	417	98	200	5.2	5.3	16.4	22.2	13.1	22.2	18.3	0.6	0.6	8.8	21	9.2	—	—	—	14.9	16.4	15.3
Avg. [s]	0.2	0.2	0.2	6.8	1.4	4.4	0.4	0.4	1.5	0.2	0.1	0.2	3	0.1	0.1	0.3	0.7	0.3	—	—	—	0.1	0.1	0.1
Satallax																								
Solved	18	19	19	58	78	51	17	17	15	113	114	112	7	7	7	35	37	33	2	2	2	138	136	138
-THM	18	19	19	54	74	47	17	17	15	104	105	103	7	7	7	31	34	30	2	2	2	138	136	138
Σ [s]	97	157	157	290	52	252	204	202	192	267	361	276	52	76	77	42	71	68	3.5	3.5	14.6	153	119	201
Avg. [s]	5.4	8.3	8.3	5	0.7	4.9	12	11.9	12.8	2.4	3.2	2.5	7.5	10.9	10.9	1.2	1.9	2.0	1.7	1.7	7.3	1.1	0.9	1.5

Results and Discussion. Table 1 displays the benchmark result summary. For the two ATP systems LEO-II and Satallax, the number of solved problems and the CPU time is shown for the original problems (denoted Org) in the respective domain and the pre-processed problem domains (denoted N_1 and N_2 respectively). The results for the remaining two normalization procedures are omitted since they are very similar to the ones shown.[3] The number of solved problems (thereof theorems) is denoted *Solved* (*-THM*). The sum (average) of CPU time spent on solving all input problems (that could be solved using the respective normalization procedure) is denoted Σ (*Avg.*).

As can be seen, in the case of the LEO-II prover, the normalization procedure N_1 results in more solved problems in benchmark domains CSR, GEG and PUZ. A decrease in solved problems can be observed in domains LCL (only 90 %solved) and PHI (83 % solved). The results of the remaining three benchmark domains only differ in the overall (and average) solving time. For the Satallax prover, the results are even better: More problems were solved in domains AGT, CSR, LCL and SET using N_1. Only in domain SET there are some problems that could not be proven anymore (see remark on domain SET below). In all cases, the normalization procedure N_2 did not improve the reasoning effectivity.

The most striking increase in solved problems can be observed in domain CSR where 8 problem were additionally solved due to N_1 (roughly 13 %) by LEO-II. Also, the overall (and average) proving time in N_1 only takes approximately a quarter of the original time (while proving more problems in that time). These observations also apply for Satallax, where 20 more problems (34 %) were solved by using N_1 while reducing the reasoning time to roughly one quarter of the original time. Detailed measurement results for problem domain CSR are shown in Table 2, where the fifteen best speed-ups are displayed for each employed ATP system. In order to provide additional evidence for the practicability of the presented normalization procedures, we included Isabelle/HOL in these measurements. The average speed-up for the Isabelle system is approx. 66 %. Here, some problems that were originally provable by Isabelle's CVC4 routine become provable by Blast after normalization with N_1, hence the speed-up.

[3] The average time results for normalization procedure N_3 are in nearly all cases within a range of 0.1 % of the results for N_1. Likewise results apply for N_4 and N_2.

Table 2. The 15 best relative time improvements with normalization N_1 in CSR domain for the respective prover (ordered by speed-up). Normalization N_2 is shown for comparison. A timeout result of a system is denoted †.

	Time [s]		
Problem	Orig.	N_1	N_2
CSR153^2	38.254	0.054	†
CSR138^1	9.858	0.029	9.875
CSR153^1	5.492	0.038	5.493
CSR126^2	31.425	0.676	31.451
CSR139^1	10.022	0.266	10.027
CSR137^2	1.351	0.039	4.270
CSR134^1	9.713	0.338	0.359
CSR122^2	19.307	0.683	19.266
CSR143^2	2.714	0.238	†
CSR153^3	7.743	0.988	†
CSR119^3	26.588	3.672	†
CSR120^3	26.609	3.678	†
CSR137^1	0.242	0.042	0.246
CSR152^3	14.960	3.671	†
CSR151^3	14.939	3.668	27.075

(a) Satallax

	Time [s]		
Problem	Orig.	N_1	N_2
CSR139^2	5.331	0.146	5.633
CSR132^2	5.331	0.283	†
CSR139^1	1.196	0.055	1.142
CSR150^1	1.633	0.091	1.604
CSR141^2	1.229	0.202	†
CSR148^1	0.295	0.057	†
CSR149^2	1.002	0.194	1.479
CSR123^2	0.930	0.192	†
CSR124^2	0.731	0.190	†
CSR122^2	0.725	0.193	†
CSR125^2	0.757	0.327	†
CSR119^2	0.355	0.173	†
CSR138^1	0.104	0.054	0.122
CSR120^2	0.388	0.202	†
CSR127^2	0.342	0.190	0.529

(b) LEO-II

	Time [s]		
Problem	Orig.	N_1	N_2
CSR128^2	52.286	14.305	49.489
CSR153^2	50.682	14.056	51.087
CSR131^2	49.519	13.983	52.484
CSR133^2	48.984	13.902	50.028
CSR148^2	43.199	14.172	42.063
CSR149^2	40.294	14.310	38.664
CSR138^2	39.387	15.305	40.481
CSR150^1	29.746	11.767	29.844
CSR130^2	49.632	21.136	46.207
CSR132^2	55.217	24.252	56.829
CSR129^2	47.401	20.979	45.682
CSR119^1	26.858	12.259	12.343
CSR141^2	52.501	26.366	44.813
CSR123^2	52.272	26.337	54.192
CSR127^2	27.227	14.248	26.311

(c) Isabelle

The above result show a significant increase in reasoning effectivity for problems of the CSR (common*s*ense *r*easoning) domain. In the investigated THF subset of that domain, a majority of problems represent HOL embeddings of SUMO [9] reasoning tasks.

Another major observation is that a TPTP rating 1.0 problem (i.e. a problem that could not be solved by any ATP system) became provable after normalization with procedure N_1. Here, the problem PUZ145^1 from the puzzles domain can be shown to be a theorem in 5.8s by Satallax.

Another interesting aspect of the normalization procedures is the impact on occurrences of defined equalities in problems. As discussed before, some problems in our benchmark suite became unprovable after normalization. This could be due to the fact that the HOL ATPs under consideration provide some special techniques for the manipulation of defined equalities, for example, Leibniz equalities. Leibniz equalities have the form $\forall P.Pa \Leftrightarrow Pb$, $\forall P.Pa \supset Pb$, $\forall P.\neg Pa \vee Pb$, etc., for arbitrary terms a and b.[4] The special techniques are aiming at a more goal directed equality handling as is possible with the above formulas, cf. [4].

However, our implemented normalization procedures do not yet support a similar detection and treatment of defined equalities. Thus, executing a normalization strategy can alter the structure of a (sub-)formula in a way that the HOL ATPs do subsequently not recognize it as an instance of a defined equality anymore. Augmenting our procedures with special techniques for defined equalities might therefore further improve the above results. Similar techniques might be useful for description and choice.

[4] The prover LEO-II, for example, is able to detect such (sub-)formulas and to replace them by primitive equalities $a = b$.

Implementation. For the above experiments, the described normalization techniques were implemented into a stand-alone pre-processing tool, called *Leonora* (for *Leo*'s *norm*a*lization*) that can be used to normalize any higher-order problem file in THF format [15], ready to employ in conjunction with any TPTP-compliant HO ATP system [14]. The selection of normalization steps to apply on the input problem can be controlled individually for each technique via flags, e.g. `-a` for enabling argument extraction or `-r` for formula renaming. A preliminary version of Leonora is freely available under MIT license and can be found at GitHub[5]. For the experiments we used version 1 of Leonora.

4 Further Work

Even though we have not yet included special techniques for defined equalities, the results show a significant improvement in some problem domains. Introducing such techniques has large potential to further improve these results.

Future work is to additionally support normalization techniques that are more suited for non-CNF based calculi. While it was indeed possible to improve Satallax's performance in both speed and proving capabilities in some problem domains, the system is using a quite different approach, i.e. a tableau method with an iterative queuing, SAT solver and special treatment for existentially quantified formulas. Applying the standard prenex algorithm could complicate the problems for Satallax. In order to deal with provers that are not suited for working with Skolem variables, we will additionally implement a procedure that moves all quantifiers outside and omits the skolemization.

Additionally there is a lot of potential for further improvements, such as finding meaningful size functions for the formula renaming procedure (cf. #CNF from Sect. 2.2). One example would be size functions that aim at minimizing the number of tableau branches created by the input problem.

5 Conclusion

In this work, we have adopted prominent first-order normalization techniques for higher-order logic. First benchmark results of these techniques on a set of HO problems indicate promising results. For each of the employed ATP systems the normalization procedures enables an improvement in both speed as well as number of solved problems for most benchmark domains. In some domains, up to 20 (34 %) more problems could be solved. Additionally, a problem that was not solved by any ATP system before could be solved by Satallax in less than six seconds after normalization.

We have observed that a straight-forward adaption of FOL techniques is not enough for the HOL case. Especially the treatment of nested formulas has great potential. Additionally, we have identified that a simple application of these techniques can interfere with occurrence of defined equalities, which is a

[5] The Leonora repository can be found at https://github.com/Ryugoron/Leonora.

problem specifically arising in HOL. By lifting nested formulas to top-level, we have on the one hand established a prenex normal form for HOL. On the other hand, the lifting contributes to the improvement of a prover's performance.

References

1. Andrews, P.: Church's type theory. In: Zalta, E.N. (ed.) The Stanford Encyclopedia of Philosophy. Stanford University, Spring (2014)
2. Barrett, C., Conway, C.L., Deters, M., Hadarean, L., Jovanović, D., King, T., Reynolds, A., Tinelli, C.: CVC4. In: Gopalakrishnan, G., Qadeer, S. (eds.) CAV 2011. LNCS, vol. 6806, pp. 171–177. Springer, Heidelberg (2011)
3. Benzmüller, C.: Higher-order automated theorem provers. In: Delahaye, D., Woltzenlogel Paleo, B. (eds.) All about Proofs, Proof for All. Mathematical Logic and Foundations, pp. 171–214. College Publications, London (2015)
4. Benzmüller, C., Brown, C., Kohlhase, M.: Cut-simulation, impredicativity. Log. Methods Comput. Sci. **5**(1:6), 1–21 (2009)
5. Benzmüller, C., Paulson, L.C., Sultana, N., Theiß, F.: The higher-order prover LEO-II. J. Autom. Reason. **55**(4), 389–404 (2015)
6. Blanchette, J.C., Nipkow, T.: Nitpick: a counterexample generator for higher-order logic based on a relational model finder. In: Kaufmann, M., Paulson, L.C. (eds.) ITP 2010. LNCS, vol. 6172, pp. 131–146. Springer, Heidelberg (2010)
7. Brown, C.E.: Satallax: an automatic higher-order prover. In: Gramlich, B., Miller, D., Sattler, U. (eds.) IJCAR 2012. LNCS, vol. 7364, pp. 111–117. Springer, Heidelberg (2012)
8. Kern, K.: Improved computation of CNF in higher-order logics. Bachelor thesis, Freie Universität Berlin (2015)
9. Niles, I., Pease, A.: Towards a standard upper ontology. In: Proceedings of the International Conference on Formal Ontology in Information Systems, pp. 2–9. ACM (2001)
10. Nipkow, T., Paulson, L.C., Wenzel, M. (eds.): Isabelle/HOL - A Proof Assistant for Higher-Order Logic. LNCS, vol. 2283. Springer, Heidelberg (2002)
11. Nonnengart, A., Rock, G., Weidenbach, C.: On generating small clause normal forms. In: Kirchner, C., Kirchner, H. (eds.) CADE 1998. LNCS (LNAI), vol. 1421, pp. 397–411. Springer, Heidelberg (1998)
12. Nonnengart, A., Weidenbach, C.: Computing small clause normal forms. In: Robinson, J., Voronkov, A. (eds.) Handbook of Automated Reasoning, vol. 1, pp. 335–367. Gulf Professional Publishing, Houston (2001)
13. Paulson, L.C.: A generic tableau prover and its integration with isabelle. J. Univers. Comput. Sci. **5**(3), 73–87 (1999)
14. Sutcliffe, G.: The TPTP problem library and associated infrastructure. J. Autom. Reason. **43**(4), 337–362 (2009)
15. Sutcliffe, G., Benzmüller, C.: Automated reasoning in higher-order logic using the TPTP THF infrastructure. J. Formaliz. Reason. **3**(1), 1–27 (2010)
16. Weidenbach, C., Gaede, B., Rock, G.: SPASS & FLOTTER version 0.42. In: McRobbie, M.A., Slaney, J.K. (eds.) Cade-13. LNCS, vol. 1104, pp. 141–145. Springer, Heidelberg (1996)
17. Wisniewski, M., Steen, A., Benzmüller, C.: The Leo-III project. In: Bolotov, A., Kerber, M. (eds.) Joint Automated Reasoning Workshop and Deduktionstreffen, p. 38 (2014)

Modal and Temporal Logics

Complexity Optimal Decision Procedure for a Propositional Dynamic Logic with Parallel Composition

Joseph Boudou(✉)

IRIT – Toulouse University, Toulouse, France
boudou@irit.fr

Abstract. $\mathsf{PPDL}^{\mathrm{det}}$ extends propositional dynamic logic (PDL) with parallel composition of programs. This new construct has separation semantics: to execute the parallel program $(\alpha \,||\, \beta)$ the initial state is separated into two substates and the programs α and β are executed on these substates. By adapting the elimination of Hintikka sets procedure, we provide a decision procedure for the satisfiability problem of $\mathsf{PPDL}^{\mathrm{det}}$. We prove that this decision procedure can be executed in deterministic exponential time, hence that the satisfiability problem of $\mathsf{PPDL}^{\mathrm{det}}$ is EXPTIME-complete.

1 Introduction

Propositional dynamic logic (PDL) is a multi-modal logic designed to reason about behaviors of programs [11,23]. A modal operator $\langle\alpha\rangle$ is associated to each program α, formulas $\langle\alpha\rangle\varphi$ being read "the program α can be executed from the current state to reach a state where the formula φ holds". The set of programs is structured by the following operators: sequential composition $(\alpha\,;\beta)$ of programs α and β executes β after α; nondeterministic choice $(\alpha \cup \beta)$ of programs α and β executes α or β; test $\varphi?$ on formula φ checks whether the current state satisfies φ; iteration α^* of program α executes α a nondeterministic number of times. The satisfiability problem of PDL is EXPTIME-complete [11,23]. Since PDL programs are abstract, this logic has been successfully adapted to many different domains like knowledge representation or linguistics [9,10,26].

A limitation of PDL is the lack of a construct to reason about concurrency. Different extensions of PDL have been devised to overcome this limitation; let us mention interleaving PDL [1], PDL with intersection [13] and the concurrent dynamic logic [22]. A noteworthy property of these logics is that whenever a parallel program is executable, some of its subprograms are executable too. But in some situations, for example when some agents are forced to cooperate, it may be the case that the parallel composition of some programs is executable while no other programs (but tests) are. The propositional dynamic logic with storing, recovering and parallel composition (PRSPDL) [4] can cope with such situations. This logic extends PDL with parallel compositions of programs and four special programs (for storing and recovering). These five new constructs are

N. Olivetti and A. Tiwari (Eds.): IJCAR 2016, LNAI 9706, pp. 373–388, 2016.
DOI: 10.1007/978-3-319-40229-1_26

inspired by fork algebras [12] and their semantics rely on a single ternary relation which intuitively models the decomposition of states into two substates. For the parallel program $(\alpha \,||\, \beta)$ to be executed at some state x, x must be decomposed into two states w_1 and w_2 by the ternary relation, then α is executed at w_1 reaching w_3, β is executed at w_2 reaching w_4 and the final state y is obtained by composing w_3 and w_4 by the ternary relation. These semantics for parallel programs are inspired by the concurrent separation logic [6,21] and the ternary relation, called the *separation relation*, is closely related to the Kripke semantics of binary normal modal logics like the Boolean logic of bunched implication [18, 24]. In contrast with unary modalities which usually express relations *between* states, a binary modality can express the internal structure of states, a formula of the form $\varphi \circ \psi$ being read "the current state can be decomposed in two substates, the first one satisfying φ, the other one satisfying ψ". In PRSPDL, the binary modality interpreted by the separation relation can be defined as $\varphi \circ \psi \doteq \langle \varphi? \,||\, \psi? \rangle \top$. Hence, PRSPDL embeds both PDL and the minimal binary normal modal logic. Since binary modalities have been used in various fields of logics (see for instance [14,19,27]), combining one with PDL's actions is promising. Despite these interesting features, little is known about PRSPDL. To our knowledge, the only complexity results to date are that the satisfiability problem of PRSPDL is in 2EXPTIME [2] and that variants of PRSPDL interpreted in classes of models where the decomposition of states is deterministic are undecidable [3].

In the present work, we study the logic $\mathsf{PPDL}^{\mathrm{det}}$ which is a variant of PRSPDL. Its language is the fragment of PRSPDL without the four store/recover programs. These special programs were designed to reason about data structures and are of little use to reason about concurrency. The formulas of $\mathsf{PPDL}^{\mathrm{det}}$ are interpreted in the class of models where the composition of states is deterministic: there is at most one way to merge two states. This semantic condition, called $\lhd$-determinism, is quite natural and has been studied in many logics with a binary modality such as Boolean logics of bunched implication [18,24], separation logics [8,25] and arrow logics [19]. Formally, $\lhd$-determinism forces the ternary relation to be a partial binary operator. The satisfiability problem for $\mathsf{PPDL}^{\mathrm{det}}$ has been shown in [5] to be in NEXPTIME, but the exact complexity of this problem was still unknown. In the present paper, we adapt Pratt's elimination of Hintikka sets decision procedure for PDL [23] to prove that the satisfiability problem of $\mathsf{PPDL}^{\mathrm{det}}$ is EXPTIME-complete. Thus, adding a $\lhd$-deterministic parallel composition to PDL does not increase the complexity of the logic. The adaptation of the elimination of Hintikka sets procedure to $\mathsf{PPDL}^{\mathrm{det}}$ is not straightforward. First, as it has been already outlined in [2,5], a comprehensive decomposition of formulas such as the Fischer-Ladner closure is not expressible in $\mathsf{PPDL}^{\mathrm{det}}$. Second, whereas in Pratt [23] states can be considered independently, for $\mathsf{PPDL}^{\mathrm{det}}$ the decomposition path leading to each state must be remembered. Third, like for the filtration [5], $\lhd$-determinism is not preserved by the elimination of Hintikka set procedure.

The paper is organized as follows. In the next section, the language and semantics of $\mathsf{PPDL}^{\mathrm{det}}$ are introduced, along with the $\mathsf{PPDL}^{\mathrm{det}}$ specific notions of threads and twines. In Sect. 3, an adaptation of the Fischer-Ladner closure to

$\mathsf{PPDL}^{\mathsf{det}}$ is proposed. Section 4 presents the optimal decision procedure, which is proved to be complete and sound in Sects. 5 and 6 respectively. Section 7 draws a conclusion and proposes perspectives for future works.

2 Language and Semantics of $\mathsf{PPDL}^{\mathsf{det}}$

Let Π_0 be a countable set of atomic programs (denoted by $a, b \ldots$) and Φ_0 a countable set of propositional variables (denoted by $p, q \ldots$). The sets Π and Φ of programs and formulas are defined by:

$$\alpha, \beta := a \mid (\alpha\,;\beta) \mid (\alpha \cup \beta) \mid \varphi? \mid \alpha^* \mid (\alpha \parallel \beta)$$
$$\varphi := p \mid \bot \mid \neg\varphi \mid \langle\alpha\rangle\varphi$$

Parentheses may be omitted for clarity. We define the dual modalities as usual: $[\alpha]\,\varphi \doteq \neg\langle\alpha\rangle\neg\varphi$. The missing Boolean operators can be defined too, starting with $\varphi \rightarrow \psi \doteq [\varphi?]\,\psi$. The syntactic operator $\sim$ is defined such that $\sim\varphi = \psi$ if $\varphi = \neg\psi$ for some ψ and $\sim\varphi = \neg\varphi$ otherwise. We write $|\alpha|$ and $|\varphi|$ for the number of occurrences of symbols in the program α and the formula φ, respectively.

A frame is a tuple $(W, R, \lhd)$ where W is a non-empty set of states (denoted by $w, x, y \ldots$), R is a function assigning a binary relation over W to each atomic program and $\lhd$ is a ternary relation over W called the separation relation. Intuitively, $x\,R\,(a)\,y$ means that the program a can be executed in state x, reaching state y. Similarly, $x \lhd (y, z)$ means that x can be split into the states y and z. We say that y and z are *substates* of x by the decomposition $(x, y, z) \in\lhd$. Equivalently, $x \lhd (y, z)$ means that the substates y and z can be merged to obtain x. When the merging of substates is functional, the frame is said to be $\lhd$-deterministic. This is a common restriction, for instance in Boolean logics of bunched implication [18]. Formally, a frame is $\lhd$-deterministic iff for all $x, y, w_1, w_2 \in W$,

$$\text{if } x \lhd (w_1, w_2) \text{ and } y \lhd (w_1, w_2) \text{ then } x = y \qquad (\lhd\text{-determinism})$$

A model is a tuple $(W, R, \lhd, V)$ where $(W, R, \lhd)$ is a frame and V is a valuation function associating a subset of W to each propositional variable. A model is $\lhd$-deterministic iff its frame is $\lhd$-deterministic. The forcing relation $\vDash$ is defined by parallel induction along with the extension of R to all programs:

$\mathcal{M}, x \vDash p$	iff $x \in V(p)$
$\mathcal{M}, x \vDash \bot$	never
$\mathcal{M}, x \vDash \neg\varphi$	iff $\mathcal{M}, x \nvDash \varphi$
$\mathcal{M}, x \vDash \langle\alpha\rangle\varphi$	iff $\exists y \in W,\ x\,R\,(\alpha)\,y$ and $\mathcal{M}, y \vDash \varphi$
$x\,R\,(\alpha\,;\beta)\,y$	iff $\exists z \in W,\ x\,R\,(\alpha)\,z$ and $z\,R\,(\beta)\,y$
$x\,R\,(\alpha \cup \beta)\,y$	iff $x\,R\,(\alpha)\,y$ or $x\,R\,(\beta)\,y$
$x\,R\,(\varphi?)\,y$	iff $x = y$ and $\mathcal{M}, x \vDash \varphi$
$x\,R\,(\alpha^*)\,y$	iff $x\,R\,(\alpha)^*\,y$ where $R\,(\alpha)^*$ is the reflexive and transitive closure of $R\,(\alpha)$
$x\,R\,(\alpha \parallel \beta)\,y$	iff $\exists w_1, w_2, w_3, w_4 \in W$, $x \lhd (w_1, w_2), w_1\,R\,(\alpha)\,w_3, w_2\,R\,(\beta)\,w_4$ and $y \lhd (w_3, w_4)$

$\mathsf{PPDL}^{\mathrm{det}}$ is the logic with language Φ interpreted in the class of $\lhd$-deterministic frames. A formula $\varphi \in \Phi$ is $\mathsf{PPDL}^{\mathrm{det}}$ *satisfiable* iff there exists a $\lhd$-deterministic model $\mathcal{M} = (W, R, \lhd, V)$ and a state $w \in W$ such that $\mathcal{M}, w \vDash \varphi$. The satisfiability problem of $\mathsf{PPDL}^{\mathrm{det}}$ is the decision problem answering whether a formula in Φ is $\mathsf{PPDL}^{\mathrm{det}}$ satisfiable.

Because of the semantics of parallel programs, $\mathsf{PPDL}^{\mathrm{det}}$ does not have the tree-like model property. To overcome this difficulty, $\mathsf{PPDL}^{\mathrm{det}}$ models can be partitioned into parts which have good properties. In [5], *threads* and *twines* were introduced for that purpose. A thread is a set of states which can be reached from each other by some program. Formally, given a model $\mathcal{M} = (W, R, \lhd, V)$, a thread is an equivalence class over W by the symmetric and transitive closure of the relation $\leadsto$ defined by: $x \leadsto y$ iff there exists some program α such that $x \; R(\alpha) \; y$. A twine is either a thread which contains no substates of another state or a pair of threads such that whenever a state in one thread is a substate by a decomposition then the other substate by this decomposition belongs to the other thread in the twine. Formally, a twine is an ordered pair $\theta = (t_\ell, t_r)$ of threads such that for all $x, y, z \in W$ if $x \lhd (y, z)$ then $y \notin t_r$, $z \notin t_\ell$ and $y \in t_\ell \Leftrightarrow z \in t_r$. We abusively identify twines with the union of their threads. It has been proved in [5] that whenever a formula is $\mathsf{PPDL}^{\mathrm{det}}$ satisfiable, it is satisfiable in a model $\mathcal{M} = (W, R, \lhd, V)$ such that the set of twines of $\mathcal{M}$ is a partition of W and for any twine θ in $\mathcal{M}$, there is at most two decompositions $(w, x, y) \in \lhd$ such that $\{x, y\} \subseteq \theta$.

3 Fischer-Ladner Closure

The Fischer-Ladner closure [11] is a decomposition of PDL formulas into a comprehensive set of subformulas. This decomposition is used in the elimination of Hintikka sets decision procedure of Pratt [23]. For $\mathsf{PPDL}^{\mathrm{det}}$ we need such a decomposition but parallel compositions of programs cause some difficulties. Firstly, the language of $\mathsf{PPDL}^{\mathrm{det}}$ is not expressive enough for a set of formulas to capture the semantics of formulas of the form $\langle \alpha \parallel \beta \rangle \varphi$. What is missing is some formulas to put after the modalities $\langle \alpha \rangle$ and $\langle \beta \rangle$. For this purpose, we add the special propositional variables L_1, L_2, R_1 and R_2. These new propositional variables identify corresponding left and right components of decompositions by the separation relation: if $x \lhd (y, z)$ then L_t and R_t identify y and z respectively, for some $t \in \{1, 2\}$. These propositional variables are special because we will not include their valuation in the model. Instead, we will allow us to modify their valuation depending on the states we consider (see Sect. 6). Two pairs of new propositional variables are needed because we will consider pairs of decompositions in Sect. 4 and we will need to distinguish them. We write $\Delta^+ = \{L_1, L_2, R_1, R_2\}$, $\Phi_0^+ = \Phi_0 \cup \Delta^+$ and Φ^+ for the formulas over Π_0 and Φ_0^+. Implicitly, Φ denotes the set of formulas over Π_0 and Φ_0, i.e. formulas which do not contain any propositional variables from Δ^+. Secondly, we need to keep track of the level of separation of each subformula. Hence we consider *localized formulas*. A *location* is a word on the alphabet $\{\ell, r\}$, the empty word being denoted by ϵ. A localized formula is a pair (μ, φ) composed of a location μ and a formula φ.

Then, given a localized formula (μ, φ) we construct the *closure* $\mathrm{Cl}(\mu, \varphi)$ of (μ, φ) as the least set of localized formulas containing (μ, φ) and closed by the rules in Fig. 1. It has to be noted that the closure presented here is an *ad hoc* decomposition devised for the needs of the decision procedure in Sect. 4. More general (and involved) closures for $\mathsf{PPDL}^{\mathrm{det}}$ have been presented in [2,5].

$$\frac{(\mu,\ \langle a\rangle\varphi)}{(\mu,\ \varphi)} \qquad \frac{(\mu,\ \varphi)}{(\mu,\ \sim\varphi)}$$

$$\frac{(\mu,\ \langle\varphi?\rangle\psi)}{(\mu,\ \varphi) \quad (\mu,\ \psi)} \qquad \frac{(\mu,\ \langle\alpha\,;\beta\rangle\varphi)}{(\mu,\ \langle\alpha\rangle\langle\beta\rangle\varphi)}$$

$$\frac{(\mu,\ \langle\alpha^*\rangle\varphi)}{(\mu,\ \langle\alpha\rangle\langle\alpha^*\rangle\varphi) \quad (\mu,\ \varphi)} \qquad \frac{(\mu,\ \langle\alpha\cup\beta\rangle\varphi)}{(\mu,\ \langle\alpha\rangle\varphi) \quad (\mu,\ \langle\beta\rangle\varphi)}$$

$$\frac{(\mu,\ \langle\alpha\parallel\beta\rangle\varphi)}{(\mu.\ell,\ \langle\alpha\rangle L_1) \quad (\mu.\ell,\ \langle\alpha\rangle L_2) \quad (\mu.r,\ \langle\beta\rangle R_1) \quad (\mu.r,\ \langle\beta\rangle R_2) \quad (\mu,\ \varphi)}$$

Fig. 1. Rules of the closure calculus

In the remainder of this paper we will be mainly interested in the closure of localized formulas of the form (ϵ, φ_0) for some formula $\varphi_0 \in \Phi$. We define the abbreviations $\mathrm{Cl}(\varphi_0) = \mathrm{Cl}(\epsilon, \varphi_0)$, $\mathrm{SP}(\varphi_0) = \{\alpha \mid \exists\mu, \exists\varphi, (\mu, \langle\alpha\rangle\varphi) \in \mathrm{Cl}(\varphi_0)\}$ and $\mathrm{Loc}(\varphi_0) = \{\mu \mid \exists\varphi, (\mu, \varphi) \in \mathrm{Cl}(\varphi_0)\}$. The cardinality of $\mathrm{Cl}(\varphi_0)$ is denoted by N_{φ_0}. It can be easily checked that the closure has the two properties expressed by Lemmas 1 and 2.

Lemma 1. *For any location μ, any program α and any formulas $\psi \in \Phi^+$ and $\varphi_0 \in \Phi$, if $(\mu, \langle\alpha\rangle\psi) \in Cl(\varphi_0)$ then $(\mu, \psi) \in Cl(\varphi_0)$.*

Lemma 2. *For any location μ and and any formulas $\psi \in \Phi^+$ and $\varphi_0 \in \Phi$, if $(\mu, \psi) \in Cl(\varphi_0)$ and $\psi? \in \mathrm{SP}(\varphi_0)$ then there are no occurrences of propositional variables from Δ^+ in ψ.*

Moreover, the proof from [11] can be adapted to prove the following lemma:

Lemma 3. *The cardinality of $Cl(\varphi_0)$ is linear in $|\varphi_0|$.*

Proof. For any localized formula (μ, φ), we define the *restricted* closure $\mathrm{rCl}(\mu, \varphi)$ of φ like the closure $\mathrm{Cl}(\mu, \varphi)$ except that the rules for negations, iterations, nondeterministic choices and parallel compositions are replaced with the rules in Fig. 2. The new propositional variables of the form Q_ψ serve the same role as in [11]. Obviously, $\mathrm{Cl}(\epsilon, \varphi_0)$ can be obtained from $\mathrm{rCl}(\epsilon, \varphi_0)$ and the cardinality of $\mathrm{Cl}(\epsilon, \varphi_0)$ is not greater than four times the cardinality of $\mathrm{rCl}(\epsilon, \varphi_0)$. Then, the function γ on programs and formulas is inductively defined by:

$$
\begin{aligned}
\gamma(p) &= 1 & \gamma(a) &= 1 \\
\gamma(L_1) &= 1 & \gamma(\varphi?) &= \gamma(\varphi) + 1 \\
\gamma(R_1) &= 1 & \gamma(\alpha\,;\beta) &= \gamma(\alpha) + \gamma(\beta) + 1 \\
\gamma(Q_\varphi) &= 1 & \gamma(\alpha^*) &= \gamma(\alpha) + 2 \\
\gamma(\neg\varphi) &= \gamma(\varphi) + 1 & \gamma(\alpha \cup \beta) &= \gamma(\alpha) + \gamma(\beta) + 3 \\
\gamma(\langle\alpha\rangle\varphi) &= \gamma(\alpha) + \gamma(\varphi) & \gamma(\alpha \parallel \beta) &= \gamma(\alpha) + \gamma(\beta) + 3
\end{aligned}
$$

The following properties can be easily proved by induction on $n > 0$:

1. For any program α, if $n = |\alpha|$ then $\gamma(\alpha) \leq 3n$.
2. For any formula φ, if $n = |\varphi|$ then $\gamma(\varphi) \leq 3n$.
3. For any localized formula (μ, φ), if $\gamma(\varphi) = n$ then the cardinality of $\mathrm{rCl}(\mu, \varphi)$ is less or equal to n. □

$$
\frac{(\mu,\ \langle\alpha^*\rangle\,\varphi)}{(\mu,\ \langle\alpha\rangle\,Q_{\langle\alpha^*\rangle\varphi}) \quad (\mu,\ \varphi)}
\qquad
\frac{(\mu,\ \langle\alpha \cup \beta\rangle\,\varphi)}{(\mu,\ \langle\alpha\rangle Q_\varphi) \quad (\mu,\ \langle\beta\rangle Q_\varphi) \quad (\mu,\ \varphi)}
$$

$$
\frac{(\mu,\ \neg\varphi)}{(\mu,\ \varphi)}
\qquad
\frac{(\mu,\ \langle\alpha \parallel \beta\rangle\,\varphi)}{(\mu.\ell,\ \langle\alpha\rangle\,L_1) \quad (\mu.r,\ \langle\beta\rangle\,R_1) \quad (\mu,\ \varphi)}
$$

Fig. 2. Replacement rules for the restricted closure calculus

4 Elimination of Hintikka Sets Procedure

In this section we describe a decision procedure for the satisfiability problem of $\mathsf{PPDL}^{\mathrm{det}}$. This decision procedure is based on the elimination of Hintikka set decision procedure devised for PDL by Pratt [23]. The principle of such decision procedures is to first construct a potential finite model for the formula φ_0 being tested for satisfiability. This initial model must somehow embed any possible model which could satisfy φ_0. Then the states of that model not fulfilling some *eventualities* are recursively removed. The procedure succeeds if the final model still contains some states satisfying φ_0. For PDL, states of the initial model are some subsets of the Fischer-Ladner closure called Hintikka sets, an eventuality is a formula of the form $\langle\alpha\rangle\psi$ belonging to a state and a state satisfies a formula if it contains this formula. There are two main difficulties in adapting this decision procedure to $\mathsf{PPDL}^{\mathrm{det}}$. Firstly, Hintikka sets are not sufficient to characterize states of $\mathsf{PPDL}^{\mathrm{det}}$ models. The decomposition path leading to each state is an essential information. Therefore, we introduce *plugs*, which correspond to decompositions by the separation relation, and *sockets*, which are sets of plugs and correspond to twines. A state of the initial model is a pair (H, S) where H is a Hintikka set and S a socket. Secondly, the resulting model is not $\lhd$-deterministic. Hence to prove that whenever the procedure succeeds the formula is satisfiable, a $\lhd$-deterministic model must be constructed from the final model. This construction is detailed in Sect. 5. In the remainder of the present section, we formally describe the elimination of Hintikka sets procedure for $\mathsf{PPDL}^{\mathrm{det}}$.

Definition 1. *Let $\varphi_0 \in \Phi$ be a formula and μ a location in* Loc(φ_0). *A* Hintikka set *H over φ_0 at μ is any maximal subset of $Cl(\varphi_0)$ verifying all the following conditions:*

1. *If $(\mu', \varphi) \in H$, then $\mu' = \mu$.*
2. *If $(\mu, \neg\varphi) \in Cl(\varphi_0)$, then $(\mu, \neg\varphi) \in H$ iff $(\mu, \varphi) \notin H$.*
3. *If $(\mu, \langle\alpha\,;\beta\rangle\varphi) \in Cl(\varphi_0)$, then $(\mu, \langle\alpha\,;\beta\rangle\varphi) \in H$ iff $(\mu, \langle\alpha\rangle\langle\beta\rangle\varphi) \in H$.*
4. *If $(\mu, \langle\alpha \cup \beta\rangle\varphi) \in Cl(\varphi_0)$, then $(\mu, \langle\alpha \cup \beta\rangle\varphi) \in H$ iff $(\mu, \langle\alpha\rangle\varphi) \in H$ or $(\mu, \langle\beta\rangle\varphi) \in H$.*
5. *If $(\mu, \langle\varphi?\rangle\psi) \in Cl(\varphi_0)$, then $(\mu, \langle\varphi?\rangle\psi) \in H$ iff $(\mu, \varphi) \in H$ and $(\mu, \psi) \in H$.*
6. *If $(\mu, \langle\alpha^*\rangle\varphi) \in Cl(\varphi_0)$, then $(\mu, \langle\alpha^*\rangle\varphi) \in H$ iff $(\mu, \langle\alpha\rangle\langle\alpha^*\rangle\varphi) \in H$ or $(\mu, \varphi) \in H$.*

μ is called the location *of H, denoted by $\lambda(H)$. The set of all Hintikka sets over φ_0 at all $\mu \in$* Loc(φ_0) *is denoted by $\mathcal{H}in\,(\varphi_0)$.*

Definition 2. *A* plug *for φ_0 is a triple $P = (H, H_1, H_2)$ of Hintikka sets from $\mathcal{H}in\,(\varphi_0)$ such that:*

1. *$\lambda(H_1) = \lambda(H).\ell$ and $\lambda(H_2) = \lambda(H).r$;*
2. *P has a* type, *which is an index $t \in \{1, 2\}$ such that $(\lambda(H_1), L_t) \in H_1$ and $(\lambda(H_2), R_t) \in H_2$.*

Notice that a plug may have more than one type. Two plugs have different types if there is no $t \in \{1, 2\}$ such that t is a type of both plugs. The *location* of the plug $P = (H, H_1, H_2)$, denoted by $\lambda(P)$, is the location of H.

Definition 3. *A* socket *for φ_0 is a set S of plugs for φ_0 such that:*

1. *S is either the empty set, a singleton or a set $\{P, P'\}$ such that P and P' have the same location but different types;*
2. *for any $(H, H_1, H_2), (H', H_3, H_4) \in S$, any type t' of (H', H_3, H_4) and any α, β, φ such that $(\lambda(H), \langle\alpha \parallel \beta\rangle\varphi) \in Cl(\varphi_0)$,*

$$\begin{aligned} &\textit{if } (\lambda(H'), \varphi) \in H' && \textit{and}\\ &\quad (\lambda(H_1), \langle\alpha\rangle L_{t'}) \in H_1 && \textit{and}\\ &\quad (\lambda(H_2), \langle\beta\rangle R_{t'}) \in H_2 &&\\ &\textit{then } (\lambda(H), \langle\alpha \parallel \beta\rangle\varphi) \in H. && \end{aligned}$$

The set of all sockets for φ_0 is denoted by $\mathcal{S}\,(\varphi_0)$. The location set *of a socket S, denoted by $\Lambda(S)$, is defined such that $\Lambda(\emptyset) = \{\epsilon\}$ and for all $S \neq \emptyset$, $\Lambda(S) = \{\lambda(P).\ell, \lambda(P).r \mid P \in S\}$.*

Given a formula $\varphi_0 \in \Phi$ we construct inductively for each $k \in \mathbb{N}$ the tuple $\mathcal{M}_k^{\varphi_0} = (W_k^{\varphi_0}, R_k^{\varphi_0}, \lhd_k^{\varphi_0}, V_k^{\varphi_0})$ where $W_k^{\varphi_0} \subseteq \mathcal{H}in\,(\varphi_0) \times \mathcal{S}\,(\varphi_0)$. Each of these tuples is a model iff $W_k^{\varphi_0} \neq \emptyset$. The *restricted accessibility relation* $\widehat{R}_k^{\varphi_0}(\alpha)$ over $W_k^{\varphi_0}$ is inductively defined for all $k \in \mathbb{N}$ and all $\alpha \in \Pi$ by:

- $(H,S)\ \widehat{R}_k^{\varphi_0}(a)\ (H',S')$ iff $(H,S)\ R_k^{\varphi_0}(a)\ (H',S')$,
- $(H,S)\ \widehat{R}_k^{\varphi_0}(\varphi?)\ (H',S')$ iff $(H,S)=(H',S')$ and $(\lambda(H),\varphi) \in H$,
- $(H,S)\ \widehat{R}_k^{\varphi_0}(\alpha\,;\beta)\ (H',S')$ iff $\exists\,(H'',S'') \in W_k^{\varphi_0}$, $(H,S)\ \widehat{R}_k^{\varphi_0}(\alpha)\ (H'',S'')$ and $(H'',S'')\ \widehat{R}_k^{\varphi_0}(\beta)\ (H',S')$,
- $(H,S)\ \widehat{R}_k^{\varphi_0}(\alpha \cup \beta)\ (H',S')$ iff $(H,S)\ \widehat{R}_k^{\varphi_0}(\alpha)\ (H',S')$ or $(H,S)\ \widehat{R}_k^{\varphi_0}(\beta)\ (H',S')$,
- $(H,S)\ \widehat{R}_k^{\varphi_0}(\alpha^*)\ (H',S')$ iff $(H,S)\ \widehat{R}_k^{\varphi_0}(\alpha)^*\ (H',S')$ where $\widehat{R}_k^{\varphi_0}(\alpha)^*$ is the reflexive and transitive closure of $\widehat{R}_k^{\varphi_0}(\alpha)$,
- $(H,S)\ \widehat{R}_k^{\varphi_0}(\alpha \parallel \beta)\ (H',S')$ iff $S = S'$ and $\exists H_1,H_2,H_3,H_4 \in \mathcal{H}in\,(\varphi_0)$, $S'' = \{(H,H_1,H_2),(H',H_3,H_4)\} \in \mathcal{S}\,(\varphi_0)$, $(H_1,S'')\ \widehat{R}_k^{\varphi_0}(\alpha)\ (H_3,S'')$ and $(H_2,S'')\ \widehat{R}_k^{\varphi_0}(\beta)\ (H_4,S'')$.

Initial Step. The initial tuple $\mathcal{M}_0^{\varphi_0} = (W_0^{\varphi_0}, R_0^{\varphi_0}, \lhd_0^{\varphi_0}, V_0^{\varphi_0})$ is constructed as follows:

- $W_0^{\varphi_0}$ is the set of pairs $(H,S) \in \mathcal{H}in\,(\varphi_0) \times \mathcal{S}\,(\varphi_0)$ such that $\lambda(H) \in \Lambda(S)$,
- for all $a \in \Pi_0$, $(H,S)\ R_0^{\varphi_0}(a)\ (H',S')$ iff $S = S'$ and $\forall(\mu,\varphi) \in H'$, if $(\mu,\langle a\rangle\varphi) \in \mathrm{Cl}\,(\varphi_0)$ then $(\mu,\langle a\rangle\varphi) \in H$,
- $(H,S) \lhd_0^{\varphi_0} ((H_1,S_1),(H_2,S_2))$ iff $S_1 = S_2$ and $(H,H_1,H_2) \in S_1$,
- for all $p \in \Phi_0^+$, $V_0^{\varphi_0}(p) = \{(H,S) \in W_0^{\varphi_0} \mid (\lambda(H),p) \in H\}$.

Inductive $(k+1)^{th}$ Step. Suppose $\mathcal{M}_k^{\varphi_0} = (W_k^{\varphi_0}, R_k^{\varphi_0}, \lhd_k^{\varphi_0}, V_k^{\varphi_0})$ has already been defined. A state $(H,S) \in W_k^{\varphi_0}$ is *demand-satisfied* in $\mathcal{M}_k^{\varphi_0}$ iff for any program α and any formula φ, if $(\lambda(H),\langle\alpha\rangle\varphi) \in H$ then there exists $(H',S') \in W_k^{\varphi_0}$ such that $(H,S)\ \widehat{R}_k^{\varphi_0}(\alpha)\ (H',S')$ and $(\lambda(H'),\varphi) \in H'$. Define $\mathcal{M}_{k+1}^{\varphi_0} = (W_{k+1}^{\varphi_0}, R_{k+1}^{\varphi_0}, \lhd_{k+1}^{\varphi_0}, V_{k+1}^{\varphi_0})$ such that :

- $W_{k+1}^{\varphi_0} = \{(H,S) \in W_k^{\varphi_0} \mid (H,S)$ is demand-satisfied in $\mathcal{M}_k^{\varphi_0}\}$,
- $R_{k+1}^{\varphi_0}(a) = R_k^{\varphi_0}(a) \cap (W_{k+1}^{\varphi_0})^2$ for all $a \in \Pi_0$,
- $\lhd_{k+1}^{\varphi_0} = \lhd_k^{\varphi_0} \cap (W_{k+1}^{\varphi_0})^3$,
- $V_{k+1}^{\varphi_0}(p) = V_k^{\varphi_0}(p) \cap W_{k+1}^{\varphi_0}$ for all $p \in \Phi_0^+$.

It can be easily proved that there is less than $2^{7N_{\varphi_0}+1}$ states in $W_0^{\varphi_0}$. Therefore, there exists $n \leq 2^{7N_{\varphi_0}+1}$ such that $\mathcal{M}_n^{\varphi_0} = \mathcal{M}_{n+k}^{\varphi_0}$ for all $k \in \mathbb{N}$. Let $\mathcal{M}^{\varphi_0} = (W^{\varphi_0}, R^{\varphi_0}, \lhd^{\varphi_0}, V^{\varphi_0}) = \mathcal{M}_n^{\varphi_0}$. Our procedure succeeds iff there is a state $(H_0,S_0) \in W^{\varphi_0}$ such that $(\epsilon,\varphi_0) \in H_0$.

Lemma 4. *Given a formula φ_0, to construct the corresponding model $\mathcal{M}^{\varphi_0}$ and to check whether there is a state $(H_0,S_0) \in W^{\varphi_0}$ such that $(\epsilon,\varphi_0) \in H_0$ can be done in deterministic exponential time.*

Proof. We have already stated that the procedure constructs at most an exponential number of models. The method from [16] can be easily adapted to prove that $\widehat{R}_k^{\varphi_0}(\alpha)$ can be computed in time polynomial in the cardinality of $W_k^{\varphi_0}$. Therefore, the whole procedure can be executed in deterministic exponential time. □

The remainder of this work is devoted to prove that this procedure is a decision procedure for the satisfiability problem of $\mathsf{PPDL}^{\det}$. We use the traditional vocabulary used for the dual problem of validity.

5 Completeness

In this section, we suppose that $\mathcal{M}^{\varphi_0} = (W^{\varphi_0}, R^{\varphi_0}, \lhd^{\varphi_0}, V^{\varphi_0})$ has been constructed, for a given formula $\varphi_0 \in \Phi$, as defined in the previous section and that there exists $(H_0, S_0) \in W^{\varphi_0}$ such that $(\epsilon, \varphi_0) \in H_0$. We will prove that φ_0 is satisfiable in the class of $\lhd$-deterministic models. Obviously, $\mathcal{M}^{\varphi_0}$ is a model. But in the general case, $\mathcal{M}^{\varphi_0}$ is not $\lhd$-deterministic. Therefore we will construct from $\mathcal{M}^{\varphi_0}$ a $\lhd$-deterministic model $\mathcal{M}^{\det} = \left(W^{\det}, R^{\det}, \lhd^{\det}, V^{\det}\right)$ satisfying φ_0. The main idea is to consider the equivalence classes by the relation $\asymp$ over W^{φ_0} defined such that $(H, S) \asymp (H', S')$ iff $S = S'$. It can be proved that, by removing from $\mathcal{M}^{\varphi_0}$ some "unreachable" states, these equivalence classes are twines and that the removed states are not needed in the proofs of the present section. Hence we will abusively call these equivalence classes *twines*. Remark that each such twine corresponds exactly to a socket. The *initial* twine θ_0 is the twine which corresponds to the empty socket $\emptyset$. The model $\mathcal{M}^{\det}$ is constructed inductively as follows. Initially, the model contains only a copy of the initial twine θ_0. Then, whenever two states in $\mathcal{M}^{\det}$ are copies of states reachable in $\mathcal{M}^{\varphi_0}$ by a parallel program, a copy of the twine linking these two states in $\mathcal{M}^{\varphi_0}$ is added to $\mathcal{M}^{\det}$. Since there are no decompositions within twines, we can ensure that $\mathcal{M}^{\det}$ is $\lhd$-deterministic, while preserving the satisfiability of φ_0.

Formally, to be able to copy twines, hence states, the states of $\mathcal{M}^{\det}$ are pairs $(i, (H, S))$ where i is a positive natural number and (H, S) is a state from $\mathcal{M}^{\varphi_0}$. We define the set $PL \subseteq \mathbb{N} \times W^{\varphi_0} \times \mathrm{SP}(\varphi_0) \times W^{\varphi_0}$ of parallel links such that $(n, (H, S), \alpha, (H', S')) \in PL$ iff $(H, S)\ \widehat{R}^{\varphi_0}(\alpha)\ (H', S')$ and there exists $\beta, \gamma \in \Pi$ such that $\alpha = \beta \parallel \gamma$. As both W^{φ_0} and $\mathrm{SP}(\varphi_0)$ are finite, PL can be totally ordered such that $(n_1, (H_1, S_1), \alpha_1, (H_1', S_1')) < (n_2, (H_2, S_2), \alpha_2, (H_2', S_2'))$ implies $n_1 \leq n_2$. If PL is not empty, such an order has a least element, hence the kth element of PL is well defined for all $k \in \mathbb{N}$. Moreover, if $(n, (H, S), \alpha, (H', S'))$ is the kth element of PL, then $n \leq k$. Now, we construct inductively the models $\left(\mathcal{M}_k^{\det}\right)_{k \in \mathbb{N}}$ as follows.

Initial Step. $\mathcal{M}_0^{\det} = \left(W_0^{\det}, R_0^{\det}, \lhd_0^{\det}, V_0^{\det}\right)$ is defined such that:

$$\begin{aligned}
W_0^{\det} &= \{(0,(H,S)) \mid (H,S) \in \theta_0\} \\
R_0^{\det}(a) &= \{((i_F,(H_F,S_F)),(i_T,(H_T,S_T))) \in W_0^{\det} \times W_0^{\det} \mid \\
&\qquad\qquad i_F = i_T \text{ and } (H_F,S_F)\ R^{\varphi_0}(a)\ (H_T,S_T)\} \\
\lhd_0^{\det} &= \emptyset \\
V_0^{\det}(p) &= \{(i,(H,S)) \in W_0^{\det} \mid (\lambda(H),p) \in H\}
\end{aligned}$$

If PL is empty, let us define $\mathcal{M}_k^{\det} = \mathcal{M}_0^{\det}$ for all $k > 0$. Otherwise, the following step is applied recursively.

Inductive $(k+1)^{th}$ *Step.* Suppose that $\mathcal{M}_k^{\det}$ has already been constructed and $(n,(H,S),\alpha \parallel \beta,(H',S'))$ is the kth tuple in PL. If $(n,(H,S)) \notin W_k^{\det}$ or $(n,(H',S')) \notin W_k^{\det}$ then $\mathcal{M}_{k+1}^{\det} = \mathcal{M}_k^{\det}$. Otherwise, since $(H,S)\ \widehat{R}^{\varphi_0}(\alpha \parallel \beta)$ (H',S'), there exists $H_1,H_2,H_3,H_4 \in \mathcal{H}in(\varphi_0)$ such that $S'' = \{(H,H_1,H_2),$ $(H',H_3,H_4)\} \in \mathcal{S}(\varphi_0)$, $(H_1,S'')\ \widehat{R}^{\varphi_0}(\alpha)\ (H_3,S'')$ and $(H_2,S'')\ \widehat{R}^{\varphi_0}(\beta)$ (H_4,S''). Let θ be the twine corresponding to S''. The model $\mathcal{M}_{k+1}^{\det}$ is defined by:

$$\begin{aligned}
W_{k+1}^{\det} &= W_k^{\det} \cup \{(i,(H''',S''')) \mid i = k+1 \text{ and } (H''',S''') \in \theta\} \\
R_{k+1}^{\det}(a) &= \{((i_F,(H_F,S_F)),(i_T,(H_T,S_T))) \in W_{k+1}^{\det} \times W_{k+1}^{\det} \mid \\
&\qquad\qquad i_F = i_T \text{ and } (H_F,S_F)\ R^{\varphi_0}(a)\ (H_T,S_T)\} \\
\lhd_{k+1}^{\det} &= \lhd_k^{\det} \cup \{((n,(H,S)),(k+1,(H_1,S'')),(k+1,(H_2,S''))), \\
&\qquad\qquad ((n,(H',S')),(k+1,(H_3,S'')),(k+1,(H_4,S'')))\} \\
v_{k+1}^{\det}(p) &= \{(i,(H''',S''')) \in W_{k+1}^{\det} \mid (\lambda(H'''),p) \in H'''\}
\end{aligned}$$

Finally, the model $\mathcal{M}^{\det}$ is defined as the union of all the models $\mathcal{M}_k^{\det}$ for $k \in \mathbb{N}$. We prove now that $\mathcal{M}^{\det}$ is a $\lhd$-deterministic model satisfying φ_0.

Lemma 5. *$\mathcal{M}^{\det}$ is $\lhd$-deterministic.*

Proof. Let us suppose that $(k,(H,S)) \lhd^{\det} ((k_1,(H_1,S_1)),(k_2,(H_2,S_2)))$ and $(k',(H',S')) \lhd^{\det} ((k_1,(H_1,S_1)),(k_2,(H_2,S_2)))$. By construction, $k_1 = k_2$ and $S_1 = S_2$. Moreover, those two tuples have been added to $\lhd^{\det}$ at the $k_1{}^{\text{th}}$ inductive step. Therefore, $k = k'$, $S = S'$ and $\{(H,H_1,H_2),(H',H_1,H_2)\} \in \mathcal{S}(\varphi_0)$. Since the types of (H,H_1,H_2) and (H',H_1,H_2) only depend on H_1 and H_2, these two plugs have the same types. Hence, by Definition 3, $H = H'$. □

To prove that $\mathcal{M}^{\det}$ satisfies φ_0 (Lemma 8), we need the following two lemmas.

Lemma 6. *For all $k \in \mathbb{N}$, all $(H,S),(H',S') \in W_k^{\varphi_0}$ and all programs α, if $(H,S)\ \widehat{R}_k^{\varphi_0}(\alpha)\ (H',S')$, then $S = S'$, $\lambda(H) = \lambda(H')$ and for all $i \leq k$, $(H,S)\ \widehat{R}_i^{\varphi_0}(\alpha)\ (H',S')$.*

Proof. By a simple induction on $|\alpha|$. We detail only the case for parallel compositions. Suppose that $(H, S)\ \widehat{R}_k^{\varphi_0}(\alpha \parallel \beta)\ (H', S')$. By definition, $S = S'$ and there exists H_1, H_2, H_3, H_4 such that $S'' = \{(H, H_1, H_2), (H', H_3, H_4)\}$ is a socket, $(H_1, S'')\ \widehat{R}_k^{\varphi_0}(\alpha)\ (H_3, S'')$ and $(H_2, S'')\ \widehat{R}_k^{\varphi_0}(\beta)\ (H_4, S'')$. Since S'' is a socket, $\lambda(H) = \lambda(H')$. By induction, for all $i \leq k$, $(H_1, S'')\ \widehat{R}_i^{\varphi_0}(\alpha)\ (H_3, S'')$ and $(H_2, S'')\ \widehat{R}_i^{\varphi_0}(\beta)\ (H_4, S'')$, hence $(H, S)\ \widehat{R}_i^{\varphi_0}(\alpha \parallel \beta)\ (H', S')$. □

Lemma 7. *For all* $(H, S), (H', S') \in W^{\varphi_0}$, *all* $\alpha \in \Pi$ *and all* $\varphi \in \Phi$, *if* $(\lambda(H), [\alpha]\varphi) \in H$ *and* $(H, S)\ \widehat{R}^{\varphi_0}(\alpha)\ (H', S')$ *then* $(\lambda(H'), \varphi) \in H'$.

Proof. The proof is by induction on $|\alpha|$. We only prove the case when α is a parallel composition. The other cases are straightforward and left to the reader. Suppose that $(\lambda(H), [\alpha \parallel \beta]\varphi) \in H$ and $(H, S)\ \widehat{R}^{\varphi_0}(\alpha \parallel \beta)\ (H', S')$. By definition, there exists $H_1, H_2, H_3, H_4 \in \mathcal{H}in(\varphi_0)$ such that, $S'' = \{(H, H_1, H_2), (H', H_3, H_4)\} \in \mathcal{S}(\varphi_0)$, $(H_1, S'')\ \widehat{R}^{\varphi_0}(\alpha)\ (H_3, S'')$ and $(H_2, S'')\ \widehat{R}^{\varphi_0}(\beta)\ (H_4, S'')$. As H is a Hintikka set, $(\lambda(H), \langle\alpha \parallel \beta\rangle\neg\varphi) \notin H$. Since (H', H_3, H_4) is a plug, there exists $t' \in \{1, 2\}$ such that $(\lambda(H_3), L_{t'}) \in H_3$ and $(\lambda(H_4), R_{t'}) \in H_4$. And since S'' is a socket, by Condition 2 of Definition 3, one of the following statements holds:

$$(\lambda(H_1), \langle\alpha\rangle L_{t'}) \notin H_1 \tag{1}$$

$$(\lambda(H_2), \langle\beta\rangle R_{t'}) \notin H_2 \tag{2}$$

$$(\lambda(H'), \neg\varphi) \notin H' \tag{3}$$

If (1) holds, then $(\lambda(H_1), [\alpha]\neg L_{t'}) \in H_1$ and by the induction hypothesis $(\lambda(H_3), \neg L_{t'}) \in H_3$ which is a contradiction. The case when (2) holds is similar. Finally, if (3) holds, then $(\lambda(H'), \varphi) \in H'$. □

We can now state the following truth lemma.

Lemma 8 (Truth lemma). *For all* $(k, (H, S)) \in W^{det}$ *and all* $(\mu, \varphi) \in Cl(\varphi_0)$,

$$(\mu, \varphi) \in H \text{ iff } \mathcal{M}^{det}, (k, (H, S)) \models \varphi \text{ and } \lambda(H) = \mu$$

Proof. The following two properties are proved by induction on n for all $n \in \mathbb{N}$ and all $(k, (H, S)) \in W^{\text{det}}$:

IH.1 for all $\alpha \in \Pi$ and all $(k', (H', S')) \in W^{\text{det}}$, if $n = |\alpha|$ and $\exists\varphi \in \Phi^+$ such that $(\lambda(H), \langle\alpha\rangle\varphi) \in \text{Cl}(\varphi_0)$, then:

$$(k, (H, S))\ R^{\text{det}}(\alpha)\ (k', (H', S')) \text{ iff } (H, S)\ \widehat{R}^{\varphi_0}(\alpha)\ (H', S') \text{ and } k = k'$$

IH.2 for all $(\mu, \varphi) \in \text{Cl}(\varphi_0)$, if $n = |\varphi|$ and $\lambda(H) = \mu$ then:

$$(\mu, \varphi) \in H \text{ iff } \mathcal{M}^{\text{det}}, (k, (H, S)) \models \varphi$$

First note that by Lemma 6 and by the construction of $\mathcal{M}^{\det}$, if $(k,(H,S)) \in W^{\det}$ and $(H,S)\ \widehat{R}^{\varphi_0}(\alpha)\ (H',S')$ then $(k,(H',S')) \in W^{\det}$. Then for IH.1, we detail only the case for parallel compositions, the other cases being straightforward. Suppose $\alpha = \beta \parallel \gamma$. For the right-to-left direction, $(k,(H,S),\alpha,(H',S')) \in PL$, hence by construction and by IH.1, $(k,(H,S))\ R^{\det}(\alpha)\ (k',(H',S'))$. For the left-to-right direction, there exists $w_i = (k_i,(H_i,S_i)) \in W^{\det}$ for each $i \in 1..4$ such that $(k,(H,S)) \lhd^{\det} (w_1,w_2)$, $w_1\ R^{\det}(\beta)\ w_3$, $w_2\ R^{\det}(\gamma)\ w_4$ and $(k',(H',S')) \lhd^{\det} (w_3,w_4)$. By IH.1, $k_1 = k_3$, $k_2 = k_4$, $(H_1,S_1)\ \widehat{R}^{\varphi_0}(\beta)\ (H_3,S_3)$ and $(H_2,S_2)\ \widehat{R}^{\varphi_0}(\gamma)\ (H_4,S_4)$. By the construction of $\mathcal{M}^{\det}$, $k = k'$, $S_1 = S_2 = S_3 = S_4$ and $\{(H,H_1,H_2),(H',H_3,H_4)\} \subseteq S_1$. And since any subset of a socket is a socket, $(H,S)\ \widehat{R}^{\varphi_0}(\alpha)\ (H',S')$. For IH.2, the cases for propositional variables and their negation are trivial. For diamond modalities, suppose $\varphi = \langle\alpha\rangle\psi$. By construction of $\mathcal{M}^{\varphi_0}$, there is $(H',S') \in W^{\varphi_0}$ such that $(H,S)\ \widehat{R}^{\varphi_0}(\alpha)\ (H',S')$ and $(\lambda(H'),\psi) \in H'$. And by IH.1 and IH.2, $\mathcal{M}^{\det},(k,(H,S)) \models \langle\alpha\rangle\psi$. The case for box modalities is managed by Lemma 7. □

By hypothesis, there exists $(H,S) \in W^{\varphi_0}$ such that $(\epsilon,\varphi_0) \in H$. And by construction, for any state $(H,S) \in W^{\varphi_0}$, if $\lambda(H) = \epsilon$ then $(0,(H,S)) \in W^{\det}$. Therefore, Lemma 8 proves $\mathcal{M}^{\det}$ satisfy φ_0.

6 Soundness

In this section we prove that for any $\mathsf{PPDL}^{\det}$ formula φ_0, if φ_0 is satisfiable then there exists $(H_0,S_0) \in W^{\varphi_0}$ such that $(\epsilon,\varphi_0) \in H_0$, where $\mathcal{M}^{\varphi_0} = (W^{\varphi_0},R^{\varphi_0},\lhd^{\varphi_0},V^{\varphi_0})$ has been obtained by the elimination of Hintikka sets procedure described in Sect. 4. The proof proceeds as follows. First, considering a $\mathsf{PPDL}^{\det}$ model $\mathcal{M}$ satisfying φ_0, a correspondence between the states of $\mathcal{M}$ and some states of $W_0^{\varphi_0}$ is constructed. Then, it is proved that the states of $W_0^{\varphi_0}$ corresponding to states in $\mathcal{M}$ can not be deleted by the procedure and that one of these states $(H_0,S_0) \in W_0^{\varphi_0}$ is such that $(\epsilon,\varphi_0) \in H_0$. The difficulties come from the involved structure of $\mathcal{M}_0^{\varphi_0}$ with locations, Hintikka sets, plugs and sockets and from the new propositional variables L_t and R_t. To overcome these difficulties, we use the following result from [5] which allows us to assume some properties about $\mathcal{M}$.

Proposition 1. *For any formula φ_0, if φ_0 is $\mathsf{PPDL}^{\det}$ satisfiable then there exists a model $\mathcal{M} = (W,R,\lhd,V)$, a state $x_0 \in W$ and a function λ from W to locations such that for all $x,y,z \in W$ and $\alpha \in \Pi$:*

$$\mathcal{M},x_0 \models \varphi_0 \tag{4}$$

$$\lambda(x_0) = \epsilon \tag{5}$$

$$\text{if } x \lhd (y,z) \text{ then } \lambda(y) = \lambda(x).\ell \text{ and } \lambda(z) = \lambda(x).r \tag{6}$$

$$\text{if } x\ R(\alpha)\ y \text{ then } \lambda(x) = \lambda(y) \tag{7}$$

From now on, we assume that $\mathcal{M}$ is as described in Proposition 1. In order to interpret the new propositional variables introduced by the closure defined in Sect. 3, extensions of the valuation V are defined as follows.

Definition 4. *For any model $\mathcal{M} = (W, R, \lhd, V)$ and any formulas φ_0, a valuation extension of $\mathcal{M}$ to φ_0 is a function $\mathcal{V}$ from the new propositional variables L_1, L_2, R_1 and R_2 to subsets of W. We write $\mathcal{M}+\mathcal{V}$ for the model $(W, R, \lhd, V')$ where V' is the function satisfying:*

$$V'(p) = V(p) \text{ iff } p \in \Phi_0$$
$$V'(p) = \mathcal{V}(p) \text{ iff } p \in \Delta^+$$

The set of all valuation extensions of $\mathcal{M}$ to φ_0 is denoted by $\mathfrak{V}(\mathcal{M}, \varphi_0)$.

It has to be noticed that by Lemma 2, there are no occurrences of propositional variables from Δ^+ in any program of $\mathrm{SP}(\varphi_0)$. Therefore, as long as only programs from $\mathrm{SP}(\varphi_0)$ are considered, the extension of R for $\mathcal{M} + \mathcal{V}$ does not depend on $\mathcal{V}$ and the notation R for this extension is not ambiguous.

Then, to define the correspondence between W and $W_0^{\varphi_0}$, we define the functions h_{hin}, h_{plug}, h_{socket} and h_{state} such that for all $x, y, z \in W$, $\mathcal{T} \subseteq \lhd$ and $\mathcal{V}, \mathcal{V}' \in \mathfrak{V}(\mathcal{M}, \varphi_0)$:

$$\begin{aligned} h_{\text{hin}}(x, \mathcal{V}) &= \{(\mu, \varphi) \in \mathrm{Cl}(\varphi_0) \mid \mu = \lambda(x) \text{ and } \mathcal{M}+\mathcal{V}, x \models \varphi\} \\ h_{\text{plug}}((x, y, z), \mathcal{V}, \mathcal{V}') &= (h_{\text{hin}}(x, \mathcal{V}'), h_{\text{hin}}(y, \mathcal{V}), h_{\text{hin}}(z, \mathcal{V})) \\ h_{\text{socket}}(\mathcal{T}, \mathcal{V}, \mathcal{V}') &= \{h_{\text{plug}}(D, \mathcal{V}, \mathcal{V}') \mid D \in \mathcal{T}\} \\ h_{\text{state}}(x, \mathcal{T}, \mathcal{V}, \mathcal{V}') &= (h_{\text{hin}}(x, \mathcal{V}), h_{\text{socket}}(\mathcal{T}, \mathcal{V}, \mathcal{V}')) \end{aligned}$$

A state $(H, S) \in W_0^{\varphi_0}$ *has a correspondence* if there exists $x \in W$, $\mathcal{T} \subseteq \lhd$ and $\mathcal{V}, \mathcal{V}' \in \mathfrak{V}(\mathcal{M}, \varphi_0)$ such that $h_{\text{state}}(x, \mathcal{T}, \mathcal{V}, \mathcal{V}') = (H, S)$. Obviously, for all $x \in W$ and all $\mathcal{V} \in \mathfrak{V}(\mathcal{M}, \varphi_0)$, $h_{\text{hin}}(x, \mathcal{V})$ is a Hintikka set. The following lemmas prove this correspondence has the desired properties.

Lemma 9. *For some $x \in W$, $\mathcal{T} \subseteq \lhd$ and $\mathcal{V}, \mathcal{V}' \in \mathfrak{V}(\mathcal{M}, \varphi_0)$, $(\epsilon, \varphi_0) \in h_{hin}(x, \mathcal{V})$ and $h_{state}(x, \mathcal{T}, \mathcal{V}, \mathcal{V}') \in W_0^{\varphi_0}$.*

Proof. Define $\mathcal{V}_\emptyset$ such that $\mathcal{V}_\emptyset(p) = \emptyset$ for all $p \in \Delta^+$. By (4) and (5), $(\epsilon, \varphi_0) \in h_{\text{hin}}(x_0, \mathcal{V}_\emptyset)$. Moreover, $h_{\text{socket}}(\emptyset, \mathcal{V}_\emptyset, \mathcal{V}_\emptyset) = \emptyset$ is trivially a socket and since $\Lambda(\emptyset) = \{\epsilon\}$, $h_{\text{state}}(x_0, \emptyset, \mathcal{V}_\emptyset, \mathcal{V}_\emptyset) \in W_0^{\varphi_0}$. □

Lemma 10. *For all $x \in W$, all $\mathcal{T} \subseteq \lhd$ and all $\mathcal{V}, \mathcal{V}' \in \mathfrak{V}(\mathcal{M}, \varphi_0)$,*

if $h_{state}(x, \mathcal{T}, \mathcal{V}, \mathcal{V}') \in W_0^{\varphi_0}$ then for all $k \in \mathbb{N}$, $h_{state}(x, \mathcal{T}, \mathcal{V}, \mathcal{V}') \in W_k^{\varphi_0}$.

Proof. We prove by induction on k that for all $k \in \mathbb{N}$, $x \in W$, $\mathcal{T} \subseteq \lhd$ and $\mathcal{V}, \mathcal{V}' \in \mathfrak{V}(\mathcal{M}, \varphi_0)$:

IH.1 if $h_{\text{state}}(x, \mathcal{T}, \mathcal{V}, \mathcal{V}') \in W_0^{\varphi_0}$ then $h_{\text{state}}(x, \mathcal{T}, \mathcal{V}, \mathcal{V}') \in W_k^{\varphi_0}$;
IH.2 for all $y \in W$ and all $\alpha \in \Pi$ such that $\exists\varphi, (\lambda(x), \langle\alpha\rangle\varphi) \in \mathrm{Cl}(\varphi_0)$, if $x \; R(\alpha) \; y$ and $h_{\text{state}}(x, \mathcal{T}, \mathcal{V}, \mathcal{V}') \in W_k^{\varphi_0}$ then $h_{\text{state}}(y, \mathcal{T}, \mathcal{V}, \mathcal{V}') \in W_k^{\varphi_0}$ and $h_{\text{state}}(x, \mathcal{T}, \mathcal{V}, \mathcal{V}') \; \widehat{R}_k^{\varphi_0}(\alpha) \; h_{\text{state}}(y, \mathcal{T}, \mathcal{V}, \mathcal{V}')$.

Base case. IH.1 is trivial. For IH.2, we first prove that $h_{\mathsf{state}}(y, \mathcal{T}, \mathcal{V}, \mathcal{V}') \in W_0^{\varphi_0}$. By hypothesis, $h_{\mathsf{socket}}(\mathcal{T}, \mathcal{V}, \mathcal{V}')$ is a socket. Hence it only remains to be proved that $\lambda(y) \in \Lambda(h_{\mathsf{socket}}(\mathcal{T}, \mathcal{V}, \mathcal{V}'))$ which is the case by (7) since $\lambda(x) \in \Lambda(h_{\mathsf{socket}}(\mathcal{T}, \mathcal{V}, \mathcal{V}'))$. The proof that $h_{\mathsf{state}}(x, \mathcal{T}, \mathcal{V}, \mathcal{V}')\ \widehat{R}_0^{\varphi_0}(\alpha)\ h_{\mathsf{state}}(y, \mathcal{T}, \mathcal{V}, \mathcal{V}')$ is by a subinduction on $|\alpha|$. We detail only the case for parallel compositions, the other cases being straightforward. Suppose $x\ R(\beta \parallel \gamma)\ y$. There exists $w_1, w_2, w_3, w_4 \in W$ such that $x \lhd (w_1, w_2)$, $w_1\ R(\beta)\ w_3$, $w_2\ R(\gamma)\ w_4$ and $y \lhd (w_3, w_4)$. Let $\mathcal{V}''$ be defined such that $\mathcal{V}''(L_1) = \{w_1\}$, $\mathcal{V}''(R_1) = \{w_2\}$, $\mathcal{V}''(L_2) = \{w_3\}$ and $\mathcal{V}''(R_2) = \{w_4\}$. Since, by hypothesis, there exists φ such that $(\lambda(x), \langle\beta \parallel \gamma\rangle\varphi) \in \mathrm{Cl}(\varphi_0)$, by Lemma 1, $(\lambda(x).\ell, L_1) \in \mathrm{Cl}(\varphi_0)$ and $(\lambda(x).r, R_1) \in \mathrm{Cl}(\varphi_0)$. Therefore, by (6), $h_{\mathsf{plug}}((x, w_1, w_2), \mathcal{V}'', \mathcal{V})$ is a plug of type 1. By a similar reasoning, $h_{\mathsf{plug}}((y, w_3, w_4), \mathcal{V}'', \mathcal{V})$ is a plug of type 2. Let $\mathcal{T}' = \{(x, w_1, w_2), (y, w_3, w_4)\}$, $S' = h_{\mathsf{socket}}(\mathcal{T}', \mathcal{V}'', \mathcal{V})$ and $H_i = h_{\mathsf{hin}}(w_i, \mathcal{V}'')$ for all $i \in 1..4$. By definition, $(H_i, S') = h_{\mathsf{state}}(w_i, \mathcal{T}', \mathcal{V}'', \mathcal{V})$ for all $i \in 1..4$. We prove that S' is a socket. For Condition 1 of Definition 3, suppose first that $h_{\mathsf{plug}}((x, w_1, w_2), \mathcal{V}'', \mathcal{V})$ has both types. Then $(\lambda(w_1), L_2) \in H_1$ and $(\lambda(w_2), R_2) \in H_2$, hence $w_1 = w_3$, $w_2 = w_4$ and $\mathcal{T}'$ is a singleton. The case is similar if $h_{\mathsf{plug}}((y, w_3, w_4), \mathcal{V}'', \mathcal{V})$ has both types. And if the plugs have different types, by (7) they have the same location. For Condition 2 of Definition 3, suppose that $(\lambda(x), \langle\alpha' \parallel \beta'\rangle\varphi') \in \mathrm{Cl}(\varphi_0)$, $(\lambda(y), \varphi') \in h_{\mathsf{hin}}(y, \mathcal{V})$, $(\lambda(w_1), \langle\alpha'\rangle L_2) \in H_1$ and $(\lambda(w_2), \langle\beta'\rangle R_2) \in H_2$, the other case being symmetric. By definition of $\mathcal{V}''$, $w_1\ R(\alpha')\ w_3$ and $w_2\ R(\beta')\ w_4$, hence $\mathcal{M} + \mathcal{V}, x \vDash \langle\alpha' \parallel \beta'\rangle\varphi'$ and $(\lambda(x), \langle\alpha' \parallel \beta'\rangle\varphi') \in h_{\mathsf{hin}}(x, \mathcal{V})$. Therefore, S' is a socket. Moreover, since $\Lambda(S') = \{\lambda(x).\ell, \lambda(x).r\}$, by (6) and (7), $\{(H_1, S'), (H_2, S'), (H_3, S'), (H_4, S')\} \subseteq W_0^{\varphi_0}$. By the subinduction hypothesis, we have $H_1\ \widehat{R}_0^{\varphi_0}(\beta)\ H_3$ and $H_2\ \widehat{R}_0^{\varphi_0}(\gamma)\ H_4$. Therefore, by definition of the restricted accessibility relation, $h_{\mathsf{state}}(x, \mathcal{T}', \mathcal{V}, \mathcal{V}'')\ \widehat{R}_0^{\varphi_0}(\beta \parallel \gamma)\ h_{\mathsf{state}}(y, \mathcal{T}', \mathcal{V}, \mathcal{V}'')$.

Inductive step. Suppose now that IH.1 and IH.2 hold for a given k. Here the order of the proofs matters since we use IH.1 for $k+1$ to prove IH.2 for $k+1$. To prove IH.1 for $k+1$, suppose that $h_{\mathsf{state}}(x, \mathcal{T}, \mathcal{V}, \mathcal{V}') \in W_k^{\varphi_0}$. Then for any formula $\langle\alpha\rangle\varphi$ such that $(\lambda(x), \langle\alpha\rangle\varphi) \in h_{\mathsf{hin}}(x, \mathcal{V})$, there exists $y \in W$ such that $x\ R(\alpha)\ y$ and $\mathcal{M} + \mathcal{V}, y \vDash \varphi$. And by IH.2, Lemma 1 and (7), $h_{\mathsf{state}}(y, \mathcal{T}, \mathcal{V}, \mathcal{V}') \in W_k^{\varphi_0}$, $h_{\mathsf{state}}(x, \mathcal{T}, \mathcal{V}, \mathcal{V}')\ \widehat{R}_k^{\varphi_0}(\alpha)\ h_{\mathsf{state}}(y, \mathcal{T}, \mathcal{V}, \mathcal{V}')$ and $(\lambda(y), \varphi) \in h_{\mathsf{hin}}(y, \mathcal{V})$. Therefore $h_{\mathsf{state}}(x, \mathcal{T}, \mathcal{V}, \mathcal{V}')$ is demand-satisfied and belongs to $W_{k+1}^{\varphi_0}$. The proof of IH.2 for $k+1$ is similar to the corresponding proof in the base case except that the hypothesis IH.1 for $k+1$ is used. For instance, in the case for parallel compositions, once it has been proved that $h_{\mathsf{state}}(w_i, \mathcal{T}', \mathcal{V}'', \mathcal{V}) \in W_0^{\varphi_0}$ for all $i \in 1..4$, we use IH.1 to state that $h_{\mathsf{state}}(w_i, \mathcal{T}', \mathcal{V}'', \mathcal{V}) \in W_{k+1}^{\varphi_0}$ for all $i \in 1..4$. Thus the subinduction hypothesis can be used to conclude. □

7 Conclusion and Perspectives

In this work, we have presented a procedure for the decision of the satisfiability problem of $\mathsf{PPDL}^{\mathsf{det}}$. This procedure is a nontrivial adaptation of Pratt's elimination of Hintikka set procedure [23]. We have proved that this decision procedure can be executed in deterministic exponential time. Since PDL can be trivially embedded into $\mathsf{PPDL}^{\det}$, this decision procedure is optimal and the satisfiability problem of $\mathsf{PPDL}^{\det}$ is EXPTIME-complete. This result extends a previous similar result for the iteration-free fragment of $\mathsf{PPDL}^{\det}$ [2]: adding a $\lhd$-deterministic separating parallel composition to PDL does not increase the theoretical complexity of the logic. This result contrasts with the 2EXPTIME complexity of the satisfiability problem of both PDL with intersection [17] and interleaving PDL [20].

Although $\lhd$-determinism is a very natural semantic condition, which turns the ternary separation relation into a partial binary operator, it would be interesting for future research to consider other classes of models. For instance, for the class of all models, only a 2EXPTIME upper bound is currently known [2]. For the class of models where the separation of states is deterministic, even though PRSPDL has been proved to be undecidable [3], it may be possible that in the absence of the four store/recover programs of PRSPDL the logic is decidable. Finally, for many concrete semantics like Petri nets or finite sets of agents, the separation relation would have to be both $\lhd$-deterministic and associative. The minimal associative binary normal modal logic is undecidable [15], therefore the variant of PDL with separating parallel composition interpreted in the class of all associative $\lhd$-deterministic models is undecidable too. But since there exists some decidable associative binary modal logics (for instance the separation logics kSL0 [7]), it would be interesting to search for decidable variants of PDL with separating parallel composition interpreted in special classes of associative $\lhd$-deterministic models.

References

1. Abrahamson, K.R.: Modal logic of concurrent nondeterministic programs. In: Kahn, G. (ed.) Semantics of Concurrent Computation. LNCS, vol. 70, pp. 21–33. Springer, Heidelberg (1979)
2. Balbiani, P., Boudou, J.: Tableaux methods for propositional dynamic logics with separating parallel composition. In: Felty, A.P., Middeldorp, A. (eds.) CADE-25. LNCS, vol. 9195, pp. 539–554. Springer, Switzerland (2015)
3. Balbiani, P., Tinchev, T.: Definability and computability for PRSPDL. In: Goré, R., Kooi, B.P., Kurucz, A. (eds.) Advances in Modal Logic - AiML, pp. 16–33. College Publications, San Diego (2014)
4. Benevides, M.R.F., de Freitas, R.P., Viana, J.P.: Propositional dynamic logic with storing, recovering and parallel composition. In: Hermann Haeusler, E., Farinas del Cerro, L. (eds.) Logical and Semantic Frameworks, with Applications - LSFA. ENTCS, vol. 269, pp. 95–107 (2011)
5. Boudou, J.: Exponential-size model property for PDL with separating parallel composition. In: Italiano, G.F., Pighizzini, G., Sannella, D.T. (eds.) MFCS 2015. LNCS, vol. 9234, pp. 129–140. Springer, Heidelberg (2015)
6. Brookes, S.: A semantics for concurrent separation logic. Theoret. Comput. Sci. **375**(1–3), 227–270 (2007)
7. Calcagno, C., Yang, H., O'Hearn, P.W.: Computability and complexity results for a spatial assertion language for data structures. In: Hariharan, R., Mukund, M., Vinay, V. (eds.) FSTTCS 2001. LNCS, vol. 2245, pp. 108–119. Springer, Heidelberg (2001)

8. Demri, S., Deters, M.: Separation logics and modalities: a survey. J. Appl. Non-Class. Logics **25**(1), 50–99 (2015)
9. van Ditmarsch, H., van der Hoek, W., Kooi, B.P.: Dynamic Epistemic Logic, vol. 337. Springer Science & Business Media, Netherlands (2007)
10. van Eijck, J., Stokhof, M.: The gamut of dynamic logics. In: Gabbay, D.M., Woods, J. (eds.) Logic and the Modalities in the Twentieth Century. Handbook of the History of Logic, vol. 7, pp. 499–600. Elsevier, Amsterdam (2006)
11. Fischer, M.J., Ladner, R.E.: Propositional dynamic logic of regular programs. J. Comput. Syst. Sci. **18**(2), 194–211 (1979)
12. Frias, M.F.: Fork Algebras in Algebra, Logic and Computer Science. Advances in Logic, vol. 2. World Scientific, Singapore (2002)
13. Harel, D.: Recurring dominoes: making the highly undecidable highly understandable (preliminary report). In: Karpinski, M. (ed.) FCT. LNCS, vol. 158, pp. 177–194. Springer, Heidelberg (1983)
14. Herzig, A.: A simple separation logic. In: Libkin, L., Kohlenbach, U., de Queiroz, R. (eds.) WoLLIC 2013. LNCS, vol. 8071, pp. 168–178. Springer, Heidelberg (2013)
15. Kurucz, Á., Németi, I., Sain, I., Simon, A.: Decidable and undecidable logics with a binary modality. J. Logic, Lang. Inf. **4**(3), 191–206 (1995)
16. Lange, M.: Model checking propositional dynamic logic with all extras. J. Appl. Logic **4**(1), 39–49 (2006)
17. Lange, M., Lutz, C.: 2-ExpTime lower bounds for propositional dynamic logics with intersection. J. Symbolic Logic **70**(4), 1072–1086 (2005)
18. Larchey-Wendling, D., Galmiche, D.: The undecidability of boolean BI through phase semantics. In: Logic in Computer Science - LICS, pp. 140–149. IEEE Computer Society (2010)
19. Marx, M., Pólos, L., Masuch, M.: Arrow Logic and Multi-modal Logic. CSLI Publications, Stanford (1996)
20. Mayer, A.J., Stockmeyer, L.J.: The complexity of PDL with interleaving. Theoret. Comput. Sci. **161**(1&2), 109–122 (1996)
21. O'Hearn, P.W.: Resources, concurrency, and local reasoning. Theoret. Comput. Sci. **375**(1–3), 271–307 (2007)
22. Peleg, D.: Concurrent dynamic logic. J. ACM **34**(2), 450–479 (1987)
23. Pratt, V.R.: Models of program logics. In: Foundations of Computer Science - FOCS, pp. 115–122. IEEE Computer Society (1979)
24. Pym, D.J.: The Semantics and Proof Theory of the Logic of Bunched Implications. Applied Logic Series, vol. 26. Springer, Netherlands (2002). Kluwer Academic Publishers
25. Reynolds, J.C.: Separation logic: a logic for shared mutable data structures. In: Logic in Computer Science - LICS, pp. 55–74. IEEE Computer Society (2002)
26. Schild, K.: A correspondence theory for terminological logics: preliminary report. In: Mylopoulos, J., Reiter, R. (eds.) International Joint Conference on Artificial Intelligence - IJCAI, pp. 466–471. Morgan Kaufmann (1991)
27. Venema, Y.: A modal logic for chopping intervals. J. Logic Comput. **1**(4), 453–476 (1991)

Interval Temporal Logic Model Checking: The Border Between Good and Bad HS Fragments

Laura Bozzelli[1], Alberto Molinari[2], Angelo Montanari[2](✉), Adriano Peron[3], and Pietro Sala[4]

[1] Technical University of Madrid (UPM), Madrid, Spain
laura.bozzelli@fi.upm.es
[2] University of Udine, Udine, Italy
molinari.alberto@gmail.com, angelo.montanari@uniud.it
[3] University of Napoli "Federico II", Napoli, Italy
adrperon@unina.it
[4] University of Verona, Verona, Italy
pietro.sala@univr.it

Abstract. The model checking problem has thoroughly been explored in the context of standard point-based temporal logics, such as LTL, CTL, and CTL*, whereas model checking for interval temporal logics has been brought to the attention only very recently.

In this paper, we prove that the model checking problem for the logic of Allen's relations *started-by* and *finished-by* is highly intractable, as it can be proved to be **EXPSPACE**-hard. Such a lower bound immediately propagates to the full Halpern and Shoham's modal logic of time intervals (HS). In contrast, we show that other noteworthy HS fragments, namely, Propositional Neighbourhood Logic extended with modalities for the Allen relation *starts* (resp., *finishes*) and its inverse *started-by* (resp., *finished-by*), turn out to have—maybe unexpectedly—the same complexity as LTL (i.e., they are **PSPACE**-complete), thus joining the group of other already studied, well-behaved albeit less expressive, HS fragments.

1 Introduction

Model checking (MC) is one of the most successful techniques in the area of formal methods. It allows one to automatically check whether some desired properties of a system, specified by a temporal logic formula, hold over a model of it. MC has proved itself to be extremely useful in formal verification [6], but it has also been successfully exploited in various areas of AI, ranging from planning to configuration and multi-agent systems (see, for instance, [8,16]). Point-based temporal logics, such as LTL [23], CTL, and CTL* [7], that allow one to predicate over computation states, are usually adopted in MC as the specification language, as they are suitable for practical purposes in many application domains. However, some relevant temporal properties, that involve, for instance, actions with duration, accomplishments, and temporal aggregations, are inherently "interval-based" and thus cannot be expressed by point-based logics. Here, we focus on MC algorithms for interval temporal logic (ITL).

N. Olivetti and A. Tiwari (Eds.): IJCAR 2016, LNAI 9706, pp. 389–405, 2016.
DOI: 10.1007/978-3-319-40229-1_27

ITLs take intervals, instead of points, as their primitive entities, providing an alternative setting for reasoning about time [10,22,28]. They have been applied in various areas of computer science, including formal verification, computational linguistics, planning, and multi-agent systems [13,22,24]. In order to check interval properties of computations, one needs to collect information about states into computation stretches: each finite path of a Kripke structure is interpreted as an interval, whose labelling is defined on the basis of the labelling of the component states. Halpern and Shoham's modal logic of time intervals HS [10] is the most famous among ITLs. It features one modality for each of the 13 possible ordering relations between pairs of intervals (the so-called Allen's relations [1]), apart from equality. Its satisfiability problem turns out to be highly undecidable for all relevant (classes of) linear orders [10]. The same holds for most fragments of it [3,12,17]. However, some meaningful exceptions exist, including the logic of temporal neighbourhood $\mathsf{A\overline{A}}$ and the logic of sub-intervals D [4,5].

In this paper, we address some open issues in the MC problem for HS, which only recently entered the research agenda [13–15,18–21]. In [18], Molinari et al. deal with MC for full HS (under the homogeneity assumption [25]). They introduce the problem and prove its non-elementary decidability and **PSPACE**-hardness. Since then, the attention was also brought to the fragments of HS, which, similarly to what happens with satisfiability, are often computationally better. Here, we focus on the border between good and bad HS fragments, showing the criticality of the combined use of modalities for interval prefixes and suffixes (modalities for Allen's relations *started-by* and *finished-by*). First, we prove that MC for the fragment BE, whose modalities can express properties of both interval prefixes and suffixes, is **EXPSPACE**-hard, and this lower bound immediately propagates to full HS. Then, we show that the complexity of MC for fragments where properties of interval prefixes and suffixes are considered separately is markedly lower. In [21], the authors proved that if we consider only properties of future and past intervals, MC is in $\mathbf{P}^{\mathbf{NP}}$; if modalities for interval extensions to the left and to the right are added, MC becomes **PSPACE**-complete [19]. Here we prove that MC for the fragment $\mathsf{A\overline{A}B\overline{B}}$ (resp., $\mathsf{A\overline{A}E\overline{E}}$), that allows one to express properties of interval prefixes (resp., suffixes), future and past intervals, and right (resp., left) interval extensions, is in **PSPACE**. Since MC for the fragment featuring only one modality for right (resp., left) interval extensions is **PSPACE**-hard [21], **PSPACE**-completeness immediately follows. Moreover, we show that if we restrict HS to modalities for either interval prefixes or suffixes (HS fragments B and E), MC turns out to be **co-NP**-complete. The MC problem for epistemic extensions of some HS fragments has been investigated by Lomuscio and Michaliszyn [13–15] (a detailed account of their results can be found in [18]). However, their semantic assumptions differ from those of [18], thus making it difficult to compare the two research lines.

In the next section, we introduce the fundamental elements of the MC problem for HS and its fragments. Then, in Sect. 3 we focus on the fragment BE, while in Sect. 4 we deal with $\mathsf{A\overline{A}E\overline{E}}$ and E (and with $\mathsf{A\overline{A}B\overline{B}}$ and B). Conclusions provide an assessment of the work done and outline future research directions.

2 Preliminaries

The interval temporal logic HS. An interval algebra to reason about intervals and their relative order was proposed by Allen in [1], while a systematic logical study of interval representation and reasoning was done a few years later by Halpern and Shoham, who introduced the interval temporal logic HS featuring one modality for each Allen relation, but equality [10]. Table 1 depicts 6 of the 13 Allen's relations, together with the corresponding HS (existential) modalities. The other 7 relations are the 6 inverse relations (given a binary relation $\mathcal{R}$, the inverse relation $\overline{\mathcal{R}}$ is such that $b\overline{\mathcal{R}}a$ if and only if $a\mathcal{R}b$) and equality.

Table 1. Allen's relations and corresponding HS modalities.

Allen relation	HS	Definition w.r.t. interval structures	Example
MEETS	$\langle A \rangle$	$[x,y]\mathcal{R}_A[v,z] \iff y = v$	
BEFORE	$\langle L \rangle$	$[x,y]\mathcal{R}_L[v,z] \iff y < v$	
STARTED-BY	$\langle B \rangle$	$[x,y]\mathcal{R}_B[v,z] \iff x = v \wedge z < y$	
FINISHED-BY	$\langle E \rangle$	$[x,y]\mathcal{R}_E[v,z] \iff y = z \wedge x < v$	
CONTAINS	$\langle D \rangle$	$[x,y]\mathcal{R}_D[v,z] \iff x < v \wedge z < y$	
OVERLAPS	$\langle O \rangle$	$[x,y]\mathcal{R}_O[v,z] \iff x < v < y < z$	

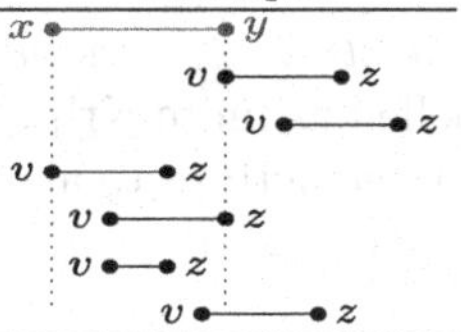

The HS language consists of a set of proposition letters $\mathcal{AP}$, the Boolean connectives $\neg$ and $\wedge$, and a temporal modality for each of the (non trivial) Allen's relations, i.e., $\langle A \rangle$, $\langle L \rangle$, $\langle B \rangle$, $\langle E \rangle$, $\langle D \rangle$, $\langle O \rangle$, $\langle \overline{A} \rangle$, $\langle \overline{L} \rangle$, $\langle \overline{B} \rangle$, $\langle \overline{E} \rangle$, $\langle \overline{D} \rangle$, and $\langle \overline{O} \rangle$. HS formulas are defined by the grammar $\psi ::= p \mid \neg\psi \mid \psi \wedge \psi \mid \langle X \rangle \psi \mid \langle \overline{X} \rangle \psi$, where $p \in \mathcal{AP}$ and $X \in \{A, L, B, E, D, O\}$. In the following, we will also exploit the standard logical connectives (disjunction $\vee$, implication $\rightarrow$, and double implication $\leftrightarrow$) as abbreviations. Furthermore, for any modality X, the dual universal modalities $[X]\psi$ and $[\overline{X}]\psi$ are defined as $\neg\langle X \rangle \neg\psi$ and $\neg\langle \overline{X} \rangle \neg\psi$, respectively.

Given any subset of Allen's relations $\{X_1, \cdots, X_n\}$, we denote by $\mathsf{X}_1 \cdots \mathsf{X}_n$ the HS fragment featuring existential (and universal) modalities for $X_1, \ldots, X_n$ only.

W.l.o.g., we assume the *non-strict semantics of HS*, which admits intervals consisting of a single point[1]. Under such an assumption, all HS modalities can be expressed in terms of modalities $\langle B \rangle, \langle E \rangle, \langle \overline{B} \rangle$, and $\langle \overline{E} \rangle$ [28]. HS can thus be viewed as a multi-modal logic with these 4 primitive modalities and its semantics can be defined over a multi-modal Kripke structure, called *abstract interval model*, where intervals are treated as atomic objects and Allen's relations as binary relations between pairs of intervals. Since later we will focus on the HS fragments $\mathsf{A\overline{A}E\overline{E}}$ and $\mathsf{A\overline{A}B\overline{B}}$—which respectively do not feature $\langle B \rangle$, $\langle \overline{B} \rangle$ and $\langle E \rangle$, $\langle \overline{E} \rangle$—we add both $\langle A \rangle$ and $\langle \overline{A} \rangle$ to the considered set of HS modalities.

Definition 1 [18]**.** *An* abstract interval model *is a tuple* $\mathcal{A} = (\mathcal{AP}, \mathbb{I}, A_{\mathbb{I}}, B_{\mathbb{I}}, E_{\mathbb{I}}, \sigma)$, *where* $\mathcal{AP}$ *is a set of proposition letters,* $\mathbb{I}$ *is a possibly infinite set of atomic objects*

[1] All the results we prove in the paper hold for the strict semantics as well.

(worlds), $A_{\mathbb{I}}$, $B_{\mathbb{I}}$, and $E_{\mathbb{I}}$ are three binary relations over $\mathbb{I}$, and $\sigma : \mathbb{I} \mapsto 2^{\mathcal{AP}}$ is a (total) labeling function, which assigns a set of proposition letters to each world.

In the interval setting, $\mathbb{I}$ is interpreted as a set of intervals and $A_{\mathbb{I}}$, $B_{\mathbb{I}}$, and $E_{\mathbb{I}}$ as Allen's relations A (*meets*), B (*started-by*), and E (*finished-by*), respectively; σ assigns to each interval in $\mathbb{I}$ the set of proposition letters that hold over it.

Given an abstract interval model $\mathcal{A} = (\mathcal{AP}, \mathbb{I}, A_{\mathbb{I}}, B_{\mathbb{I}}, E_{\mathbb{I}}, \sigma)$ and an interval $I \in \mathbb{I}$, the truth of an HS formula over I is inductively defined as follows:

- $\mathcal{A}, I \models p$ iff $p \in \sigma(I)$, for any $p \in \mathcal{AP}$;
- $\mathcal{A}, I \models \neg\psi$ iff it is not true that $\mathcal{A}, I \models \psi$ (also denoted as $\mathcal{A}, I \not\models \psi$);
- $\mathcal{A}, I \models \psi \wedge \phi$ iff $\mathcal{A}, I \models \psi$ and $\mathcal{A}, I \models \phi$;
- $\mathcal{A}, I \models \langle X \rangle \psi$, for $X \in \{A, B, E\}$, iff there is $J \in \mathbb{I}$ s.t. $I \, X_{\mathbb{I}} \, J$ and $\mathcal{A}, J \models \psi$;
- $\mathcal{A}, I \models \langle \overline{X} \rangle \psi$, for $\overline{X} \in \{\overline{A}, \overline{B}, \overline{E}\}$, iff there is $J \in \mathbb{I}$ s.t. $J \, X_{\mathbb{I}} \, I$ and $\mathcal{A}, J \models \psi$.

Kripke structures and abstract interval models. Finite state systems are usually modelled as finite Kripke structures. In [18], the authors define a mapping from Kripke structures to abstract interval models, that allows one to specify interval properties of computations by means of HS formulas.

Definition 2. *A* finite Kripke structure *is a tuple $\mathcal{K} = (\mathcal{AP}, W, \delta, \mu, w_0)$, where $\mathcal{AP}$ is a set of proposition letters, W is a finite set of states, $\delta \subseteq W \times W$ is a left-total relation between pairs of states, $\mu : W \mapsto 2^{\mathcal{AP}}$ is a total labelling function, and $w_0 \in W$ is the initial state.*

For all $w \in W$, $\mu(w)$ is the set of proposition letters that hold at w, while δ is the transition relation that describes the evolution of the system over time.

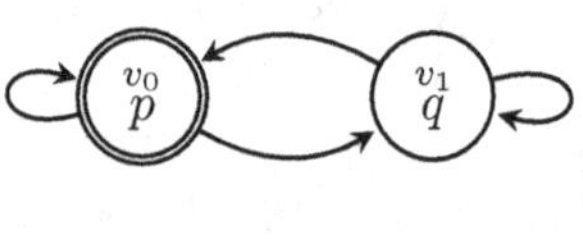

Fig. 1. The Kripke structure $\mathcal{K}_2$.

Figure 1 depicts the finite Kripke structure $\mathcal{K}_2 = (\{p, q\}, \{v_0, v_1\}, \delta, \mu, v_0)$, where $\delta = \{(v_0, v_0), (v_0, v_1), (v_1, v_0), (v_1, v_1)\}$, $\mu(v_0) = \{p\}$, and $\mu(v_1) = \{q\}$. The initial state v_0 is identified by a double circle.

Definition 3. *A* track ρ *over a finite Kripke structure $\mathcal{K} = (\mathcal{AP}, W, \delta, \mu, w_0)$ is a finite sequence of states $v_1 \cdots v_n$, with $n \geq 1$, such that $(v_i, v_{i+1}) \in \delta$ for $i = 1, \ldots, n-1$.*

Let $\text{Trk}_{\mathcal{K}}$ be the (infinite) set of all tracks over a finite Kripke structure $\mathcal{K}$. For any track $\rho = v_1 \cdots v_n \in \text{Trk}_{\mathcal{K}}$, we define:

- $|\rho| = n$, $\text{fst}(\rho) = v_1$, and $\text{lst}(\rho) = v_n$;
- any index $i \in [1, |\rho|]$ is called a *ρ-position* and $\rho(i) = v_i$;
- $\text{states}(\rho) = \{v_1, \cdots, v_n\} \subseteq W$;
- $\rho(i, j) = v_i \cdots v_j$, for $1 \leq i \leq j \leq |\rho|$, is the subtrack of ρ bounded by the ρ-positions i and j (we write ρ^i for $\rho(i, |\rho|)$, for $1 \leq i \leq |\rho|$);
- $\text{Pref}(\rho) = \{\rho(1, i) \mid 1 \leq i \leq |\rho| - 1\}$ and $\text{Suff}(\rho) = \{\rho(i, |\rho|) \mid 2 \leq i \leq |\rho|\}$ are the sets of all proper prefixes and suffixes of ρ, respectively.

Given $\rho, \rho' \in \mathrm{Trk}_{\mathcal{K}}$, we denote by $\rho \cdot \rho'$ the concatenation of the tracks ρ and ρ'. Moreover, if $\mathrm{lst}(\rho) = \mathrm{fst}(\rho')$, we denote by $\rho \star \rho'$ the track $\rho(1, |\rho| - 1) \cdot \rho'$. In particular, when $|\rho| = 1$, $\rho \star \rho' = \rho'$. In the following, when we write $\rho \star \rho'$, we implicitly assume that $\mathrm{lst}(\rho) = \mathrm{fst}(\rho')$. Finally, if $\mathrm{fst}(\rho) = w_0$ (the initial state of $\mathcal{K}$), ρ is called an *initial track.*

An abstract interval model (over $\mathrm{Trk}_{\mathcal{K}}$) can be naturally associated with a finite Kripke structure $\mathcal{K}$ by considering the set of intervals as the set of tracks of $\mathcal{K}$. Since $\mathcal{K}$ has loops (δ is left-total), the number of tracks in $\mathrm{Trk}_{\mathcal{K}}$, and thus the number of intervals, is infinite.

Definition 4. *The* abstract interval model induced by a finite Kripke structure $\mathcal{K} = (\mathcal{AP}, W, \delta, \mu, w_0)$ *is* $\mathcal{A}_{\mathcal{K}} = (\mathcal{AP}, \mathbb{I}, A_{\mathbb{I}}, B_{\mathbb{I}}, E_{\mathbb{I}}, \sigma)$, *where* $\mathbb{I} = \mathrm{Trk}_{\mathcal{K}}$, $A_{\mathbb{I}} = \{(\rho, \rho') \in \mathbb{I} \times \mathbb{I} \mid \mathrm{lst}(\rho) = \mathrm{fst}(\rho')\}$, $B_{\mathbb{I}} = \{(\rho, \rho') \in \mathbb{I} \times \mathbb{I} \mid \rho' \in \mathrm{Pref}(\rho)\}$, $E_{\mathbb{I}} = \{(\rho, \rho') \in \mathbb{I} \times \mathbb{I} \mid \rho' \in \mathrm{Suff}(\rho)\}$, *and* $\sigma : \mathbb{I} \mapsto 2^{\mathcal{AP}}$ *is such that* $\sigma(\rho) = \bigcap_{w \in \mathrm{states}(\rho)} \mu(w)$, *for all* $\rho \in \mathbb{I}$.

Relations $A_{\mathbb{I}}, B_{\mathbb{I}}$, and $E_{\mathbb{I}}$ are interpreted as the Allen's relations A, B, and E, respectively. Moreover, according to the definition of σ, $p \in \mathcal{AP}$ holds over $\rho = v_1 \cdots v_n$ iff it holds over all the states $v_1, \cdots, v_n$ of ρ. This conforms to the *homogeneity principle*, according to which a proposition letter holds over an interval if and only if it holds over all its subintervals.

Definition 5. *Let $\mathcal{K}$ be a finite Kripke structure and ψ be an HS formula; we say that a track $\rho \in \mathrm{Trk}_{\mathcal{K}}$ satisfies ψ, denoted as $\mathcal{K}, \rho \models \psi$, iff it holds that $\mathcal{A}_{\mathcal{K}}, \rho \models \psi$. Moreover, we say that $\mathcal{K}$ models ψ, denoted as $\mathcal{K} \models \psi$, iff for all* initial *tracks $\rho' \in \mathrm{Trk}_{\mathcal{K}}$ it holds that $\mathcal{K}, \rho' \models \psi$. The* model checking problem *for HS over finite Kripke structures is the problem of deciding whether $\mathcal{K} \models \psi$.*

We conclude with a simple example (a simplified version of the one given in [18]), showing that the fragments considered in this paper can express meaningful properties of state-transition systems.

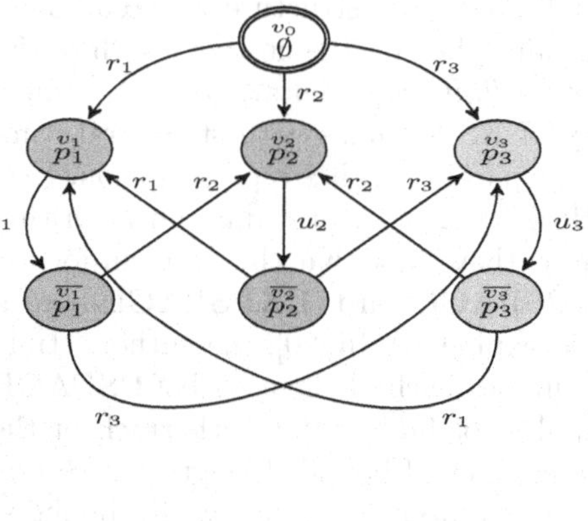

Fig. 2. The Kripke structure $\mathcal{K}_{Sched}$.

In Fig. 2, we provide an example of a finite Kripke structure $\mathcal{K}_{Sched}$ that models the behaviour of a scheduler serving three processes which are continuously requesting the use of a common resource. The initial state is v_0: no process is served in that state. In any other state v_i and $\overline{v}_i$, with $i \in \{1, 2, 3\}$, the i-th process is served (this is denoted by the fact that p_i holds in those states). For the sake of readability, edges are marked either by r_i, for $request(i)$, or by u_i, for $unlock(i)$. Edge labels do not have a semantic value, that is, they are neither part of the structure definition, nor proposition letters; they are simply used to ease reference to edges. Process i is served in state v_i, then, after "some time", a transition u_i from v_i to $\overline{v}_i$ is taken; subsequently, process i cannot be served again immediately, as v_i is not directly reachable from $\overline{v}_i$ (the scheduler cannot serve the same process twice in two successive rounds). A transition r_j, with $j \neq i$, from $\overline{v}_i$ to v_j is then taken and process j is served. This structure can easily be generalised to a higher number of processes.

We show how some meaningful properties to be checked over $\mathcal{K}_{Sched}$ can be expressed in HS, and, in particular, with formulas of $\mathsf{A\overline{A}E\overline{E}}$. In all formulas, we force the validity of the considered property over all legal computation sub-intervals by using modality $[E]$ (all computation sub-intervals are suffixes of at least one initial track). The truth of the next statements can be easily checked:

- $\mathcal{K}_{Sched} \models [E]\big(\langle \mathrm{E}\rangle^3 \top \rightarrow (\chi(p_1,p_2) \vee \chi(p_1,p_3) \vee \chi(p_2,p_3))\big)$, where $\chi(p,q) := \langle \mathrm{E}\rangle \langle \overline{\mathrm{A}}\rangle\, p \wedge \langle \mathrm{E}\rangle \langle \overline{\mathrm{A}}\rangle\, q$;
- $\mathcal{K}_{Sched} \not\models [E](\langle \mathrm{E}\rangle^{10} \top \rightarrow \langle \mathrm{E}\rangle \langle \overline{\mathrm{A}}\rangle\, p_3)$;
- $\mathcal{K}_{Sched} \not\models [E](\langle \mathrm{E}\rangle^{5} \rightarrow (\langle \mathrm{E}\rangle \langle \overline{\mathrm{A}}\rangle\, p_1 \wedge \langle \mathrm{E}\rangle \langle \overline{\mathrm{A}}\rangle\, p_2 \wedge \langle \mathrm{E}\rangle \langle \overline{\mathrm{A}}\rangle\, p_3))$.

The first formula states that in any suffix of length at least 4 of an initial track, at least 2 proposition letters are witnessed. $\mathcal{K}_{Sched}$ satisfies the formula since a process cannot be executed twice in a row. The second formula states that in any suffix of length at least 11 of an initial track, process 3 is executed at least once in some internal states (*non starvation*). $\mathcal{K}_{Sched}$ does not satisfy the formula since the scheduler can avoid executing a process ad libitum. The third formula states that in any suffix of length at least 6 of an initial track, p_1, p_2, p_3 are all witnessed. The only way to satisfy this property is to constrain the scheduler to execute the three processes in a strictly periodic manner (*strict alternation*), i.e., $p_i p_j p_k p_i p_j p_k p_i p_j p_k \cdots$, $i,j,k \in \{1,2,3\}, i \neq j \neq k \neq i$, but this is not the case.

The general picture. We now describe known and new complexity results about the model checking problem for HS fragments (see Fig. 3 for a graphical account).

In [18], the authors show that, given a finite Kripke structure $\mathcal{K}$ and a bound k on the structural complexity of HS formulas, i.e., on the nesting depth of $\langle \mathrm{E}\rangle$ and $\langle \mathrm{B}\rangle$ modalities, it is possible to obtain a *finite* representation for $\mathcal{A}_{\mathcal{K}}$, which is equivalent to $\mathcal{A}_{\mathcal{K}}$ w. r. to satisfiability of HS formulas with structural complexity less than or equal to k. Then, by exploiting such a representation, they prove that the MC problem for (full) HS is decidable, providing an algorithm with non-elementary complexity. Moreover, they show that the problem for the fragment $\mathsf{A\overline{A}BE}$, and thus for full HS, is **PSPACE**-hard (**EXPSPACE**-hard if a suitable succinct encoding of formulas is exploited). In [20], the authors study the fragments $\mathsf{A\overline{A}B\overline{B}\overline{E}}$ and $\mathsf{A\overline{A}E\overline{B}\overline{E}}$, devising for each of them an **EXPSPACE** MC algorithm which exploits the possibility of finding, for each track of the Kripke structure, a satisfiability-preserving track of bounded length (*track representative*). In this way, the algorithm needs to check only tracks with a bounded maximum length. Later [19], they prove that the problem for $\mathsf{A\overline{A}B\overline{B}\overline{E}}$ and $\mathsf{A\overline{A}E\overline{B}\overline{E}}$ is **PSPACE**-hard (if a succinct encoding of formulas is exploited, the algorithm

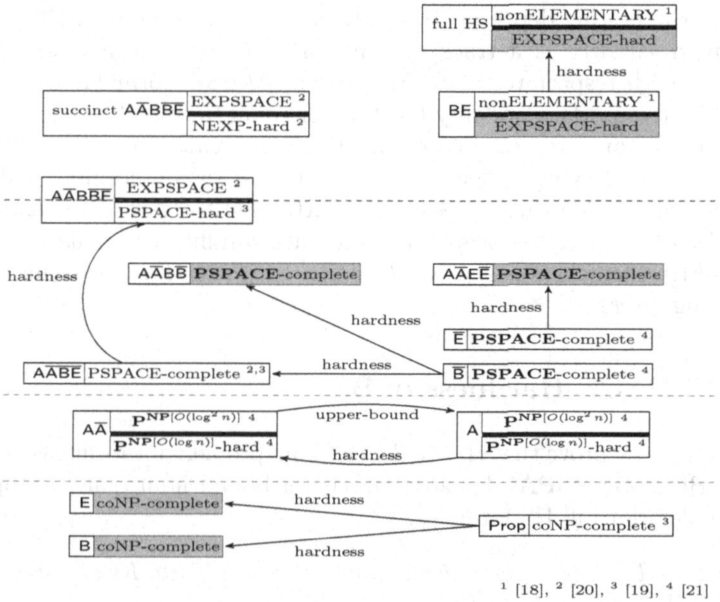

Fig. 3. Complexity of the model checking problem for HS fragments: known results are depicted in white boxes, new ones in gray boxes.

remains in **EXPSPACE**, but a **NEXPTIME** lower bound can be given [20]). Finally, they show that formulas satisfying a constant bound on the nesting depth of $\langle B \rangle$ (resp., $\langle E \rangle$) can be checked in polynomial working space [20].

In [19,21] the authors identify some well-behaved HS fragments, namely, $A\overline{A}B\overline{E}$, $\overline{B}$, $\overline{E}$, $A\overline{A}$, A, and $\overline{A}$, which are still expressive enough to capture meaningful interval properties of state-transition systems and whose model checking problem has a computational complexity markedly lower than that of full HS. In particular, they prove that the problem is **PSPACE**-complete for the first three fragments, and in between $\mathbf{P}^{\mathbf{NP}[O(\log n)]}$ and $\mathbf{P}^{\mathbf{NP}[O(\log^2 n)]}$ [9,26] for the last three. In all cases, the complexity of the problem turns out be comparable to or lower than that of LTL, which is known to be **PSPACE**-complete [27]. In this paper, we first strengthen the lower bound to the complexity of the model checking problem for full HS by proving **EXPSPACE**-hardness of the fragment BE. Then, we study two more well-behaved fragments, namely, $A\overline{A}B\overline{B}$ and $A\overline{A}E\overline{E}$, and we prove that their model checking problem is **PSPACE**-complete (the previously known upper bound was **EXPSPACE** [20]). This is somehow surprising, as their expressive power seems to be really higher than that of the fragments analyzed in [19,21], but their complexity turns out to be the same. Finally, we prove that B and E are in **co-NP**, and thus **co-NP**-complete, as the purely propositional fragment of HS, Prop, is **co-NP**-complete [19].

It is worth noticing that, to determine the complexity of $A\overline{A}B\overline{B}$ and $A\overline{A}E\overline{E}$, we exploit the structure of the specific input formula, rather than considering generically the nesting depth of ⟨B⟩ or ⟨E⟩ modalities (as done in [20]). In [20], a *track representative* is a track of *exponential length*, which is satisfiability equivalent—with respect to *all* $\mathsf{A\overline{A}B\overline{B}\overline{E}}$ (resp., $\mathsf{A\overline{A}E\overline{B}\overline{E}}$) formulas with nesting depth of ⟨B⟩ (resp., ⟨E⟩) modality less than or equal to some k—to all the (possibly infinitely many) represented tracks. Here, we weaken such a constraint by requiring satisfiability equivalence only w. r. to the *specific* formula under consideration, which allows us to restrict our attention to tracks of *polynomially-bounded length*, that is, we prove that if a track ρ fulfils a formula ψ of $\mathsf{A\overline{A}B\overline{B}}$ (resp., $\mathsf{A\overline{A}E\overline{E}}$), then there is also a *polynomial-length* track ρ' satisfying ψ (with ρ' depending on ψ).

3 EXPSPACE-Hardness of BE

In this section, we prove that the model checking problem for formulas of the HS fragment BE is **EXPSPACE**-hard. This lower-bound immediately propagates to the problem for full HS formulas.

Theorem 6. *The MC problem for* BE *formulas over finite Kripke structures is* **EXPSPACE**-*hard (under polynomial-time reductions).*

Proof. The claim is proved by a polynomial-time reduction from a domino-tiling problem for grids with rows of single exponential length [11]. An instance $\mathcal{I}$ of such problem is a tuple $\mathcal{I} = (C, \Delta, n, d_{init}, d_{final})$, where C is a finite set of colors, $\Delta \subseteq C^4$ is a set of tuples $(c_{down}, c_{left}, c_{up}, c_{right})$ of four colors, called *domino-types*, $n > 0$ is a natural number encoded in *unary*, and $d_{init}, d_{final} \in \Delta$ are domino-types. A *tiling of* $\mathcal{I}$ is a mapping $f : [0, k] \times [0, 2^n - 1] \rightarrow \Delta$, for some $k \geq 0$, satisfying the following constraints:

- two adjacent cells in a row have the same color on the shared edge: for all $(i, j) \in [0, k] \times [0, 2^n - 2]$, $[f(i, j)]_{right} = [f(i, j + 1)]_{left}$;
- two adjacent cells in a column have the same color on the shared edge: for all $(i, j) \in [0, k - 1] \times [0, 2^n - 1]$, $[f(i, j)]_{up} = [f(i + 1, j)]_{down}$;
- $f(0, 0) = d_{init}$ (*initialization*) and $f(k, 2^n - 1) = d_{final}$ (*acceptance*).

It is well-known that checking the existence (resp., the non-existence) of a tiling for $\mathcal{I}$ is **EXPSPACE**-complete [11]. We now show how to build in polynomial time a Kripke structure $\mathcal{K}_{\mathcal{I}}$ and a BE formula $\varphi_{\mathcal{I}}$ such that there exists an initial track of $\mathcal{K}_{\mathcal{I}}$ satisfying $\varphi_{\mathcal{I}}$ if and only if there exists a tiling of $\mathcal{I}$. Hence $\mathcal{K}_{\mathcal{I}} \models \neg\varphi_{\mathcal{I}}$ iff there does not exist a tiling of $\mathcal{I}$, and Theorem 6 follows.

We use the following set $\mathcal{AP}$ of proposition letters to encode tilings of $\mathcal{I}$: $\mathcal{AP} = \Delta \cup \{\$\} \cup \{0, 1\}$. Proposition letters in $\{0, 1\}$ are used to encode the value of an n-bits counter numbering the cells of one row of a tiling. In particular, a cell with content $d \in \Delta$ and column number $j \in [0, 2^n - 1]$ is encoded by the word of length $n+1$ over $\mathcal{AP}$ given by $d\, b_1 \ldots b_n$, where $b_1 \ldots b_n$ is the binary encoding of the column number j (b_n is the most significant bit). A row is encoded by the word listing the encodings of cells from left to right, and a tiling f with $k + 1$

rows is encoded by the finite word $r_0\$r_1\ldots\$r_k$, where r_i is the encoding of the i-th row of f for all $i \in [0,k]$.

The Kripke structure $\mathcal{K}_\mathcal{I}$ is defined as $\mathcal{K}_\mathcal{I} = (\mathcal{AP}, \mathcal{AP}, \mathcal{AP} \times \mathcal{AP}, \mu, d_{init})$, where $\mu(p) = \{p\}$, for any $p \in \mathcal{AP}$. Thus, the initial tracks of $\mathcal{K}_\mathcal{I}$ correspond to the finite words over $\mathcal{AP}$ which start with the initial domino type d_{init}.

In order to build the BE formula $\varphi_\mathcal{I}$, we use some auxiliary formulas, namely, length_i, $\mathit{beg}(p)$, $\mathit{end}(p)$, ϕ_{cell}, and $\theta_j(b,b')$ where $i \in [1, 2n+2]$, $j \in [2, n+1]$, $p \in \mathcal{AP}$, and $b, b' \in \{0,1\}$. The formula length_i has size linear in i and characterizes the tracks of length i. It can be expressed as follows:

$$\mathit{length}_i := (\underbrace{\langle B\rangle \ldots \langle B\rangle}_{i-1} \top) \wedge (\underbrace{[B] \ldots [B]}_{i} \bot).$$

The formula $\mathit{beg}(p)$ (resp., $\mathit{end}(p)$) captures the tracks of $\mathcal{K}$ which start (resp., end) in state p:

$$\mathit{beg}(p) := (p \wedge \mathit{length}_1) \vee \langle B\rangle (p \wedge \mathit{length}_1), \quad \mathit{end}(p) := (p \wedge \mathit{length}_1) \vee \langle E\rangle (p \wedge \mathit{length}_1).$$

The formula ϕ_{cell} captures the tracks of $\mathcal{K}_\mathcal{I}$ which encode cells:

$$\phi_{cell} := \mathit{length}_{n+1} \wedge \Big(\bigvee_{d \in \Delta} \mathit{beg}(d) \Big) \wedge [E](\mathit{beg}(0) \vee \mathit{beg}(1)).$$

Finally, for all $j \in [2, n+1]$ and $b, b' \in \{0,1\}$, the formula $\theta_j(b,b')$ is defined as $\theta_j(b,b') := \langle B\rangle(\mathit{length}_j \wedge \mathit{end}(b)) \wedge \langle E\rangle(\mathit{length}_{n-j+2} \wedge \mathit{beg}(b'))$. The formula $\theta_j(b,b')$ is satisfied by a track ρ if $|\rho| \geq j+1$, $|\rho| \geq n-j+3$, $\rho(j) = b$, and $\rho(|\rho| - n + j - 1) = b'$. In particular, for a track ρ starting with a cell c and ending with a cell c', $\theta_j(b,b')$ is satisfied by ρ if the jth bit of c is b and the jth bit of c' is b'.

Additionally, we use the derived operator $\langle G\rangle$ and its dual $[G]$, which allow us to select arbitrary subtracks of the given track, including the track itself:

$$\langle G\rangle \psi := \psi \vee \langle B\rangle \psi \vee \langle E\rangle \psi \vee \langle B\rangle \langle E\rangle \psi.$$

Then, the formula $\varphi_\mathcal{I}$ is defined as $\varphi_\mathcal{I} := \varphi_b \wedge \varphi_{req} \wedge \varphi_{inc} \wedge \varphi_{rr} \wedge \varphi_{rc}$.

φ_b checks that the given track starts with a cell with content d_{init} and column number 0, and ends with a cell with content d_{final} and column number $2^n - 1$:

$$\varphi_b := \langle B\rangle \phi_{cell} \wedge \mathit{beg}(d_{init}) \wedge \langle E\rangle(\phi_{cell} \wedge \mathit{beg}(d_{final})) \wedge \bigwedge_{j=2}^{n+1} \theta_j(0,1).$$

The conjunct φ_{req} ensures the following two requirements: (i) each occurrence of \$ in the given track is followed by a cell with column number 0 and (ii) each cell c in the given track is followed either by another cell, or by the separator \$, and in the latter case c has column number $2^n - 1$. The first requirement is encoded by the formula: $[G]\big((length_{n+2} \wedge beg(\$)) \rightarrow \langle \mathrm{E}\rangle(\phi_{cell} \wedge [E]beg(0))\big)$; the second one by the formula:

$$[G]\Big\{(length_{n+2} \wedge \bigvee_{d\in\Delta} beg(d)) \rightarrow \Big(\langle \mathrm{B}\rangle\, \phi_{cell} \wedge (end(\$) \vee \bigvee_{d\in\Delta} end(d)) \wedge (end(\$) \rightarrow [E](beg(\$) \vee beg(1)))\Big)\Big\}.$$

The conjunct φ_{inc} checks that adjacent cells along the given track have consecutive columns numbers:

$$\varphi_{inc} = [G]\Big(\phi_{two\_cells} \rightarrow \bigvee_{j=2}^{n+1} \big[\theta_j(0,1) \wedge \bigwedge_{h=2}^{j-1} \theta_h(1,0) \wedge \bigwedge_{h=j+1}^{n+1} \bigvee_{b\in\{0,1\}} \theta_h(b,b)\big]\Big),$$

where ϕ_{two_cells} is given by $length_{2n+2} \wedge \langle \mathrm{B}\rangle\, \phi_{cell} \wedge \langle \mathrm{E}\rangle\, \phi_{cell}$. Note that φ_{req} and φ_{inc} ensure that the column numbers are correctly encoded.

The conjunct φ_{rr} checks that adjacent cells in a row have the same color on the shared edge:

$$\varphi_{rr} = [G]\Big(\phi_{two\_cells} \rightarrow \bigvee_{(d,d')\in\Delta\times\Delta \mid d_{right}=d'_{left}} (beg(d) \wedge \langle \mathrm{E}\rangle(length_{n+1} \wedge beg(d')))\Big).$$

Finally, the conjunct φ_{rc} checks that adjacent cells in a column have the same color on the shared edge. For this, it suffices to require that for each subtrack of the given one containing exactly one occurrence of \$, starting with a cell c, and ending with a cell c', if c and c' have the same column number, then $d_{up} = d'_{down}$, where d (resp., d') is the content of c (resp., c'). Thus, formula φ_{rc} is defined as follows, where we use the formulas $\theta_j(b,b)$, with $j \in [2, n+1]$ and $b \in \{0,1\}$, for expressing that c and c' have the same column number:

$$\varphi_{rc} = [G]\Big\{\Big(\phi_{one}(\$) \wedge \langle \mathrm{B}\rangle\, \phi_{cell} \wedge \langle \mathrm{E}\rangle\, \phi_{cell} \wedge \bigwedge_{j=2}^{n+1} \bigvee_{b\in\{0,1\}} \theta_j(b,b)\Big) \rightarrow \bigvee_{(d,d')\in\Delta\times\Delta \mid d_{up}=d'_{down}} (beg(d) \wedge \langle \mathrm{E}\rangle(length_{n+1} \wedge beg(d')))\Big\},$$

where $\phi_{one}(\$)$ is defined as $(\langle \mathrm{B}\rangle\, end(\$)) \wedge \neg(\langle \mathrm{B}\rangle(end(\$) \wedge \langle \mathrm{B}\rangle\, end(\$)))$.

Note that $\varphi_{\mathcal{I}}$ has size polynomial in the size of $\mathcal{I}$. By construction, a track ρ of $\mathcal{K}_{\mathcal{I}}$ satisfies $\varphi_{\mathcal{I}}$ if and only if ρ encodes a tiling. Since the initial tracks of $\mathcal{K}_{\mathcal{I}}$ are the finite words over $\mathcal{AP}$ starting with d_{init}, it follows that there exists a tiling of $\mathcal{I}$ if and only if there exists an initial track of $\mathcal{K}_{\mathcal{I}}$ which satisfies $\varphi_{\mathcal{I}}$. Hence, the result follows, which concludes the proof. □

4 The Fragments $\mathsf{A\overline{A}E\overline{E}}$ and $\mathsf{A\overline{A}B\overline{B}}$: Polynomial-Size Model-Track Property

In this section, we show that the MC problem for the fragments $\mathsf{A\overline{A}E\overline{E}}$ and $\mathsf{A\overline{A}B\overline{B}}$ is in **PSPACE** by proving a *polynomial size model-track property*, that is, we show that if a track ρ of a Kripke structure $\mathcal{K}$ satisfies a formula φ of $\mathsf{A\overline{A}E\overline{E}}$ or $\mathsf{A\overline{A}B\overline{B}}$, then there is a track π, whose length is polynomial in the sizes of φ and $\mathcal{K}$, starting from and leading to the same states as ρ, that satisfies φ. Moreover, we show that the problem is in **co-NP** for the smaller fragments B and E. We conclude the section by providing two model checking procedures, one for $\mathsf{A\overline{A}E\overline{E}}$ formulas and one for E formulas. In the following, we focus on $\mathsf{A\overline{A}E\overline{E}}$ and the smaller fragment E, being the cases of $\mathsf{A\overline{A}B\overline{B}}$ and B completely symmetric.

Let $\mathcal{K} = (\mathcal{AP}, W, \delta, \mu, w_0)$ be a Kripke structure. We start by introducing the notions of *induced track* and *well-formed track*, which will be exploited to prove the polynomial size model-track property.

Definition 7. *Let $\rho \in \mathrm{Trk}_{\mathcal{K}}$ be a track of length n. A* track induced by ρ *is a track $\pi \in \mathrm{Trk}_{\mathcal{K}}$ such that there is an increasing sequence of ρ-positions $i_1 < \ldots < i_k$, with $i_1 = 1$, $i_k = n$, and $\pi = \rho(i_1)\cdots\rho(i_k)$. Moreover, we say that the π-position j and the ρ-position i_j are corresponding. The induced track π is* well-formed with respect to ρ *if, for all π-positions j, with corresponding ρ-positions i_j, and all proposition letters $p \in \mathcal{AP}$, it holds that $\mathcal{K}, \pi^j \models p \iff \mathcal{K}, \rho^{i_j} \models p$.*

Note that if π is induced by ρ, then $\mathrm{fst}(\pi) = \mathrm{fst}(\rho)$, $\mathrm{lst}(\pi) = \mathrm{lst}(\rho)$, and $|\pi| \leq |\rho|$ (in particular, $|\pi| = |\rho|$ iff $\pi = \rho$). Intuitively, a track induced by ρ is obtained by contracting ρ, namely, by concatenating some subtracks of ρ, provided that the resulting sequence is a track of $\mathcal{K}$ as well. Well-formedness implies that the suffix of π starting from position j and the suffix of ρ starting from the corresponding position i_j agree over all the proposition letters in $\mathcal{AP}$, i.e., they have the same satisfiability pattern of proposition letters. In particular, $\mathcal{K}, \pi \models p$ iff $\mathcal{K}, \rho \models p$, for all $p \in \mathcal{AP}$. It can be easily seen that *the well-formedness relation is transitive.*

The following proposition shows how it is possible to contract a track, preserving the same satisfiability of proposition letters with respect to suffixes. Such a criterion represents a "basic step" in a contraction process which will allow us to prove the polynomial size model-track property.

Proposition 8. *For any track ρ of $\mathcal{K} = (\mathcal{AP}, W, \delta, \mu, w_0)$, there exists a track π of $\mathcal{K}$, which is well-formed with respect to ρ, such that $|\pi| \leq |W| \cdot (|\mathcal{AP}| + 1)$.*

Proof. Let $\rho \in \mathrm{Trk}_{\mathcal{K}}$ be a track of length n. If $n \leq |W| \cdot (|\mathcal{AP}| + 1)$, the thesis trivially holds. Let us assume $n > |W| \cdot (|\mathcal{AP}| + 1)$. We show that there exists a track of $\mathcal{K}$ which is well-formed with respect to ρ and whose length is smaller than n. Since $n > |W| \cdot (|\mathcal{AP}| + 1)$, there is some state $w \in W$ occurring in ρ at least $|\mathcal{AP}| + 2$ times. Assume that for all ρ-positions i and j, with $j > i$, if $\rho(i) = \rho(j) = w$, then there exists some $p \in \mathcal{AP}$ such that $\mathcal{K}, \rho^j \models p$ and $\mathcal{K}, \rho^i \not\models p$. This assumption leads to a contradiction, as the suffixes of ρ may feature at most $|\mathcal{AP}| + 1$ distinct satisfiability patterns of proposition letters (due to the homogeneity principle in Definition 4), while there are at least $|\mathcal{AP}| + 2$ occurrences of w. As a consequence, there are two ρ-positions i and j, with $j > i$, such that $\rho(i) = \rho(j) = w$ and, for all $p \in \mathcal{AP}$, $\mathcal{K}, \rho^j \models p$ iff $\mathcal{K}, \rho^i \models p$. It is easy to see that $\pi = \rho(1, i) \star \rho(j, n) \in \mathrm{Trk}_{\mathcal{K}}$ is well-formed with respect to ρ and $|\pi| < n$. Now, if $|\pi| \leq |W| \cdot (|\mathcal{AP}| + 1)$, the thesis is proved; otherwise, the same basic step can be iterated a finite number of times, and the thesis follows by transitivity of the well-formedness relation. □

The next definition introduces some distinguished positions in a track. The intuition is that—as we will see in the proof of Theorem 10—if we perform a contraction (as we did in the proof of Proposition 8) between a pair of such positions, we get an equivalent track with respect to satisfiability of the considered $\mathsf{A\overline{A}E\overline{E}}$ formula. In the following, we restrict ourselves to formulas in *negation normal form* (NNF), namely, formulas where negation is applied only to proposition letters. By using De Morgan's laws and the dual modalities $[E]$, $[\overline{E}]$, $[A]$, and $[\overline{A}]$ of $\langle \mathrm{E} \rangle$, $\langle \overline{\mathrm{E}} \rangle$, $\langle \mathrm{A} \rangle$, and $\langle \overline{\mathrm{A}} \rangle$, we can trivially convert in linear time a formula into an equivalent one in NNF, of at most double length.

Definition 9 (Witness Positions). *Let ρ be a track of $\mathcal{K}$ and φ be a formula of $\mathsf{A\overline{A}E\overline{E}}$. Let us denote by $E(\varphi, \rho)$ the set of subformulas $\langle \mathrm{E} \rangle \psi$ of φ such that $\mathcal{K}, \rho \models \langle \mathrm{E} \rangle \psi$. The* set $Wt(\varphi, \rho)$ of witness positions of ρ for φ *is the* minimal *set of ρ-positions satisfying the following constraint: for each $\langle \mathrm{E} \rangle \psi \in E(\varphi, \rho)$, the greatest ρ-position $i > 1$ such that $\mathcal{K}, \rho^i \models \psi$ belongs to $Wt(\varphi, \rho)$*[2]*.*

It is immediate to see that the cardinalities of $E(\varphi, \rho)$ and of $Wt(\varphi, \rho)$ are at most $|\varphi| - 1$. We are now ready to prove the polynomial-size model-track property.

Theorem 10 (Polynomial-Size Model-Track Property). *Let $= (\mathcal{AP}, W, \delta, \mu, w_0)$, $\rho, \sigma \in \mathrm{Trk}_{\mathcal{K}}$, and φ be an $\mathsf{A\overline{A}E\overline{E}}$ formula in NNF such that $\mathcal{K}, \rho \star \sigma \models \varphi$. Then, there is $\pi \in \mathrm{Trk}_{\mathcal{K}}$, induced by ρ, such that $\mathcal{K}, \pi \star \sigma \models \varphi$ and $|\pi| \leq |W| \cdot (|\varphi| + 1)^2$.*

Notice that Theorem 10 holds in particular if $|\sigma| = 1$, and thus $\rho \star \sigma = \rho$ and $\pi \star \sigma = \pi$. In this case, if $\mathcal{K}, \rho \models \varphi$, then $\mathcal{K}, \pi \models \varphi$, where π is induced by ρ and $|\pi| \leq |W| \cdot (|\varphi| + 1)^2$. The more general statement of Theorem 10 is needed for technical reasons in the soundness/completeness proof of the next algorithms.

Proof. W.l.o.g., we can restrict ourselves to set of proposition letters occurring in φ, thus assuming $|\mathcal{AP}| \leq |\varphi|$. Let $Wt(\varphi, \rho \star \sigma)$ be the set of witness positions of $\rho \star \sigma$ for φ. Let $\{i_1, \ldots, i_k\}$ be the ordering of $Wt(\varphi, \rho \star \sigma)$ such that $i_1 < \ldots < i_k$. Let $i_0 = 1$ and $i_{k+1} = |\rho \star \sigma|$. Hence, $1 = i_0 < i_1 < \ldots < i_k \leq i_{k+1} = |\rho \star \sigma|$.

If the length of ρ is at most $|W| \cdot (|\varphi| + 1)^2$, the thesis trivially holds. Let us assume that $|\rho| > |W| \cdot (|\varphi| + 1)^2$. We show that there exists a track π induced by ρ, with $|\pi| < |\rho|$, such that $\mathcal{K}, \pi \star \sigma \models \varphi$.

W.l.o.g., we can assume that $i_0 < i_1 < \ldots < i_j$, for some $j \geq 0$, are ρ-positions (while $i_{j+1} < \ldots < i_{k+1}$ are $(\rho \star \sigma)$-positions not in ρ). We claim that either (i) there exists $t \in [0, j-1]$ such that $i_{t+1} - i_t > |W| \cdot (|\varphi| + 1)$ or (ii) $|\rho(i_j, |\rho|)| > |W| \cdot (|\varphi| + 1)$. By way of contradiction, suppose that neither (i) nor (ii) holds. We need to distinguish two cases. If $\rho \star \sigma = \rho$, then $|\rho| = (i_{k+1} - i_0) + 1 \leq (k+1) \cdot |W| \cdot (|\varphi| + 1) + 1$; otherwise ($|\rho| < |\rho \star \sigma|$), $|\rho| = (i_j - i_0) + |\rho(i_j, |\rho|)| \leq j \cdot |W| \cdot (|\varphi| + 1) + |W| \cdot (|\varphi| + 1) \leq (k+1) \cdot |W| \cdot (|\varphi| + 1)$. The contradiction follows since $(k+1) \cdot |W| \cdot (|\varphi| + 1) + 1 \leq |\varphi| \cdot |W| \cdot (|\varphi| + 1) + 1 \leq |W| \cdot (|\varphi| + 1)^2$.

[2] Note that such a ρ-position exists by definition of $E(\varphi, \rho)$.

Let us define $(\alpha, \beta) = (i_t, i_{t+1})$ in case (i), and $(\alpha, \beta) = (i_j, |\rho|)$ in case (ii). Moreover let $\rho' = \rho(\alpha, \beta)$. In both cases, we have $|\rho'| > |W| \cdot (|\varphi| + 1) \geq |W| \cdot (|\mathcal{AP}| + 1)$, being $|\mathcal{AP}| \leq |\varphi|$. By Proposition 8, there exists a track π' of $\mathcal{K}$, well-formed with respect to ρ', such that $|\pi'| \leq |W| \cdot (|\mathcal{AP}|+1) < |\rho'|$. Let π be the track induced by ρ obtained by replacing the subtrack ρ' of ρ with π'. Since $|\pi| < |\rho|$, it remains to prove that $\mathcal{K}, \pi \star \sigma \models \varphi$.

Let us denote $\pi \star \sigma$ by $\overline{\pi}$ and $\rho \star \sigma$ by $\overline{\rho}$. Moreover, let $H : [1, |\overline{\pi}|] \to [1, |\overline{\rho}|]$ be the function mapping positions of $\overline{\pi}$ into positions of $\overline{\rho}$ in this way: positions "outside" π' (i.e., outside the interval $[\alpha, \alpha + |\pi'| - 1]$) are mapped into their original position in $\overline{\rho}$; positions "inside" π' (i.e., in $[\alpha, \alpha + |\pi'| - 1]$) are mapped to the corresponding position in ρ' (exploiting well-formedness of π' w.r.t. to ρ').

$$H(m) = \begin{cases} m & \text{if } \ m < \alpha \\ \alpha + \ell_{m-\alpha+1} - 1 & \text{if } \ \alpha \leq m < \alpha + |\pi'| \\ m + (|\rho'| - |\pi'|) & \text{if } \ m \geq \alpha + |\pi'| \end{cases} \tag{1}$$

where ℓ_m is the ρ'-position corresponding to the π'-position m. It is easy to check that H satisfies the following properties: (1) H is strictly monotonic, i.e., for all $j, j' \in [1, |\overline{\pi}|]$, $j < j'$ iff $H(j) < H(j')$; (2) for all $j \in [1, |\overline{\pi}|]$, $\overline{\pi}(j) = \overline{\rho}(H(j))$; (3) $H(1) = 1$ and $H(|\overline{\pi}|) = |\overline{\rho}|$; (4) $Wt(\varphi, \overline{\rho}) \subseteq \{H(j) \mid j \in [1, |\overline{\pi}|]\}$; (5) for each $j \in [1, |\overline{\pi}|]$ and $p \in \mathcal{AP}$, $\mathcal{K}, \overline{\pi}^j \models p$ iff $\mathcal{K}, \overline{\rho}^{H(j)} \models p$.

The statement $\mathcal{K}, \overline{\pi} \models \varphi$ is an immediate consequence of the following claim, considering that $H(1) = 1$, $\mathcal{K}, \overline{\rho} \models \varphi$, $\overline{\rho}^1 = \overline{\rho}$, and $\overline{\pi}^1 = \overline{\pi}$.

Claim. For all $j \in [1, |\overline{\pi}|]$, all subformulas ψ of φ, and all $u \in \text{Trk}_{\mathcal{K}}$, it holds that if $\mathcal{K}, u \star \overline{\rho}^{H(j)} \models \psi$, then $\mathcal{K}, u \star \overline{\pi}^j \models \psi$.

Proof. Assume that $\mathcal{K}, u \star \overline{\rho}^{H(j)} \models \psi$. Note that $u \star \overline{\rho}^{H(j)}$ is defined iff $u \star \overline{\pi}^j$ is defined. We prove by induction on the structure of φ that $\mathcal{K}, u \star \overline{\pi}^j \models \psi$. Since φ is in NNF, only the following cases occur:

- $\psi = p$ or $\psi = \neg p$ for some $p \in \mathcal{AP}$. By Property 5 of H, $\mathcal{K}, \overline{\pi}^j \models p$ iff $\mathcal{K}, \overline{\rho}^{H(j)} \models p$. Hence, $\mathcal{K}, u \star \overline{\pi}^j \models p$ iff $\mathcal{K}, u \star \overline{\rho}^{H(j)} \models p$, and the result holds.
- $\psi = \theta_1 \wedge \theta_2$ or $\psi = \theta_1 \vee \theta_2$, for some $\mathsf{A\overline{A}E\overline{E}}$ formulas θ_1 and θ_2: the result directly follows from the inductive hypothesis.
- $\psi = [E]\theta$. We need to show that for each proper suffix η of $u \star \overline{\pi}^j$, $\mathcal{K}, \eta \models \theta$. We distinguish two cases:
  - η is *not* a proper suffix of $\overline{\pi}^j$. Hence, η is of the form $u^h \star \overline{\pi}^j$ for some $h \in [2, |u|]$. Since $\mathcal{K}, u \star \overline{\rho}^{H(j)} \models [E]\theta$, then $\mathcal{K}, u^h \star \overline{\rho}^{H(j)} \models \theta$. By induction, $\mathcal{K}, u^h \star \overline{\pi}^j \models \theta$.
  - η is a proper suffix of $\overline{\pi}^j$. Hence, $\eta = \overline{\pi}^h$ for some $h \in [j + 1, |\overline{\pi}|]$. By Property 1 of H, $H(h) > H(j)$, and since $\mathcal{K}, u \star \overline{\rho}^{H(j)} \models \psi$, we have that $\mathcal{K}, \overline{\rho}^{H(h)} \models \theta$. By induction, $\mathcal{K}, \overline{\pi}^h \models \theta$.

  Therefore, $\mathcal{K}, u \star \overline{\pi}^j \models [E]\theta$.
- $\psi = \langle E \rangle\, \theta$. We need to show that there exists a proper suffix of $u \star \overline{\pi}^j$ satisfying θ. Since $\mathcal{K}, u \star \overline{\rho}^{H(j)} \models \psi$, there exists a proper suffix η' of $u \star \overline{\rho}^{H(j)}$ such that $\mathcal{K}, \eta' \models \theta$. We distinguish two cases:

- η' is *not* a proper suffix of $\overline{\rho}^{H(j)}$. Hence, η' is of the form $u^h \star \overline{\rho}^{H(j)}$ for some $h \in [2, |u|]$. By induction, $\mathcal{K}, u^h \star \overline{\pi}^j \models \theta$, and $\mathcal{K}, u \star \overline{\pi}^j \models \langle \mathrm{E} \rangle\, \theta$.
- η' is a proper suffix of $\overline{\rho}^{H(j)}$. Hence, $\eta' = \overline{\rho}^i$ for some $i \in [H(j)+1, |\overline{\rho}|]$, and $\mathcal{K}, \overline{\rho}^i \models \theta$. Let i' be the greatest position of $\overline{\rho}$ such that $\mathcal{K}, \overline{\rho}^{i'} \models \theta$. Hence $i' \geq i$ and, by Definition 9, $i' \in Wt(\varphi, \overline{\rho})$. By Property 4 of H, $i' = H(h)$ for some $\overline{\pi}$-position h. Since $H(h) > H(j)$, it holds that $h > j$ (Property 1). By induction, $\mathcal{K}, \overline{\pi}^h \models \theta$, and $\mathcal{K}, u \star \overline{\pi}^j \models \langle \mathrm{E} \rangle\, \theta$.

- $\psi = [\overline{E}]\theta$ or $\psi = \langle \overline{\mathrm{E}} \rangle\, \theta$: a direct consequence of the inductive hypothesis.
- $\psi = [A]\theta$, $\psi = \langle \mathrm{A} \rangle\, \theta$, $\psi = [\overline{A}]\theta$ or $\psi = \langle \overline{\mathrm{A}} \rangle\, \theta$. Since $u \star \overline{\pi}^j$ and $u \star \overline{\rho}^{H(j)}$ start at the same state and lead to the same state (by Properties 2 and 3 of H), the result trivially follows. This concludes the proof of the claim.

We have proved that $\mathcal{K}, \overline{\pi} \models \varphi$, with $|\pi| < |\rho|$. If $|\pi| \leq |W| \cdot (|\varphi| + 1)^2$, the thesis is proved; otherwise, we can iterate the above contraction (a finite number of times) until the bound is achieved. □

Now, by exploiting the polynomial-size model-track property stated by Theorem 10, it is easy to define a **PSPACE** MC algorithm for $\mathsf{A}\overline{\mathsf{A}}\mathsf{E}\overline{\mathsf{E}}$ formulas, and a **co-NP** MC algorithm for E formulas. The main MC procedure for $\mathsf{A}\overline{\mathsf{A}}\mathsf{E}\overline{\mathsf{E}}$ formulas is `ModCheck`($\mathcal{K}, \psi$) (Algorithm 1). All the initial tracks $\tilde{\rho}$, obtained by visiting the unravelling of $\mathcal{K}$ from w_0 up to depth $|W| \cdot (2|\psi| + 3)^2$, are checked w.r. to ψ by the function `Check`($\mathcal{K}, \psi, \tilde{\rho}$) (Algorithm 2)—which decides whether $\mathcal{K}, \tilde{\rho} \models \psi$ by basically calling itself recursively on the subformulas of ψ and unravelling again $\mathcal{K}$—until either some initial track is found that does not model ψ or all of them model ψ (and thus $\mathcal{K} \models \psi$). Notice that the for-loop at the first line considers all initial tracks of length at most $|W| \cdot (2|\psi| + 3)^2 \geq |W| \cdot (|NNF(\neg\psi)| + 1)^2$.

Algorithm 1. `ModCheck`($\mathcal{K}, \psi$)

1: **for** all initial $\tilde{\rho} \in \mathrm{Trk}_{\mathcal{K}}$ s.t. $|\tilde{\rho}| \leq |W| \cdot (2|\psi| + 3)^2$ **do**
2:    **if** `Check`($\mathcal{K}, \psi, \tilde{\rho}$) $= 0$ **then**
3:       **return** 0: "$\mathcal{K}, \tilde{\rho} \not\models \psi$" ◁ *Counterexample found*
4: **return** 1: "$\mathcal{K} \models \psi$"

The reason is that in the soundness and completeness proof of the algorithm, we need to consider the NNF of $\neg\psi$ and to apply the polynomial bound of Theorem 10 to such a form. The next theorem states soundness and completeness of the procedures.

Theorem 11 (See [2] for the proof). *Let ψ be an $\mathsf{A}\overline{\mathsf{A}}\mathsf{E}\overline{\mathsf{E}}$ formula and $\mathcal{K}$ be a Kripke structure. Then, (i) `ModCheck`($\mathcal{K}, \psi$) $= 1$ if and only if $\mathcal{K} \models \psi$; (ii) for any track $\tilde{\rho} \in \mathrm{Trk}_{\mathcal{K}}$, `Check`($\mathcal{K}, \psi, \tilde{\rho}$) $= 1$ if and only if $\mathcal{K}, \tilde{\rho} \models \psi$.*

The given procedures require *polynomial working space*, since (*i*) `ModCheck` needs to store only a track no longer than $|W| \cdot (2|\psi| + 3)^2$ (many tracks are

Algorithm 2. $\mathtt{Check}(\mathcal{K}, \psi, \tilde{\rho})$

if $\psi = p$, for $p \in \mathcal{AP}$ **then**
  if $p \in \bigcap_{s \in \mathtt{states}(\tilde{\rho})} \mu(s)$ **then**
    return 1 **else return** 0
else if $\psi = \varphi_1 \wedge \varphi_2$ **then**
  if $\mathtt{Check}(\mathcal{K}, \varphi_1, \tilde{\rho}) = 0$ **then**
    return 0
  else
    return $\mathtt{Check}(\mathcal{K}, \varphi_2, \tilde{\rho})$
else if $\psi = \langle A \rangle \varphi$ **then**
  for all $\rho \in \mathrm{Trk}_{\mathcal{K}}$ such that $\mathrm{fst}(\rho) = \mathrm{lst}(\tilde{\rho})$, and $|\rho| \leq |W| \cdot (2|\varphi| + 1)^2$ **do**
    if $\mathtt{Check}(\mathcal{K}, \varphi, \rho) = 1$ **then**
      return 1
  return 0
else if $\psi = \langle E \rangle \varphi$ **then**
  for each $\overline{\rho}$ suffix of $\tilde{\rho}$ **do**
    if $\mathtt{Check}(\mathcal{K}, \varphi, \overline{\rho}) = 1$ **then**
      return 1
  return 0
else if $\psi = \langle \overline{E} \rangle \varphi$ **then**
  for all $\rho \in \mathrm{Trk}_{\mathcal{K}}$ such that $\mathrm{lst}(\rho) = \mathrm{fst}(\tilde{\rho})$, and $2 \leq |\rho| \leq |W| \cdot (2|\varphi| + 1)^2$ **do**
    if $\mathtt{Check}(\mathcal{K}, \varphi, \rho \star \tilde{\rho}) = 1$ **then**
      return 1
  return 0
else if $\psi = \neg\varphi$ **then**
  return $1 - \mathtt{Check}(\mathcal{K}, \varphi, \tilde{\rho})$
... ◁ *$\psi = \langle \overline{A} \rangle \varphi$ is analogous to $\psi = \langle A \rangle \varphi$*

generated while visiting the unravelling of $\mathcal{K}$, but only one at a time needs to be stored), (*ii*) each recursive call to `Check` (possibly) needs space for a track no longer than $|W| \cdot (2|\varphi| + 1)^2$, where φ is a subformula of ψ with $|\varphi| = |\psi| - 1$, and (*iii*) at most 1 call to `ModCheck` and $|\psi|$ calls to `Check` are jointly active. Thus, the maximum space needed by the algorithms is $(|\psi| + 1) \cdot O(\log |W|) \cdot (|W| \cdot (2|\psi| + 3)^2)$ bits, where $O(\log |W|)$ bits are needed to represent a state of $\mathcal{K}$.

Corollary 12. *The MC problem for* $\mathsf{A\overline{A}E\overline{E}}$ *formulas over finite Kripke structures is* **PSPACE**-*complete.*

Proof. **PSPACE**-hardness immediately follows from that of $\mathsf{\overline{E}}$ [21]. □

By means of simple modifications to the proposed procedures, it is possible to prove the following corollary. For a more detailed explanation, we refer to [2].

Corollary 13. *The MC problem for* E *formulas over finite Kripke structures is* **co-NP**-*complete.*

Proof (Sketch). First, checking an E formula φ over a track ρ can be done in deterministic polynomial time in $|\rho|$ and $|\varphi|$. Then, by Theorem 10, one can restrict to non-deterministically guessing a possible counterexample (i.e., an initial track *not* satisfying the input formula ψ) of length at most $|W| \cdot (|NNF(\neg\psi)| + 1)^2$. If a counterexample can be found, $\mathcal{K} \not\models \psi$. It follows that the MC problem for E is in **co-NP**. **co-NP**-hardness immediately follows from that of Prop [19]. □

5 Conclusions

In this paper, we have sharpened the border between good and bad HS fragments with respect to model checking. On the one hand, we have shown that the presence of both modalities $\langle B \rangle$ and $\langle E \rangle$ suffices for a fragment to be **EXPSPACE**-hard. This lower bound immediately propagates to full HS. On the other hand, we have studied two well-behaved, **PSPACE**-complete fragments, $\mathsf{A\overline{A}E\overline{E}}$ and $\mathsf{A\overline{A}B\overline{B}}$, which are quite promising from the point of view of applications.

The fragment $\mathsf{A\overline{A}B\overline{B}\overline{E}}$ (as well as the symmetric fragment $\mathsf{A\overline{A}E\overline{B}\overline{E}}$), investigated in [20], still lies somehow across the border between good and bad fragments, as it is situated in between **EXPSPACE** and **PSPACE**. One possibility for $\mathsf{A\overline{A}B\overline{B}\overline{E}}$ is to be **PSPACE**-complete—which would mean that $\langle \overline{\mathrm{E}} \rangle$ does not add complexity to $\mathsf{A\overline{A}B\overline{B}}$, and analogously $\langle \mathrm{B} \rangle$ to $\mathsf{A\overline{A}\overline{B}\overline{E}}$. Another possibility is that the presence of both $\langle \mathrm{B} \rangle$ and $\langle \overline{\mathrm{E}} \rangle$ causes a significant blow-up in complexity. A larger complexity gap is the one for full HS: we have shown it to be **EXPSPACE**-hard, but the only known upper bound is non-elementary. In our future work, we will definitely come back to both $\mathsf{A\overline{A}B\overline{B}\overline{E}}$ and full HS.

References

1. Allen, J.F.: Maintaining knowledge about temporal intervals. Commun. ACM **26**(11), 832–843 (1983)
2. Bozzelli, L., Molinari, A., Montanari, A., Peron, A., Sala, P.: Interval Temporal Logic Model Checking: the Border Between Good and Bad HS Fragments (2016). https://www.dimi.uniud.it/la-ricerca/pubblicazioni/preprints/1.2016
3. Bresolin, D., Della Monica, D., Goranko, V., Montanari, A., Sciavicco, G.: The dark side of interval temporal logic: marking the undecidability border. Ann. Math. Artif. Intell. **71**(1–3), 41–83 (2014)
4. Bresolin, D., Goranko, V., Montanari, A., Sala, P.: Tableau-based decision procedures for the logics of subinterval structures over dense orderings. J. Logic Comput. **20**(1), 133–166 (2010)
5. Bresolin, D., Goranko, V., Montanari, A., Sciavicco, G.: Propositional interval neighborhood logics: expressiveness, decidability, and undecidable extensions. Ann. Pure Appl. Logic **161**(3), 289–304 (2009)
6. Clarke, E.M., Grumberg, O., Peled, D.A.: Model Checking. MIT Press, Cambridge (2002)
7. Emerson, E.A., Halpern, J.Y.: "Sometimes" and "not never" revisited: on branching versus linear time temporal logic. J. ACM **33**(1), 151–178 (1986)
8. Giunchiglia, F., Traverso, P.: Planning as model checking. In: Biundo, S., Fox, M. (eds.) ECP 1999. LNCS, vol. 1809, pp. 1–20. Springer, Heidelberg (2000)
9. Gottlob, G.: NP trees and Carnap's modal logic. J. ACM **42**(2), 421–457 (1995)
10. Halpern, J.Y., Shoham, Y.: A propositional modal logic of time intervals. J. ACM **38**(4), 935–962 (1991)
11. Harel, D.: Algorithmics: The Spirit of Computing. Wesley, Reading (1992)
12. Lodaya, K.: Sharpening the undecidability of interval temporal logic. In: Kleinberg, R.D., Sato, M. (eds.) ASIAN 2000. LNCS, vol. 1961, pp. 290–298. Springer, Heidelberg (2000)
13. Lomuscio, A., Michaliszyn, J.: An epistemic Halpern-Shoham logic. In: IJCAI, pp. 1010–1016 (2013)
14. Lomuscio, A., Michaliszyn, J.: Decidability of model checking multi-agent systems against a class of EHS specifications. In: ECAI, pp. 543–548 (2014)
15. Lomuscio, A., Michaliszyn, J.: Model checking epistemic Halpern-Shoham logic extended with regular expressions. CoRR abs/1509.00608 (2015)

16. Lomuscio, A., Raimondi, F.: MCMAS: a model checker for multi-agent systems. In: Hermanns, H., Palsberg, J. (eds.) TACAS 2006. LNCS, vol. 3920, pp. 450–454. Springer, Heidelberg (2006)
17. Marcinkowski, J., Michaliszyn, J.: The undecidability of the logic of subintervals. Fundamenta Informaticae **131**(2), 217–240 (2014)
18. Molinari, A., Montanari, A., Murano, A., Perelli, G., Peron, A.: Checking interval properties of computations. Acta Informatica (2015, accepted for publication)
19. Molinari, A., Montanari, A., Peron, A.: Complexity of ITL model checking: some well-behaved fragments of the interval logic HS. In: TIME, pp. 90–100 (2015)
20. Molinari, A., Montanari, A., Peron, A.: A model checking procedure for interval temporal logics based on track representatives. In: CSL, pp. 193–210 (2015)
21. Molinari, A., Montanari, A., Peron, A., Sala, P.: Model checking well-behaved fragments of HS: the (almost) final picture. In: KR (2016)
22. Moszkowski, B.: Reasoning about digital circuits. Ph.D. thesis, Dept. of Computer Science, Stanford University, Stanford, CA (1983)
23. Pnueli, A.: The temporal logic of programs. In: FOCS, pp. 46–57. IEEE (1977)
24. Pratt-Hartmann, I.: Temporal prepositions and their logic. Artif. Intell. **166**(1–2), 1–36 (2005)
25. Roeper, P.: Intervals and tenses. J. Philos. Logic **9**, 451–469 (1980)
26. Schnoebelen, P.: Oracle circuits for branching-time model checking. In: Baeten, J.C.M., Lenstra, J.K., Parrow, J., Woeginger, G.J. (eds.) ICALP 2003. LNCS, vol. 2719, pp. 790–801. Springer, Heidelberg (2003)
27. Sistla, A.P., Clarke, E.M.: The complexity of propositional linear temporal logics. J. ACM **32**(3), 733–749 (1985)
28. Venema, Y.: Expressiveness and completeness of an interval tense logic. Notre Dame J. Formal Logic **31**(4), 529–547 (1990)

K_SP: A Resolution-Based Prover for Multimodal K

Cláudia Nalon[1(✉)], Ullrich Hustadt[2], and Clare Dixon[2]

[1] Department of Computer Science, University of Brasília, C.P. 4466 - CEP:70.910-090, Brasília, DF, Brazil
nalon@unb.br

[2] Department of Computer Science, University of Liverpool, Liverpool L69 3BX, UK
{U.Hustadt,CLDixon}@liverpool.ac.uk

Abstract. In this paper, we describe an implementation of a hyper-resolution-based calculus for the propositional basic multimodal logic, K_n. The prover was designed to support experimentation with different combinations of refinements for its basic calculus: it is primarily based on the set of support strategy, which can then be combined with other refinements, simplification techniques and different choices for the underlying normal form and clause selection. The prover allows for both local and global reasoning. We show experimental results for different combinations of strategies and comparison with existing tools.

1 Introduction

In this paper, we present K_SP, a theorem prover for the basic multimodal logic K_n which implements a variation of the set of support strategy [21] for the modal resolution-based procedure described in [14]. The prover also implements several other refinements and simplification techniques in order to reduce the search space for a proof. Besides the set of support strategy, all other refinements of the calculus are implemented as independent modules, allowing for a better evaluation of how effective they are.

The paper is organised as follows. We introduce the syntax and semantics of K_n in Sect. 2. In Sect. 3 we briefly describe the normal form and the calculus presented in [14]. Section 4 describes the available strategies and their implementations. Evaluation of strategies and of the performance of the prover compared to existing tools are given in Sect. 5. We summarise our results in Sect. 6.

2 Language

Let $\mathsf{A} = \{1, \ldots, n\}$, $n \in \mathbb{N}$, be a finite fixed set of indexes and $\mathsf{P} = \{p, q, s, t, p', q', \ldots\}$ be a denumerable set of propositional symbols. The *set of modal formulae*, $\mathsf{WFF_K}$, is the least set such that every $p \in \mathsf{P}$ is in $\mathsf{WFF_K}$; if φ and ψ are in $\mathsf{WFF_K}$, then so are $\neg\varphi$, $(\varphi \wedge \psi)$, and $\boxed{a}\varphi$ for each $a \in \mathsf{A}$. The formulae

N. Olivetti and A. Tiwari (Eds.): IJCAR 2016, LNAI 9706, pp. 406–415, 2016.
DOI: 10.1007/978-3-319-40229-1_28

false, **true**, $(\varphi \vee \psi)$, $(\varphi \Rightarrow \psi)$, and $\diamondsuit_a \varphi$ are introduced as the usual abbreviations for $(\varphi \wedge \neg\varphi)$, $\neg$**false**, $\neg(\neg\varphi \wedge \neg\psi)$, $(\neg\varphi \vee \psi)$, and $\neg \boxed{a} \neg\varphi$, respectively (where $\varphi, \psi \in \mathsf{WFF}_\mathsf{K}$). A *literal* is either a propositional symbol or its negation; the set of literals is denoted by L. A *modal literal* is either $\boxed{a} l$ or $\diamondsuit_a l$, where $l \in \mathsf{L}$ and $a \in \mathsf{A}$. The *modal depth* of a formula is given by the maximal number of nested occurrences of modal operators in that formula. The *modal level* of a formula is the maximal number of nested occurrences of modal operators in which scope the formula occurs. For instance, in $\boxed{a} \diamondsuit_a p$, the modal depth of p is 0 and its modal level is 2. Formal definitions can be found at [14].

As our calculus operates on a labelled clausal normal form that is closely linked to the tree model property of Kripke models for K_n, we briefly overview of the semantics of K_n. A *tree-like Kripke model M for n agents over* P is given by a tuple $(W, w_0, R_1, \ldots, R_n, \pi)$, where W is a set of possible *worlds* with a distinguished world w_0, each *accessibility relation* R_a is a binary relation on W such that their union is a tree with *root* w_0, and $\pi : W \to (\mathsf{P} \to \{\mathit{true}, \mathit{false}\})$ is a function which associates with each world $w \in W$ a truth-assignment to propositional symbols. Satisfaction of a formula at a world w of a model M is defined by:

- $\langle M, w \rangle \models p$ if, and only if, $\pi(w)(p) = \mathit{true}$, where $p \in \mathsf{P}$;
- $\langle M, w \rangle \models \neg\varphi$ if, and only if, $\langle M, w \rangle \not\models \varphi$;
- $\langle M, w \rangle \models (\varphi \wedge \psi)$ if, and only if, $\langle M, w \rangle \models \varphi$ and $\langle M, w \rangle \models \psi$;
- $\langle M, w \rangle \models \boxed{a} \varphi$ if, and only if, for all w', $w R_a w'$ implies $\langle M, w' \rangle \models \varphi$.

Let $M = (W, w_0, R_1, \ldots, R_n, \pi)$ be a model. A formula φ is *locally satisfied in* M, denoted by $M \models_L \varphi$, if $\langle M, w_0 \rangle \models \varphi$. The formula φ is *locally satisfiable* if there is a model M such that $\langle M, w_0 \rangle \models \varphi$. A formula φ is *globally satisfied in* M, if for all $w \in W$, $\langle M, w \rangle \models \varphi$. We denote by $\mathsf{depth}(w)$ the length of the unique path from w_0 to w through the union of the accessibility relations in M. We call a *modal layer* the equivalence class of worlds at the same depth in a model.

We note that checking the local satisfiability of a formula φ can be reduced to the problem of checking the local satisfiability of its subformulae at the modal layer of a model which corresponds to the modal level where those subformulae occur (see [1]). Due to this close correspondence of modal layer and modal level we use the terms interchangeably. Also, checking the global satisfiability of φ can be reduced to checking the local satisfiability of φ at all modal layers (up to an exponential distance from the root) of a model [3,19]. Thus, an uniform approach based on modal levels can be used to deal with both problems, as we show in the next section.

3 A Calculus for K_n

The calculus for K_n presented in [14] is clausal, where clauses are labelled by the modal level at which they occur. In order to refer explicitly to modal levels,

Table 1. Inference rules, where $ml = \sigma(\{ml_1, \ldots, ml_{m+1}, ml_{m+2} - 1\})$ in GEN1, GEN3; $ml = \sigma(\{ml, ml'\})$ in LRES, MRES; and $ml = \sigma(\{ml_1, ml_2, ml_3\})$ in GEN2.

[LRES]
$$\begin{array}{r} ml : D \vee l \\ ml' : D' \vee \neg l \\ \hline ml : D \vee D' \end{array}$$

[MRES]
$$\begin{array}{r} ml : l_1 \Rightarrow \boxed{a} l \\ ml' : l_2 \Rightarrow \Diamond_{a} \neg l \\ \hline ml : \neg l_1 \vee \neg l_2 \end{array}$$

[GEN2]
$$\begin{array}{r} ml_1 : l'_1 \Rightarrow \boxed{a} l_1 \\ ml_2 : l'_2 \Rightarrow \boxed{a} \neg l_1 \\ ml_3 : l'_3 \Rightarrow \Diamond_{a} l_2 \\ \hline ml : \neg l'_1 \vee \neg l'_2 \vee \neg l'_3 \end{array}$$

[GEN1]
$$\begin{array}{r} ml_1 : l'_1 \Rightarrow \boxed{a} \neg l_1 \\ \vdots \\ ml_m : l'_m \Rightarrow \boxed{a} \neg l_m \\ ml_{m+1} : l' \Rightarrow \Diamond_{a} \neg l \\ ml_{m+2} : l_1 \vee \ldots \vee l_m \vee l \\ \hline ml : \neg l'_1 \vee \ldots \vee \neg l'_m \vee \neg l' \end{array}$$

[GEN3]
$$\begin{array}{r} ml_1 : l'_1 \Rightarrow \boxed{a} \neg l_1 \\ \vdots \\ ml_m : l'_m \Rightarrow \boxed{a} \neg l_m \\ ml_{m+1} : l' \Rightarrow \Diamond_{a} l \\ ml_{m+2} : l_1 \vee \ldots \vee l_m \\ \hline ml : \neg l'_1 \vee \ldots \vee \neg l'_m \vee \neg l' \end{array}$$

the modal language is extended with labels. We write $ml : \varphi$ to denote that φ is true at the modal layer ml in a Kripke model, where $ml \in \mathbb{N} \cup \{*\}$. By $* : \varphi$ we mean that φ is true at all modal layers in a Kripke model. The notion of local satisfiability is extended as expected: for a model M, $M \models_L ml : \varphi$ if, and only if, for all worlds $w \in W$ such that $\mathsf{depth}(w) = ml$, we have $\langle M, w \rangle \models_L \varphi$. Then, the layered normal form, called SNF_{ml}, is given by a conjunction of *literal clauses* of the form $ml : D$, where D is a disjunction of literals, and *modal clauses* of the form $ml : l \Rightarrow l'$, where $l \in \mathsf{L}$ and l' is a modal literal. Transformation into SNF_{ml} uses *renaming* and preserves satisfiability.

The motivation for the use of this labelled clausal normal form is that inference rules can then be guided by the semantic information given by the labels and applied to smaller sets of clauses, reducing the number of unnecessary inferences, and therefore improving the efficiency of the proof procedure. The calculus comprises the set of inference rules given in Table 1. Unification on sets of labels is defined by $\sigma(\{ml, *\}) = ml$; and $\sigma(\{ml\}) = ml$; otherwise, σ is undefined. The inference rules can only be applied if the unification of their labels is defined (where $* - 1 = *$). This calculus has been shown to be sound, complete, and terminating [14].

4 Implementation

K_SP is an implementation, written in C, of the calculus given in [14]. The prover was designed to support experimentation with different combinations of refinements of its basic calculus. Refinements and options for (pre)processing the input are coded as independently as possible in order to allow for the easy addition and testing of new features. This might not lead to optimal performance (e.g. one technique needs to be applied after the other, whereas most tools would apply them together), but it helps to evaluate how the different options independently contribute to achieve efficiency. In the following we give a brief overview of the main aspects of the implementation.

Transformation to Clausal Form: If the input is a set of formulae, then these formulae are first transformed into their prenex or antiprenex normal form (or one after the other) [12] and then into Negation Normal Form (NNF) or into Box Normal Form (BNF) [16]. With options *nnfsimp* (resp. *bnfsimp*), simplification is applied to formulae in NNF (resp. BNF); with option *early_mlple*, pure literal elimination is applied at every modal level. There are then four different options that determine the normal form. In SNF^{+}_{ml}, negative literals in the scope of modal operators are renamed by propositional symbols; in SNF^{++}_{ml} all literals in the scope of modal operators are renamed by propositional symbols. SNF^{-}_{ml} and SNF^{--}_{ml} are defined analogously, with positive literals being renamed by negative ones. The reuse of propositional symbols in renaming can also be controlled. In our evaluation, given in Sect. 5, the same propositional symbol is used for all occurrences of a formula being renamed.

Preprocessing of Clauses: Self-subsumption is applied at this step if the options for forward and/or backward subsumption are set [9]. The inference rules MRES and GEN2 are also exhaustively applied at this step, that is, before the prover enters the main loop.

Main Loop: The main loop is based on the given-clause algorithm implemented in Otter [11], a variation of the set of support strategy [21], a refinement which restricts the set of choices of clauses participating in a derivation step. For the classical case, a set of clauses Δ is partitioned in two sets Γ and $\Lambda = \Delta \setminus \Gamma$, where Λ must be satisfiable. McCune refers to Γ as the *set of support* (the sos, aka *passive* or *unprocessed* set); and Λ is called the *usable* (aka as *active* or *processed* set). The *given clause* is chosen from Γ, resolved with clauses in Λ, and moved from Γ to Λ. Resolvents are added to Γ. For the modal calculus, the set of clauses is further partitioned according to the modal layer at which clauses are true. That is, for each modal layer ml there are three sets: Γ^{lit}_{ml}, Λ^{lit}_{ml} and Λ^{mod}_{ml}, where the first two sets contain literal clauses while the latter contains modal clauses. As the calculus does not generate new modal clauses and because the set of modal clauses by itself is satisfiable, there is no need for a set for unprocessed modal clauses. Attempts to apply an inference rule are guided by the choice, for each modal layer ml, of a literal clause in Γ^{lit}_{ml}, which can be resolved with either a literal clause in Λ^{lit}_{ml} or with a set of modal clauses in Λ^{mod}_{ml-1}. There are six options for automatically populating the usable: all negative clauses, all positive clauses, all non-negative clauses, all non-positive clauses, all clauses whose maximal literal is positive, and all clauses whose maximal literal is negative. The prover can either perform *local* or *global* reasoning.

Refinements: Besides the basic calculus with a set-of-support strategy, the user can further restrict LRES by choosing ordered (clauses can only be resolved on their maximal literals with respect to an ordering chosen by the prover in such a way to preserve completeness), negative (one of the premises is a negative clause, i.e. a clause where all literals are of the form $\neg p$ for some $p \in \mathsf{P}$), positive (one of the premises is a positive clause), or negative + ordered resolution (both negative and ordered resolution inferences are performed).

The completeness of some of these refinements depends on the particular normal form chosen. For instance, negative resolution is incomplete without SNF^{+}_{ml} or SNF^{++}_{ml}. For example, the set $\{p, p \Rightarrow \Box\neg q, p \Rightarrow \Diamond s, \neg s \vee q\}$ is unsatisfiable, but as there is no negative literal clause in the set, no refutation can be found. By renaming $\neg q$ in the scope of $\Box$, we obtain the set $\{p, p \Rightarrow \Box t, p \Rightarrow \Diamond s, \neg s \vee q, \neg t \vee \neg q\}$ in SNF^{+}_{ml}, from which a refutation using negative resolution can be found. Similarly, ordered resolution requires SNF^{++}_{ml} for completeness, while positive resolution requires SNF^{-}_{ml} or SNF^{--}_{ml}.

Inference Rules: Besides the inference rules given in Table 1, three more inference rules are also implemented: *unit resolution*, which propagates unit clauses through all literal clauses and the right-hand side of modal clauses; *lhs unit resolution*, which propagates unit clauses through the left-hand side of modal clauses; and *ires*, which together with the *global* option, implements initial resolution and, thus, the calculus given in [13].

Redundancy Elimination: Pure literal elimination can be applied globally or by modal layer. Both forward and backward subsumption are implemented. Subsumption is applied in lazy mode: a clause is tested for subsumption only when it is selected from the sos and only against clauses in the usable. As pointed out in [18], this avoids expensive checks for clauses that might never be selected during the search of a proof.

Clause Selection: There are five different heuristics for choosing a literal clause in the sos: shortest, newest, oldest, greatest maximal literal, and the smallest maximal literal.

For a comprehensive list of options, see [15], where the sources and instructions on how to install and use K$_S$P can be found.

5 Evaluation

We have compared K$_S$P with BDDTab [4], FaCT++ 1.6.3 [20], InKreSAT 1.0 [7], Spartacus 1.0 [5], and a combination of the optimised functional translation [6] with Vampire 3.0 [8][1]. In this context, FaCT++ represents the previous generation of reasoners while the remaining systems have all been developed in recent years. Unless stated otherwise, the reasoners were used with their default options.

Our benchmarks [15] consist of three collections of modal formulae:

1. The complete set of TANCS-2000 modalised random QBF (MQBF) formulae [10] complemented by the additional MQBF formulae provided by Tebbi and

[1] We have excluded *SAT from the comparison as it produced incorrect results on a number of benchmark formulae.

Kaminski [7]. This collection consists of five classes, called qbf, qbfL, qbfS, qbfML, and qbfMS in the following, with a total of 1016 formulae, of which 617 are known to be satisfiable and 399 are known to be unsatisfiable (due to at least one of the provers being able to solve the formula). The minimum modal depth of formulae in this collection is 19, the maximum 225, average 69.2 with a standard deviation of 47.5.

2. LWB basic modal logic benchmark formulae [2], with 56 formulae chosen from each of the 18 parameterised classes. In most previous uses of these benchmarks, only parameter values 1 to 21 were used for each class, with the result that provers were able to solve all benchmark formulae for most of the classes. Instead we have chosen the 56 parameter values so that the best current prover will not be able to solve all the formulae within a time limit of 1000 CPU second. The median value of the maximal parameter value used for the 18 classes is 1880, far beyond what has ever been tested before. Of the resulting 1008 formulae, half are satisfiable and half are unsatisfiable by construction of the benchmark classes.
3. Randomly generated 3CNF$_K$ formulae [17] over 3 to 10 propositional symbols with modal depth 1 or 2. We have chosen formulae from each of the 11 parameter settings given in the table on page 372 of [17]. For the number of conjuncts we have focused on a range around the critical region where about half of the generated formulae are satisfiable and half are unsatisfiable. The resulting collection contains 1000 formulae, of which 457 are known to be satisfiable and 464 are known to be unsatisfiable. Note that this collection is quite distinct to the one used in [7] which consisted of 135 3CNF$_K$ formulae over 3 propositional symbols with modal depth 2, 4 or 6, all of which were satisfiable.

Benchmarking was performed on PCs with an Intel i7-2600 CPU @ 3.40 GHz and 16 GB main memory. For each formula and each prover we have determined the median run time over five runs with a time limit of 1000 CPU seconds for each run.

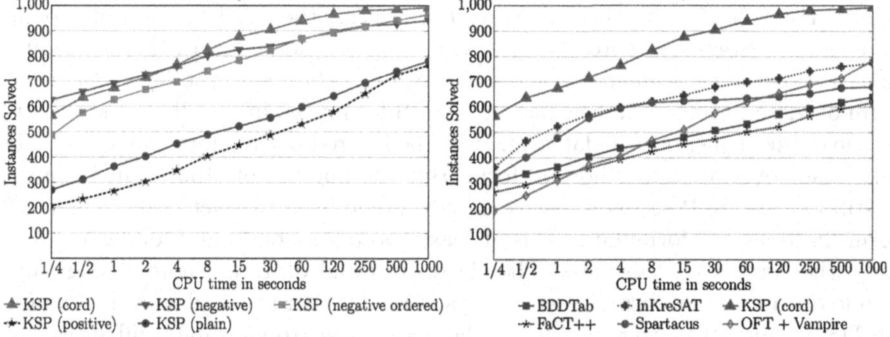

(a) K_SP with different refinements (b) All provers

Fig. 1. Benchmarking results for MQBF (Color figure online)

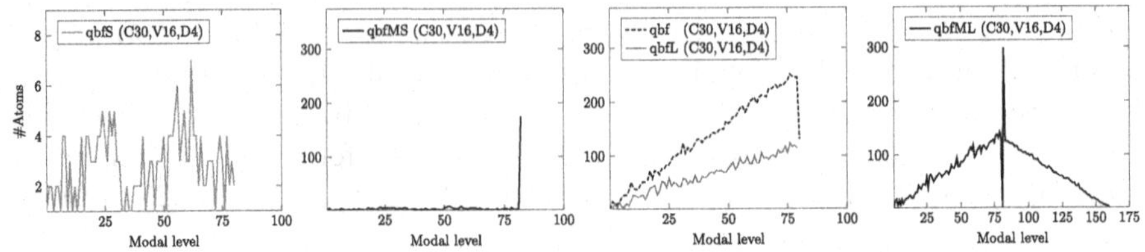

Fig. 2. Modal structure of MQBF formulae (Color figure online)

Figure 1a compares the impact of different refinements on the performance of K$_S$P on the MQBF collection. With *plain* K$_S$P uses the rules shown in Table 1, without additional refinement, on a set of SNF_{ml}^{++} clauses. With *cord* and *negative ordered*, K$_S$P applies ordered resolution and negative + ordered resolution, respectively, again on a set of SNF_{ml}^{++} clauses. The configuration *negative* uses negative resolution on a set of SNF_{ml}^{+} clauses, while with *positive*, K$_S$P applies positive resolution on SNF_{ml}^{-} clauses. Irrespective of the refinement, the shortest clause is selected to perform inferences; both forward and backward subsumption are used; the lhs-unit resolution rule is applied; prenex is set; and no simplification steps are applied. The usable is populated with clauses whose maximal literal is positive, except for positive resolution where it is populated with clauses whose maximal literal is negative. The *cord* configuration offers the best performance. Ordered resolution restricts the applicability of the rules further than the other refinements. Not only is this an advantage on satisfiable formulae in that a saturation can be found more quickly, but also unsatisfiable formulae where with this refinement K$_S$P finds refutations much more quickly than with any of the other refinements.

Figure 1b compares the performance of all the provers on the MQBF collection. It shows that K$_S$P performs better than any of the other provers. The graphs in Fig. 2 offer some insight into why K$_S$P performs well on these formulae. Each of the four graphs shows for one formula from each class how many atomic subformulae occur at each modal level, the formulae originate from MQBF formulae with the same number of propositional symbols, conjuncts and QBF quantifier depth. Formulae in the class qbfS are the easiest, the total number of atomic subformulae is low and spread over a wide range of modal levels, thereby reducing the possibility of inference steps between the clauses in the layered normal form of these formulae. In contrast, in qbfMS formulae almost all atomic subformulae occur at just one modal level. Here the layered normal form can offer little advantage over a simpler normal form. But the number of atomic subformulae is still low and K$_S$P seems to derive an advantage from the fact that the normal form 'flattens' the formula: K$_S$P is at least two orders of magnitude faster than any other prover on this class. The classes qbf and qbfL are more challenging. While the atomic subformulae are more spread out over the modal levels than for qbfMS, at a lot of these modal levels there are more atomic subformulae than in a qbfMS formula in total. The layered modal translation is effective at reducing the number of inferences for these classes, but more inference possibilities remain

Table 2. Detailed benchmarking results on LWB

	BDDTab		FaCT++		InKreSAT		K$_S$P (cord)		Spartacus		OFT + Vampire	
k_branch_n	22	*22*	12	*12*	15	*15*	18	*18*	12	*12*	**50**	*70*
k_branch_p	22	*22*	12	*12*	22	*22*	23	*23*	14	*14*	**50**	*70*
k_d4_n	20	*440*	6	*40*	34		**48**	*1560*	28	*760*	14	*200*
k_d4_p	26	*640*	24	*600*	18	*360*	**54**	*1800*	32	*920*	21	*960*
k_dum_n	39	*2400*	42	*2640*	23	*1120*	**49**	*3200*	44	*2800*	17	*640*
k_dum_p	42	*2640*	38	*2320*	28	*1520*	**50**	*3280*	46	*2960*	18	*720*
k_grz_n	35	*2600*	27	*1800*	50	*4500*	5	*50*	**52**	*5500*	24	*1500*
k_grz_p	35	*2600*	27	*1800*	51	*5000*	29	*2000*	**52**	*5500*	27	*1800*
k_lin_n	46	*4000*	43	*3400*	33	*2500*	1	*10*	**50**	*4800*	40	*3100*
k_lin_p	14	*500*	28	*10000*	**56**	*500000*	23	*5000*	55	*400000*	28	*10000*
k_path_n	37	*290*	48	*400*	7	*14*	**54**	*1000*	48	*400*	41	*330*
k_path_p	35	*270*	48	*400*	5	*12*	**54**	*1000*	48	*400*	41	*330*
k_ph_n	10	*10*	8	*16*	24	*90*	3	*6*	**21**	*75*	15	*45*
k_ph_p	**11**	*11*	9	*8*	10	*10*	5	*5*	9	*9*	10	*10*
k_poly_n	39	*600*	34	*500*	30		36	*540*	**44**	*720*	20	*220*
k_poly_p	38	*580*	34	*500*	28	*400*	36	*540*	**44**	*700*	20	*220*
k_t4p_n	40	*3500*	24	*1500*	17	*800*	39	*3000*	**45**	*6000*	11	*200*
k_t4p_p	48	*7500*	49	*8000*	28		49	*8000*	**53**	*12000*	14	*500*

than for qbfMS. Finally, qbfML combines the worst aspects of qbfL and qbfMS, the number of atomic subformulae is higher than for any other class and there is a 'peak' at one particular modal level. This is the only MQBF class containing formulae that K$_S$P cannot solve.

Figure 3 shows the benchmarking results on the LWB and 3CNF$_K$ collections. On the LWB collection K$_S$P performs about as well as BDDTab, FaCT++ and InKreSAT, while Spartacus performs best and the combination of the optimised functional translation with Vampire (OFT + Vampire) performs worst. Table 2 provides more detailed results. For each prover it shows in the left column how many of the 56 formulae in a class have been solved and in the right column the parameter value of the most difficult formula solved. For InKreSAT we are not reporting this parameter value for three classes on which the prover's runtime does not increase monotonically with the parameter value but fluctuates instead. As is indicated, BDDTab and InKreSAT are the best performing provers on one class each, OFT + Vampire on two, K$_S$P on six, and Spartacus on eight classes. A characteristic of the classes on which K$_S$P performs best is again that atomic subformulae are evenly spread over a wide range of modal levels.

It is worth pointing out that simplification alone is sufficient to detect that formulae in k_lin_p are unsatisfiable. For k_grz_p, pure literal elimination can be used to reduce all formulae in this class to the same simple formula; the same is true for k_grz_n and k_lin_n. Thus, these classes are tests of how effectively and efficiently, if at all, a prover uses these techniques and Spartacus does best

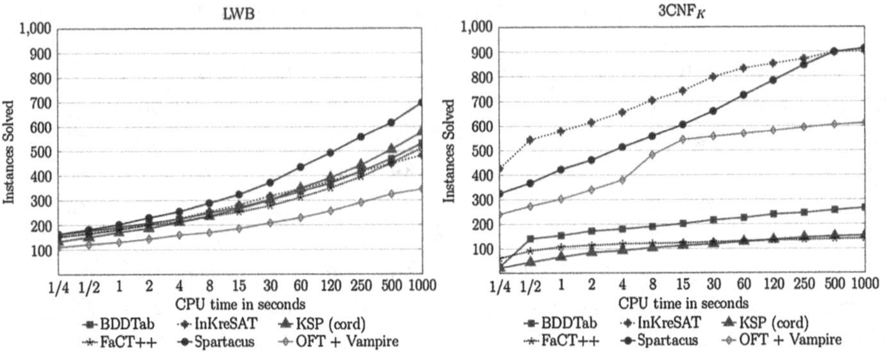

Fig. 3. Benchmarking results for LWB and 3CNF$_K$ (Color figure online)

on these classes. Note that pure literal elimination has been disabled in the *cord* configuration we have used for K_SP. With it K_SP would perform better on k_grz_p and k_grz_n, but worse on other classes where this simplification has no beneficial effects.

Finally, on the 3CNF$_K$ collection, InKreSAT is the best performing prover and K_SP the worst performing one. This should now not come as a surprise. For 3CNF$_K$ we specifically restricted ourselves to formulae with low modal depth which in turn means that the layered normal form has little positive effect.

6 Conclusions and Future Work

The evaluation indicates that K_SP works well on problems with high modal depth where the separation of modal layers can be exploited to improve the efficiency of reasoning.

As with all provers that provide a variety of strategies and optimisations, to get the best performance for a particular formula or class of formulae it is important to choose the right strategy and optimisations. K_SP currently leaves that choice to the user and the development of an "auto mode" in which the prover makes a choice of its own, based on an analysis of the given formula, is future work.

The same applies to the transformation to the layered normal form. Again, K_SP offers a number of ways in which this can be done as well as a number of simplifications that can be applied during the process. It is clear that this affects the performance of the prover, but we have yet to investigate the effects on the benchmark collections introduced in this paper.

References

1. Areces, C., Gennari, R., Heguiabehere, J., Rijke, M.D.: Tree-based heuristics in modal theorem proving. In: Proceedings of ECAI 2000, pp. 199–203. IOS Press (2000)

2. Balsiger, P., Heuerding, A., Schwendimann, S.: A benchmark method for the propositional modal logics K, KT, S4. J. Autom. Reasoning **24**(3), 297–317 (2000)
3. Goranko, V., Passy, S.: Using the universal modality: gains and questions. J. Logic and Comput. **2**(1), 5–30 (1992)
4. Goré, R., Olesen, K., Thomson, J.: Implementing tableau calculi using BDDs: BDDTab system description. In: Demri, S., Kapur, D., Weidenbach, C. (eds.) IJCAR 2014. LNCS, vol. 8562, pp. 337–343. Springer, Heidelberg (2014)
5. Götzmann, D., Kaminski, M., Smolka, G.: Spartacus: a tableau prover for hybrid logic. Electr. Notes Theor. Comput. Sci. **262**, 127–139 (2010)
6. Horrocks, I.R., Hustadt, U., Sattler, U., Schmidt, R.: Computational modal logic. In: Handbook of Modal Logic, pp. 181–245. Elsevier, Amsterdam (2006)
7. Kaminski, M., Tebbi, T.: InKreSAT: modal reasoning via incremental reduction to SAT. In: Bonacina, M.P. (ed.) CADE 2013. LNCS, vol. 7898, pp. 436–442. Springer, Heidelberg (2013)
8. Kovács, L., Voronkov, A.: First-order theorem proving and Vampire. In: Sharygina, N., Veith, H. (eds.) CAV 2013. LNCS, vol. 8044, pp. 1–35. Springer, Heidelberg (2013)
9. Lee, R.C.T.: A completeness theorem and computer program for finding theorems derivable from given axioms. Ph.D. thesis, Berkeley (1967)
10. Massacci, F., Donini, F.M.: Design and results of TANCS- non-classical (modal) systems comparison. TABLEAUX 2000. LNCS, vol. 1847, pp. 52–56. Springer, Heidelberg (2000)
11. McCune, W.W.: OTTER 3.0 reference manual and guide, 07 May 2007
12. Nalon, C., Dixon, C.: Anti-prenexing and prenexing for modal logics. In: Fisher, M., van der Hoek, W., Konev, B., Lisitsa, A. (eds.) JELIA 2006. LNCS (LNAI), vol. 4160, pp. 333–345. Springer, Heidelberg (2006)
13. Nalon, C., Dixon, C.: Clausal resolution for normal modal logics. J. Algorithms **62**, 117–134 (2007)
14. Nalon, C., Hustadt, U., Dixon, C.: A modal-layered resolution calculus for K. In: De Nivelle, H. (ed.) TABLEAUX 2015. LNCS, vol. 9323, pp. 185–200. Springer, Heidelberg (2015). doi:10.1007/978-3-319-24312-2_13
15. Nalon, C., Hustadt, U., Dixon, C.: KSP sources and benchmarks (2016). http://www.cic.unb.br/~nalon/#software
16. Pan, G., Sattler, U., Vardi, M.Y.: BDD-based decision procedures for the modal logic K. J. Appl. Non-Class. Logics **16**(1–2), 169–208 (2006)
17. Patel-Schneider, P.F., Sebastiani, R.: A new general method to generate random modal formulae for testing decision procedures. J. Artif. Intell. Res. (JAIR) **18**, 351–389 (2003)
18. Schulz, S.: Simple and efficient clause subsumption with feature vector indexing. In: Bonacina, M.P., Stickel, M.E. (eds.) Automated Reasoning and Mathematics. LNCS, vol. 7788, pp. 45–67. Springer, Heidelberg (2013)
19. Spaan, E.: Complexity of modal logics. Ph.D. thesis, University of Amsterdam (1993)
20. Tsarkov, D., Horrocks, I.: FaCT++ description logic reasoner: system description. In: Furbach, U., Shankar, N. (eds.) IJCAR 2006. LNCS (LNAI), vol. 4130, pp. 292–297. Springer, Heidelberg (2006)
21. Wos, L., Robinson, G., Carson, D.: Efficiency and completeness of the set of support strategy in theorem proving. J. ACM **12**, 536–541 (1965)

Inducing Syntactic Cut-Elimination for Indexed Nested Sequents

Revantha Ramanayake(✉)

Technische Universität Wien, Vienna, Austria
revantha@logic.at

Abstract. The key to the proof-theoretical study of a logic is a cut-free proof calculus. Unfortunately there are many logics of interest lacking suitable proof calculi. The proof formalism of nested sequents was recently generalised to indexed nested sequents in order to yield cutfree proof calculi for extensions of the modal logic K by Geach (Lemmon-Scott) axioms. The proofs of completeness and cut-elimination therein were semantical and intricate. Here we identify a subclass of the labelled sequent formalism and show that it corresponds to the indexed nested sequent formalism. This correspondence is then exploited to induce syntactic proofs for indexed nested sequents using the elegant existing proofs in the labelled sequent formalism. A larger goal of this work is to demonstrate how specialising existing proof-theoretical transformations (adapting these as required to remain within the subclass) is an alternative proof method which can alleviate the need for independent proofs from 'scratch' in each formalism. Moreover, such coercion can be used to induce new cutfree calculi. We demonstrate by presenting the first indexed nested sequent calculi for intermediate logics.

Keywords: Proof theory · Cut-elimination · Nested labelled sequents · Modal logic

1 Introduction

Gentzen [9] introduced the *sequent calculus* as an elegant formal proof system for classical and intuitionistic logics. The building blocks of the sequent calculus are *traditional sequents* of the form $X \vdash Y$ where X and Y are formula multisets (formula lists in the original formulation). To simulate the rule of *modus ponens*—a rule that is present in many logics—while preserving the nice properties of his calculus, Gentzen was led to introduce the cut-rule below right.

$$\frac{A \qquad A \to B}{B}\,modus\,ponens \qquad\qquad \frac{X \vdash Y, A \qquad A, U \vdash V}{X, U \vdash Y, V}\,\text{cut}$$

Supported by the Austrian Science Fund (FWF), START project Y544.

N. Olivetti and A. Tiwari (Eds.): IJCAR 2016, LNAI 9706, pp. 416–432, 2016.
DOI: 10.1007/978-3-319-40229-1_29

Unfortunately the presence of the cut-rule has a great cost: reading the rule from the conclusion to the premises—as in backward proof search—the cut-rule can introduce in the premises an arbitrary formula A that might not even occur in the conclusion. Gentzen's response was the cut-elimination theorem (*Hauptsatz*) which shows how to eliminate the cut-rule from any proof (derivation) in the sequent calculus, in effect showing that the cut-rule is redundant. This is the central result for the sequent calculus since the resulting *cutfree* calculi possess the *subformula property*: only subformulae of the formula to be proved may appear in its derivation. This property places a strong restriction on the set of possible derivations of a given formula, in turn enabling elegant and constructive proof-theoretic arguments of logical properties such as consistency, decidability, complexity and interpolation. Gentzen himself utilised the cut-elimination result to give a proof of consistency of arithmetic with a suitable induction principle. Subsequently, the elimination of the cut-rule from a derivation has even been given a computational interpretation.

Following the seminal work of Gentzen, efforts were made to obtain cutfree sequent calculi (i.e. to give a proof-theory) for the many logics of interest but the sequent calculus is often not expressive enough. This has led to various generalisations of the Gentzen sequent calculus in order to obtain a proof formalism capable of presenting these logics.

Nested sequent calculus [2,12] is a popular proof formalism that has been used to present intuitionistic logic [7], conditional logics [16], logics in the classical and intuitionistic modal cube [12,13,19] and path axiom extensions of classical modal logic [10]. The idea is to use a *tree* of traditional sequents as the basic building block rather than just a single traditional sequent. The tree structure is encoded using the nesting of $[\cdots]$ and comma. Here are some examples:

$$X \vdash Y, [P \vdash Q, [U \vdash V], [L \vdash M]], [S \vdash T] \quad (1)$$

$$X \vdash Y, [U \vdash V, [L \vdash M], [S \vdash T]] \quad (2)$$

The preceeding two nested sequents can thus be depicted graphically as a (finite directed rooted) tree whose nodes are decorated by traditional sequents.

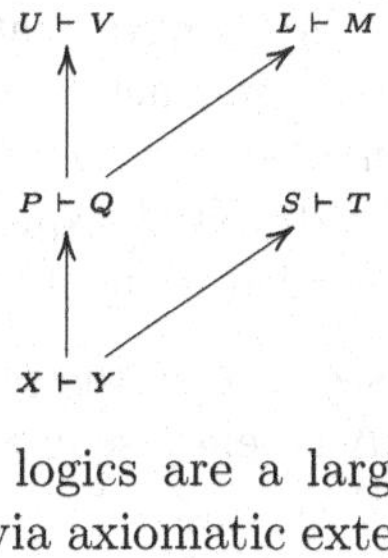

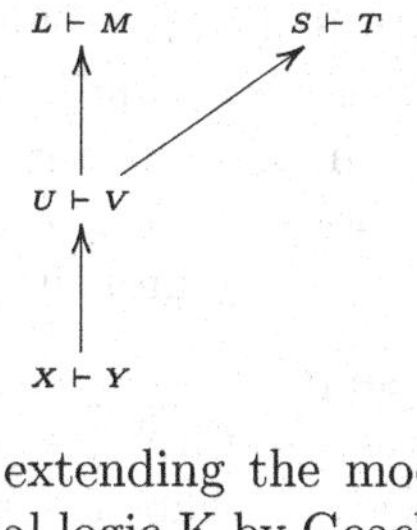

Geach logics are a large class of modal logics extending the modal cube, obtained via axiomatic extension of the normal modal logic K by Geach axioms:

$$G(h, i, j, k) := \Diamond^h \Box^i p \to \Box^j \Diamond^k p \quad (h, i, j, k \geq 0) \quad (3)$$

There are no known nested sequent calculi for the general class of Geach logics.

Indexed nested sequents are a recent extension of nested sequents introduced by Fitting [5] to obtain calculi for the Geach logics. An indexed nested sequent is obtained by assigning an index to each traditional sequent in the nested sequent and permitting multiple traditional sequents to possess the same index. The following are examples of indexed nested sequents:

$$X \vdash^0 Y, [P \vdash^1 Q, [U \vdash^2 V], [L \vdash^3 M]], [S \vdash^3 T] \tag{4}$$

$$X \vdash^0 Y, [P \vdash^1 Q, [U \vdash^0 V], [L \vdash^1 M]], [S \vdash^2 T] \tag{5}$$

As before we can present an indexed nested sequent as a tree whose nodes are decorated with a traditional sequent and an index (first two graphs below).

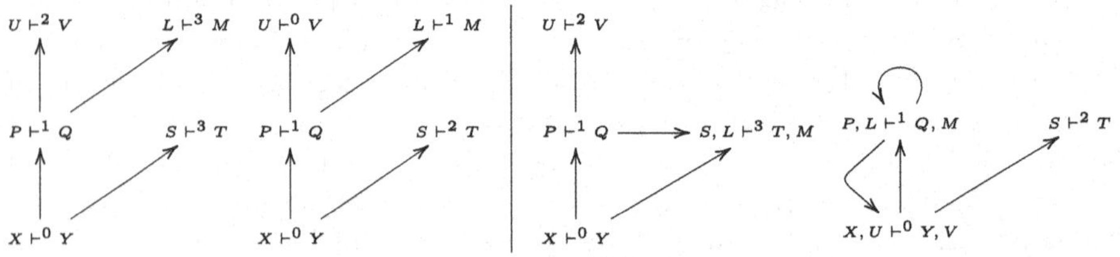

Viewing an indexed nested sequent as above does not bring us anything new. The key to obtaining more expressivity is to interpret those nodes of the tree with the same index ($\Gamma_1 \vdash^n \Delta_1, \ldots, \Gamma_{N+1} \vdash^n \Delta_{N+1}$, say) as a *single* node decorated with $\Gamma_1, \ldots, \Gamma_{N+1} \vdash \Delta_1, \ldots, \Delta_{N+1}$.[1] Thus indexed nested sequents correspond to *directed graphs* obtained by conflating certain nodes of a tree. The indexed nested sequents (4) and (5) are depicted by the two rightmost graphs above.

Fitting does not prove syntactic cut-elimination for his indexed nested sequent calculi for Geach logics, but instead establishes that the calculus minus the cut-rule is complete with respect to the corresponding logic's semantics. Such a proof is called a *semantic proof of cut-elimination* and contrasts with the syntactic proofs à la Gentzen where the elimination of cuts is constructive.

Apart from the technical interest in syntactic proofs of cut-elimination (after all, proof-theory is concerned primarily with the syntax), such proofs yield a constructive procedure and the cutfree derivation is related in a formal sense to the original derivation. However, the downside of syntactic proofs is that they tend to be highly technical and difficult to verify. We believe that the best response is to reuse and adapt whenever possible those syntactic proofs that are already in existence rather than presenting new proofs from scratch. In this paper, we show syntactic cut-elimination for the indexed nested sequent calculi for Geach logics by inducing the existing results for labelled sequent calculi.

Labelled sequents [6,14] generalise the traditional sequent by the prefixing of *state variables* to formulae occurring in the sequent. A labelled sequent has the form $\mathcal{R}, X \vdash Y$ where the *relation mset* (multiset) $\mathcal{R}$ consists of terms of the form Rxy. Meanwhile X and Y are multisets of labelled formulae (e.g. $x : A \rightarrow$

[1] Of course, this interpretation needs to be justified. This is shown in Sects. 3–5. Here we want to provide an intuition for these objects.

B, $y : p$). A labelled sequent can be viewed as a directed graph (defined using the set $\mathcal{R}$) with sequents at each node [18].

Negri [15] has presented a method for generating cutfree and contraction-free labelled sequent calculi for the large family of modal logics whose Kripke semantics are defined by geometric frame conditions. The proof of cut-elimination is general in the sense that it applies uniformly to every modal logic defined by geometric frame conditions. This result has been extended to labelled sequent calculi for intermediate and other non-classical logics [3] and indeed to arbitrary first-order formulae [4].

It is well-known that every modal logic obtained by the addition of Geach axioms to K can be defined semantically using the geometric frame conditions. For example, the logic $K + \Diamond\Box p \to \Box\Diamond p$ is the logic whose Kripke frames satisfy the geometric condition $\forall xyz(Rxy \wedge Rxz \longrightarrow \exists u(Ryu \wedge Rzu))$.

Labelled tree sequents (LTS) are a special instance of the labelled sequent that are *isomorphic up to state variable names* [11] to the nested sequent (an isomorphism with prefixed tableaux has also been shown [7]). The idea is to impose restrictions on the relation mset of a labelled sequent $\mathcal{R}, X \vdash Y$ to ensure that the directed graph defined by $\mathcal{R}$ is a tree. For example, the nested sequents (1) and (2) correspond to the LTS

$$\underbrace{Rxy, Rxz, Ryu, Ryv}_{\text{relation mset}}, x : X, y : P, z : S, u : U, v : L \vdash x : Y, y : Q, z : T, u : V, v : M$$

$$\overbrace{Rxy, Ryu, Ryv}, x : X, y : U, u : L, v : S \vdash x : Y, y : V, u : M, v : T$$

The isomorphism will be more transparent to the reader if he/she consults the trees that we presented following (1) and (2). In particular, the relation msets above define those trees and the labelled formulae multisets specify where to place the multisets on the tree. The extension of the isomorphism to nested sequent and labelled tree sequent calculi was used [11] to answer a question [17] concerning the relationship between two distinct proof calculi.

In this work we use a technical extension[2] introduced in Negri [15] to write a labelled sequent as $\mathcal{R}, \mathcal{E}, X \vdash Y$ where $\mathcal{E}$ consists of terms of the form $x = y$. In this formulation of a labelled sequent, a labelled tree sequent has $\mathcal{R}$ defining a tree and $\mathcal{E} = \emptyset$. We then characterise indexed nested sequents as labelled sequents where $\mathcal{R}$ is a tree but $\mathcal{E}$ is not forced to be empty. We call such sequents *labelled tree sequents with equality* (LTSE). We then lift the map between indexed nested sequents and LTSE to calculi built from these sequents. In this way we develop the technical machinery mentioned by Fitting in [5] as 'a significant different direction' in the study of indexed nested sequents.

[2] The equality relation facilitates the mapping with indexed nested sequents. In any case, the equality terms can be 'compiled away' to obtain a standard labelled sequent.

The results of [11] relating nested sequent and labelled tree sequent calculi was restricted by the fact that the crucial substitution lemma does not hold for the latter. We show here that a nuanced substitution lemma does hold for LTSE calculi. Using this result we induce general results from labelled sequent calculi to LTSE calculi. The situation is delicate as it is necessary to remain within the LTSE-fragment at all times. In this way we efficiently obtain the first syntactic proof of cut-elimination (and hence completeness) for the indexed nested sequent calculi, reusing existing results and alleviating the need for new, independent proofs. We then extend these results to introduce indexed nested sequent calculi for propositional intermediate logics i.e. logics extending intuitionistic logic. We are not aware of any existing nested sequent calculi for intermediate logics.

2 Preliminaries

The set of natural numbers is denoted by $\mathbb{N}$. We assume a set $\{p, q, r, \ldots\}$ of propositional variables. A formula in the language of classical or intermediate logic is either a propositional variable or the logical constants $\bot, \top$ or has the form $A \star B$ where A and B are formulae and $\star \in \{\vee, \wedge, \rightarrow\}$. The language of modal logic has formula $\Box A$ whenever A is a formula. The size of a formula is the sum of the number of connectives and propositional variables and logical constants it contains.

Assume that we have at our disposal an infinite set $\mathbb{SV} = \{x_1, x_2, \ldots\}$ of *state variables* disjoint from the set of propositional variables. A *labelled formula* has the form $x : A$ where $x \in \mathbb{SV}$ and A is a formula. If $X = \{A_1, \ldots A_n\}$ is a formula multiset, then $x : X$ denotes the multiset $\{x : A_1, \ldots, x : A_n\}$ of labelled formulae. Notice that if the formula multiset X is empty, then the labelled formula multiset $x : X$ is also empty.

In this paper we discuss several different types of sequents. The following definitions are standard. A *rule* is a sequence of sequents of some type, typically written as $(s_1, \ldots, s_N/s_{N+1})$. The sequent s_{N+1} is called the *conclusion* of the rule, the remaining sequents are called the *premises* of the rule. If $N = 0$ then the rule is called an *initial sequent.* A calculus consists of a finite set of rules. A derivation in the calculus is defined recursively as either an initial sequent or the object obtained by applying a rule ρ in the calculus to smaller derivations whose bottommost sequents (*endsequents*) are legal premises of ρ. The *height* of a derivation is the number of rules on its longest branch (viewing the derivation as a tree whose nodes are sequents and the root is the endsequent).

A *relation mset* $\mathcal{R}$ is a multiset of relation terms Rxy $(x, y \in \mathbb{SV})$. An *equality mset* $\mathcal{E}$ is a multiset of equality terms $x = y$ $(x, y \in \mathbb{SV})$. Let $\mathcal{E}$ be an equality mset. Then for a relation mset $\mathcal{R}$, let $\mathcal{R}[\mathcal{E}]$ denote the relation mset obtained by replacing every x_j in $\mathcal{R}$ with x_i where i is the least number such that $\mathcal{E} \models x_i = x_j$. Here $\models$ is the usual consequence relation for the theory of equality. Define $X[\mathcal{E}]$ analogously for a labelled formula multiset X.

Definition 1 (Labelled Sequent LS). *A labelled sequent has the form* $\mathcal{R}, \mathcal{E}, X \vdash Y$ *where* $\mathcal{R}$ *is a relation mset,* $\mathcal{E}$ *is an equality mset and* X *and* Y *are multisets of labelled formulae.*

The equality-free LS corresponding to $\mathcal{R}, \mathcal{E}, X \vdash Y$ is $\mathcal{R}[\mathcal{E}], X[\mathcal{E}] \vdash Y[\mathcal{E}]$.

The labelled sequent calculus **LSEq-K** is given in Fig. 1. Some rules in [15]—e.g. reflexivity and transitivity rule for equality—do not appear here. These rules are admissible for sequents containing no equality terms in the succedent. We will be unable to derive $\vdash x = x$ and $x = y \wedge y = z \vdash x = z$ but this is fine because **LSEq-K** is intended for deriving modal formulae. We introduce a new rule (ls-sc) to simulate the (sc) rule of **INS-K** that we will introduce later. ($\Box$r) has a side condition stating that variable y (the *eigenvariable*) does not appear in the conclusion. An *atomic term* has the form Rxy or $x = y$ for $x, y \in \mathbb{SV}$.

$$\frac{}{\mathcal{R}, \mathcal{E}, x : \bot, \Gamma \vdash \Delta}\ (\text{init-}\bot) \qquad \frac{}{\mathcal{R}, \mathcal{E}, x : p, \Gamma \vdash \Delta, x : p}\ (\text{init})$$

$$\frac{\mathcal{R}, \mathcal{E}, x : A, \Gamma \vdash \Delta \qquad \mathcal{R}, \mathcal{E}, x : B, \Gamma \vdash \Delta}{\mathcal{R}, \mathcal{E}, x : A \vee B, \Gamma \vdash \Delta}\ (\vee\text{l}) \qquad \frac{\mathcal{R}, \mathcal{E}, \Gamma \vdash \Delta, x : A, x : B}{\mathcal{R}, \mathcal{E}, \Gamma \vdash \Delta, x : A \vee B}\ (\vee\text{r})$$

$$\frac{\mathcal{R}, \mathcal{E}, x : A, x : B, \Gamma \vdash \Delta}{\mathcal{R}, \mathcal{E}, x : A \wedge B, \Gamma \vdash \Delta}\ (\wedge\text{l}) \qquad \frac{\mathcal{R}, \mathcal{E}, \Gamma \vdash \Delta, x : A \qquad \mathcal{R}, \mathcal{E}, \Gamma \vdash \Delta, x : B}{\mathcal{R}, \mathcal{E}, \Gamma \vdash \Delta, x : A \wedge B}\ (\wedge\text{r})$$

$$\frac{\mathcal{R}, \mathcal{E}, \Gamma \vdash \Delta, x : A \qquad \mathcal{R}, \mathcal{E}, x : B, \Gamma \vdash \Delta}{\mathcal{R}, \mathcal{E}, x : A \to B, \Gamma \vdash \Delta}\ (\to\text{l}) \qquad \frac{\mathcal{R}, \mathcal{E}, x : A, \Gamma \vdash \Delta, x : B}{\mathcal{R}, \mathcal{E}, \Gamma \vdash \Delta, x : A \to B}\ (\to\text{r})$$

$$\frac{\mathcal{R}, Rxy, \mathcal{E}, x : \Box A, y : A, \Gamma \vdash \Delta}{\mathcal{R}, Rxy, \mathcal{E}, x : \Box A, \Gamma \vdash \Delta}\ (\Box\text{l}) \qquad \frac{\mathcal{R}, Rxy, \mathcal{E}, \Gamma \vdash \Delta, y : A}{\mathcal{R}, \mathcal{E}, \Gamma \vdash \Delta, x : \Box A}\ (\Box\text{r})$$

$$\frac{\mathcal{R}, \mathcal{E}, x = y, x : A, y : A, \Gamma \vdash \Delta}{\mathcal{R}, \mathcal{E}, x = y, x : A, \Gamma \vdash \Delta}\ (\text{rep-l}) \qquad \frac{\mathcal{R}, \mathcal{E}, x = y, \Gamma \vdash \Delta, x : A, y : A}{\mathcal{R}, \mathcal{E}, x = y, \Gamma \vdash \Delta, x : A}\ (\text{rep-r})$$

$$\frac{\mathcal{R}Rxz, Ryz, \mathcal{E}, x = y, \Gamma \vdash \Delta}{\mathcal{R}, Rxz, \mathcal{E}, x = y, \Gamma \vdash \Delta}\ (\text{rep-R1}) \qquad \frac{\mathcal{R}, Rzx, Rzy, \mathcal{E}, x = y, \Gamma \vdash \Delta}{\mathcal{R}, Rzx, \mathcal{E}, x = y, \Gamma \vdash \Delta}\ (\text{rep-R2})$$

$$\frac{\mathcal{R}, Rxy, Ruv, \mathcal{E}, x = u, y = v, \Gamma \vdash \Delta}{\mathcal{R}, Rxy, \mathcal{E}, x = u, y = v, \Gamma \vdash \Delta}\ (\text{ls-sc})\ x \text{ is not } u;\ v \text{ not in conclusion}$$

Fig. 1. The labelled sequent calculus **LSEq-K**. ($\Box$r) has the side condition: y does not appear in the conclusion.

Definition 2 (Geometric Axiom). *A* geometric axiom *is a formula in the first-order language (binary relations R, $=$) of the following form where the P_i are atomic formulae and $\hat{Q}_j$ is a conjunction $Q_{j1} \wedge \ldots \wedge Q_{jk_j}$ of atomic formulae.*

$$\forall \bar{z}(P_1 \wedge \ldots \wedge P_m \to \exists \bar{x}(\hat{Q}_1 \vee \ldots \vee \hat{Q}_n)) \tag{6}$$

A modal logic extending the basic normal modal logic K can be defined by addition of axioms to the Hilbert calculus for K. Alternatively, we can consider the logic K as the set of formulae valid on all Kripke frames. Modal logics extending K can be obtained as the set of formulae valid on various subclasses of Kripke frames. For example, the axiomatic extension $K + \Box p \to p$ ($K + \Box p \to \Box\Box p$) corresponds to the set of formulae valid on reflexive (resp. transitive) Kripke frames. A modal logic is *defined* by a set of geometric axioms to mean that the logic consists exactly of the formulae valid on Kripke frames satisfying such geometric axiom (first-order frame conditions). See [1] for further details.

Theorem 3 (Negri). *Let L be a modal logic defined by the geometric axioms $\{\alpha_i\}_{i \in I}$. Then **LSEq-K**+$\{\rho_i\}_{i \in I}$ is a LS calculus for L where ρ_i is a structural rule of the form GRS below corresponding to the geometric axiom (6).*

$$\frac{Q_1\{y_1/x_1\}, P, \mathcal{R}, \mathcal{E}, \mathcal{E}, \Gamma \vdash \Delta \qquad \dots \qquad Q_N\{y_N/x_N\}, P, \mathcal{R}, \mathcal{E}, \mathcal{E}, \Gamma \vdash \Delta}{P, \Gamma \vdash \Delta} GRS$$

Here $Q_j = Q_{j1}, \dots, Q_{jk_j}$ and $P = P_1, \dots, P_m$. The rule ρ_i has the side condition that the eigenvariables $y_1 \dots, y_n$ *do not appear in the conclusion.*

We write **LSEq-K*** to denote some extension of **LSEq-K** by GRS rules.

Example 4.

$$\Diamond\Box p \to \Box\Diamond p \quad \forall xyz\,(Rxy \wedge Rxz \to \exists uv\,(Ryu \wedge Rzv \wedge u = v))$$

$$\frac{\mathcal{R}, Rxy, Rxz, Ryu, Rzv, \mathcal{E}, u = v, \Gamma \vdash \Delta}{R, Rxy, Rxz, \mathcal{E}\Gamma \vdash \Delta} u, v \text{ not in conclusion} \qquad (7)$$

The graph $G(\mathcal{R})$ defined by relation mset $\mathcal{R}$ is the directed graph whose nodes are the state variables in $\mathcal{R}$ and $x \to y$ is a directed edge in $G(\mathcal{R})$ iff $Rxy \in \mathcal{R}$.

Definition 5 (Treelike). *A non-empty relation mset $\mathcal{R}$ is* treelike *if the directed graph defined by $\mathcal{R}$ is a tree (i.e. it is rooted, irreflexive and its underlying undirected graph has no cycle).*

Example 6. Consider the following relation msets: $\{Rxx\}$, $\{Rxy, Ruv\}$, $\{Rxy, Rzy\}$, and $\{Rxy, Rxz, Ryu, Rzu\}$. The graphs defined by these sets are, respectively,

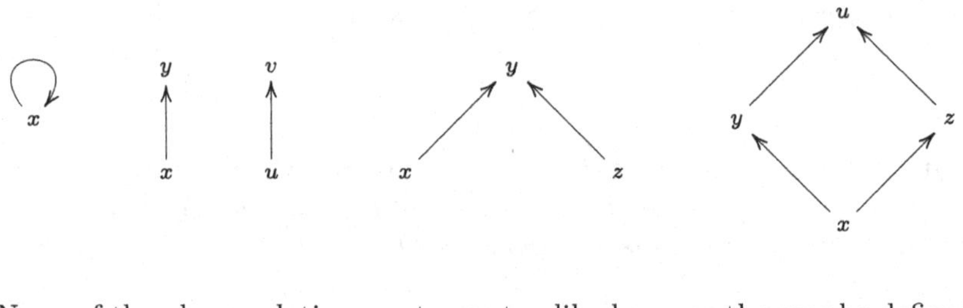

None of the above relation msets are treelike because the graphs defined by their relation msets are not trees. From left-to-right, graph 1 contains a reflexive state; graph 2 and graph 3 are not rooted. Frame 4 is not a tree because the underlying undirected graph contains a cycle.

Definition 7 (Labelled Tree Sequent LTS). *A* labelled tree-sequent *is a labelled sequent of the form $\mathcal{R}, \mathcal{E}, X \vdash Y$ where $\mathcal{E} = \emptyset$ and:*

(i) if $\mathcal{R} \neq \emptyset$ then $\mathcal{R}$ is treelike and every state variable x that occurs in $X \cup Y$ occurs in $\mathcal{R}$.
(ii) if $\mathcal{R} = \emptyset$ then every label in X and Y is the same.

Some examples of LTS: $x : A \vdash x : B$ $\quad \vdash y : A$ $\quad Rxy, Rxz, x : A \vdash y : B$

A state variable may occur in the relation mset and not in the X, Y multisets (e.g. z above far right). Below are *not* LTS (assume no two in x, y, z identical).

$$x : A \vdash x : B, z : C \qquad Rxy, x : A \vdash z : B \qquad Rxy, Ryz, Rxz \vdash$$

From left-to-right above, the first labelled sequent is not an LTS because the relation mset is empty and yet two distinct state variables x and z occur in the sequent, violating condition Definition 7(ii). The next sequent violates Definition 7(i) because z does not appear in the relation mset. The final sequent is not an LTS because the relation mset is not treelike.

Definition 8 (Labelled Tree Sequent with Equality). *A* labelled tree-sequent with equality *(LTSE) is a labelled sequent of the form* $\mathcal{R}, \mathcal{E}, X \vdash Y$ *where:*

(i) if $\mathcal{R} \neq \emptyset$ *then* $\mathcal{R}$ *is treelike and every state variable in* X, Y *and* $\mathcal{E}$ *occurs in* $\mathcal{R}$.
(ii) if $\mathcal{R} = \emptyset$ *then every label in* X, Y *and* $\mathcal{E}$ *is the same.*

Clearly every LTS is an LTSE. Each of the following is an LTSE:

$$Rxy, x = y, x : A \vdash y : B \qquad y = y \vdash y : A \qquad Rxy, Rxz, y = z, x : A \vdash z : B$$

The following are *not* LTSE (assume that no two in x, y and z are identical).

$$x = z, x : A \vdash x : B \qquad Rxy, y = z, x : A \vdash y : B \qquad Rxy, Ryz, Rxz, x = y, x = z \vdash$$

From left-to-right above, the first labelled sequent is not an LTSE because the relation mset is empty and yet the sequent contains more than one label. The next sequent violates Definition 8(ii) because z does not appear in the relation mset. The final sequent is not an LTSE because the relation mset is not treelike.

Definition 9 (Nested Sequent NS). *A* nested sequent *is a finite object defined recursively as follows:*

$$NS := X \vdash Y \text{ where } X \text{ and } Y \text{ are formula multisets}$$
$$NS := NS, [NS], \ldots, [NS]$$

The underlying structure of a nested sequent is a tree decorated with sequents. In Sect. 1 we presented the decorated trees defined by (1) and (2).

Definition 10 (Indexed Nested Sequent INS). *An* indexed nested sequent *is a finite object defined recursively as follows:*

$$INS := X \vdash^n Y \text{ where } X \text{ and } Y \text{ are formula multisets and } n \in \mathbb{N}$$
$$INS := INS, [INS], \ldots, [INS]$$

There is nothing to prevent two traditional sequents in an INS from being given the same index $n \in \mathbb{N}$.

Notation. We write $\Gamma\{X \vdash^n Y, \Delta\}$ to mean the INS Γ containing the occurrence $X \vdash^n Y, \Delta$. Also $\Gamma\{X \vdash^n Y, \Delta\}\{U \vdash^m Y, \Sigma\}$ denotes an INS Γ containing two distinct occurrences: $X \vdash^n Y, \Delta$ and $U \vdash^m Y, \Sigma$. Here Δ and Σ have the form [INS], ..., [INS] (possibly empty).

The indexed nested sequent calculus **INS-K** [5] for K is given in Fig. 2. The usual nested sequent calculus [7,13] for K can be obtained by ignoring the indices and deleting the rules (fc-l), (fc-r) and (sc). Fitting's Geach scheme [5, Sect. 8] yields an INS rule corresponding to $G(h, i, j, k)$ when $i, j > 0$. E.g. here is the rule corresponding to $\Diamond\Box p \to \Box\Diamond p$ (*index c does not appear in conclusion*).

$$\frac{\Gamma\{[X]\}\vdash^a Y, \Delta, [\vdash^c]], [U \vdash^b V, \Sigma, [\vdash^c]]\}}{\Gamma\{[X \vdash^a Y, \Delta], [U \vdash^b V, \Sigma]\}} \tag{8}$$

In contrast to [5], in this work we give INS calculi for *all* Geach axioms.

$$\frac{}{\Gamma\{\bot, X \vdash^n Y, \Delta\}}\ (\text{init-}\bot) \qquad \frac{}{\Gamma\{p, X \vdash^n Y, p, \Delta\}}\ (\text{init}) \qquad \frac{\Gamma\{\Box A, X \vdash^n Y, \Delta, [A, U \vdash^m V, \Sigma]\}}{\Gamma\{\Box A, X \vdash^n Y, \Delta, [U \vdash^m V, \Sigma]\}}\ (\Box\text{l})$$

$$\frac{\Gamma\{X \vdash^n Y, A, B, \Delta\}}{\Gamma\{X \vdash^n Y, A \vee B, \Delta\}}\ (\vee\text{r}) \qquad \frac{\Gamma\{X, A, B \vdash^n Y, \Delta\}}{\Gamma\{X, A \wedge B \vdash^n Y, \Delta\}}\ (\wedge\text{l}) \qquad \frac{\Gamma\{X \vdash^n Y, A, \Delta\} \qquad \Gamma\{X \vdash^n Y, B, \Delta\}}{\Gamma\{X \vdash^n Y, A \wedge B, \Delta\}}\ (\wedge\text{r})$$

$$\frac{\Gamma\{A, X \vdash^n Y, \Delta][A, U \vdash^n V, \Sigma\}}{\Gamma\{A, X \vdash^n Y, \Delta][U \vdash^n V, \Sigma\}}\ (\text{fc-l}) \qquad \frac{\Gamma\{A, X \vdash^n Y, B, \Delta\}}{\Gamma\{X \vdash^n Y, A \to B, \Delta\}}\ (\to\text{r}) \qquad \frac{\Gamma\{X \vdash^n Y, \Delta, [\ \vdash^m A]\}}{\Gamma\{X \vdash^n Y, \Delta, \Box A\}}\ (\Box\text{r})$$

$$\frac{\Gamma\{A, X \vdash^n Y, \Delta\} \qquad \Gamma\{B, X \vdash^n Y, \Delta\}}{\Gamma\{A \vee B, X \vdash^n Y, \Delta\}}\ (\vee\text{l}) \qquad \frac{\Gamma\{X \vdash^n Y, A, \Delta\} \qquad \Gamma\{B, X \vdash^n Y, \Delta\}}{\Gamma\{A \to B, X \vdash^n Y, \Delta\}}\ (\to\text{l})$$

$$\frac{\Gamma\{X \vdash^n Y, A, \Delta\}\{U \vdash^n V, A, \Sigma\}}{\Gamma\{X \vdash^n Y, A, \Delta\}\{U \vdash^n V, \Sigma\}}\ (\text{fc-r}) \qquad \frac{\Gamma\{X \vdash^n Y, \Delta, [P \vdash^m Q, \Pi]\}\{U \vdash^n V, \Sigma, [\ \vdash^m\ ]\}}{\Gamma\{X \vdash^n Y, \Delta, [P \vdash^m Q, \Pi]\}\{U \vdash^n V, \Sigma\}}\ (\text{sc})$$

Fig. 2. The indexed nested sequent calculus **INS-K** for modal logic K. The rule ($\Box$r) has the side condition that index m does not appear in the conclusion.

3 Syntactic Cut-Elimination for LTSE-Derivations

Definition 11 (LTSE-Derivation). *A derivation δ in **LSEq-K** is called an LTSE-derivation if every sequent in δ is an LTSE.*

We now present a syntactic proof of cut-elimination simultaneously for the labelled sequent calculus **LSEq-K*** and for LTSE-derivations in **LSEq-K***. Of course, the proof for **LSEq-K*** appears in [15] so our focus is on LTSE-derivations. We aim to reuse as much as we can from that proof. The main adaptation for LTSE-derivations is a nuanced version of the substitution lemma. Note that certain rule instances—e.g. (rep-R1) and non-trivial instances of (rep-R2)—*cannot* occur in an LTSE-derivation as they do not preserve LTSE. The following is by induction on the size of A.

Lemma 12. *The sequent $\mathcal{R}, \mathcal{E}, x : A, \Gamma \vdash x : A, \Delta$ for arbitrary formula A is derivable in **LSEq-K***. Moreover, if the sequent is LTSE then it has an LTSE-derivation in **LSEq-K***.*

Let $\bar{x} = (x_1, \ldots, x_{N+1})$ and $\bar{y} = (y_1, \ldots, y_{N+1})$. Then $\mathcal{R}\{\bar{y}/\bar{x}\}$ is obtained by replacing every occurrence of x_i in relation mset $\mathcal{R}$ with y_i $(1 \leq i \leq N+1)$. For an equality mset $\mathcal{E}$ and a labelled formula multiset X define $\mathcal{E}\{y/x\}$ and $X\{y/x\}$ analogously. When $\bar{x} = (x)$ and $\bar{y} = (y)$ we simply write $\{y/x\}$. In words, $\{y/x\}$ is the substitution of every occurrence of x with y. The following is trivial.

Lemma 13. *Let δ be a derivation of $s = \mathcal{R}, \mathcal{E}, \Gamma \vdash \Delta$ in **LSEq-K***. For $x, y \in \mathbb{SV}$ s.t. y is not in s, there is a derivation δ' of $\mathcal{R}\{y/x\}, \mathcal{E}\{y/x\}, \Gamma\{y/x\} \vdash \Delta\{y/x\}$ of the same height. Moreover, if δ is an LTSE-derivation then so is δ'.*

The labelled sequent calculus **LSEq-K*** has a stronger substitution property: for all $x, y \in \mathbb{SV}$, if $\mathcal{R}, \mathcal{E}, \Gamma \vdash \Delta$ is derivable, then there is a derivation of $\mathcal{R}\{y/x\}, \mathcal{E}\{y/x\}, \Gamma\{y/x\} \vdash \Delta\{y/x\}$ of the same height. However this property does *not* preserve LTSE-derivations. We need the following nuanced property.

Lemma 14 (Substitution). *Let δ be a derivation of $\mathcal{R}, Rxy, Rxz, \mathcal{E}, y = z, \Gamma \vdash \Delta$ in **LSEq-K*** where $y = z \notin \mathcal{E}$ and $Rxy \notin \mathcal{R}$.[3] Then there is a derivation δ' of $\mathcal{R}\{z/y\}, Rxz, \mathcal{E}\{z/y\}, X\{z/y\} \vdash Y\{z/y\}$ of the same height. Moreover, if δ is an LTSE-derivation then so is δ'.*

Proof. Induction on the height of δ. Consider the last rule ρ in δ. For most ρ it suffices to apply the induction hypothesis to its premises and then reapply ρ.

Suppose ρ is ($\Box$R) below left (α not in conclusion; $x \neq \alpha$). For fresh w:

$$\frac{\mathcal{R}, Rx\alpha, \mathcal{E}, \Gamma \vdash \Delta, \alpha : A}{\mathcal{R}, \mathcal{E}, \Gamma \vdash \Delta, x : \Box A}\,\rho \qquad \dfrac{\dfrac{\dfrac{\mathcal{R}, Rx\alpha, \mathcal{E}, \Gamma \vdash \Delta, \alpha : A}{\mathcal{R}, Rxw, \mathcal{E}, \Gamma \vdash \Delta, w : A}\,\text{IH}}{(\mathcal{R}, Rxw)\{\alpha/\beta\}, \mathcal{E}\{\alpha/\beta\}, \Gamma\{\alpha/\beta\} \vdash \Delta\{\alpha/\beta\}, w : A}\,\text{IH}}{\mathcal{R}\{\alpha/\beta\}, \mathcal{E}\{\alpha/\beta\}, \Gamma\{\alpha/\beta\} \vdash \Delta\{\alpha/\beta\}, (x : \Box A)\{\alpha/\beta\}}\,\rho$$

The case of a GRS rule is similar. Now suppose that the last rule in δ is (ls-sc). By inspection the Ruv term in the premise of (ls-sc) in Fig. 1 cannot be active in the rule for then the premise would have the form $\mathcal{R}', Rus, Ruv, s = v, \ldots$ $(s \in \mathbb{SV})$. Since u is not x by the side condition, $u = v$ is not $s = v$, and hence the latter must occur in the conclusion contradicting the side condition of (ls-sc).

By inspection, δ' is an LTSE-derivation whenever δ is an LTSE-derivation.

Lemma 15 (Weakening). *Let δ be a derivation of $\mathcal{R}, \mathcal{E}, \Gamma \vdash \Delta$ in **LSEq-K***. Then there is a derivation δ' of $s = \mathcal{R}, \mathcal{R}', \mathcal{E}, \mathcal{E}', \Gamma, \Gamma' \vdash \Delta, \Delta'$ of the same height. Moreovoer, if s is an LTSE and δ is an LTSE-derivation, then so is δ'.*

Proof. Induction on the height of δ. Consider the last rule ρ in δ. Use the substitution of eigenvariables with new variables to avoid a clash with the variables in the weakening terms. The claim that δ' is an LTSE-derivation whenever s is an LTSE and δ is an LTSE-derivation can be verified by inspection.

[3] Note: $Rxy \notin \mathcal{R}$ and $y = z \notin \mathcal{E}$ ensure that $\mathcal{R}\{z/y\}, Rxz, \mathcal{E}\{z/y\}, X\{z/y\} \vdash Y\{z/y\}$ is an LTSE whenever s is an LTSE and prevents a residual term $z = z$, respectively.

The proof of the following lemma follows the proof in [15].

Lemma 16 (Invertible). *Let δ be a derivation of the conclusion of a rule instance in* ***LSEq-K**** *. Then there is a derivation δ' of the premise of that rule instance of the same height. Moreover, if δ is an LTSE-derivation then so is δ'.*

There are four rules of contraction.

$$\frac{R, Rxy, Rxy, \mathcal{E}, \Gamma \vdash \Delta}{R, Rxy, \mathcal{E}, \Gamma \vdash \Delta} \quad \frac{R, \mathcal{E}, x = y, x = y, \Gamma \vdash \Delta}{R, \mathcal{E}, x = y, \Gamma \vdash \Delta} \quad \frac{R, \mathcal{E}, x : A, x : A, \Gamma \vdash \Delta}{R, \mathcal{E}, x : A, \Gamma \vdash \Delta} \quad \frac{R, \mathcal{E}, \Gamma \vdash \Delta, x : A, x : A}{R, \mathcal{E}, x : A, \Gamma \vdash \Delta, x : A}$$

Lemma 17 (Contraction). *Let δ be a derivation in* ***LSEq-K**** *of the premise of any of the above contraction rules. Then there is a derivation δ' of its conclusion with the same height. Moreover, if δ is an LTSE-derivation then so is δ'.*

Proof. Induction on the height of δ. In the general case (δ not necessarily an LTSE-derivation): applying height-preserving invertibility to $\mathcal{R}, Rxy, \mathcal{E}, \Gamma \vdash \Delta, y : A, x : \Box A$ yields $\mathcal{R}, Rxy, Rxz, \mathcal{E}, \Gamma \vdash \Delta, y : A, z : A$ and then the stronger substitution property yields height-preserving $\mathcal{R}, Rxz, Rxz, \mathcal{E}, \Gamma \vdash \Delta, z : A, z : A$ and the required sequent follows from the IH.

Consider the case when δ is an LTSE-derivation whose last rule is ($\Box$r). Then the required LTSE-derivation is below right (y and z are eigenvariables).

$$\frac{\mathcal{R}, Rxy, \mathcal{E}, \Gamma \vdash \Delta, y : A, x : \Box A}{\mathcal{R}, \mathcal{E}, \Gamma \vdash \Delta, x : \Box A, x : \Box A}(\Box\text{r}) \qquad \cfrac{\cfrac{\cfrac{\cfrac{\cfrac{\mathcal{R}, Rxy, \mathcal{E}, \Gamma \vdash \Delta, y : A, x : \Box A}{\mathcal{R}, Rxy, Rxz, \mathcal{E}, \Gamma \vdash \Delta, y : A, z : A}\text{inv}}{\mathcal{R}, Rxy, Rxz, \mathcal{E}, y = z, \Gamma \vdash \Delta, y : A, z : A}\text{weak}}{\mathcal{R}, Rxz, \mathcal{E}, \Gamma \vdash \Delta, z : A, z : A}\text{subs.}}{\mathcal{R}, Rxz, \mathcal{E}, \Gamma \vdash \Delta, z : A}\text{IH}}{\mathcal{R}, \mathcal{E}, \Gamma \vdash \Delta, x : \Box A}(\Box\text{r})$$

The usual cut-rule is below left. We use the cut-rule below right which has the property that the conclusion is an LTSE whenever the premises are LTSE.

$$\frac{R_1, \mathcal{E}_1, \Gamma \vdash \Delta, x : A \qquad R_2, \mathcal{E}_2, x : A, \Sigma \vdash \Pi}{R_1, R_2, \mathcal{E}_1, \varepsilon_2, \Sigma \vdash \Delta, \Pi} \quad \frac{R, \mathcal{E}_1, \Gamma \vdash \Delta, x : A \qquad R_2, \mathcal{E}_2, x : A, \Sigma \vdash \Pi}{R_1, \mathcal{E}_1, \varepsilon_2, \Gamma, \Sigma \vdash \Delta, \Pi}$$

Theorem 18. *The cut-rule is eliminable in* ***LSEq-K**** *. Moreover, if the original derivation is an LTSE-derivation, then so is the transformed derivation.*

Proof. We will show how to eliminate a derivation ending with the cut-rule whose premises are cutfree. Primary induction on the size of the cutformula and secondary induction on the sum of the heights of the derivations of the premises. We focus on the case of an LTSE-derivation.

First suppose that the cut-formula is *not* principal in the left premise (the argument is analogous if the cut-formula is not principal in the right premise). When the last rule in the left premise is a unary rule we have the following situation (the case of a n-ary rule is similar).

$$\dfrac{\dfrac{\mathcal{R},\mathcal{R}',\mathcal{E}_1,\mathcal{E}',\Gamma_1 \vdash \Delta_1, x:B}{\mathcal{R},\mathcal{E}_1,\Gamma_1 \vdash \Delta_1, x:B}\,\rho \qquad \mathcal{R},\mathcal{E}_2, x:B,\Gamma_2\vdash\Delta_2}{\mathcal{R},\mathcal{E}_1,\mathcal{E}_2,\Gamma_1,\Gamma_2\vdash \Delta_1,\Delta_2}\,\text{cut}$$

Even if ρ has a side condition, since the eigenvariable(s) $\bar{y}$ do not appear in its conclusion, it follows that no variable in $\bar{y}$ appears in $\mathcal{R}$ and hence not in the right premise of cut either, by Definition 8(i). So we may proceed:

$$\dfrac{\dfrac{\mathcal{R},\mathcal{R}',\mathcal{E}_1,\mathcal{E}',\Gamma_1 \vdash \Delta_1, x:B \qquad \mathcal{R},\mathcal{E}_2,x:B,\Gamma_2\vdash\Delta_2}{\mathcal{R},\mathcal{R}',\mathcal{E}_1,\mathcal{E}',\mathcal{E}_2,\Gamma_1,\Gamma_2\vdash\Delta_1,\Delta_2}\,\text{cut}}{\mathcal{R},\mathcal{E}_1,\mathcal{E}_2,\Gamma_1\Gamma_2\vdash\Delta_1,\Delta_2}\,\rho$$

The remaining case to consider is when the cut-formula is principal in both left and right premise. Once again the interesting case is when the original derivation is an LTSE-derivation and last rule in the left premise is ($\Box$r).

$$\dfrac{\dfrac{\mathcal{R},Rxy,Rxz,\mathcal{E}_1,\Gamma_1\vdash\Delta_1,y:A}{\mathcal{R},Rxz,\mathcal{E}_1,\Gamma_1\vdash\Delta_1,x:\Box A}\,(\Box\text{r}) \qquad \dfrac{\mathcal{R},Rxz,\mathcal{E}_2,x:\Box A,z:A\vdash\Gamma_2\vdash\Delta_2}{\mathcal{R},Rxz,\mathcal{E}_2,x:\Box A,\Gamma_2\vdash\Delta_2}\,(\Box\text{l})}{\mathcal{R},Rxz,\mathcal{E}_1,\mathcal{E}_2,\Gamma_1,\Gamma_2\vdash\Delta_1,\Delta_2}\,\text{cut}$$

Then proceed:

$$\dfrac{\dfrac{\dfrac{\mathcal{R},Rxy,Rxz,\mathcal{E}_1,\Gamma_1\vdash\Delta_1,y:A}{\mathcal{R},Rxy,Rxz,y=z,\mathcal{E}_1,\Gamma_1\vdash\Delta_1,y:A}\,\text{weak}}{\mathcal{R},Rxz,\mathcal{E}_1,\Gamma_1\vdash\Delta_1,y:A}\,\text{subs.} \qquad \text{cut}\left\{\dfrac{\begin{array}{c}\mathcal{R},Rxz,\mathcal{E}_1,\Gamma_1\vdash\Delta_1,x:\Box A\\ \mathcal{R},Rxz,\mathcal{E}_2,x:\Box A,z:A\vdash\Gamma_2\vdash\Delta_2\end{array}}{\mathcal{R},Rxz,\mathcal{E}_1,\mathcal{E}_2,z:A\vdash\Gamma_1,\Gamma_2\vdash\Delta_1,\Delta_2}\right.}{\dfrac{\mathcal{R},Rxz,\mathcal{E}_1,\mathcal{E}_1,\mathcal{E}_2,\Gamma_1,\Gamma_1,\Gamma_2\vdash\Delta_1,\Delta_1,\Delta_2}{\mathcal{R},Rxz,\mathcal{E}_1,\mathcal{E}_2,\Gamma_1,\Gamma_2\vdash\Delta_1,\Delta_2}\,\text{ctr}}\,\text{cut}$$

4 LTSE Derivations: Sound and Complete for Geach Logics

We already know that there are labelled sequent calculi for Geach logics since the corresponding frame conditions are geometric. However it remains to show that these labelled sequent calculi are complete for the Geach logic when we restrict to LTSE derivations. That is the content of this section. First define:

$$\mathsf{R}^0xy := \emptyset \qquad \mathsf{R}^{n+1}xy := \{Rxy_1, Ry_1y_2, \ldots, Ry_{n-1}y_n, Ry_ny\}$$

$$\hat{\mathsf{R}}^0xy := \top \qquad \hat{\mathsf{R}}^{n+1}xy := Rxy_1 \wedge Ry_1y_2 \wedge \ldots \wedge Ry_{n-1}y_n \wedge Ry_ny$$

E.g. $\mathsf{R}^2xy = \{Rxy_1, Ry_1y\}$ and $\hat{\mathsf{R}}^2 = Rxy_1 \wedge Ry_1y$.

Let $\bar{y} = y_1, \ldots, y_{h-1}, y$; $\bar{z} = z_1, \ldots, z_{j-1}, z$; $\bar{u} = u_1, \ldots, u_{i-1}, u$ and $\bar{v} = v_1, \ldots, v_{k-1}, v$ and let λ be the function that returns the last element of a non-empty sequence. It is well-known that the first-order frame condition $f(h,i,j,k)$ below corresponds to the Geach formula $G(h,i,j,k)$ given in (3).

$$\forall x\bar{y}\bar{z}\left(\hat{\mathsf{R}}^h xy \wedge \hat{\mathsf{R}}^j xz \longrightarrow \exists\bar{u}\bar{v}\left(\hat{\mathsf{R}}^i\lambda(x\bar{y})u \wedge \hat{\mathsf{R}}^k\lambda(x\bar{z})v \wedge \lambda(x\bar{y}\bar{u}) = \lambda(x\bar{z}\bar{v})\right)\right)$$

Some examples of Geach formulae and their corresponding frame conditions:

(ref) $\Box p \to p$	$\forall x(\top \to \exists u(Rxu \wedge u = x))$	so $i = 1$ and others are 0
(trans) $\Box p \to \Box\Box p$	$\forall xz_1z(Rxz_1 \wedge Rz_1z \to \exists u(Rxu \wedge u = x))$	so $i = 1$, $j = 2$

From the Sahlqvist correspondence and completeness theorems (see [1]) we have that the modal logic defined by the set $\{f(h_s, i_s, j_s, k_s)\}_{s \in S}$ is precisely the modal logic $K + \{G(h_s, i_s, j_s, k_s)\}_{s \in S}$. Here is the corresponding structural rule where u and u' do not appear in the conclusion.

$$\frac{\mathcal{R}, \mathsf{R}^h xy, \mathsf{R}^j xz, \mathsf{R}^i \lambda(x\bar{y})u, \mathsf{R}^k \lambda(x\bar{z})v, \lambda(x\bar{y}\bar{u}) = \lambda(x\bar{z}\bar{v}), \mathcal{E}, \Gamma \vdash \Delta}{\mathcal{R}, \mathsf{R}^h xy, \mathsf{R}^j xz, \mathcal{E}, \Gamma \vdash \Delta} \rho(h, i, j, k) \quad (9)$$

Here is the LTSE derivation of $G(h, i, j, k)$ using $\rho(h, i, j, k)$:

$$\dfrac{\dfrac{\dfrac{\dfrac{\dfrac{\dfrac{\dfrac{\mathsf{R}^h xy, \mathsf{R}^j xz, \mathsf{R}^i yu, \mathsf{R}^k zu, u = u', y : \Box^i p, \ldots, u' : p, u : p \vdash z : \Diamond^k p, \ldots, u' : p}{\mathsf{R}^h xy, \mathsf{R}^j xz, \mathsf{R}^i yu, \mathsf{R}^k zu, u = u', y : \Box^i p, \ldots, u : p \vdash z : \Diamond^k p, \ldots, u' : p}\text{Repl}}{\mathsf{R}^h xy, \mathsf{R}^j xz, \mathsf{R}^i yu, \mathsf{R}^k zu, u = u', y : \Box^i p, \ldots, u : p \vdash z : \Diamond^k p}\Diamond\text{r}}{\mathsf{R}^h xy, \mathsf{R}^j xz, \mathsf{R}^i yu, \mathsf{R}^k zu', u = u', y : \Box^i p \vdash z : \Diamond^k p}\Box\text{l*}}{\mathsf{R}^h xy, \mathsf{R}^j xz, y : \Box^i p \vdash z : \Diamond^k p}\rho(h, i, j, k)}{\mathsf{R}^h xy, y : \Box^i p \vdash x : \Box^j \Diamond^k p}\Box\text{r}}{x : \Diamond^h \Box^i p \vdash x : \Box^j \Diamond^k p}\Diamond\text{l*}}{\vdash x : \Diamond^h \Box^i p \to \Box^j \Diamond^k p}$$

Theorem 19. *$A \in K + \{G(h_s, i_s, j_s, k_s)\}_{s \in S}$ iff there is an LTSE-derivation of $\vdash x : A$ in **LSEq-K**$+\{f(h_s, i_s, j_s, k_s)\}_{s \in S}$.*

Proof. The direction ($\Rightarrow$) is completeness. We need to show that every axiom and rule of the logic is LTSE-derivable. We saw that $G(h_s, i_s, j_s, k_s)$ is LTSE-derivable. Here is the normal axiom:

$$\dfrac{\dfrac{\dfrac{\dfrac{\dfrac{Rxy, x : \Box A, y : A \vdash y : B, y : A \qquad Rxy, y : B, x : \Box A, y : A \vdash y : B}{Rxy, x : \Box A, y : A, x : \Box(A \to B), y : A \to B \vdash y : B}(\to\text{l})}{Rxy, x : \Box A, x : \Box(A \to B) \vdash y : B}(\Box\text{l}), (\Box\text{l})}{x : \Box A, x : \Box(A \to B) \vdash x : \Box B}(\Box\text{r})}{x : \Box(A \to B) \vdash x : (\Box A \to \Box B)}(\to\text{r})}{\vdash x : \Box(A \to B) \to (\Box A \to \Box B)}(\to\text{r})$$

Simulating *modus ponens* and necessitation is as usual. Recall that the former requires the cut-rule. From the cut-elimination theorem it follows that we can restrict our attention to LTSE-derivations.

To show soundness ($\Leftarrow$) it suffices to show that every rule instance (restricted to LTSE-derivations) of the calculus is sound for $K + \{G(h_s, i_s, j_s, k_s)\}_{s \in S}$. In fact, there is a shortcut for proving soundness: since we already know [15] that Negri's labelled sequent calculus with equality is sound for this logic (interpreting the sequents under the Kripke semantics), derive each of the **LSEq-K** rules in that calculus. Indeed the only new rule is (ls-sc). Given the premises of (ls-sc), apply the full substitution lemma with $\{x/u\}$ and $\{y/v\}$ together with appropriate contraction and weakening to get the conclusion.

5 Maps Between LSEq-K* and INS-K* Calculi

It has already been shown [11] that a nested sequent is isomorphic to a labelled tree sequent up to labelling of state variables. This means that the rules in one formalism immediately induce rules in the other formalism under this mapping. Now let us extend this map to an isomorphism between INS and LTSE.

INS to LTSE: (i) Rewrite an INS s as the tree τ_s (each node will thus be decorated with a sequent $X \vdash^n Y$ from s). Describe the structure of the tree τ_s as a relation mset $\mathcal{R}$ (each node in τ_s corresponds to a distinct variable in $\mathbb{SV}$).

(ii) For $y_1, \ldots, y_{N+1} \in \mathbb{SV}$ occurring in τ_s with corresponding sequents $U_1 \vdash^i V_1, \ldots, U_{N+1} \vdash^i V_{N+1}$ i.e. each of these has the same index i, define $\mathcal{E}_i = \{u = v | u \neq v \,\text{and}\, u, v \in \{y_1, \ldots y_{N+1}\}\}$. Set $\mathcal{E}$ as the union over all indices in s.

(iii) The LTSE corresponding to INS is $\mathcal{R}, \mathcal{E}, X \vdash Y$ where $X = \cup_i x_i : U_i$ and $Y = \cup_i x_i : V_i$. Here the union is over all $\mathbb{SV}$ variables in τ_s and $U_i \vdash V_i$ is the traditional sequent decorating τ_s at x_i.

LTSE to INS: Given the LTSE $\mathcal{R}, \mathcal{E}, X \vdash Y$, construct the tree τ_s defined by $\mathcal{R}$ (so each node in τ_s corresponds to a distinct variable in $\mathbb{SV}$). Decorate node u in τ_s with the traditional sequent $X_u \vdash Y_u$. Here X_u (Y_u) denotes the subset of X (resp. Y) of formulae with label u. It remains to assign an index to each traditional sequent decorating τ_s. Assign the index so that two traditional sequents have the same index iff their corresponding nodes $u, v \in \mathbb{S}$ in τ_s have the property that $\mathcal{E} \models u = v$.

Extending to maps between LSEq-K* and INS-K* calculi. We now demonstrate how to simulate the rules of one calculus in the other. For a rule (ρ) ($\rho \in \{\text{init}, \text{init}{-}\bot, \vee l, \vee r, \wedge l, \wedge r, \to l, \to r, \Box l, \Box r\}$) in **INS-K*** (**LSEq-K***) the corresponding rule in **LSEq-K***(resp. **INS-K***) under the isomorphism is precisely the rule with the same name. By inspection, (fc-l) corresponds to (rep-l) and (fc-r) corresponds to (rep-r).

Rule (rep-R1) and non-trivial instances of (rep-R2)—the trivial case contracts a relation term and hence does not change the INS—cannot occur in LTSE-derivations. The (sc) rule is simulated by (ls-sc).

Each Geach rule in **INS-K*** maps to the GRS of the corresponding rule in **LSEq-K*** e.g. (7) $\leftrightarrow$ (8). The variable restrictions are translated as follows: suppose the GRS premise has the term $u = v$ and the corresponding INS index is s. If both u and v are eigenvariables in the GRS then s cannot appear in the conclusion—e.g. (8). If only u is an eigenvariable then the restriction is weaker: the INS premise must contain an occurrence $[\vdash^s]$ in the appropriate position which does not appear in the conclusion—e.g. (ref) and (trans).

Note that [5] does not give rules for $\{\Diamond^h p \to \Box^j p | h, j \geq 0\}$. We are able to handle these logics. E.g. here are the INS and LTSE rules for $\Diamond p \to \Box p$.

$$\frac{\Gamma\,[[X \vdash^a Y, \Delta], [U \vdash^a V, \Sigma]]}{\Gamma\,[[X \vdash^a Y, \Delta], [U \vdash^b V, \Sigma]]} \qquad \frac{\mathcal{R}, \mathcal{E}, Rxy, Rxz, \mathcal{E}, y = z, \Gamma \vdash \Delta}{\mathcal{R}, \mathcal{E}, Rxy, Rxz, \mathcal{E}, \Gamma \vdash \Delta}$$

From Theorem 19 and the above translation we get:

Corollary 20. *$K + \{G(h_s, i_s, j_s, k_s)\}_{s \in S}$ has an indexed nested sequent calculus.*

6 Intermediate Logics

Nested sequent calculi have been presented for intuitionistic logic [8] and logics in the intuitionistic modal cube [13,19]. Logics between classical and intuitionistic logic are called *intermediate logics.* We are not aware of any nested sequent calculi for intermediate logics. A labelled sequent calculus **LSEq-Ip** (Fig. 3) has been presented [3] for propositional intuitionistic logic and extended via structural rules to capture those intermediate logics whose Kripke semantics are defined by geometric axioms. In this section we show how to use **LSEq-Ip** to obtain an INS calculus for suitable intermediate logics.

$$\text{(init-}\bot\text{)} \qquad \frac{\mathcal{R}, Rxu, u = x, \mathcal{E}, \Gamma \vdash \Delta}{\mathcal{R}, \mathcal{E}, \Gamma \vdash \Delta}\ \text{(ref)} \qquad \frac{\mathcal{R}, Rxy, Ryz, Rxu, u = z, \mathcal{E}, \Gamma \vdash \Delta}{\mathcal{R}, Rxy, Ryz, \mathcal{E}, \Gamma \vdash \Delta}\ \text{(trans)}$$

$$\mathcal{R}, Rxy, \mathcal{E}, x : p, \Gamma \vdash \Delta, y : p \qquad (\vee\text{l}) \qquad (\vee\text{r}) \qquad (\wedge\text{l}) \qquad (\wedge\text{r})$$

$$\frac{\mathcal{R}, Rxy, \mathcal{E}, x : A \to B, \Gamma \vdash y : A, \Delta \qquad \mathcal{R}, Rxy, \mathcal{E}, x : A \to B, y : B, \Gamma \vdash \Delta}{\mathcal{R}, Rxy, \mathcal{E}, x : A \to B, y : B, \Gamma \vdash \Delta}\ (\to\text{l})$$

$$\frac{\mathcal{R}, Rxy, \mathcal{E}, y : A, \Gamma \vdash \Delta, y : B}{\mathcal{R}, \mathcal{E}, \Gamma \vdash \Delta, x : A \to B}\ (\to\text{r}) \qquad \text{(rep-l)} \qquad \text{(rep-r)} \qquad \text{(rep-R1)} \qquad \text{(rep-R2)} \qquad \text{(ls-sc)}$$

Fig. 3. The labelled sequent calculus **LSEq-Ip**. In (→r), y does not appear in the conclusion. In (ref) and (trans), u does not appear in the conclusion.

$$\Gamma\{p, X \vdash^n Y, [U \vdash^m V, p, \Delta]\} \qquad \frac{\Gamma\{X \vdash^n Y, [\vdash^n], \Delta\}}{\Gamma\{X \vdash^n Y, \Delta\}}\ \text{(ref)}$$

$$\frac{\Gamma\{X \vdash^n Y, \Delta, [U \vdash^m V, \Sigma, [P \vdash^s Q, \Pi]], [\vdash^s]\}}{\Gamma\{X \vdash^n Y, \Delta, [U \vdash^m V, \Sigma, [P \vdash^s Q, \Pi]], \}}\ \text{(trans)} \qquad \frac{\Gamma\{X \vdash^n Y, [A \vdash^m B], \Delta\}}{\Gamma\{X \vdash^n Y, A \to B, \Delta\}}\ (\to\text{r})$$

$$\text{(init-}\bot\text{)} \qquad (\vee\text{l}) \qquad (\vee\text{r}) \qquad (\wedge\text{l}) \qquad (\wedge\text{r})$$

$$\frac{\Gamma\{A \to B, X \vdash^n Y, \Delta, [U \vdash^m V, A, \Sigma]\} \qquad \Gamma\{A \to B, X \vdash^n Y, \Delta, [B, U \vdash^m V, \Sigma]\}}{\Gamma\{A \to B, X \vdash^n Y, \Delta, [U \vdash^m V, \Sigma]\}}\ (\to\text{l})$$

$$\text{(fc-l)} \qquad \text{(fc-r)} \qquad \text{(sc)}$$

Fig. 4. The INS calculus **INS-Ip**.

The INS calculus **INS-Ip** (Fig. 4) is obtained from the rules of **LSEq-Ip** by translating each LTSE into an INS. By similar argument to Sect. 3 we can show that the cut-rule is eliminable from **LSEq-Ip** while preserving LTSE-derivations. Analogous to Sect. 4: intermediate logic obtained by axiomatic extension via axioms which (i) correspond to a geometric axiom and (ii) have an LTSE-derivation in the corresponding structural rule extension of **LSEq-Ip** *have* an indexed nested sequent calculus. For example, for the logic Ip+$(p \to \bot) \vee ((p \to \bot) \to \bot)$, the frame condition and corresponding GRS is given in (7). Here is the LTSE-derivation of the axiom. We leave it to the reader to construct the corresponding INS derivation in the calculus extending **INS-Ip** with rule (8).

$$
\begin{array}{c}
\dfrac{Rxy, Rxz, Ryu, Rzv, \mathbf{y} = \mathbf{v}, \mathbf{y} : \mathbf{p}, \mathbf{v} : \mathbf{p}, z : p \to \bot \vdash y : \bot, z : \bot, \mathbf{v} : \mathbf{p}}{Rxy, Rxz, Ryu, Rzv, \mathbf{y} = \mathbf{v}, \mathbf{y} : \mathbf{p}, z : p \to \bot \vdash y : \bot, z : \bot, \mathbf{v} : \mathbf{p}}\ (\text{rep-l}) \\
(\to\text{l})\left\{ \quad \right. \\
\dfrac{\dfrac{Rxy, Rxz, Ryu, Rzv, y = v, y : p, z : p \to \bot, \mathbf{v} : \bot \vdash y : \bot, z : \bot}{Rxy, Rxz, Ryu, \mathbf{Rzv}, y = v, y : p, \mathbf{z} : \mathbf{p} \to \bot \vdash y : \bot, z : \bot}}{\dfrac{Rxy, Rxz, y : p, z : p \to \bot \vdash y : \bot, z : \bot}{\vdash x : (p \to \bot) \vee ((p \to \bot) \to \bot)}}\ \text{GRS}
\end{array}
$$

Finally, if we ignore the indices in **INS-Ip** and delete the rules (fc-l),(fc-r) and (sc) then we obtain a nested sequent calculus for intuitionistic logic. It is instructive to compare this nested sequent calculus with [8].

7 Conclusion

This work can be seen as part of a larger program to classify various proof-systems as subsystems of the labelled sequent calculus. Understanding the relationships between the various formalisms is not only of theoretical importance, it will also help to simplify the proofs, and avoid the need for independent proofs in each formalism. It is not our intention to suggest that only one of {NS, LTS} and only one of {INS, LTSE} is worthy of consideration. Aside from notational preference, there are distinct advantages to each approach:

Recall that (I)NS were obtained by generalising the traditional sequent. Meanwhile LTS(E) can be seen as specific cases of the labelled sequents. Extending a formalism by generalisation has the advantage of intuition: extend just enough to capture the logic of interest without losing nice syntactic properties. Obtaining a formalism by specialisation opens the possibility of coercing existing results to the new situation as we have done here.

Any modal logic whose corresponding frame condition is geometric has a labelled sequent calculus (not just those frame conditions corresponding to Geach logics). Which of these logics has a INS calculus? It needs to be checked if the axiom corresponding to the frame condition has an LTSE-derivation. Using the tools of correspondence theory we can envisage the construction of a 'general' LTSE derivation, perhaps starting from the first-order Kracht formulae [1].

References

1. Blackburn, P., de Rijke, M., Venema, Y.: Modal Logic. Cambridge Tracts in Theoretical Computer Science, vol. 53. Cambridge University Press, Cambridge (2001)
2. Brünnler, K.: Deep sequent systems for modal logic. In: Advances in modal logic, vol. 6, pp. 107–119. Coll. Publ., London (2006)
3. Dyckhoff, R., Negri, S.: Proof analysis in intermediate logics. Arch. Math. Log. **51**(1–2), 71–92 (2012)
4. Dyckhoff, R., Negri, S.: Geometrization of first-order logic. Bull. Symbolic Logic **21**, 123–163 (2015)
5. Fitting, M.: Cut-free proof systems for Geach logics. IfCoLog J. Logics Appl. **2**, 17–64 (2015)
6. Fitting, M.: Proof Methods for Modal and Intuitionistic Logics, Synthese Library, vol. 169. D. Reidel Publishing Co., Dordrecht (1983)

7. Fitting, M.: Prefixed tableaus and nested sequents. Ann. Pure Appl. Logic **163**(3), 291–313 (2012)
8. Fitting, M.: Nested sequents for intuitionistic logics. Notre Dame J. Formal Logic **55**(1), 41–61 (2014)
9. Gentzen, G.: The collected papers of Gerhard Gentzen. In: Szabo, M.E. (ed.) Studies in Logic and the Foundations of Mathematics. North-Holland Publishing Co., Amsterdam (1969)
10. Goré, R., Postniece, L., Tiu, A.: On the correspondence between display postulates and deep inference in nested sequent calculi for tense logics. Log. Methods Comput. Sci. **7**(2), 2:8–3:8 (2011)
11. Goré, R., Ramanayake, R.: Labelled tree sequents, tree hypersequents and nested (deep) sequents. In: Advances in modal logic, vol. 9. College Publications, London (2012)
12. Kashima, R.: Cut-free sequent calculi for some tense logics. Stud. Logica. **53**(1), 119–135 (1994)
13. Marin, S., Straßburger, L.: Label-free modular systems for classical and intuitionistic modal logics. In: Advances in Modal Logic 10, Invited and Contributed papers from the Tenth Conference on "Advances in Modal Logic," held in Groningen, The Netherlands, 5–8 August 2014, pp. 387–406 (2014)
14. Mints, G.: Indexed systems of sequents and cut-elimination. J. Philos. Logic **26**(6), 671–696 (1997)
15. Negri, S.: Proof analysis in modal logic. J. Philos. Logic **34**(5–6), 507–544 (2005)
16. Olivetti, N., Pozzato, G.L.: Nested sequent calculi and theorem proving for normal conditional logics: the theorem prover NESCOND. Intelligenza Artificiale **9**(2), 109–125 (2015)
17. Poggiolesi, F.: A purely syntactic and cut-free sequent calculus for the modal logic of provability. Rev. Symbolic Logic **2**(4), 593–611 (2009)
18. Restall, G.: Comparing modal sequent systems. http://consequently.org/papers/comparingmodal.pdf
19. Straßburger, L.: Cut elimination in nested sequents for intuitionistic modal logics. In: Pfenning, F. (ed.) FOSSACS 2013 (ETAPS 2013). LNCS, vol. 7794, pp. 209–224. Springer, Heidelberg (2013)

Non-classical Logics

A Tableau System for Quasi-Hybrid Logic

Diana Costa and Manuel A. Martins(✉)

CIDMA - Department Mathematics, University Aveiro, Aveiro, Portugal
{dianafcosta,martins}@ua.pt

Abstract. Hybrid logic is a valuable tool for specifying relational structures, at the same time that allows defining accessibility relations between states, it provides a way to nominate and make mention to what happens at each specific state. However, due to the many sources nowadays available, we may need to deal with contradictory information. This is the reason why we came with the idea of Quasi-hybrid logic, which is a paraconsistent version of hybrid logic capable of dealing with inconsistencies in the information, written as hybrid formulas.

In [5] we have already developed a semantics for this paraconsistent logic. In this paper we go a step forward, namely we study its proof-theoretical aspects. We present a complete tableau system for Quasi-hybrid logic, by combining both tableaux for Quasi-classical and Hybrid logics.

1 Introduction

Hybrid logic [1] is the simplest tool for the description of relational structures: it allows establishing accessibility relations between states and furthermore, nominating and making mention to what happens at specific states.

Unfortunately, we may collect contradictory information due to the many sources nowadays available. This is the reason why we came with the idea of Quasi-hybrid logic [5], which is a paraconsistent version of hybrid logic, thus capable of dealing with the inconsistencies in the information, written as formulas in hybrid logic. This kind of logic is useful for comparing the amount of inconsistency among databases, and has proved to be applicable in a wide range of real-life applications, namely we have studied how can inconsistencies relate to the health care flow of a patient [4], and we are currently working in robotics in order to create a robot which uses a paraconsistent reasoning to determine its movements and actions.

This work proposes to introduce proof-theoretical aspects of QH logic. We aimed to combine both tableaux for quasi-classical and hybrid logics, [3,8] respectively, which resulted in a new tableau system as desired.

Classically, tableau systems rely on a backwards reasoning where we start with a formula whose validity we want to prove. A tableau, *i.e.*, a tree, is created using some predefined rules, and whose starting point is the negation of the formula we are investigating. If we come to a point where each branch of the tree contains both a formula of the form φ and a formula of the form $\neg\varphi$, we

N. Olivetti and A. Tiwari (Eds.): IJCAR 2016, LNAI 9706, pp. 435–451, 2016.
DOI: 10.1007/978-3-319-40229-1_30

say that the tableau is closed and verify that there are no counter-models for the original formula, thus it is proved that the formula is valid.

In our paraconsistent setting, we will consider a database Δ, and a query ψ whose satisfiability will be verified in the bistructures (introduced in Sect. 3) that satisfy the formulas in the database. Analogously to the classical case, we start with Δ and ψ^*, where ψ^* will be defined later (in particular it will be a satisfaction statement). Our tableau is constructed using strong rules (which yield disjunctive syllogism) for formulas in Δ and a weaker version (which rejects DS) for ψ^*. If we end up with a tableau which is closed, *i.e.*, in which every branch has a formula of the form φ and φ^*, we can conclude that φ is true in every bistructure that satisfies Δ.

2 The Basic Hybrid Language

We start by presenting the simplest form of hybrid logic: the *basic hybrid language*, $\mathcal{H}(@)$. The basic hybrid language introduces nominals and the satisfaction operator into the propositional modal logic. Although being a simple extension, it carries great power in terms of expressivity.

Definition 1. *Let* $L = \langle \text{Prop}, \text{Nom} \rangle$ *be a* hybrid similarity type *where* Prop *is a set of* propositional symbols *and* Nom *is a set disjoint from* Prop. *We use* p, q, r, *etc. to refer to the elements in* Prop. *The elements in* Nom *are called* nominals *and we typically write them as* i, j, k, *etc.. The set of well-formed formulas over* L, $\text{Form}_@(L)$, *is defined by the following grammar:*

$$WFF := i \mid p \mid \neg\varphi \mid \varphi \vee \psi \mid \varphi \wedge \psi \mid \Diamond\varphi \mid \Box\varphi \mid @_i\varphi$$

For any nominal i, $@_i$ *is called a* satisfaction operator, *and for a formula* φ, $@_i\varphi$ *is called a* satisfaction statement.

Given a hybrid similarity type $L = \langle \text{Prop}, \text{Nom} \rangle$, a *hybrid structure* $\mathcal{H}$ over L is a tuple (W, R, N, V) such that:

- W is a non-empty set called *domain*, whose elements are called *states* or *worlds*.
- R is a binary relation on W and is called the *accessibility relation*.
- $N : \text{Nom} \to W$ is a function called *hybrid nomination* that assigns nominals to elements in W. For any nominal i, $N(i)$ is the element of W named by i.
- V is a *hybrid valuation*, which means that V is a function with domain Prop and range $Pow(W)$ such that $V(p)$ tells us at which states (if any) each propositional symbol is true.

The pair (W, R) is called the *frame* underlying $\mathcal{H}$ and $\mathcal{H}$ is said to be a structure based on this frame.

The satisfaction relation, which is defined as follows, is a generalization of Kripke-style satisfaction.

Definition 2 (Satisfaction). *The relation of local satisfaction* $\models$ *between a hybrid structure* $\mathcal{H} = (W, R, N, V)$*, a state* $w \in W$ *and a hybrid formula is recursively defined by:*

1. $\mathcal{H}, w \models i$ *iff* $w = N(i)$;
2. $\mathcal{H}, w \models p$ *iff* $w \in V(p)$;
3. $\mathcal{H}, w \models \neg\varphi$ *iff it is false that* $\mathcal{H}, w \models \varphi$;
4. $\mathcal{H}, w \models \varphi \wedge \psi$ *iff* $\mathcal{H}, w \models \varphi$ *and* $\mathcal{H}, w \models \psi$;
5. $\mathcal{H}, w \models \varphi \vee \psi$ *iff* $\mathcal{H}, w \models \varphi$ *or* $\mathcal{H}, w \models \psi$;
6. $\mathcal{H}, w \models \Diamond\varphi$ *iff* $\exists w' \in W(wRw'$ *and* $\mathcal{H}, w' \models \varphi)$;
7. $\mathcal{H}, w \models \Box\varphi$ *iff* $\forall w' \in W(wRw' \Rightarrow \mathcal{H}, w' \models \varphi)$;
8. $\mathcal{H}, w \models @_i\varphi$ *iff* $\mathcal{H}, w' \models \varphi$, *where* $w' = N(i)$;

If $\mathcal{H}, w \models \varphi$ *we say that* φ *is* satisfied in $\mathcal{H}$ *at* w. *If* φ *is satisfied at all states in a structure* $\mathcal{H}$, *we write* $\mathcal{H} \models \varphi$. *If* φ *is satisfied at all states in all structures based on a frame* $\mathcal{F}$, *then we say that* φ *is* valid on $\mathcal{F}$ *and we write* $\mathcal{F} \models \varphi$. *If* φ *is valid on all frames, then we simply say that* φ *is valid and we write* $\models \varphi$. *For* $\Delta \subseteq \text{Form}_@(L)$, *we say that* $\mathcal{H}$ *is a* model of Δ *iff for all* $\theta \in \Delta, \mathcal{H} \models \theta$.

Definition 3. *A formula* $\bar{\varphi} \in \text{Form}_@(L)$ *is said to be* (logically) equivalent to $\varphi \in \text{Form}_@(L)$ *iff for every hybrid structure* $\mathcal{H} = (W, R, N, V)$, *for all* $w \in W$,

$$\mathcal{H}, w \models \varphi \Leftrightarrow \mathcal{H}, w \models \bar{\varphi}.$$

It is easy to see that boolean connectives have the usual properties, and that $\Box\varphi$ is equivalent to $\neg\Diamond\neg\varphi$.

We define the notion of negation normal form of a formula (*i.e.*, formulas in which the negation symbol occurs immediately before propositional symbols and/or nominals) for hybrid logic and we establish an analogous result to the one in [2] for classical propositional logic that states that any modal formula is logically equivalent to one in the negation normal form.

Definition 4. *Let* $L = \langle \text{Prop}, \text{Nom} \rangle$ *be a hybrid similarity type. A formula is said to be in* negation normal form, *for short* NNF, *if negation only appears directly before propositional variables and/or nominals. The set of* NNF *formulas over* L, $\text{Form}_{\text{NNF}(@)}(L)$, *is recursively defined as follows:*

For $p \in \text{Prop}$, $i \in \text{Nom}$,

1. p, i, $\neg p$, $\neg i$ *are in* NNF;
2. *If* φ, ψ *are formulas in* NNF, *then* $\varphi \vee \psi$, $\varphi \wedge \psi$ *are in* NNF;
3. *If* φ *is in* NNF, *then* $\Box\varphi$, $\Diamond\varphi$ *are in* NNF;
4. *If* φ *is in* NNF, *then* $@_i\varphi$ *is in* NNF.

The next proposition shows that we do not lose generality by considering just formulas in negation normal form.

Proposition 1 [5]. *Every formula* $\varphi \in \mathrm{Form}_{@}(L)$ *is logically equivalent to a formula* $\bar{\varphi} \in \mathrm{Form}_{\mathrm{NNF}(@)}(L)$.

The *negation normal form* of a formula is defined just as in classical propositional logic. A recursive procedure that puts formulas in negation normal form $nnf : \mathrm{Form}_{@}(L) \to \mathrm{Form}_{\mathrm{NNF}(@)}(L)$, is set as usual. For example:

$$nnf(l) \stackrel{def}{=} l, \text{if } l \text{ is a literal,}$$

$$nnf(\neg(\psi_1 \wedge \psi_2)) \stackrel{def}{=} nnf(\neg\psi_1) \vee nnf(\neg\psi_2) \quad \text{and} \quad nnf(\neg\Box\psi) \stackrel{def}{=} \Diamond nnf(\neg\psi)$$

3 Paraconsistency in Hybrid Logic

In this section we study paraconsistency in Hybrid logic following an approach inspired by the work of Grant and Hunter ([6,7]).

First of all, we define a *Quasi-hybrid (*QH*) Basic Logic.* The assumption in [6] is that all formulas are in *Prenex Conjunctive Normal Form*; in QH logic we will assume henceforth that all formulas are in *Negation Normal Form.* This assumption does not lead to loss of generality since any hybrid formula is equivalent to a formula in negation normal form (cf. Proposition 1).

Next, concepts of bistructure, decoupled and strong satisfaction and QH model will be presented. We define the paraconsistent diagram of a bistructure.

3.1 Quasi-Hybrid Basic Logic

As already mentioned, we will assume that all formulas are in negation normal form, *i.e.*, given a hybrid similarity type $L = \langle \mathrm{Prop}, \mathrm{Nom} \rangle$, the set of formulas is $\mathrm{Form}_{\mathrm{NNF}(@)}(L)$.

Definition 5. *Let* θ *be a formula in* NNF*. We define the* complementation operation $\sim$ *from* $\sim \theta := nnf(\neg\theta)$.

The $\sim$ operator is not part of the object hybrid similarity type but it makes some definitions clearer.

Recall that a *hybrid structure* for a hybrid similarity type L is a tuple (W, R, N, V). However, in order to accommodate contradictions in a model, we will use two valuations for propositions: V^+ and V^-.

Definition 6. *A* hybrid bistructure *is a tuple* (W, R, N, V^+, V^-)*, where* (W, R, N, V^+) *and* (W, R, N, V^-) *are hybrid structures.*

The map V^+ is interpreted as the acceptance of a propositional symbol, and V^- as the rejection. This is formalized in the definition for decoupled satisfaction.

Definition 7. *For a hybrid bistructure* $E = (W, R, N, V^+, V^-)$ *we define a satisfiability relation* $\models_d$ *called* decoupled satisfaction *at* $w \in W$ *for propositional symbols and nominals as follows:*

1. $E, w \models_d p$ *iff* $w \in V^+(p)$;
2. $E, w \models_d i$ *iff* $w = N(i)$;
3. $E, w \models_d \neg p$ *iff* $w \in V^-(p)$;
4. $E, w \models_d \neg i$ *iff* $w \neq N(i)$;

Since we allow both a propositional symbol and its negation to be simultaneously satisfied and also allow both to be non-satisfied, we have decoupled, at the level of the structure, the link between a formula and its complement. In contrast, if a classical hybrid structure satisfies a propositional symbol at some world, it is forced to not satisfy its complement at that world.

This decoupling gives us the basis for a semantics for paraconsistent reasoning. Paraconsistency involves a tradeoff; in order to allow contradictions, one of the following three principles must be abandoned: disjunction introduction, disjunctive syllogism, and transitivity. In this approach, we chose to keep the disjunctive syllogism and transitivity and discard disjunction introduction.

In Quasi-hybrid logic, "*or*" statements involve an intensional disjunction. Such a disjunction is one whose satisfaction entails not merely that at least one of the disjuncts is the case, but also that if one of the disjuncts were not the case, then the other one would be the case.

Definition 8. *A satisfiability relation* $\models_s$ *called* strong satisfaction, *is defined as follows:*

1. $E, w \models_s p$ *iff* $E, w \models_d p$;
2. $E, w \models_s \neg p$ *iff* $E, w \models_d \neg p$;
3. $E, w \models_s i$ *iff* $E, w \models_d i$;
4. $E, w \models_s \neg i$ *iff* $E, w \models_d \neg i$;
5. $E, w \models_s \theta_1 \vee \theta_2$ *iff* $[E, w \models_s \theta_1$ *or* $E, w \models_s \theta_2]$ *and* $[E, w \models_s \sim \theta_1 \Rightarrow E, w \models_s \theta_2]$ *and* $[E, w \models_s \sim \theta_2 \Rightarrow E, w \models_s \theta_1]$;
6. $E, w \models_s \theta_1 \wedge \theta_2$ *iff* $E, w \models_s \theta_1$ *and* $E, w \models_s \theta_2$;
7. $E, w \models_s \Diamond\theta$ *iff* $\exists w'(wRw'$ & $E, w' \models_s \theta)$;
8. $E, w \models_s \Box\theta$ *iff* $\forall w'(wRw' \Rightarrow E, w' \models_s \theta)$;
9. $E, w \models_s @_i\theta$ *iff* $E, w' \models_s \theta$ *where* $w' = N(i)$;

We define *strong validity* as follows: $E \models_s \theta$ iff for all $w \in W$, $E, w \models_s \theta$.

We say that E is a *quasi-hybrid model* of Δ iff for all $\theta \in \Delta, E \models_s \theta$ and we write $E \models_s \Delta$.

4 A Tableau for Quasi-Hybrid Logic

In this section we discuss a decision procedure for Quasi-hybrid logic, based on a tableau system. This new tableau system is a fusion between the tableau system for Quasi-classical logic introduced in [8], and the tableau system for Hybrid logic proposed in [3].

We will consider a database Δ of hybrid formulas that express real situations where inconsistencies may appear at some states, and we will check if a query φ is a consequence of the database, *i.e.*, we will want to check if every bistructure that strongly validates all formulas in Δ also validates φ weakly.

We will restrict our attention to formulas which are satisfaction statements.

4.1 Tableau Rules for QH Logic

We will start by introducing some definitions that will be useful later when we explain the construction of the tableau. We present the rules and a theorem for checking soundness.

Definition 9. *We define* weak satisfaction $\models_{\mathrm{w}}$ *as strong satisfaction* ($\models_s$), *except for the case of disjunction, which we will consider as a classical disjunction:*

$$E, w \models_{\mathrm{w}} \theta_1 \vee \theta_2 \textit{ iff } E, w \models_{\mathrm{w}} \theta_1 \textit{ or } E, w \models_{\mathrm{w}} \theta_2$$

The reader may observe that for any $\theta \in \mathrm{Form}_{\mathrm{NNF}(@)}(L)$, $E, w \models_s \theta$ implies $E, w \models_{\mathrm{w}} \theta$. And that, by contraposition, $\not\models_{\mathrm{w}} \subseteq \not\models_s$.

Similarly to the definition of strong validity, we define *weak validity* as follows: $E \models_{\mathrm{w}} \theta$ iff for all $w \in W$, $E, w \models_{\mathrm{w}} \theta$.

From now on, we will restrict our attention to satisfaction statements.

Definition 10 (Quasi-Hybrid Consequence Relation). *Let Δ be a set of satisfaction statements called* database, *and φ be a satisfaction statement, called* query. *We say that φ is a* consequence of Δ in quasi-hybrid logic *if and only if, for all bistructures E which are quasi-hybrid models of Δ, φ is weakly valid.*

Formally,

$$\Delta \models_{\mathrm{QH}} \varphi \textit{ iff } \forall E\, (E \models_s \Delta \Rightarrow E \models_{\mathrm{w}} \varphi)$$

Before introducing the tableau-based proof procedure, some definitions are required, namely:

Definition 11. *Given a hybrid similarity type $L = \langle \mathrm{Prop}, \mathrm{Nom} \rangle$, we denote the set of satisfaction statements over L as $L_@$.*

We duplicate the set of satisfaction statements by considering starred copies. The extended set is denoted by $L_@^*$ and is defined as: $L_@^* = L_@ \cup \{\varphi^* \mid \varphi \in L_@\}$.

The satisfaction of the new formulas φ^* is in some sense the complementary of the satisfaction of φ.

Definition 12. *We extend both weak and strong satisfaction relations to starred formulas as follows:*

$E, w \models_s \varphi^*$ *iff* $E, w \not\models_s \varphi$

$E, w \models_{\mathrm{w}} \varphi^*$ *iff* $E, w \not\models_{\mathrm{w}} \varphi$

Weak and strong validity of starred formulas are defined in the natural way.

We can now introduce two types of decomposition rules to be used in our QH semantic tableau: *strong rules* and *weak rules*, strong rules are applied to non-starred formulas, and weak rules are applied to starred formulas. The tableau system will be denoted T_{QH}.

Tableau Rules:

Strong rules (S-rules)

- For connectives and operators:

$$\frac{@_i(\alpha \vee \beta)}{(@_i(\sim \alpha))^* \mid @_i\beta}\ (\vee_1) \qquad \frac{@_i(\alpha \vee \beta)}{(@_i(\sim \beta))^* \mid @_i\alpha}\ (\vee_2) \qquad \frac{@_i(\alpha \vee \beta)}{@_i\alpha \mid @_i\beta}\ (\vee_3)$$

$$\frac{@_i(\alpha \wedge \beta)}{@_i\alpha, @_i\beta}\ (\wedge) \qquad \frac{@_i@_j\alpha}{@_j\alpha}\ (@) \qquad \frac{@_i\Box\alpha, @_i\Diamond t}{@_t\alpha}\ (\Box)$$

$$\frac{@_i\Diamond\alpha}{@_i\Diamond t, @_t\alpha}\ (\Diamond)(\text{i})$$

- For nominals:

$$\frac{}{@_i i}\ (Ref)(\text{ii}) \qquad \frac{@_a c, @_a\varphi}{@_c\varphi}\ (Nom_1)(\text{iii}) \qquad \frac{@_a c, @_a\Diamond b}{@_c\Diamond b}\ (Nom_2)$$

Weak rules (W-rules)

- For connectives and operators:

$$\frac{(@_i(\alpha \vee \beta))^*}{(@_i\alpha)^*, (@_i\beta)^*}\ (\vee^*) \qquad \frac{(@_i(\alpha \wedge \beta))^*}{(@_i\alpha)^* | (@_i\beta)^*}\ (\wedge^*)$$

$$\frac{(@_i@_j\alpha)^*}{(@_j\alpha)^*}\ (@^*) \qquad \frac{(@_i\Box\alpha)^*}{@_i\Diamond t, (@_t\alpha)^*}\ (\Box^*)(\text{iv}) \qquad \frac{(@_i\Diamond\alpha)^*, @_i\Diamond t}{(@_t\alpha)^*}\ (\Diamond^*)$$

$$\frac{(@_i\Box\neg t)^*}{@_i\Diamond t}\ (\Box^*_{\neg i})$$

- For nominals:

$$\frac{@_a c, (@_a\varphi)^*}{(@_c\varphi)^*}\ (Nom_1^*)(\text{iii}) \qquad \frac{@_a c, (@_a\Box b)^*}{(@_c\Box b)^*}\ (Nom_2^*)$$

(i) t is a new nominal, α is not a nominal.
(ii) for i in the branch.
(iii) for φ a propositional variable/nominal.
(iv) t is a new nominal, α is not of the form $\neg j$, for j a nominal.

The strong and weak rules for nominals, together with $(\vee_1), (\vee_2), (\Box)$ and $(\Diamond^*)$ are called *non-destructive rules.* The remaining are called *destructive.*

The star in the formulas can be seen as a kind of meta-negation; the weak rules, which involve starred formulas, can thus be viewed as duals of the strong ones, except for the case where we obtain the classical disjunction.

Next theorem states that T_{QH} is sound:

Theorem 1 (Soundness). *The tableau rules are sound in the following sense:*

- *for any r-rule* $\frac{\Lambda}{\Sigma}$, *any bistructure* E *and any state* $w \in W$, $E, w \models_r \Lambda$ *implies* $E, w \models_r \Sigma$.
- *for any r-rule* $\frac{\Lambda}{\Sigma \mid \Gamma}$, *any bistructure* E *and any state* $w \in W$, $E, w \models_r \Lambda$ *implies* $E, w \models_r \Sigma$ *or* $E, w \models_r \Gamma$,

for Λ; Σ *and* Γ *lists of formulas in* $L^*_@$ *and* $r \in \{s, \mathrm{w}\}$.

Proof. The proof can be obtained by checking each rule.

4.2 Properties of the Tableau System and its Construction

The idea of a tableau is that we apply the rules previously introduced to root formulas and to formulas which occur in the tableau after the application of a rule. This indiscriminate way of applying rules leads to infinite tableaux, where there may be repeated formulas and loops, thus we must find a systematic construction that terminates a tableau and allows us to take some conclusions from it.

Definition 13. *We say that a formula* $\chi \in L^*_@$ *is a* strong occurrence/s-occurs *if it is the result of applying a strong rule. Analogously we say that* χ *is a* weak occurrence/w-occurs *if it is the result of applying a weak rule. A formula occurs if it s-occurs or w-occurs.*

Definition 14. *The notion of a* subformula *is defined by the following conditions:*

- φ *is a subformula of* φ;
- *if* $\psi \wedge \theta$ *or* $\psi \vee \theta$ *is a subformula of* φ, *then so are* ψ *and* θ;
- *if* $@_a\psi$, $\Box\psi$, *or* $\Diamond\psi$ *is a subformula of* φ, *then so is* ψ.

The tableau system T_{QH} satisfies the following quasi-subformula property:

Theorem 2 (Quasi-Subformula Property). *If a formula* $@_a\varphi$ *s-occurs in a tableau where* φ *is not a nominal and* φ *is not of the form* $\Diamond b$, *then* φ *is a subformula of a root formula. If a formula* $(@_a\varphi)^*$ *w-occurs in a tableau, then* φ *is a subformula of the premise in the applied rule.*

Proof. The proof can be obtained by checking each rule.

Definition 15. *Let* Θ *be a branch of a tableau and let* Nom^Θ *be the set of nominals occurring in the formulas of* Θ. *Define a binary relation* $\sim_\Theta$ *on* Nom^Θ *by* $a \sim_\Theta b$ *if and only if the formula* $@_ab$ *occurs on* Θ.

Definition 16. *Let* b *and* a *be nominals occurring on a branch* Θ *of a tableau in* T_{QH}. *The nominal* a *is said to be included in the nominal* b *with respect to* Θ *if the following holds:*

- *for any subformula φ of a root formula, if the $@_a\varphi$ s-occurs on Θ, then $@_b\varphi$ also s-occurs on Θ; and*
- *if $(@_a\varphi)^*$ w-occurs on Θ, then $(@_b\varphi)^*$ also w-occurs on Θ.*

If a is included in b with respect to Θ, and the first occurrence of b on Θ is before the first occurrence of a, then we write $a \subseteq_\Theta b$.

Definition 17 (Tableau Construction). *Given a database Δ of satisfaction statements and a query $@_a\varphi$ of QH, one wants to verify if $@_a\varphi$ is a consequence of Δ. In order to do so, we define by induction a sequence $\tau_0, \tau_1, \tau_2, \cdots$ of finite tableaux in T_{QH}, each of which is embedded in its successor.*

Let τ_0 be the finite tableau constituted by the formulas in Δ and $(@_a\varphi)^*$. τ_{n+1} is obtained from τ_n if it is possible to apply an arbitrary rule to τ_n with the following three restrictions:

1. *If a formula to be added to a branch by applying a rule already occurs on the branch, then the addition of the formula is simply omitted.*
2. *After the application of a destructive rule to a formula occurrence φ on a branch, it is recorded that the rule was applied to φ with respect to the branch and the rule will not again be applied to φ with respect to the branch or any extension of it.*
3. *The existential rules $(\Diamond, \Box^*)$ are not applied to a formula occurrence $@_a\Diamond\varphi$ or $(@_a\Box\varphi)^*$ on a branch Θ if there exists a nominal b such that $a \subseteq_\Theta b$.*

Note that due to the first restriction, a formula cannot occur more than once on a branch. Also note that no information is recorded about applications of nondestructive rules. The conditions on applications of the existential rules $(\Diamond, \Box^*)$ in the third restriction are the loop-check conditions. The intuition behind loopchecks is that an existential rule is not applied in a world if the information in that world can be found already in an ancestor world. Hence, the introduction of a new world by the existential rule is blocked.

A branch is *closed* iff there is a formula ψ for which ψ and ψ^* are in that branch; we use the symbol $\times$ to mark a closed branch. A T_{QH} tableau is *closed* iff every branch is closed. A branch is *open* if it is not closed and there are no more rules to apply; we use the symbol $\odot$ to mark an open branch. A tableau is *open* if it has an open branch.

A *terminal tableau* is a tableau where the rules have been exhaustively used *i.e.*, there are no more rules applicable to the tableau obeying the restrictions in Definition 17.

Henceforth, θ is a branch of a terminal tableau.

Definition 18. *Let U be the subset of Nom^Θ containing any nominal a having the property that there is no nominal b such that $a \subseteq_\Theta b$. Let $\approx$ be the restriction of $\sim_\Theta$ to U.*

Note that U contains all nominals present in the root formulas since they are the first formulas of the branch Θ. Observe that Θ is closed under the rules (Ref) and (Nom1), so both $\sim_\Theta$ and $\approx$ are equivalence relations.

Given a nominal a in U, we let $[a]_{\approx}$ denote the equivalence class of a with respect to $\approx$ and we let $U/\approx$ denote the set of equivalence classes.

Definition 19. *Let R be the binary relation on U defined by aRc if and only if there exists a nominal $c' \approx c$ such that one of the following two conditions is satisfied:*

1. *The formula $@_a\Diamond c'$ occurs on Θ.*
2. *There exists a nominal d in Nom^{Θ} such that the formula $@_a\Diamond d$ occurs on Θ and $d \subseteq_{\Theta} c'$.*

Note that the nominal d referred to in the second item in the definition is not an element of U. It follows from Θ being closed under the rule (Nom2) that R is compatible with $\approx$ in the first argument and it is trivial that R is compatible with $\approx$ in the second argument. We let $\bar{R}$ be the binary relation on $U/\approx$ defined by $[a]_{\approx}\,\bar{R}\,[c]_{\approx}$ if and only if aRc.

Definition 20. *Let $\bar{N} : U \to U/\approx$ be defined as $\bar{N}(a) = [a]_{\approx}$.*

Definition 21. *Let V^+ be the function that to each ordinary propositional symbol assigns the set of elements of U where that propositional variable occurs, i.e., $a \in V^+(p)$ iff $@_a p$ occurs on Θ. Analogously, let V^- be the function that to each ordinary propositional symbol assigns the set of elements of U where the negation of that propositional variable occurs, i.e., $a \in V^-(p)$ iff $@_a\neg p$ occurs on Θ.*

We let $V^+_{\approx}$ be defined by $V^+_{\approx}(p) = \{[a]_{\approx} \mid a \in V^+(p)\}$. We define $V^-_{\approx}$ analogously: $V^-_{\approx}(p) = \{[a]_{\approx} \mid a \in V^-(p)\}$.

Given a branch Θ, let $M^{\Theta} = \left(U/\approx, \bar{R}, \bar{N}, V^+_{\approx}, V^-_{\approx}\right)$. We will omit the reference to the branch in M^{Θ} if it is clear from the context.

Theorem 3 (Model Existence). *Assume that the branch Θ is open. For any satisfaction statement $@_a\varphi$ which contains only nominals from U, the following conditions hold:*

(i) *If $@_a\varphi$ s-occurs on Θ, then $M, [a]_{\approx} \models_s \varphi$*
(ii) *If $@_a\varphi$ w-occurs on Θ, then $M, [a]_{\approx} \models_{\mathrm{w}} \varphi$*
(iii) *If $(@_a\varphi)^*$ s-occurs on Θ, then $M, [a]_{\approx} \not\models_s \varphi$.*
(iv) *If $(@_a\varphi)^*$ w-occurs on Θ, then $M, [a]_{\approx} \not\models_{\mathrm{w}} \varphi$.*

Proof. The proof is by induction on the structure of φ.

- $\varphi = i$, i a nominal
  (i) $@_a i$ s-occurs on Θ, then $[a]_{\approx} = [i]_{\approx}$, hence $M, [a]_{\approx} \models_s i$.
  (ii) $@_a i$ never w-occurs.
  (iii) $(@_a i)^*$ s-occurs on Θ, then, since the branch is open, $@_a i$ does not occur on Θ, in particular, it does not s-occur. So, $[a]_{\approx} \neq [i]_{\approx}$, then $M, [a]_{\approx} \not\models_s i$.
  (iv) $(@_a i)^*$ w-occurs on Θ, then analogously to the previous case, $@_a i$ does not occur on Θ, in particular, it does not w-occur. So, $[a]_{\approx} \neq [i]_{\approx}$, then $M, [a]_{\approx} \not\models_{\mathrm{w}} i$.

- $\varphi = \neg i$, i a nominal
  The proof is analogous to the case $\varphi = i$.
- $\varphi = p$, p a propositional variable
  (i) $@_a p$ s-occurs on Θ, then $a \in V_{\approx}^{+}(p)$, thus $M, [a]_{\approx} \models_s p$.
  (ii) $@_a p$ never w-occurs.
  (iii) $(@_a p)^*$ s-occurs on Θ, then, since the branch is open, $@_a p$ does not occur on Θ, in particular, it does not s-occur. It also means that $a \notin V_{\approx}^{+}(p)$ so, $M, [a]_{\approx} \not\models_s p$.
  (iv) $(@_a p)^*$ w-occurs on Θ; analogously to the previous case, considering that $@_a p$ does not w-occur we get that $M, [a]_{\approx} \not\models_{\mathrm{w}} p$.
- $\varphi = \neg p$, p a propositional variable
  Each case will be analogous to the one in $\varphi = p$, only with the difference that we now consider $V_{\approx}^{-}(p)$ instead of $V_{\approx}^{+}(p)$.
- $\varphi = \phi \wedge \psi$, ϕ, ψ formulas
  (i) $@_a(\phi \wedge \psi)$ s-occurs on Θ, then, by applying the S-rule $(\wedge)$, $@_a\phi$ and $@_a\psi$ s-occur on Θ. By the induction hypothesis, $M, [a]_{\approx} \models_s \phi$ and $M, [a]_{\approx} \models_s \psi$. Thus, $M, [a]_{\approx} \models_s \phi \wedge \psi$.
  (ii) $@_a(\phi \wedge \psi)$ never w-occurs.
  (iii) $(@_a(\phi \wedge \psi))^*$ s-occurs on Θ, then, by applying the W-rule $(\wedge^*)$, $(@_a\phi)^*$ or $(@_a\psi)^*$ w-occur on Θ.Hence, by induction hypothesis, $M, [a]_{\approx} \not\models_{\mathrm{w}} \phi$ or $M, [a]_{\approx} \not\models_{\mathrm{w}} \psi$. Thus, $M, [a]_{\approx} \not\models_{\mathrm{w}} \phi \wedge \psi$. Therefore, $M, [a]_{\approx} \not\models_s \phi \wedge \psi$.
  (iv) $(@_a(\phi \wedge \psi))^*$ w-occurs on Θ; follow an analogous approach to (iii).
- $\varphi = @_i\phi$, ϕ a formula
  (i) $@_a @_i \phi$ s-occurs on Θ, then, by applying the S-rule $(@)$, $@_i\phi$ s-occurs on Θ. By the induction hypothesis, $M, [i]_{\approx} \models_s \phi$; thus, by the definition of satisfiability, $M, [a]_{\approx} \models_s @_i\phi$.
  (ii) $@_a @_i \phi$ never w-occurs.
  (iii) $(@_a @_i \phi)^*$ s-occurs on Θ, then, by applying the W-rule $(@^*)$, $(@_i\phi)^*$ w-occurs on Θ. By the induction hypothesis, $M, [i]_{\approx} \not\models_{\mathrm{w}} \phi$. Thus, $M, [i]_{\approx} \not\models_s \phi$ and it follows that $M, [a]_{\approx} \not\models_s @_i\phi$.
  (iv) $(@_a @_i \phi)^*$ w-occurs on Θ; follow an analogous approach to (iii).
- $\varphi = \phi \vee \psi$, ϕ, ψ formulas
  (i) $@_a(\phi \vee \psi)$ s-occurs on Θ, then, since we can apply three S-rules, namely $(\vee_1, \vee_2, \vee_3)$, one may obtain 8 new branches, represented as follows:

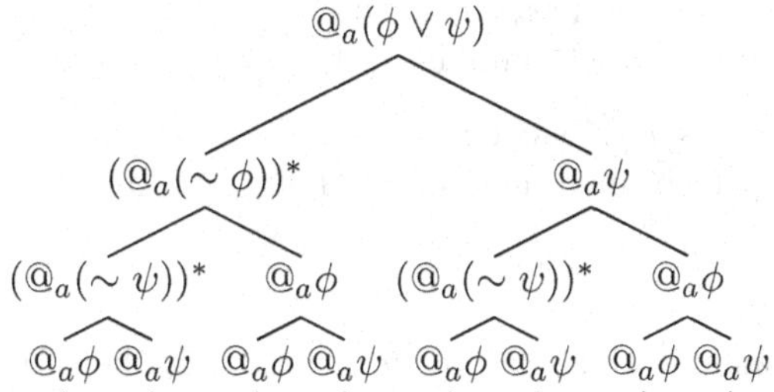

Let us check what happens at each branch:

1. $(@_a(\sim \phi))^*$, $(@_a(\sim \psi))^*$ and $@_a\phi$ s-occur.
   By induction hypothesis:
   $M, [a]_\approx \not\models_s \sim \phi$,
   $M, [a]_\approx \not\models_s \sim \psi$, and
   $M, [a]_\approx \models_s \phi$.

   Recall that $M, [a]_\approx \models_s \phi \vee \psi$
   $\Leftrightarrow (M, [a]_\approx \models_s \phi$ or $M, [a]_\approx \models_s \psi)$
   and $(M, [a]_\approx \models_s \sim \phi \Rightarrow M, [a]_\approx \models_s \psi)$
   and $(M, [a]_\approx \models_s \sim \psi \Rightarrow M, [a]_\approx \models_s \phi)$
   Thus, under the conditions in branch number 1, it is verified that $M, [a]_\approx \models_s \phi \vee \psi$.
2. $(@_a(\sim \phi))^*$, $(@_a(\sim \psi))^*$ and $@_a\psi$ s-occur.
   Thus,
   $M, [a]_\approx \not\models_s \sim \phi$,
   $M, [a]_\approx \not\models_s \sim \psi$, and
   $M, [a]_\approx \models_s \psi$.
   Using the same approach as before, one has that $M, [a]_\approx \models_s \phi \vee \psi$.
3. $(@_a(\sim \phi))^*$, and $@_a\phi$ s-occur.
   Thus,
   $M, [a]_\approx \not\models_s \sim \phi$, and
   $M, [a]_\approx \models_s \phi$.

   Then, $M, [a]_\approx \models_s \phi \vee \psi$.
4. $(@_a(\sim \phi))^*$, $@_a\phi$ and $@_a\psi$ s-occur.
   Thus,
   $M, [a]_\approx \not\models_s \sim \phi$,
   $M, [a]_\approx \models_s \phi$, and
   $M, [a]_\approx \models_s \psi$.

   So, $M, [a]_\approx \models_s \phi \vee \psi$.
5. $@_a\psi$, $(@_a(\sim \psi))^*$ and $@_a\phi$ s-occur.
   Analogously to branch number 4, $M, [a]_\approx \models_s \phi \vee \psi$.
6. $@_a\psi$ and $(@_a(\sim \psi))^*$ s-occur.
   Analogously to branch number 3, $M, [a]_\approx \models_s \phi \vee \psi$.
7. and 8. $@_a\phi$ and $@_a\psi$ s-occur.
   Thus, $M, [a]_\approx \models_s \phi \vee \psi$.

We can thus conclude that, if $@_a(\phi \vee \psi)$ s-occurs, then $M, [a]_\approx \models_s \phi \vee \psi$.

(ii) $@_a(\phi \vee \psi)$ never w-occurs.
(iii) $(@_a(\phi \vee \psi))^*$ s-occurs on Θ, then by applying rule $(\vee^*)$, we obtain that $(@_a\phi)^*$ and $(@_a\psi)^*$ w-occur. By induction hypothesis, $M, [a]_\approx \not\models_w \phi$ and $M, [a]_\approx \not\models_w \psi$; thus, $M, [a]_\approx \not\models_w \phi \vee \psi$. Therefore $M, [a]_\approx \not\models_s \phi \vee \psi$.
(iv) $(@_a(\phi \vee \psi))^*$ w-occurs on Θ, analogous to the previous case.

- $\varphi = \Diamond\psi$, where:
  - ψ is a nominal t
    For the first two cases we have that:
    (i) $@_a\Diamond t$ s-occurs on Θ, then $[a]_\approx \bar{R}[t]_\approx$. By definition of satisfiability, $M, [a]_\approx \models_s \Diamond t$.
    (ii) $@_a\Diamond t$ w-occurs on Θ; it follows from the previous explanation that $M, [a]_\approx \models_w \Diamond t$.
  - ψ is not a nominal
    For the first two cases we have that:
    (i) $@_a\Diamond\psi$ s-occurs on Θ, then by the application of the S-rule $(\Diamond)$, $@_a\Diamond t$ and $@_t\psi$ s-occur, for t a new nominal.
    * if $t \in U$, $[a]_\approx \bar{R}[t]_\approx$ and $M, [a]_\approx \models_s \Diamond\psi$.
    * if $t \notin U$, $\exists d$ such that $t \subseteq_\Theta d$. Assume that there is no e such that $d \subseteq_\Theta e$, *i.e.*, $d \in U$. By Theorem 2, the formula ψ is a subformula of a root formula, so $@_d\psi$ s-occurs on Θ. By induction hypothesis, $M, [d]_\approx \models_s \psi$ and $[a]_\approx \bar{R}[d]_\approx$. Thus yielding that $M, [a]_\approx \models_s \Diamond\psi$.
    (ii) $@_a\Diamond\psi$ never w-occurs.

  - The last two cases are applicable for either ψ a nominal or not:
    (iii) $(@_a\Diamond\psi)^*$ s-occurs on Θ.
    We want to prove that $M, [a]_\approx \not\models_s \Diamond\psi$, *i.e.*, that for all $[c]_\approx$ such that $[a]_\approx \bar{R}[c]_\approx$, $M, [c]_\approx \not\models_s \psi$.
    By definition, $[a]_\approx \bar{R}[c]_\approx$ implies that $\exists c' : c' \approx c$ that satisfies one of the following two conditions:
    * $@_a\Diamond c'$ occurs.
    Then, by the W-rule $(\Diamond^*)$, $(@_{c'}\psi)^*$ w-occurs on Θ. By induction, $M, [c']_\approx \not\models_w \psi$.
    Since $[c']_\approx = [c]_\approx$, then $[a]_\approx \bar{R}[c']_\approx$. Thus, $M, [a]_\approx \not\models_w \Diamond\psi$, which implies that $M, [a]_\approx \not\models_s \Diamond\psi$.
    * $\exists d \in \mathrm{Nom}^\Theta$ such that $@_a\Diamond d$ occurs and $d \subseteq_\Theta c'$.
    By the application of the W-rule $(\Diamond^*)$, $(@_d\psi)^*$ occurs on Θ. By Theorem 2, ψ is a subformula of the premise in the applied rule.
    Since $d \subseteq_\Theta c'$, $(@_{c'}\psi)^*$ occurs on Θ.
    By induction we conclude that $M, [c']_\approx \not\models_w \psi$, thus $M, [a]_\approx \not\models_w \Diamond\psi$ and finally, $M, [a]_\approx \not\models_s \Diamond\psi$.
    (iv) $(@_a\Diamond\psi)^*$ w-occurs on Θ. The proof is the same as the previous one, stopping at the point where $M, [a]_\approx \not\models_w \Diamond\psi$.

- $\varphi = \Box\psi$
  In order to prove the first two cases, consider ψ an arbitrary formula.
  (i) $@_a\Box\psi$ s-occurs on Θ.
  We want to prove that $M, [a]_\approx \models_s \Box\psi$, *i.e.*, that for all $[c]_\approx$ such that $[a]_\approx\bar{R}[c]_\approx$, $M, [c]_\approx \models_s \psi$.
  As you may verify, this is a similar proof to the one made for $(@_a\Diamond\psi)^*$:
  By definition, $[a]_\approx\bar{R}[c]_\approx$ implies that $\exists c' : c' \approx c$ that satisfies one of the following two conditions:

  - $@_a\Diamond c'$ occurs.
    Then, by the S-rule $(\Box)$, $@_{c'}\psi$ s-occurs on Θ. By induction, $M, [c']_\approx \models_s \psi$. Since $[c']_\approx = [c]_\approx$, then $[a]_\approx\bar{R}[c']_\approx$. Thus, $M, [a]_\approx \models_s \Box\psi$.
  - $\exists d \in \text{Nom}^\Theta$ such that ${}_a\Diamond d$ occurs and $d \subseteq_\Theta c'$.
    By the application of the S-rule $(\Box)$, $@_d\psi$ s-occurs on Θ. By Theorem 2, ψ is a subformula.
    Since $d \subseteq_\Theta c'$, $@_{c'}\psi$ occurs on Θ.
    By induction we conclude that $M, [c']_\approx \models_s \psi$, thus $M, [a]_\approx \models_s \Box\psi$.

  (ii) $@_a\Box\psi$ never w-occurs.

  The last two cases require a separation between ψ of the form $\neg i$, with i a nominal, and not of the form described.
  Let us consider:

  - ψ of the form $\neg i$, with i a nominal.
    (iii) $(@_a\Box\neg i)^*$ s-occurs on Θ, thus, by the rule $(\Box^*_{\neg i})$ which implies that $(@_a\Diamond i)$ w-occurs.
    An occurrence of $(@_a\Diamond i)$ means that $[a]_\approx\bar{R}[i]_\approx$. So, by definition, $M, [a]_\approx \models_w \Diamond i$, which entails that $M, [a]_\approx \not\models_w \Box\neg i$. Therefore, $M, [a]_\approx \not\models_s \Box\neg i$.
    (iv) $(@_a\Box\neg i)^*$ w-occurs on Θ follows the same approach.
  - ψ otherwise.
    (iii) $(@_a\Box\psi)^*$ s-occurs on Θ, thus by application of the W-rule $(\Box^*)$, $@_a\Diamond t$ and $(@_t\psi)^*$ w-occur, for t a new nominal.
    * if $t \in U$, $[a]_\approx\bar{R}[t]_\approx$. By induction hypothesis, $M, [t]_\approx \not\models_w \psi$. Thus $M, [a]_\approx \not\models_w \Box\psi$. To conclude, $M, [a]_\approx \not\models_s \Box\psi$.
    * if $t \notin U$, $\exists d$ such that $t \subseteq_\Theta d$. Assume that there is no e such that $d \subseteq_\Theta e$, *i.e.*, $d \in U$. By Theorem 2, the formula ψ is a subformula of a root formula, so $(@_d\psi)^*$ w-occurs on Θ. By induction hypothesis, $M, [d]_\approx \not\models_w \psi$ and $[a]_\approx\bar{R}[d]_\approx$. Thus yielding that $M, [a]_\approx \not\models_w \Box\psi$, which, in its turn, yields $M, [a]_\approx \not\models_s \Box\psi$.

    (iv) $(@_a\Box\psi)^*$ s-occurs on Θ; the proof is the same as before, stopping at $M, [a]_\approx \not\models_w \Box\psi$, which already gives the desired result.

From this theorem together with the soundness theorem we have the following decision procedure:

Decision Procedure: *Given a database Δ and a query $@_a\varphi$ whose consequence from Δ we want to decide, let τ_n be a terminal tableau generated by the tableau construction algorithm. If the tableau is closed, then $@_a\varphi$ is a consequence of Δ. Analogously, if the tableau is open, then $@_a\varphi$ is not a consequence of Δ.*

Example 1. *Let $\Delta = \{@_i(p \lor q), @_j\Diamond i, @_j q, @_j\neg q\}$ be a database and consider a query $\varphi = @_j\Diamond p$. Let us decide if φ is a consequence of Δ using the tableau procedure described:*

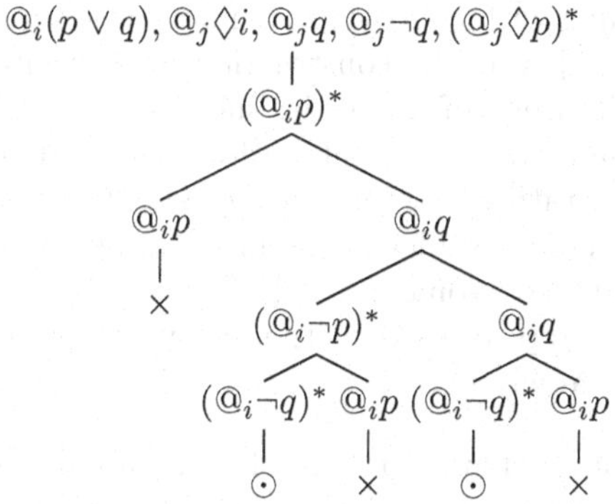

Since the tableau has some open branches, it means that the tableau is open, thus φ is not a consequence of Δ. A counter-model is the bistructure with two worlds i and j where only $@_i q$, $@_j q$, $@_j\neg q$ and $@_j\Diamond i$, $@_i\Box\neg i$, $@_j\Box\neg j$ are satisfied, which clearly is a bistructure that satisfies Δ, but where φ is not true.

We can easily verify that the database has an inconsistency, as it contains both $@_j q$ and $@_j\neg q$. If we were to evaluate φ in hybrid logic (usual version, non-paraconsistent), then φ would trivially follow.

Example 2. *Let $\Delta = \{@_t(p \land q \land r), @_i\Box\neg p, @_i\Diamond t\}$ be a database and consider a query $\varphi = (@_t(p \land \neg p))$. Let us decide if φ is a consequence of Δ using the tableau procedure described:*

$$@_t(p \land q \land r),\ @_i\Box\neg p,\ @_i\Diamond t,\ (@_t(p \land \neg p))^*$$
$$|$$
$$@_t p,\ @_t q,\ @_t r$$
$$|$$
$$@_t\neg p$$
$$(@_t p)^*\quad (@_t\neg p)^*$$
$$\quad|\qquad\quad|$$
$$\quad\times\qquad\quad\times$$

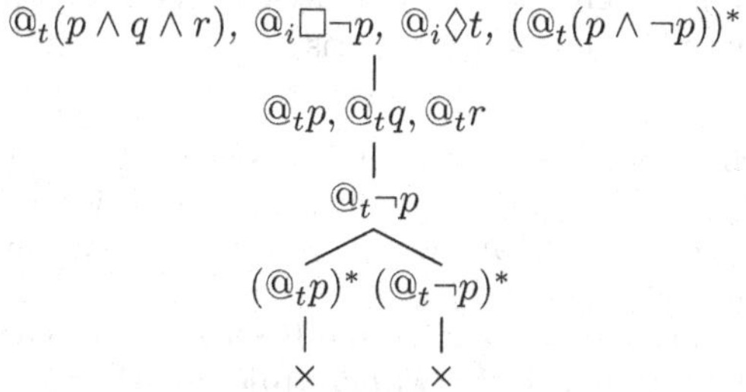

Note that the database has an inconsistency and the query itself is inconsistent. However, from the tableau procedure we verify that, since it is closed, φ is a consequence of Δ.

These examples show the mechanism of the tableau construction in action. Both examples yield an inconsistency in the database, but it is possible to still reach sensible conclusions about the validity of the query, thanks to the paraconsistent aspect of the tableau system presented.

5 Conclusion

Paraconsistent logics have been around for some decades, and their importance is justified by the unavoidable occurrence of inconsistencies in data and knowledge management. Inconsistent information can appear everywhere, and for many reasons. Contradictory information may arise in systems which are safety critical, such as health systems, aviation systems and many others.

After introducing Quasi-hybrid logic [5], a paraconsistent version of hybrid logic capable of dealing with contradictory information under the form of hybrid formulas, the next step was the construction of a proof-theoretical system for this new logic and the study of properties such as completeness, decidability, *etc.*

The challenge here was the combination of two different tableau systems already existent in literature, one for Quasi-classical logic and another for Hybrid logic. We ended up with a tableau system for Quasi-hybrid logic, described in this work, which ends with some examples.

This is clearly a step forward in the study of QH logic, which we aim to continue exploring, namely by:

- investigating inconsistency at the level of nominals, by allowing two nominations: a positive one for nominals, and a negative one for the negation of nominals. This way, we are able to handle the possibility of receiving information of the form $@_i j$ and $@_i \neg j$;
- accounting for inconsistency in modalities, for example by allowing both $@_i \Diamond j$ and $@_i \Box \neg j$ to be true;
- studying paraconsistency in the context of strong Priorean logic;
- use a combination of all of the above to study reactive deontic logics and switch graphs, where the base accessibility relation changes when its edges are traversed, in the context of QH logic.

Paraconsistency has already been considered in the context of modal logic (e.g. [9,10]). It will be interesting to compare our system with the hybrid systems obtained by adding nominals and the satisfaction operator to those systems.

Acknowledgements. We are very grateful to J. Marcos for productive discussions, in several related topics, that were very important for achieving the reported results. We would also like to mention that the careful work of the anonymous reviewers have improved the quality of the present paper.

This work was supported in part by the Portuguese Foundation for Science and Technology (FCT) through CIDMA within project UID/MAT/04106/2013 and the project EU FP7 Marie Curie PIRSES-GA-2012-318986 GeTFun: Generalizing Truth-Functionality. Diana Costa also thanks the support of FCT via the Ph.D. scholarship PD/BD/105730/2014, and the Calouste Gulbenkian Foundation through the Research Stimulus Program 2015 (*Programa de Estímulo à Investigação 2015*).

References

1. Blackburn, P.: Representation, reasoning, and relational structures: a hybrid logic manifesto. Log. J. IGPL **8**(3), 339–365 (2000)
2. Boolos, G.S., Burgess, J.P., Jeffrey, R.C.: Computability and Logic, 5th edn. Cambridge University Press, Cambridge (2007)
3. Braüner, T.: Hybrid Logic and Its Proof-Theory. Springer, Heidelberg (2010)
4. Costa, D., Martins, M.: Inconsistencies in health care knowledge. In: 2014 IEEE 16th International Conference on e-Health Networking, Applications and Services (Healthcom), pp. 37–42, October 2014
5. Costa, D., Martins, M.: Paraconsistency in hybrid logic. Accepted in the Journal of Logic and Computation (2016). http://sweet.ua.pt/martins/documentos/preprint_2014/phl14.pdf
6. Grant, J., Hunter, A.: Measuring inconsistency in knowledge bases. J. Intell. Inf. Syst. **27**(2), 159–184 (2006)
7. Grant, J., Hunter, A.: Analysing inconsistent first-order knowledge bases. Artif. Intell. **172**(8–9), 1064–1093 (2008)
8. Hunter, A.: A semantic tableau version of first-order quasi-classical logic. In: Benferhat, S., Besnard, P. (eds.) ECSQARU 2001. LNCS (LNAI), vol. 2143, pp. 544–555. Springer, Heidelberg (2001)
9. Odintsov, S.P., Wansing, H.: Modal logics with Belnapian truth values. J. Appl. Non-Class. Logics **20**(3), 279–304 (2010)
10. Rivieccio, U., Jung, A., Jansana, R.: Four-valued modal logic: Kripke semantics and duality. J. Logic Comput. (2015). http://logcom.oxfordjournals.org/citmgr?gca=logcom%3Bexv038v1. doi:10.1093/logcom/exv038

Machine-Checked Interpolation Theorems for Substructural Logics Using Display Calculi

Jeremy E. Dawson[1(✉)], James Brotherston[2], and Rajeev Goré[1]

[1] Research School of Computer Science, Australian National University, Canberra, Australia
Jeremy.Dawson@anu.edu.au
[2] University College London, London, UK

Abstract. We present a mechanised formalisation, in Isabelle/HOL, of Brotherston and Goré's proof of Craig interpolation for a large of class display calculi for various propositional substructural logics. Along the way, we discuss the particular difficulties associated with the local interpolation property for various rules, and some important differences between our proofs and those of Brotherston and Goré, which are motivated by the ease of mechanising the development. Finally, we discuss the value for this work of using a prover with a programmable user interface (here, Isabelle with its Standard ML interface).

Keywords: Craig interpolation · Display logic · Interactive theorem proving

1 Introduction

In calculi for logical entailment, *Craig interpolation* is the property that for any entailment $A \vdash B$ between formulae, there exists an *interpolant* formula F such that $A \vdash F$ and $F \vdash B$ are both entailments of the calculus, while F contains only those variables or nonlogical constants that are common to both A and B. It has long been known that there are close connections between interpolation and other central logical concerns (see e.g. [9]); indeed, one of Craig's original applications of interpolation was to give a new proof of Beth's Definability Theorem [6]. Significant applications of interpolation have been found recently in program verification in the inference of *loop invariants* [11], and in model checking [5].

Recently, Brotherston and Goré [3] gave a modular proof of interpolation for a class of propositional substructural logics, based on Belnap's *display logic* [2] (here we prefer the term *display calculi*). Roughly speaking, display calculi are two-sided sequent calculi equipped with a richer-than-usual notion of sequent structure and a principle by which sequents can be rearranged so as to "display" any chosen substructure as the *entire* left or right hand side (much like

J.E. Dawson—Supported by Australian Research Council Discovery Grant DP120101244.

J. Brotherston—Supported by an EPSRC Career Acceleration Fellowship.

N. Olivetti and A. Tiwari (Eds.): IJCAR 2016, LNAI 9706, pp. 452–468, 2016.
DOI: 10.1007/978-3-319-40229-1_31

rearranging a mathematical equation for a chosen variable). The main attraction of display calculi is Belnap's general cut-elimination result, which says that cut-elimination holds for any display calculus whose rules satisfy eight easily verifiable syntactic conditions. Cut-elimination is generally essential to the standard proof-theoretic approach to interpolation, which is to proceed by induction on cut-free derivations (see e.g. [4]). Despite the availability of a general cut-elimination result, however, there seem to have been no proofs on interpolation based on display calculi prior to [3], probably due to the inherent complexity of their sequent structure and display principles. Indeed, in line with this general expectation, Brotherston and Goré's proof is very technical, involving many case distinctions, and many intricate properties of substitutions. Moreover, due to space restrictions, most of the proofs are only sketched, leaving the potential for errors. Thus we believe it is vital to verify these intricate details using an interactive theorem prover, to give us greater confidence in their very general interpolation theorems.

In this paper, we describe the Isabelle/HOL formalisation of their results, discuss the difficulties in formalising their proofs and describe the differences between their proofs and ours. We also highlight the usefulness of a programmable user interface. Our Isabelle mechanisation, comprising over 8000 lines of Isabelle theory and proof code, can be found at [1].

2 Display Calculi for (Some) Substructural Logics

Here, we briefly describe display calculi, and recall those display calculi for which Brotherston and Goré proved interpolation [3].

We assume a fixed infinite set of propositional variables. *Formulae* F and *structures* X are then given by the following grammars, where P ranges over propositional variables:

$$F ::= P \mid \top \mid \bot \mid \neg F \mid F \mathbin{\&} F \mid F \vee F \mid F \to F \mid \top_a \mid \bot_a \mid F \mathbin{\&_a} F \mid F \vee_a F$$
$$X ::= F \mid \emptyset \mid \sharp X \mid X \,;\, X$$

Formula connectives with an "a" subscript stand for an *additive* version of that connective, while connectives without a subscript are construed as *multiplicative*. However, in the Isabelle formulation we do not duplicate the connectives in this way — rather, we identify the logical rules for which our various results apply. We write F, G etc. to range over formulae and W, X, Y, Z etc. to range over structures. If X and Y are structures then $X \vdash Y$ is a *consecution* (sequent).

The complete set of proof rules for our display calculi is given in Fig. 1. As usual, we begin by giving a set of *display postulates*, and taking the least equivalence closed under the postulates to be our notion of *display-equivalence*. We then have the usual *display theorem*, which says that for any structure occurrence Z in a consecution $X \vdash Y$, one has either $X \vdash Y \equiv_D Z \vdash W$ or $X \vdash Y \equiv_D W \vdash Z$ for some W, depending on whether Z occurs positively or negatively in $X \vdash Y$. Rearranging $X \vdash Y$ into $Z \vdash W$ or $W \vdash Z$ in this way is called *displaying* Z.

We remark that the display postulates "build in" commutativity of the structural semi-colon, so that we consider only calculi for commutative logics.

Brotherston and Goré [3] consider the additive rules, collectively, and each structural rule, individually, to be *optional* inclusions in their calculi. At present, our mechanisation assumes the presence of the unit rules ($\emptyset$W$_L$), ($\emptyset$W$_R$), ($\emptyset$C$_L$), ($\emptyset$C$_R$) and the associativity rule (α). Thus the smallest display calculus we consider gives (classical) multiplicative linear logic MLL. By adding the additive logical rules we obtain multiplicative-additive linear logic MALL, and by adding the full weakening rule (W) or the full contraction rule (C) we obtain affine or strict variants of these logics, respectively. Note that rules for weakening and contraction on the right can be derived using the display postulates from the corresponding left rules. Of course, if we add *both* weakening and contraction then we obtain standard classical propositional logic.

No matter which variant of these display calculi we consider, we have the standard *cut-elimination* result due to Belnap. Since we omit the cut rule from our presentation of the display calculi in Fig. 1, we state it here in the weaker form of *cut admissibility*:

Theorem 1 (cf. [3]). *If $X \vdash F$ and $F \vdash Y$ are both provable then so is $X \vdash Y$. Moreover, this property is not affected by the presence or otherwise of the additive logical rules (collectively), or of any of the structural rules.*

3 Interpolation for Display Calculi

In many traditional sequent calculi, it is fairly straightforward to decorate each rule with interpolants by building up the interpolant for the conclusion sequent from the interpolants for the premise sequents. In multi-conclusion intuitionistic logic it is not trivial to make it work since we may have to consider all possible splittings of the sequent. Our approach can be seen as a generalisation of this one, where "splittings" are replaced by "display-and-associativity rearrangements" since the sequent $X \vdash Y$ goes through many transformations while displaying some substructure Z. Brotherston and Goré therefore consider the following "LADI" property [3, Definition 3.4], where $\equiv_{AD}$ is the equivalence obtained by combining display equivalence with applications of associativity (α) if present in the calculus:

LADI: a rule with premises $\mathcal{C}_i$ and conclusion $\mathcal{C}$ satisfies the local AD display interpolation property (LADI) if for all premises $\mathcal{C}_i$, all sequents $\mathcal{C}_i'$ such that $\mathcal{C}_i' \equiv_{AD} \mathcal{C}_i$ satisfy the interpolation property, then all sequents $\mathcal{C}'$ such that $\mathcal{C}' \equiv_{AD} \mathcal{C}$ satisfy the interpolation property.

Although Brotherston and Goré [3] give the separate variants of the logical connectives $\top, \bot, \wedge$ and $\vee$ for the additive and multiplicative forms of the logical introduction rules, we just use one connective for each of $\top, \bot, \wedge$ and $\vee$. Although the additive and multiplicative forms are equivalent in the presence of

Display postulates:

$$X;Y \vdash Z \quad \rightleftarrows_D \quad X \vdash \sharp Y;Z \quad \rightleftarrows_D \quad Y;X \vdash Z$$

$$X \vdash Y;Z \quad \rightleftarrows_D \quad X;\sharp Y \vdash Z \quad \rightleftarrows_D \quad X \vdash Z;Y$$

$$X \vdash Y \quad \rightleftarrows_D \quad \sharp Y \vdash \sharp X \quad \rightleftarrows_D \quad \sharp\sharp X \vdash Y$$

Identity rules:

$$\frac{}{P \vdash P}\,(\mathrm{Id}) \qquad \frac{X' \vdash Y'}{X \vdash Y}\; X \vdash Y \equiv_D X' \vdash Y' \;(\equiv_D)$$

Multiplicative logical rules:

$$\frac{\emptyset \vdash X}{\top \vdash X}\,(\top\mathrm{L}) \qquad \frac{}{\emptyset \vdash \top}\,(\top\mathrm{R}) \qquad \frac{F;G \vdash X}{F \,\&\, G \vdash X}\,(\&\mathrm{L}) \qquad \frac{X \vdash F \quad Y \vdash G}{X\,;Y \vdash F \,\&\, G}\,(\&\mathrm{R})$$

$$\frac{}{\bot \vdash \emptyset}\,(\bot\mathrm{L}) \qquad \frac{X \vdash \emptyset}{X \vdash \bot}\,(\bot\mathrm{R}) \qquad \frac{F \vdash X \quad G \vdash Y}{F \vee G \vdash X\,;Y}\,(\vee\mathrm{L}) \qquad \frac{X \vdash F;G}{X \vdash F \vee G}\,(\vee\mathrm{R})$$

$$\frac{\sharp F \vdash X}{\neg F \vdash X}\,(\neg\mathrm{L}) \qquad \frac{X \vdash \sharp F}{X \vdash \neg F}\,(\neg\mathrm{R}) \qquad \frac{X \vdash F \quad G \vdash Y}{F \to G \vdash \sharp X\,;Y}\,(\to\mathrm{L}) \qquad \frac{X\,;F \vdash G}{X \vdash F \to G}\,(\to\mathrm{R})$$

Additive logical rules:

$$\frac{}{\bot_a \vdash X}\,(\bot_a\mathrm{L}) \qquad \frac{F_i \vdash X}{F_1 \,\&_a\, F_2 \vdash X}\; i \in \{1,2\}\,(\&_a\mathrm{L}) \qquad \frac{F \vdash X \quad G \vdash X}{F \vee_a G \vdash X}\,(\vee_a\mathrm{L})$$

$$\frac{}{X \vdash \top_a}\,(\top_a\mathrm{R}) \qquad \frac{X \vdash F \quad X \vdash G}{X \vdash F \,\&_a\, G}\,(\&_a\mathrm{R}) \qquad \frac{X \vdash F_i}{X \vdash F_1 \vee_a F_2}\; i \in \{1,2\}\,(\vee_a\mathrm{R})$$

Structural rules:

$$\frac{\emptyset;X \vdash Y}{X \vdash Y}\,(\emptyset\mathrm{C_L}) \qquad \frac{X \vdash Y;\emptyset}{X \vdash Y}\,(\emptyset\mathrm{C_R}) \qquad \frac{X \vdash Y}{\emptyset;X \vdash Y}\,(\emptyset\mathrm{W_L}) \qquad \frac{X \vdash Y}{X \vdash Y;\emptyset}\,(\emptyset\mathrm{W_R})$$

$$\frac{(W;X);Y \vdash Z}{W;(X;Y) \vdash Z}\,(\alpha) \qquad \frac{X \vdash Z}{X;Y \vdash Z}\,(\mathrm{W}) \qquad \frac{X;X \vdash Y}{X \vdash Y}\,(\mathrm{C})$$

Fig. 1. Display calculus proof rules. In the display rule ($\equiv_D$), the relation $\equiv_D$ is the least equivalence containing the relation $\rightleftarrows_D$ given by the display postulates. Note that all our formalisation, and all our results, omit the $\to$-connective and its rules.

contraction and weakening, [3] contains results which are relevant to the situation where not all structural rules are included. Thus they prove results for both the rules shown below, even though the second rule is much easier to deal with.

$$\frac{X \vdash A \quad Y \vdash B}{X, Y \vdash A \wedge B} \qquad \frac{X \vdash A \quad X \vdash B}{X \vdash A \wedge B}$$

We first considered the second (additive) rule shown; we subsequently developed a proof dealing with the first (multiplicative) rule directly.

4 The Isabelle Mechanisation

Our mechanisation builds on the work of Dawson and Goré [7] in formalising Display Logic. Some of our notation and choices of properties, lemmas, etc., are attributable to this. In particular, we use `Comma`, `Star` and `I` for ';', '#' and '∅'.

The work in [7] is a deep embedding of rules and of the variables in them, and we have followed that approach here (see [8] for our understanding of what this means, and for more details). That is, we define a language of formulae and structures, which contains explicit structure and formula variables, for which we define explicit substitution functions. We also define the rules as specific data structures (of which there is a small finite number, such as those in Fig. 1), and infinitely many substitution instances of these rules.

4.1 Formalising Display Logic in Isabelle

An actual derivation in a Display Calculus involves structures containing formulae which are composed of primitive propositions (which we typically represent by p, q, r). It uses rules which are *expressed* using structure and formula variables, typically X, Y, Z and A, B, C respectively, to represent structures and formulae made up from primitive propositions. We are using a "deep embedding" of variables, so our Isabelle formulation explicitly represents variables such as X, Y, Z and A, B, C, and defines substitution for them of given structures and formulae, which may themselves contain variables.

Thus, in our Isabelle formulation we use `PP` *name*, `FV` *name* and `SV` *name* to represent propositional, formula and structure variables respectively. The operator `Structform` "casts" a formula into a structure. We can then give recursive datatypes `formula` and `structr` for formulas and structures respectively (possibly parameterised by formula or structure variables), in the obvious way.

Thus the datatype `formula` for formulae has constructors `FV`, `PP` and the logical operators & $\vee$ $\neg$ $\top$ $\bot$ whereas the datatype `structr` for structures has constructors `SV`, `Structform` and the structure operators ; $\sharp$ $\emptyset$.

A rule (type) is represented as a list of premises and a conclusion, and a sequent by a Isabelle/HOL datatype:

```
types 'a psc ="'a list * 'a"
datatype sequent = Sequent structr structr
```

A sequent `(Sequent X Y)` can also be represented as `$X |- $Y`.

Since a "deep" embedding requires handling substitution explicitly, we defined functions to substitute for structure and formula variables, in structures, sequents and rules. In particular, we have an operator `rulefs`, where `rulefs` *rules* is the set of substitution instances of rules in the set *rules*. Also, when we refer to derivability using a set of rules, this allows inferences using substitution instances of these rules, and `derivableR` *rules sequents* means the set of sequents which can be derived from *sequents* using *rules*, instantiated.

```
derivableR  :: "rule set => sequent set => sequent set
rulefs :: "rule set => rule set"
```

We also use some general functions to describe derivability. An inference rule of type `'a psc` is a list `ps` of premises and a conclusion `c`. Then `derl rls` is the set of rules derivable from the rule set `rls` while `derrec rls prems` is the set of sequents derivable using rules `rls` from the set `prems` of premises. We defined these using Isabelle's package for inductively defined sets. A more detailed expository account of these, with many useful lemmas, is given in [10].

```
derl       :: "'a psc set => 'a psc set"
derrec     :: "'a psc set => 'a set => 'a set"
```

Note that these functions do not envisage instantiation of rules. Thus we have the following relationship between `derivableR` and `derrec`.

```
"derivableR ?rules == derrec (rulefs ?rules)"
```

The "deep embedding" approach to rules enables us to express properties of rules, such as that "no structure variable appears in both antecedent and succedent positions". Such lemmas apply to all display postulates satisfying conditions of this sort. We used this in [7] in showing that cut-admissibility applies whenever the structural rules were all of a particular form (as in Belnap's cut elimination theorem). In regards to interpolation, possible future work may include showing that interpolation results hold whenever rules are of a particular form, but our present work (except for some lemmas) does not do this.

The work in [7] is also a deep embedding of proofs (where we took proof objects and explicitly manipulated them) but we have *not* done that here.

4.2 Definitions Relating to Interpolation

We define the following sets of rules:

`dps`: is the set of six display postulates shown in Fig. 1 [3, Definition 2.4];
`aidps`: is `dps`, their inverses, and the associativity rule (i.e., 13 rules);
`ilrules`: is the unit-contraction and unit-weakening rules;
`rlscf`: is the set of all rules of the logic as shown in [3, Figures 1 and 3], plus `aidps`; that is, the rules of Fig. 1, except the additive logical rules (and we omit throughout this work the derivable rules for implication $\rightarrow$);

`rlscf_nw`: is as `rlscf`, but excluding the weakening rule.

Definition 1. *We define several predicates to do with interpolation:*

```
interp :: "rule set => sequent => formula => bool"
edi    :: "rule set => rule set => sequent => bool"
ldi    :: "rule set => rule set => sequent list * sequent => bool"
cldi   :: "rule set => rule set => sequent list * sequent => bool"
```

`interp rules` $(X \vdash Y)$ `intp`: *iff* `intp` *is an interpolant for* $X \vdash Y$*. Thus* $X \vdash$ `intp` *and* `intp` $\vdash Y$ *are derivable using* `rules` *and the (formula) variables in* `intp` *are among the formula variables of the structures* X *and* Y*;*

`edi lrules drules` $(X \vdash Y)$: *(*Extended Display Interpolation*) iff for all sequents* $X' \vdash Y'$ *from which* $X \vdash Y$ *is derivable using* `lrules`*, the sequent* $X' \vdash Y'$ *has an interpolant defined in terms of derivability using* `drules` *where* `lrules` *would typically be a set of display postulates;*

`ldi lrules drules` (ps, c): *(*Local Display Interpolation*) iff the rule* (ps, c) *preserves the property* `edi`*: that is, if, for all* $p \in ps$*,* `edi lrules drules` p *holds, then* `edi lrules drules` c *holds. Thus, if* `lrules` *is the set* AD *of rules (our* `aidps`*), and* `drules` *is the set of rules of the logic, then the LADI-property [3, Definition 3.4] as shown before on page 3 for rule* (ps, c) *is* `ldi aidps drules` (ps, c).

Note that none of these definitions involves a condition that $X \vdash Y$ be derivable. Of course, cut-admissibility would imply that if $X \vdash Y$ has an interpolant then $X \vdash Y$ is derivable, but we avoid proving or using cut-admissibility. Even so, in most cases we do not need such a condition. However we do need the derivability of $X \vdash Y$ in the case of a sequent $I \vdash Y, \#X$ produced by weakening and displaying the I structural connective. Thus we need a predicate with that condition:

Definition 2 (Conditional Local Display Interpolation)
`cldi lrules drules` (ps, c) *holds iff:*
if c *is is derivable using* `drules`*, then* `ldi lrules drules` (ps, c) *holds.*

We also need variants `interpn`, `edin`, `ldin` and `cldin`, of these predicates, where the derivation of interpolated sequents is from a given set of rules, rather than from given rules and their substitution instances. We use these in lemmas which involve rule sets which are not closed under substitution.

We mention here that many of our lemmas about these properties assume, although we do not specifically say so, that AD (rule set `aidps`) is used as to instantiate *lrules* in the above definitions, and that the derivation rules, *drules* in the above definitions, contain the AD rules.

Lemma 3.5 of [3] says that if all rules satisfy the local AD-interpolation property, then the calculus has the interpolation property. In fact the stronger result, Lemma 1(a) (below) is true, that LADI is preserved under derivation. But for the conditional local display interpolation property, a result analogous to the first-mentioned, only, of these results holds: see Lemma 1(b)

Lemma 1 (`ldi_derl, cldi_ex_interp`).

(a) if each rule from a set of rules satisfies the local display interpolation property, then so does a rule derived from them;
(b) if all the derivation rules satisfy the conditional local AD-interpolation property, then the calculus has the interpolation property.

4.3 Substitution of Congruent Occurrences

In [3, Definitions 3.6 and 3.7] the concept of congruent occurrences of some structure Z is used, with substitution for such congruent occurrences. Where two sequents $\mathcal{C}$ and $\mathcal{C}'$ are related by a display postulate, or sequence of them, a particular occurrence of Z in $\mathcal{C}$ will correspond to a particular occurrence of Z in $\mathcal{C}'$, according to the sequence of display postulates used to obtain $\mathcal{C}'$ from $\mathcal{C}$.

This concept looked rather difficult to define and express precisely and formally: note that in the notation in [3], $\mathcal{C}[Z/A] \equiv_{AD} \mathcal{C}'[Z/A]$, the meanings of $\mathcal{C}[Z/A]$ and $\mathcal{C}'[Z/A]$ depend on each other, because they refer to particular, corresponding, instances of A in $\mathcal{C}$ and $\mathcal{C}'$.

So we adopted the alternative approach, used successfully in [7]: rather than trying to define $\mathcal{C}'[Z/A]$ we would prove that there exists a sequent (call it $\mathcal{C}'_{Z/A}$) satisfying $\mathcal{C}[Z/A] \equiv_{AD} \mathcal{C}'_{Z/A}$ and satisfying the property that some occurrences of A in $\mathcal{C}'$ are replaced by Z in $\mathcal{C}'_{Z/A}$. This approach turned out to be sufficient for all the proofs discussed here.

In previous work [7], we defined and used a relation `seqrep`, defined as follows.

Definition 3 (`seqrep`).

```
seqrep : "bool => structr => structr => (sequent * sequent) set"
```

$(U, V) \in$ `seqrep` $b\ X\ Y$ means that some (or all or none) of the occurrences of X in U are replaced by Y, to give V; otherwise U and V are the same; the occurrences of X which are replaced by Y must all be in succedent or antecedent position according to whether b is true or false.

For this we write $U \ {}^{X}\!\!\leadsto^{Y} V$, where the appropriate value of b is understood.
Analogous to [3, Lemma 3.9] we proved the following result

Lemma 2 (`SF_some_sub`). *For formula F, structure Z, and rule set `rules`,*[1] *if*

(a) the conclusions of `rules` do not contain formulae; and
(b) the conclusion of a rule in `rules` does not contain more than one occurrence of any structure variable; and
(c) the `rules` obeys Belnap's C4 condition: when the conclusion and a premise of a rule both contain a structure variable, then both occurrences are in antecedent or both are in succedent positions; and
(d) `concl` is derivable from `prems` using `rules`; and

[1] In Lemma 3.9 [3] this set of rules is the set of AD rules.

(e) `concl` $^F\!\rightsquigarrow^Z$ `sconcl`

then there exists a list `sprems` *(of the same length as* `prems`*) such that*

(1) `sconcl` *is derivable from* `sprems` *using* `rules`*; and*
(2) $\texttt{prem}_n$ $^F\!\rightsquigarrow^Z$ $\texttt{sprem}_n$ *holds for corresponding members* $\texttt{prem}_n$ *of* `prems` *and* $\texttt{sprem}_n$ *of* `sprems`*.*

4.4 LADI Property for Unary Logical Rules

Proposition 3.10 of [3] covers the display postulates, the associativity rule, and the nullary or unary logical introduction rules.

The first case (($\equiv_D$), that is, any sequence of display postulates) of [3, Proposition 3.10] is covered by the following result (which holds independently of the choice of set of derivation rules).

Lemma 3 (bi_lrule_ldi_lem). *Let rule ρ be a substitution instance of a rule in AD. Then ρ has the LADI property.*

With the next lemma we can handle the rules (Id), ($\top R$) and ($\bot L$).

Lemma 4 (non_bin_lem_gen). *Assume the derivation rules include ($\neg L$) and ($\neg R$). Let ρ be a substitution instance of a rule in AD whose premise does not contain any ';'. If the premise of ρ has an interpolant then so does its conclusion.*

Since the conclusions of the three nullary rules (Id), ($\top R$) and ($\bot L$) clearly themselves have interpolants, Lemma 4 shows they satisfy the extended display interpolation property, and so the rules have the LADI property.

Proposition 1. *The rules (Id), ($\top R$) and ($\bot L$) satisfy the LADI property.*

The remaining cases of Proposition 3.10 are the logical introduction rules with a single premise. For these we use the four lemmas (of which one is shown)

Lemma 5 (sdA1). *If the rule shown below left is a logical introduction rule, and the condition in the middle holds, then the rule shown below right is derivable (i.e., using AD and the logical introduction rules)*

$$\frac{Y' \vdash U}{Y \vdash U} \qquad W \ {}^{Y}\!\rightsquigarrow^{Y'} \ W' \qquad \frac{W' \vdash Z}{W \vdash Z}$$

Then from these lemmas we get

Lemma 6 (seqrep_interpA). *For the logical introduction rule shown below left, if formula variables in Y' also appear in Y, the condition on the right holds, and I is an interpolant for $W' \vdash Z'$, then I is also an interpolant for $W \vdash Z$:*

$$\frac{Y' \vdash U}{Y \vdash U} \qquad W \vdash Z \ {}^{Y}\!\rightsquigarrow^{Y'} \ W' \vdash Z' \ \textit{(in antecedent positions)}$$

Finally we get the following result which gives Proposition 3.10 for single premise logical introduction rules (additive or multiplicative).

Proposition 2 (`logA_ldi`). *If F is a formula, and the rule $(F \vdash)$ below is a logical introduction rule, and the formula variables in Y are also in F, then $(F \vdash)$ satisfies the LADI property:*

$$\frac{Y \vdash U}{F \vdash U}(F \vdash)$$

This last result requires Lemma 2. We have analogous results for a logical introduction rule for a formula on the right.

Remark 1. At this stage, we have a general method for proving local display interpolation for a given rule ρ, with premises ps_ρ and conclusion c_ρ: identify a relation rel such that

(a) $(ps_\rho, c_\rho) \in rel$
(b) whenever $c \equiv_{AD} c_\rho$, we can find a list ps (often got from sequents in ps_ρ using the same sequence of display postulates which get c from c_ρ) such that $(ps, c) \in rel$, and $p \equiv_{AD} p_\rho$ for each $p \in ps$ and corresponding $p_\rho \in ps_\rho$
(c) whenever $(ps, c) \in rel$, c is derivable from ps (not used except to prove (d))
(d) whenever $(ps, c) \in rel$, and each $p \in ps$ has an interpolant, then c has an interpolant (proof of this will normally use (c)).

4.5 LADI Property for (Unit) Contraction

This is relatively easy for the unit-contraction rule: the relation rel is given by: $(p, c) \in rel$ if p is obtained from c by deleting, somewhere in c, some $\#^n\emptyset$, and we get (b) using roughly the same sequence of display postulates.

Lemma 7 (`ex_box_uc`). *If sequent Cd is obtained from C by deleting one occurrence of some $\#^n\emptyset$, and if $Cd' \rightarrow^*_{AD} Cd$, then there exists C', such that $C' \rightarrow^*_{AD} C$, and Cd' is obtained from C' by deleting one occurrence of $\#^n\emptyset$.*

The proof of this required a good deal of programming repetitive use of complex tactics similar to (but less complex than) those described in Sect. 4.6.

The following lemma gives (c) of the general proof method above.

Lemma 8 (`delI_der`). *If $(p, c) \in rel$ (defined above), and if the derivation rules include AD and the unit contraction rules, then c is derivable from p*

Proposition 3 (`ldi_ila, ldi_ils`). *The unit contraction rules satisfy LADI.*

For the case of contraction, we defined a relation `mseqctr`: $(C, C') \in$ `mseqctr` means that C' is obtained from C, by contraction of substructures (X, X) to X. Contractions may occur (of different substructures) in several places or none.

Lemma 9 (`ex_box_ctr`). *If sequent Cd is obtained from C by contraction(s) of substructure(s), and if $Cd' \rightarrow^*_{AD} Cd$, then there exists C', such that $C' \rightarrow^*_{AD} C$, and Cd' is obtained from C' by substructure contraction(s).*

Proof. The proof of `ex_box_ctr` is a little more complex than that for unit-contraction, because (for example) when $X;Y \vdash Z \equiv_{AD} X \vdash Z;\#Y$, and $X;Y \vdash Z$ is obtained by contracting $(X;Y);(X;Y) \vdash Z$, we need to show $(X;Y);(X;Y) \vdash Z \equiv_{AD} X;X \vdash Z;\#(Y;Y)$

Lemma 10 (`ctr_der`). *If $(p,c) \in$ `mseqctr` (defined above), and if the derivation rules include AD and the left contraction rule, then c is derivable from p.*

Proposition 4 (`ldi_cA`). *The left contraction rule satisfies the LADI property.*

4.6 Deletion Lemma ([3], Lemma 4.2)

For weakening or unit-weakening, it is more difficult: a sequence of display postulates applied to the conclusion $X;\emptyset \vdash Y$ may give $\emptyset \vdash Y;\#X$, so the same or similar sequence cannot be applied to the premise $X \vdash Y$.

For this situation we need Lemma 4.2 (Deletion Lemma) of [3]: this result says that for F a formula sub-structure occurrence in C, or $F = \emptyset$, and $C \to^*_{AD} C'$, then (in the usual case) $C \setminus F \to^*_{AD} C' \setminus F$, where $C \setminus F$ and $C' \setminus F$ mean deleting only particular occurrence(s) of F in C, and deleting the *congruent* (corresponding) occurrence(s) of F in C', where congruence is determined by the course of the derivation of C' from C.

We did not define congruent occurrences in this sense: see the general discussion of this issue in Sect. 4.3. It seemed easier to define and use the relation `seqdel`:

Definition 4 (`seqdel`). *Define $(C,C') \in$ `seqdel` Fs to mean that C' is obtained from C by deleting one occurrence in C of a structure in the set Fs.*

Then we proved the following result about deletion of a formula:

Lemma 11 (`deletion`). *Let F be a formula or $F = \emptyset$. If sequent Cd is obtained from C by deleting an occurrence of some $\#^i F$, and if $C \to^*_{AD} C'$, then either*

(a) there exists Cd′, such that $Cd \to^*_{AD} Cd'$, and Cd′ is obtained from C′ by deleting an occurrence of some $\#^j F$, or
(b) C′ is of the form $\#^n F \vdash \#^m(Z_1;Z_2)$ or $\#^m(Z_1;Z_2) \vdash \#^n F$, where $Cd \to^*_{AD} (Z_1 \vdash \#Z_2)$, or $Cd \to^*_{AD} (\#Z_1 \vdash Z_2)$

Proof. Thus the premise is that Cd is got from C by deleting instance(s) of the substructure formula F, possibly with some # symbols. The main clause of the result says that there exists Cd' (this corresponds to $C' \setminus F$ in [3]) which is got from Cd by deleting instance(s) of $\#^n F$ (for some n), but there is also an exceptional case where $\#^n F$ is alone on one side of the sequent.

The proof of this result required considerable ML programming of proof tactics.

When we get cases as to the last rule used in the derivation $C \to^*_{AD} C'$, this gives 13 possibilities. For each rule there are two main cases for the shape of the sequent after the preceding rule applications: in the first, $\#^n F$ appears in

$\#^n F, Z$ or $Z, \#^n F$ and so could be deleted (F is "delible"), and in the second, the relevant occurrence of $\#^n F$ is the whole of one side of the sequent.

Then where, in the case of the associativity rule for example, the sequent which is $(X;Y);Z \vdash W$ (instantiated) has F delible, $\#^n F$ may be equal to X, Y or Z, or may be delible from X, Y, Z or W. Without the possibility of programming a tactic in Standard ML to deal with all these possibilities, each of these seven cases, and a similar (less numerous) set of cases for each of the other 12 rules, would require its own separate proof.

For the second case, where $\#^n F$ is equal to one side of the sequent (W in the above example), a variety of tactics is required: for those display rules which move the comma from one side to the other one function works for all, but the other cases have to be proved individually. $\dashv$

We then proved this result for $F = \emptyset$ instead of a formula, to give a theorem `deletion_I`; the changes required in the proof were trivial.

4.7 LADI Property for (Unit) Weakening Rules

To handle weakening in a similar way, we considered two separate rules, one to weaken with instances of $\#^n\emptyset$ and one to change any instance of $\emptyset$ to any formula. Thus, where $Y_\emptyset$ means a structure like Y but with every formula or structure variable in it changed to $\emptyset$, a weakening is produced as shown:

$$X \vdash Z \Longrightarrow X, Y_\emptyset \vdash Z \Longrightarrow X, Y \vdash Z$$

We first consider the second of these, replacing any instance of $\emptyset$ with a structural atom, that is, a formula or a structure variable which are atomic so far as the structure language is concerned.

We use the relation `seqrepI str_atoms`: $(c, p) \in$ `seqrepI str_atoms` means that some occurrences of $\emptyset$ in p are changed to structural atoms in c.

Lemma 12 (`ex_box_repI_atoms`). *If sequent C is obtained from Cd by replacing $\emptyset$ by structural atoms, and if $C' \rightarrow^*_{AD} C$, then there exists Cd', such that $Cd' \rightarrow^*_{AD} Cd$, and C' is obtained from Cd' by replacing $\emptyset$ by structural atoms.*

For this relation, property (b) was quite easy to prove, since exactly the same sequence of AD-rules can be used.

We proved that there are derived rules permitting replacing instances of $\emptyset$ by anything, and this gave us that such rules, where the replacement structure is a formula or structure variable, have the the local display interpolation property.

Lemma 13 (`seqrepI_der`). *If the derivation rules include weakening and unit-contraction, and $(c, p) \in$ `seqrepI` Fs, i.e. some occurrences of $\emptyset$ in p are replaced by anything to give c, then c is derivable from p.*

The next lemma gives the LADI property, not for a rule of the system, but for inferences $([p], c)$ where $(c, p) \in$ `seqrepI str_atoms`.

Proposition 5 (`ldi_repI_atoms`). *Where* $(c, p) \in$ *seqrepI str_atoms, i.e. some occurrences of* $\emptyset$ *in* p *are replaced by structural atoms to give* c, $([p], c)$ *has the LADI property.*

Next we consider the structural rules allowing insertion of $\#^n\emptyset$.

We use the variant of the theorem `deletion` (see Sect. 4.6) which applies to deletion of $\emptyset$ rather than of a formula.

Then we show that inserting occurrences of anything preserves derivability.

Lemma 14 (`seqwk_der`). *If the derivation rules include weakening, and* $(c, p) \in$ *seqdel Fs, i.e.,* c *is obtained from* p *by weakening, then* c *is derivable from* p.

Then we need the result that such rules satisfy the local display interpolation property. In this case, though, where a sequent containing $\emptyset$ is rearranged by the display postulates such that the $\emptyset$ is alone on one side (such as where $X \vdash Y; \emptyset$ is rearranged to $X; \#Y \vdash \emptyset$), then to prove the LADI property requires using the derivability of $X \vdash Y$ rather than the fact that $X \vdash Y$ satisfies LADI. Thus we can prove only the conditional local display interpolation property.

Proposition 6 (`ldi_wkI_alt`). *If the derivation rules include unit weakening, unit contraction, and the left and right introduction rules for* $\top$ *and* $\bot$, *then a rule for* $\#^n\emptyset$*-weakening (i.e., inserting* $\#^n\emptyset$*) satisfies the conditional LADI property.*

At this point we also proved that the additive forms of the binary logical introduction rules satisfy the LADI property. The proofs are conceptually similar to those for the unary logical introduction rules — but more complex where single structures/sequents become lists of these entities. For reasons of space, and because we proceed to deal with the more difficult multiplicative forms of the binary introduction rules, we omit details.

Now we can give the result for a system which contains weakening, contraction, and the binary rules in either additive or multiplicative form. To get this we define a set of rules called `ldi_rules_a`, which contains the additive binary logical rules, and does not contain the weakening rules but does contain the relations of Propositions 6 and 5. We have that all of its rules satisfy the conditional LADI property, so the system has interpolants. We show this gives a deductive system equivalent to the given set of rules `rlscf`, which system therefore also has interpolants. Details are similar to the derivation of Theorem 3.

Theorem 2 (`rlscf_interp`). *The system of substitutable rules* `rlscf` *satisfies display interpolation*

4.8 LADI Property for Binary Multiplicative Logical Rules

Here, we just consider the multiplicative version of these rules; the case of their additive analogues is similar, but easier.

We deal with these rules in two stages — firstly, weakening in occurrences of $\#^n\emptyset$, then changing any occurrence of $\emptyset$ to any structural atom, as shown below.

$$\dfrac{\dfrac{X \vdash A}{X, Y_\emptyset \vdash A}\ wk^*_\emptyset \qquad \dfrac{Y \vdash B}{X_\emptyset, Y \vdash B}\ wk^*_\emptyset}{X, Y \vdash A \wedge B}$$

Here $X_\emptyset$ and $Y_\emptyset$ mean the structures X and Y, with each structural atom (formula or uninterpreted structure variable) replaced by $\emptyset$. The first stage, the inferences labelled $wk^*_\emptyset$ above, are obtained by repeatedly weakening by occurrences of $\#^n\emptyset$ in some substructure. The second stage (for which we define the relation **ands_rep**), consists of changing the $X_\emptyset$ of one premise, and the $Y_\emptyset$ of the other premise, to X and Y respectively. For the second of these stages, then, when any sequence of display postulates is applied to $X, Y \vdash A \wedge B$, the same sequence can be applied to the two premises, $X, Y_\emptyset \vdash A$ and $X_\emptyset, Y \vdash B$. This simplifies the proof of local display interpolation for these rules.

For the first stage we proceed as described for Sect. 4.7, using Proposition 6 to show that the inferences labelled $wk^*_\emptyset$ satisfy the conditional LADI property.

The second stage consists of the rule shown as **ands_rep** in the diagram. Considering the four points in Remark 1, since any display postulate applied to the conclusion can be applied to the premises, we need to define a suitable relation between conclusion and premises which is preserved by applying any display postulate to them. For this we define a relation `lseqrepm` between sequents, analogous to `seqrep` (Sect. 4.3, Definition 3) and `lseqrep`:

```
lseqrepm       :: "(structr * structr list) set =>
    bool => [structr, structr list] => (sequent * sequent list) set"
```

Definition 5 (`lseqrepm, repnI_atoms`)

(a) $(U, Us) \in$ ***lseqrepm*** *orel b Y Ys means that there is one occurrence of Y in U which is replaced by the nth member of Ys in the nth member of Us; this occurrence is at an antecedent or succedent position, according to whether b is* ***True*** *or* ***False****. However elsewhere in U, each structural atom A in U is replaced by the nth member of As in the nth member of Us, where* $(A, As) \in orel$;

(b) $(A, As) \in$ ***repnI_atoms*** *iff one of the As is A and the rest of the As are* $\emptyset$.

We use `lseqrepm` only with $orel =$ **repnI_atoms**. For example, for the ($\wedge$R) rule, we use `lseqrepm repnI_atoms True` $(A \wedge B)$ $[A, B]$. as the relation *rel* of the four points in Remark 1. We get the following lemmas.

Lemma 15 (**repm_some1sub**). *Whenever Y is a formula, and* $(U, Us) \in$ ***lseqrepm*** *orel b Y Ys, and U is manipulated by a display postulate (or sequence of them) to give V, then the Us can be manipulated by the same display postulate(s) to give Vs (respectively), where* $(V, Vs) \in rel$.

The following lemmas refer to derivation in the system containing the **ands_rep** rule (not the regular ($\wedge$R) rule), and also unit-weakening and unit-contraction.

Lemma 16 (**ands_mix_gen**). *Whenever* $(V, Vs) \in rel$*, then V can be derived from the Vs.*

This lemma relies on taking the conjunction or disjunction of interpolants of premises. So the next two results require a deductive system containing the `ands_rep` and `ora_rep` rules, and also the (∨R) and (∧L) rules.

Lemma 17 (`lseqrepm_interp_andT`). *Whenever* $(V, Vs) \in rel$*, then we can construct an interpolant for* V *from interpolants for the* Vs.

Proposition 7 (`ldin_ands_rep`). *The rule* `ands_rep` *satisfies LADI.*

We recall from Sect. 4.2 the set `rlscf_nw` of substitutable rules, the rules of Fig. 1, except the additive logical rules and weakening. From this set we define a set of rules called `ldi_add` by omitting from `rlscf_nw` the binary logical rules $(\vee L)$ and $(\wedge R)$, but including the rule `ands_rep` (see diagram above) and a corresponding rule `ora_rep`. Note that these latter rules, and therefore `ldi_add`, are not closed under substitution. (Note that, as mentioned earlier, our formalisation had not included the connective →, or any rules for it).

Lemma 18 (`ldi_add_equiv`). *The calculi* `ldi_add` *and* `rlscf_nw` *(defined above) are deductively equivalent.*

Theorem 3 (`ldi_add_interp, rlscf_nw_interp`)

(a) the system `ldi_add` *satisfies display interpolation*
(b) the system of substitutable rules `rlscf_nw` *satisfies display interpolation*

Proof. We have all rules in `ldi_add` satisfying at least the conditional local display interpolation property (`ldi_add_cldin`). By `cldin_ex_interp`, this gives us that the system `ldi_add` satisfies display interpolation `ldi_add_interp`, and so therefore does the equivalent system of substitutable rules `rlscf_nw`.

5 Discussion and Further Work

Our formalisation does not include implication connectives and rules since we assume that implication is a defined connective via the involutive negation. Thus we have only captured substructural "classical" logics.

The presence of the involutive negation is not necessary. If one has intuitionistic-style logics then the display postulates typically capture residuation and everything goes through in the same way. But then we have to re-work the formalisation to include explicit rules for implication as it is no longer a defined connective.

Commutativity of conjunction is also assumed because it makes life easier (e.g. in reducing the number of connectives and rules). One could imagine including commutativity as an optional structural rule, but this would then cause implication to split into a left and right implication (slash). However, we are not sure whether one could prove the LADI property for this rule directly, or whether one would have to build it into display-equivalence as we currently do for associativity.

Another avenue to explore to be more explicit about when derivability is needed for the proof of interpolation. This is an interesting point concerning structural rules.

As the proof-sketch of Lemma 9 indicates, it might be possible to replace "$\rightarrow^*_{AD}$" by an $\equiv_{AD}$, where it occurs in the paper.

6 Conclusions

As we have seen, interpolation proofs for display calculi are very technical, due to the inherent complexity of mixing the display principle with the definition of interpolation. As a consequence of this, the proof of Brotherston and Goré [3] is very technical, and most of the proofs were only sketched, leaving the potential for errors. Consequently it is valuable to have confirmed the correctness of their result using a mechanised theorem prover. And while the detailed proofs have only been sketched in this paper too, the files containing the Isabelle proofs enable the proofs to be examined to any desired level of detail.

This work has illustrated some interesting issues in the use of a mechanised prover. We have indicated where we found it necessary to follow a (slightly) different line of proof. This arose where their proof involved looking at corresponding parts of two display-equivalent sequents — an intuitively clear notion, but one which seemed so difficult to formalise that a different approach seemed easier. The two-stage approach used in Sect. 4.8 is also somewhat different from the proof in [3].

This work illustrated the enormous value of having a prover with a programmable user interface. Isabelle is written in Standard ML, and (for its older versions) the user interacts with it using that language. This proved invaluable in the work described in Sects. 4.5 and 4.6, where we were able to code up sequences of attempted proof steps which handled enormous numbers of cases efficiently.

Acknowledgements. We are grateful for the many comments from the IJCAR reviewers, which have improved the paper considerably.

References

1. Isabelle/HOL mechanisation of our interpolation proofs. http://users.cecs.anu.edu.au/jeremy/isabelle/2005/interp/
2. Belnap, N.D.: Display logic. J. Philos. Logic **11**, 375–417 (1982)
3. Brotherston, J., Goré, R.: Craig interpolation in displayable logics. In: Brünnler, K., Metcalfe, G. (eds.) TABLEAUX 2011. LNCS, vol. 6793, pp. 88–103. Springer, Heidelberg (2011)
4. Buss, S.R.: Introduction to proof theory. In: Handbook of Proof Theory, chap. I. Elsevier Science (1998)
5. Caniart, N.: Merit: an interpolating model-checker. In: Touili, T., Cook, B., Jackson, P. (eds.) CAV 2010. LNCS, vol. 6174, pp. 162–166. Springer, Heidelberg (2010)

6. Craig, W.: Three uses of the Herbrand-Gentzen theorem in relating model theory and proof theory. J. Symbolic Logic **22**(3), 269–285 (1957)
7. Dawson, J.E., Goré, R.P.: Formalised cut admissibility for display logic. In: Carreño, V.A., Muñoz, C.A., Tahar, S. (eds.) TPHOLs 2002. LNCS, vol. 2410, pp. 131–147. Springer, Heidelberg (2002)
8. Dawson, J.E., Goré, R.: Generic methods for formalising sequent calculi applied to provability logic. In: Fermüller, C.G., Voronkov, A. (eds.) LPAR-17. LNCS, vol. 6397, pp. 263–277. Springer, Heidelberg (2010)
9. Feferman, S.: Harmonious logic: craigs interpolation theorem and its descendants. Synthese **164**, 341–357 (2008)
10. Goré, R.P.: Machine checking proof theory: an application of logic to logic. In: Ramanujam, R., Sarukkai, S. (eds.) ICLA 2009. LNCS (LNAI), vol. 5378, pp. 23–35. Springer, Heidelberg (2009)
11. McMillan, K.L.: Quantified invariant generation using an interpolating saturation prover. In: Ramakrishnan, C.R., Rehof, J. (eds.) TACAS 2008. LNCS, vol. 4963, pp. 413–427. Springer, Heidelberg (2008)

Intuitionistic Layered Graph Logic

Simon Docherty(✉) and David Pym

University College London, London, UK
{simon.docherty14,d.pym}@ucl.ac.uk

Abstract. Models of complex systems are widely used in the physical and social sciences, and the concept of layering, typically building upon graph-theoretic structure, is a common feature. We describe an intuitionistic substructural logic that gives an account of layering. As in bunched systems, the logic includes the usual intuitionistic connectives, together with a non-commutative, non-associative conjunction (used to capture layering) and its associated implications. We give soundness and completeness theorems for labelled tableaux and Hilbert-type systems with respect to a Kripke semantics on graphs. To demonstrate the utility of the logic, we show how to represent a range of systems and security examples, illuminating the relationship between services/policies and the infrastructures/architectures to which they are applied.

1 Introduction

Complex systems can be defined as the field of science that studies, on the one hand, how it is that the behaviour of a system, be it natural or synthetic, derives from the behaviours of its constituent parts and, on the other, how the system interacts with its environment. A commonly employed and highly effective concept that helps to manage the difficulty in conceptualizing and reasoning about complex systems is that of *layering*: the system is considered to consist of a collection of interconnected layers each of which has a distinct, identifiable role in the system's operations. Layers can be informational or physical and both kinds may be present in a specific system. In [3,13], multiple layers are given by multiple relations over a single set of nodes.

We employ three illustrative examples. First, a transport network that uses buses to move people. It has an infrastructure layer (i.e., roads, together with their markings, traffic signals, etc., and buses running to a timetable), and a social layer (i.e., the groupings and movements of people enabled by the bus services). Second, a simple example of the relationship between a security policy and its underlying system architecture. Finally, we consider the security architecture of an organization that operates high- and low-security internal systems as well as providing access to its systems from external mobile devices. These examples illustrate the interplay between services/policies and the architectures/infrastructures to which they are intended to apply.

We give a graph-theoretic definition of layering and provide an associated logic for reasoning about layers. There is very little work in the literature on

N. Olivetti and A. Tiwari (Eds.): IJCAR 2016, LNAI 9706, pp. 469–486, 2016.
DOI: 10.1007/978-3-319-40229-1_32

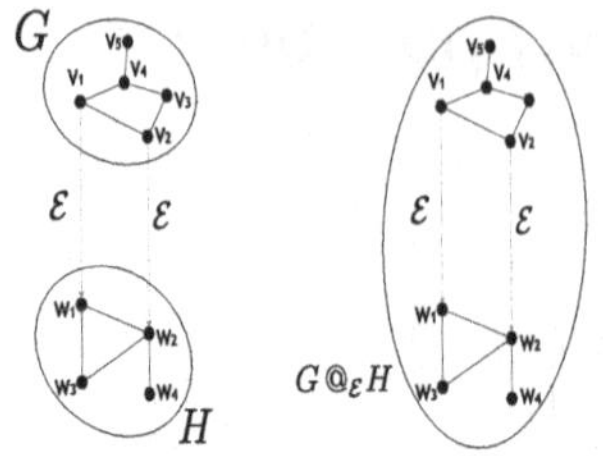

Fig. 1. A graph for which $G @_{\mathcal{E}} H$ is defined, and the resulting composition

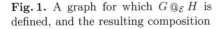

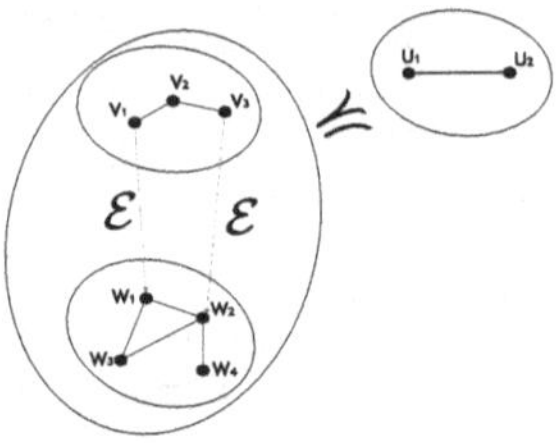

Fig. 2. Preordered scaffold

layering in graphs. Notable exceptions are [10,18,19]. Layered graphs are an instance of a general algebraic semantics for the logic. Our approach stands in contrast to our previous work in this area [6,7] in that the additive component of the bunched logic [12,17] we employ is intuitionistic, with the consequence that we are able to obtain a tableaux system for the logic together with a completeness theorem for the layered graph semantics. In Sect. 2, we introduce layered graph semantics and ILGL, the associated intuitionistic layered graph logic. In Sect. 3, we establish its basic metatheory — the soundness and completeness of ILGL's tableaux system with respect to layered graph semantics — and, in Sect. 4, we give an algebraic semantics and a (sound and complete) Hilbert-type proof system for ILGL. In Sect. 5, we sketch a modal extension of ILGL that is convenient for practical modelling, explaining its theoretical status and developing the three examples mentioned above.

2 Intuitionistic Layered Graph Logic

Layered Graph Semantics. We begin with a formal, graph-theoretic account of the notion of layering that, we claim, captures the concept as used in complex systems. In this notion, two layers in a directed graph are connected by a specified set of edges, each element of which starts in the upper layer and ends in the lower layer.

Given a directed graph, $\mathcal{G}$, we refer to its *vertex set* and its *edge set* by $V(\mathcal{G})$ and $E(\mathcal{G})$ respectively, while its set of subgraphs is denoted $Sg(\mathcal{G})$ with $H \subseteq \mathcal{G}$ iff $H \in Sg(\mathcal{G})$. For a distinguished edge set $\mathcal{E} \subseteq E(\mathcal{G})$, the reachability relation $\rightsquigarrow_{\mathcal{E}}$ on subgraphs of $\mathcal{G}$ is $H \rightsquigarrow_{\mathcal{E}} K$ iff a vertex of K can be reached from a vertex of H by an $\mathcal{E}$-edge.

We then have a composition $@_{\mathcal{E}}$ on subgraphs where $G @_{\mathcal{E}} H \downarrow$ iff $V(G) \cap V(H) = \emptyset, G \rightsquigarrow_{\mathcal{E}} H$ and $H \not\rightsquigarrow_{\mathcal{E}} G$ (where $\downarrow$ denotes definedness) with output given by the graph union of the two subgraphs and the $\mathcal{E}$-edges between them. For a graph G, we say it is *layered* (with respect to $\mathcal{E}$) if there exist H, K such that $H @_{\mathcal{E}} K \downarrow$ and $G = H @_{\mathcal{E}} K$ (see Fig. 1). Layering is evidently neither commutative nor associative.

Within a given ambient graph, $\mathcal{G}$, we can identify a specific form of layered structure, called a *preordered scaffold*, that will facilitate our definition of a model of intuitionistic layered graph logic. Properties of graphs that are inherited by their subgraphs are naturally captured in an intuitionistic logic. This idea is generalized by the structure carried by a preordered scaffold. To set this up, we begin by defining an *admissible subgraph set* is a subset $X \subseteq Sg(\mathcal{G})$ such that, for all $G, H \in Sg(\mathcal{G})$, if $G @_{\mathcal{E}} H\downarrow$, then $G, H \in X$ iff $G @_{\mathcal{E}} H \in X$. Then, a *preordered scaffold* (see Fig. 2) is a structure $\mathcal{X} = (\mathcal{G}, \mathcal{E}, X, \preccurlyeq)$ such that $\mathcal{G}$ is a graph, $\mathcal{E} \subseteq E(\mathcal{G})$, X an admissible subgraph set, $\preccurlyeq$ a preorder on X. Layers are present if $G @_{\mathcal{E}} H\downarrow$ for at least one pair $G, H \in X$.

Note that the scaffold is preordered and we choose a subset of the subgraph set. There are several reasons for these choices. From a modelling perspective, we can look closely at the precise layering structure of the graph that is of interest. In particular, we can avoid degenerate cases of layering. (Note that this is a more general definition of scaffold than that taken in [6,7], where the structure was less tightly defined.) Technical considerations also come into play. When we restrict to interpreting ILGL on the full subgraph set, it is impossible to perform any composition of models without the worlds (states) proliferating wildly. A similar issue arises during the construction of countermodels from the tableaux system of Sect. 3, a procedure that is impossible when we are forced to take the full subgraph set as the set of worlds.

Having established the basic semantic structures that are required, we can now set up ILGL. Let Prop be a set of atomic propositions, ranged over by p. The set Form of all propositional formulae is generated by the following grammar:

$$\phi ::= \mathrm{p} \mid \top \mid \bot \mid \phi \wedge \phi \mid \phi \vee \phi \mid \phi \rightarrow \phi \mid \phi \blacktriangleright \phi \mid \phi \mathrel{-\!\!\blacktriangleright} \phi \mid \phi \mathrel{\blacktriangleright\!\!-} \phi$$

The familiar connectives will be interpreted intuitionistically. The non-commutative, non-associative conjunction, $\blacktriangleright$, which will be used to capture layering, is interpreted intuitionistically, as in BI [12,17], and has associated right ($\mathrel{\blacktriangleright\!\!-}$) and left ($\mathrel{-\!\!\blacktriangleright}$) implications. We define intuitionistic negation in terms of the connectives: $\neg\phi ::= \phi \rightarrow \bot$.

Definition 1 (Layered Graph Model). *A layered graph model, $\mathcal{M}$, of* ILGL *is a pair $(\mathcal{X}, \mathcal{V})$, where $\mathcal{X}$ is a preordered scaffold and $\mathcal{V}$:* Prop $\rightarrow \wp(X)$ *is a* persistent *valuation; that is, $G \preccurlyeq H$ and $G \in \mathcal{V}(\mathrm{p})$ implies $H \in \mathcal{V}(\mathrm{p})$.* □

Satisfaction in layered graph models is then defined in a familiar way.

Definition 2 (Satisfaction in Layered Graph Models). *Given a layered graph model $\mathcal{M} = (\mathcal{X}, \mathcal{V})$, we generate the satisfaction relation $\models_{\mathcal{M}} \subseteq X \times$ Form as follows:*

$$G \models_{\mathcal{M}} \top \text{ always} \qquad G \models_{\mathcal{M}} \bot \text{ never} \qquad G \models_{\mathcal{M}} \mathrm{p} \text{ iff } G \in \mathcal{V}(\mathrm{p})$$
$$G \models_{\mathcal{M}} \varphi \wedge \psi \text{ iff } G \models_{\mathcal{M}} \varphi \text{ and } G \models_{\mathcal{M}} \psi \qquad G \models_{\mathcal{M}} \varphi \vee \psi \text{ iff } G \models_{\mathcal{M}} \varphi \text{ or } G \models_{\mathcal{M}} \psi$$
$$G \models_{\mathcal{M}} \varphi \rightarrow \psi \text{ iff, for all } G' \text{ such that } G \preccurlyeq G', G' \models_{\mathcal{M}} \varphi \text{ implies } G' \models_{\mathcal{M}} \psi$$

$$G \models_{\mathcal{M}} \varphi \blacktriangleright \psi \text{ iff there exist } H, K \text{ such that } H @_{\mathcal{E}} K\downarrow, H @_{\mathcal{E}} K \preccurlyeq G, \text{and } H \models_{\mathcal{M}} \varphi \text{ and } K \models_{\mathcal{M}} \psi$$
$$G \models_{\mathcal{M}} \varphi \mathrel{-\!\!\blacktriangleright} \psi \text{ iff for all } G \preccurlyeq H \text{ and all } K \text{ such that } H @_{\mathcal{E}} K\downarrow, K \models_{\mathcal{M}} \varphi \text{ implies } H @_{\mathcal{E}} K \models_{\mathcal{M}} \psi$$
$$G \models_{\mathcal{M}} \varphi \mathrel{\blacktriangleright\!\!-} \psi \text{ iff for all } G \preccurlyeq H \text{ and all } K \text{ such that } K @_{\mathcal{E}} H\downarrow, K \models_{\mathcal{M}} \varphi \text{ implies } K @_{\mathcal{E}} H \models_{\mathcal{M}} \psi \;\square$$

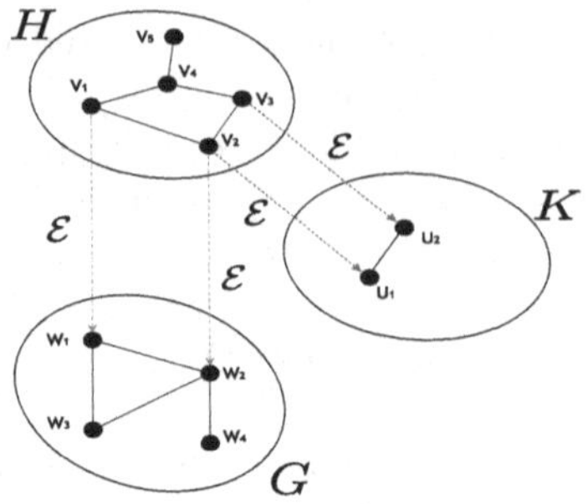

Fig. 3. The $\mathcal{E}$-reachability preorder

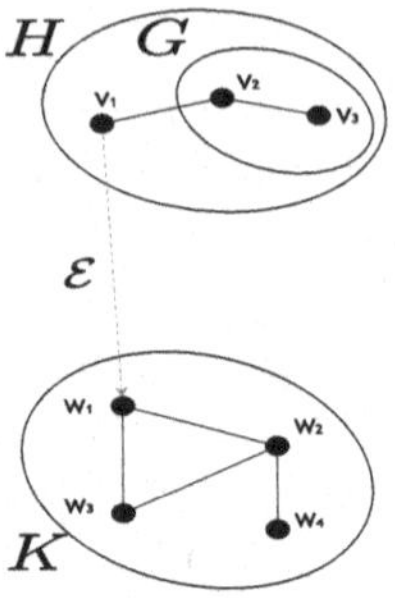

Fig. 4. The subgraph order

Definition 3 (Validity). *A formula ϕ is valid in a layered graph model $\mathcal{M}$ ($\models_{\mathcal{M}} \phi$) iff, for all $G \in X$, $G \models_{\mathcal{M}} \phi$. A formula ϕ is valid ($\models \phi$) iff, for all layered graph models $\mathcal{M}$, $\models_{\mathcal{M}} \phi$.* □

Lemma 1 (Persistence). *Persistence extends to all formulae with respect to the layered graph semantics. That is, for all $\varphi \in Form$, $G \preccurlyeq H$ and $G \models_{\mathcal{M}} \varphi$ implies $H \models_{\mathcal{M}} \varphi$.*

Proof. By induction on the complexity of formulae. The additive fragment, corresponding to intuitionistic propositional logic (IPL), is standard and we restrict attention to two examples of the multiplicative connectives.

Suppose $G \models_{\mathcal{M}} \varphi \blacktriangleright \psi$ and $G \preccurlyeq H$. There are K, K' s.t. $K @_{\mathcal{E}} K' \downarrow$ and $K @_{\mathcal{E}} K' \preccurlyeq G$, with $K \models_{\mathcal{M}} \varphi$ and $K' \models_{\mathcal{M}} \psi$. By transitivity of $\preccurlyeq$, $K @_{\mathcal{E}} K' \preccurlyeq H$, so $H \models_{\mathcal{M}} \varphi \blacktriangleright \psi$.

Suppose $G \models_{\mathcal{M}} \varphi \mathrel{-\!\!\blacktriangleright} \psi$. Then, for all K such that $G \preccurlyeq K$ and all K' s.t. $K @_{\mathcal{E}} K' \downarrow$, if $K' \models_{\mathcal{M}} \varphi$, then $K @_{\mathcal{E}} K' \models_{\mathcal{M}} \psi$. Let $G \preccurlyeq H$ and suppose $H \preccurlyeq K$ and K' are s.t. $K @_{\mathcal{E}} K' \downarrow$ and $K' \models_{\mathcal{M}} \varphi$. So, since $G \preccurlyeq H \preccurlyeq K$, it follows that $K @_{\mathcal{E}} K' \models_{\mathcal{M}} \varphi$. So $H \models_{\mathcal{M}} \varphi \mathrel{-\!\!\blacktriangleright} \psi$. The case for $\varphi \blacktriangleright\!\!- \psi$ is similar. □

Note that, unlike in BI, we require the restriction 'for all H, $G \preccurlyeq H \ldots$' in the semantic clauses for the multiplicative implications. Without this we cannot prove persistence because we cannot proceed with the inductive step in those cases. The reason for this is that we put no restriction on the interaction between $\preccurlyeq$ and @ in the definition of preordered scaffold. This is unlike the analogous case for BI, where the monoidal composition is required to be bifunctorial with respect to the ordering. One might resolve this issue with the following addendum to the definition of preordered scaffold: if $G \preccurlyeq H$ and $H @_{\mathcal{E}} K \downarrow$, then $G @_{\mathcal{E}} K \downarrow$ and $G @_{\mathcal{E}} K \preccurlyeq H @_{\mathcal{E}} K$.

Two natural examples of subgraph preorderings show that this would be undesirable. First, consider the layering preorder. Let $\preccurlyeq$ be the reflexive, transitive closure of the relation $R(G, H)$ iff $H @_{\mathcal{E}} G \downarrow$, restricted to the admissible subgraph set X. Figure 3 shows a subgraph H with $G \preccurlyeq H$ and $H @_{\mathcal{E}} K \downarrow$ but $G @_{\mathcal{E}} K \uparrow$ (we write $\uparrow$ for undefinedness). Second, consider the subgraph relation. In Fig. 4, we have $G \subseteq H$ and $H @_{\mathcal{E}} K \downarrow$ but $G @_{\mathcal{E}} K \uparrow$. It is, however, the case that, with this ordering, if $G \subseteq H, H @_{\mathcal{E}} K$ and $G @_{\mathcal{E}} K \downarrow$, then $G @_{\mathcal{E}} K \subseteq H @_{\mathcal{E}} K$.

Labelled Tableaux. We define a labelled tableaux system for ILGL, utilising a method first showcased on tableaux systems for BBI and DMBI [8,16] and in the spirit of previous work for BI [12].

Definition 4 (Graph Labels). *Let* $\Sigma = \{c_i \mid i \in \mathbb{N}\}$ *be a countable set of* atomic labels. *We define the set* $\mathbb{L} = \{x \in \Sigma^\star \mid 0 < |x| \leq 2\} \setminus \{c_i c_i \mid c_i \in \Sigma\}$ *to be the set of* graph labels. *A* sub-label y *of a label* x *is a non-empty sub-word of* x*, and we denote the set of sub-labels of* x *by* $\mathcal{S}(x)$. □

The graph labels are a syntactic representation of the subgraphs of a model, with labels of length 2 representing a graph that can be decomposed into two layers. We exclude the possibility $c_i c_i$ as layering is anti-reflexive. In much the same way we give a syntactic representation of preorder.

Definition 5 (Constraints). *A* constraint *is an expression of the form* $x \preccurlyeq y$*, where* x *and* y *are graph labels.* □

Let $\mathcal{C}$ be a set of constraints. The *domain* of $\mathcal{C}$ is the set of all non-empty sub-labels appearing in $\mathcal{C}$. In particular, $\mathcal{D}(\mathcal{C}) = \bigcup_{x \preccurlyeq y \in \mathcal{C}} (\mathcal{S}(x) \cup \mathcal{S}(y))$ The *alphabet* of $\mathcal{C}$ is the set of atomic labels appearing in $\mathcal{C}$. In particular, we have $\mathcal{A}(\mathcal{C}) = \Sigma \cap \mathcal{D}(\mathcal{C})$.

$$\frac{x \preccurlyeq y}{x \preccurlyeq x}\langle R_1 \rangle \qquad \frac{x \preccurlyeq y}{y \preccurlyeq y}\langle R_2 \rangle \qquad \frac{x \preccurlyeq yz}{y \preccurlyeq y}\langle R_3 \rangle \qquad \frac{x \preccurlyeq yz}{z \preccurlyeq z}\langle R_4 \rangle$$

$$\frac{xy \preccurlyeq z}{x \preccurlyeq x}\langle R_5 \rangle \qquad \frac{xy \preccurlyeq z}{y \preccurlyeq y}\langle R_6 \rangle \qquad \frac{x \preccurlyeq y \quad y \preccurlyeq z}{x \preccurlyeq z}\langle Tr \rangle$$

Fig. 5. Rules for closure of constraints

Definition 6 (Closure of Constraints). *Let* $\mathcal{C}$ *be a set of constraints. The closure of* $\mathcal{C}$*, denoted* $\overline{\mathcal{C}}$*, is the least relation closed under the rules of Fig. 5 such that* $\mathcal{C} \subseteq \overline{\mathcal{C}}$. □

This closure yields a preorder on $\mathcal{D}(\mathcal{C})$, with $\langle R_1 \rangle - \langle R_6 \rangle$ generating reflexivity and $\langle Tr \rangle$ yielding transitivity. Crucially, taking the closure of the constraint set does not cause labels to proliferate and the generation of any particular constraint from an arbitrary constraint set $\mathcal{C}$ is fundamentally a finite process.

Proposition 1. *Let $\mathcal{C}$ be a set of constraints. (1) $x \in \mathcal{D}(\overline{\mathcal{C}})$ iff $x \preccurlyeq x \in \overline{\mathcal{C}}$. (2) $\mathcal{D}(\mathcal{C}) = \mathcal{D}(\overline{\mathcal{C}})$ and $\mathcal{A}(\mathcal{C}) = \mathcal{A}(\overline{\mathcal{C}})$.* □

Lemma 2 (Compactness). *Let $\mathcal{C}$ be a (possibly countably infinite) set of constraints. If $x \preccurlyeq y \in \overline{\mathcal{C}}$, then there is a finite set of constraints $\mathcal{C}_f \subseteq \mathcal{C}$ such that $x \preccurlyeq y \in \overline{\mathcal{C}_f}$.* □

Definition 7. *A* labelled formula *is a triple $(\mathbb{S}, \varphi, x) \in \{\mathbb{T}, \mathbb{F}\} \times \mathrm{Form} \times \mathbb{L}$, written $\mathbb{S}\varphi : x$. A* constrained set of statements *(CSS) is a pair $\langle \mathcal{F}, \mathcal{C} \rangle$, where $\mathcal{F}$ is a set of labelled formulæ and $\mathcal{C}$ is a set of constraints, satisfying the following properties: for all $x \in \mathbb{L}$ and distinct $c_i, c_j, c_k \in \Sigma$, (1) (Ref) if $\mathbb{S}\varphi : x \in \mathcal{F}$, then $x \preccurlyeq x \in \overline{\mathcal{C}}$, (2) (Contra) if $c_i c_j \in \mathcal{D}(\mathcal{C})$, then $c_j c_i \notin \mathcal{D}(\mathcal{C})$, and (3) (Freshness) if $c_i c_j \in \mathcal{D}(\mathcal{C})$, then $c_i c_k, c_k c_i, c_j c_k, c_k c_j \notin \mathcal{D}(\mathcal{C})$. A CSS $\langle \mathcal{F}, \mathcal{C} \rangle$ is* finite *if $\mathcal{F}$ and $\mathcal{C}$ are finite. The relation $\subseteq$ is defined on CSSs by $\langle \mathcal{F}, \mathcal{C} \rangle \subseteq \langle \mathcal{F}', \mathcal{C}' \rangle$ iff $\mathcal{F} \subseteq \mathcal{F}'$ and $\mathcal{C} \subseteq \mathcal{C}'$. We denote by $\langle \mathcal{F}_f, \mathcal{C}_f \rangle \subseteq_f \langle \mathcal{F}, \mathcal{C} \rangle$ when $\langle \mathcal{F}_f, \mathcal{C}_f \rangle \subseteq \langle \mathcal{F}, \mathcal{C} \rangle$ holds and $\langle \mathcal{F}_f, \mathcal{C}_f \rangle$ is finite.* □

The CSS properties ensure models can be built from the labels: (Ref) ensures we have enough data for the closure rules to generate a preorder, (Contra) ensures the contra-commutativity of graph layering is respected, and (Freshness) ensures the layering structure of the models we construct is exactly that specified by the labels and constraints in the CSS. As with constraint closure, CSSs have a finite character.

Proposition 2. *For any CSS $\langle \mathcal{F}_f, \mathcal{C} \rangle$ in which $\mathcal{F}_f$ is finite, there exists $\mathcal{C}_f \subseteq \mathcal{C}$ such that $\mathcal{C}_f$ is finite and $\langle \mathcal{F}_f, \mathcal{C}_f \rangle$ is a CSS.* □

Figure 6 presents the rules of the tableaux system for ILGL. That 'c_i and c_j are fresh atomic labels' means $c_i \neq c_j \in \Sigma \setminus \mathcal{A}(\mathcal{C})$. We denote by $\oplus$ the concatenation of lists.

Definition 8 (Tableaux). *Let $\langle \mathcal{F}_0, \mathcal{C}_0 \rangle$ be a finite CSS. A* tableau *for this CSS is a list of CSS, called* branches, *built inductively according the following rules:*

1. *The one branch list $[\langle \mathcal{F}_0, \mathcal{C}_0 \rangle]$ is a tableau for $\langle \mathcal{F}_0, \mathcal{C}_0 \rangle$;*
2. *If the list $\mathcal{T}_m \oplus [\langle \mathcal{F}, \mathcal{C} \rangle] \oplus \mathcal{T}_n$ is a tableau for $\langle \mathcal{F}_0, \mathcal{C}_0 \rangle$ and*

$$\frac{cond\langle \mathcal{F}, \mathcal{C} \rangle}{\langle \mathcal{F}_1, \mathcal{C}_1 \rangle \mid \ldots \mid \langle \mathcal{F}_k, \mathcal{C}_k \rangle}$$

is an instance of a rule of Fig. 6 for which $cond\langle \mathcal{F}, \mathcal{C} \rangle$ is fulfilled, then the list $\mathcal{T}_m \oplus [\langle \mathcal{F} \cup \mathcal{F}_1, \mathcal{C} \cup \mathcal{C}_1 \rangle; \ldots; \langle \mathcal{F} \cup \mathcal{F}_k, \mathcal{C} \cup \mathcal{C}_k \rangle] \oplus \mathcal{T}_n$ is a tableau for $\langle \mathcal{F}_0, \mathcal{C}_0 \rangle$.

A tableau *for the formula φ is a* tableau *for $\langle \{\mathbb{F}\varphi : c_0\}, \{c_0 \preccurlyeq c_0\} \rangle$.* □

It is a simple but tedious exercise to show that the rules of Fig. 6 preserve the CSS properties of Definition 7. We now give the notion of proof for our labelled tableaux.

$$\frac{\mathbb{T}\varphi \wedge \psi : x \in \mathcal{F}}{\langle\{\mathbb{T}\varphi : x, \mathbb{T}\psi : x\}, \emptyset\rangle}\langle\mathbb{T}\wedge\rangle \qquad \frac{\mathbb{F}\varphi \wedge \psi : x \in \mathcal{F}}{\langle\{\mathbb{F}\varphi : x\}, \emptyset\rangle \mid \langle\{\mathbb{F}\psi : x\}, \emptyset\rangle}\langle\mathbb{F}\wedge\rangle$$

$$\frac{\mathbb{T}\varphi \vee \psi : x \in \mathcal{F}}{\langle\{\mathbb{T}\varphi : x\}, \emptyset\rangle \mid \langle\{\mathbb{T}\psi : x\}, \emptyset\rangle}\langle\mathbb{T}\vee\rangle \qquad \frac{\mathbb{F}\varphi \vee \psi : x \in \mathcal{F}}{\langle\{\mathbb{F}\varphi : x, \mathbb{F}\psi : x\}, \emptyset\rangle}\langle\mathbb{F}\vee\rangle$$

$$\frac{\mathbb{T}\varphi \to \psi : x \in \mathcal{F} \text{ and } x \preccurlyeq y \in \overline{C}}{\langle\{\mathbb{F}\varphi : y\}, \emptyset\rangle \mid \langle\{\mathbb{T}\psi : y\}, \emptyset\rangle}\langle\mathbb{T}\to\rangle \qquad \frac{\mathbb{F}\varphi \to \psi : x \in \mathcal{F}}{\langle\{\mathbb{T}\varphi : c_i, \mathbb{F}\psi : c_i\}, \{x \preccurlyeq c_i\}\rangle}\langle\mathbb{F}\to\rangle$$

$$\frac{\mathbb{T}\varphi \blacktriangleright \psi : x \in \mathcal{F}}{\langle\{\mathbb{T}\varphi : c_i, \mathbb{T}\psi : c_j\}, \{c_i c_j \preccurlyeq x\}\rangle}\langle\mathbb{T}\blacktriangleright\rangle \qquad \frac{\mathbb{F}\varphi \blacktriangleright \psi : x \in \mathcal{F} \text{ and } yz \preccurlyeq x \in \overline{C}}{\langle\{\mathbb{F}\varphi : y\}, \emptyset\rangle \mid \langle\{\mathbb{F}\psi : z\}, \emptyset\rangle}\langle\mathbb{F}\blacktriangleright\rangle$$

$$\frac{\mathbb{T}\varphi \multimap\!\!\!\blacktriangleright \psi : x \in \mathcal{F} \text{ and } x \preccurlyeq y, yz \preccurlyeq yz \in \overline{C}}{\langle\{\mathbb{F}\varphi : z\}, \emptyset\rangle \mid \langle\{\mathbb{T}\psi : yz\}, \emptyset\rangle}\langle\mathbb{T}\multimap\!\!\!\blacktriangleright\rangle \qquad \frac{\mathbb{F}\varphi \multimap\!\!\!\blacktriangleright \psi : x \in \mathcal{F}}{\langle\{\mathbb{T}\varphi : c_j, \mathbb{F}\psi : c_i c_j\}, \{x \preccurlyeq c_i, c_i c_j \preccurlyeq c_i c_j\}\rangle}\langle\mathbb{F}\multimap\!\!\!\blacktriangleright\rangle$$

$$\frac{\mathbb{T}\varphi \blacktriangleright\!\!\!- \psi : x \in \mathcal{F} \text{ and } x \preccurlyeq y, zy \preccurlyeq zy \in \overline{C}}{\langle\{\mathbb{F}\varphi : z\}, \emptyset\rangle \mid \langle\{\mathbb{T}\psi : zy\}, \emptyset\rangle}\langle\mathbb{T}\blacktriangleright\!\!\!-\rangle \qquad \frac{\mathbb{F}\varphi \blacktriangleright\!\!\!- \psi : x \in \mathcal{F}}{\langle\{\mathbb{T}\varphi : c_j, \mathbb{F}\psi : c_j c_i\}, \{x \preccurlyeq c_i, c_j c_i \preccurlyeq c_j c_i\}\rangle}\langle\mathbb{F}\blacktriangleright\!\!\!-\rangle$$

with c_i and c_j being fresh atomic labels

Fig. 6. Tableaux rules for ILGL

Definition 9 (Closed Tableau/Proof). *A CSS $\langle \mathcal{F}, \mathcal{C}\rangle$ is* closed *if one of the following conditions holds: (1) $\mathbb{T}\varphi : x \in \mathcal{F}$, $\mathbb{F}\varphi : y \in \mathcal{F}$ and $x \preccurlyeq y \in \overline{\mathcal{C}}$; (2) $\mathbb{F}\top : x \in \mathcal{F}$; and (3) $\mathbb{T}\bot : x \in \mathcal{F}$. A CSS is* open *iff it is not closed. A tableau is closed iff all its branches are closed. A* proof *for a formula φ is a closed tableau for φ.* □

CSSs are related back to the graph semantics via the notion of realization.

Definition 10 (Realization). *Let $\langle \mathcal{F}, \mathcal{C}\rangle$ be a CSS. A* realization *of $\langle \mathcal{F}, \mathcal{C}\rangle$ is a triple $\mathfrak{R} = (\mathcal{X}, \mathcal{V}, \lfloor.\rfloor)$ where $\mathcal{M} = (\mathcal{X}, \mathcal{V})$ is a layered graph model and $\lfloor.\rfloor : \mathcal{D}(\mathcal{C}) \to X$ is such that (1) $\lfloor.\rfloor$ is total: for all $x \in \mathcal{D}(\mathcal{C})$, $\lfloor x\rfloor \downarrow$, (2) for all $x \in \mathcal{D}(\mathcal{C})$, if $x = c_i c_j$, then $\lfloor c_i\rfloor @_{\mathcal{E}} \lfloor c_j\rfloor \downarrow$ and $\lfloor x\rfloor = \lfloor c_i\rfloor @_{\mathcal{E}} \lfloor c_j\rfloor)$, (3) if $x \preccurlyeq y \in \mathcal{C}$, then $\lfloor x\rfloor \preccurlyeq_{\mathcal{M}} \lfloor y\rfloor$, (4) if $\mathbb{T}\varphi : x \in \mathcal{F}$, then $\lfloor x\rfloor \vDash_{\mathcal{M}} \varphi$, and (5) if $\mathbb{F}\varphi : x \in \mathcal{F}$, then $\lfloor x\rfloor \nvDash_{\mathcal{M}} \varphi$.* □

We say that a CSS is *realizable* is there exists a realization of it. We say that a tableau is *realizable* if at least one of its branches is realizable. We can also show that the relevant clauses of the definition extend to the closure of the constraint set automatically.

Proposition 3. *Let $\langle \mathcal{F}, \mathcal{C}\rangle$ be a CSS and $\mathfrak{R} = (\mathcal{X}, \mathcal{V}, \lfloor.\rfloor)$ a realization of it. Then: (1) for all $x \in \mathcal{D}(\overline{\mathcal{C}})$, $\lfloor x\rfloor$ is defined; (2) if $x \preccurlyeq y \in \overline{\mathcal{C}}$, then $\lfloor x\rfloor \preccurlyeq_{\mathcal{M}} \lfloor y\rfloor$.* □

3 Metatheory

We now establish the soundness and, via countermodel extraction, the completeness of ILGL's tableaux system with respect to layered graph semantics. The proof of soundness is straightforward (cf. [8,11,12,16]). We begin with two key lemmas about realizability and closure. Their proofs proceed by simple case analysis.

Lemma 3. *The tableaux rules for ILGL preserve realizability.* □

Lemma 4. *Closed branches are not realizable.* □

Theorem 1 (Soundness). *If there exists a closed tableau for the formula φ, then φ is valid in layered graph models.*

Proof. Suppose that there exists a proof for φ. Then there is a closed tableau $\mathcal{T}_\varphi$ for the CSS $\mathfrak{C} = \langle\{\mathbb{F}\varphi : c_0\}, \{c_0 \preccurlyeq c_0\}\rangle$. Now suppose that φ is not valid. Then there is a countermodel $\mathcal{M} = (\mathcal{X}, \mathcal{V})$ and a subgraph $G \in X$ such that $G \not\models_{\mathcal{M}} \varphi$. Define $\mathfrak{R} = (\mathcal{M}, \mathcal{V}, \lfloor . \rfloor)$ with $\lfloor c_0 \rfloor = G$. Note that $\mathfrak{R}$ is a realization of $\mathfrak{C}$, hence by Lemma 3, $\mathcal{T}_\varphi$ is realizable. By Lemma 4, $\mathcal{T}_\varphi$ cannot be closed. But, this contradicts the fact that $\mathcal{T}_\varphi$ is a proof and therefore a closed tableau. It follows that φ is valid. □

We now proceed to establish the completeness of the labelled tableaux with respect to layered graph semantics. We begin with the notion of a Hintikka CSS, which will facilitate the construction of countermodels. All remaining proofs omitted from this section are provided in the research note [9].

Definition 11 (Hintikka CSS). *A CSS $\langle \mathcal{F}, \mathcal{C} \rangle$ is a* Hintikka CSS *iff, for any formulas $\varphi, \psi \in$ Form and any graph labels $x, y \in \mathbb{L}$, we have the following:*

1. $\mathbb{T}\varphi : x \notin \mathcal{F}$ or $\mathbb{F}\varphi : y \notin \mathcal{F}$ or $x \preccurlyeq y \notin \overline{\mathcal{C}}$ 2.$\mathbb{F}\top : x \notin \mathcal{F}$ 3.$\mathbb{T}\bot : x \notin \mathcal{F}$
4. if $\mathbb{T}\varphi \wedge \psi : x \in \mathcal{F}$, then $\mathbb{T}\varphi : x \in \mathcal{F}$ and $\mathbb{T}\psi : x \in \mathcal{F}$
5. if $\mathbb{F}\varphi \wedge \psi : x \in \mathcal{F}$, then $\mathbb{F}\varphi : x \in \mathcal{F}$ or $\mathbb{F}\psi : x \in \mathcal{F}$
6. if $\mathbb{T}\varphi \vee \psi : x \in \mathcal{F}$, then $\mathbb{T}\varphi : x \in \mathcal{F}$ or $\mathbb{T}\psi : x \in \mathcal{F}$
7. if $\mathbb{F}\varphi \vee \psi : x \in \mathcal{F}$, then $\mathbb{F}\varphi : x \in \mathcal{F}$ and $\mathbb{F}\psi : x \in \mathcal{F}$
8. if $\mathbb{T}\varphi \rightarrow \psi : x \in \mathcal{F}$, then, for all $y \in \mathbb{L}$, if $x \preccurlyeq y \in \overline{\mathcal{C}}$, then $\mathbb{F}\varphi : y \in \mathcal{F}$ or $\mathbb{T}\psi : y \in \mathcal{F}$
9. if $\mathbb{F}\varphi \rightarrow \psi : x \in \mathcal{F}$, then there exists $y \in \mathbb{L}$ such that $x \preccurlyeq y \in \overline{\mathcal{C}}$ and $\mathbb{T}\varphi : y \in \mathcal{F}$ and $\mathbb{F}\psi : y \in \mathcal{F}$
10. if $\mathbb{T}\varphi \blacktriangleright \psi : x \in \mathcal{F}$, then there are $c_i, c_j \in \Sigma$ such that $c_i c_j \preccurlyeq x \in \overline{\mathcal{C}}$ and $\mathbb{T}\varphi : c_i \in \mathcal{F}$ and $\mathbb{T}\psi : c_j \in \mathcal{F}$
11. if $\mathbb{F}\varphi \blacktriangleright \psi : x \in \mathcal{F}$, then, for all $c_i, c_j \in \Sigma$, if $c_i c_j \preccurlyeq x \in \overline{\mathcal{C}}$, then $\mathbb{F}\varphi : c_i \in \mathcal{F}$ or $\mathbb{F}\psi : c_j \in \mathcal{F}$
12. if $\mathbb{T}\varphi \multimap\!\!\blacktriangleright \psi : x \in \mathcal{F}$, then, for all $c_i, c_j \in \Sigma$, if $x \preccurlyeq c_i \in \overline{\mathcal{C}}$ and $c_i c_j \in \mathcal{D}(\overline{\mathcal{C}})$, then $\mathbb{F}\varphi : c_j \in \mathcal{F}$ or $\mathbb{T}\psi : c_i c_j \in \mathcal{F}$
13. if $\mathbb{F}\varphi \multimap\!\!\blacktriangleright \psi : x \in \mathcal{F}$, then there are $c_i, c_j \in \Sigma$ such that $x \preccurlyeq c_i \in \overline{\mathcal{C}}$ and $c_i c_j \in \mathcal{D}(\overline{\mathcal{C}})$ and $\mathbb{T}\varphi : c_j \in \mathcal{F}$ and $\mathbb{F}\psi : c_i c_j \in \mathcal{F}$
14. if $\mathbb{T}\varphi \blacktriangleright\!\!- \psi : x \in \mathcal{F}$, then, for all $c_i, c_j \in \Sigma$, if $x \preccurlyeq c_i \in \overline{\mathcal{C}}$ and $c_j c_i \in \mathcal{D}(\overline{\mathcal{C}})$, then $\mathbb{F}\varphi : c_j \in \mathcal{F}$ or $\mathbb{T}\psi : c_j c_i \in \mathcal{F}$
15. if $\mathbb{F}\varphi \multimap\!\!\blacktriangleright \psi : x \in \mathcal{F}$, then there are $c_i, c_j \in \Sigma$ such that $x \preccurlyeq c_i \in \overline{\mathcal{C}}$ and $c_j c_i \in \mathcal{D}(\overline{\mathcal{C}})$ and $\mathbb{T}\varphi : c_j \in \mathcal{F}$ and $\mathbb{F}\psi : c_j c_i \in \mathcal{F}$. □

We now give the definition of a function Ω that extracts a countermodel from a Hintikka CSS. A Hintikka CSS can thus be seen as the *labelled* tableaux counterpart of Hintikka sets, which are maximally consistent sets satisfying a subformula property.

Definition 12 (Function Ω). *Let $\langle \mathcal{F}, \mathcal{C}\rangle$ be a Hintikka CSS. The function Ω associates to $\langle \mathcal{F}, \mathcal{C}\rangle$ a tuple $\Omega(\langle \mathcal{F}, \mathcal{C}\rangle) = (\mathcal{G}, \mathcal{E}, X, \preccurlyeq, \mathcal{V})$, such that (1) $V(\mathcal{G}) = \mathcal{A}(\mathcal{C})$, (2) $E(\mathcal{G}) = \{(c_i, c_j) \mid c_i c_j \in \mathcal{D}(\mathcal{C})\} = \mathcal{E}$, $X = \{x^\Omega \mid x \in \mathcal{D}(\mathcal{C})\}$, where $V(c_i^\Omega) = \{c_i\}$, $E(c_i^\Omega) = \emptyset$, $V((c_i c_j)^\Omega) = \{c_i c_j\}$, and $E((c_i c_j)^\Omega) = \{(c_i, c_j)\}$, (3) $x^\Omega \preccurlyeq y^\Omega$ iff $x \preccurlyeq y \in \overline{\mathcal{C}}$, and (4) $x^\Omega \in \mathcal{V}(p)$ iff there exists $y \in \mathcal{D}(\mathcal{C})$ such that $y \preccurlyeq x \in \overline{\mathcal{C}}$ and $\mathbb{T}p : y \in \mathcal{F}$.* □

The next lemma shows that there is a precise correspondence between the structure that the Hintikka CSS properties impose on the labels and the layered structure specified by the construction of the model.

Lemma 5. *Let $\langle \mathcal{F}, \mathcal{C}\rangle$ be a Hintikka CSS and $\Omega(\langle \mathcal{F}, \mathcal{C}\rangle) = (\mathcal{G}, \mathcal{E}, X, \preccurlyeq, \mathcal{V})$. (1) If $c_i, c_j \in \mathcal{A}(\mathcal{C})$, then $c_i c_j \in \mathcal{D}(\mathcal{C})$ iff $c_i^\Omega \mathbin{@_\mathcal{E}} c_j^\Omega \downarrow$. (2) If $c_i c_j \in \mathcal{D}(\mathcal{C})$, then $(c_i c_j)^\Omega = c_i^\Omega \mathbin{@_\mathcal{E}} c_j^\Omega$. 3. $x^\Omega \mathbin{@_\mathcal{E}} y^\Omega \downarrow$ iff there exist $c_i, c_j \in \mathcal{A}(\mathcal{C})$ s.t. $x = c_i$, $y = c_j$ and $c_i c_j \in \mathcal{D}(\mathcal{C})$.* □

Lemma 6. *Let $\langle \mathcal{F}, \mathcal{C}\rangle$ be a Hintikka CSS. $\Omega(\langle \mathcal{F}, \mathcal{C}\rangle)$ is a layered graph model.* □

Lemma 7. *Let $\langle \mathcal{F}, \mathcal{C}\rangle$ be a Hintikka CSS and $\mathcal{M} = \Omega(\langle \mathcal{F}, \mathcal{C}\rangle) = (\mathcal{G}, \mathcal{E}, X, \preccurlyeq, \mathcal{V})$. For all formulas $\varphi \in$ Form, and all $x \in \mathcal{D}(\mathcal{C})$. we have (1) if $\mathbb{F}\varphi : x \in \mathcal{F}$, then $x^\Omega \not\models_\mathcal{M} \varphi$, and (2) if $\mathbb{T}\varphi : x \in \mathcal{F}$, then $x^\Omega \models_\mathcal{M} \varphi$. Hence, if $\mathbb{F}\varphi : x \in \mathcal{F}$, then φ is not valid and $\Omega(\langle \mathcal{F}, \mathcal{C}\rangle)$ is a countermodel of φ.* □

This construction of a countermodel would fail in a labelled tableaux system for LGL (i.e., the layered graph logic with classical additives [6]). This is because it is impossible to construct the internal structure of each subgraph in the model systematically, as the classical semantics for ► demands strict equality between the graph under interpretation and the decomposition into layers. This issue is sidestepped for ILGL since each time the tableaux rules require a decomposition of a subgraph into layers we can move to a 'fresh' layered subgraph further down the ordering. Thus we can safely turn each graph label into the simplest instantiation of the kind of graph it represents: either a single vertex (indecomposable) or two vertices and an edge (layered).

We now show how to construct such a CSS. We first require a listing of all labelled formulae that may need to be added to the CSS in order to satisfy properties 4–15. We require a particularly strong condition on the listing to make this procedure work: that every labelled formula appears infinitely often to be tested.

Definition 13 (Fair Strategy). *A* fair strategy *for a language $\mathcal{L}$ is a labelled sequence of formulæ $(\mathbb{S}_i \chi_i : (x_i))_{i \in \mathbb{N}}$ in $\{\mathbb{T}, \mathbb{F}\} \times$ Form $\times\, \mathbb{L}$ such that $\{i \in \mathbb{N} \mid \mathbb{S}_i \chi_i : (x_i) \equiv \mathbb{S}\chi : x\}$ is infinite for any $\mathbb{S}\chi : x \in \{\mathbb{T}, \mathbb{F}\} \times$ Form $\times\, \mathbb{L}$.* □

Proposition 4. *There exists a fair strategy for the language of ILGL.* □

Next we need the concept of an oracle. Here an oracle allows Hintikka sets to be constructed inductively, testing the required consistency properties at each stage.

Definition 14. *Let $\mathcal{P}$ be a set of CSSs. (1) $\mathcal{P}$ is* $\subseteq$-closed *if $\langle \mathcal{F}, \mathcal{C} \rangle \in \mathcal{P}$ holds whenever $\langle \mathcal{F}, \mathcal{C} \rangle \subseteq \langle \mathcal{F}', \mathcal{C}' \rangle$ and $\langle \mathcal{F}', \mathcal{C}' \rangle \in \mathcal{P}$ holds. (2) $\mathcal{P}$ is of* finite character *if $\langle \mathcal{F}, \mathcal{C} \rangle \in \mathcal{P}$ holds whenever $\langle \mathcal{F}_f, \mathcal{C}_f \rangle \in \mathcal{P}$ holds for every $\langle \mathcal{F}_f, \mathcal{C}_f \rangle \subseteq_f \langle \mathcal{F}, \mathcal{C} \rangle$. (3) $\mathcal{P}$ is* saturated *if, for any $\langle \mathcal{F}, \mathcal{C} \rangle \in \mathcal{P}$ and any instance*

$$\frac{cond(\mathcal{F}, \mathcal{C})}{\langle \mathcal{F}_1, \mathcal{C}_1 \rangle \mid \ldots \mid \langle \mathcal{F}_k, \mathcal{C}_k \rangle}$$

of a rule of Fig. 6 if $cond(\mathcal{F}, \mathcal{C})$ is fulfilled, then $\langle \mathcal{F} \cup \mathcal{F}_i, \mathcal{C} \cup \mathcal{C}_i \rangle \in \mathcal{P}$, for at least one $i \in \{1, \ldots, k\}$. □

Definition 15 (Oracle). *An* oracle *is a set of open CSSs which is $\subseteq$-closed, of finite character, and saturated.* □

Definition 16 (Consistency/Finite Consistency). *Let $\langle \mathcal{F}, \mathcal{C} \rangle$ be a CSS. We say $\langle \mathcal{F}, \mathcal{C} \rangle$ is consistent if it is finite and has no closed tableau. We say $\langle \mathcal{F}, \mathcal{C} \rangle$ is finitely consistent if every finite sub-CSS $\langle \mathcal{F}_f, \mathcal{C}_f \rangle$ is consistent.* □

Proposition 5. *(1) Consistency is $\subseteq$-closed. (2) A finite CSS is consistent iff it is finitely consistent.* □

Lemma 8. *The set of finitely consistent CSS, $\mathcal{P}$, is an oracle.* □

We can now show completeness of our tableaux system. Consider a formula φ for which there exists no closed tableau. We show there is a countermodel to φ. We start with the initial tableau $\mathcal{T}_0$ for φ. Then, we have (1) $\mathcal{T}_0 = [\langle \{\mathbb{F}\varphi : c_0\}, \{c_0 \preccurlyeq c_0\} \rangle]$ and (2) $\mathcal{T}_0$ cannot be closed. Let $\mathcal{P}$ be as in Lemma 8. By Proposition 4, there exists a fair strategy, which we denote by $\mathcal{S}$, with $\mathbb{S}_i X_i : (x_i)$ the ith formula of $\mathcal{S}$. As $\mathcal{T}_0$ cannot be closed, $\langle \{\mathbb{F}\varphi : c_0\}, \{c_0 \preccurlyeq c_0\} \rangle \in \mathcal{P}$. We build a sequence $\langle \mathcal{F}_i, \mathcal{C}_i \rangle_{i \geqslant 0}$ as follows:

- $\langle \mathcal{F}_0, \mathcal{C}_0 \rangle = \langle \{\mathbb{F}\varphi : c_0\}, \{c_0 \preccurlyeq c_0\} \rangle$;
- if $\langle \mathcal{F}_i \cup \{\mathbb{S}_i X_i : (x_i)\}, \mathcal{C}_i \rangle \notin \mathcal{P}$, then we have $\langle \mathcal{F}_{i+1}, \mathcal{C}_{i+1} \rangle = \langle \mathcal{F}_i, \mathcal{C}_i \rangle$; and
- if $\langle \mathcal{F}_i \cup \{\mathbb{S}_i X_i : (x_i)\}, \mathcal{C}_i \rangle \in \mathcal{P}$, then we have $\langle \mathcal{F}_{i+1}, \mathcal{C}_{i+1} \rangle = \langle \mathcal{F}_i \cup \{\mathbb{S}_i X_i : (x_i)\} \cup F_e, \mathcal{C}_i \cup \mathcal{C}_e \rangle$ such that F_e and $\mathcal{C}_e$ are determined by

$\mathbb{S}_i$	X_i	F_e	$\mathcal{C}_e$
$\mathbb{F}$	$\varphi \rightarrow \psi$	$\{\mathbb{T}\varphi : c_{\mathfrak{I}+1}, \mathbb{F}\psi : c_{\mathfrak{I}+1}\}$	$\{x_i \preccurlyeq c_{\mathfrak{I}+1}\}$
$\mathbb{T}$	$\varphi \blacktriangleright \psi$	$\{\mathbb{T}\varphi : c_{\mathfrak{I}+1}, \mathbb{T}\psi : c_{\mathfrak{I}+2}\}$	$\{c_{\mathfrak{I}+1}c_{\mathfrak{I}+2} \preccurlyeq x_i\}$
$\mathbb{F}$	$\varphi \mathrel{-\!\!\blacktriangleright} \psi$	$\{\mathbb{T}\varphi : c_{\mathfrak{I}+2}, \mathbb{F}\psi : c_{\mathfrak{I}+1}c_{\mathfrak{I}+2}\}$	$\{x_i \preccurlyeq c_{\mathfrak{I}+1}, c_{\mathfrak{I}+1}c_{\mathfrak{I}+2} \preccurlyeq c_{\mathfrak{I}+1}c_{\mathfrak{I}+2}\}$
$\mathbb{F}$	$\varphi \mathrel{\blacktriangleright\!\!-} \psi$	$\{\mathbb{T}\varphi : c_{\mathfrak{I}+2}, \mathbb{F}\psi : c_{\mathfrak{I}+2}c_{\mathfrak{I}+1}\}$	$\{x_i \preccurlyeq c_{\mathfrak{I}+1}, c_{\mathfrak{I}+2}c_{\mathfrak{I}+1} \preccurlyeq c_{\mathfrak{I}+2}c_{\mathfrak{I}+1}\}$
Otherwise		$\emptyset$	$\emptyset$

with $\mathfrak{I} = \max\{j \mid c_j \in \mathcal{A}(\mathcal{C}_i) \cup \mathcal{S}(x_i)\}$.

Proposition 6. *For any $i \in \mathbb{N}$, the following properties hold: (1) $\mathcal{F}_i \subseteq \mathcal{F}_{i+1}$ and $\mathcal{C}_i \subseteq \mathcal{C}_{i+1}$; (2) $\langle \mathcal{F}_i, \mathcal{C}_i \rangle \in \mathcal{P}$.* □

We now define the limit $\langle \mathcal{F}_\infty, \mathcal{C}_\infty \rangle = \langle \bigcup_{i \geqslant 0} \mathcal{F}_i, \bigcup_{i \geqslant 0} \mathcal{C}_i \rangle$ of the sequence $\langle \mathcal{F}_i, \mathcal{C}_i \rangle_{i \geqslant 0}$.

Proposition 7. *The following properties hold: (1) $\langle \mathcal{F}_\infty, \mathcal{C}_\infty \rangle \in \mathcal{P}$; (2) For all labelled formulæ $\mathbb{S}\varphi : x$, if $\langle \mathcal{F}_\infty \cup \{\mathbb{S}\varphi : x\}, \mathcal{C}_\infty \rangle \in \mathcal{P}$, then $\mathbb{S}\varphi : x \in \mathcal{F}_\infty$.* □

Lemma 9. *The limit CSS is a Hintikka CSS.* □

Theorem 2 (Completeness). *If φ is valid, then there exists a closed tableau for φ.* □

4 A Hilbert System and an Algebraic Semantics

We give a Hilbert-type proof system, $\mathrm{ILGL_H}$, for ILGL in Fig. 7. The additive fragment, corresponding to intuitionistic propositional logic, is standard (e.g., [2]). The presentation of the multiplicative fragment is similar to that for BI's multiplicatives [20], but for the non-commutative and non-associative (following from the absence of a multiplicative counterpart to $\wedge_2$) conjunction, $\blacktriangleright$, together with its associated left and right implications (cf. [14,15]).

$$\frac{}{\varphi \vdash \varphi}\,(\mathrm{Ax}) \qquad \frac{\varphi \vdash \psi \qquad \psi \vdash \chi}{\varphi \vdash \chi}\,(\mathrm{Cut}) \qquad \frac{}{\varphi \vdash \top}\,(\top) \qquad \frac{}{\bot \vdash \varphi}\,(\bot)$$

$$\frac{\varphi \vdash \psi \qquad \varphi \vdash \chi}{\varphi \vdash \psi \wedge \chi}\,(\wedge_1) \qquad \frac{}{\varphi_1 \wedge \varphi_2 \vdash \varphi_i}\,(\wedge_2) \qquad \frac{}{\varphi_i \vdash \varphi_1 \vee \varphi_2}\,(\vee_1) \qquad \frac{\varphi \vdash \chi \qquad \psi \vdash \chi}{\varphi \vee \psi \vdash \chi}\,(\vee_2)$$

$$\frac{\varphi \vdash \psi \to \chi \qquad \nu \vdash \psi}{\varphi \wedge \nu \vdash \chi}\,(\to_1) \qquad \frac{\varphi \wedge \psi \vdash \chi}{\varphi \vdash \psi \to \chi}\,(\to_2) \qquad \frac{\varphi \vdash \psi \qquad \chi \vdash \upsilon}{\varphi \blacktriangleright \chi \vdash \psi \blacktriangleright \upsilon}\,(\blacktriangleright)$$

$$\frac{\varphi \vdash \psi \mathrel{-\!\!\blacktriangleright} \chi \qquad \upsilon \vdash \psi}{\varphi \blacktriangleright \upsilon \vdash \chi}\,(\mathrel{-\!\!\blacktriangleright}_1) \qquad \frac{\varphi \blacktriangleright \psi \vdash \chi}{\varphi \vdash \psi \mathrel{-\!\!\blacktriangleright} \chi}\,(\mathrel{-\!\!\blacktriangleright}_2) \qquad \frac{\varphi \vdash \psi \mathrel{\blacktriangleright\!\!-} \chi \qquad \upsilon \vdash \psi}{\upsilon \blacktriangleright \varphi \vdash \chi}\,(\mathrel{\blacktriangleright\!\!-}_1) \qquad \frac{\varphi \blacktriangleright \psi \vdash \chi}{\psi \vdash \varphi \mathrel{\blacktriangleright\!\!-} \chi}\,(\mathrel{\blacktriangleright\!\!-}_2)$$

Fig. 7. Rules of the Hilbert system, $\mathrm{ILGL_H}$, for ILGL

This section concludes with equivalence of $\mathrm{ILGL_H}$ and ILGL's tableaux system.

Definition 17 (Layered Heyting Algebra). *A layered Heyting algebra is a structure $\mathbb{A} = (A, \wedge, \vee, \to, \bot, \top, \blacktriangleright, \mathrel{-\!\!\blacktriangleright}, \mathrel{\blacktriangleright\!\!-})$ such that $(A, \wedge, \vee, \to, \bot, \top)$ is a Heyting algebra, $\blacktriangleright$, $\mathrel{-\!\!\blacktriangleright}$, and $\mathrel{\blacktriangleright\!\!-}$ are binary operations on A satisfying $a \leq a'$ and $b \leq b'$ implies $a \blacktriangleright b \leq a' \blacktriangleright b'$ and $a \blacktriangleright b \leq c$ iff $a \leq b \mathrel{-\!\!\blacktriangleright} c$ iff $b \leq a \mathrel{\blacktriangleright\!\!-} c$.* □

We interpret ILGL on layered Heyting algebras. Let $\mathcal{V}$: Prop $\to A$ be a valuation on the layered Heyting algebra $(A, \wedge_A, \vee_A, \to_A, \bot_A, \top_A, \blacktriangleright_A, \twoheadrightarrow_A, \blacktriangleright\!\!-_A)$. We maintain the subscripts to distinguish the operations of the algebra from the connectives of ILGL. We uniquely define an interpretation function $[\![-]\!]$: Form $\to A$ by extending with respect to the connectives in the usual fashion: $[\![\top]\!] = \top_A$, $[\![\bot]\!] = \bot_A$, $[\![p]\!] = V(p)$, and $[\![\varphi \circ \psi]\!] = [\![\varphi]\!] \circ_A [\![\psi]\!]$ for $\circ \in \{\wedge, \vee, \to, \blacktriangleright, \twoheadrightarrow, \blacktriangleright\!\!- \}$.

Proposition 8 (Soundness). *For any layered Heyting algebra $\mathbb{A}$ and any interpretation $[\![-]\!]$: Form $\to \mathbb{A}$: if $\varphi \vdash \psi$ then $[\![\varphi]\!] \leq [\![\psi]\!]$.*

Proof. By induction on the derivation rules of $\mathrm{ILGL_H}$. The cases for the additive fragment are standard. For rule $(\blacktriangleright)$, we use the property $a \leq_A a'$ and $b \leq_A b'$ implies $a \blacktriangleright_A b \leq_A a' \blacktriangleright_A b'$ and for the remaining rules pertaining to the multiplicative implications we use the adjointness property $a \blacktriangleright_A b \leq_A c$ iff $a \leq_A b \twoheadrightarrow_A c$ iff $b \leq_A a \blacktriangleright\!\!-_A c$. □

Lemma 10. *There is a layered Heyting algebra $\mathcal{T}$ and an interpretation $[\![-]\!]_{\mathcal{T}}$: Prop $\to \mathcal{T}$ such that if $\varphi \not\vdash \psi$ then $[\![\varphi]\!]_{\mathcal{T}} \not\leq [\![\psi]\!]_{\mathcal{T}}$.*

Proof. We give a Lindenbaum term-algebra construction on the syntax of ILGL with the equivalence relation $\varphi \equiv \psi$ iff $\varphi \vdash \psi$ and $\psi \vdash \varphi$. The set of all such equivalence classes $[\varphi]$ gives the underlying set of the layered Heyting algebra, $\mathcal{T}$: $\top_{\mathcal{T}} := [\top]$, $\bot_{\mathcal{T}} := [\bot]$, and $[\varphi] \circ_{\mathcal{T}} [\psi] := [\varphi \circ \psi]$ for $\circ \in \{\wedge, \vee, \to, \blacktriangleright, \twoheadrightarrow, \blacktriangleright\!\!- \}$.

The fragment $(\mathcal{T}, \wedge_{\mathcal{T}}, \vee_{\mathcal{T}}, \top_{\mathcal{T}}, \bot_{\mathcal{T}})$ forms a bounded distributive lattice with order $[\varphi] \leq_{\mathcal{T}} [\psi]$ iff $[\varphi] \wedge_{\mathcal{T}} [\psi] = [\varphi]$. It is straightforward to use rules (Ax), $(\wedge_1)$ and $(\wedge_2)$ to show that the right hand condition holds iff $\varphi \vdash \psi$. We then obtain adjointness of $\wedge_{\mathcal{T}}$ and $\to_{\mathcal{T}}$ from rules $(\to_1)$ and $(\to_2)$, monotonicity of $\blacktriangleright_{\mathcal{T}}$ from rule $(\blacktriangleright)$ and the adjointness of $\blacktriangleright_{\mathcal{T}}, \twoheadrightarrow_{\mathcal{T}}$ and $\blacktriangleright\!\!-_{\mathcal{T}}$ from rules $(\twoheadrightarrow_1), (\twoheadrightarrow_2), (\blacktriangleright\!\!-_1)$, and $(\blacktriangleright\!\!-_2)$. Thus $\mathcal{T}$ is a layered Heyting algebra with an interpretation given by $[\![\varphi]\!] = [\varphi]$. By the definition of the ordering, $\varphi \not\vdash \psi$ implies $[\![\varphi]\!] \not\leq_{\mathcal{T}} [\![\psi]\!]$, as required. □

We now standardly obtain completeness.

Theorem 3 (Completeness). *For any propositions φ, ψ of ILGL, if $[\![\varphi]\!] \leq [\![\psi]\!]$ for all interpretations $[\![-]\!]$ on layered Heyting algebras then $\varphi \vdash \psi$ in $\mathrm{ILGL_H}$.* □

We now show that the layered graph semantics is a special case of the algebraic semantics.

Definition 18 (Preordered Layered Magma). *A preordered layered magma is a tuple $(X, \preccurlyeq, \circ)$, with X a set, $\preccurlyeq$ a preorder on X, and $\circ$ a binary partial operation on X.* □

It is clear that, given a preordered scaffold $(\mathcal{G}, \mathcal{E}, X, \preccurlyeq)$, the structure $(X, \preccurlyeq, @_{\mathcal{E}})$ is a preordered layered magma. Analogously to the classical case [6], we can generate a layered Heyting algebra.

Proposition 9. *Every preordered layered magma generates a layered Heyting algebra.*

Proof. Let $(X, \preccurlyeq, \circ)$ be a preordered layered magma. An up-set of the preorder $(X, \preccurlyeq)$ is a set $U \subseteq X$ such that $x \in U$ and $x \preccurlyeq y$ implies $y \in U$. Denote the set of all up-sets of X by $\mathrm{Up}(X)$. The structure $(\mathrm{Up}(X), \cup, \cap, \to, \emptyset, X)$ is a Heyting algebra, where $\to$ is defined as follows: $U \to V := \{x \in X \mid \text{ for all } y\,(x \preccurlyeq y \text{ and } y \in U \text{ implies } y \in V)\}$ We define the operators $\blacktriangleright, \mathrel{-\!\!\blacktriangleright}, \mathrel{\blacktriangleright\!\!-}$ as follows:

$$\begin{aligned}
U \blacktriangleright V &:= \{x \in X \mid \text{ there exists } y \in U, z \in V\,(y \circ z{\downarrow} \text{ and } y \circ z \preccurlyeq x)\}\\
U \mathrel{-\!\!\blacktriangleright} V &:= \{x \in X \mid \text{ for all } y, z\,(x \preccurlyeq y \text{ and } y \circ z{\downarrow} \text{ and } z \in U \text{ implies } y \circ z \in V)\}\\
U \mathrel{\blacktriangleright\!\!-} V &:= \{x \in X \mid \text{ for all } y, z\,(x \preccurlyeq y \text{ and } z \circ y{\downarrow} \text{ and } z \in U \text{ implies } z \circ y \in V)\}
\end{aligned}$$

It is straightforward that these all define up-sets, and are thus well-defined. It remains to prove monotonicity of $\blacktriangleright$ and adjointness of the operators. For monotonicity, let $U \subseteq U'$, $V \subseteq V'$ and $x \in U \blacktriangleright V$. Then there exist $y \in U \subseteq U'$ and $z \in V \subseteq V'$ such that $y \circ z{\downarrow}$ and $y \circ z \leq x$. It follows immediately that $x \in U' \blacktriangleright V'$.

Next, adjointness. We give just one case, for $\mathrel{\blacktriangleright\!\!-}$. The others are similar. Suppose $V \subseteq U \mathrel{\blacktriangleright\!\!-} W$. We must show $U \blacktriangleright V \subseteq W$, so assume $x \in U \blacktriangleright V$. It follows that there exist $x_0 \in U$ and $x_1 \in V$ such that $x_0 \circ x_1{\downarrow}$ and $x_0 \circ x_1 \preccurlyeq x$. By assumption, $x_1 \in U \mathrel{\blacktriangleright\!\!-} W$ and we have $x_1 \preccurlyeq x_1$, $x_0 \circ x_1{\downarrow}$ and $x_0 \in U$, so it follows that $x_0 \circ x_1 \in W$. Finally, W is an up-set, so $x_0 \circ x_1 \preccurlyeq x$ entails $x \in W$, and the verification is complete. □

We can now get the soundness and completeness of the layered graph semantics with respect to $\mathrm{ILGL_H}$ as a special case of the algebraic semantics. Note that a *persistent* valuation $\mathcal{V}\colon \mathrm{Prop} \to \wp(X)$ corresponds uniquely to a valuation $\mathcal{V}: \mathrm{Prop} \to \mathrm{Up}(X)$. By definition, for each propositional variable p, $\mathcal{V}(\mathrm{p})$ is an up-set of the preorder $(X, \preccurlyeq)$ and trivially an up-set of $(X, \preccurlyeq)$ is an element of $\wp(X)$. We can thus use a persistent valuation to generate an interpretation $[\![-]\!]_{\mathcal{V}}$ on the layered Heyting algebra generated by $(X, \preccurlyeq, @_{\mathcal{E}})$.

Proposition 10. *For any layered graph model $\mathcal{M}$ with valuation $\mathcal{V}\colon \mathrm{Prop} \to \wp(X)$ and every formula φ of* ILGL, *we have $[\![\varphi]\!]_{\mathcal{V}} = \{G \in X \mid G \models_{\mathcal{M}} \varphi\} \in \mathrm{Up}(X)$.* □

Hence the layered graph semantics of ILGL is a special case of the algebraic semantics and $\mathrm{ILGL_H}$ is sound and complete with respect to the layered graph semantics.

Proposition 11 (Equivalence of the Hilbert and Tableaux Systems). $\vdash \varphi$ *is provable in* $\mathrm{ILGL_H}$ *iff there is closed tableau for φ.* □

5 Extension to Resources and Actions: Examples

To express the examples mentioned in Sect. 1 conveniently and efficiently, we consider an extension of layered graph semantics and ILGL in which we label the ambient graph with resources and consider action modalities (cf. Stirling's intuitionistic Hennessy–Milner logic [22]) that express resource manipulations.

This extension introduces a degree of statefulness to ILGL without changing the underlying semantics.

This extension is based on an assignment of a set of resources R to the vertices of the graph G. That is, each $r \in R$ is situated at vertices of G. Such assignments are denoted $G[R]$, where we think of G as the (directed) graph of locations in a system model. Resources should also carry sufficient structure to allow some basic operations on resource elements. In [4,5,17], resources are required to form pre-ordered partial monoids, such as the natural numbers $(\mathbb{N}, \leq, +, 0)$, and we use this approach here. Let $(\mathcal{R}, \sqsubseteq, \circ, e)$ be a resource monoid, where $\mathcal{R}$ is a collection of sets of resources and $\circ : \mathcal{R} \times \mathcal{R} \to \mathcal{R}$ is a commutative and associative binary operation. It is easy to see that assignments of resources can be composed and that the algebraic semantics can be easily extended (cf. [6]).

Lemma 11. *Consider @ and $\circ$. Both are binary operations with @ non-commutative and non-associative while $\circ$ is commutative and associative. A non-commutative, non-associative operation on graphs labelled with resources can be defined.*

Proof. We have $@_{\mathcal{E}} : \mathcal{G} \times \mathcal{G} \to \mathcal{G}$ and $\circ : \mathcal{R} \times \mathcal{R} \to \mathcal{R}$. Define $\bullet_{\mathcal{E}} : (\mathcal{G} \times \mathcal{R}) \times (\mathcal{G} \times \mathcal{R}) \to (\mathcal{G} \times \mathcal{R})$ as $(G_1, R_1) \bullet_{\mathcal{E}} (G_2, R_2) = (G_1 \, @_{\mathcal{E}} \, G_2, R_1 \circ R_2)$. It is clear that $\bullet_{\mathcal{E}}$ is both non-commutative and non-associative. □

We write $G[R] \preccurlyeq G'[R']$ to denote the evident containment ordering on labelled graphs and resources (i.e., G' is a subgraph of G and $R \sqsubseteq R'$). We assume also a countable set *Act* of actions, with elements a, etc. Action modalities, $\langle a \rangle$ and $[a]$ manipulate (e.g., add to, remove from) the resources assigned to the vertices of the graph.

Definition 19 (Satisfaction in Resource-Labelled Models). *We extend layered graph models to graphs labelled with resources and extend the interpretation of formulae to the action modalities. For a resource monoid $\mathcal{R}$, a countable set of actions, Act, and a layered graph model $\mathcal{M} = (\mathcal{X}, \mathcal{V})$ over labelled graphs, with the containment ordering on labelled graphs, we generate the satisfaction relation* $\models_{\mathcal{M}} \subseteq X[R] \times \mathrm{Form}$ *as*

$$G[R] \models_{\mathcal{M}} \top \text{ always} \qquad G[R] \models_{\mathcal{M}} \bot \text{ never} \qquad G[R] \models_{\mathcal{M}} \mathrm{p} \text{ iff } G[R] \in \mathcal{V}(\mathrm{p})$$
$$G[R] \models_{\mathcal{M}} \varphi \wedge \psi \text{ iff } G[R] \models_{\mathcal{M}} \varphi \text{ and } G[R] \models_{\mathcal{M}} \psi \quad G[R] \models_{\mathcal{M}} \varphi \vee \psi \text{ iff } G[R] \models_{\mathcal{M}} \varphi \text{ or } G[R] \models_{\mathcal{M}} \psi$$
$$G[R] \models_{\mathcal{M}} \varphi \to \psi \text{ iff, for all } G'[R'] \text{ such that } G[R] \preccurlyeq G'[R'], G'[R'] \models_{\mathcal{M}} \varphi \text{ implies } G'[R'] \models_{\mathcal{M}} \psi$$

$$\begin{aligned} G[R] \models_{\mathcal{M}} \varphi_1 \blacktriangleright \varphi_2 \text{ iff for some } & G_1[R_1], G_2[R_2] \text{ such that } G_1[R_1] \bullet_{\mathcal{E}} G_2[R_2] \preccurlyeq G[R], \\ & G_1[R_1] \models_{\mathcal{M}} \varphi_1 \text{ and } G_2[R_2] \models_{\mathcal{M}} \varphi_2 \\ G[R] \models_{\mathcal{M}} \varphi \mathrel{-\!\!\blacktriangleright} \psi \text{ iff for all } & G[R] \preccurlyeq H[S] \text{ and all } K[T] \text{ such that } H[S] \bullet_{\mathcal{E}} K[T] \downarrow, \\ & K[T] \models_{\mathcal{M}} \varphi \text{ implies } (H[S] \bullet_{\mathcal{E}} K[T]) \models_{\mathcal{M}} \psi \\ G[R] \models_{\mathcal{M}} \varphi \mathrel{\blacktriangleright\!\!-} \psi \text{ iff for all } & G[R] \preccurlyeq H[S] \text{ and all } K[T] \text{ with } K[T] \bullet_{\mathcal{E}} H[S] \downarrow, \\ & K[T] \models_{\mathcal{M}} \varphi \text{ implies } (K[T] \bullet_{\mathcal{E}} H[S]) \models_{\mathcal{M}} \psi \end{aligned}$$

$$G[R] \models_{\mathcal{M}} \langle a \rangle \varphi \text{ iff } \textit{for some well-formed } G[R'] \text{ such that} G[R] \xrightarrow{a} G[R'], G[R'] \models_{\mathcal{M}} \varphi$$
$$G[R] \models_{\mathcal{M}} [a] \varphi \text{ iff } \textit{for all well-formed } G[R'] \text{ such that } G[R] \xrightarrow{a} G[R'], G[R'] \models_{\mathcal{M}} \varphi \quad \square$$

We defer the presentation of the metatheory to account for this extension, including proof systems and completeness results, to another occasion. To do so

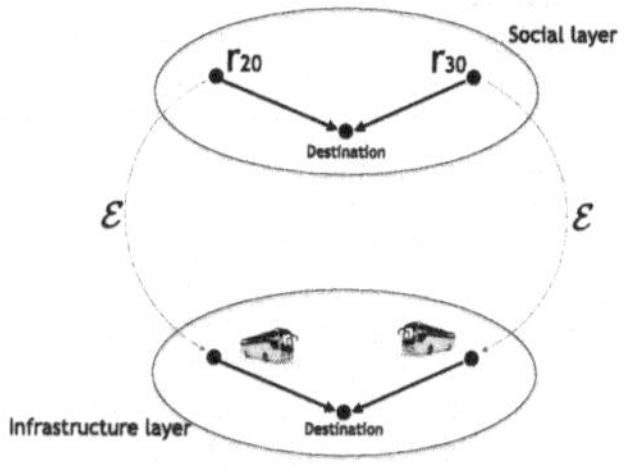

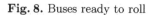

Fig. 8. Buses ready to roll

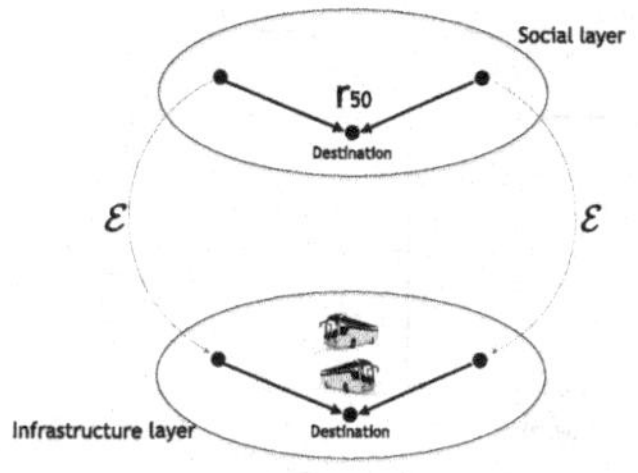

Fig. 9. Buses arrive at meeting

we follow the approach of dynamic epistemic logics [23], wherein the transitions underlying the action modalities correspond to maps between models rather than states. It is clear persistence will not (and should not) hold for action modalities, but at any given model persistence will hold. To extend the tableaux system we should instead take sequences of CSSs, together with a history of actions following similar approaches in the proof theory of Public Announcement Logic [1].

Example 1 (A Transportation Network). Here we abstract a public transportation network into social and infrastructure layers. For a meeting in the social layer to be quorate, sufficient people (say 50) must attend. To achieve this, there must be buses of sufficient capacity to transport 50 people, represented as resources, to the meeting hall, in the infrastructure layer (see Figs. 8 and 9). The formula ϕ_{quorum} denotes a quorate meeting, ϕ_x denotes that x number of people are picked up at bus stops, and the arrival of buses of capacity x in the infrastructure layer is denoted by the action modality $\langle \mathrm{bus}_x \rangle$. These actions move x amount of people from the bus stops to the meeting hall in the social layer. Let $\phi_{meeting}$ assert the existence of a meeting in the social layer, G_1. Then, if G_2 denotes the graph of the infrastructure layer, we have the formulae

$$G_2[R] \models_{\mathcal{M}} \langle \mathrm{bus}_{25} \rangle \langle \mathrm{bus}_{35} \rangle ((\phi_{meeting} \blacktriangleright \phi_{50}) \mathrel{-\!\!\blacktriangleright} \phi_{quorum})$$
$$G_2[R] \models_{\mathcal{M}} \langle \mathrm{bus}_{40} \rangle ((\phi_{meeting} \blacktriangleright \phi_{40}) \mathrel{-\!\!\blacktriangleright} \neg \phi_{quorum})$$

which assert that having two buses available with a total capacity of more than 50 will allow the meeting to proceed, but that a single bus with capacity 40 will not.

Example 2 (A Security Barrier). This example (see Fig. 10) is a situation highlighted by Schneier [21], wherein a security system is ineffective because of the existence of a side-channel that allows a control to be circumvented. The security policy, as expressed in the security layer, with graph G_1, requires that a token be possessed in order to pass from the outside to the inside; that is, $\langle \mathrm{pass} \rangle (\phi_{\mathrm{inside}} \rightarrow \phi_{\mathrm{token}})$. However, in the routes layer, with graph G_2, it is possible to perform an action $\langle \mathrm{swerve} \rangle$ to drive around the gate, as shown in the Fig. 11; that is,

$$G_1 \mathbin{@_{\mathcal{E}}} G_2 \models_{\mathcal{M}} (\langle \mathrm{pass} \rangle (\phi_{\mathrm{inside}} \rightarrow \phi_{\mathrm{token}}) \blacktriangleright \langle \mathrm{swerve} \rangle (\phi_{\mathrm{inside}} \wedge \neg \phi_{\mathrm{token}}))$$

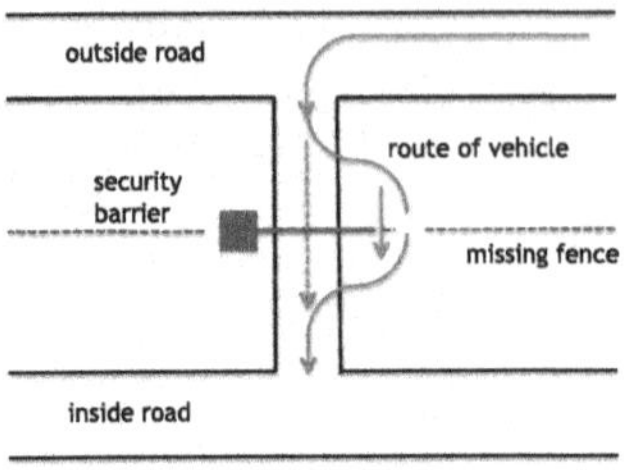

Fig. 10. The security barrier and side channel

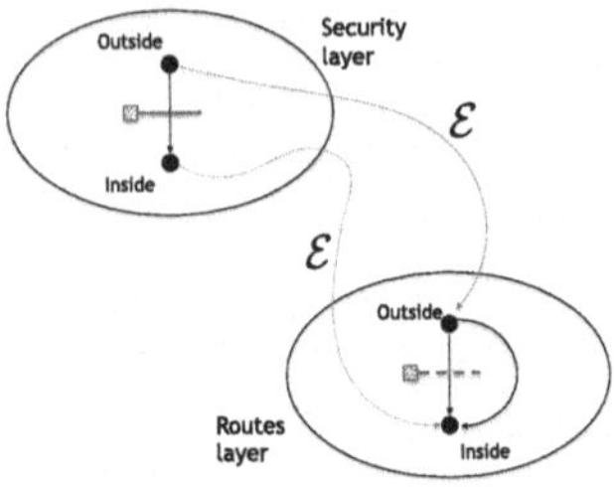

Fig. 11. The layered graph model

Thus we can express the mismatch between the security policy and architecture to which it is intended to apply.

Example 3 (An Organizational Security Architecture). Our final example concerns an organization which internally has high- and low-security parts of its network. It also operates mobile devices that are outside of its internal network but able to connect to it. Figure 12 illustrates our layered graph model of this set-up. We can give a characterization in ILGL of a side channel that allows a resource from the high-security part of the internal network to transfer to the low-security part via the external mobile connection. Associated with the mobile layer are actions that allow the transference of data We have two local compliance properties, in the high- and low-security parts of the network, respectively: $\chi_{\mathrm{high}}(r)$ describes compliance with a policy allowing resource in the high-security network and $\chi_{\mathrm{sec}}(r)$ is a correctness condition that if a resource r is not permitted in the low-security network, then it is not in it. We take actions copy, download, upload associated with the mobile layer G_2, allowing data to be copied to another location as well as moved down and up $\mathcal{E}$-edges respectively, with $\theta(r)$ a compliance property such that $G_2[R] \models_{\mathcal{M}} \langle \mathrm{copy} \rangle \theta(r)$ in order to copy data r. Now we have that

$$G_2[R] \models_{\mathcal{M}} \langle \mathrm{download} \rangle ((\chi_{\mathrm{high}}(r) \blacktriangleright \theta(r)) \wedge \langle \mathrm{copy} \rangle \langle \mathrm{upload} \rangle (\theta(r) \blacktriangleright \neg \chi_{\mathrm{sec}}(r)))$$

showing that the mobile layer is a side channel that can undermine the policy χ_{sec}.

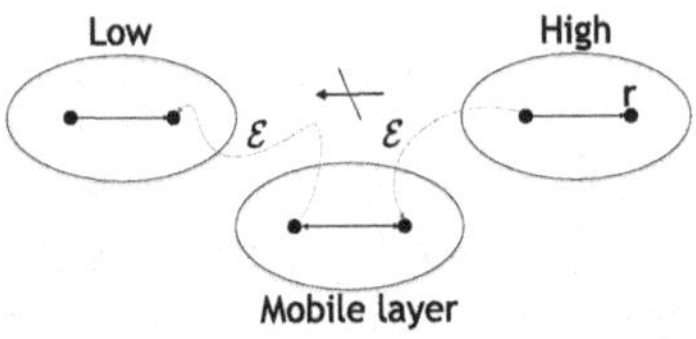

Fig. 12. Organizational security architecture

References

1. Balbiani, P., Ditmarsch, H., Herzig, A., de Lima, T.: Tableaux for public announcement logic. J. Logic Comput. **20**(1), 55–76 (2008)
2. Bezhanishvili, N., de Jongh, D.: Intuitionistic logic. Technical report PP-2006-25, Institute for Logic, Language and Computation Universiteit van Amsterdam (2006)
3. Bródka, P., Skibicki, K., Kazienko, P., Musiał, K.: A degree centrality in multi-layered social network. In: International Conference on Computational Aspects of Social Networks (2011)
4. Collinson, M., Monahan, B., Pym, D.: A Discipline of Mathematical Systems Modelling. College Publications, London (2012)
5. Collinson, M., Pym, D.: Algebra and logic for resource-based systems modelling. Math. Struct. Comput. Sci. **19**(5), 959–1027 (2009)
6. Collinson, M., McDonald, K., Pym, D.: A substructural logic for layered graphs. J. Logic Comput. **24**(4), 953–988 (2014). doi:10.1093/logcom/exv019. http://logcom.oxfordjournals.org/content/early/2015/06/04/logcom.exv019.full.pdf+html
7. Collinson, M., McDonald, K., Pym, D.: Layered graph logic as an assertion language for access control policy models. J. Logic Comput. (2015). doi:10.1093/logcom/exv020
8. Courtault, J.-R., Galmiche, D.: A modal separation logic for resource dynamics. J. Logic Comput. (2015). doi:10.1093/logcom/exv031
9. Docherty, S., Pym, D.: Intuitionistic layered graph logic. Research note RN 16/03, Department of Computer Science, UCL. http://www.cs.ucl.ac.uk/fileadmin/UCL-CS/research/Research_Notes/RN_16_03.pdf
10. Fiat, A., Foster, D., Karloff, H., Rabani, Y., Ravid, Y., Vishwanathan, S.: Competitive algorithms for layered graph traversal. SIAM J. Comput. **28**(2), 447–462 (1998)
11. Fitting, M.: Tableau methods of proof for modal logics. Notre Dame J. Formal Logic **13**(2), 237–247 (1972)
12. Galmiche, D., Méry, D., Pym, D.: The semantics of BI and resource tableaux. Math. Struct. Comput. Sci. **15**(06), 1033–1088 (2005)
13. Kurant, M., Thiran, P.: Layered complex networks. Phys. Rev. Lett. **96**, 138701 (2006)
14. Lambek, J.: On the calculus of syntactic types. In: Proceedings of the 12th Symposia on Applied Mathematics, Studies of Language and Its Mathematical Aspects, Providence, pp. 166–178 (1961)
15. Lambek, J.: From categorical grammar to bilinear logic. In: Schroeder-Heister, P., Došen, K. (eds.) Substructural Logics, pp. 207–237. Oxford University Press, Oxford (1993)
16. Larchey-Wendling, D.: The formal proof of the strong completeness of partial monoidal boolean BI. J. Logic Comput. (2014). doi:10.1093/logcom/exu031
17. O'Hearn, P., Pym, D.: The logic of bunched implications. Bull. Symbolic Logic **5**(2), 215–244 (1999)
18. Papadimitriou, C., Yannakakis, M.: Shortest paths without a map. Theoret. Comput. Sci. **84**(1), 127–150 (1991)
19. Paz, A.: A theory of decomposition into prime factors of layered interconnection networks. Discrete Appl. Math. **159**(7), 628–646 (2011)
20. Pym, D., O'Hearn, P., Yang, H.: Possible worlds and resources: the semantics of BI. Theoret. Comput. Sci. **315**(1), 257–305 (2004)

21. Schneier, B.: The weakest link (2005). (https://www.schneier.com/blog/archives/2005/02/the_weakest_lin.html). Schneier on Security (https://www.schneier.com)
22. Stirling, C.: Modal logics for communication systems. Theoret. Comput. Sci. **49**, 311–347 (1987)
23. van Ditmarsch, H., van der Hoek, W., Kooi, B.: Dynamic epistemic logic. Synthese Library (2008)

Gen2sat: An Automated Tool for Deciding Derivability in Analytic Pure Sequent Calculi

Yoni Zohar[1(✉)] and Anna Zamansky[2]

[1] Tel Aviv University, Tel Aviv, Israel
yoni.zohar@cs.tau.ac.il
[2] Haifa University, Haifa, Israel
annazam@is.haifa.ac.il

Abstract. Gen2sat [1] is an efficient and generic tool that can decide derivability for a wide variety of propositional non-classical logics given in terms of a sequent calculus. It contributes to the line of research on computer-supported tools for investigation of logics in the spirit of the "logic engineering" paradigm. Its generality and efficiency are made possible by a reduction of derivability in analytic pure sequent calculi to SAT. This also makes Gen2sat a "plug-and-play" tool so it is compatible with any standard off-the-shelf SAT solver and does not require any additional logic-specific resources. We describe the implementation details of Gen2sat and an evaluation of its performance, as well as a pilot study for using it in a "hands on" assignment for teaching the concept of sequent calculi in a logic class for engineering practitioners.

1 Introduction

Logic Engineering [2] is a quickly developing field which studies ways to investigate and construct new logical formalisms with "nice" properties (such as decidability, appropriate expressive power, effective reasoning methods, etc.), for a particular need or application. Handling whole families (or "product lines") of non-classical logics calls for automatic methods for their construction and investigation, as well as for new approaches in teaching these topics to future logicians and practitioners. Such ideas of automatic support for the investigation of logics, first presented in [21], were realized in a variety of tools, such as MultLog [5], TINC [8], NESCOND [22], LoTREC [13], MetTeL [24], and many others.

The tool Gen2sat is a contribution to the above paradigm, which particularly aims to support investigators who use *sequent calculi* for the specification of logics. Sequent calculi are a prominent proof-theoretic framework, suitable for performing proof search in a wide variety of different logics. However, a great deal of ingenuity is required for developing efficient proof-search algorithms for sequent calculi (see, e.g., [12]). Aiming to support users with minimal background in programming and automated reasoning techniques, Gen2sat uses a *uniform* method for deciding derivability of a sequent in a given calculus using

This research was supported by The Israel Science Foundation (grant no. 817-15).

N. Olivetti and A. Tiwari (Eds.): IJCAR 2016, LNAI 9706, pp. 487–495, 2016.
DOI: 10.1007/978-3-319-40229-1_33

the polynomial reduction of [16] to SAT. Shifting the intricacies of implementation and heuristic considerations to the realm of off-the-shelf SAT solvers, the tool is lightweight and focuses solely on the transformation of derivability to a SAT instance. As such, it also has the potential to serve as a tool that can enhance learning and research of concepts related to proof theory and semantics of non-classical logics, in particular those of sequent calculi. To demonstrate the educational potential of the tool, in this paper we report on a pilot on using it to enhance learning sequent calculi by graduate Information Systems students at the University of Haifa.

2 Analytic Pure Sequent Calculi

While this paper's focus is on the implementation and usage of Gen2sat, the theoretical background can be found in [16]. Below we briefly review the relevant results.

The variety of sequent calculi which can be handled by Gen2sat includes the family of *pure analytic calculi.* A derivation rule is called *pure* if it does not enforce any limitations on the context formulas. For example, the right introduction rule of implication in classical logic is pure, but not in intuitionistic logic, where only left context formulas are allowed. A sequent calculus is called pure if it includes all the standard structural rules:[1] weakening, identity and cut; and all its derivation rules are pure.

For a finite set $\circledcirc$ of unary connectives, we say that a formula φ is a $\circledcirc$-subformula of a formula ψ if either φ is a subformula of ψ, or $\varphi = \circ\psi'$ for some $\circ \in \circledcirc$ and proper subformula ψ' of ψ. A pure calculus is $\circledcirc$-*analytic* if whenever a sequent s is provable in it, s can be proven using only formulas from $sub^{\circledcirc}(s)$, the set of $\circledcirc$-subformulas of s. We call a calculus *analytic* if it is $\circledcirc$-analytic for some set $\circledcirc$. Note that $\emptyset$-analyticity amounts to the usual subformula property. Many well-known logics can be represented by analytic pure sequent calculi, including three and four-valued logics, various paraconsistent logics, and extensions of primal infon logic ([16] presents some examples).

In order to decide derivability in sequent calculi, Gen2sat adopts the following semantic view:

Definition 1. Let $\mathbf{G}$ be an analytic pure sequent calculus. A $\mathbf{G}$-*legal bivaluation* is a function v from some set of formulas to $\{0, 1\}$ that respects each rule of $\mathbf{G}$, that is, for every instance of a rule, if v assigns 1 to all premises, it also assigns 1 to the conclusion. This definition relies on the following extension of bivaluations to sequents: $v(\Gamma \Rightarrow \Delta) = 1$ iff $v(\psi) = 0$ for some $\psi \in \Gamma$ or $v(\psi) = 1$ for some $\psi \in \Delta$.

Theorem 1 ([16]). *Let $\circledcirc$ be a set of unary connectives, $\mathbf{G}$ a $\circledcirc$-analytic pure sequent calculus, and s a sequent. s is provable in $\mathbf{G}$ if and only if there is no $\mathbf{G}$-legal bivaluation v with domain $sub^{\circledcirc}(s)$ such that $v(s) = 0$.*

[1] We take sequents to be pairs of *sets* of formulas, and therefore exchange and contraction are built in.

Thus, given a ⓒ-analytic calculus **G** and a sequent s as its input, Gen2sat does not search for a proof. Instead, it searches for a countermodel of the sequent, by encoding in a SAT instance the following properties of the countermodel: (*1*) Assigning 0 to s; and (*2*) Being **G**-legal with domain $sub^{©}(s)$.

Gen2sat is capable of handling also impure rules of the form $(*i)\ \dfrac{\Gamma \Rightarrow \Delta}{*\Gamma \Rightarrow *\Delta}$ for *Next*-operators. This requires some adaptations of the above reduction, that are described in [16]. $(*i)$ is the usual rule for *Next* in LTL (see, e.g., [15]). It is also used as $\Box$ (and $\Diamond$) in the modal logic $KD!$ of functional Kripke frames (also known as KF and $KDalt1$). In primal infon logic [10] *Next* operators play the role of quotations.

3 Features and Usage

There is a variety of tools developed in the spirit of logic engineering, such as MultLog [5], TINC [8], NESCOND [22], LoTREC [13], and finally MetTeL [24], which generates a theorem prover for a given logic, as well as a source code for the prover, that can be further optimized. The aim of Gen2sat is similar, allowing the user to specify the logic and automatically obtain a decision procedure. In contrast to MetTeL which uses tableaux, in Gen2sat the logic is given by a sequent calculus. Moreover, the core of Gen2sat is a reduction to SAT, thus it leaves the"hard work" and heuristic considerations of optimizations to state of the art SAT solvers, allowing the user to focus solely on the *logical* considerations.

Gen2sat can be run both via a web interface and from the command line. In the web-based version the user fills in a form; in the command line a property file is passed as an argument. From the command line, Gen2sat is called by: `java -jar gen2sat.jar <path>`. The form has the following fields:

Connectives. A comma separated list of connectives, each specified by its symbol and arity, separated by a colon.
Next **operators.** A comma separated list of the symbols for the next operators.
Rules. Each rule is specified in a separate line that starts with "rule:". The rule itself has two parts separated by "/": the premises, which is a semicolon separated list of sequents, and the conclusion, which is a sequent.
Analyticity. For the usual subformula property this field is left empty. For other forms of analyticity, it contains a comma separated list of unary connectives.
Input sequent. The sequent whose derivability should be decided.

The web-based version includes predefined forms for some propositional logics (e.g. classical logic, primal infon logic and more). In addition, it allows the user to import sequent calculi from *Paralyzer*.[2]

[2] Paralyzer is a tool that transforms Hilbert calculi of a certain general form into equivalent analytic sequent calculi. It was described in [7] and can be found at http://www.logic.at/people/lara/paralyzer.html.

```
Input file
connectives: P:2, E:2
rule: =>a; =>b / =>aPb
rule: a=> / aPb=>
rule: b=> / aPb=>
rule: =>a; =>b / =>aEb
rule: =>b; a=> / aEb=>
analyticity:
inputSequent: (((m1 P m2 ) E k) E k),k=>m1

Output
provable
There's a proof that uses only these rules:
[=>b; a=> / a E b=>, a=> / a P b=>]
```

Fig. 1. An provable instance

```
Input file
connectives: AND:2,OR:2,IMPLIES:2,TOP:0
nextOperators: q1 said, q2 said, q3 said
rule: =>p1; =>p2 / =>p1 AND p2
rule: p1,p2=> / p1 AND p2=>
rule: =>p1,p2 / =>p1 OR p2
rule: =>p2 / =>p1 IMPLIES p2
rule: =>p1; p2=> / p1 IMPLIES p2=>
rule: / => TOP
analyticity:
inputSequent: =>q1said (p IMPLIES p)

Output
unprovable
Countermodel:
q1said p=false, q1said(p IMPLIES p)=false
```

Fig. 2. An unprovable instance

If the sequent is unprovable, Gen2sat outputs a countermodel. If it is provable, Gen2sat recovers a sub-calculus in which the sequent is already provable (naturally, the full proof is unobtainable due to the semantic approach of Gen2sat). Thus, for a provable sequent Gen2sat outputs a subset of rules that suffice to prove the sequent.

Figures 1 and 2 present examples for the usage of Gen2sat. In Fig. 1, the input contains a sequent calculus for the Dolev-Yao intruder model [9]. The connectives E and P correspond to encryption and pairing. The sequent is provable, meaning that given two messages m_1 and m_2 that are paired and encrypted twice with k, the intruder can discover m_1 if it knows k. The output also contains the only two rules that are needed in order to prove the sequent. In Fig. 2, the input file contains a sequent calculus for primal infon logic, where the implication connective is not reflexive, and hence the input sequent is unprovable. Note that the rules for the next operators are fixed, and therefore they are not included in the input file. Both calculi are $\emptyset$-analytic, and hence the analyticity field is left empty. Gen2sat supports analyticity w.r.t. any number of unary connectives, and hence this field may include a list of unary connectives.

4 Implementation Details

Gen2sat is implemented in Java and uses sat4j [17] as its underlying SAT solver. Since the algorithm from [16] is a "one-shot" reduction to SAT, no changes are needed in the SAT solver itself. In particular, sat4j can be easily replaced by other available solvers. Figure 3 includes a partial class diagram of Gen2sat. The two main modules of sat4j that we use are `specs`, which provides the solver itself, and `xplain`, which searches for an unsat core. The main class of Gen2sat is `DecisionProcedure`, that is instantiated with a specific `SequentCalculus`. Its main method `decide` checks whether the input sequent is provable. Given a `Sequent` s, `decide` generates a `SatInstance` stating that s has a countermodel, by applying the rules of the calculus on the relevant formulas, as described above.

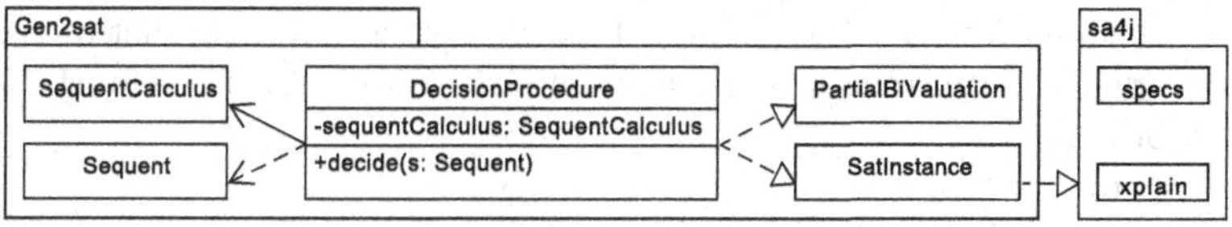

Fig. 3. A partial class diagram of Gen2sat

`SatInstance` is the only class that uses sat4j directly, and thus it is the only class that will change if another SAT solver is used.

For satisfiable instances, the `specs` module returns a satisfying assignment, which is directly translated to a countermodel in the form of a `PartialBivaluation`. For unsatisfiable instances, the `xplain` module generates a subset of clauses that is itself unsatisfiable. Tracking back to the rules that induced these clauses, we are able recover a smaller sequent calculus in which s is already provable. Note however, that the smaller calculus need not be analytic, and then the correctness, that relies on Theorem 1 might fail. Nevertheless, correctness is preserved in this case, as the "if" part of Theorem 1 holds even for non-analytic calculi. Thus, although Gen2sat does not provide a proof of the sequent, we do obtain useful information about the rules that were used in it.

5 Performance

Gen2sat can be used in applications based on reasoning in non-classical logics, and especially paraconsistent logics [6], as there exist analytic pure sequent calculi for many of them [4]. For evaluating its performance, we have considered the well-known paraconsistent logic C_1 [11], using the $\{\neg\}$-analytic calculus from [4], and compared its running time with KEMS [18–20], a theorem prover that implements several KE tableau calculi for classical logic and paraconsistent logics.[3] For this comparison, a lighter version of Gen2sat was compiled, called $Gen2sat_m$, that only decides whether the input sequent is provable, without providing a countermodel or a smaller calculus. The experiments were made on a dedicated Linux machine with four dual-core 2.53 GHz AMD Opteron 285 processors and 8 GB RAM. We have used the problem families of *provable* sequents for benchmarking paraconsistent logics provers from [18, 20], and extended them by creating similar problem families of *unprovable* sequents. For example, problems of the first family from [18] have the form:

$$\Phi_n^1 = \bigwedge_{i=1}^{n}(\neg A_i), \bigwedge_{i=1}^{n}(\circ A_i \to A_i), (\bigvee_{i=1}^{n} \circ A_i) \vee (\neg A_n \to C) \Rightarrow C$$

[3] The tableau calculus for C_1 in KEMS can be translated to a sequent calculus (the connection between the two frameworks was discussed e.g. in [3]). However, the translated sequent calculus is non-analytic,and thus cannot be used with Gen2sat.

where $\circ$ is defined by $\circ A = \neg(A \wedge \neg A)$. Dismissing the first conjunct of the first formula leads to an unprovable sequent. Similarly, problems of the fourth family have the form:

$$\Phi_n^4 = \bigwedge_{i=1}^{n} A_i, \bigwedge_{j=1}^{n}((A_j \vee B_j) \rightarrow \circ A_{j+1}), (\bigwedge_{k=2}^{n} \circ A_k) \rightarrow A_{n+1} \Rightarrow \neg\neg A_{n+1}$$

Replacing $\neg\neg A_{n+1}$ with $\neg A_{n+1}$ leads to an unprovable sequent.

Table 1 includes the running times in miliseconds for the first and fourth problem families.[4] Similar results were obtained on the other families.

In our tests, KEMS performed better on smaller inputs, while Gen2sat performed better on larger ones. Also, provable sequents are easier than unprovable ones for KEMS (which searches for a proof), while the opposite holds for Gen2sat (which searches for a countermodel). For provable sequents, *Gen2sat*$_m$ performs much faster than the full version, as it does not call the `xplain` module of sat4j. In contrast, for unprovable sequents, the difference between the two versions is negligible. The table also contains the number of variables and clauses in the SAT instances that were generated by Gen2sat. In the case of C_1, the set of variables corresponds to the set of subformulas of the sequent and their negations.

Table 1. Benchmark results for provable and unprovable sequents in C_1

	provable					unprovable				
	KEMS	Gen2sat	*Gen2sat*$_m$	#vars	#clauses	KEMS	Gen2sat	*Gen2sat*$_m$	#vars	#clauses
Φ_{10}^1	133	342	213	137	344	153	224	215	135	339
Φ_{20}^1	675	252	73	277	694	686	70	70	275	689
Φ_{50}^1	13934	747	143	697	1744	14247	146	159	695	1739
Φ_{80}^1	75578	1393	148	1117	2794	78212	175	203	1115	2789
Φ_{100}^1	175716	2178	235	1397	3494	182904	226	284	1395	3489
Φ_{10}^4	124	291	212	173	410	205	220	219	173	410
Φ_{20}^4	502	207	78	353	840	1416	83	78	353	840
Φ_{50}^4	8723	444	158	893	2130	36282	137	159	893	2130
Φ_{80}^4	45130	744	178	1433	3420	226422	194	190	1433	3420
Φ_{100}^4	123619	908	220	1793	4280	661078	227	227	1793	4280

6 Gen2sat for Education: A Pilot

There is an ongoing debate on the appropriate way of teaching logic and formal methods to future software engineering practitioners, in particular on ways to *bridge* between the taught material and the software domain (see, e.g., [23]). As a contribution to the latter question, we have initiated a pilot of integrating a "hands-on" assignment based on Gen2sat into a logic course for Information Systems graduate students[5], exploring its potential to enhance learning of the

[4] Out of 11 possible formula comparator choices of KEMS, Table 1 presents the results for the best performing one in each problem.

[5] The second author has been teaching the course for several years at the University of Haifa; see [25] for further details on the course design.

concept of sequent calculi. The assignment aimed to allow them to experiment with different sequent calculi, discovering "a whole new world" of non-classical logics. To increase their engagement, the assignment had the "look and feel" of a software engineering assignment whose domain is non-classical logics.

After a two hour lecture on sequent calculi and the system LK, we introduced Gen2sat in class and explained its functionality and features. The students were then requested to play the role of *testers* of the tool. More concretely, they were requested to provide a test plan (as small as possible) which would cover all possible scenarios the tool could encounter. For a quantifiable measure for success we used a standard approach of measuring code coverage, instructing them to install the Eclemma plug-in for Eclipse [14] for determining the percentage of code activated for a given input. Thus, basically the students' assignment was producing a minimal test plan that would achieve maximal code coverage. When analyzing different inputs to the tool, the students would potentially gain insights into the wide variety of non-classical logics defined in terms of sequent calculi.

The results of our pilot were encouraging: of eight students who participated in the assignment, all ended up submitting[6] test plans which achieved between 70 % – 85 % coverage, and included non-trivial sequent calculi for different languages. An anonymous feedback questionnaire showed that the students found the assignment helpful for understanding sequent calculi, as well as engaging and fun. This seems to us an indication of the potential of integrating "hands on" assignments in the spirit of logic engineering in illuminating educational logical content through experimenting with software tools.

7 Conclusions and Future Work

We have introduced Gen2sat, an efficient tool that decides derivability for a wide family of non-classical logics via the reduction to SAT given in [16]. In the spirit of the "logic engineering" paradigm, Gen2sat is *generic*: it receives as input the language and rules of a sequent calculus of a very general form. In addition, Gen2sat works on top of standard off-the-shelf SAT solvers, without requiring any additional logic-specific resources. Our preliminary experimental results show that the generality of Gen2sat does not come at the expence of its performance, making it appropriate for practical automated reasoning in non-classical logics. We plan to extend these experiments to include more provers for logics with analytic pure sequent calculi.

As a result of its semantic approach, Gen2sat currently does not provide actual proofs of provable sequents. This can be overcome by integrating Gen2sat with other existing theorem provers so that for unprovable sequents, the theorem prover will not have to search for a proof, while for provable sequents, the search space can be potentially reduced by exploiting gensat's capability of supplying a sufficient subset of rules. Our experience with teaching the concept of sequent

[6] Interestingly, seven students employed new connectives with arity greater than 2 and three employed also 0-ary connectives (which indeed increased coverage), although they have not seen any such example in class.

calculi in a "hands-on" assignment on test design for Gen2sat shows its potential in educational settings. Another direction for further research is developing an educational version of the tool, in which learning concepts from non-classical logics could be achieved via interacting with the software.

References

1. Gen2sat website. http://www.cs.tau.ac.il/research/yoni.zohar/gen2sat
2. Areces, C.E.: Logic engineering: the case of description and hybrid logics. Institute for Logic, Language and Computation (2000)
3. Avron, A.: Gentzen-type systems, resolution, tableaux. J. Autom. Reasoning **10**(2), 265–281 (1993)
4. Avron, A., Konikowska, B., Zamansky, A.: Efficient reasoning with inconsistent information using C-systems. Inf. Sci. **296**, 219–236 (2015)
5. Baaz, M., Fermüller, C.G., Salzer, G., Zach, R.: Multlog 1.0: towards an expert system for many-valued logics. In: McRobbie, M.A., Slaney, J.K. (eds.) CADE 1996. LNCS, vol. 1104, pp. 226–230. Springer, Heidelberg (1996)
6. Carnielli, W., Coniglio, M.E., Marcos, J.: Logics of formal inconsistency. In: Gabbay, D.M., Guenthner, F. (eds.) Handbook of Philosophical Logic, vol. 14, pp. 1–93. Springer, New York (2007)
7. Ciabattoni, A., Lahav, O., Spendier, L., Zamansky, A.: Automated support for the investigation of paraconsistent and other logics. In: Artemov, S., Nerode, A. (eds.) LFCS 2013. LNCS, vol. 7734, pp. 119–133. Springer, Heidelberg (2013)
8. Ciabattoni, A., Spendier, L.: Tools for the investigation of substructural and paraconsistent logics. In: Fermé, E., Leite, J. (eds.) JELIA 2014. LNCS, vol. 8761, pp. 18–32. Springer, Heidelberg (2014)
9. Comon-Lundh, H., Shmatikov, V.: Intruder deductions, constraint solving and insecurity decision in presence of exclusive OR. In: 2003 Proceedings of the 18th Annual IEEE Symposium on Logic in Computer Science, pp. 271–280, June 2003
10. Cotrini, C., Gurevich, Y.: Basic primal infon logic. J. Logic Comput. **26**(1), 117–141 (2013)
11. da Costa, N.C.: Sistemas formais inconsistentes, vol. 3. Editora UFPR (1993)
12. Degtyarev, A., Voronkov, A.: The inverse method. In: Robinson, A., Voronkov, A. (eds.) Handbook of Automated Reasoning, vol. 1, pp. 179–272. MIT Press, Cambridge (2001)
13. Gasquet, O., Herzig, A., Longin, D., Sahade, M.: LoTREC: logical tableaux research engineering companion. In: Beckert, B. (ed.) TABLEAUX 2005. LNCS (LNAI), vol. 3702, pp. 318–322. Springer, Heidelberg (2005)
14. Hoffmann, M., Iachelini, G.: Code coverage analysis for eclipse. In: Eclipse Summit Europe (2007)
15. Kawai, H.: Sequential calculus for a first order infinitary temporal logic. Math. Logic Q. **33**(5), 423–432 (1987)
16. Lahav, O., Zohar, Y.: SAT-based decision procedure for analytic pure sequent calculi. In: Demri, S., Kapur, D., Weidenbach, C. (eds.) IJCAR 2014. LNCS, vol. 8562, pp. 76–90. Springer, Heidelberg (2014)
17. Le Berre, D., Parrain, A.: The Sat4j library, release 2.2. J. Satisfiability Boolean Mode. Comput. **7**, 59–64 (2010)
18. Neto, A., Finger, M.: Effective prover for minimal inconsistency logic. In: Bramer, M. (ed.) Artificial Intelligence in Theory and Practice. IFIP, vol. 217, pp. 465–474. Springer US, London (2006)

19. Neto, A., Finger, M.: Kems-a multi-strategy tableau prover. In: Proceedings of the VI Best MSc Dissertation/PhD Thesis Contest (CTDIA 2008), Salvador (2008)
20. Neto, A., Kaestner, C.A.A., Finger, M.: Towards an efficient prover for the paraconsistent logic C1. Electron. Notes Theoret. Comput. Sci. **256**, 87–102 (2009)
21. Ohlbach, H.J.: Computer support for the development and investigation of logics. Logic J. IGPL **4**(1), 109–127 (1996)
22. Olivetti, N., Pozzato, G.L.: NESCOND: an implementation of nested sequent calculi for conditional logics. In: Demri, S., Kapur, D., Weidenbach, C. (eds.) IJCAR 2014. LNCS, vol. 8562, pp. 511–518. Springer, Heidelberg (2014)
23. Page, R.L.: Software is discrete mathematics. ACM SIGPLAN Not. **38**, 79–86 (2003). ACM
24. Tishkovsky, D., Schmidt, R.A., Khodadadi, M.: Mettel2: towards a tableau prover generation platform. In: PAAR@ IJCAR, pp. 149–162 (2012)
25. Zamansky, A., Farchi, E.: Teaching logic to information systems students: challenges and opportunities. In: Fourth International Conference on Tools for Teaching Logic, TTL (2015)

Verification

Model Checking Parameterised Multi-token Systems via the Composition Method

Benjamin Aminof[1(✉)] and Sasha Rubin[2]

[1] Technische Universität Wien, Vienna, Austria
benj@forsyte.at
[2] Università di Napoli "Federico II", Naples, Italy
sasha.rubin@unina.it

Abstract. We study the model checking problem of parameterised systems with an arbitrary number of processes, on arbitrary network-graphs, communicating using multiple multi-valued tokens, and specifications from indexed-branching temporal logic. We prove a composition theorem, in the spirit of Feferman-Vaught [21] and Shelah [31], and a finiteness theorem, and use these to decide the model checking problem. Our results assume two constraints on the process templates, one of which is the standard fairness assumption introduced in the cornerstone paper of Emerson and Namjoshi [18]. We prove that lifting any of these constraints results in undecidability. The importance of our work is three-fold: (i) it demonstrates that the composition method can be fruitfully applied to model checking complex parameterised systems; (ii) it identifies the most powerful model, to date, of parameterised systems for which model checking indexed *branching-time* specifications is decidable; (iii) it tightly marks the borders of decidability of this model.

1 Introduction

Many concurrent systems consist of identical processes running in parallel, such as peer-to-peer systems, sensor networks, multi-agent systems, etc. [14,27,29]. Model checking is a successful technique for establishing correctness of such systems: model a system as the product transition system $\mathbf{P}^G$, where $\mathbf{P}$ is a transition system representing the process, and G is a network-graph describing the communication lines [9]. If the number of processes is not known, or too large for model-checking tools, it is appropriate to express correctness as a parameterised model checking (PMC) problem: decide if a given temporal logic specification holds irrespective of the number of processes [8,22]. That is, for a fixed infinite set $\mathcal{G}$ of network-graphs (e.g., $\mathcal{G}$ may be the set of all ring network-graphs), decide, given process $\mathbf{P}$ and specification φ if $\mathbf{P}^G \models \varphi$ for all $G \in \mathcal{G}$. Not surprisingly, PMC is a hard problem, i.e., even for a given $\mathbf{P}$, PMC consists

Benjamin Aminof is supported by the Vienna Science and Technology Fund (WWTF) through grant ICT12-059. Sasha Rubin is a Marie Curie fellow of the Istituto Nazionale di Alta Matematica.

N. Olivetti and A. Tiwari (Eds.): IJCAR 2016, LNAI 9706, pp. 499–515, 2016.
DOI: 10.1007/978-3-319-40229-1_34

of model-checking infinitely many systems; in other words, it can be thought of as model checking infinite-state systems [15,23]. It quickly becomes undecidable [8], even if the participating processes are finite-state [32], and even if they do not communicate with each other at all [25]. Thus, much work has focused on proving decidability for restricted systems, i.e., by limiting both the communication mechanism and the specification logic [1–3,6,7,10,13,17,20].

We consider specifications in indexed branching temporal logic without the "next-time" operator X (formulas without X are stuttering-insensitive, and are thus natural for specifying asynchronous concurrent systems [18]). More specifically, we use formulas of prenex indexed-$\mathsf{CTL}^*_d\backslash\mathsf{X}$ (CTL^* without X in which there at most $d \in \mathbb{N}$ nested path quantifiers). These are formulas of the form $\forall x_1 \exists x_2 \cdots \forall x_k \phi$ where the variables x_i vary over processes, and ϕ is a $\mathsf{CTL}^*_d\backslash\mathsf{X}$ formula where atomic propositions are paired with the index variables x_i [30]. This specification language allows one to express many natural properties, e.g. mutual-exclusion. Non-prenex temporal logic is so powerful that its parameterised model checking is undecidable already for indexed $\mathsf{LTL}\backslash\mathsf{X}$ specifications, even for non-communicating processes [25]. We consider systems with an arbitrary number of processes, on arbitrary network-graphs, communicating using multiple multi-valued tokens. Such systems arise in various contexts: multiple tokens are a means to resolve conflicts over multiple shared resources such as in the drinking philosophers problem [12], they can represent mobile finite-state agents [4,5,29], and tokens are used in self-stabilisation algorithms [24]. We further allow the edges of the network-graph to carry directions, called local port-numberings, along which the processes may send and receive tokens. Such network-graphs are typical in the distributed computing literature, for instance in mobile finite-state agents, e.g., [14,26,27]. Note that even slightly more powerful communication primitives such as pairwise-rendezvous have undecidable PMC for expressive logics such as prenex indexed $\mathsf{CTL}^*\backslash\mathsf{X}$ [3].

The Compositional Method for Parameterised Model-Checking. Composition theorems, pioneered in the seminal work of Feferman and Vaught [21] and Shelah [31], are tools that reduce reasoning about compound structures to reasoning about their component parts. Unfortunately, composition theorems for product systems are not easy to come by [28]. Nonetheless, we successfully apply the composition method to multi-token systems and prenex indexed-$\mathsf{CTL}^*_d\backslash\mathsf{X}$ specification languages. Our composition result (Theorem 3) states, roughly, that if two processes $\mathbf{X}, \mathbf{Y}$ are bisimilar, and if two network-graphs G, H with k visible vertices $\bar{g}, \bar{h}$ (i.e., vertices that formulas can talk about) are $\mathsf{CTL}^*_d\backslash\mathsf{X}$-equivalent, then the product systems $\mathbf{X}^G, \mathbf{Y}^H$ with visible processes at $\bar{g}, \bar{h}$ are $\mathsf{CTL}^*_d\backslash\mathsf{X}$-equivalent. We complement this with a finiteness result (Theorem 5) that states, roughly, that for every $d, k \in \mathbb{N}$, there are only finitely many $\mathsf{CTL}^*_d\backslash\mathsf{X}$-types of network-graphs G with k visible vertices (even though, over all graphs, there are infinitely many logically inequivalent $\mathsf{CTL}^*_d\backslash\mathsf{X}$ formulas, already for $d = 1$). Combining the composition and finiteness we reduce reasoning about $\mathbf{P}^G$ for all $G \in \mathcal{G}$ to reasoning about finitely many $G \in \mathcal{G}$, and thus decide the PMC.

Our systems employ two fairness conditions: the standard assumption (introduced in [18]) that processes that make infinitely many transitions must make infinitely many token passing transitions; and the assumption that from every state from which a process can send (resp. receive) a token, it can also reach a state in which it can send (resp. receive) the token in any other given direction and value. We show that if either of the fairness conditions is removed PMC becomes undecidable; furthermore, it remains undecidable even if other very restrictive assumptions are added. It is notable that until now it was not known if the standard fairness assumption was necessary for decidability. Thus, our results answer this question in the affirmative. Due to space constraints, some proofs are only sketched or omitted.

2 Definitions

A *labeled transition system (LTS)* is a tuple $\langle \mathsf{AP}, \Sigma, Q, Q_0, \delta, \lambda \rangle$ where AP is a finite set of *atomic propositions* (also called *atoms*), Σ is a finite set of *actions*, Q is a finite set of *states*, $Q_0 \subseteq Q$ is a set of *initial states*, $\delta \subseteq Q \times \Sigma \times Q$ is a *transition relation*, and $\lambda : Q \to 2^{\mathsf{AP}}$ is a *labeling* function. We write $q \xrightarrow{\sigma} q'$ if $(q, \sigma, q') \in \delta$, and write $q \to q'$ if $(q, \sigma, q') \in \delta$ for some $\sigma \in \Sigma$. An LTS is *total* if for every $q \in Q$ there exists $q' \in Q$ such that $q \to q'$. A *transition system (TS)* is a tuple $\langle \mathsf{AP}, Q, Q_0, \delta, \lambda \rangle$ like an LTS, except that $\delta \subseteq Q \times Q$. A *path* of an LTS is a finite string $q_0 q_1 \dots q_n \in Q^+$ or an infinite string $q_0 q_1 \dots \in Q^\omega$ such that $q_i \to q_{i+1}$ for all i. An *edge-path* of an LTS is a (finite or infinite) sequence of transitions $(q_0, \sigma_0, q_1)(q_1, \sigma_1, q_2) \dots$ of δ. Every edge-path $(q_0, \sigma_0, q_1)(q_1, \sigma_1, q_2) \dots$ *induces* the path $q_0 q_1 q_2 \dots$. A path is *simple* if no vertex repeats, and it is a *simple cycle* if the (only) two equal vertices are the first and last. An edge-path is *simple (resp. simple cycle)* if the induced path is. A *run* of an LTS is a maximal path starting in an initial state. An LTS can be translated into a TS by simply removing the actions from transitions. We will implicitly use this translation.

System Model. Informally, a token-passing system is an LTS obtained by taking some finite edge-labeled graph (called a *topology* or *network-graph*), placing one process at each of its vertices, and having all processes execute the same code (given in the form of a finite-state *process template*). Processes synchronize by sending one of finitely many tokens along the edges of the topology, which are labeled with a send direction, a receive direction, and a token-value.[1] In the most general model, processes can choose the direction to send the token, from which direction to receive a token, and the value of the token. In case there is more than one possible recipient for a token, one is chosen nondeterministically.

In what follows, we use a finite non-empty set of *token values* Σ_{val}, finite disjoint non-empty sets Σ_{snd} of *send directions* and Σ_{rcv} of *receive directions*,

[1] The direction-labels on the edges (also called a local orientation) represent network port numbers [14,27]. All of our results also hold for the case that each edge has a single direction-label that combines send and receive directions, e.g., "clockwise".

and an integer $T > 0$ (the number of tokens in the system). Since these data are usually fixed, we do not mention them if they are clear from the context.

Process Template. Fix a countable set $\mathcal{AP}$ of atomic propositions for use by all process templates. We assume that $\mathcal{AP}$ also contains, for every integer $i \geq 0$, the special proposition tok^i. A *process template* (w.r.t. $T \in \mathbb{N}, \Sigma_{\mathsf{val}}, \Sigma_{\mathsf{snd}}, \Sigma_{\mathsf{rcv}}$) is a total LTS $\mathbf{P} = \langle \mathsf{AP}_{\mathsf{pr}}, \Sigma_{\mathsf{pr}}, Q, Q_0, \delta, \lambda \rangle$ such that: (i) $\mathsf{AP}_{\mathsf{pr}} \subset \mathcal{AP}$ is a finite set containing tok^i for $0 \leq i \leq T$; (ii) for every $q \in Q$ there is exactly one i such that $\mathsf{tok}^i \in \lambda(q)$ (and we say that q *has* i *tokens*); (ii) $\Sigma_{\mathsf{pr}} = \{\mathsf{int}\} \cup [(\Sigma_{\mathsf{snd}} \cup \Sigma_{\mathsf{rcv}}) \times \Sigma_{\mathsf{val}}]$; (iv) $Q_0 = \{\iota_T, \iota_0\}$ where ι_T has T tokens, and ι_0 has 0 tokens; (v) For every transition $q \xrightarrow{\sigma} q'$: if $\sigma \in \Sigma_{\mathsf{snd}} \times \Sigma_{\mathsf{val}}$ then $\exists i > 0$ such that q has i tokens and q' has $(i-1)$ tokens; if $\sigma \in \Sigma_{\mathsf{rcv}} \times \Sigma_{\mathsf{val}}$ then $\exists i < T$ such that q has i tokens and q' has $(i+1)$ tokens; and if $\sigma = \mathsf{int}$ then q and q' have the same number of tokens.

Notation. We say that the *initial states* of two templates $\mathbf{X}, \mathbf{Y}$ *are bisimilar* if, writing $\iota_0^{\mathbf{Z}}, \iota_T^{\mathbf{Z}}$ for the initial states of template $\mathbf{Z}$, we have that $\iota_\epsilon^{\mathbf{X}} \sim \iota_\epsilon^{\mathbf{Y}}$ for $\epsilon \in \{0, T\}$, where $\sim$ is a bisimulation relation between $\mathbf{X}$ and $\mathbf{Y}$. The elements of Q are called *local states* and the transitions in δ are called *local transitions*. A transition (q, σ, q') is called a *local send transition* if $\sigma \in \Sigma_{\mathsf{snd}} \times \Sigma_{\mathsf{val}}$; it is called a *local receive transition* if $\sigma \in \Sigma_{\mathsf{rcv}} \times \Sigma_{\mathsf{val}}$; and it is called a *local internal transition* if $\sigma = \mathsf{int}$. The local send/receive transitions are collectively known as *local token-passing transitions.* A local state q for which there exists a local send-transition (resp. receive-transition) $(q, (d, m), q')$ is called *ready to send (resp. receive) in direction* d *and value* m; it is also called *ready to send (resp. receive) in direction* d, *ready to send (resp. receive) value* m, or simply *ready to send (resp. receive).*

Fairness Notions. A template $\mathbf{P}$ is *fair* if every infinite path $q_1 q_2 \cdots$ in $\mathbf{P}$ satisfies that for infinitely many i the transition from q_i to q_{i+1} is a local token passing transition [18]. Other restrictions that we consider involve treating different directions and/or different token values in an unbiased way, and thus "fairly". Formally, a state q of $\mathbf{P}$ having i tokens that is ready to send (resp. receive) is called an *i-sending* (resp. *i-receiving*) state. A path in $\mathbf{P}$ is an i-path if it only mentions states having i tokens. A template $\mathbf{P}$ is *direction/value-fair* if for every $\mathsf{d} \in \Sigma_{\mathsf{rcv}}$, $\mathsf{e} \in \Sigma_{\mathsf{snd}}$ and $\mathsf{m} \in \Sigma_{\mathsf{val}}$, for every i-receiving (resp. i-sending) state q there is a finite i-path from q ending in a state that is ready to receive (resp. send) in direction d (resp. from direction e) and value m. We denote by $\mathcal{P}^{\mathsf{FDV}}$ the set of fair and direction/value-fair process-templates; and by $\mathcal{P}^{\mathsf{FD}}$ the set of fair, direction-fair, and valueless (i.e., with $|\Sigma_{\mathsf{val}}| = 1$) process templates. As noted in the introduction, the undecidability results (Sect. 4) show that the limitations of $\mathcal{P}^{\mathsf{FDV}}$ are, in a strong sense, minimal limitations one can impose and still obtain a decidable parameterized model checking problem.

Topology/Network-Graph. LTS $G = \langle \emptyset, \Sigma_{\mathsf{snd}} \times \Sigma_{\mathsf{rcv}}, V, \{\mathsf{init}\}, E, \lambda \rangle$ is a *topology* (w.r.t. $\Sigma_{\mathsf{snd}}, \Sigma_{\mathsf{rcv}}$) if: (i) $V = [n]$ for some $n \in \mathbb{N}$ is a set of *vertices* (or *process indices*); (ii) $\mathsf{init} \in V$ is an *initial vertex*; (iii) $E \subseteq V \times (\Sigma_{\mathsf{snd}} \times \Sigma_{\mathsf{rcv}}) \times V$ is called the *edge relation*, (iv) and λ is the constant function $\lambda(v) = \emptyset$. We abbreviate and write $G = \langle V, E, \mathsf{init} \rangle$ or $G = \langle V_G, E_G, \mathsf{init}_G \rangle$. The *underlying graph of* G has vertex set V and edge (v, w) iff $\exists \mathsf{d}, \mathsf{e}.(v, (\mathsf{d}, \mathsf{e}), w) \in E$. We assume that the underlying graph is irreflexive, contains no vertices without outgoing edges, and that every vertex $v \in V$ is reachable from init. These are natural assumptions since paths in the topology represent the paths along which the tokens can move.

Parameterized Topology $\mathcal{G}$. Let $\mathcal{G}$ denote a countable set of topologies. For example, the set of all pipelines (see Fig. 1), or the set of all rings.

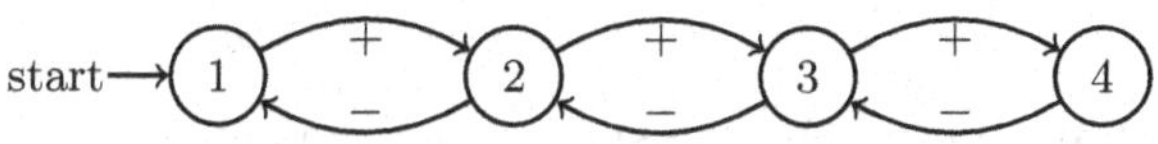

Fig. 1. Example pipeline topology: + signifies a (snd_E, rcv_W) label, and – (snd_W, rcv_E).

Token-Passing System. Given a process template $\mathbf{P} = \langle \mathsf{AP}_{\mathsf{pr}}, \Sigma_{\mathsf{pr}}, Q, Q_0, \delta, \lambda \rangle$ and a topology $G = \langle V, E, \mathsf{init} \rangle$, we define the *token-passing system* (or TPS for short) to be the LTS $\mathbf{P}^G = \langle \mathsf{AP}_{\mathsf{sys}}, \Sigma_{\mathsf{sys}}, S, S_0, \Delta, \Lambda \rangle$. Informally, the system $\mathbf{P}^G$ can be thought of as the interleaving parallel composition of $\mathbf{P}$ over G. The tokens start with process init. Time is discrete: at each step either exactly one process makes a local internal transition; or exactly two processes, say at vertices v, w, simultaneously make local token-passing transitions as the process at v sends a token in direction d with value m, and the one at w receives the token from direction e with value m. Such a transition can occur only if $(v, (\mathsf{d}, \mathsf{e}), w) \in E$.

Formally: (1) $\mathsf{AP}_{\mathsf{sys}} := \mathsf{AP}_{\mathsf{pr}} \times V$ is the set of *indexed atomic propositions* (because it is standard notation, we sometimes write p_i instead of (p, i)); (2) $\Sigma_{\mathsf{sys}} := \{\mathsf{int}\} \cup (\Sigma_{\mathsf{snd}} \times \Sigma_{\mathsf{rcv}} \times \Sigma_{\mathsf{val}})$ is the set of actions; (3) The set of *global states* is $S := Q^V$, i.e., all functions from V to Q (informally, if $s \in Q^V$ is a global state then $s(i)$ denotes the local state of the process with index i); (4) The set of *global initial states* S_0 consists of the unique global state $s \in Q_0^V$ such that $s(\mathsf{init}) = \iota_T$, and for all $i \neq \mathsf{init}$, $s(i) = \iota_0$; (5) The labeling $\Lambda(s) \subset \mathsf{AP}_{\mathsf{sys}}$ for $s \in S$ is defined as follows: $p_i \in \Lambda(s)$ if and only if $p \in \lambda(s(i))$, for $p \in \mathsf{AP}_{\mathsf{pr}}$ and $i \in V$ (informally, p_i is true at s if and only if p is true at the corresponding local state of the process with index i); (6) The *global transition relation* $\Delta \subseteq S \times \Sigma_{\mathsf{sys}} \times S$ consists of global internal transitions and global token-passing transitions (collectively called *global transitions*), defined as follows: *(6a)* The *global internal transitions* are elements of the form (s, int, s') for which there exists a process index $v \in V$ such that $s(v) \stackrel{\mathsf{int}}{\to} s'(v)$ is a local internal transition of $\mathbf{P}$, and for all $w \in V \setminus \{v\}$, $s(w) = s'(w)$. Such a global transition is said to *involve* v. *(6b)* The *global token-passing transitions* are elements of the form $(s, (\mathsf{d}, \mathsf{e}, \mathsf{m}), s')$ for which there exist process indices $v, w \in V$ such that: $(v, (\mathsf{d}, \mathsf{e}), w) \in E$; there is a local send transition $s(v) \stackrel{(\mathsf{d},\mathsf{m})}{\to} s'(v)$ and a local receive transition $s(w) \stackrel{(\mathsf{e},\mathsf{m})}{\to} s'(w)$ of $\mathbf{P}$; and for every $u \in V \setminus \{v, w\}$, $s'(u) = s(u)$. Such a transition is said to have v *sending in direction* d *an* m*-valued token to* w *from direction* e.

Observe that the definition above specifies that at the beginning all tokens are at the initial state of the topology. This is not a real restriction since a system which allows the tokens to start already distributed in a nondeterministic way among multiple vertices can be simulated by adding to the topology a new initial state that starts with all the tokens and from which they are later distributed. One can even accommodate many fixed initial distributions by modifying the specification formula to remove from consideration unwanted distributions.

For a global state s, let tokens(s) be the set of v such that $s(v)$ has one or more tokens. If $T = 1$, we define tokens(s) to be the vertex that has the token.

Specification Language. For the syntax and semantics of CTL^* see [9]. In this work, TL denotes a syntactic fragment of CTL^*, such as $\mathsf{CTL}^*\backslash\mathsf{X}$ (i.e., CTL^* without the "next" operator), or the fragment $\mathsf{CTL}^*_d\backslash\mathsf{X}$ of $\mathsf{CTL}^*\backslash\mathsf{X}$ in which the nesting-depth of the path quantifiers E, A is at most $d \in \mathbb{N}_0$ (see [30]).

A *partition* of an infinite path $\pi = \pi_1\pi_2\dots$ is an infinite sequence $B_1, B_2, \dots$ of finite intervals of $\mathbb{N}$ such that there exist integers $m_1 < m_2 < \dots$ with $m_1 = 1$ and for all $i \in \mathbb{N}$, $B_i = [m_i, m_{i+1} - 1]$. The intervals B_i are called *blocks.*

Definition 1. *[30] For TSs $M = \langle AP, S, S_0, \Delta, \Lambda\rangle$, $M' = \langle AP, S', S_0', \Delta', \Lambda'\rangle$ (over the same set of atomic propositions AP), and non-negative integer d, define relations $\equiv_d \subseteq S \times S'$ as follows:* (i) *$s \equiv_0 s'$ if $\Lambda(s) = \Lambda'(s')$; and* (ii) *$s \equiv_{d+1} s'$ if for every infinite path π in M from s there exists an infinite path π' in M' from s' (and vice versa) and a partition $B_1B_2\dots$ of π and a partition $B_1'B_2'\dots$ of π' such that for every $i \in \mathbb{N}$ and every $b \in B_i, b' \in B_i'$ we have that $\pi_b \equiv_d \pi'_{b'}$.*

In case we need to stress the LTSs, we write $\equiv_d^{M,M'}$ instead of $\equiv_d$. Say that M is TL*-equivalent* to M', denoted by $M \equiv_{\mathsf{TL}} M'$, if they agree on all TL formulas, i.e., for every TL formula φ over AP it holds that $M \models \varphi$ iff $M' \models \varphi$. The next proposition characterizes $\mathsf{CTL}^*_d\backslash\mathsf{X}$-equivalence, denoted $\equiv_{\mathsf{CTL}^*_d\backslash\mathsf{X}}$.

Proposition 1. *[30] For every integer d, TS M with a single initial state s, and TS M' with a single initial state s': $M \equiv_{\mathsf{CTL}^*_d\backslash\mathsf{X}} M'$ if and only if $s \equiv_d s'$.*

Indexed Temporal Logics (ITLs) were introduced in [11,18,19] to model specifications of systems with multiple processes. ITL formulas are built from TL formulas by adding the ability to quantify over process indices using the universal and existential process quantifiers $\forall x_{cond}$ and $\exists x_{cond}$ (generally written as Qx). Accordingly, the atoms are $\mathcal{AP} \times \mathsf{Vars}$, where $\mathsf{Vars} = \{x, y, z, \dots\}$ is some fixed infinite set of index variables (we write p_x instead of $(p, x) \in \mathcal{AP} \times \mathsf{Vars}$). For example, the formula $\forall x \forall y_{x \neq y}. \mathsf{A} \neg \mathsf{F}\, c_x \wedge c_y$ specifies mutual exclusion, i.e., that it is never the case that two different processes simultaneously satisfy atom c. Syntactically, *Indexed-*CTL^* *formulas* are formed by adding the following to the syntax of CTL^* formulas over atomic propositions $\mathsf{AP} \times \mathsf{Vars}$: if φ is an indexed-CTL^* state (resp. path) formula then so are the formulas $\forall x_{cond}.\varphi$ and $\exists x_{cond}.\varphi$, where $x, y \in \mathsf{Vars}$, and *cond* is Boolean combination over predicates of the form true, $(x, y) \in E$, $(y, x) \in E$, and $x = y$.

Semantics. Indexed-CTL^* formulas over variables Vars and atomic propositions $\mathsf{AP}_{\mathsf{pr}}$ are interpreted over a token-passing system $\mathbf{P}^G$, where $\mathbf{P}$ has atomic propositions $\mathsf{AP}_{\mathsf{pr}}$, and $G = \langle V, E, \mathsf{init}\rangle$. A *valuation* is a function $e : \mathsf{Vars} \to V_G$. An *$x$-variant* of e is a valuation e' with $e'(y) = e(y)$ for all $y \in \mathsf{Vars} \setminus x$. First we inductively define what it means for *valuation e to satisfy cond*, written $e \models cond$: $e \models \mathsf{true}$ (for all e); $e \models x = y$ iff $e(x) = e(y)$; $e \models (x, y) \in E$ iff $(e(x), e(y)) \in E$; $e \models \neg cond$ iff $e \not\models cond$; $e \models cond \wedge cond'$ iff $e \models cond$ and $e \models cond'$.

For a TPS $\mathbf{P}^G = \langle \mathsf{AP}_{\mathsf{sys}}, \Sigma_{\mathsf{sys}}, S, S_0, \Delta, \Lambda\rangle$, a global state s, a state formula φ, and a valuation e, define $(\mathbf{P}^G, s) \models \varphi[e]$ inductively:

- $(\mathbf{P}^G, s) \models p_x[e]$ iff $p_{e(x)} \in \Lambda(s)$,
- $(\mathbf{P}^G, s) \models \mathsf{E}\,\psi[e]$ iff $(\mathbf{P}^G, \pi) \models \psi[e]$ for some infinite path π form s in $\mathbf{P}^G$,
- $(\mathbf{P}^G, s) \models \forall x_{cond}.\varphi[e]$ (resp. $(\mathbf{P}^G, s) \models \exists x_{cond}.\varphi[e]$) iff for all (resp. for some) x-variants e' of e that satisfy *cond*, it holds that $(\mathbf{P}^G, s) \models \varphi[e']$,
- $(\mathbf{P}^G, s) \models \varphi \wedge \varphi'[e]$ iff $(\mathbf{P}^G, s) \models \varphi[e]$ and $(\mathbf{P}^G, s) \models \varphi'[e]$, and
- $(\mathbf{P}^G, s) \models \neg\varphi[e]$ iff it is not the case that $(\mathbf{P}^G, s) \models \varphi[e]$.

Path formulas are interpreted similarly, but over $(\mathbf{P}^G, \pi)$, where π is an infinite path. An indexed CTL^* formula is a *sentence* if every atom is in the scope of a process quantifier. Let φ be an indexed-CTL^* state formula. For a valuation e, define $\mathbf{P}^G \models \varphi[e]$ if $(\mathbf{P}^G, s_0) \models \varphi[e]$, where s_0 is the initial state of $\mathbf{P}^G$. If φ is also a sentence, define $\mathbf{P}^G \models \varphi$ if for all valuations e (equivalently, for some valuation) it holds that $(\mathbf{P}^G, s_0) \models \varphi[e]$. Similarly, define $(\mathbf{P}^G, s) \models \varphi$ iff for all valuations (equivalently, for some valuation) $e : \mathsf{Vars} \to V_G$ it holds that $(\mathbf{P}^G, s) \models \varphi[e]$. We use the usual shorthands, e.g., $\forall x.\varphi$ is shorthand for $\forall x_{\mathsf{true}}.\varphi$.

Prenex indexed-TL is a syntactic fragment of indexed-TL in which all the processes' index quantifiers are at the front of the formula, e.g., prenex indexed $\mathsf{CTL}^*\backslash\mathsf{X}$ consists of formulas of the form $(Q_1 x_1) \ldots (Q_k x_k)\,\varphi$ where φ is a $\mathsf{CTL}^*\backslash\mathsf{X}$ formula over atoms $\mathsf{AP} \times \{x_1, \ldots, x_k\}$, and the $Q_i x_i$s are index quantifiers. Such formulas with k quantifiers are called *k-indexed*, collectively written $\{\forall, \exists\}^k$-TL. The union of $\{\forall, \exists\}^k$-TL for $k \in \mathbb{N}$ is written $\{\forall, \exists\}^*$-TL and called (full) prenex indexed TL. The remainder of this paper deals with prenex indexed-$\mathsf{CTL}^*_d\backslash\mathsf{X}$.

Parameterized Model Checking Problem $\mathsf{PMCP}_{\mathcal{G}}(\mathcal{P}, \mathcal{F})$. The *parameterized model checking (PMC) problem* is to decide, given $\mathbf{P} \in \mathcal{P}$ and $\varphi \in \mathcal{F}$, whether or not for all $G \in \mathcal{G}$, $\mathbf{P}^G \models \varphi$. Here $\mathcal{P}$ is a set of process templates, and $\mathcal{F}$ is a set of ITL formulas.

Cutoffs and Decidability. A *cutoff* for $\mathsf{PMCP}_{\mathcal{G}}(\mathcal{P}, \mathcal{F})$ is a natural number c such that for every $\mathbf{P} \in \mathcal{P}$ and $\varphi \in \mathcal{F}$, if $\mathbf{P}^G \models \varphi$ for all $G \in \mathcal{G}$ with $|V_G| \leq c$ then $\mathbf{P}^G \models \varphi$ for all $G \in \mathcal{G}$. Note: if $\mathsf{PMCP}_{\mathcal{G}}(\mathcal{P}, \mathcal{F})$ has a cutoff, then it is decidable. Note that the existence of a cutoff only implies the existence of a decision procedure. For instance, the statement "for every $k \in \mathbb{N}$, $\mathsf{PMCP}_{\mathcal{G}}(\mathcal{P}, \{\forall, \exists\}^k\text{-}\mathsf{TL})$ has a cutoff" does not imply, a priori, that $\mathsf{PMCP}_{\mathcal{G}}(\mathcal{P}, \{\forall, \exists\}^*\text{-}\mathsf{TL})$ is decidable.

3 Decidability Results

In this section we prove that token-passing systems have decidable PMC problem for specifications from k-indexed $\mathsf{CTL}_d^*\backslash\mathsf{X}$ for fair and direction/value-fair process templates. We begin with some definitions.

Notation. A *k-tuple over V_G*, written $\bar{g}$, denotes a tuple $(g_1, \dots, g_k)$ of elements of V_G. We write $v \in \bar{g}$ if $v = g_i$ for some i. Given a valuation $e : \mathsf{Vars} \to V_G$, the relevant part of e for a $\mathsf{CTL}_d^*\backslash\mathsf{X}$ formula with k free variables (w.l.o.g. called $x_1, \dots x_k$) can be described by a k-tuple $\bar{g}$ over V_G (with $g_i = e(x_i)$ for $1 \leq i \leq k$).

The Restriction $\mathbf{P}^G|\bar{g}$. Fix process template $\mathbf{P}$, topology G, and nodes $\bar{g} \in V_G^k$. Define the *restriction of* $\mathbf{P}^G = \langle \mathsf{AP}_{\mathsf{sys}}, S, S_0, \Delta, \Lambda \rangle$ *onto* $\bar{g}$, written $\mathbf{P}^G|\bar{g}$, as the LTS $(\mathsf{AP}_{@}, S, S_0, \Delta, L)$ over atomic propositions $\mathsf{AP}_{@} = \{p@i : p \in \mathsf{AP}_{\mathsf{pr}}, i \in [k]\}$, where for all $s \in S$ the labeling $L(s)$ is defined as follows: $L(s) := \{p@i : p_{g_i} \in \Lambda(s), i \in [k]\}$. Informally, $\mathbf{P}^G|\bar{g}$ is the LTS $\mathbf{P}^G$ with a modified labeling that, for every $g_i \in \bar{g}$, replaces the indexed atom p_{g_i} by the atom $p@i$ (i.e., process indices are replaced by their *positions* in $\bar{g}$); all other atoms are removed. Intuitively, $p@i$ means that the atom $p \in \mathsf{AP}_{\mathsf{pr}}$ holds in the process with index (i.e., at the vertex) g_i. Note that $\mathbf{P}^G$ and $\mathbf{P}^G|\bar{g}$ only differ in their labelling. It is not hard to see that given a k-indexed formula $\theta := Q_1 x_1 \dots Q_k x_k.\ \varphi$, the truth value of φ in $\mathbf{P}^G$, with respect to a valuation for $x_1, \dots x_k$ described by a k-tuple $\bar{g}$, can be deduced by reasoning instead on $\mathbf{P}^G|\bar{g}$ (since for this evaluation φ only "sees" the atomic propositions of processes in vertices in $\bar{g}$).

The Valuation TS $G[\![\bar{g}]\!]$. The idea is to annotate the topology G by atoms that allow logical formulae to talk about the movement of tokens in and out of vertices in $\bar{g}$. In order to capture the directions involved in such movements, we insert new nodes in the middle of any edge of G that is incident with a vertex in $\bar{g}$. Thus, $G[\![\bar{g}]\!]$ is a TS formed as follows: (i) the atoms true at v are the positions that v appears in $\bar{g}$, if any; (ii) split each edge labeled (d, e) involving (one or two) vertices from $\bar{g}$ by inserting a state whose atoms label the directions to or from the vertices from $\bar{g}$ that are involved; (iii) remove all edge labels.

Formally, let $G = \langle V, E, \mathsf{init} \rangle$ be a topology, and let $\bar{g}$ be a k-tuple over V. Define the *valuation TS* $G[\![\bar{g}]\!]$ as the TS $\langle \mathsf{AP}, Q, Q_0, \delta, \lambda \rangle$ where

- $\mathsf{AP} = [k] \cup \Sigma_{\mathsf{snd}} \cup \Sigma_{\mathsf{rcv}}$,
- $Q = V \cup \{[v, \mathsf{d}, \mathsf{e}, w] \mid (v, (\mathsf{d}, \mathsf{e}), w) \in E$ and either $v \in \bar{g}$ or $w \in \bar{g}\}$,
- $Q_0 = \{\mathsf{init}\}$,
- $\delta \subset Q \times Q$ is the union of $\{(v, v') : \exists \mathsf{d}, \mathsf{e}.(v, (\mathsf{d}, \mathsf{e}), v') \in E, v \notin \bar{g} \wedge v' \notin \bar{g}\}$ and $\{(v, [v, \mathsf{d}, \mathsf{e}, w]) : [v, \mathsf{d}, \mathsf{e}, w] \in Q\}$ and $\{([v, \mathsf{d}, \mathsf{e}, w], w) : [v, \mathsf{d}, \mathsf{e}, w] \in Q\}$.
- $\lambda(v) := \{i \in [k] : v = g_i\}$ (for $v \in V$); and $\lambda([v, \mathsf{d}, \mathsf{e}, w])$ is $\{\mathsf{d}\}$ if $v \in \bar{g}, w \notin \bar{g}$, is $\{\mathsf{e}\}$ if $w \in \bar{g}, v \notin \bar{g}$, and is $\{\mathsf{d}, \mathsf{e}\}$ if $v \in \bar{g}, w \in \bar{g}$.

Since Σ_{snd} and Σ_{rcv} are disjoint, the label of $(v, \mathsf{d}, \mathsf{e}, w)$ determines which of v and w is in $\bar{g}$. A valuation TS $G[\![\bar{g}]\!]$ with $|\bar{g}| = k$ is called a *k-valuation TS*. Every edge-path $\xi \in G$ naturally induces a path $\mathsf{map}(\xi)$ in $G[\![\bar{g}]\!]$. Observe that $\mathsf{map}(\xi)$ starts in a node of V_G, and if ξ is finite also ends in a node of V_G. Formally:

$\mathsf{map}((v,\sigma,v'))$ is defined to be vv' if $v \notin \bar{g}$ and $v' \notin \bar{g}$, and $v \cdot [v,\sigma,v'] \cdot v'$ otherwise; and $\mathsf{map}(\xi \cdot (v,\sigma,v'))$ is defined to be $\mathsf{map}(\xi) \cdot v'$ if $v \notin \bar{g}$ and $v' \notin \bar{g}$, and is $\mathsf{map}(\xi) \cdot [v,\sigma,v'] \cdot v'$, otherwise. Note that for a path ρ in $G[\![\bar{g}]\!]$ that begins in a node of V_G (and, if ρ is finite, also ends in V_G), the set $\mathsf{map}^{-1}(\rho)$ is non-empty.

We can now define the COMPOSITION and FINITENESS properties.

- COMPOSITION Property for $\langle \mathsf{TL}, \mathcal{P}, \mathcal{G} \rangle$: For every $k \in \mathbb{N}$, processes $\mathbf{X}, \mathbf{Y} \in \mathcal{P}$, topologies $G, H \in \mathcal{G}$, and k-tuples $\bar{g} \in V_G^k$, $\bar{h} \in V_H^k$: if $G[\![\bar{g}]\!] \equiv_{\mathsf{TL}} H[\![\bar{h}]\!]$ and the initial states of $\mathbf{X}$ and $\mathbf{Y}$ are bisimilar, then $\mathbf{X}^G|\bar{g} \equiv_{\mathsf{TL}} \mathbf{Y}^H|\bar{h}$. In words, the COMPOSITION property states that if the initial states of $\mathbf{X}$ and $\mathbf{Y}$ are bisimilar then one can deduce the logical equivalence of the restrictions $\mathbf{X}^G|\bar{g}, \mathbf{Y}^H|\bar{h}$ from the logical equivalence of the valuation TSs $G[\![\bar{g}]\!], H[\![\bar{h}]\!]$.
- FINITENESS Property for $\langle \mathsf{TL}, \mathcal{G} \rangle$: For every $k \in \mathbb{N}$, the set $\mathcal{M} := \{G[\![\bar{g}]\!] : G \in \mathcal{G}, \bar{g} \in V_G^k\}$ has only finitely many $\equiv_{\mathsf{TL}}$ equivalence classes.

Later in this section we will prove the COMPOSITION and FINITENESS properties with $\mathsf{TL} = \mathsf{CTL}_d^*\backslash\mathsf{X}$ (for fixed $d \in \mathbb{N}$), $\mathcal{P} = \mathcal{P}^{\mathsf{FD}}$. Note that if instead of using valuation TSs one uses arbitrary TSs then the finiteness property does not hold even for $\mathsf{TL} = \mathsf{CTL}_1^*\backslash\mathsf{X}$.[2] Thus, the proof of the finiteness property for $\mathsf{CTL}_d^*\backslash\mathsf{X}$ must, and does, exploit properties of valuation TSs; in particular, the fact that the number of atoms is bounded and no atom is true in more than one state of $G[\![\bar{g}]\!]$ in every path between two vertices in $\bar{g}$.

We now state the main theorem of this section.

Theorem 1. *$PMCP_{\mathcal{G}}(\mathcal{P}^{\mathsf{FDV}}, \{\forall,\exists\}^k\text{-}CTL_d^*\backslash X)$ is decidable for every parameterised topology $\mathcal{G}$ and every $d, k \in \mathbb{N}$.*

The proof is in two steps. First, one removes the token values by encoding them in the directions. Thus, in the statement of Theorem 1, we may replace $\mathcal{P}^{\mathsf{FDV}}$ by $\mathcal{P}^{\mathsf{FD}}$. In step two, we show that (for every $\mathcal{G}, k, d$) the PMC problem has a cutoff using the composition method, following the recipe from [2]:

Theorem 2. *If $\langle TL, \mathcal{P}, \mathcal{G} \rangle$ has the* COMPOSITION *property and $\langle TL, \mathcal{G} \rangle$ has the* FINITENESS *property, then for all $k \in \mathbb{N}$, $PMCP_{\mathcal{G}}(\mathcal{P}, \{\forall,\exists\}^k\text{-}TL)$ has a cutoff.*

Proof (sketch). The truth value of a $\{\forall,\exists\}^k$-TL formula $\theta := Q_1 x_1 \dots Q_k x_k.\, \varphi$ in a system $\mathbf{P}^G$ is a Boolean combination of the truth values of the (non-indexed) TL formula φ, resulting from different valuations of the variables $x_1, \dots, x_k$. By the COMPOSITION property, two different topologies G, H, with corresponding valuations $\bar{g}, \bar{h}$, that yield TL-equivalent valuation TSs will admit the same truth values of φ in $\mathbf{P}^G$, $\mathbf{P}^H$. By the FINITENESS property, all the valuation TSs fall into finitely many TL-equivalence classes. Hence, given G, evaluating θ in $\mathbf{P}^G$ amounts to evaluating a Boolean function (that depends only on $G, Q_1, \dots Q_k$) over finitely many variables (one variable for each representative valuation TS);

[2] Indeed, there are infinitely many $\mathsf{CTL}_1^*\backslash\mathsf{X}$ formulas that are pairwise logically-inequivalent. E.g., every finite word over $\{0,1\}$ can be represented as an LTS, which itself can be axiomatised by a $\mathsf{CTL}_1^*\backslash\mathsf{X}$ formula that uses the U operator.

and evaluating θ with respect to $\mathcal{G}$ amounts to evaluating a set of such functions (all using the same variables). Since there are only finitely many Boolean functions over a finite set of variables we obtain a cutoff.[3] □

3.1 The Composition Theorem

Theorem 3 ***(Composition).*** *For all $d, k \in \mathbb{N}$, topologies G, H, processes $\mathbf{X}, \mathbf{Y} \in \mathcal{P}^{\mathsf{FD}}$, $\bar{g} \in V_G^k$ and $\bar{h} \in V_H^k$: if $G[\![\bar{g}]\!] \equiv_{\mathsf{CTL}_d^*\backslash\mathsf{X}} H[\![\bar{h}]\!]$ and the initial states of $\mathbf{X}$ and $\mathbf{Y}$ are bisimilar, then $\mathbf{X}^G|\bar{g} \equiv_{\mathsf{CTL}_d^*\backslash\mathsf{X}} \mathbf{Y}^H|\bar{h}$.*

Proof (sketch). The proof has the following outline. Let s_0 and s_0' be the initial states of $\mathbf{X}^G|\bar{g}$ and $\mathbf{Y}^H|\bar{h}$, respectively. By Proposition 1, it is enough to show that if the assumption of the theorem holds then $s_0 \equiv_d s_0'$. This is done by induction on d. For stating the inductive hypothesis we first need the following definition: given a system $\mathbf{P}^G$, a function $\beta : [T] \to \text{tokens}(s)$ that maps token numbers to vertices in G is a *token assignment at s* if, for every $v \in \text{tokens}(s)$, the number of tokens mapped to v is equal to the number of tokens at v according to s, i.e., $(\mathsf{tok}^{|\beta^{-1}(v)|}, v) \in \Lambda(s)$, where Λ is the labelling of $\mathbf{P}^G$.

The dth Inductive Hypothesis. *For every global state s of $\mathbf{X}^G|\bar{g}$ and global state s' of $\mathbf{Y}^H|\bar{h}$, conclude that $s \equiv_d s'$ if the following two conditions hold:*

1. *$s(g_i) \sim s'(h_i)$ for all $i \in [k]$, and*
2. *there exists a token assignment β at s, and there exists a token assignment β' at s', such that for all $i \in [T]$ we have that $\beta(i) \equiv_d \beta'(i)$.*

The first condition says that s and s' assign bisimilar states to matching processes in $\bar{g}$ and $\bar{h}$; the second condition says that s and s' have their tokens in nodes of G, H (respectively) that are equivalent according to $\equiv_d^{G[\![\bar{g}]\!],H[\![\bar{h}]\!]}$. The theorem follows by showing that s_0, s_0' satisfy these assumptions.

For the induction base, observe that (by the first assumption in the inductive hypothesis) s and s' assign bisimilar local states to matching processes in $\bar{g}, \bar{h}$ and thus, s, s' have the same labelling and are indistinguishable by a $\mathsf{CTL}_0^*\backslash\mathsf{X}$ formula; now apply Proposition 1.

The main work in proving the inductive step is to satisfy the second condition in the definition of $\equiv_d$ (Definition 1). This requires that, for every path π in $\mathbf{X}^G|\bar{g}$ starting in s, one can find a $(d-1)$-matching path π' in $\mathbf{Y}^H|\bar{h}$ starting at s' (and vice versa). Note that since our setup is symmetric we can ignore the "vice-versa" and only find π' given π. We construct π' using the general scheme graphically depicted in Fig. 2.

[3] The existence of a cutoff is independent of whether $\mathcal{G}$ is computable. However, deciding whether a given number is a cutoff may not be easy. Consider for example the limited setting of [2]: there exists a computable $\mathcal{G}$ and a fixed $\mathbf{P}$ such that it is impossible, given $k, d \in \mathbb{N}$ (even fixing $d = 1$), to compute a cutoff [2]. Nonetheless, by [3], in the same setting (and we believe that also in our broader setting) one can compute a cutoff for many natural parameterized topologies $\mathcal{G}$.

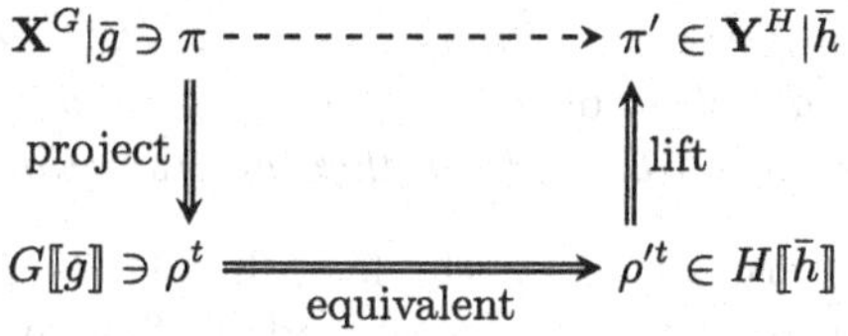

Fig. 2. Proving the COMPOSITION property via Definition 1.

First, given π, we assign to each token a unique number from 1 through T. This allows us to track the movements of individual tokens in G (according to the token-passing transitions of π) which we arbitrarily assume obey the following rule: all processes start in the initial vertex, and during a global token-passing transition, the smallest numbered token that the sending process has is the one being sent. Using this rule, for $i \in \mathbb{N}$ and $t \in [T]$ we can define the function $\mathsf{token}_i : [T] \to V_G$ such that $\mathsf{token}_i(t)$ is the vertex in which the token numbered t is located in the global state π_i. However, in order to construct π' so that it mimics π, we also need to know the directions the tokens take when entering and leaving nodes (which is information that is not explicitly present in π). The reason this is needed is that processes in $\mathbf{X}^G|\bar{g}$ can change state based on the direction a token is sent to or received from (this is possible even if the process is direction-fair). To solve this, we arbitrarily choose some edge-path ξ in $\mathbf{X}^G|\bar{g}$ that induces π — being an edge-path, ξ contains the directions as part of each edge. As it turns out, we only need to know the directions of token-passing transitions affecting processes in $\bar{g}$. These are the send (resp. receive) directions of edges in G that start (resp. end) in a state in $\bar{g}$, and are captured by the extra nodes added to G to construct $G[\![\bar{g}]\!]$. Thus, for each $t \in [T]$ we obtain from ξ a path ρ^t in $G[\![\bar{g}]\!]$ that records the movement of token t along π.

For $t \in [T]$, by the assumption in the inductive hypothesis, $\beta(t) \equiv_d \beta'(t)$. Apply Definition 1 to the path ρ^t (that starts in $\beta(t)$) of $G[\![\bar{g}]\!]$ to get a $(d-1)$-matching path ρ'^t that starts in $\beta'(t)$ of $H[\![\bar{h}]\!]$. The paths $\rho'^1, \ldots, \rho^T$ will serve as the paths of tokens' movements for the path π'.

We construct π' by mimicking the transitions of π: an internal transition in $\mathbf{Y}^H|\bar{h}$ that involves a process $g_i \in \bar{g}$ is mimicked by a bisimilar internal transition using h_i; and a token passing transition[4] of the token number t (for $t \in [T]$) is mimicked as follows: we take the portion w of ρ^t this transition corresponds to, match it to a portion w' of ρ'^t (using the partitioning of ρ^t and ρ'^t into matching blocks), and then push the token t along an edge-path in H that induces w'.

The following lemma says that this last "pushing" step is possible.

Lemma.[5] *Let* $\mathbf{P} \in \mathcal{P}^{\mathsf{FD}}$, *let* p, q *be states of* $\mathbf{P}$, *and let* G *be a topology. Let* $\rho = (v_1, (\mathsf{d}_1, e_1), v_2)(v_2, (\mathsf{d}_2, e_2), v_3) \ldots (v_{m-1}, (\mathsf{d}_{m-1}, e_{m-1}), v_m)$ *be an edge-path that is a simple path (or a simple cycle) in* G, *and let* s *be a state of* $\mathbf{P}^G$ *with* $v_1 \in \text{tokens}(s)$. *There exists a finite edge-path* $\alpha = (f_0, \sigma_0, f_1)(f_1, \sigma_1, f_2) \ldots (f_{h-1}, \sigma_{h-1}, f_h)$ *in* $\mathbf{P}^G$, *with* $f_0 = s$, *such that:*

[4] Fortunately, we only have to mimic such transitions that cross blocks in ρ^t.

[5] The full version of this lemma contains two more conclusions.

1. *α has $m-1$ token-passing transitions and in the ith token-passing transition v_i sends a token in direction* $\mathsf{d_i}$ *to v_{i+1} from direction* $\mathsf{e_i}$ *(for $0 \leq i < m$);*
2. *If vertex $x \in V_G$ is not on the path ρ, then no transition of α involves x.*

This lemma makes crucial use of the fact that **P** is fair and direction-fair. Fairness ensures that tokens can always be made to flow in and out of a process, and direction-fairness ensures that tokens can always flow in any given direction. Indeed, the lemma is not true without both assumptions which is the main reason that without them the composition theorem does not hold and, as we show in Sect. 4, the PMC problem becomes undecidable. □

3.2 FINITENESS Property for $\mathsf{CTL}_d^*\backslash\mathsf{X}$

Our aim in this section is to prove the FINITENESS property for $\mathsf{CTL}_d^*\backslash\mathsf{X}$. We begin by recursively defining, given positive integers k, d and a k-valuation TS $G[\![\bar{g}]\!] = \langle \mathsf{AP}, Q, \{\mathsf{init}\}, \delta, \lambda \rangle$, a marking function Ξ_d^k. This function associates with each vertex $v \in Q$ a $k+1$-dimensional vector $\Xi_d^k(v)$ whose ith coordinate $\Xi_d^k(v)[i]$ is a set of strings over the alphabet $\cup_{u \in Q}\{\Xi_{d-1}^k(u)\}$. The marking function Ξ_d^k will help us later in defining the $\mathsf{CTL}_d^*\backslash\mathsf{X}$-*character* of a valuation TS, which succinctly captures the $\mathsf{CTL}_d^*\backslash\mathsf{X}$-equivalence class of this valuation TS.

Notation. The *positions* of a string v is the set $\{1, \ldots, |v|\}$ if v is finite, and $\mathbb{N}$ otherwise. A string w is *ultimately constant* if $|w| = \infty$ and $w_i = w_j$ for all $j \geq i$, for some i. Recall that the destuttering of a string is formed by removing identical consecutive letters. Define a mapping $pos_v : [|v|] \to [|v|]$ as follows: $pos_v(1) = 1$ and; for $i > 1$, $pos_v(i) := pos_v(i-1)$ if $v_{i-1} = v_i$, and otherwise $pos_v(i) := pos_v(i-1) + 1$. Intuitively, pos_v maps a position i of the string v to its corresponding position in $\mathsf{destut}(v)$. Note that the image of pos_v is of the form $[L]$ for some $L \leq |v|$. Formally, $\mathsf{destut}(v)$ is the string w of length L such that for all $i \leq L$, $w_i = v_{\min\{j : pos_v(j) = i\}}$. Thus, $v_i = w_{pos_v(i)}$ for all $i \leq L$. **The Marking Ξ_d^k.** Fix $k, d \in \mathbb{N}$, topology G, and k-tuple $\bar{g}$ over V_G. Let $G[\![\bar{g}]\!]$ be $\langle \mathsf{AP}, Q, Q_0, \delta, \lambda \rangle$. For every vertex $v \in Q$, let $v^{\rightsquigarrow}$ be the set of maximal paths in G starting in v that have no intermediate nodes in $\bar{g}$. Formally, a (finite or infinite) path $\pi = \pi_1\pi_2 \ldots$ is in $v^{\rightsquigarrow}$ iff: $\pi_1 = v$ and, for all $1 < i < |\pi|$ we have $\pi_i \notin \bar{g}$ and, if π is finite then $|\pi| \geq 2$ and $\pi_{|\pi|} \in \bar{g}$. We write $v^{\rightsquigarrow 0} := \{\pi \in v^{\rightsquigarrow} \mid |\pi| = \infty\}$ for the infinite paths in $v^{\rightsquigarrow}$; also, for every $i \in [k]$, we write $v^{\rightsquigarrow i} := \{\pi \in v^{\rightsquigarrow} \mid \pi_{|\pi|} = g_i\}$ for the set of paths in $v^{\rightsquigarrow}$ that end in g_i.

In the definition below, for a (finite or infinite) path π, we write $\Xi_{d-1}^k(\pi) := \Xi_{d-1}^k(\pi_1)\Xi_{d-1}^k(\pi_2)\ldots$ for the concatenation of the $d-1$ markings of the nodes of π. We define the marking Ξ_d^k of a node inductively (on d) as follows: $\Xi_0^k(v) := \lambda(v)$ and, for $d > 0$, $\Xi_d^k(v)$ is the vector $(\Xi_d^k(v)[0], \ldots, \Xi_d^k(v)[k])$, where $\Xi_d^k(v)[i] := \cup_{\pi \in v^{\rightsquigarrow 0}}\{\mathsf{destut}(\Xi_{d-1}^k(\pi)\}$ if $i = 0$; and $\Xi_d^k(v)[i] := \cup_{\pi \in v^{\rightsquigarrow i}}\{ \mathsf{destut}(\Xi_{d-1}^k(\pi')) \mid \pi' = \pi_1 \ldots \pi_{|\pi|-1}\}$, for $1 \leq i \leq k$. That is, for $d = 0$, the marking $\Xi_d^k(v)$ is the label $\lambda(v)$; and for $d > 1$ the marking $\Xi_d^k(v)$ is a vector of sets of strings, where the ith coordinate of the vector contains the set of strings obtained by de-stuttering the Ξ^k_{d-1} markings of the nodes of paths in $v^{\rightsquigarrow}$ (excluding the last node if $i > 0$) that end in g_i (if $i > 0$), or never visit any node in $\bar{g}$ (if $i = 0$). Observe that, for every $0 \le i \le k$ and every $d > 0$, the marking $\Xi^k_d(v)[i]$ is a set of strings over the alphabet[6] $\cup_{u \in Q}\{\Xi^k_{d-1}(u)\}$, and that all strings in $\Xi^k_d(v)[i]$ start with the letter $\Xi^k_{d-1}(v)$.

Since, for all $0 \le i \le k$ and $d > 0$, all strings in $\Xi^k_d(v)[i]$ start with the letter $\Xi^k_{d-1}(v)$, and since $\Xi^k_d(v)[i] = \emptyset$ iff $v^{\rightsquigarrow i} = \emptyset$, we get the following lemma:

Lemma 1. *For every $d > 0$, if v, u are nodes (of possibly different k-valuation TSs) such that $\Xi^k_d(v) = \Xi^k_d(u)$, then for all $0 < j \le d$ we have that $\Xi^k_j(v) = \Xi^k_j(u)$. If, in addition, $v^{\rightsquigarrow} \neq \emptyset$ then also $\Xi^k_0(v) = \Xi^k_0(u)$.*

The $\mathsf{CTL}^*_d\backslash\mathsf{X}$-*character* of a Valuation TS. Given a k-valuation TS $G[\![\bar{g}]\!] = \langle \mathsf{AP}, Q, \{\mathsf{init}\}, \delta, \lambda \rangle$, the $\mathsf{CTL}^*_d\backslash\mathsf{X}$-*character* of $G[\![\bar{g}]\!]$ is defined as the following vector $(\langle \lambda(\mathsf{init}), \Xi^k_d(\mathsf{init})\rangle, \langle \lambda(g_1), \Xi^k_d(g_1)\rangle, \ldots, \langle \lambda(g_k), \Xi^k_d(g_k)\rangle)$ of the pairs of labels and Ξ^k_d markings of the initial state and the states in $\bar{g}$.

The following theorem relates the $\mathsf{CTL}^*_d\backslash\mathsf{X}$-character of a valuation TS and its $\mathsf{CTL}^*_d\backslash\mathsf{X}$-equivalence class.

Theorem 4. *For every $k, d \in \mathbb{N}$, if $G[\![\bar{g}]\!], H[\![\bar{h}]\!]$ are two k-valuation TSs with the same $\mathsf{CTL}^*_d\backslash\mathsf{X}$-character, then $G[\![\bar{g}]\!] \equiv_{\mathsf{CTL}^*_d\backslash\mathsf{X}} H[\![\bar{h}]\!]$.*

Our next goal is to prove that there are (for given $k, d \in \mathbb{N}$) only finitely many $\mathsf{CTL}^*_d\backslash\mathsf{X}$-characters for all k-valuation TSs. We do this by showing that (for fixed alphabets $\Sigma_{\mathsf{snd}}, \Sigma_{\mathsf{rcv}}$) for all k, d, all k-valuation TSs $G[\![\bar{g}]\!]$, and all $v \in G$, we have that $\Xi^k_d(v)$ ranges over finitely many values. This is clearly true for $d = 0$. For $d > 0$, we prove this by defining a finite poset Υ^k_d, which depends only on k and d, and showing that $\Xi^k_d(v) \in \Upsilon^k_d$. We begin by defining a relation $\preceq$ between sets of strings.

Definition of $\preceq$. For sets of strings $X, Y \subseteq (\Sigma^+ \cup \Sigma^\omega)$, define $X \preceq Y$ if for all $x \in X$ there exists $y \in Y$ such that x is a (not necessarily proper) suffix of y.

It is easy to verify that the relation $\preceq$ is reflexive and transitive, but that it may not be antisymmetric (consider for example $X = \{b, ab\}$ and $Y = \{ab\}$).

Lemma 2. *Given a k-valuation TS $G[\![\bar{g}]\!]$, and a path $\pi_1 \ldots \pi_t$ in it satisfying $\pi_l \notin \bar{g}$ for all $1 < l \le t$, we have that: $\Xi^k_d(\pi_j)[i] \preceq \Xi^k_d(\pi_h)[i]$ for every $0 \le i \le k$, and $d > 0$, and $1 \le h < j \le t$.*

The relation $\preceq$ is antisymmetric when restricted to the domain consisting of sets of strings Z such that: *(i)* all strings in Z start with the same letter $first(Z)$ (i.e., there exists $first(Z) \in \Sigma$ such that for all $w \in Z$, $w_1 = first(Z)$); *(ii)* in every string in Z the letter $first(Z)$ appears only once (i.e., for all $w \in Z$, $i > 1$ implies $w_i \neq first(Z)$). Given an alphabet Σ, let $\mathbb{P}_\Sigma \subset 2^{\Sigma^+ \cup \Sigma^\omega}$ denote the set of all sets of strings Z (over Σ) satisfying the above two conditions. We have:

---

[6] Here, the empty set $\emptyset$ is a letter in $2^{[k]}$, not to be confused with the empty string ϵ.

Lemma 3. *$(\mathbb{P}_\Sigma, \preceq)$ is a partially ordered set.*

Definition of $(\Upsilon_d^k, \preceq_d)$. The definition is by induction on d: for $d = 0$ we have $\Upsilon_0^k := 2^{\mathsf{AP}}$ (recall that $\mathsf{AP} = [k] \cup \Sigma_{\mathsf{snd}} \cup \Sigma_{\mathsf{rcv}}$); and $\preceq_0$ is the transitive closure of the relation obtained by having, for every $X \in 2^{[k]}$, every $\mathsf{d} \in \Sigma_{\mathsf{snd}}$, and every $\mathsf{e} \in \Sigma_{\mathsf{rcv}}$, that: $\{\mathsf{d}\} \preceq_0 X, \{\mathsf{d}, \mathsf{e}\} \preceq_0 X, \emptyset \preceq_0 \{\mathsf{d}\}$, and $\{\mathsf{e}\} \preceq_0 \emptyset$. For $d > 0$, let: $\Upsilon_d^k = \{X \in (\mathbb{P}_{\Upsilon_{d-1}^k})^{k+1} \mid w \in X[i]$ implies $w_{j+1} \prec_{d-1} w_j$ for all $0 \leq i \leq k$ and $1 \leq j < |w|\}$ and take $\preceq_d$ to be the point-wise ordering of vectors, i.e., $X \prec_d Y$ iff $X[i] \preceq Y[i]$ for every $0 \leq i \leq k$, where $\preceq$ is the ordering defined earlier for sets of strings. Intuitively, $X \in \Upsilon_d^k$ iff every coordinate of X contains strings over the alphabet Υ_{d-1}^k that all start with the same letter and are all strictly decreasing chains of the poset $(\Upsilon_{d-1}^k, \preceq_{d-1})$. Observe that if Υ_{d-1}^k is a finite set then there are finitely many strictly decreasing chains (each of finite length) in $(\Upsilon_{d-1}^k, \preceq_{d-1})$, implying that Υ_d^k is also finite. Since Υ_0^k is finite, we can conclude, for every $d \geq 0$, that Υ_d^k is a finite set of finite strings.

The following lemma states that for fixed k, d (recall that we assume fixed alphabets $\Sigma_{\mathsf{snd}}, \Sigma_{\mathsf{rcv}}$) the domain of Ξ_d^k is contained in Υ_d^k (and is thus finite). Note that this also implies that even though the strings in $\Xi_d^k(v)[0]$ are obtained by de-stuttering markings of infinite paths in $v^{\rightsquigarrow 0}$ they are all finite strings.

Lemma 4. *For all k, d, if v is a vertex of a k-valuation TS then $\Xi_d^k(v) \in \Upsilon_d^k$.*

We conclude with the finiteness theorem for $\mathsf{CTL}_d^*\backslash\mathsf{X}$.

Theorem 5 (Finiteness). *For every $k, d \in \mathbb{N}$, the set $\{G[\![\bar{g}]\!] : G$ is a topology, $\bar{g} \in V_G^k\}$ has only finitely many $\equiv_{\mathsf{CTL}_d^*\backslash\mathsf{X}}$ equivalence classes.*

Proof. The theorem follows immediately from the fact that the $\mathsf{CTL}_d^*\backslash\mathsf{X}$-character of a valuation TS is a finite vector, Lemma 4, and Theorem 4.

4 Undecidability

The positive decidability results appearing in Sect. 3 are the strongest one can hope for. Indeed, we prove that if one drops any of the restrictions that were imposed on the process template, namely of fairness and direction/value-fairness, then PMC becomes undecidable. Furthermore, these undecidability results hold even if multiple other strong restrictions are put instead (such as having a single token, having no values, having one send or one receive direction, etc.)

Our proofs reduce the non-halting problem for counter-machines (CMs) to the PMC problem. The basic encoding uses one process (the *controller*) to orchestrate the simulation and store the line number of the CM, and many *memory* processes, each having one bit for each counter. The main difficulty we face, compared to other reductions that follow this basic encoding (e.g., in [2,18,20,32]), is how to make sure that the controller's commands are executed by the memory processes given that the restrictions imposed in the theorems prevent the controller from communicating its commands to the memory processes.

Theorem 6. *Let $\mathcal{P}^{\mathsf{DV}}$ denote the set of process templates that are direction/value-fair but not necessarily fair. There exists $\mathcal{G}$ such that $\mathsf{PMCP}_{\mathcal{G}}$ $(\mathcal{P}^{\mathsf{DV}}, \{\forall\}^5\text{-}\mathsf{LTL}\backslash\mathsf{X})$ is undecidable, even if one limits the processes to have a single valueless token (i.e. $T = 1$ and $|\Sigma_{\mathsf{val}}| = 1$), and with a single receive direction (i.e., $|\Sigma_{\mathsf{rcv}}| = 1$). The same holds replacing "receive" by "send"; furthermore, $\mathcal{G}$ is computable.*

A template **P** is *receive-direction fair* if for every i-sending state q and for every $\mathsf{d} \in \Sigma_{\mathsf{rcv}}$, there is a finite i-path from q ending in a state that is ready to receive in direction d; it is *send-direction fair* if the previous condition holds with "send(ing)" replacing "receive(ing)" and Σ_{snd} replacing Σ_{rcv}; it is *direction-fair* if it is both receive- and send-direction fair. A template **P** is *value-fair* if for every i-receiving (resp. i-sending) state q, and for every token-value $\mathsf{m} \in \Sigma_{\mathsf{val}}$, there is a finite i-path from q ending in a state that is ready to receive (resp. send) value m. It is important to note that a template that is both direction-fair and value-fair is *not*, in general, direction/value-fair. The difference is that while the former can correlate values with directions, the latter cannot. For example, it may be that from every state it can only receive/send in direction a if the value of the token is 0, and receive/send in direction b only if the token value is 1. This kind of behaviour is not allowed if the template is direction/value-fair.

Theorem 7. *Let $\mathcal{P}^{\mathsf{F}}$ be the set of process templates that are fair but not necessarily direction/value-fair. There exists $\mathcal{G}$ such that $\mathsf{PMCP}_{\mathcal{G}}(\mathcal{P}^{\mathsf{F}}, \{\forall\}^5\text{-}\mathsf{LTL}\backslash\mathsf{X})$ is undecidable, even for direction fair and value fair templates with $|\Sigma_{\mathsf{rcv}}| = 1$; furthermore, $\mathcal{G}$ is computable.*

5 Discussion

The literature contains PMC decidability results of token-passing systems with a single token [2,3,13,16,18], and with multiple tokens [16,22].[7] However, the results on multiple tokens (and their proofs) only apply to linear-time specifications, and only to ring or clique network-graphs. In contrast, our results apply to branching-time specifications and to general network-graphs.

The proof of our decidability result follows the framework outlined in [2] (inspired by [13,18]) which suggests combining composition and finiteness results.[8] Rabinovich [28] also uses the composition method for solving PMC. He considers the PMC problem for propositional modal logic assuming the parameterized network-graphs $\mathcal{G}$ have a decidable monadic-second order validity problem. While the systems in [28] are very general, the specification language, i.e., modal logic, is orthogonal to ours (e.g., it can not express liveness properties).

To the best of our knowledge, [2,28] are the only other works that use composition to establish decidability of PMC of distributed systems. While proving

[7] Communication in [22] is by rendezvous, powerful enough to express token-passing.

[8] Moreover, our work inherits from [2,13] the non-uniformity of the decision problem. We leave for future work the problem of calculating explicit cutoffs for concrete classes of network-graphs, as was done in [3].

composition and finiteness may not be easy, we find the methodology to be elegant and powerful. Indeed, in all of these cases, no other method is known (e.g., automata, tableaux) for proving decidability. We leave for future work the intriguing problem of applying this methodology to other problems.

References

1. Abdulla, P.A., Delzanno, G., Rezine, O., Sangnier, A., Traverso, R.: On the verification of timed ad hoc networks. In: Fahrenberg, U., Tripakis, S. (eds.) FORMATS 2011. LNCS, vol. 6919, pp. 256–270. Springer, Heidelberg (2011)
2. Aminof, B., Jacobs, S., Khalimov, A., Rubin, S.: Parameterized model checking of token-passing systems. In: McMillan, K.L., Rival, X. (eds.) VMCAI 2014. LNCS, vol. 8318, pp. 262–281. Springer, Heidelberg (2014)
3. Aminof, B., Kotek, T., Rubin, S., Spegni, F., Veith, H.: Parameterized model checking of rendezvous systems. In: Baldan, P., Gorla, D. (eds.) CONCUR 2014. LNCS, vol. 8704, pp. 109–124. Springer, Heidelberg (2014)
4. Aminof, B., Murano, A., Rubin, S., Zuleger, F.: Verification of asynchronous mobile-robots in partially-known environments. In: Chen, Q., Torroni, P., Villata, S., Hsu, J., Omicini, A. (eds.) PRIMA 2015. LNCS, vol. 9387, pp. 185–200. Springer, Heidelberg (2015). doi:10.1007/978-3-319-25524-8_12
5. Aminof, B., Murano, A., Rubin, S., Zuleger, F.: Automatic verification of multi-agent systems in parameterised grid-environments. In: AAMAS (2016)
6. Aminof, B., Rubin, S., Zuleger, F., Spegni, F.: Liveness of parameterized timed networks. In: Halldórsson, M.M., Iwama, K., Kobayashi, N., Speckmann, B. (eds.) ICALP 2015. LNCS, vol. 9135, pp. 375–387. Springer, Heidelberg (2015)
7. Aminof, B., Rubin, S., Zuleger, F.: On the expressive power of communication primitives in parameterised systems. In: Davis, M., Fehnker, A., McIver, A., Voronkov, A. (eds.) LPAR-20 2015. LNCS, vol. 9450, pp. 313–328. Springer, Heidelberg (2015). doi:10.1007/978-3-662-48899-7_22
8. Apt, K., Kozen, D.: Limits for automatic verification of finite-state concurrent systems. Inf. Process. Lett. **22**, 307–309 (1986)
9. Baier, C., Katoen, J.-P.: Principles of Model Checking. MIT Press, Cambridge (2008)
10. Bloem, R., Jacobs, S., Khalimov, A., Konnov, I., Rubin, S., Veith, H., Widder, J.: Decidability of parameterized verification. Synth. Lect. Distrib. Comput. Theory **6**(1), 1–170 (2015). M&C
11. Browne, M.C., Clarke, E.M., Grumberg, O.: Reasoning about networks with many identical finite state processes. Inf. Comput. **81**, 13–31 (1989)
12. Chandy, K.M., Misra, J.: The drinking philosophers problem. ACM TOPLAS **6**(4), 632–646 (1984)
13. Clarke, E., Talupur, M., Touili, T., Veith, H.: Verification by network decomposition. In: Gardner, P., Yoshida, N. (eds.) CONCUR 2004. LNCS, vol. 3170, pp. 276–291. Springer, Heidelberg (2004)
14. Das, S.: Mobile agents in distributed computing: network exploration. Bull. EATCS **109**, 54–69 (2013)
15. Demri, S., Poitrenaud, D.: Verification of infinite-state systems. In: Haddad, S., Kordon, F., Pautet, L., Petrucci, L. (eds.) Models and Analysis in Distributed Systems, Chap. 8, pp. 221–269. Wiley (2011)

16. Emerson, E.A., Kahlon, V.: Parameterized model checking of ring-based message passing systems. In: Marcinkowski, J., Tarlecki, A. (eds.) CSL 2004. LNCS, vol. 3210, pp. 325–339. Springer, Heidelberg (2004)
17. Emerson, E.A., Kahlon, V.: Model checking guarded protocols. In: LICS, pp. 361–370. IEEE (2003)
18. Emerson, E.A., Namjoshi, K.S.: Reasoning about rings. In: POPL, pp. 85–94 (1995). Journal version: Int. J. Found. Comp. Sci. **14**(4) (2003)
19. Emerson, E.A., Sistla, A.: Symmetry and model checking. In: CAV, pp. 463–478 (1993)
20. Esparza, J., Finkel, A., Mayr, R.: On the verification of broadcast protocols. In: LICS, pp. 352–359. IEEE (1999)
21. Feferman, S., Vaught, R.L.: The first-order properties of algebraic systems. Fund. Math. **47**, 57–103 (1959)
22. German, S., Sistla, A.: Reasoning about systems with many processes. JACM **39**(3), 675–735 (1992)
23. Ghilardi, S., Nicolini, E., Ranise, S., Zucchelli, D.: Combination methods for satisfiability and model-checking of infinite-state systems. In: Pfenning, F. (ed.) CADE 2007. LNCS (LNAI), vol. 4603, pp. 362–378. Springer, Heidelberg (2007)
24. Herman, T.: Probabilistic self-stabilization. Inf. Process. Lett. **35**(2), 63–67 (1990)
25. John, A., Konnov, I., Schmid, U., Veith, H., Widder, J.: Parameterized model checking of fault-tolerant distributed algorithms by abstraction. In: FMCAD, pp. 201–209 (2013)
26. Kosowski, A.: Time and Space-Efficient Algorithms for Mobile Agents in an Anonymous Network. Habilitation, U. Sciences et Technologies - Bordeaux I (2013)
27. Kranakis, E., Krizanc, D., Rajsbaum, S.: Computing with mobile agents in distributed networks. In: Rajasekaran, S., Reif, J. (eds.) Handbook of Parallel Computing: Models, Algorithms, and Applications. CRC Press (2007)
28. Rabinovich, A.: On compositionality and its limitations. ACM TOCL **8**(1), 4 (2007)
29. Rubin, S.: Parameterised verification of autonomous mobile-agents in static but unknown environments. In: AAMAS, pp. 199–208 (2015)
30. Shamir, S., Kupferman, O., Shamir, E.: Branching-depth hierarchies. ENTCS **39**(1), 65–78 (2003)
31. Shelah, S.: The monadic theory of order. Ann. Math. **102**, 379–419 (1975)
32. Suzuki, I.: Proving properties of a ring of finite-state machines. Inf. Process. Lett. **28**(4), 213–214 (1988)

Unbounded-Thread Program Verification using Thread-State Equations

Konstantinos Athanasiou(✉), Peizun Liu, and Thomas Wahl

Northeastern University, Boston, USA
konathan@ccs.neu.edu

Abstract. Infinite-state reachability problems arising from unbounded-thread program verification are of great practical importance, yet algorithmically hard. Despite the remarkable success of explicit-state exploration methods to solve such problems, there is a sense that SMT technology can be beneficial to speed up the decision making. This vision was pioneered in recent work by Esparza et al. on SMT-based coverability analysis of Petri nets. We present here an approximate coverability method that operates on *thread-transition systems*, a model naturally derived from predicate abstractions of multi-threaded programs. In addition to successfully proving uncoverability for *all* our safe benchmark programs, our approach extends previous work by the ability to decide the *unsafety* of many unsafe programs, and to provide a witness path. We also demonstrate experimentally that our method beats all leading explicit-state techniques on safe benchmarks and is competitive on unsafe ones, promising to be a very accurate and fast coverability analyzer.

1 Introduction

Unbounded-thread program verification continues to attract the attention it deserves: it targets programs designed to run on multi-user platforms and web servers, where concurrent software threads respond to service requests of a number of clients that can usually neither be predicted nor meaningfully bounded from above a priori. To account for these circumstances, such programs are designed for an unspecified and unbounded number of parallel threads.

We target in this paper unbounded-thread shared-memory programs where each thread executes a non-recursive, finite-data procedure. This model is popular, as it connects to multi-threaded C programs via predicate abstraction, a technique that has enjoyed progress for concurrent programs in recent years [5]. The model is also popular since basic program state reachability questions are decidable, although of high complexity: the corresponding *coverability problem* for Petri nets was shown to be EXPSPACE-complete [4,21].

Owing to the importance of this problem, much effort has since been invested into finding practically viable algorithms [1,3,10,11,15–17]. The vast majority of these are flavors of explicit-state exploration tools. Given the impressive advances

This work is supported by NSF grant no. CCF-1253331.

N. Olivetti and A. Tiwari (Eds.): IJCAR 2016, LNAI 9706, pp. 516–531, 2016.
DOI: 10.1007/978-3-319-40229-1_35

that SMT technology has made, and its widespread "infiltration" of program verification, an obvious question is to what extent such technology can assist in solving the coverability problem.

An encouraging answer to this question was given in a recent symbolic implementation of the Petri net *marking equations* technique for coverability checking [6]. The equations are expressed as integer linear arithmetic constraints and passed to an SMT solver. While the constraints overapproximate the coverability condition, causing the technique to produce false positives, its success rate was very convincing.

Building on the promise of this technique, in this paper

1. we develop a similar approach that applies to a computational model more fitting for software verification, called *thread-transition systems* (TTS). This model makes shared and local thread storage explicit and is designed for encodings of shared-variable concurrent programs. It enjoys a one-to-one correspondence with multi-threaded *Boolean programs*. The latter in turn is a widely used software abstraction employed in concurrency-capable tools such as SatAbs [5] and Bfc [15]. Naturally, we dub our constraint sets *thread-state equations*;
2. we equip our approach with a straightforward but effective component to detect spurious assignments, and to refine the constraints if needed. This component enables the approach to prove systems *unsafe* and generate counterexamples; a feature that was not addressed in [6].

Our method is sound but theoretically incomplete. We implemented it in a tool called Tse; Sect. 5 contains an extensive evaluation on a large number of Boolean program benchmarks. We give a preview of our findings here:

- Notwithstanding said incompleteness, Tse was able to correctly decide 98 % of all TTS instances; this includes safe and unsafe ones.
- Comparing to the most competitive *complete* coverability checker for replicated Boolean programs, Bfc [15], Tse proves to be very close in efficiency on unsafe benchmarks, and *much more efficient* than Bfc on safe ones. (The gap is even larger with other explicit-state explorers.)

In summary, we envision our work to introduce the power of constraint-based coverability analysis to the world of unbounded-thread program verification. Our results showcase Tse as a very capable and highly successful replicated Boolean program verifier.

2 Thread-Transition Systems

We assume multi-threaded programs are given in the form of an abstract state machine called *thread-transition system* (TTS) [15]. Such a system reflects the replicated nature of programs we consider: programs consisting of threads executing a given procedure defined over shared and (thread-)local variables. A thread-transition system is defined over a set of *thread states* $T = S \times L$, where S and L are the finite sets of *shared* and *local* states respectively. $R \subseteq T \times T$ is the transition relation on T, partitioned into $R = \mapsto \cup \looparrowright$; the two partitions intuitively represent *thread transitions* and *spawn transitions*, respectively (semantics below). We refer to elements of R as *edges*. A TTS can now be defined as $\mathcal{P} = (T, R)$. Figure 1 shows an example.

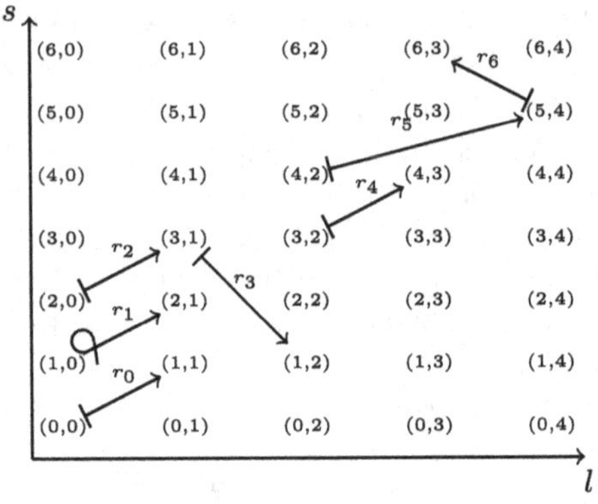

Fig. 1. A thread-transition system with thread ($\mapsto$) and spawn ($\looparrowright$) transitions.

A TTS induces an infinite-state transition system $\mathcal{P}_\infty = (V_\infty, R_\infty)$, as follows. For a positive integer n, let $V_n = S \times L^n$ and $V_\infty = \cup_{i=1}^{\infty} V_i$. We write $v = (s|l_1, \ldots, l_n)$ to denote a (global) system state with a shared component s, and n threads in local states l_i ($i \in \{1, ..., n\}$).

A transition, written as $(s|l_1, \ldots, l_n) \rightarrowtail (s'|l'_1, \ldots, l'_{n'})$, belongs to relation R_∞ exactly if one of the following conditions holds:

Thread Transition: $n' = n$ and there exists $(s, l) \mapsto (s', l') \in R$ and i such that $l_i = l$, $l'_i = l'$, and for all $j \neq i$, $l'_j = l_j$.
Spawn Transition: $n' = n + 1$ and there exists $(s, l) \looparrowright (s', l') \in R$ and i such that $l_i = l$, $l'_{n'} = l'$, and for all $j < n'$, $l'_j = l_j$.

Thus, a transition in R_∞ affects the shared state, and the local state of at most one thread. It may fire only if one thread—the *active* thread—is currently at the corresponding TTS edge's source thread state. We denote by $w \rightarrowtail_{(\mapsto)} w'$ the fact that the thread active in $w \rightarrowtail w'$ fires a $\mapsto$ edge; similarly for $\rightarrowtail_{(\looparrowright)}$.

Let $L_I \subseteq L$ be a set of initial local states and s_I be the unique initial shared state; initial states of $\mathcal{P}_\infty$ hence have the form $v_I = (s_I|l_1, \ldots, l_n)$ where $l_i \in L_I$ for all i. A *path* in $\mathcal{P}_\infty$ is a finite sequence of states in V_∞ starting from any v_I whose adjacent elements are related by R_∞.

In order to state the problem we are tackling, define the *covers* relation $\succeq$ over V_∞ as $(s|l_1, \ldots, l_n) \succeq (s'|l'_1, \ldots, l'_{n'})$ if $s = s'$ and $[l_1, \ldots, l_n] \supseteq [l'_1, \ldots, l'_{n'}]$, where $[\cdot]$ denotes a *multi-set*. We are solving in this paper the *coverability* problem for a given (final) state $v_F \in V_\infty$: is v_F coverable, i.e. does there exists a path in $\mathcal{P}_\infty$ leading to a state v that covers $v_F : v \succeq v_F$? We denote the final shared state by s_F, i.e. $v_F = (s_F|\ldots)$. As an example, state $(1|0)$ is coverable in the 2-thread system derived from the TTS in Fig. 1 with the unique initial thread state $(0, 0)$; the path consists of one thread firing the edge $(0, 0) \mapsto (1, 1)$.

The coverability problem is decidable: relation $\succeq$ is a well-quasi order with respect to which the system $\mathcal{P}_\infty$ is *monotone* [15]. Algorithms for deciding coverability over such systems exist [2,11] but are of high complexity, e.g. EXPSPACE-complete for standard Petri nets [4,21], which are equivalent in expressiveness to infinite-state transition systems obtained from TTS [15].

3 Safety Proofs via Thread-State Equations

In this section we describe how, given a coverability problem, we derive a set of equations whose inconsistency (unsatisfiability of their conjunction) implies the unreachability of any global state covering the final state v_F, and hence the safety of the infinite-state system. We do so by determining constraints on the number of threads in each local state when a global state is reached, as well as constraints that encode the synchronization that shared states enforce among the threads.

3.1 Thread and Transition Counting

Given an initial global state, a finite path p in $\mathcal{P}_\infty$ can be succinctly and unambiguously represented as a sequence of pairs (r, i), where $r \in R$ is a TTS edge and i is a thread index. An abstraction of such a sequence is given by the number of times each edge in R fires along p. This "counting abstraction", which can be seen as simplifying an edge sequence to a multi-set, is rather crude, as it ignores the order of edges fired along p. On the other hand, it allows us to express the coverability condition: from the numbers of times each edge fires, we can obtain the number of threads per local state in the final global state of p. We now require that they match or exceed the thread counts in v_F. Along with the obvious non-negativity constraints for counters, we obtain a first approximation of our thread-state equations, as follows.

Given a TTS $\mathcal{P} = (T, R)$ and a final state v_F, we fix a total order on all edges, and a total order on all local states. We further introduce:

- an integer vector $\mathbf{r}$ of $|R|$ variables, representing the number of occurrences of each edge along p (the edges appear in $\mathbf{r}$ in the given total order);
- an integer vector $\mathbf{l}_I$ of $|L|$ variables, representing the number of threads per local state in the initial state of p (the local states appear in $\mathbf{l}_I$ in the given total order);
- an integer vector $\mathbf{l}_F$ of $|L|$ variables, representing the number of threads per local state in the final state of p (the local states appear in $\mathbf{l}_F$ in the given total order);
- an $|L| \times |R|$ integer matrix $\mathbf{c}$ (a constant) that captures the effect of each edge on each local state, as follows:

$$\mathbf{c}(l, r) = \begin{cases} +1 & \text{if edge } r \text{ } \textit{ends} \text{ in local state } l \\ -1 & \text{if } r \in \mapsto \text{ and } r \text{ } \textit{starts} \text{ in local state } l \\ 0 & \text{otherwise.} \end{cases} \qquad (1)$$

(For simplicity, we identify l with the local state with number l in the total order, similarly for r.) We assume R has no self-loops (which are irrelevant for coverability), hence the quantity $\mathbf{c}(l, r)$ is well-defined. Note that the -1 case only applies to standard thread transition edges ("$\mapsto$"), not to spawns: the latter affect only the number of threads in the target local state. Also note that $\mathbf{c}$ does not capture shared-state changes.

With these variables, we define the following system of *local-state constraints* $\mathcal{C}_L$:

$$\mathcal{C}_L = \bigwedge \begin{cases} \mathbf{r} \geq 0 & \text{non-negative edge counters} \\ \mathbf{l}_I \geq 0 & \\ \mathbf{l}_F \geq 0 & \Big\} \text{ non-neg. local state ctrs.} \\ \bigwedge_{l \notin L_I} \mathbf{l}_I(l) = 0 & \text{initial state condition} \\ \mathbf{l}_F = \mathbf{l}_I + \mathbf{c} \cdot \mathbf{r} & \text{final state condition} \\ \bigwedge_{l \in L} \mathbf{l}_F(l) \geq |\{i : v_F(i) = l\}| & \text{coverability condition} \end{cases} \tag{2}$$

The notation $\mathbf{r} \geq 0$ means "pointwise non-negative"; similarly for $\mathbf{l}_I$ and $\mathbf{l}_F$. Symbol $\mathbf{l}_I(l)$ refers to the component of $\mathbf{l}_I$ corresponding to local state l; similarly for $\mathbf{l}_F(l)$. Operator $\cdot$ denotes matrix multiplication, $|\{\ldots\}|$ is set cardinality, and $v_F(i)$ stands for the local state of thread i in state v_F. These constraints stipulate that all edge and local state counters be natural numbers; that no thread start out in a non-initial local state; that the final local state counters account for the effect of all edges; and that the final global state covers v_F.

3.2 Shared State Synchronization

The thread and transition counting constraints reflected in $\mathcal{C}_L$ ignore the order in which edges fire along a path p, since distinguishing ordered edge sequences symbolically is prohibitively expensive. Some of the ordering information can, however, be recovered, by taking shared state changes into account (which have also been ignored so far): consecutive edges along p must synchronize on the shared state "in the middle".

This requirement can be formalized as follows. Consider an assignment to $(\mathbf{r}, \mathbf{l}_I, \mathbf{l}_F)$ satisfying the constraints $\mathcal{C}_L$. We call an edge $r \in R$ *active* if $\mathbf{r}(r) > 0$, and a shared state *active* if at least one of its adjacent edges is active.

Observation 1. *Let $G_{\mathbf{r}}|_S$ be the directed* ***multi****-graph with node set S and edge* ***multi****-set $[r \in R : \mathbf{r}(r) > 0]|_S$. That is, $G_{\mathbf{r}}|_S$ is defined over the active edges in the multiplicity given by $\mathbf{r}$, projected to S. An edge sequence p*

1. *uses exactly the edges in the multiplicity given by $\mathbf{r}$, and*
2. *has consecutive edges that synchronize on the shared state,*

exactly if p is an ***Euler path*** *in $G_{\mathbf{r}}|_S$.*

This observation is easily seen to hold: the Euler criterion guarantees that exactly all edges in $G_{\mathbf{r}}|_S$ (= the active edges, in the given multiplicity) are used. The "pathness" in the S-projection guarantees the synchronization condition.

We are thus looking for an Euler path in $G_{\mathbf{r}}|_S$. To formalize its existence, we use the following standard adjacency notions from graph theory:

$$\begin{aligned} in(s) &= \{r \in R \mid r \text{ ends in shared state } s\}, \qquad adj(s) = in(s) \cup out(s), \\ out(s) &= \{r \in R \mid r \text{ starts in shared state } s\}. \end{aligned}$$

Note that edges that leave the shared state invariant (denoting thread-internal transitions) are contained in both the *in* and *out* sets.

The existence of an Euler path from s_I to s_F in $G_{\mathbf{r}}|_S$ is known to be equivalent to the conjunction of the following two conditions (see, e.g., [8]):

Flow: each shared state except s_I and s_F is entered and exited the same number of times (along with some special conditions on s_I and s_F),

Connectivity: the *undirected* subgraph of $G_{\mathbf{r}}|_S$ induced by the active shared states is *connected* (has a path between any two nodes).

We now describe how we formalize these conditions as symbolic constraints.

Flow Condition. We write shared state s's *flow constraints* as

$$\mathit{flow}(s) \quad :: \quad \sum_{r \in in(s)} \mathbf{r}(r) \; - \; \sum_{r \in out(s)} \mathbf{r}(r) \; = \; N \tag{3}$$

where N is defined depending on the relationship between s, s_I, and s_F:

$$N \; = \; \begin{cases} \;\;\,0 \;\text{ if } s \notin \{s_I, s_F\} \textbf{ or } s = s_I = s_F \\ -1 \;\text{ if } s = s_I \neq s_F \\ +1 \;\text{ if } s = s_F \neq s_I \end{cases} \tag{4}$$

Our overall flow condition enforces flow constraints (3) for all shared states: $\mathcal{C}_F = \bigwedge_{s \in S} \mathit{flow}(s)$.

Connectivity Condition. For an Euler path to exist in $G_{\mathbf{r}}|_S$, the undirected graph induced by its active shared state nodes must be connected. This is equivalent to the existence of a simple *undirected* path between the initial shared state s_I and s, for each shared state s. To this end we introduce, for each $s \in S$,

- a vector $\mathbf{e}_s$ of $|R|$ integer variables. These variables, later constrained to be in $\{0, 1\}$, encode, in unary, the set of undirected edges of $G_{\mathbf{r}}|_S$ participating in the simple path between s_I and s.
- a predicate for the existence of such a path to s:

$$\begin{aligned} p(s) \quad :: \quad & \textstyle\sum_{r \in adj(s_I)} \mathbf{e}_s(r) = 1 \;\wedge\; \sum_{r \in adj(s)} \mathbf{e}_s(r) = 1 \\ & \wedge\; \forall s' \in S \setminus \{s_I, s\} \;\; \textstyle\sum_{r \in adj(s')} \mathbf{e}_s(r) \;\in\; \{0, 2\} \end{aligned} \tag{5}$$

  The first two sums ensure that the initial (s_I) and target (s) shared states of the simple path have exactly one adjacent transition (and thus degree 1). The last two ensure that each other shared state is either part of the simple path (and has degree 2) or it is not (and has degree 0).
- a predicate characterizing active shared states: $act(s) :: \sum_{r \in adj(s)} \mathbf{r}(r) > 0$.

We now formulate the following system of *connectivity constraints* $\mathcal{C}_C$:

$$\mathcal{C}_C = \bigwedge \begin{cases} \bigwedge_{s\in S} \bigwedge_{r\in R} \mathbf{e}_s(r) \in \{0,1\} \\ \bigwedge_{r\in R} (\mathbf{r}(r) = 0 \implies \bigwedge_{s\in S} \mathbf{e}_s(r) = 0) \\ \bigwedge_{s\in S\setminus\{s_I,s_F\}} act(s) \implies p(s) \\ s_I \neq s_F \wedge act(s_F) \implies p(s_F) \end{cases} \quad (6)$$

These constraints state that the $\mathbf{e}_s$ are bitvectors (used to encode the edge set of $G_{\mathbf{r}}|_S$ in unary); that inactive edges are excluded from the connected subgraph; and that each active shared state except s_I and s_F is connected by a simple path to the initial shared state; the last line requires the same of s_F unless $s_I = s_F$.

Just like $\mathcal{C}_L$, constraints $\mathcal{C}_F$ and $\mathcal{C}_C$ are expressible in the decidable theory of integer linear arithmetic (ILA). Formulas $\mathcal{C}_L$ and $\mathcal{C}_F$ require a number of variables linear in the size of the input TTS, namely $|R|+2|L|$, while $\mathcal{C}_C$ requires a quadratic number of variables, namely $|S|\times|R|$. This larger number of variables has consequences for deciding the $\mathcal{C}_C$ constraints; a fact that is taken into account by the coverability algorithm proposed in Sect. 4.1.

We finally remark that satisfiability of all conditions together, i.e. $\mathcal{C}_L \wedge \mathcal{C}_F \wedge \mathcal{C}_C$, does not guarantee that the edges given by $\mathbf{r}$ can be sequenced to a proper path through $\mathcal{P}_\infty$. Figure 2 shows a TTS and a satisfying assignment to $(\mathbf{r}, \mathbf{l}_I, \mathbf{l}_F)$ that suggests to form a path consisting of exactly one occurrence of each edge in the TTS. The S-projection of these edges is connected. However, it is easy to see that no permutation of the three edges constitutes a valid firing sequence.

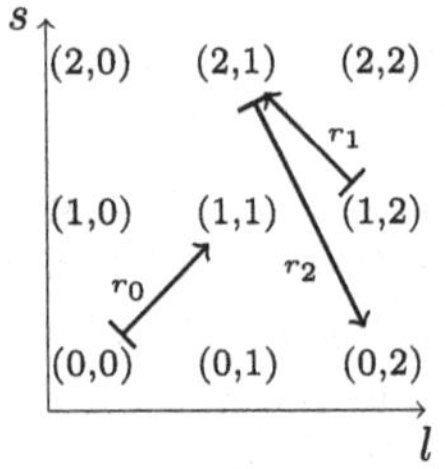

$$\mathbf{r} = \begin{bmatrix}1\\1\\1\end{bmatrix}, \; \mathbf{l}_I = \begin{bmatrix}1\\0\\0\end{bmatrix}, \; \mathbf{l}_F = \begin{bmatrix}0\\1\\0\end{bmatrix}$$

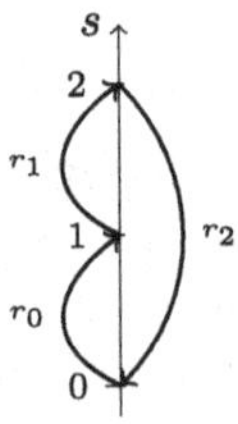

Fig. 2. A TTS (left) with $s_I = 0$, $L_I = \{0\}$, and $v_F = (0|1)$, an assignment satisfying $\mathcal{C}_L \wedge \mathcal{C}_F \wedge \mathcal{C}_C$ (middle), and its S-projection $G_{\mathbf{r}}|_S$ (right)

3.3 Thread-State Equations by Example

We use the TTS of Fig. 3 to showcase how our approach attempts to symbolically solve the coverability problem, by reducing it to a conjunction of integer linear constraints. We consider the case where $s_I = 0$, $L_I = \{0\}$ and therefore the initial state of $\mathcal{P}_\infty$ is of the form (0|0,...,0). We would like to confirm safety with respect to the "bad" final thread state $t_F = (1,1)$. It is not coverable, i.e. there exists no state reachable from any initial state that covers the final state $v_F = (1|1)$.

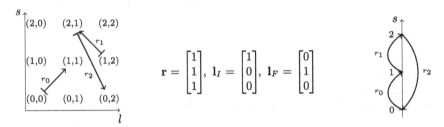

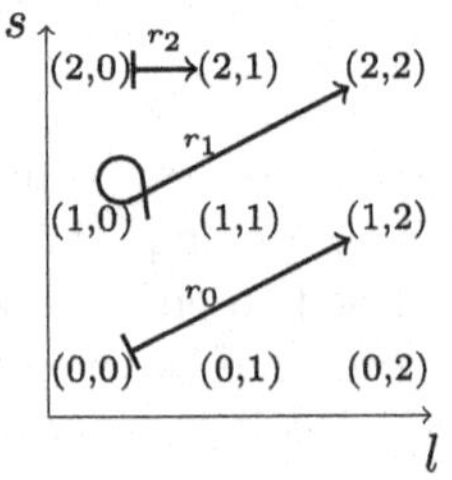

Fig. 3. A TTS

We start by formulating the $\mathcal{C}_L$ constraints, described in Eq. (2). The first equation on the left column of Fig. 4 is the final state condition for local state 0. No edges enter local state 0 but two edges exit it: r_0 and r_2; hence counters $\mathbf{r}(0)$ and $\mathbf{r}(2)$ are subtracted from $\mathbf{l}_I(0)$ to yield $\mathbf{l}_F(0)$. (Recall that spawn edges leave the active thread's local state intact; hence r_1 does not affect $\mathbf{l}_F(0)$.) We derive the final state constraints for the remaining local states similarly; for these the entries of $\mathbf{l}_I$ are 0 by the initial state condition of (2). The coverability condition of (2) for final state $v_F = (1 \mid 1)$ translates into $\mathbf{l}_F(1) \geq 1$.

$$
\begin{aligned}
\mathbf{l}_I(0) - \mathbf{r}(0) - \mathbf{r}(2) &= \mathbf{l}_F(0) & -\mathbf{r}(0) &= -1 \\
\mathbf{r}(2) &= \mathbf{l}_F(1) & \mathbf{r}(0) - \mathbf{r}(1) &= 1 \\
\mathbf{r}(0) + \mathbf{r}(1) &= \mathbf{l}_F(2) & \mathbf{r}(1) &= 0
\end{aligned}
$$

$$
\mathbf{l}_F(1) \geq 1 \qquad \mathbf{r}(1) = 0 \implies \mathbf{e}_2(1) = 0 \wedge \mathbf{e}_1(1) = 0
$$

$$
p(2) \; :: \; \mathbf{e}_2(0) = 1 \wedge \mathbf{e}_2(1) = 1 \wedge \big(\mathbf{e}_2(0) + \mathbf{e}_2(1) = 0 \vee \mathbf{e}_2(0) + \mathbf{e}_2(1) = 2\big)
$$

Fig. 4. Thread-state equations for the TTS of Fig. 3. Notation $\mathbf{l}_F(i)$ stands for the counter variable for local state $l_i \in \{0, 1, 2\}$; $\mathbf{r}(i)$ for the counter variable for edge r_i. The left column shows the local state constraints $\mathcal{C}_L$, the right column the synchronization constraints $\mathcal{C}_F$ and the inactive edge condition for r_1; the last row shows the path predicate $p(2)$.

Next we show in Fig. 4 (right) the flow constraints as defined by in (3). The first equation deals with shared state 0, which has only one adjacent transition: r_0. Since it exists 0 and 0 is initial, we obtain $-\mathbf{r}(0) = -1$. The next equation deals with shared state 1, which has two adjacent transitions: r_0 (entering) and r_1 (exiting); since 1 is final, we obtain $\mathbf{r}(0) - \mathbf{r}(1) = 1$. Regarding shared state 2, edge r_2 leaves it invariant, while r_1 enters it; we obtain $\mathbf{r}(1) = 0$.

Finally we write the constraints for the active edge condition for transition r_1, and the predicate $p(2)$. If r_1 occurs 0 times then the values $\mathbf{e}_1(1)$ and $\mathbf{e}_2(1)$ are set to 0 so that the undirected edges they encode cannot be part of simple paths between the initial shared state and shared states 1 and 2 respectively. $p(2)$ checks for existence of a simple, undirected path between shared states 0 (initial) and 2. The values encoding their adjacent edges, $\mathbf{e}_2(0)$ and $\mathbf{e}_2(1)$, are set to 1 so that shared states 0 and 2 serve as source and target of the path. For shared state 1, the sum of their adjacent edges is set to either 0 or 2 to allow it to either be part of the simple path or not.

The above TSE are unsatisfiable, confirming the uncoverability of t_F. The $\mathcal{C}_F$ constraints enforce that $\mathbf{r}(1)$ is 0, implying that $\mathbf{e}_2(1)$ is 0, which prevents a path between shared states 0 and 2. It turns out that *without* the connectivity condition, the TSE permit the spurious two-thread solution $\mathbf{r}(2) = 1$, $\mathbf{r}(0) = 1$: firing these edges in some order would cover local state 1 (local-state constraints), and the flow constraints are satisfied as well; note that edge r_2, once projected to S, is a self-loop and thus irrelevant for Eq. (3). The two edges do not, however, synchronize on the shared state, no matter which order they fire (the S-projection permits no Euler path). This failure is caught by Eq. (6).

4 Coverability Analysis via Thread-State Equations

We are now ready to incorporate our thread-state equations into an algorithm for establishing system safety. We also present a simple refinement scheme that, very often in practice, enables our algorithm to prove unsafety.

4.1 Coverability via TSE: The Algorithm

Overview. Our algorithm employs the local-state, flow, and connectivity constraints given by $\mathcal{C}_L$, $\mathcal{C}_F$, and $\mathcal{C}_C$, respectively. Constraints $\mathcal{C}_C$, formulating the (non-trivial) connectedness condition for graph $G_{\mathbf{r}}\big|_S$, use a number of variables quadratic in the size $|S|+|L|+|R|$ of the TTS (see Sect. 3.2). As we have determined empirically, they tend to be more expensive to check for satisfiability than $\mathcal{C}_L$ and $\mathcal{C}_F$. Our algorithm is therefore composed of two sub-processes, as follows. Process **A** implements the main algorithm and is described in detail below. Process **B** runs in parallel with **A** and attempts to prove safety using the full set of constraints $\psi = \mathcal{C}_L \wedge \mathcal{C}_F \wedge \mathcal{C}_C$, including $\mathcal{C}_C$. If it proves ψ unsatisfiable, it kills **A** and returns "uncoverable" as the overall answer. If ψ is satisfiable, or **B** runs out of memory, it exits without returning an answer, and process **A** continues alone.

The composition of processes **A** and **B** is shown in Algorithm 1, which attempts to decide the reachability, in $\mathcal{P}_\infty$, of a global state covering v_F. We describe in the following the implementation of process **A**, which uses the (more lightweight) counting and flow constraints to prove safety, and a witness generation scheme to prove unsafety. Process **A** begins by building the thread-state equations $\varphi = \mathcal{C}_L \wedge \mathcal{C}_F$ for the given $\mathcal{P}$, and passing it to a model-building SMT solver capable of deciding integer linear arithmetic formulas (Line **A**1). If the solver decides φ is unsatisfiable, the algorithm returns "uncoverable".

Otherwise let m be a model, i.e. an assignment to $(\mathbf{r}, \mathbf{l}_I, \mathbf{l}_F)$ (Line **A**2). From these assignments we can extract the number n_m of threads that exist at the beginning of the path to be built as is the sum of all $\mathbf{l}_I$ variables, and the number s_m of threads spawned along the path as the sum of all $\mathbf{r}$ variables that correspond to spawn edges (Lines **A**3 and **A**4).

Process **A** now needs to check whether the assignment obtained in m is spurious, or whether it can be turned into a proper witness path in $\mathcal{P}_\infty$. To do this efficiently, we generalize this task and ask whether v_F is coverable along *any* path, but given limited resources, namely n_m initial threads and at most s_m spawns. The key is that this is a finite-state search problem. We have built our own, reasonably efficient and complete, counterexample-producing explorer for this purpose; it is invoked in Line **A**5. If this search is successful, we have a solution to the infinite-state search problem as well: we return the witness path generated by $\text{Fss}(\mathcal{P}, n_m, s_m)$ as the answer produced by Algorithm 1.

If the finite-state search is unsuccessful, it shows that, if a state covering v_F is reachable, then only along a path that starts with more than n_m initial threads ("$n > n_m$") **or** spawns more than s_m threads along the way ("$s > s_m$"). This condition is enforced in Line **A**7, thus strengthening φ. In contrast to Lines

Algorithm 1. Coverability($\mathcal{P}, s_I, L_I, v_F$).
The **return** statements kill off the respective other process before returning

Input: TTS $\mathcal{P}$; initial shared state s_I; initial local states set L_I; final state v_F
Output: "uncoverable", or "coverable" + witness path

Process A

1: $\varphi := \mathcal{C}_L \wedge \mathcal{C}_F$
2: **while** $\exists m : m \models \varphi$
3: $\quad n_m := \sum_{l \in L} \mathbf{l}_I(l)(m)$
4: $\quad s_m := \sum_{r \in \looparrowright} \mathbf{r}(r)(m)$
5: $\quad$ **if** Fss($\mathcal{P}, n_m, s_m$) = "coverable" + witness p
6: $\quad\quad$ **return** "coverable" + p
7: $\quad \varphi := \varphi \wedge (n > n_m \vee s > s_m)$
8: **return** "uncoverable"

||

Process B

1: $\psi := \mathcal{C}_L \wedge \mathcal{C}_F \wedge \mathcal{C}_C$
2: **if** ψ is unsat
3: $\quad$ **return** "uncoverable"

A3 and **A**4, the strengthening is expressed *symbolically* over the variables in $\mathbf{l}_I$ and $\mathbf{r}$. More precisely, $n > n_m$ abbreviates the formula

$$\mathbf{l}_I(0) + \quad \ldots \quad + \mathbf{l}_I(|L| - 1) > n_m ,$$

where the $\mathbf{l}_I(i)$ are variables, and n_m is the constant computed in Line **A**3. The formula abbreviated by $s > s_m$ is built similarly; here the sum expression for s is formed over the variables in $\mathbf{r}$ that correspond to spawn edges.

Given the strengthening to φ computed in Line **A**7, process **A** returns to the beginning of the loop and checks φ for satisfiability.

Finite-State Search. A breadth-first style algorithm for routine Fss is shown on the right. It maintains a worklist W and an *explored* set E, both initialized to the of initial states I_n, which covers all combinations of initial threads with size n. Each state w maintains a counter s to record the remaining number of spawns that can be fired from w; for $w \in I$, $w.s = s$. In each step, Fss removes a state w from W and expands it to w' if $w.s$ allows so. It returns coverable if $w' \succeq v_F$; otherwise steps forward. w' decreases the value of s inherited from w if the transition is due to a spawn.

Algorithm 2. Fss($\mathcal{P}, n, s$)

$W := I_n$; $E := I_n$
while $\exists w \in W$
$\quad W := W \setminus \{w\}$
$\quad$ **for each** $w' \notin E$: $w \rightarrowtail_{(\mapsto)} w'$ $\vee (w \rightarrowtail_{(\looparrowright)} w' \wedge w.s > 0)$
$\quad\quad$ **if** $w' \succeq v_F$ **then**
$\quad\quad\quad$ **return** "coverable"
$\quad\quad$ **if** $w \rightarrowtail_{(\looparrowright)} w'$ **then**
$\quad\quad\quad w'.s{-}{-}$
$\quad\quad W := W \cup \{w'\}$; $E := E \cup \{w\}$
return "uncoverable"

4.2 Coverability via TSE: Analysis

We first prove the soundness (partial correctness) of Algorithm 1, and then discuss its termination. We assume that Lines **A**2 and **B**2 use a sound, complete, and model-building ILA solver; we use Z3 [20] in our experiments.

Partial Correctness. We begin our analysis with the following property.

Lemma 2. *If v_F is coverable in $\mathcal{P}_\infty$, then φ built in Lines **A1** and **A7** and ψ built in Line **B1** are satisfiable.*

Theorem 3 (Soundness). *If Algorithm 1 returns "coverable", v_F is coverable in $\mathcal{P}_\infty$. If Algorithm 1 returns "uncoverable", v_F is uncoverable in $\mathcal{P}_\infty$.*

Proof. If Algorithm 1 returns "coverable", v_F is coverable, as procedure FSS running on a finite state space is sound and complete. If Algorithm 1 returns "uncoverable", triggered by the unsatisfiability of φ in Line **A2** or ψ in Line **B2**, then v_F is uncoverable in $\mathcal{P}_\infty$ by Lemma 2. □

Termination. In general, Algorithm 1 is not guaranteed to terminate: neither of the two processes **A** and **B** may return. Two different scenarios can lead to non-termination. The first is that despite an uncoverable final state, **A** keeps finding spurious assignments, and **B** does the same or times out. Consider again the scenario and the assignment $(\mathbf{r}, \mathbf{l}_I, \mathbf{l}_F)$ shown in Fig. 2. As discussed in Sect. 3.2, this assignment is spurious, as will be confirmed by the invocation of FSS$(\mathcal{P}, 1, 0)$, which fails to reach a state covering v_F. φ is strengthened by $\mathbf{l}_I(0) > 1$. The result is again satisfiable, this time with a model that sets all of $\mathbf{r}(0)$, $\mathbf{r}(1)$, $\mathbf{r}(2)$ and $\mathbf{l}_F(1)$ to 2. We see that, for any n_m, there exists a model of φ satisfying $\mathbf{r}(i) = n_m$ for $i \in \{0 \ldots 2\}$, $\mathbf{l}_I(0) = n_m$, $\mathbf{l}_F(1) = n_m$, which never translates to a genuine path in $\mathcal{P}_\infty$. Therefore Algorithm 1 will not terminate.

The other non-termination scenario is that of a *coverable* final state that is overlooked as the search diverges in the wrong direction. The problem is that increments applied to the initial thread count n and the spawn count s by the solver may not be *fair*: Line **A7** only requires one of them to go up. As a special case, if the TTS has no spawn transitions ($\rightarrowtail = \emptyset$), we can tighten Line **A7** to $\varphi := \varphi \wedge n > n_m$, in which case the algorithm is (in principle) *complete for unsafe instances.*

5 Empirical Evaluation

The technique presented in this paper is implemented in a coverability checker named TSE (for "Thread-State Equation"). TSE is written in C++ and uses Z3 (v4.3.1) as the back-end ILA solver. It takes as input coverability problems for TTS. We used a benchmark suite of concurrent Boolean programs to evaluate TSE. We ran TSE on Boolean programs in order to compare with the following state-of-the-art checkers[1]:

Petrinizer: An SMT-based coverability checker described in [6] (v1.0)
BFC: A coverability checker with forward oracle presented in [15] (v2.0)
BFC-KM: A generalized Karp-Miller procedure presented in [15] (v1.0)
IIC: Incremental, inductive coverability algorithm [17]
MIST-AR: An abstraction refinement method presented in [10] (v1.1)
MIST-EEC: Forward analysis with enumerative refinement [11] (v1.1)

[1] Available at www.cprover.org/bfc/; github.com/pierreganty/mist; and http://www.mpi-sws.org/~fniksic/cav2014/repository.tgz.

Benchmarks. Our benchmark set contains 339 concurrent Boolean programs generated from concurrent C programs (taken from [15,19]), 135 of which are safe. For each example, we consider a reachability property that is specified via an assertion. The table on the right shows the size ranges of the BPs.

BP	min.	max.
$\|S\|$	5	32769
$\|L\|$	17	55
$\|R\|$	18	584384

To apply TSE to C programs, we use SATABS to transform those programs to TTS (option `--build-tts`) via intermediate Boolean programs [5]. When SATABS requires several CEGAR iterations over the C programs until the abstraction permits a decision, the same C source program gives rise to several Boolean programs and TTSs.

Experimental Setup. The main objective of our experiments with BPs is to measure the competitiveness of TSE against state-of-the-art infinite-thread BP checkers; this is mostly variants of the BFC tool. We also investigated how TSE fares against tools targeting Petri nets, of which there are many; most interesting for us is the Petrinizer tool, as it implements an idea similar to the one used in (and inspirational for) TSE. Petri net tools can be used for BP verification by converting those programs to Petri nets. We have experimented with two translators: one used in [6,15][2], and one by Pierre Ganty et al. github.com/pevalme/bfc_fork, which tries to alleviate the blowup incurred by shared state conversion. As different tools accept different translations, we used both translators in our experiments. The running times we report in the results *ignore* translation time, which ranges from almost nothing up to dozens of seconds.

All experiments are performed on a 2.3 GHz Intel Xeon machine with 64 GB memory, running 64-bit Linux. The timeout is set to 30 min and the memory limit to 4 GB. All benchmarks and our tool are available online [18].

Precision. Table 1 compares the results of precision on BPs for all tools. TSE successfully decides *all* BPs except 5 unsafe instances, where the SMT solver runs out of memory. Both BFC and BFC-KM prove 4 out of these 5 instances. As for the safe instances, it was interesting to observe that the connectivity constraints $\mathcal{C}_C$ were never required to conclude unsatisfiability, i.e. the constraints $\mathcal{C}_L$ and $\mathcal{C}_F$ were already inconsistent. This means that process **B** in Algorithm 1 never ran to completion.

Table 1. Precision results for all tools. Note that Petrinizer decides only safe benchmarks

suite \ tools	TSE	Petrinizer	BFC	BFC-KM	IIC	MIST-AR	EEC	# instances
safe BP (%)	100	100	57.04	2.22	81.48	94.07	34.81	135
unsafe BP (%)	97.55	–	99.02	98.04	62.75	12.75	18.63	204
total (%)	98.53	–	82.60	59.88	70.21	45.13	25.07	339

[2] www.cprover.org/bfc/.

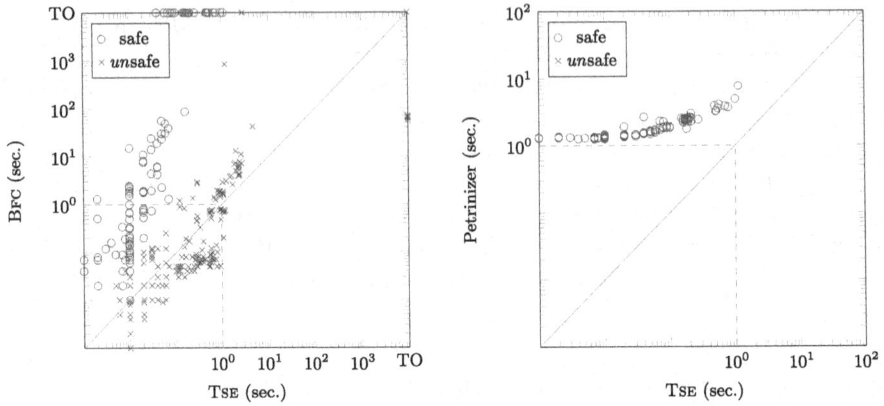

Fig. 5. Performance comparison for Boolean Programs: TSE vs. BFC (left) and vs. Petrinizer (right). Each dot represents execution time for one program (TO = timeout) (Color figure online)

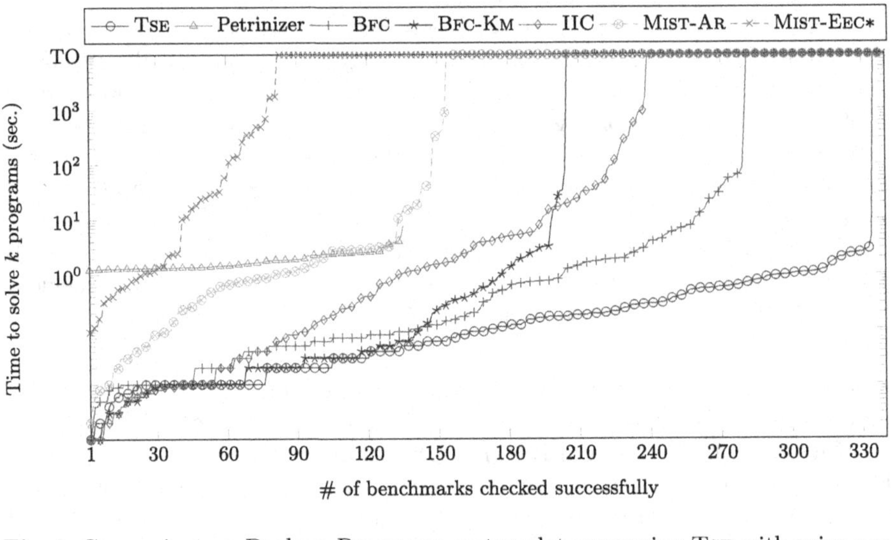

Fig. 6. Comparison on Boolean Programs: cactus plot comparing TSE with prior coverability tools. An entry of the form (k, t) for some curve shows the time t it took to solve the k easiest — for the method associate with that curve — benchmarks (order varies across methods). $*$ indicates that inputs to this tool are Petri nets from Pierre Ganty's translator. (Color figure online)

Efficiency. Figure 5 (left) plots the detailed comparison against BFC (the most efficient of the competing tools, according to [15]) over each benchmark. TSE clearly beats BFC on safe instances and remains competitive on unsafe ones. In general, we observe that BFC outperforms TSE on very small benchmarks which are solved within one second, an effect that can be attributed to the overhead added by the solver. Figure 5 (right) plots the comparison with Petrinizer[3] [6]. Since Petrinizer does not handle unsafe instances, we focus on safe ones. TSE is invariably faster. Figure 6 is a cumulative plot showing the total time (log-scale) taken to solve the k, for $1 \leq k \leq 339$, easiest of our benchmark problems, for all tools. The results demonstrate that in most cases TSE terminates within 5 seconds. BFC is the most competitive among other tools.

Summary. We summarize the precision and efficiency results as follows. Given that our tool is sound (it never gives an incorrect answer), and that it does give an answer in the vast majority of the benchmarks we have used, it is prudent to base the comparison on the efficiency results, even against exact tools. Here we observe the strength of TSE especially as a safety prover, i.e. on uncoverable instances. The aggressive search for counterexamples used in BFC gives that tool a nominal advantage for coverable instances, which is, however, hardly decisive as the running times on those instances tend to be very small.

6 Related Work and Discussion

Groundbreaking results in infinite-state system analysis include the decidability of coverability in *vector addition systems* (VAS) [16], and the work by German and Sistla on modeling communicating finite-state threads as VAS [13]. Numerous results have since improved on the original procedure in [16] in practice [11,12,22,23]. Others extend it to more general computational models, including *well-structured* [9] or *well-quasi-ordered* (wqo) transition systems [2,3].

Explicit-state techniques that combine forward and backward exploration (IIC,BFC) [15,17] or apply abstraction refinement (MIST-EEC,MIST-AR) [10,11] have been shown to efficiently decide the coverability problem for large instances, like the ones we consider in our work.

Contrary to the above mentioned complete methods for coverability, [6] follows the direction of trading completeness for performance, by reducing the coverability problem to linear constraint solving and discharging it to a SMT solver, and serves as the inspiration of our work. The thread-state equations we present can be viewed as an instantiation of the marking equation – a classical Petri net technique – in the domain of TTS. In addition to TSE, we extend the approach of [6] by equipping our method with a refinement scheme and a straightforward finite-state search in order to efficiently discover unsafe instances and provide coverability witnesses.

Another incomplete symbolic approach for coverability analysis utilizing the marking equation is presented in [24]. CEGAR is applied on top of the marking equation and is used to guide the solution space of the integer linear constraints. More complex strategies for guiding the solution space were recently introduced in [14]. Such schemes differ from ours, as the solutions to TSE are used as the starting point of the finite state space exploration. If the latter is unsuccessful, TSE are strengthened to allow a simple but efficient refinement scheme.

[3] Petrinizer offers four methods; we use the most powerful: refinement over integers.

Conclusions. Our experimental results demonstrate the trade-off between complete, explicit state exploration and incomplete, symbolic approaches. Verifying safe instances often becomes infeasible when trying to retain completeness, but is shown to be very efficient when posed as a constraint solving problem, as also pointed out in [6]. Our approach aims at *continuing* this trend of devising incomplete yet practical methods for problems of high computational cost [7], by providing an algorithm that fills in the gap of verification of unsafe instances, and efficiently solves the coverability problem in software verification for almost all of our instances.

References

1. Abdulla, P.A., Haziza, F., Holík, L.: All for the price of few (parameterized verification through view abstraction). In: Giacobazzi, R., Berdine, J., Mastroeni, I. (eds.) VMCAI 2013. LNCS, vol. 7737, pp. 476–495. Springer, Heidelberg (2013)
2. Abdulla, P.A.: Well (and better) quasi-ordered transition systems. Bull. Symbolic Logic **16**(4), 457–515 (2010)
3. Abdulla, P.A., Cerans, K., Jonsson, B., Tsay, Y.K.: General decidability theorems for infinite-state systems. In: LICS, pp. 313–321 (1996)
4. Cardoza, E., Lipton, R.J., Meyer, A.R.: Exponential space complete problems for petri nets and commutative semigroups: preliminary report. In: STOC, pp. 50–54 (1976)
5. Donaldson, A., Kaiser, A., Kroening, D., Wahl, T.: Symmetry-aware predicate abstraction for shared-variable concurrent programs. In: Gopalakrishnan, G., Qadeer, S. (eds.) CAV 2011. LNCS, vol. 6806, pp. 356–371. Springer, Heidelberg (2011)
6. Esparza, J., Ledesma-Garza, R., Majumdar, R., Meyer, P., Niksic, F.: An SMT-based approach to coverability analysis. In: Biere, A., Bloem, R. (eds.) CAV 2014. LNCS, vol. 8559, pp. 603–619. Springer, Heidelberg (2014)
7. Esparza, J., Meyer, P.J.: An SMT-based approach to fair termination analysis. In: FMCAD, pp. 49–56 (2015)
8. Even, S.: Graph Algorithms. W. H. Freeman & Co., New York (1979)
9. Finkel, A., Schnoebelen, P.: Well-structured transition systems everywhere!. Theor. Comput. Sci. **256**(1–2), 63–92 (2001)
10. Ganty, P., Raskin, J.F., Van Begin, L.: From many places to few: automatic abstraction refinement for petri nets. Fundam. Inf. **88**(3), 275–305 (2008)
11. Geeraerts, G., Raskin, J.F., Begin, L.V.: Expand, Enlarge and Check: New algorithms for the coverability problem of WSTS. J. Comput. Syst. Sci. **72**(1), 180–203 (2006)
12. Geeraerts, G., Raskin, J.-F., Van Begin, L.: On the efficient computation of the minimal coverability set for petri nets. In: Namjoshi, K.S., Yoneda, T., Higashino, T., Okamura, Y. (eds.) ATVA 2007. LNCS, vol. 4762, pp. 98–113. Springer, Heidelberg (2007)
13. German, S.M., Sistla, A.P.: Reasoning about systems with many processes. J. ACM **39**(3), 675–735 (1992)
14. Hajdu, Á., Vörös, A., Bartha, T.: New search strategies for the petri net CEGAR approach. In: Devillers, R., Valmari, A. (eds.) PETRI NETS 2015. LNCS, vol. 9115, pp. 309–328. Springer, Heidelberg (2015)

15. Kaiser, A., Kroening, D., Wahl, T.: A widening approach to multithreaded program verification. ACM Trans. Program. Lang. Syst. **36**(4), 14 (2014)
16. Karp, R.M., Miller, R.E.: Parallel program schemata. J. Comput. Syst. Sci. **3**(2), 147–195 (1969)
17. Kloos, J., Majumdar, R., Niksic, F., Piskac, R.: Incremental, inductive coverability. In: Sharygina, N., Veith, H. (eds.) CAV 2013. LNCS, vol. 8044, pp. 158–173. Springer, Heidelberg (2013)
18. Liu, P.: www.ccs.neu.edu/home/lpzun/tse/
19. Liu, P., Wahl, T.: Infinite-state backward exploration of Boolean broadcast programs. In: FMCAD, pp. 155–162 (2014)
20. de Moura, L., Bjørner, N.S.: Z3: an efficient SMT solver. In: Ramakrishnan, C.R., Rehof, J. (eds.) TACAS 2008. LNCS, vol. 4963, pp. 337–340. Springer, Heidelberg (2008)
21. Rackoff, C.: The covering and boundedness problems for vector addition systems. Theor. Comput. Sci. **6**, 223–231 (1978)
22. Reynier, P.-A., Servais, F.: Minimal coverability set for petri nets: Karp and Miller algorithm with pruning. In: Kristensen, L.M., Petrucci, L. (eds.) PETRI NETS 2011. LNCS, vol. 6709, pp. 69–88. Springer, Heidelberg (2011)
23. Valmari, A., Hansen, H.: Old and new algorithms for minimal coverability sets. In: Haddad, S., Pomello, L. (eds.) PETRI NETS 2012. LNCS, vol. 7347, pp. 208–227. Springer, Heidelberg (2012)
24. Wimmel, H., Wolf, K.: Applying CEGAR to the petri net state equation. Log. Methods Comput. Sci. **8**(3), 827–846 (2012)

A Complete Decision Procedure for Linearly Compositional Separation Logic with Data Constraints

Xincai Gu[1,2], Taolue Chen[3], and Zhilin Wu[1(✉)]

[1] State Key Laboratory of Computer Science, Institute of Software, Chinese Academy of Sciences, Beijing, China
wuzl@ios.ac.cn

[2] University of Chinese Academy of Sciences, Beijing, China

[3] Department of Computer Science, Middlesex University, London, UK

Abstract. Separation logic is a widely adopted formalism to verify programs manipulating dynamic data structures. Entailment checking of separation logic constitutes a crucial step for the verification of such programs. In general this problem is undecidable, hence only incomplete decision procedures are provided in most state-of-the-art tools. In this paper, we define a linearly compositional fragment of separation logic with inductive definitions, where traditional shape properties for linear data structures, as well as data constraints, e.g., the sortedness property and size constraints, can be specified in a unified framework. We provide complete decision procedures for both the satisfiability and the entailment problem, which are in NP and Π_3^{P} respectively.

1 Introduction

Program verification requires reasoning about complex, unbounded size data structures that may carry data ranging over infinite domains. Examples of such data structures are multi-linked lists, nested lists, trees, etc. Programs manipulating these data structures may modify their shape (due to dynamic creation and destructive updates) as well as the data attached to their elements.

Separation Logic (SL) is a well-established approach for deductive verification of programs that manipulate dynamic data structures [18,24]. Typically, SL is used in combination with inductive definitions, which provide a natural description of the data structures manipulated by a program.

In program verification, SL is normally used to express assertions about program configurations, for example in the style of Hoare logic. Checking the validity

Taolue Chen is partially supported by the ARC Discovery Project (DP160101652), the Singapore Ministry of Education AcRF Tier 2 grant (MOE2015-T2-1-137), and an oversea grant from the State Key Laboratory of Novel Software Technology, Nanjing Unviersity. Zhilin Wu is partially supported by the NSFC grants (No. 61100062, 61272135, 61472474, and 61572478).

N. Olivetti and A. Tiwari (Eds.): IJCAR 2016, LNAI 9706, pp. 532–549, 2016.
DOI: 10.1007/978-3-319-40229-1_36

of these assertions is naturally reduced to the *entailment* problem of the logic, i.e., given two SL formulae φ and ψ, to check whether $\varphi \models \psi$ holds.

Because of its importance, entailment checking has been explored extensively (see, e.g., [1,9,16]). In general, it is an undecidable problem, hence only *incomplete* decision procedures can be expected. This is especially the case when both shape properties and data (size) constraints are taken into consideration. Indeed, various separation logic based tools, e.g., INFER [8], SLEEK/HIP [9], DRYAD [19,23], and SPEN [13], only provide incomplete decision procedures.

Undoubtedly *complete* decision procedures are highly desirable: besides being theoretically appealing, they also have practical importance, for instance in tasks such as debugging of specification, counterexample generation, etc. The challenge is thus to find fragments of SL which are sufficiently expressive for writing program assertions while still feature a complete decision procedure for the entailment checking. This would enable efficient automated validation of the verification conditions.

Contributions. In this paper, we define a *linearly compositional* fragment of SL with inductive definitions (abbreviated as $\mathsf{SLID_{LC}}$), where both shape properties, e.g., singly and doubly linked lists, linked lists with tail pointers, and data constraints, e.g., sortedness property and size constraints, can be expressed. The basic idea of $\mathsf{SLID_{LC}}$ is to focus on the compositional predicates introduced in [14], while restricting to linear shapes (e.g., singly and doubly linked lists, or linked lists with tail pointers), and data constraints in the form of difference bound relations (which are sufficient to express sortedness properties and size constraints). Our main contribution is to provide complete decision procedures for the satisfiability and entailment problem of $\mathsf{SLID_{LC}}$.

For the satisfiability problem, from each $\mathsf{SLID_{LC}}$ formula φ we define an abstraction of φ, i.e., $\mathsf{Abs}(\varphi)$, where Boolean variables are introduced to encode the spatial part of φ, together with quantifier free Presburger formulae to represent the transitive closure of the data constraint in the inductive definitions. The satisfiability of φ is then reduced to the satisfiability of $\mathsf{Abs}(\varphi)$, which can be solved by the start-of-the-art SMT solvers (e.g., Z3 [25]), with an NP upper-bound.

For the entailment problem, from each $\mathsf{SLID_{LC}}$ formula φ we first construct a graph representation $\mathcal{G}_\varphi$. We then demonstrate some nice properties of $\mathcal{G}_\varphi$, which enable us to extend and adapt the concept of homomorphisms introduced in [11], to obtain a decision procedure to perform entailment checking with a Π_3^{P} upper-bound. Compared to the logic in [11], the logic $\mathsf{SLID_{LC}}$ is different in the following sense: (1) we adopt the classical semantics whereas [11] adopted the intuitionistic semantics, which can be considered as a special case, and is arguably less meaningful for program verification. (2) the logic in [11] only addresses singly linked list segments, the logic $\mathsf{SLID_{LC}}$ is much more expressive: $\mathsf{SLID_{LC}}$ allows specifying data constraints, as well as defining more shapes, e.g., doubly linked lists, linked lists with tail pointers; in addition, we allow different predicates to occur in φ and ψ for the entailment problem $\varphi \models \psi$. Because of these differences, we are not able to repeat the approach in [11] to transform the graphs into normal forms and then check graph homomorphism between the normal forms.

Instead our decision procedure introduces some new concepts e.g. allocating plans for φ and is considerably more involved than that in [11].

Related Work. We first discuss work on separation logic with inductive definitions where both shape properties and data constraints can be expressed. Various fragments have been explored and we focus on decision procedures for the entailment problem.

The most relevant work is [3], where data constraints, specified by universal quantifiers over index variables, were added to a fragment of separation logic with the *lseg* predicate (where *lseg* denotes list segments). Compared with the work in [3]: For the shape constraints, the logic there focused on singly linked lists, while in $\mathsf{SLID}_{\mathsf{LC}}$, various linear data structures can be specified. For the data constraints, the logic there can specify set and multiset constraints, while $\mathsf{SLID}_{\mathsf{LC}}$ does not. On the other hand, when restricted to arithmetic constraints over integer variables, the decision procedure in [3] is incomplete for fragments that can express list segments where the data values are consecutive, which can be easily expressed in $\mathsf{SLID}_{\mathsf{LC}}$ (cf. *plseg* predicate in Example 1).

The tool SLEEK/HIP [9] provides a decision procedure which is incomplete in general and relies on the invariants of the inductive definitions. These invariants are essentially the transitive closures of the data constraints in the inductive definitions, and are supposed to be provided by the user. In comparison, we focus on a less expressive logic $\mathsf{SLID}_{\mathsf{LC}}$, and our decision procedure can *automatically* compute the *precise* invariants of the inductive definitions.

The tool GRASSHOPPER [20–22] encoded separation logic with inductive definitions into a fragment of first-order logic with reachability predicates, whose satisfiability problem was shown in NP. The logic considered there includes both shape and data constraints and the decision procedure is complete. However the logic is unable to encode the size or multiset constraints. In contrast, our approach can fully handle the size constraints, and the multiset constraints on condition that their transitive closure can be computed (or provided as an oracle).

The tool DRYAD [19,23] reduces to the satisfiability problem in the theory of uninterpreted functions, which is sound, but incomplete. In addition, the decision procedure is *not* fully automatic since it relies on the users to provide lemmas, e.g., $lseg(E_1; E_2) * lseg(E_2; E_3) \models lseg(E_1; E_3)$.

Other work includes the cyclic-proof approach [6,10] which is based on induction on the paths of proof trees. The approach can deal with data constraints but the decision procedures there are incomplete. The work [14] considered the automated lemma generation, where the concept of compositional predicates was introduced. However, the decision procedure provided there is incomplete.

There have also been much work on the decision procedures for the fragments of SL with inductive definitions that contain no data constraints. To cite a few, the work [2,15] focused on the symbolic heap fragments where the shape constraints for list segments and binary trees can be specified and complete proof systems were given, the tool SLIDE [16,17] considered separation logic with general inductive definitions and reduced the entailment problem to the language inclusion problem of tree automata, tool SPEN [13] provided an incomplete decision procedure for a compositional fragment of separation logic with inductive definitions, and the paper [7] designed a complete decision procedure for the satisfiability problem of separation logic with general inductive definitions.

There are also other works on separation logic. The work [4] considered first-order separation logic over linked lists extended with length constraints where the decidability frontier was identified. However, neither data structures other than singly linked lists nor other forms of data constraints (e.g. sortedness) were addressed. The work [5,12] considered the fragments of first-order separation logic (without inductive definitions). The authors identified the decidability frontier and resolved some long-standing expressibility issues.

2 Linearly Compositional Separation Logic with Inductive Definitions

In this section, we introduce the *linearly compositional* fragment of separation logic with inductive definitions, denoted by $\mathsf{SLID_{LC}}[\mathcal{P}]$, where $\mathcal{P}$ is a finite set of *inductive predicates*. In $\mathsf{SLID_{LC}}[\mathcal{P}]$, both shape properties (e.g. doubly linked lists) and data constraints (e.g. sortedness and size constraints) can be specified.

We consider two data types, i.e., the *location* type $\mathbb{L}$ and the *integer* type $\mathbb{Z}$. As a convention, $l, l', \cdots \in \mathbb{L}$ denote locations and $n, n', \cdots \in \mathbb{Z}$ denote integers. Accordingly, variables in $\mathsf{SLID_{LC}}[\mathcal{P}]$ comprise *location variables* of the location type and *data variables* of the integer type. Namely, we assume a set of location variables LVars ranged over by uppercase letters $E, F, X, Y, \cdots$ and a set of data variables DVars ranged over by lowercase letters $x, y, \cdots$. Note that in literature sometimes locations are treated simply as a subset of integers, which is not adopted here for the sake of clarity. We consider two kinds of *fields*, i.e., location fields from $\mathcal{F}$ and data fields from $\mathcal{D}$. Each field $f \in \mathcal{F}$ (resp. $d \in \mathcal{D}$) is associated with $\mathbb{L}$ (resp. $\mathbb{Z}$).

$\mathsf{SLID_{LC}}[\mathcal{P}]$ formulae may contain inductive predicates, each of which is of the form $P(E, \boldsymbol{\alpha}; F, \boldsymbol{\beta}; \boldsymbol{\xi})$ and has an associated inductive definition. The parameters of an inductive predicate are classified into three groups: *source parameters* $\boldsymbol{\alpha}$, *destination parameters* $\boldsymbol{\beta}$, and *static parameters* $\boldsymbol{\xi}$. We require that the source parameters $\boldsymbol{\alpha}$ and the destination parameters $\boldsymbol{\beta}$ are *matched* in type, namely, the two tuples have the same length $\ell > 0$ and for each $i : 1 \leqslant i \leqslant \ell$, α_i and β_i have the same data type. Without loss of generality, it is assumed that the first components of $\boldsymbol{\alpha}$ and $\boldsymbol{\beta}$ are a location variable. In the sequel, for clarity, we explicitly identify the first parameters of $\boldsymbol{\alpha}$ and $\boldsymbol{\beta}$, and write $E, \boldsymbol{\alpha}$ and $F, \boldsymbol{\beta}$.

$\mathsf{SLID_{LC}}[\mathcal{P}]$ formulae comprise three types of formulae: *pure formulae* Π, *data formulae* Δ, and *spatial formulae* Σ, which are defined by the following rules,

$$\begin{array}{ll}
\Pi ::= E = F \mid E \neq F \mid \Pi \wedge \Pi & \text{(pure formulae)} \\
\Delta ::= \mathtt{true} \mid x \ \mathtt{o} \ c \mid x \ \mathtt{o} \ y + c \mid \Delta \wedge \Delta & \text{(data formulae)} \\
\Sigma ::= \mathtt{emp} \mid E \mapsto \rho \mid P(E, \boldsymbol{\alpha}; F, \boldsymbol{\beta}; \boldsymbol{\xi}) \mid \Sigma * \Sigma & \text{(spatial formulae)} \\
\rho ::= (f, X) \mid (d, x) \mid \rho, \rho &
\end{array}$$

where $\circ \in \{=, \leqslant, \geqslant\}$, c is an integer constant, $P \in \mathcal{P}$, $f \in \mathcal{F}$, and $d \in \mathcal{D}$. For spatial formulae Σ, formulae of the form $\mathtt{emp}$, $E \mapsto \rho$, or $P(E, \boldsymbol{\alpha}; F, \boldsymbol{\beta}; \boldsymbol{\xi})$ are called *spatial atoms*. In particular, formulae of the form $E \mapsto \rho$ and $P(E, \boldsymbol{\alpha}; F, \boldsymbol{\beta}; \boldsymbol{\xi})$ are called *points-to atoms* and *predicate atoms* respectively. Moreover, we call E as *the root* of these points-to or predicate atoms.

We are now in a position to introduce the *linearly compositional* predicates, which are the main focus of the current paper. A predicate $P \in \mathcal{P}$ is *linearly compositional* if the inductive definition of P is given by the following two rules,

- base rule $R_0 : P(E, \boldsymbol{\alpha}; F, \boldsymbol{\beta}; \boldsymbol{\xi}) ::= E = F \wedge \boldsymbol{\alpha} = \boldsymbol{\beta} \wedge \mathtt{emp}$,
- inductive rule $R_1 : P(E, \boldsymbol{\alpha}; F, \boldsymbol{\beta}; \boldsymbol{\xi}) ::= \exists \boldsymbol{X} \exists \boldsymbol{x}.\ \Delta \wedge E \mapsto \rho * P(Y, \boldsymbol{\gamma}; F, \boldsymbol{\beta}; \boldsymbol{\xi})$.

The left-hand (resp. right-hand) side of a rule is called the *head* (resp. *body*) of the rule. We note that the body of R_1 does not contain pure formulae.

In the sequel, we specify some constraints on the inductive rule R_1 which enable us to obtain *complete* decision procedures for the satisfiability and entailment problem later.

The first constraint (**C1**) is from [14] which guarantees that $P(E, \boldsymbol{\alpha}; F, \boldsymbol{\beta}; \boldsymbol{\xi})$ enjoys the composition lemma (cf. Proposition 1). This lemma is the basis of our decision procedure for the entailment problem (cf. Sect. 4.2).

C1. None of the variables from $F, \boldsymbol{\beta}$ occur elsewhere in the body of R_1, that is, in Δ, or $E \mapsto \rho$.

The second (**C2**) and third (**C3**) constraint address the data constraint Δ in the body of R_1. Intuitively, the two constraints require that different data parameters of $P(E, \boldsymbol{\alpha}; F, \boldsymbol{\beta}; \boldsymbol{\xi})$ do not interfere with each other and the value of each data source parameter α_i is determined either by ρ, or γ_i.

C2. Each conjunct of Δ is of the form $\alpha_i \circ c$, $\alpha_i \circ \xi_j$, or $\alpha_i \circ \gamma_i + c$ for $\circ \in \{=, \leqslant, \geqslant\}$, $1 \leqslant i \leqslant |\boldsymbol{\alpha}| = |\boldsymbol{\gamma}|$, $1 \leqslant j \leqslant |\boldsymbol{\xi}|$, and $c \in \mathbb{Z}$.

C3. For each $1 \leqslant i \leqslant |\boldsymbol{\alpha}|$ such that α_i is a data variable, either α_i occurs in ρ, or Δ contains $\alpha_i = \gamma_i + c$ for some $c \in \mathbb{Z}$.

Furthermore, we have **C4**–**C6**, which are self-explained.

C4. Each variable occurs in $P(Y, \boldsymbol{\gamma}; F, \boldsymbol{\beta}; \boldsymbol{\xi})$ (resp. ρ) at most once.

C5. All location variables from $\boldsymbol{\alpha} \cup \boldsymbol{\xi} \cup \boldsymbol{X}$ occur in ρ.

C6. $Y \in \boldsymbol{X}$ and $\boldsymbol{\gamma} \subseteq \{E\} \cup \boldsymbol{X} \cup \boldsymbol{x}$.

Note that according to the constraint **C6**, none of the variables from $\boldsymbol{\alpha} \cup \boldsymbol{\xi}$ occur in $\boldsymbol{\gamma}$. Moreover, from the constraint **C5** and **C6**, we know that Y occurs in ρ. By the semantics defined later, this would guarantee that in each model of $P(E, \boldsymbol{\alpha}; F, \boldsymbol{\beta}; \boldsymbol{\xi})$, the sub-heap represented by $P(E, \boldsymbol{\alpha}; F, \boldsymbol{\beta}; \boldsymbol{\xi})$, seen as a directed graph, is connected.

We remark that these constraints are technical, and we leave as future work to make them as general as possible. However, in practice, inductive predicates satisfying these constraints are sufficient to model linear data structures with data and size constraints, cf. Example 1.

For a linearly compositional predicate $P \in \mathcal{P}$, let $\text{Flds}(P)$ (resp. $\text{LFlds}(P)$) denote the set of fields (resp. location fields) occurring in the inductive rules of P. Moreover, define the *principal* location field of P, denoted by PLFld(P), as the location field $f \in \text{LFlds}(P)$ such that (f, Y) occurs in ρ. Note that the principal location field is unique. For a spatial atom a, let Flds(a) denote the set of fields that a refers to: if $a = E \mapsto \rho$, then Flds(a) is the set of fields occurring in ρ; if $a = P(-)$, then $\text{Flds}(a) := \text{Flds}(P)$.

We write $\mathsf{SLID_{LC}}[\mathcal{P}]$ for the collection of separation logic formulae $\varphi = \Pi \wedge \Delta \wedge \Sigma$ satisfying the following constraints,

- **linearly compositional predicates**: all predicates from $\mathcal{P}$ are linearly compositional,
- **domination of principal location field**: for each pair of predicates $P_1, P_2 \in \mathcal{P}$, if $\text{Flds}(P_1) = \text{Flds}(P_2)$, then $\text{PLFld}(P_1) = \text{PLFld}(P_2)$,
- **uniqueness of predicates**: there is $P \in \mathcal{P}$ such that each predicate atom of Σ is of the form $P(-)$, and for each points-to atom occurring in Σ, the set of fields of this atom is Flds(P).

For an $\mathsf{SLID_{LC}}[\mathcal{P}]$ formula φ, let $\mathsf{Vars}(\varphi)$ (resp. $\mathsf{LVars}(\varphi)$, resp. $\mathsf{DVars}(\varphi)$) denote the set of (resp. location, resp. data) variables occurring in φ. Moreover, we use $\varphi[\boldsymbol{\mu}/\boldsymbol{\alpha}]$ to denote the simultaneous replacement of the variables α_j by μ_j in φ.

For the semantics of $\mathsf{SLID_{LC}}[\mathcal{P}]$, each formula is interpreted on the states. Formally, a *state* is a pair (s, h), where

- s is an assignment function which is a partial function from $\mathsf{LVars} \cup \mathsf{DVars}$ to $\mathbb{L} \cup \mathbb{Z}$ such that $dom(s)$ is finite and s respects the data type,
- h is a *heap* which is a partial function from $\mathbb{L} \times (\mathcal{F} \cup \mathcal{D})$ to $\mathbb{L} \cup \mathbb{D}$ such that
  - h respects the data type of fields, that is, for each $l \in \mathbb{L}$ and $f \in \mathcal{F}$ (resp. $l \in \mathbb{L}$ and $d \in \mathcal{D}$), if $h(l, f)$ (resp. $h(l, d)$) is defined, then $h(l, f) \in \mathbb{L}$ (resp. $h(l, d) \in \mathbb{Z}$); and
  - h is field-consistent, i.e. every location in h possess the same set of fields.

For a heap h, we use $\mathsf{ldom}(h)$ to denote the set of locations $l \in \mathbb{L}$ such that $h(l, f)$ or $h(l, d)$ is defined for some $f \in \mathcal{F}$ and $d \in \mathcal{D}$. Moreover, we use Flds(h) to denote the set of fields $f \in \mathcal{F}$ or $d \in \mathcal{D}$ such that $h(l, f)$ or $h(l, d)$ is defined for some $l \in \mathbb{L}$.

Two heaps h_1 and h_2 are said to be *field-compatible* if $\text{Flds}(h_1) = \text{Flds}(h_2)$. We write $h_1 \# h_2$ if $\mathsf{ldom}(h_1) \cap \mathsf{ldom}(h_2) = \emptyset$. Moreover, we write $h_1 \uplus h_2$ for the disjoint union of two field-compatible fields h_1 and h_2 (this implies that $h_1 \# h_2$).

Let (s, h) be a state and φ be an $\mathsf{SLID_{LC}}[\mathcal{P}]$ formula. The semantics of $\mathsf{SLID_{LC}}[\mathcal{P}]$ formulae is defined as follows,

- $(s, h) \models E = F$ (resp. $(s, h) \models E \neq F$) if $s(E) = s(F)$ (resp. $s(E) \neq s(F)$),
- $(s, h) \models \Pi_1 \wedge \Pi_2$ if $(s, h) \models \Pi_1$ and $(s, h) \models \Pi_2$,
- $(s, h) \models x \ o\ c$ (resp. $(s, h) \models x \ o\ y + c$) if $s(x) \ o\ c$ (resp. $s(x) \ o\ s(y) + c$),
- $(s, h) \models \Delta_1 \wedge \Delta_2$ if $(s, h) \models \Delta_1$ and $(s, h) \models \Delta_2$,
- $(s, h) \models \mathtt{emp}$ if $\mathsf{ldom}(h) = \emptyset$,
- $(s, h) \models E \mapsto \rho$ if $\mathsf{ldom}(h) = s(E)$, and for each $(f, X) \in \rho$ (resp. $(d, x) \in \rho$), $h(s(E), f) = s(X)$ (resp. $h(s(E), d) = s(x)$),

- $(s,h) \models P(E,\boldsymbol{\alpha};F,\boldsymbol{\beta};\boldsymbol{\xi})$ if $(s,h) \in [\![P(E,\boldsymbol{\alpha};F,\boldsymbol{\beta};\boldsymbol{\xi})]\!]$,
- $(s,h) \models \Sigma_1 * \Sigma_2$ if there are h_1, h_2 such that $h = h_1 \uplus h_2$, $(s,h_1) \models \Sigma_1$ and $(s,h_2) \models \Sigma_2$.

where the semantics of predicates $[\![P(E,\boldsymbol{\alpha};F,\boldsymbol{\beta};\boldsymbol{\xi})]\!]$ is given by the least fixed point of a monotone operator constructed from the body of rules for P in a standard way as in [7].

Example 1. Below are a few examples of the data structures definable in $\mathsf{SLID_{LC}}[\mathcal{P}]$: *slseg* for sorted list segments, *dllseg* for doubly linked list segments, *tlseg* for list segments with tail pointers, *plseg* for list segments where the data values are consecutive, and *ldllseg* for doubly list segments with lengths.

$$
\begin{aligned}
&slseg(E,x;F,x') ::= E = F \wedge x = x' \wedge \mathtt{emp},\\
&slseg(E,x;F,x') ::= \exists X, x''.\ x \leqslant x'' \wedge\\
&\qquad\qquad E \mapsto ((\mathtt{next},X),(\mathtt{data},x)) * slseg(X,x'';F,x').\\
&dllseg(E,P;F,L) ::= E = F \wedge P = L \wedge \mathtt{emp},\\
&dllseg(E,P;F,L) ::= \exists X.\ E \mapsto ((\mathtt{next},X),(\mathtt{prev},P)) * dllseg(X,E;F,L).\\
&tlseg(E;F;B) ::= E = F \wedge \mathtt{emp},\\
&tlseg(E;F;B) ::= \exists X.\ E \mapsto ((\mathtt{next},X),(\mathtt{tail},B)) * tlseg(X;F;B).\\
&plseg(E,x;F,x') ::= E = F \wedge x = x' \wedge \mathtt{emp},\\
&plseg(E,x;F,x') ::= \exists X, x''.\ x'' = x + 1 \wedge\\
&\qquad\qquad E \mapsto ((\mathtt{next},X),(\mathtt{data},x)) * plseg(X,x'';F,x').\\
&ldllseg(E,P,x;F,L,x') ::= E = F \wedge P = L \wedge x = x' \wedge \mathtt{emp},\\
&ldllseg(E,P,x;F,L,x') ::= \exists X, x''.\ x = x'' + 1 \wedge E \mapsto ((\mathtt{next},X),(\mathtt{prev},P))\\
&\qquad\qquad * ldllseg(X,E,x'';F,L,x').
\end{aligned}
$$

On the other hand, the predicate *tlseg*2 defined below is *not* linearly compositional, since F occurs twice in the body of the inductive rule.

$$
\begin{aligned}
&tlseg2(E;F) ::= E = F \wedge \mathtt{emp},\\
&tlseg2(E;F) ::= \exists X.\ E \mapsto ((\mathtt{next},X),(\mathtt{tail},F)) * tlseg2(X;F).
\end{aligned}
$$

For a formula φ, let $[\![\varphi]\!]$ denote the set of states (s,h) such that $(s,h) \models \varphi$. Let φ, ψ be $\mathsf{SLID_{LC}}[\mathcal{P}]$ formulae, then define $\varphi \models \psi$ as $[\![\varphi]\!] \subseteq [\![\psi]\!]$.

Proposition 1 [14]. *For each linearly compositional predicate* $P \in \mathcal{P}$, *it holds that* $P(E,\boldsymbol{\alpha};F,\boldsymbol{\beta};\boldsymbol{\xi}) * P(F,\boldsymbol{\beta};G,\boldsymbol{\gamma};\boldsymbol{\xi}) \models P(E,\boldsymbol{\alpha};G,\boldsymbol{\gamma};\boldsymbol{\xi})$.

We focus on the following two decision problems.

- Satisfiability: Given an $\mathsf{SLID_{LC}}[\mathcal{P}]$ formula φ, decide whether $[\![\varphi]\!]$ is empty.
- Entailment: Given two $\mathsf{SLID_{LC}}[\mathcal{P}]$ formulae φ, ψ such that $\mathsf{Vars}(\psi) \subseteq \mathsf{Vars}(\varphi)$, decide whether $\varphi \models \psi$ holds.

The rest of this paper is devoted to sound and complete decision procedures for the satisfiability and entailment problem of $\mathsf{SLID_{LC}}[\mathcal{P}]$.

3 Satisfiability

To decide the satisfiability of a separation logic formula φ, in [13], a *Boolean abstraction* $\mathsf{BoolAbs}(\varphi)$ of φ was constructed such that φ is satisfiable iff $\mathsf{BoolAbs}(\varphi)$ is satisfiable. Our decision procedure for $\mathsf{SLID_{LC}}[\mathcal{P}]$ follows this general approach. However, $\mathsf{SLID_{LC}}[\mathcal{P}]$ admits data constrains (viz. difference bound constraints specified in the data formulae) which are considerably more involved. The following example shows these data constraints are somehow intertwined with the "shape" part of the logic and they should be taken into account simultaneously when the satisfiability is concerned.

Example 2. Suppose $\varphi = E_1 = E_4 \wedge x_1 > x_2+1 \wedge \mathit{ldllseg}(E_1, E_3, x_1; E_2, E_4, x_2)$. From the inductive definition of *ldllseg* and $x_1 > x_2 + 1$, we know that if φ is satisfiable, then for any state (s, h) such that $(s, h) \models \varphi$, it holds that $|\mathsf{ldom}(h)| \geqslant 2$. On the other hand, in any heap (s, h) such that $(s, h) \models \mathit{ldllseg}(E_1, E_3, x_1; E_2, E_4, x_2)$ and $|\mathsf{ldom}(h)| \geqslant 2$, we know that both $s(E_1)$ and $s(E_4)$ are allocated and $s(E_1) \neq s(E_4)$. This contradicts to the fact that $E_1 = E_4$ is a conjunct in φ. Therefore, φ is unsatisfiable.

In the rest of this section, we will show how to extend the abstraction of formulae in [13] to obtain an abstraction in the presence of data constraints. In this case, the abstraction is *not* a Boolean formula, but a formula involving Boolean variables, (in)equality constraints over location variables, and difference bounded constraints over data variables. The satisfiability of these formulae can be decided by off-the-shelf SMT solvers. We also remark that, compared to the logic in [13], predicates in $\mathsf{SLID_{LC}}[\mathcal{P}]$ may have more than one source or destination parameter which gives rises to further technical difficulties.

Let $\varphi = \Pi \wedge \Delta \wedge \Sigma$ be an $\mathsf{SLID_{LC}}[\mathcal{P}]$ formula. Suppose $\Sigma = a_1 * \cdots * a_n$, where each a_i is either a points-to atom or a predicate atom.

Assume $a_i = P(Z_1, \boldsymbol{\mu}; Z_2, \boldsymbol{\nu}; \boldsymbol{\chi})$ where the inductive rule for P is

$$R_1 : P(E, \boldsymbol{\alpha}; F, \boldsymbol{\beta}; \boldsymbol{\xi}) ::= \exists \boldsymbol{X} \exists \boldsymbol{x}.\ \Delta' \wedge E \mapsto \rho * P(Y, \boldsymbol{\gamma}; F, \boldsymbol{\beta}; \boldsymbol{\xi}).$$

We extract the data constraint $\Delta_P(\boldsymbol{\alpha}', \boldsymbol{\beta}')$ out of R_1. Formally, $\Delta_P(\boldsymbol{\alpha}', \boldsymbol{\beta}') := \Delta'[\boldsymbol{\beta}'/\boldsymbol{\gamma}']$, where $\boldsymbol{\alpha}'$ (resp. $\boldsymbol{\gamma}'$, $\boldsymbol{\beta}'$) is the projection of $\boldsymbol{\alpha}$ (resp. $\boldsymbol{\gamma}$, $\boldsymbol{\beta}$) to data variables. For instance, $\Delta_{ldllseg}(x, x') := (x = x'' + 1)[x'/x''] = (x = x' + 1)$. Note that $\Delta_P(\boldsymbol{\alpha}', \boldsymbol{\beta}')$ may contain data variables from $\boldsymbol{\xi}$.

Furthermore, by Proposition 2, a Presburger formula $\psi_P(k, \boldsymbol{\alpha}', \boldsymbol{\beta}')$ where k occurs as a free variable, can be constructed to describe the composition of the relation corresponding to $\Delta_P(\boldsymbol{\alpha}', \boldsymbol{\beta}')$ for k times. In the running example, $\psi_{ldllseg}(k, x, x') := x = x' + k$.

Proposition 2. *Suppose $P(E, \boldsymbol{\alpha}; F, \boldsymbol{\beta}; \boldsymbol{\xi}) \in \mathcal{P}$. Then a quantifier free Presburger formula $\psi_P(k, \boldsymbol{\alpha}', \boldsymbol{\beta}')$ where k occurs as a free variable, can be constructed in linear time to define, for each $k \geqslant 1$, the composition of the relation corresponding to $\Delta_P(\boldsymbol{\alpha}', \boldsymbol{\beta}')$ for k times.*

As the next step, we define two formulae $\mathsf{Ufld}_1(a_i)$ and $\mathsf{Ufld}_{\geqslant 2}(a_i)$ obtained by unfolding the rule R_1 once and at least twice respectively. For each a_i, we introduce a fresh integer variable k_i. Before the definition of the two formulae, we introduce a notation first.

Definition 1 ($\mathsf{idx}_{(P,\gamma,E)}$). *Let $P \in \mathcal{P}$ and R_1 be the inductive rule in the definition of P. If in the body of R_1, E occurs in γ, then we use $\mathsf{idx}_{(P,\gamma,E)}$ to denote the unique index j such that $\gamma_j = E$ (The uniqueness follows from* **C4***).*

We define $\mathsf{Ufld}_1(a_i)$ and $\mathsf{Ufld}_{\geqslant 2}(a_i)$ by distinguishing the following two cases.

- If in the body of R_1, E occurs in γ, then let

$$\begin{aligned}&\mathsf{Ufld}_1(a_i) := \\ &\quad (E = \beta_{\mathsf{idx}_{(P,\gamma,E)}} \wedge k_i = 1 \wedge \psi_P(k_i, \boldsymbol{\alpha}', \boldsymbol{\beta}'))[Z_1/E, \boldsymbol{\mu}/\boldsymbol{\alpha}, Z_2/F, \boldsymbol{\nu}/\boldsymbol{\beta}, \boldsymbol{\chi}/\boldsymbol{\xi}],\end{aligned}$$

  and

$$\begin{aligned}&\mathsf{Ufld}_{\geqslant 2}(a_i) := \\ &\quad (E \neq \beta_{\mathsf{idx}_{(P,\gamma,E)}} \wedge k_i \geqslant 2 \wedge \psi_P(k_i, \boldsymbol{\alpha}', \boldsymbol{\beta}'))[Z_1/E, \boldsymbol{\mu}/\boldsymbol{\alpha}, Z_2/F, \boldsymbol{\nu}/\boldsymbol{\beta}, \boldsymbol{\chi}/\boldsymbol{\xi}].\end{aligned}$$

- Otherwise, let

$$\mathsf{Ufld}_1(a_i) := (k_i = 1 \wedge \psi_P(k_i, \boldsymbol{\alpha}', \boldsymbol{\beta}'))[Z_1/E, \boldsymbol{\mu}/\boldsymbol{\alpha}, Z_2/F, \boldsymbol{\nu}/\boldsymbol{\beta}, \boldsymbol{\chi}/\boldsymbol{\xi}],$$

  and

$$\mathsf{Ufld}_{\geqslant 2}(a_i) := (k_i \geqslant 2 \wedge \psi_P(k_i, \boldsymbol{\alpha}', \boldsymbol{\beta}'))[Z_1/E, \boldsymbol{\mu}/\boldsymbol{\alpha}, Z_2/F, \boldsymbol{\nu}/\boldsymbol{\beta}, \boldsymbol{\chi}/\boldsymbol{\xi}].$$

Example 3. Let φ be the formula in Example 2 and a_1 be the (unique) spatial atom in φ. Since the atom $P(X, E, x''; F, L, x')$ occurs in body of the inductive rule of *ldllseg* (where we have $E = \gamma_1$), we deduce that $\mathsf{Ufld}_1(a_1) := E_1 = E_4 \wedge k_1 = 1 \wedge x_1 = x_2 + k_1$ and $\mathsf{Ufld}_{\geqslant 2}(a_1) := E_1 \neq E_4 \wedge k_1 \geqslant 2 \wedge x_1 = x_2 + k_1$.

For each atom $a_i = P(Z_1, \boldsymbol{\mu}; Z_2, \boldsymbol{\nu}; \boldsymbol{\chi})$ in Σ, we introduce a Boolean variable $[Z_1, i]$. Moreover, if in the body of the inductive rule of P, E occurs in γ, then introduce a Boolean variable $[\nu_{\mathsf{idx}_{(P,\gamma,E)}}, i]$. Let $\mathsf{BVars}(\varphi)$ denote the set of introduced Boolean variables. We define *the abstraction of* φ to be $\mathsf{Abs}(\varphi) ::= \Pi \wedge \Delta \wedge \phi_\Sigma \wedge \phi_*$ over $\mathsf{BVars}(\varphi) \cup \{k_i \mid 1 \leqslant i \leqslant n\} \cup \mathsf{Vars}(\varphi)$, where ϕ_Σ and ϕ_* are defined as follows.

- $\phi_\Sigma = \bigwedge_{1 \leqslant i \leqslant n} \mathsf{Abs}(a_i)$ is an abstraction of Σ where
  - if $a_i = E \mapsto \rho$, then $\mathsf{Abs}(a_i) = [E, i]$,
  - if $a_i = P(Z_1, \boldsymbol{\mu}; Z_2, \boldsymbol{\nu}; \boldsymbol{\chi})$ and in the body of the inductive rule of P, E occurs in γ, then

$$\begin{aligned}\mathsf{Abs}(a_i) = &(\neg[Z_1, i] \wedge \neg[\nu_{\mathsf{idx}_{(P,\gamma,E)}}, i] \wedge Z_1 = Z_2 \wedge \boldsymbol{\mu} = \boldsymbol{\nu} \wedge k_i = 0) \vee \\ &([Z_1, i] \wedge [\nu_{\mathsf{idx}_{(P,\gamma,E)}}, i] \wedge \mathsf{Ufld}_1(P(Z_1, \boldsymbol{\mu}; Z_2, \boldsymbol{\nu}; \boldsymbol{\chi}))) \vee \\ &([Z_1, i] \wedge [\nu_{\mathsf{idx}_{(P,\gamma,E)}}, i] \wedge \mathsf{Ufld}_{\geqslant 2}(P(Z_1, \boldsymbol{\mu}; Z_2, \boldsymbol{\nu}; \boldsymbol{\chi}))),\end{aligned}$$

- if $a_i = P(Z_1, \boldsymbol{\mu}; Z_2, \boldsymbol{\nu}; \boldsymbol{\chi})$ and in the body of the inductive rule of P, E does not occur in $\boldsymbol{\gamma}$, then

$$\begin{aligned}\mathsf{Abs}(a_i) = &(\neg[Z_1, i] \wedge Z_1 = Z_2 \wedge \boldsymbol{\mu} = \boldsymbol{\nu} \wedge k_i = 0) \vee \\ &([Z_1, i] \wedge \mathsf{Ufld}_1(P(Z_1, \boldsymbol{\mu}; Z_2, \boldsymbol{\nu}; \boldsymbol{\chi}))) \vee \\ &([Z_1, i] \wedge \mathsf{Ufld}_{\geqslant 2}(P(Z_1, \boldsymbol{\mu}; Z_2, \boldsymbol{\nu}; \boldsymbol{\chi}))),\end{aligned}$$

– ϕ_* states the separation constraint of spatial atoms,

$$\phi_* = \bigwedge_{[Z_1,i],[Z_1',j] \in \mathsf{BVars}(\varphi), i \neq j} (Z_1 = Z_1' \wedge [Z_1, i]) \rightarrow \neg[Z_1', j].$$

Example 4. Suppose φ is the formula in Example 3. Then

$$\begin{aligned}\mathsf{Abs}(\varphi) = &E_1 = E_4 \wedge x_1 > x_2 + 1 \wedge \\ &((E_1 = E_2 \wedge E_3 = E_4 \wedge x_1 = x_2 \wedge k_1 = 0) \vee \\ &([E_1, 1] \wedge [E_4, 1] \wedge E_1 = E_4 \wedge k_1 = 1 \wedge x_1 = x_2 + k_1) \vee \\ &([E_1, 1] \wedge [E_4, 1] \wedge E_1 \neq E_4 \wedge k_1 \geqslant 2 \wedge x_1 = x_2 + k_1)).\end{aligned}$$

It is easy to see that $\mathsf{Abs}(\varphi)$ is unsatisfiable.

Proposition 3. *For each* $\mathsf{SLID_{LC}}[\mathcal{P}]$ *formula* φ, φ *is satisfiable iff* $\mathsf{Abs}(\varphi)$ *is satisfiable.*

The satisfiability of $\mathsf{Abs}(\varphi)$ can be discharged by the state-of-the-art SMT solvers, e.g., Z3. It is well known that the satisfiability of the quantifier-free presburger arithmetic formulae can be decided in NP. Hence we have:

Theorem 1. *The satisfiability problem of* $\mathsf{SLID_{LC}}[\mathcal{P}]$ *is in* NP.

Note that the problem whether the satisfiability problem of $\mathsf{SLID_{LC}}[\mathcal{P}]$ is NP-hard is open.

4 Entailment

In this section, we present a complete decision procedure for the entailment problem $\varphi \models \psi$, where φ, ψ are two $\mathsf{SLID_{LC}}[\mathcal{P}]$ formulae. We assume, without loss of generality, that $\mathsf{Vars}(\psi) \subseteq \mathsf{Vars}(\varphi)$, both φ and ψ are satisfiable, and $\mathrm{Flds}(\varphi) = \mathrm{Flds}(\psi)$.

On a high level, the decision procedure is similar to that in [11]. Loosely speaking, we construct graph representations $\mathcal{G}_\varphi$ and $\mathcal{G}_\psi$ of φ and ψ respectively and reduce the entailment problem to (a variant of) the graph homomorphism problem from $\mathcal{G}_\psi$ to $\mathcal{G}_\varphi$. However, our decision procedure is considerably more involved due to the additional expressibility of the logic and the non-intuitionistic semantics.

Recall that, in the previous section, from an $\mathsf{SLID_{LC}}[\mathcal{P}]$ formula φ one can construct an abstraction $\mathsf{Abs}(\varphi)$. Let $\sim_\varphi$ denote the equivalence relation defined over $\mathsf{LVars}(\varphi)$ as follows: For $X, Y \in \mathsf{LVars}(\varphi)$, $X \sim_\varphi Y$ iff $\mathsf{Abs}(\varphi) \models X = Y$. For $X \in \mathsf{LVars}(\varphi)$, let $[X]_\varphi$ denote the equivalence class of X under $\sim_\varphi$.

4.1 Graph Representations of $\mathsf{SLID_{LC}}[\mathcal{P}]$ Formulae

For a satisfiable $\mathsf{SLID_{LC}}[\mathcal{P}]$ formula φ, we will construct a graph $\mathcal{G}_\varphi$ from φ. Without loss of generality, we assume that φ contains at least one points-to atom or predicate atom.

Assume $\varphi = \Pi \wedge \Delta \wedge \Sigma$ with $\Sigma = a_1 * \ldots * a_n$ $(n \geqslant 1)$, and f_0 denotes the principal location field of φ. (Recall the "uniqueness of predicates" assumption for $\mathsf{SLID_{LC}}[\mathcal{P}]$ formulae in Sect. 2.)

We construct a directed *multigraph* (i.e., a directed graph with parallel arcs) $\mathcal{G}_\varphi = (\mathcal{V}_\varphi, \mathcal{R}_\varphi, \mathcal{L}_\varphi)$:

- $\mathcal{V}_\varphi = \{[E] \mid E \in \mathsf{LVars}(\varphi)\}$, where we use $[E]$ as an abbreviation of $[E]_\varphi$, that is, the equivalence class of $\sim_\varphi$ containing E.
- $\mathcal{R}_\varphi$ is the set of arcs and $\mathcal{L}_\varphi$ is the arc-labeling function, defined as follows:
  - for each pair of location variables (E, F) such that Σ contains a *points-to atom* $a_i = E \mapsto \rho$ and (f_0, F) occurs in ρ for $f_0 \in \mathbb{L}$, there is an arc from $[E]$ to $[F]$ labeled by $f_0[\rho']$, where ρ' is obtained by removing (f_0, F) from ρ — this arc e is said to be *field-labeled* and we write $\mathcal{L}_\varphi(e) = f_0[\rho']$;
  - for each pair of location variables (E, F) such that Σ contains a *predicate atom* $a_i = P(E, \boldsymbol{\alpha}; F, \boldsymbol{\beta}; \boldsymbol{\xi})$ and $\mathsf{Abs}(\varphi) \not\models \neg[E, i]$, there is an arc from $[E]$ to $[F]$ labeled by $P(\boldsymbol{\alpha}; \boldsymbol{\beta}; \boldsymbol{\xi})$ — this arc e is said to be *predicate-labeled* and we write $\mathcal{L}_\varphi(e) = P(\boldsymbol{\alpha}; \boldsymbol{\beta}; \boldsymbol{\xi})$.

From the construction, each field-labeled or predicate-labeled arc e corresponds to an unique atom a_i in Σ. Let $i(e)$ denote the index i of the atom.

Example 5. Let

$$\begin{aligned}\varphi = \; & \underbrace{ldllseg(E_1, E_1', x_1; E_3, E_3', x_3)}_{a_1} * \underbrace{ldllseg(E_2, E_2', x_2; E_4, E_4', x_4)}_{a_2} * \\ & \underbrace{ldllseg(E_3, E_3', x_3; E_4, E_4', x_4)}_{a_3} * \underbrace{ldllseg(E_4, E_4', x_4'; E_3, E_3', x_3')}_{a_4} * \\ & \underbrace{ldllseg(E_3, E_3', x_3; E_5, E_5', x_5)}_{a_5} * \underbrace{ldllseg(E_5, E_5', x_5'; E_3, E_3', x_3')}_{a_6} * \\ & \underbrace{ldllseg(E_4, E_4', x_5; E_6, E_6', x_6)}_{a_7}.\end{aligned}$$

The graph $\mathcal{G}_\varphi$ is as illustrated in Fig. 1, where each equivalence class of $\sim_\varphi$ is a singleton and $\mathcal{V}_\varphi = \{[E_1], \ldots, [E_6], [E_1'], \ldots, [E_6']\}$. Note that there are no arcs between the nodes $[E_1'], \ldots, [E_6']$.

We use standard graph-theoretic notions, for instance, paths, connected components (CCs) and strongly connected components (SCCs). In particular, a path in $\mathcal{G}_\varphi$ is a (possibly empty) sequence of consecutive arcs in $\mathcal{G}_\varphi$. If there is a path from $[E]$ to $[F]$, then $[F]$ is said to be *reachable* from $[E]$ and $[E]$ is said to be an *ancestor* of $[F]$. For a node $[E]$ and an arc e with source node $[E']$, e is said

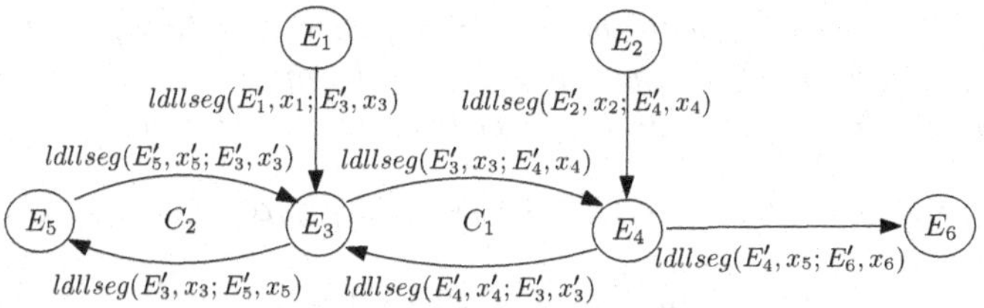

Fig. 1. The graph $\mathcal{G}_\varphi$

to be *reachable* from $[E]$ if $[E']$ is reachable from $[E]$. A CC or SCC $\mathcal{C}$ of $\mathcal{G}_\varphi$ is said to be *nontrivial* if $\mathcal{C}$ contains at least one arc.

We shall reveal some structural properties of the graph $\mathcal{G}_\varphi$.

Proposition 4. *The graph $\mathcal{G}_\varphi$ satisfies the following properties:*

1. *If there is a field-labeled arc out of $[E]$, then there are no predicate-labeled arcs out of $[E]$.*
2. *For each pair of distinct nodes $[E]$ and $[F]$ in $\mathcal{G}_\varphi$, there is at most one simple path from $[E]$ to $[F]$ in $\mathcal{G}_\varphi$.*

Proposition 5. *Each nontrivial SCC $\mathcal{S}$ satisfies the following constraints.*

- *Each pair of different simple cycles in $\mathcal{S}$ share at most one node — The set of shared nodes is called the set of cut nodes of $\mathcal{S}$, denoted by* $\mathsf{Cut}(\mathcal{S})$. *Here by "different", we mean that the two sets of arcs in the two cycles are different.*
- *The collection of simple cycles in $\mathcal{S}$ is organised into a tree. More precisely, let $\{C_1, \ldots, C_n\}$ be the set of all the simple cycles in $\mathcal{S}$ and $\mathcal{T}_\mathcal{S} = (\{C_1, \ldots, C_n\}, \mathsf{Cut}(\mathcal{S}), \mathcal{R})$ be the undirected bipartite graph such that for each $i : 1 \leqslant i \leqslant n$, $\{C_i, [E]\} \in \mathcal{R}$ iff $[E] \in \mathsf{Cut}(\mathcal{S}) \cap C_i$. Then $\mathcal{T}_\mathcal{S}$ is a tree.*

Example 6. The graph $\mathcal{G}_\varphi$ in Fig. 1 has just one nontrivial SCC $\mathcal{S}$ comprising the nodes $[E_3], [E_4], [E_5]$. The graph $\mathcal{T}_\mathcal{S} = (\{C_1, C_2\}, \{[E_3]\}, \{\{C_1, [E_3]\}, \{C_2, [E_3]\}\})$ is a tree.

4.2 Entailment Checking by Graph Homomorphisms

As a starting point, we illustrate how a path in $\mathcal{G}_\varphi$ is matchable to an arc in $\mathcal{G}_\psi$, which is the basis of our decision procedure.

Definition 2. *Given an arc e from $[E]_\psi$ to $[F]_\psi$ with label $P'(\boldsymbol{\alpha}'; \boldsymbol{\beta}'; \boldsymbol{\xi}')$ in $\mathcal{G}_\psi$, a (possibly empty) path $\pi = [E_0]_\varphi [E_1]_\varphi \ldots [E_n]_\varphi$ from $[E]_\varphi$ to $[F]_\varphi$ in $\mathcal{G}_\varphi$ is said to be* matchable *to e wrt.* $\mathsf{Abs}(\varphi)$ *if (1) either π is empty and* $\mathsf{Abs}(\varphi) \models E = F \wedge \boldsymbol{\alpha}' = \boldsymbol{\beta}'$, *(2) or π is nonempty and there are $\boldsymbol{\alpha}'_0, \boldsymbol{\alpha}'_1, \ldots, \boldsymbol{\alpha}'_n$ such that $\boldsymbol{\alpha}'_0 = \boldsymbol{\alpha}'$, $\boldsymbol{\alpha}'_n = \boldsymbol{\beta}'$, and for each $i : 1 \leqslant i \leqslant n$, the arc from $[E_{i-1}]_\varphi$ to $[E_i]_\varphi$ in π is*

– *either a field-labeled arc with the label* $f_0[\rho']$ *such that* $\mathsf{Abs}(\varphi) \wedge E_{i-1} \mapsto \rho \models P'(E_{i-1}, \boldsymbol{\alpha}'_{i-1}; E_i, \boldsymbol{\alpha}'_i; \boldsymbol{\xi}')$*, where* ρ *is obtained from* ρ' *by adding* (f_0, E_i)*;*
– *or a predicate-labeled arc with the label* $P(\boldsymbol{\alpha}; \boldsymbol{\beta}; \boldsymbol{\xi})$ *such that*

$$\mathsf{Abs}(\varphi) \wedge P(E_{i-1}, \boldsymbol{\alpha}; E_i, \boldsymbol{\beta}; \boldsymbol{\xi}) \models P'(E_{i-1}, \boldsymbol{\alpha}'_{i-1}; E_i, \boldsymbol{\alpha}'_i; \boldsymbol{\xi}').$$

Note that in the above definition, we abuse the notation slightly, since $\mathsf{Abs}(\varphi)$ may contain Boolean variables $[E', j]$, the integer variables k_j, and disjunctions, thus strictly speaking, $\mathsf{Abs}(\varphi) \wedge E_{i-1} \mapsto \rho$ and $\mathsf{Abs}(\varphi) \wedge P(E_{i-1}, \boldsymbol{\alpha}; E_i, \boldsymbol{\beta}; \boldsymbol{\xi})$ are not $\mathsf{SLID_{LC}}[\mathcal{P}]$ formulae.

Example 7. Let φ be the formula in Example 5 and $\psi = dllseg(E_1, E'_1; E_6, E'_6) * dllseg(E_2, E'_2; E_4, E'_4)$. Then the path $[E_1]_\varphi[E_3]_\varphi[E_4]_\varphi[E_6]_\varphi$ in $\mathcal{G}_\varphi$ is matchable to the arc e from $[E_1]_\psi$ to $[E_6]_\psi$ with the label $dllseg(E'_1; E'_6)$ in $\mathcal{G}_\psi$. More specifically, there are $\boldsymbol{\alpha}'_0 = E'_1$, $\boldsymbol{\alpha}'_1 = E'_3$, $\boldsymbol{\alpha}'_2 = E'_4$, and $\boldsymbol{\alpha}'_3 = E'_6$ such that

$$\begin{aligned}
\mathsf{Abs}(\varphi) \wedge ldllseg(E_1, E'_1, x_1; E_3, E'_3, x_3) &\models dllseg(E_1, E'_1; E_3, E'_3),\\
\mathsf{Abs}(\varphi) \wedge ldllseg(E_3, E'_3, x_3; E_4, E'_4, x_4) &\models dllseg(E_3, E'_3; E_4, E'_4),\\
\mathsf{Abs}(\varphi) \wedge ldllseg(E_4, E'_4, x_5; E_6, E'_6, x_6) &\models dllseg(E_4, E'_4; E_6, E'_6).
\end{aligned}$$

Proposition 6. *Suppose* φ *is an* $\mathsf{SLID_{LC}}[\mathcal{P}]$ *formula,* $a = E \mapsto \rho$ *or* $a = P(E, \boldsymbol{\alpha}; F, \boldsymbol{\beta}; \boldsymbol{\xi})$ *is a spatial atom in* φ*, and* $P'(E, \boldsymbol{\alpha}'; F, \boldsymbol{\beta}'; \boldsymbol{\xi}')$ *is a predicate atom (not necessarily in* φ*) such that* $\mathsf{Vars}(P'(E, \boldsymbol{\alpha}'; F, \boldsymbol{\beta}'; \boldsymbol{\xi}')) \subseteq \mathsf{Vars}(\varphi)$*. Then (1) the entailment problem* $\mathsf{Abs}(\varphi) \wedge a \models P'(E, \boldsymbol{\alpha}'; F, \boldsymbol{\beta}'; \boldsymbol{\xi}')$ *is in* Δ_2^p*; and (2) if there exist* $\boldsymbol{\alpha}', \boldsymbol{\alpha}''$ *such that* $\mathsf{Abs}(\varphi) \wedge a \models P'(E, \boldsymbol{\alpha}'; F, \boldsymbol{\beta}'; \boldsymbol{\xi}')$ *and* $\mathsf{Abs}(\varphi) \wedge a \models P'(E, \boldsymbol{\alpha}''; F, \boldsymbol{\beta}'; \boldsymbol{\xi}')$*, then* $\mathsf{Abs}(\varphi) \models \boldsymbol{\alpha}' = \boldsymbol{\alpha}''$*. Such an unique* $\boldsymbol{\alpha}'$ *can be computed effectively from* $\mathsf{Abs}(\varphi)$*, the atom* a*,* $P'(E, -; F, \boldsymbol{\beta}'; \boldsymbol{\xi}')$*, and the inductive definition of* P *and* P'*.*

The complexity upper bound Δ_2^p in Proposition 6 follows from the fact that, to solve $\mathsf{Abs}(\varphi) \wedge a \models P'(E, \boldsymbol{\alpha}'; F, \boldsymbol{\beta}'; \boldsymbol{\xi}')$, it is necessary to use an oracle to decide the satisfiability of quantifier-free Presburger formulae, which is in NP. The uniqueness of $\boldsymbol{\alpha}'$ in Proposition 6 is guaranteed by the constraints **C2**, **C3**, and **C5** in the inductive definition of predicates.

Proposition 6 shows that Definition 2 is effective, namely,

Proposition 7. *Check whether a path* π *in* $\mathcal{G}_\varphi$ *is matchable to a predicate-labeled arc* e *in* $\mathcal{G}_\psi$ *can be done in* Δ_2^p*.*

We are ready to present the decision procedure. We will introduce a concept of allocating plans $\mathcal{AP}$ (cf. Definition 5), which are the pairs $(\mathsf{Abs}_{\mathcal{AP}}[\varphi], \mathcal{G}_{\mathcal{AP}}[\varphi])$, where $\mathsf{Abs}_{\mathcal{AP}}[\varphi]$ is a formula obtained from $\mathsf{Abs}(\varphi)$, and $\mathcal{G}_{\mathcal{AP}}[\varphi]$ is a simplification of $\mathcal{G}_\varphi$. The entailment problem is reduced to checking the existence of a homomorphism from $(\mathsf{Abs}(\psi), \mathcal{G}_\psi)$ to $(\mathsf{Abs}_{\mathcal{AP}}[\varphi], \mathcal{G}_{\mathcal{AP}}[\varphi])$, for each allocating plan $\mathcal{AP}$. For each CC $\mathcal{C}$ of $\mathcal{G}_\varphi$, $\mathsf{Cyc}_\mathcal{C}$ denotes the set of simple cycles in $\mathcal{C}$ and $\mathsf{NScc}_\mathcal{C}$ denotes the set of nontrivial SCCs in $\mathcal{C}$. For $i \in \mathbb{N}$, let $[i] = \{1, \ldots, i\}$.

Definition 3 (Allocating pseudo-plans). *Let $\mathcal{C}_1, \ldots, \mathcal{C}_k$ be an enumeration of the nontrivial CCs of $\mathcal{G}_\varphi$, and for each $i \in [k]$, $\mathsf{Cyc}_{\mathcal{C}_i} = \{C_{i,1}, \ldots, C_{i,l_i}\}$ (where $l_i \geqslant 0$). Then an allocating pseudo-plan Ω for $\mathcal{G}_\varphi$ is a function such that $\Omega(i) \in \{0\} \cup [l_i]$ for each $i \in [k]$.*

Intuitively, $\Omega(i) \in [l_i]$ means that some arc in the simple cycle $C_{i,\Omega(i)}$ is assigned to be an nonempty heap, and accordingly, $\Omega(i) = 0$ means that all arcs in nontrivial SCCs of $\mathcal{C}_i$ are assigned to be empty heaps (cf. Definition 4).

For each arc e with $a_{i(e)} = P(E, \boldsymbol{\alpha}; F, \boldsymbol{\beta}; \boldsymbol{\xi})$, we use ϕ_e to denote $[E, i(e)]$.

Definition 4 ($\Omega[\mathsf{Abs}(\varphi)]$ and feasible allocating pseudo-plans). *Let Ω be an allocating pseudo-plan of $\mathcal{G}_\varphi$. We define $\Omega[\mathsf{Abs}(\varphi)] := \mathsf{Abs}(\varphi) \wedge \bigwedge_{i \in [k]} \zeta_i$, where for each $i \in [k]$, $\zeta_i := \bigvee\limits_{e \in C_{i,\Omega(i)}} \phi_e$ if $\Omega(i) \neq 0$; and $\zeta_i := \bigwedge\limits_{\mathcal{S} \in \mathsf{NScc}_{\mathcal{C}_i}} \bigwedge\limits_{e \in \mathcal{S}} \neg \phi_e$ if $\Omega(i) = 0$. An allocating pseudo-plan Ω is* feasible *if $\Omega[\mathsf{Abs}(\varphi)]$ is satisfiable.*

For an allocating pseudo-plan Ω of $\mathcal{G}_\varphi$, we construct a graph $\Omega[\mathcal{G}_\varphi] = (\mathcal{V}_\Omega, \mathcal{R}_\Omega, \mathcal{L}_\Omega)$ from φ, similarly to $\mathcal{G}_\varphi$, with $\sim_\varphi$ replaced by $\sim_\Omega$ (on $\mathsf{LVars}(\varphi)$) defined as follows: $E \sim_\Omega F$ iff $\Omega[\mathsf{Abs}(\varphi)] \models E = F$.

A directed graph $\mathcal{G}$ is said to be *DAG-like* (DAG: directed acyclic graph) if for each CC $\mathcal{C}$ of $\mathcal{G}$, either $\mathcal{C}$ is a DAG, or $\mathcal{C}$ contains exactly one simple cycle C which is reachable from every node in $\mathcal{C} \setminus C$.

Definition 5 (Allocating plans $\mathcal{AP}$). *Given a formula φ, an allocating plan $\mathcal{AP} = (\mathsf{Abs}_{\mathcal{AP}}[\varphi], \mathcal{G}_{\mathcal{AP}}[\varphi])$ of φ is obtained from $\mathcal{G}_\varphi$ by a sequence of allocating pseudo-plans $\Omega_1, \ldots, \Omega_n$ ($n \geqslant 0$) such that: (1) $\phi_0 = \mathsf{Abs}(\varphi)$, $\mathcal{G}_0 = \mathcal{G}_\varphi$; for each $i : 1 \leqslant i \leqslant n$, (2) Ω_i is a feasible allocating pseudo-plan of $\mathcal{G}_{i-1}$, $\phi_i = \Omega_i[\phi_{i-1}]$, $\mathcal{G}_i = \Omega_i[\mathcal{G}_{i-1}]$; (3) $\mathsf{Abs}_{\mathcal{AP}}[\varphi] = \phi_n$, $\mathcal{G}_{\mathcal{AP}}[\varphi] = \mathcal{G}_n$, and $\mathcal{G}_{\mathcal{AP}}[\varphi]$ is DAG-like.*

For an allocating plan $\mathcal{AP}$ of φ, we use $\Sigma_{\mathcal{AP}}[\varphi]$ to denote the spatial formula corresponding to $\mathcal{G}_{\mathcal{AP}}[\varphi]$. In addition, let $\varphi_{\mathcal{AP}} = \mathsf{Abs}_{\mathcal{AP}}[\varphi] \wedge \Sigma_{\mathcal{AP}}[\varphi]$.

Example 8. Let φ be the formula in Example 5. The graph $\mathcal{G}_\varphi$ contains exactly one nontrivial connected component $\mathcal{C}_1$ (cf. Fig. 1). In addition, suppose Ω_1 and Ω_2 are the allocating pseudo-plans such that $\Omega_1(1) = 1$ and $\Omega_2(1) = 0$. Then $(\Omega_1[\mathsf{Abs}(\varphi)], \Omega_1[\mathcal{G}_\varphi])$ and $(\Omega_2[\mathsf{Abs}(\varphi)], \Omega_2[\mathcal{G}_\varphi])$ are illustrated in Fig. 2. Since both $\Omega_1[\mathcal{G}_\varphi]$ and $\Omega_2[\mathcal{G}_\varphi]$ are DAG-like, we know that $(\Omega_1[\mathsf{Abs}(\varphi)], \Omega_1[\mathcal{G}_\varphi])$ and $(\Omega_2[\mathsf{Abs}(\varphi)], \Omega_2[\mathcal{G}_\varphi])$ are both allocating plans.

Lemma 1. *Let φ, ψ be two $\mathsf{SLID_{LC}}[\mathcal{P}]$ formulae such that $\mathsf{Vars}(\psi) \subseteq \mathsf{Vars}(\varphi)$. Then $\varphi \models \psi$ iff the following two conditions hold.*

- $\mathsf{Abs}(\varphi) \models \exists \boldsymbol{Z}.\mathsf{Abs}(\psi)$, *where $\boldsymbol{Z} = \mathsf{Vars}(\mathsf{Abs}(\psi)) \setminus \mathsf{Var}(\psi)$, i.e., the set of additional variables introduced when constructing $\mathsf{Abs}(\psi)$ from ψ.*
- *For each allocating plan $\mathcal{AP}$ of φ, $\varphi_{\mathcal{AP}} \models \psi$.*

By Lemma 1, the entailment problem $\varphi \models \psi$ can be reduced to checking $\varphi_{\mathcal{AP}} \models \psi$ for each allocating plan $\mathcal{AP}$, which we now show that can be further reduced to checking the existence of a (graph) homomorphism from $(\mathsf{Abs}(\psi), \mathcal{G}_\psi)$ to $(\mathsf{Abs}_{\mathcal{AP}}[\varphi], \mathcal{G}_{\mathcal{AP}}[\varphi])$.

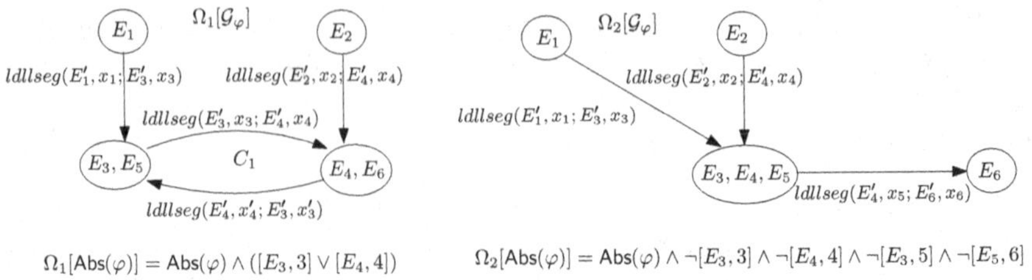

Fig. 2. $(\Omega_1[\mathsf{Abs}(\varphi)], \Omega_1[\mathcal{G}_\varphi])$ and $(\Omega_2[\mathsf{Abs}(\varphi)], \Omega_2[\mathcal{G}_\varphi])$

Definition 6 (Homomorphisms). *Let $\mathcal{AP}$ be an allocating plan of φ, $\mathcal{G}_{\mathcal{AP}}[\varphi] = (\mathcal{V}_{\mathcal{AP}}, \mathcal{R}_{\mathcal{AP}}, \mathcal{L}_{\mathcal{AP}})$, and $\mathcal{G}_\psi = (\mathcal{V}_\psi, \mathcal{R}_\psi, \mathcal{L}_\psi)$. Then a* homomorphism *from $(\mathsf{Abs}(\psi), \mathcal{G}_\psi)$ to $(\mathsf{Abs}_{\mathcal{AP}}[\varphi], \mathcal{G}_{\mathcal{AP}}[\varphi])$ is a pair of functions (θ, η) where θ is from $\mathcal{V}_\psi$ to $\mathcal{V}_{\mathcal{AP}}$ and η is from $\mathcal{R}_\psi$ to the set of paths in $\mathcal{G}_{\mathcal{AP}}[\varphi]$ satisfying the following constraints.*

- **Variable subsumption:** *For each node $[E] \in \mathcal{V}_\psi$, $[E] \subseteq \theta([E])$.*
- **Field-labeled arcs:** *For each field-labeled arc e from $[E]$ to $[F]$ in $\mathcal{G}_\psi$, $\eta(e)$ is a field-labeled arc from $\theta([E])$ to $\theta([F])$ in $\mathcal{G}_{\mathcal{AP}}[\varphi]$.*
- **Predicate-labeled arcs:** *For each predicate-labeled arc e from $[E]$ to $[F]$ in $\mathcal{G}_\psi$, both $\theta([E])$ and $\theta([F])$ must be in some CC $\mathcal{C}$, and the following conditions are satisfied.*
  - *If $\mathcal{C}$ is a DAG, then*
    - *if $\theta([E]) \neq \theta([F])$, then $\eta(e)$ is the unique simple path from $\theta([E])$ to $\theta([F])$ in $\mathcal{G}_{\mathcal{AP}}[\varphi]$,*
    - *otherwise, $\eta(e)$ is the empty path from $\theta([E])$ to $\theta([F])$.*
  - *Otherwise, let C be the unique simple cycle in $\mathcal{C}$.*
    - *If $\theta([E]) \neq \theta([F])$, moreover, the unique simple path from $\theta([E])$ to $\theta([F])$ in $\mathcal{C}$ is either node-disjoint from C, or contains at least two nodes in C, then $\eta(e)$ is the unique simple path from $\theta([E])$ to $\theta([F])$ in $\mathcal{C}$.*
    - *If $\theta([E]) \neq \theta([F])$, moreover, the unique simple path from $\theta([E])$ to $\theta([F])$ in $\mathcal{C}$ contains exactly one node in C (i.e. $\theta([F])$), then $\eta(e)$ is either the unique simple path from $\theta([E])$ to $\theta([F])$ or the composition of the unique simple path from $\theta([E])$ to $\theta([F])$ and the cycle C.*
    - *If $\theta([E]) = \theta([F])$ and $\theta([F])$ belongs to C, then $\eta(e)$ is either the empty path or the simple cycle C from $\theta([E])$ to $\theta([F])$.*
    - *If $\theta([E]) = \theta([F])$ and $\theta([F])$ does not belong to C, then $\eta(e)$ is the empty path.*
- **Matching of paths to arcs:** *For each arc e in $\mathcal{G}_\psi$, $\eta(e)$ is matchable to e wrt. $\mathsf{Abs}_{\mathcal{AP}}[\varphi]$.*
- **Separation constraint:** *For each pair of distinct arcs e_1, e_2 in $\mathcal{G}_\psi$, $\eta(e_1)$ and $\eta(e_2)$ are arc-disjoint.*
- **Coverage of allarcs in $\mathcal{G}_{\mathcal{AP}}[\varphi]$:** *Each arc of $\mathcal{G}_{\mathcal{AP}}[\varphi]$ occurs in $\eta(e)$ for some arc e in $\mathcal{G}_\psi$.*

Lemma 2. *Let φ, ψ be two formulae satisfying the premise of Lemma 1. Then for each allocating plan $\mathcal{AP}$ of $\mathcal{G}_\varphi$, $\varphi_{\mathcal{AP}} \models \psi$ iff there is a homomorphism from $(\mathsf{Abs}(\psi), \mathcal{G}_\psi)$ to $(\mathsf{Abs}_{\mathcal{AP}}[\varphi], \mathcal{G}_{\mathcal{AP}}[\varphi])$.*

Theorem 2. *The entailment problem of $\mathsf{SLID_{LC}}[\mathcal{P}]$ formulae is in Π_3^{P}.*

Complexity Analysis: Deciding whether there exists a homomorphism from $(\mathsf{Abs}(\psi), \mathcal{G}_\psi)$ to $(\mathsf{Abs}_{\mathcal{AP}}[\varphi], \mathcal{G}_{\mathcal{AP}}[\varphi])$ can be done in Σ_2^{P}, by Proposition 7 and guessing a homomorphism (θ, η) in Definition 6. Furthermore, by Lemma 2, $\varphi \not\models \psi$ iff either $\mathsf{Abs}(\varphi) \neq \exists \boldsymbol{Z}.\mathsf{Abs}(\psi)$ (cf. Lemma 1), or there is an allocating plan $\mathcal{AP}$ such that there is no homomorphism from $(\mathsf{Abs}(\psi), \mathcal{G}_\psi)$ to $(\mathsf{Abs}_{\mathcal{AP}}[\varphi], \mathcal{G}_{\mathcal{AP}}[\varphi])$. Hence, deciding $\varphi \not\models \psi$ is in $\mathrm{NP}^{\Pi_2^{\mathrm{P}}} = \Sigma_3^{\mathrm{P}}$. We conclude that the entailment problem is in Π_3^{P}.

5 Conclusion

In this paper, we have defined $\mathsf{SLID_{LC}}[\mathcal{P}]$, a linearly compositional fragment of separation logic with inductive definitions, where both linear shapes, e.g., singly or doubly linked lists, lists with tail pointers, and data constraints, e.g., sortedness and size constraints, are expressible. We have provided complete decision procedures for both the satisfiability and the entailment problem, with complexity upper-bounds NP and Π_3^{P} respectively. For the satisfiability problem, it turned out that computing the transitive closure of data constraints is critical to the completeness of the decision procedure. For the entailment checking, a novel concept of allocating plans was introduced. Note that we made no efforts to tighten the $\mathrm{NP}/\Pi_3^{\mathrm{P}}$ upper-bound or to provide lower-bounds, which might be interesting subjects of further research. More importantly, we believe that the approach introduced in this paper is amenable to implementations and can be extended to handle non-linear shapes (e.g., nested lists, binary search trees) as well as other kinds of data constraints (e.g., set or multiset constraints). These are left as future work.

References

1. Antonopoulos, T., Gorogiannis, N., Haase, C., Kanovich, M., Ouaknine, J.: Foundations for decision problems in separation logic with general inductive predicates. In: Muscholl, A. (ed.) FOSSACS 2014 (ETAPS). LNCS, vol. 8412, pp. 411–425. Springer, Heidelberg (2014)
2. Berdine, J., Calcagno, C., O'Hearn, P.W.: Symbolic execution with separation logic. In: Yi, K. (ed.) APLAS 2005. LNCS, vol. 3780, pp. 52–68. Springer, Heidelberg (2005)
3. Bouajjani, A., Drăgoi, C., Enea, C., Sighireanu, M.: Accurate invariant checking for programs manipulating lists and arrays with infinite data. In: Chakraborty, S., Mukund, M. (eds.) ATVA 2012. LNCS, vol. 7561, pp. 167–182. Springer, Heidelberg (2012)

4. Bozga, M., Iosif, R., Perarnau, S.: Quantitative separation logic and programs with lists. J. Autom. Reasoning **45**(2), 131–156 (2010)
5. Brochenin, R., Demri, S., Lozes, É.: On the almighty wand. Inf. Comput. **211**, 106–137 (2012)
6. Brotherston, J., Distefano, D., Petersen, R.L.: Automated cyclic entailment proofs in separation logic. In: Bjørner, N., Sofronie-Stokkermans, V. (eds.) CADE 2011. LNCS, vol. 6803, pp. 131–146. Springer, Heidelberg (2011)
7. Brotherston, J., Fuhs, C., Perez, J.A.N., Gorogiannis, N.: A decision procedure for satisfiability in separation logic with inductive predicates. In: LICS (2014)
8. Calcagno, C., Distefano, D.: Infer: an automatic program verifier for memory safety of C programs. In: Bobaru, M., Havelund, K., Holzmann, G.J., Joshi, R. (eds.) NFM 2011. LNCS, vol. 6617, pp. 459–465. Springer, Heidelberg (2011)
9. Chin, W.-N., David, C., Nguyen, H.H., Qin, S.: Automated verification of shape, size and bag properties via user-defined predicates in separation logic. Sci. Comput. Program. **77**(9), 1006–1036 (2012)
10. Chu, D., Jaffar, J., Trinh, M.: Automating proofs of data-structure properties in imperative programs. CoRR, abs/1407.6124 (2014)
11. Cook, B., Haase, C., Ouaknine, J., Parkinson, M., Worrell, J.: Tractable reasoning in a fragment of separation logic. In: Katoen, J.-P., König, B. (eds.) CONCUR 2011. LNCS, vol. 6901, pp. 235–249. Springer, Heidelberg (2011)
12. Demri, S., Deters, M.: Expressive completeness of separation logic with two variables and no separating conjunction. In: CSL-LICS, p. 37 (2014)
13. Enea, C., Lengál, O., Sighireanu, M., Vojnar, T.: Compositional entailment checking for a fragment of separation logic. In: Garrigue, J. (ed.) APLAS 2014. LNCS, vol. 8858, pp. 314–333. Springer, Heidelberg (2014)
14. Enea, C., Sighireanu, M., Wu, Z.: On automated lemma generation for separation logic with inductive definitions. ATVA 2015. LNCS, vol. 9364, pp. 80–96. Springer, Heidelberg (2015). doi:10.1007/978-3-319-24953-7_7
15. Hou, Z., Goré, R., Tiu, A.: Automated theorem proving for assertions in separation logic with all connectives. In: Felty, A.P., Middeldorp, A. (eds.) CADE-25. LNCS(LNAI), vol. 9195, pp. 501–516. Springer, Heidelberg (2015)
16. Iosif, R., Rogalewicz, A., Simacek, J.: The tree width of separation logic with recursive definitions. In: Bonacina, M.P. (ed.) CADE 2013. lncs, vol. 7898, pp. 21–38. Springer, Heidelberg (2013)
17. Iosif, R., Rogalewicz, A., Vojnar, T.: Deciding entailments in inductive separation logic with tree automata. In: Cassez, F., Raskin, J.-F. (eds.) ATVA 2014. LNCS, vol. 8837, pp. 201–218. Springer, Heidelberg (2014)
18. O'Hearn, P.W., Reynolds, J.C., Yang, H.: Local reasoning about programs that alter data structures. In: Fribourg, L. (ed.) CSL 2001 and EACSL 2001. LNCS, vol. 2142, p. 1. Springer, Heidelberg (2001)
19. Pek, E., Qiu, X., Madhusudan, P.: Natural proofs for data structure manipulation in C using separation logic. In: PLDI, pp. 440–451. ACM (2014)
20. Piskac, R., Wies, T., Zufferey, D.: Automating separation logic using SMT. In: Sharygina, N., Veith, H. (eds.) CAV 2013. LNCS, vol. 8044, pp. 773–789. Springer, Heidelberg (2013)
21. Piskac, R., Wies, T., Zufferey, D.: Automating separation logic with trees and data. In: Biere, A., Bloem, R. (eds.) CAV 2014. LNCS, vol. 8559, pp. 711–728. Springer, Heidelberg (2014)
22. Piskac, R., Wies, T., Zufferey, D.: GRASShopper - complete heap verification with mixed specifications. In: Ábrahám, E., Havelund, K. (eds.) TACAS 2014 (ETAPS). LNCS, vol. 8413, pp. 124–139. Springer, Heidelberg (2014)

23. Qiu, X., Garg, P., Stefănescu, A., Madhusudan, P.: Naturalproofs for structure, data, and separation. In: PLDI, pp. 231–242. ACM (2013)
24. Reynolds, J.C.: Separation logic: a logic for shared mutable datastructures. In: LICS, pp. 55–74. ACM (2002)
25. Z3. http://rise4fun.com/z3

Lower Runtime Bounds for Integer Programs

F. Frohn[1], M. Naaf[1], J. Hensel[1], M. Brockschmidt[2], and J. Giesl[1(✉)]

[1] LuFG Informatik 2, RWTH Aachen University, Aachen, Germany
{florian.frohn,hensel,giesl}@informatik.rwth-aachen.de,
matthias.naaf@rwth-aachen.de
[2] Microsoft Research, Cambridge, UK
mabrocks@microsoft.com

Abstract. We present a technique to infer *lower* bounds on the worst-case runtime complexity of integer programs. To this end, we construct symbolic representations of program executions using a framework for iterative, under-approximating program simplification. The core of this simplification is a method for (under-approximating) program acceleration based on recurrence solving and a variation of ranking functions. Afterwards, we deduce *asymptotic* lower bounds from the resulting simplified programs. We implemented our technique in our tool LoAT and show that it infers non-trivial lower bounds for a large number of examples.

1 Introduction

Recent advances in program analysis yield efficient methods to find *upper* bounds on the complexity of sequential integer programs. Here, one usually considers "worst-case complexity", i.e., for any variable valuation, one analyzes the length of the longest execution starting from that valuation. But in many cases, in addition to upper bounds, it is also important to find *lower* bounds for this notion of complexity. Together with an analysis for upper bounds, this can be used to infer *tight* complexity bounds. Lower bounds also have important applications in security analysis, e.g., to detect possible denial-of-service or side-channel attacks, as programs whose runtime depends on a secret parameter "leak" information about that parameter. In general, *concrete* lower bounds that hold for arbitrary variable valuations can hardly be expressed concisely. In contrast, *asymptotic* bounds are easily understood by humans and witness possible attacks in a convenient way.

We first introduce our program model in Sect. 2. In Sect. 3, we show how to use a variation of classical ranking functions which we call *metering functions* to under-estimate the number of iterations of a simple loop (i.e., a single transition t looping on a location ℓ). Then, we present a framework for repeated program simplifications in Sect. 4. It simplifies full programs (with branching and sequences of possibly nested loops) to programs with only simple loops. Moreover, it eliminates simple loops by (under-)approximating their effect using

Supported by the DFG grant GI 274/6-1 and the Air Force Research Laboratory (AFRL).

N. Olivetti and A. Tiwari (Eds.): IJCAR 2016, LNAI 9706, pp. 550–567, 2016.
DOI: 10.1007/978-3-319-40229-1_37

a combination of metering functions and recurrence solving. In this way, programs are transformed to *simplified programs* without loops. In Sect. 5, we then show how to extract asymptotic lower bounds and variables that influence the runtime from simplified programs. Finally, we conclude with an experimental evaluation of our implementation LoAT in Sect. 6. For all proofs, we refer to [16].

Related Work. While there are many techniques to infer *upper bounds* on the worst-case complexity of integer programs (e.g., [1–4,8,9,14,19,26]), there is little work on *lower bounds*. In [3], it is briefly mentioned that their technique could also be adapted to infer lower instead of upper bounds for *abstract cost rules*, i.e., integer procedures with (possibly multiple) outputs. However, this only considers *best-case* lower bounds instead of worst-case lower bounds as in our technique. Upper and lower bounds for *cost relations* are inferred in [1]. Cost relations extend recurrence equations such that, e.g., non-determinism can be modeled. However, this technique also considers best-case lower bounds only.

A method for best-case lower bounds for logic programs is presented in [11]. Moreover, we recently introduced a technique to infer worst-case lower bounds for term rewrite systems (TRSs) [15]. However, TRSs differ fundamentally from the programs considered here, since they do not allow integers and have no notion of a "program start". Thus, the technique of [15], based on synthesizing families of reductions by automatic induction proofs, is very different to the present paper.

To simplify programs, we use a variant of *loop acceleration* to summarize the effect of applying a loop repeatedly. Acceleration is mostly used in over-approximating settings (e.g., [13,17,21,24]), where handling non-determinism is challenging, as loop summaries have to cover *all* possible non-deterministic choices. However, our technique is under-approximating, i.e., we can instantiate non-deterministic values arbitrarily. In contrast to the under-approximating acceleration technique in [22], instead of quantifier elimination we use an adaptation of ranking functions to under-estimate the number of loop iterations symbolically.

2 Preliminaries

We consider sequential non-recursive imperative integer programs, allowing non-linear arithmetic and non-determinism, whereas heap usage and concurrency are not supported. While most existing abstractions that transform heap programs to integer programs are "over-approximations", we would need an under-approximating abstraction to ensure that the inference of worst-case lower bounds is sound. As in most related work, we treat numbers as mathematical integers $\mathbb{Z}$. However, the transformation from [12] can be used to handle machine integers correctly by inserting explicit normalization steps at possible overflows.

$\mathcal{A}(\mathcal{V})$ is the set of arithmetic terms[1] over the variables $\mathcal{V}$ and $\mathcal{F}(\mathcal{V})$ is the set of conjunctions[2] of (in)equations over $\mathcal{A}(\mathcal{V})$. So for $x, y \in \mathcal{V}$, $\mathcal{A}(\mathcal{V})$ contains terms like $x \cdot y + 2^y$ and $\mathcal{F}(\mathcal{V})$ contains formulas such as $x \cdot y \leq 2^y \wedge y > 0$.

We fix a finite set of *program variables* $\mathcal{PV}$ and represent integer programs as directed graphs. Nodes are program *locations* $\mathcal{L}$ and edges are program *transitions* $\mathcal{T}$ where $\mathcal{L}$ contains a *canonical start location* ℓ_0. W.l.o.g., no transition leads back to ℓ_0 and all transitions $\mathcal{T}$ are reachable from ℓ_0. To model non-deterministic program data, we introduce pairwise disjoint finite sets of *temporary variables* $\mathcal{TV}_\ell$ with $\mathcal{PV} \cap \mathcal{TV}_\ell = \varnothing$ and define $\mathcal{V}_\ell = \mathcal{PV} \cup \mathcal{TV}_\ell$ for all locations $\ell \in \mathcal{L}$.

Definition 1 (Programs). Configurations $(\ell, \boldsymbol{v})$ *consist of a location* $\ell \in \mathcal{L}$ *and a* valuation $\boldsymbol{v} : \mathcal{V}_\ell \to \mathbb{Z}$. $\mathcal{V}al_\ell = \mathcal{V}_\ell \to \mathbb{Z}$ *is the set of all valuations for* $\ell \in \mathcal{L}$ *and valuations are lifted to terms* $\mathcal{A}(\mathcal{V}_\ell)$ *and formulas* $\mathcal{F}(\mathcal{V}_\ell)$ *as usual. A* transition $t = (\ell, \gamma, \eta, c, \ell')$ *can evaluate a configuration* $(\ell, \boldsymbol{v})$ *if the* guard $\gamma \in \mathcal{F}(\mathcal{V}_\ell)$ *is satisfied (i.e.,* $\boldsymbol{v}(\gamma)$ *holds) to a new configuration* $(\ell', \boldsymbol{v}')$*. The* update $\eta : \mathcal{PV} \to \mathcal{A}(\mathcal{V}_\ell)$ *maps any* $x \in \mathcal{PV}$ *to a term* $\eta(x)$ *where* $\boldsymbol{v}(\eta(x)) \in \mathbb{Z}$ *for all* $\boldsymbol{v} \in \mathcal{V}al_\ell$*. It determines* $\boldsymbol{v}'$ *by setting* $\boldsymbol{v}'(x) = \boldsymbol{v}(\eta(x))$ *for* $x \in \mathcal{PV}$*, while* $\boldsymbol{v}'(x)$ *for* $x \in \mathcal{TV}_{\ell'}$ *is arbitrary. Such an* evaluation step *has* cost $k = \boldsymbol{v}(c)$ *for* $c \in \mathcal{A}(\mathcal{V}_\ell)$ *and is written* $(\ell, \boldsymbol{v}) \to_{t,k} (\ell', \boldsymbol{v}')$*. We use* $\mathsf{src}(t) = \ell$, $\mathsf{guard}(t) = \gamma$, $\mathsf{cost}(t) = c$, *and* $\mathsf{dest}(t) = \ell'$*. We sometimes drop the indices* t, k *and write* $(\ell, \boldsymbol{v}) \to_k^* (\ell', \boldsymbol{v}')$ *if* $(\ell, \boldsymbol{v}) \to_{k_1} \cdots \to_{k_m} (\ell', \boldsymbol{v}')$ *and* $\sum_{1 \leq i \leq m} k_i = k$*. A* program *is a set of transitions* $\mathcal{T}$*.*

Figure 1 shows an example, where the pseudo-code on the left corresponds to the program on the right. Here, `random`(x, y) returns a random integer m with $x < m < y$ and we fix $-\omega < m < \omega$ for all numbers m. The loop at location ℓ_1 sets y to a value that is quadratic in x. Thus, the loop at ℓ_2 is executed quadratically often where in each iteration, the inner loop at ℓ_3 may also be repeated quadratically often. Thus, the length of the program's worst-case execution is a polynomial of degree 4 in x. Our technique can infer such lower bounds automatically.

In the graph of Fig. 1, we write the costs of a transition in [ ] next to its name and represent the updates by imperative commands. We use x to refer to the value of the variable x before the update and x' to refer to x's value after the update. Here, $\mathcal{PV} = \{x, y, z, u\}$, $\mathcal{TV}_{\ell_3} = \{tv\}$, and $\mathcal{TV}_\ell = \varnothing$ for all locations $\ell \neq \ell_3$. We have $(\ell_3, \boldsymbol{v}) \to_{t_4} (\ell_3, \boldsymbol{v}')$ for all valuations $\boldsymbol{v}$ where $\boldsymbol{v}(u) > 0$, $\boldsymbol{v}(tv) > 0$, $\boldsymbol{v}'(u) = \boldsymbol{v}(u) - \boldsymbol{v}(tv)$, and $\boldsymbol{v}'(v) = \boldsymbol{v}(v)$ for all $v \in \{x, y, z\}$.

Our goal is to find a lower bound on the worst-case runtime of a program $\mathcal{T}$. To this end, we define its *derivation height* [18] by a function $\mathsf{dh}_\mathcal{T}$ that

[1] Our implementation only supports addition, subtraction, multiplication, division, and exponentiation. Since we consider integer programs, we only allow programs where all variable values are integers (so in contrast to $x = \frac{1}{2}x$, the assignment $x = \frac{1}{2}x + \frac{1}{2}x^2$ is permitted). While our program simplification technique preserves this property, we do not allow division or exponentiation in the *initial* program to ensure its validity.

[2] Note that negations can be expressed by negating (in)equations directly, and disjunctions in programs can be expressed using multiple transitions.

```
ℓ0: y = 0
ℓ1: while x > 0 do
        y = y + x
        x = x − 1
    done
    z = y
ℓ2: while z > 0 do
        u = z − 1
ℓ3:     while u > 0 do
            u = u − random(0, ω)
        done
        z = z − 1
    done
```

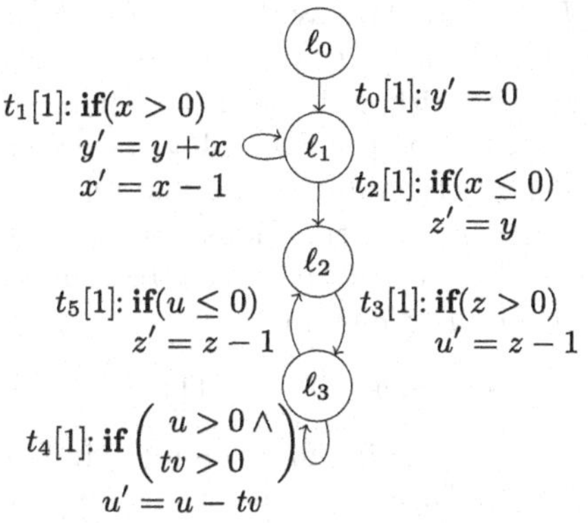

Fig. 1. Example integer program

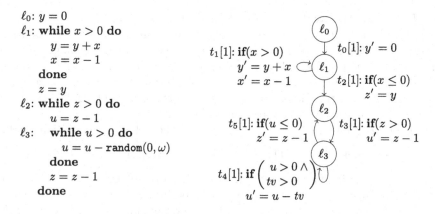

operates on valuations $\boldsymbol{v}$ of the program variables (i.e., $\boldsymbol{v}$ is not defined for temporary variables). The function $\mathsf{dh}_{\mathcal{T}}$ maps $\boldsymbol{v}$ to the maximum of the costs of all evaluation sequences starting in configurations $(\ell_0, \boldsymbol{v}_{\ell_0})$ where $\boldsymbol{v}_{\ell_0}$ is an extension of $\boldsymbol{v}$ to $\mathcal{V}_{\ell_0}$. So in our example we have $\mathsf{dh}_{\mathcal{T}}(\boldsymbol{v}) = 2$ for all valuations $\boldsymbol{v}$ where $\boldsymbol{v}(x) = 0$, since then we can only apply the transitions t_0 and t_2 once. For all valuations $\boldsymbol{v}$ with $\boldsymbol{v}(x) > 1$, our method will detect that the worst-case runtime of our program is at least $\frac{1}{8}\boldsymbol{v}(x)^4 + \frac{1}{4}\boldsymbol{v}(x)^3 + \frac{7}{8}\boldsymbol{v}(x)^2 + \frac{7}{4}\boldsymbol{v}(x)$. From this concrete lower bound, our approach will infer that the asymptotic runtime of the program is in $\Omega(x^4)$. In particular, the runtime of the program depends on x. Hence, if x is "secret", then the program is vulnerable to side-channel attacks.

Definition 2 (Derivation Height). *Let* $\mathcal{V}al = \mathcal{PV} \to \mathbb{Z}$. *The* derivation height $\mathsf{dh}_{\mathcal{T}} : \mathcal{V}al \to \mathbb{R}_{\geq 0} \cup \{\omega\}$ *of a program* $\mathcal{T}$ *is defined as* $\mathsf{dh}_{\mathcal{T}}(\boldsymbol{v}) = \sup\{k \in \mathbb{R} \mid \exists \boldsymbol{v}_{\ell_0} \in \mathcal{V}al_{\ell_0}, \ell \in \mathcal{L}, \boldsymbol{v}_\ell \in \mathcal{V}al_\ell \,.\; \boldsymbol{v}_{\ell_0}|_{\mathcal{PV}} = \boldsymbol{v} \wedge (\ell_0, \boldsymbol{v}_{\ell_0}) \to_k^* (\ell, \boldsymbol{v}_\ell)\}$.

Since $\to_k^*$ also permits evaluations with 0 steps, we always have $\mathsf{dh}_{\mathcal{T}}(\boldsymbol{v}) \geq 0$. Obviously, $\mathsf{dh}_{\mathcal{T}}$ is not computable in general, and thus our goal is to compute a lower bound that is as precise as possible (i.e., a lower bound which is, e.g., unbounded,[3] exponential, or a polynomial of a degree as high as possible).

3 Estimating the Number of Iterations of Simple Loops

We now show how to under-estimate the number of possible iterations of a *simple loop* $t = (\ell, \gamma, \eta, c, \ell)$. More precisely, we infer a term $b \in \mathcal{A}(\mathcal{V}_\ell)$ such that for all $\boldsymbol{v} \in \mathcal{V}al_\ell$ with $\boldsymbol{v} \models \gamma$, there is a $\boldsymbol{v}' \in \mathcal{V}al_\ell$ with $(\ell, \boldsymbol{v}) \to_t^{\lceil \boldsymbol{v}(b) \rceil} (\ell, \boldsymbol{v}')$. Here, $\lceil k \rceil = \min\{m \in \mathbb{N} \mid m \geq k\}$ for all $k \in \mathbb{R}$. Moreover, $(\ell, \boldsymbol{v}) \to_t^m (\ell, \boldsymbol{v}')$ means that $(\ell, \boldsymbol{v}) = (\ell, \boldsymbol{v}_0) \to_{t,k_1} (\ell, \boldsymbol{v}_1) \to_{t,k_2} \cdots \to_{t,k_m} (\ell, \boldsymbol{v}_m) = (\ell, \boldsymbol{v}')$ for some

[3] Programs with $\mathsf{dh}_{\mathcal{T}}(\boldsymbol{v}) = \omega$ result from non-termination or non-determinism. As an example, consider the program $x = \mathtt{random}(0, \omega)$; **while** $x > 0$ **do** $x = x - 1$ **done**.

costs $k_1, \ldots, k_m$. We say that $(\ell, \boldsymbol{v}) \to_t^m (\ell, \boldsymbol{v}')$ *preserves* $\mathcal{TV}_\ell$ iff $\boldsymbol{v}(tv) = \boldsymbol{v}_i(tv) = \boldsymbol{v}'(tv)$ for all $tv \in \mathcal{TV}_\ell$ and all $0 \leq i \leq m$. Accordingly, we lift the update η to arbitrary arithmetic terms by leaving temporary variables unchanged (i.e., if $\mathcal{PV} = \{x_1, \ldots, x_n\}$ and $b \in \mathcal{A}(\mathcal{V}_\ell)$, then $\eta(b) = b[x_1/\eta(x_1), \ldots, x_n/\eta(x_n)]$, where $[x/a]$ denotes the substitution that replaces all occurrences of the variable x by a).

To find such estimations, we use an adaptation of ranking functions [2, 6,25] which we call *metering functions.* We say that a term $b \in \mathcal{A}(\mathcal{V}_\ell)$ is a *ranking function*[4] for $t = (\ell, \gamma, \eta, c, \ell)$ iff the following conditions hold.

$$\gamma \implies b > 0 \qquad (1) \qquad\qquad \gamma \implies \eta(b) \leq b - 1 \qquad (2)$$

So e.g., x is a ranking function for t_1 in Fig. 1. If $\mathcal{TV}_\ell = \varnothing$, then for any valuation $\boldsymbol{v} \in \mathcal{V}al$, $\boldsymbol{v}(b)$ *over-estimates* the number of repetitions of the loop t: (2) ensures that $\boldsymbol{v}(b)$ decreases at least by 1 in each loop iteration, and (1) requires that $\boldsymbol{v}(b)$ is positive whenever the loop can be executed. In contrast, metering functions are *under-estimations* for the maximal number of repetitions of a simple loop.

Definition 3 (Metering Function). *Let $t = (\ell, \gamma, \eta, c, \ell)$ be a transition. We call $b \in \mathcal{A}(\mathcal{V}_\ell)$ a* metering function *for t iff the following conditions hold:*

$$\neg\gamma \implies b \leq 0 \qquad (3) \qquad\qquad \gamma \implies \eta(b) \geq b - 1 \qquad (4)$$

Here, (4) ensures that $\boldsymbol{v}(b)$ decreases at most by 1 in each loop iteration, and (3) requires that $\boldsymbol{v}(b)$ is non-positive if the loop cannot be executed. Thus, the loop can be executed *at least* $\boldsymbol{v}(b)$ times (i.e., $\boldsymbol{v}(b)$ is an under-estimation).

For the transition t_1 in the example of Fig. 1, x is also a valid metering function. Condition (3) requires $\neg x > 0 \implies x \leq 0$ and (4) requires $x > 0 \implies x - 1 \geq x - 1$. While x is a metering *and* a ranking function, $\frac{x}{2}$ is a metering, but not a ranking function for t_1. Similarly, x^2 is a ranking, but not a metering function for t_1. Theorem 4 states that a simple loop t with a metering function b can be executed at least $\lceil \boldsymbol{v}(b) \rceil$ times when starting with the valuation $\boldsymbol{v}$.

Theorem 4 (Metering Functions are Under-Estimations). *Let b be a metering function for $t = (\ell, \gamma, \eta, c, \ell)$. Then b* under-estimates *t, i.e., for all $\boldsymbol{v} \in \mathcal{V}al_\ell$ with $\boldsymbol{v} \models \gamma$ there is an evaluation $(\ell, \boldsymbol{v}) \to_t^{\lceil \boldsymbol{v}(b) \rceil} (\ell, \boldsymbol{v}')$ that preserves $\mathcal{TV}_\ell$.*

Our implementation builds upon a well-known transformation based on Farkas' Lemma [6,25] to find *linear* metering functions. The basic idea is to search for coefficients of a linear template polynomial b such that (3) and (4) hold for all possible instantiations of the variables $\mathcal{V}_\ell$. In addition to (3) and (4), we also require (1) to avoid trivial solutions like $b = 0$. Here, the coefficients of b are existentially quantified, while the variables from $\mathcal{V}_\ell$ are universally quantified. As in [6,25], eliminating the universal quantifiers using Farkas' Lemma allows us to use standard SMT solvers to search for b's coefficients efficiently.

When searching for a metering function for $t = (\ell, \gamma, \eta, c, \ell)$, one can omit constraints from γ that are irrelevant for t's termination. So if γ is $\varphi \wedge \psi$, $\psi \in \mathcal{F}(\mathcal{PV})$, and $\gamma \implies \eta(\psi)$, then it suffices to find a metering function b for $t' = (\ell, \varphi, \eta, c, \ell)$.

---

[4] In the following, we often use arithmetic terms $\mathcal{A}(\mathcal{V}_\ell)$ to denote functions $\mathcal{V}_\ell \to \mathbb{R}$.

The reason is that if $\boldsymbol{v} \models \gamma$ and $(\ell, \boldsymbol{v}) \to_{t'} (\ell, \boldsymbol{v}')$, then $\boldsymbol{v}' \models \psi$ (since $\boldsymbol{v} \models \gamma$ entails $\boldsymbol{v} \models \eta(\psi)$). Hence, if $\boldsymbol{v} \models \gamma$ then $(\ell, \boldsymbol{v}) \to_{t'}^{\lceil \boldsymbol{v}(b) \rceil} (\ell, \boldsymbol{v}')$ implies $(\ell, \boldsymbol{v}) \to_{t}^{\lceil \boldsymbol{v}(b) \rceil} (\ell, \boldsymbol{v}')$, i.e., b under-estimates t. So if $t = (\ell,\ x < y \wedge 0 < y,\ x' = x + 1, c, \ell)$, we can consider $t' = (\ell,\ x < y,\ x' = x+1, c, \ell)$ instead. While t only has complex metering functions like $\min(y - x, y)$, t' has the metering function $y - x$.

Example 5 (Unbounded Loops). Loops $t = (\ell, \gamma, \eta, c, \ell)$ where the *whole* guard can be omitted (since $\gamma \implies \eta(\gamma)$) do not terminate. Here, we also allow ω as under-estimation. So for $\mathcal{T} = \{(\ell_0, \mathit{true}, \mathsf{id}, 1, \ell),\ t\}$ with $t = (\ell,\ 0 < x,\ x' = x+1,\ y,\ \ell)\}$, we can omit $0 < x$ since $0 < x \implies 0 < x + 1$. Hence, ω under-estimates the resulting loop $(\ell,\ \mathit{true},\ x' = x + 1,\ y,\ \ell)$ and thus, ω also under-estimates t.

4 Simplifying Programs to Compute Lower Bounds

We now define *processors* mapping programs to simpler programs. Processors are applied repeatedly to transform the program until extraction of a (concrete) lower bound is straightforward. For this, processors should be sound, i.e., any lower-bound for the derivation height of $\mathsf{proc}(\mathcal{T})$ should also be a lower bound for $\mathcal{T}$.

Definition 6 (Sound Processor). *A mapping* proc *from programs to programs is* sound *iff* $\mathsf{dh}_{\mathcal{T}}(\boldsymbol{v}) \geq \mathsf{dh}_{\mathsf{proc}(\mathcal{T})}(\boldsymbol{v})$ *holds for all programs* $\mathcal{T}$ *and all* $\boldsymbol{v} \in \mathcal{V}al$.

In Sect. 4.1, we show how to *accelerate* a simple loop t to a transition which is equivalent to applying t multiple times (according to a metering function for t). The resulting program can be simplified by *chaining* subsequent transitions which may result in new simple loops, cf. Sect. 4.2. We describe a simplification strategy which alternates these steps repeatedly. In this way, we eventually obtain a *simplified* program without loops which directly gives rise to a concrete lower bound.

4.1 Accelerating Simple Loops

Consider a simple loop $t = (\ell, \gamma, \eta, c, \ell)$. For $m \in \mathbb{N}$, let η^m denote m applications of η. To accelerate t, we compute its *iterated* update and costs, i.e., a closed form η_{it} of η^{tv} and an under-approximation $c_{\mathrm{it}} \in \mathcal{A}(\mathcal{V}_\ell)$ of $\sum_{0 \leq i < tv} \eta^i(c)$ for a fresh temporary variable tv. If b under-estimates t, then we add the transition $(\ell,\ \gamma \wedge 0 < tv < b + 1,\ \eta_{\mathrm{it}},\ c_{\mathrm{it}},\ \ell)$ to the program. It summarizes tv iterations of t, where tv is bounded by $\lceil b \rceil$. Here, η_{it} and c_{it} may also contain exponentiation (i.e., we can also infer exponential bounds).

For $\mathcal{PV} = \{x_1, \ldots, x_n\}$, the iterated update is computed by solving the recurrence equations $x^{(1)} = \eta(x)$ and $x^{(tv+1)} = \eta(x)[x_1/x_1^{(tv)}, \ldots, x_n/x_n^{(tv)}]$ for all $x \in \mathcal{PV}$ and $tv \geq 1$. So for the transition t_1 from Fig. 1 we get the recurrence equations $x^{(1)} = x - 1$, $x^{(tv_1+1)} = x^{(tv_1)} - 1$, $y^{(1)} = y + x$, and $y^{(tv_1+1)} = y^{(tv_1)} + x^{(tv_1)}$. Usually, they can easily be solved using state-of-the-art recurrence solvers [4]. In our example, we obtain the closed forms $\eta_{\mathrm{it}}(x) = x^{(tv_1)} = x - tv_1$ and

$\eta_{\text{it}}(y) = y^{(tv_1)} = y + tv_1 \cdot x - \frac{1}{2}tv_1^2 + \frac{1}{2}tv_1$. While $\eta_{\text{it}}(y)$ contains rational coefficients, our approach ensures that η_{it} always maps integers to integers. Thus, we again obtain an integer program. We proceed similarly for the iterated cost of a transition, where we may under-approximate the solution of the recurrence equations $c^{(1)} = c$ and $c^{(tv+1)} = c^{(tv)} + c[x_1/x_1^{(tv)}, \ldots, x_n/x_n^{(tv)}]$. For t_1 in Fig. 1, we get $c^{(1)} = 1$ and $c^{(tv_1+1)} = c^{(tv_1)} + 1$ which leads to the closed form $c_{\text{it}} = c^{(tv_1)} = tv_1$.

Theorem 7 (Loop Acceleration). *Let $t = (\ell, \gamma, \eta, c, \ell) \in \mathcal{T}$ and let tv be a fresh temporary variable. Moreover, let $\eta_{\text{it}}(x) = \eta^{tv}(x)$ for all $x \in \mathcal{PV}$ and let $c_{\text{it}} \leq \sum_{0 \leq i < tv} \eta^i(c)$. If b under-estimates t, then the processor mapping $\mathcal{T}$ to $\mathcal{T} \cup \{(\ell, \gamma \wedge 0 < tv < b+1, \eta_{\text{it}}, c_{\text{it}}, \ell)\}$ is sound.*

We say that the resulting new simple loop is *accelerated* and we refer to all simple loops which were not introduced by Theorem 7 as *non-accelerated.*

Example 8 (Non-Integer Metering Functions). Theorem 7 also allows metering functions that do not map to the integers. Let $\mathcal{T} = \{(\ell_0, \mathit{true}, \mathsf{id}, 1, \ell), t\}$ with $t = (\ell, 0 < x, x' = x - 2, 1, \ell)$. Accelerating t with the metering function $\frac{x}{2}$ yields $(\ell, 0 < tv < \frac{x}{2} + 1, x' = x - 2\,tv, tv, \ell)$. Note that $0 < tv < \frac{x}{2} + 1$ implies $0 < x$ as tv and x range over $\mathbb{Z}$. Hence, $0 < x$ can be omitted in the resulting guard.

Example 9 (Unbounded Loops Continued). In Example 5, ω under-estimates $t = (\ell, 0 < x, x' = x + 1, y, \ell)$. The accelerated transition is $\overline{t} = (\ell, 0 < x \wedge \gamma', x' = x + tv, tv \cdot y, \ell)$, where γ' corresponds to $0 < tv < \omega + 1 = \omega$, i.e., tv has no upper bound.

If we cannot find a metering function or fail to obtain the closed form η_{it} or c_{it} for a simple loop t, then we can simplify t by eliminating temporary variables. To do so, we fix their values by adding suitable constraints to $\mathsf{guard}(t)$. As we are interested in witnesses for maximal computations, we use a heuristic that adds constraints $tv = a$ for temporary variables tv, where $a \in \mathcal{A}(\mathcal{V}_\ell)$ is a suitable upper or lower bound on tv's values, i.e., $\mathsf{guard}(t)$ implies $tv \leq a$ or $tv \geq a$. This is repeated until we find constraints which allow us to apply loop acceleration. Note that adding additional constraints to $\mathsf{guard}(t)$ is *always* sound in our setting.

Theorem 10 (Strengthening). *Let $t = (\ell, \gamma, \eta, c, \ell') \in \mathcal{T}$ and $\varphi \in \mathcal{F}(\mathcal{V}_\ell)$. Then the processor mapping $\mathcal{T}$ to $\mathcal{T} \setminus \{t\} \cup \{(\ell, \gamma \wedge \varphi, \eta, c, \ell')\}$ is sound.*

In t_4 from Fig. 1, γ contains $tv > 0$. So γ implies the bound $tv \geq 1$ since tv must be instantiated by integers. Hence, we add the constraint $tv = 1$. Now the update $u' = u - tv$ of the transition t_4 becomes $u' = u - 1$, and thus, u is a metering function. So after fixing $tv = 1$, t_4 can be accelerated similarly to t_1.

To simplify the program, we delete a simple loop t after trying to accelerate it. So we just keep the accelerated loop (or none, if acceleration of t still fails after eliminating all temporary variables by strengthening t's guard). For our example, we obtain the program in Fig. 2 with the accelerated transitions $t_{\overline{1}}$, $t_{\overline{4}}$.

Theorem 11 (Deletion). *For $t \in \mathcal{T}$, the processor mapping $\mathcal{T}$ to $\mathcal{T} \setminus \{t\}$ is sound.*

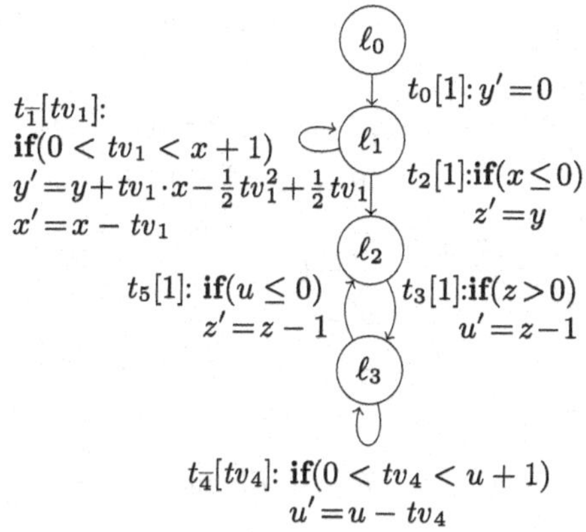

Fig. 2. Accelerating t_1 and t_4

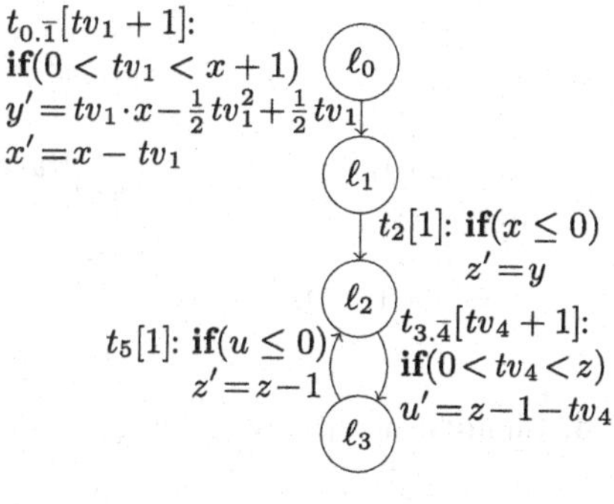

Fig. 3. Eliminating $t_{\overline{1}}$ and $t_{\overline{4}}$

4.2 Chaining Transitions

After trying to accelerate all simple loops of a program, we can *chain* subsequent transitions t_1, t_2 by adding a new transition $t_{1.2}$ that simulates their combination. Afterwards, the transitions t_1 and t_2 can (but need not) be deleted with Theorem 11.

Theorem 12 (Chaining). *Let* $t_1 = (\ell_1, \gamma_1, \eta_1, c_1, \ell_2)$ *and* $t_2 = (\ell_2, \gamma_2, \eta_2, c_2, \ell_3)$ *with* $t_1, t_2 \in \mathcal{T}$. *Let* ren *be an injective function renaming the variables in* $\mathcal{TV}_{\ell_2}$ *to fresh ones and let*[5] $t_{1.2} = (\ell_1,\ \gamma_1 \wedge \mathrm{ren}(\eta_1(\gamma_2)),\ \mathrm{ren} \circ \eta_1 \circ \eta_2,\ c_1 + \mathrm{ren}(\eta_1(c_2)), \ell_3)$. *Then the processor mapping* $\mathcal{T}$ *to* $\mathcal{T} \cup \{t_{1.2}\}$ *is sound. In the new program* $\mathcal{T} \cup \{t_{1.2}\}$, *the temporary variables of* ℓ_1 *are defined to be* $\mathcal{TV}_{\ell_1} \cup \mathrm{ren}(\mathcal{TV}_{\ell_2})$.

One goal of chaining is *loop elimination* of all accelerated simple loops. To this end, we chain all subsequent transitions t', t where t is a simple loop and t' is no simple loop. Afterwards, we delete t. Moreover, once t' has been chained with all subsequent simple loops, then we also remove t', since its effect is now covered by the newly introduced (chained) transitions. So in our example from Fig. 2, we chain t_0 with $\overline{t}_1$ and t_3 with $\overline{t}_4$. The resulting program is depicted in Fig. 3, where we always simplify arithmetic terms and formulas to ease readability.

Chaining also allows *location elimination* by chaining all pairs of incoming and outgoing transitions for a location ℓ and removing them afterwards. It is advantageous to eliminate locations with just a single incoming transition first. This heuristic takes into account which locations were the entry points of loops. So for the example in Fig. 3, it would avoid chaining t_5 and $t_{3.\overline{4}}$ in order to eliminate ℓ_2. In this way, we avoid constructing chained transitions that correspond to a run from the "middle" of a loop to the "middle" of the next loop iteration.

[5] For all $x \in \mathcal{PV}$, $\mathrm{ren} \circ \eta_1 \circ \eta_2(x) = \mathrm{ren}(\eta_1(\eta_2(x))) = \eta_2(x)[x_1/\eta_1(x_1), \ldots, x_n/\eta_1(x_n), tv_1/\mathrm{ren}(tv_1), \ldots, tv_m/\mathrm{ren}(tv_m)]$ if $\mathcal{PV} = \{x_1, \ldots, x_n\}$ and $\mathcal{TV}_{\ell_2} = \{tv_1, \ldots, tv_m\}$.

$t_{0.\bar{1}.2}[x+2]$:
if($x > 0$)
$y' = \frac{1}{2}x^2 + \frac{1}{2}x$
$x' = 0$
$z' = \frac{1}{2}x^2 + \frac{1}{2}x$

ℓ_0 ℓ_2

$t_{3.\bar{4}.5}[z+1]$:
if($z > 1$)
$u' = 0$
$z' = z - 1$

$t_{0.\bar{1}.2}[x+2]$:
if($x > 0$)
$y' = \frac{1}{2}x^2 + \frac{1}{2}x$
$x' = 0$
$z' = \frac{1}{2}x^2 + \frac{1}{2}x$

ℓ_0 ℓ_2

$t_{\overline{3.\bar{4}.5}}[tv \cdot z - \frac{1}{2}tv^2 + \frac{3}{2}tv]$:
if($0 < tv < z$)
$u' = 0$
$z' = z - tv$

$t[\frac{x^2 \cdot tv + x \cdot tv - tv^2 + 3tv + 2x + 4}{2}]$:
if($0 < tv < \frac{1}{2}x^2 + \frac{1}{2}x$)
$y' = \frac{1}{2}x^2 + \frac{1}{2}x$
$x' = 0$
$u' = 0$
$z' = \frac{1}{2}x^2 + \frac{1}{2}x - tv$

ℓ_0 ℓ_2

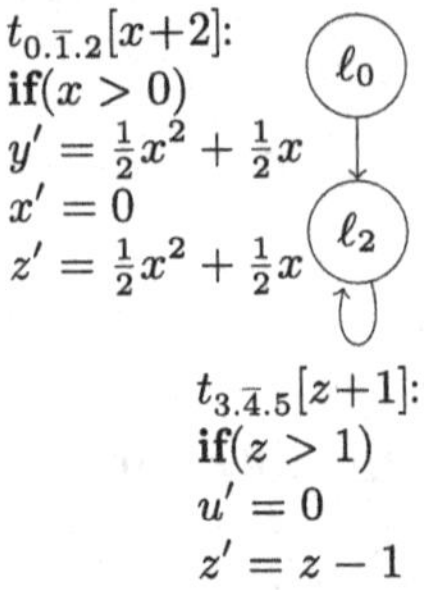

Fig. 4. Eliminating ℓ_1 and ℓ_3 **Fig. 5.** Accelerating $t_{3.\bar{4}.5}$ **Fig. 6.** Eliminating $t_{\overline{3.\bar{4}.5}}$

So instead of eliminating ℓ_2, we chain $t_{0.\bar{1}}$ and t_2 as well as $t_{3.\bar{4}}$ and t_5 to eliminate the locations ℓ_1 and ℓ_3, leading to the program in Fig. 4. Here, the temporary variables tv_1 and tv_4 vanish since, before applying arithmetic simplifications, the guards of $t_{0.\bar{1}.2}$ resp. $t_{3.\bar{4}.5}$ imply $tv_1 = x$ resp. $tv_4 = z - 1$.

Our overall approach for program simplification is shown in Algorithm 1. Of course, this algorithm is a heuristic and other strategies for the application of the processors would also be possible. The set S in Steps 3–5 is needed to handle locations ℓ with multiple simple loops. The reason is that each transition t' with $\text{dest}(t') = \ell$ should be chained with *each* of ℓ's simple loops before removing t'.

Algorithm 1 terminates: In the loop 2.1–2.2, each iteration decreases the number of temporary variables in t. The loop 2 terminates since each iteration reduces the number of non-accelerated simple loops. In loop 4, the number of simple loops is decreasing and for loop 6, the number of reachable locations decreases. The overall loop terminates as it reduces the number of reachable locations. The reason is that the program does not have simple loops anymore when the algorithm reaches Step 6. Thus, at this point there is either a location ℓ which can be eliminated or the program does not have a path of length 2.

According to Algorithm 1, in our example we go back to Step 1 and 2 and apply *Loop Acceleration* to transition $t_{3.\bar{4}.5}$. This transition has the metering function $z - 1$ and its iterated update sets u to 0 and z to $z - tv$ for a fresh temporary variable tv. To compute $t_{3.\bar{4}.5}$'s iterated costs, we have to find an under-approximation for the solution of the recurrence equations $c^{(1)} = z + 1$ and $c^{(tv+1)} = c^{(tv)} + z^{(tv)} + 1$. After computing the closed form $z - tv$ of $z^{(tv)}$, the second equation simplifies to $c^{(tv+1)} = c^{(tv)} + z - tv + 1$, which results in the closed form $c_{it} = c^{(tv)} = tv \cdot z - \frac{1}{2}tv^2 + \frac{3}{2}tv$. In this way, we obtain the program in Fig. 5. A final chaining step and deletion of the only simple loop yields the program in Fig. 6.

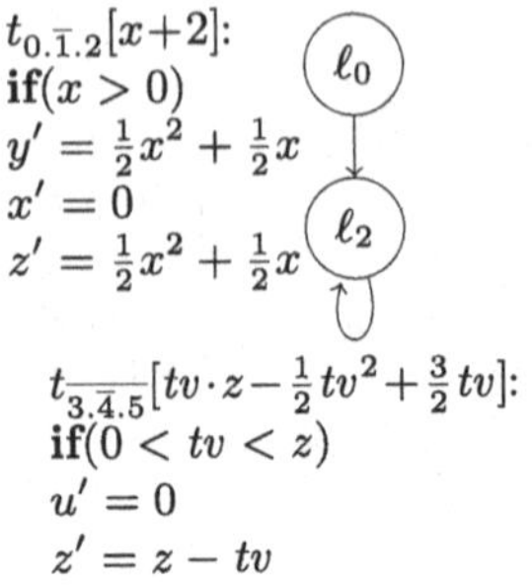

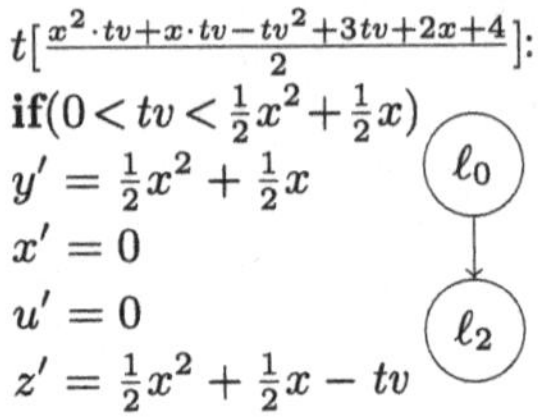

Algorithm 1. Program Simplification

While there is a path of length 2:

1. Apply *Deletion* to transitions whose guard is proved unsatisfiable.
2. While there is a non-accelerated simple loop t:
   2.1 Try to apply *Loop Acceleration* to t.
   2.2 If 2.1 failed and t uses temporary variables:
   Apply *Strengthening* to t to eliminate a temporary variable and go to 2.1
   2.3 Apply *Deletion* to t.
3. Let $S = \varnothing$.
4. While there is a simple loop t:
   4.1 Apply *Chaining* to each pair t', t where $\mathsf{src}(t') \neq \mathsf{dest}(t') = \mathsf{src}(t)$.
   4.2 Add all these transitions t' to S and apply *Deletion* to t.
5. Apply *Deletion* to each transition in S.
6. While there is a location ℓ without simple loops but with incoming and outgoing transitions (starting with locations ℓ with just one incoming transition):
   6.1 Apply *Chaining* to each pair t', t where $\mathsf{dest}(t') = \mathsf{src}(t) = \ell$.
   6.2 Apply *Deletion* to each t where $\mathsf{src}(t) = \ell$ or $\mathsf{dest}(t) = \ell$.

5 Asymptotic Lower Bounds for Simplified Programs

After Algorithm 1, all program paths have length 1. We call such programs *simplified* and let $\mathcal{T}$ be a simplified program throughout this section. Now for any $\boldsymbol{v} \in \mathcal{V}al_{\ell_0}$,

$$\max\{\boldsymbol{v}(\mathsf{cost}(t)) \mid t \in \mathcal{T}, \boldsymbol{v} \models \mathsf{guard}(t)\}, \tag{5}$$

is a lower bound on $\mathcal{T}$'s derivation height $\mathsf{dh}_{\mathcal{T}}(\boldsymbol{v}|_{\mathcal{PV}})$, i.e., (5) is the maximal cost of those transitions whose guard is satisfied by $\boldsymbol{v}$. So for the program in Fig. 6, we obtain the bound $\frac{x^2 \cdot tv + x \cdot tv - tv^2 + 3tv + 2x + 4}{2}$ for all valuations with $\boldsymbol{v} \models 0 < tv < \frac{1}{2}x^2 + \frac{1}{2}x$. However, such bounds do not provide an intuitive understanding of the program's complexity and are also not suitable to detect possible attacks. Hence, we now show how to derive *asymptotic* lower bounds for simplified programs. These asymptotic bounds can easily be understood (i.e., a high lower bound can help programmers to improve their program to make it more efficient) and they identify potential attacks. After introducing our notion of asymptotic bounds in Sect. 5.1, we present a technique to derive them automatically in Sect. 5.2.

5.1 Asymptotic Bounds and Limit Problems

While $\mathsf{dh}_{\mathcal{T}}$ is defined on valuations, asymptotic bounds are usually defined for functions on $\mathbb{N}$. To bridge this gap, we use the common definition of complexity as a function of the size of the input. So the *runtime complexity* $\mathsf{rc}_{\mathcal{T}}(n)$ is the maximal cost of any evaluation that starts with a configuration where the sum of the absolute values of all program variables is at most n.

Definition 13 (Runtime Complexity). *Let* $|\boldsymbol{v}| = \sum_{x \in \mathcal{PV}} |\boldsymbol{v}(x)|$ *for all valuations* $\boldsymbol{v}$. *The* runtime complexity $\mathsf{rc}_{\mathcal{T}} : \mathbb{N} \to \mathbb{R}_{\geq 0} \cup \{\omega\}$ *is defined as* $\mathsf{rc}_{\mathcal{T}}(n) = \sup\{\mathsf{dh}_{\mathcal{T}}(\boldsymbol{v}) \mid \boldsymbol{v} \in \mathcal{V}al, |\boldsymbol{v}| \leq n\}$.

Our goal is to derive an asymptotic lower bound for $\mathsf{rc}_{\mathcal{T}}$ from a simplified program $\mathcal{T}$. So for the program $\mathcal{T}$ in Fig. 6, we would like to derive $\mathsf{rc}_{\mathcal{T}}(n) \in \Omega(n^4)$. As usual, $f(n) \in \Omega(g(n))$ means that there is an $m > 0$ and an $n_0 \in \mathbb{N}$ such that $f(n) \geq m \cdot g(n)$ holds for all $n \geq n_0$. However, in general, the costs of a transition do not directly give rise to the desired asymptotic lower bound. For instance, in Fig. 6, the costs of the only transition are cubic, but the complexity of the program is a polynomial of degree 4 (since tv may be quadratic in x).

To infer an asymptotic lower bound from a transition $t \in \mathcal{T}$, we try to find an infinite family of valuations $\boldsymbol{v}_n \in \mathcal{V}al_{\ell_0}$ (parameterized by $n \in \mathbb{N}$) where there is an $n_0 \in \mathbb{N}$ such that $\boldsymbol{v}_n \models \mathsf{guard}(t)$ holds for all $n \geq n_0$. This implies $\mathsf{rc}_{\mathcal{T}}(|\boldsymbol{v}_n|) \in \Omega(\boldsymbol{v}_n(\mathsf{cost}(t)))$, since for all $n \geq n_0$ we have:

$$\begin{aligned} \mathsf{rc}_{\mathcal{T}}(|\boldsymbol{v}_n|) &\geq \mathsf{dh}_{\mathcal{T}}(\boldsymbol{v}_n|_{\mathcal{PV}}) && \text{as } |\,\boldsymbol{v}_n|_{\mathcal{PV}}\,| = |\boldsymbol{v}_n| \\ &\geq \boldsymbol{v}_n(\mathsf{cost}(t)) && \text{by (5)} \end{aligned}$$

We first normalize all constraints in $\mathsf{guard}(t)$ such that they have the form $a > 0$. Now our goal is to find infinitely many models $\boldsymbol{v}_n$ for a formula of the form $\bigwedge_{1\leq i\leq k}(a_i > 0)$. Obviously, such a formula is satisfied if all terms a_i are positive constants or increase infinitely towards ω. Thus, we introduce a technique which tries to find out whether fixing the valuations of some variables and increasing or decreasing the valuations of others results in positive resp. increasing valuations of $a_1, \ldots, a_k$. Our technique operates on so-called *limit problems* $\{a_1^{\bullet_1}, \ldots, a_k^{\bullet_k}\}$ where $a_i \in \mathcal{A}(\mathcal{V}_{\ell_0})$ and $\bullet_i \in \{+, -, +_!, -_!\}$. Here, a^+ (resp. a^-) means that a grows towards ω (resp. $-\omega$) and $a^{+_!}$ (resp. $a^{-_!}$) means that a has to be a positive (resp. negative) constant. So we represent $\mathsf{guard}(t)$ by an *initial limit problem* $\{a_1^{\bullet_1}, \ldots, a_k^{\bullet_k}\}$ where $\bullet_i \in \{+, +_!\}$ for all $1 \leq i \leq k$. We say that a family of valuations $\boldsymbol{v}_n$ is a *solution* to a limit problem S iff $\boldsymbol{v}_n$ "satisfies" S for large n.

To define this notion formally, for any function $f : \mathbb{N} \to \mathbb{R}$ we say that $\lim_{n\mapsto\omega} f(n) = \omega$ (resp. $\lim_{n\mapsto\omega} f(n) = -\omega$) iff for every $m \in \mathbb{Z}$ there is an $n_0 \in \mathbb{N}$ such that $f(n) \geq m$ (resp. $f(n) \leq m$) holds for all $n \geq n_0$. Similarly, $\lim_{n\mapsto\omega} f(n) = m$ iff there is an n_0 such that $f(n) = m$ holds for all $n \geq n_0$.

Definition 14 (Solutions of Limit Problems). *For any function $f : \mathbb{N} \to \mathbb{R}$ and any $\bullet \in \{+, -, +_!, -_!\}$, we say that f* satisfies $\bullet$ *iff*

$$\begin{array}{ll} \lim_{n\mapsto\omega} f(n) = \omega, \ \textit{if} \bullet = + & \exists m \in \mathbb{Z}.\ \lim_{n\mapsto\omega} f(n) = m > 0, \textit{if} \bullet = +_! \\ \lim_{n\mapsto\omega} f(n) = -\omega, \textit{if} \bullet = - & \exists m \in \mathbb{Z}.\ \lim_{n\mapsto\omega} f(n) = m < 0, \textit{if} \bullet = -_! \end{array}$$

A family $\boldsymbol{v}_n$ of valuations is a solution *of a limit problem S iff for every $a^\bullet \in S$, the function $\lambda n.\ \boldsymbol{v}_n(a)$ satisfies $\bullet$. Here, "$\lambda n.\ \boldsymbol{v}_n(a)$" is the function from $\mathbb{N} \to \mathbb{R}$ that maps any $n \in \mathbb{N}$ to $\boldsymbol{v}_n(a)$.*

Example 15 (Bound for Fig. 6). In Fig. 6 where $\mathsf{guard}(t)$ is $0 < tv < \frac{1}{2}x^2 + \frac{1}{2}x$, the family $\boldsymbol{v}_n$ with $\boldsymbol{v}_n(tv) = \frac{1}{2}n^2 + \frac{1}{2}n - 1, \boldsymbol{v}_n(x) = n$, and $\boldsymbol{v}_n(y) = \boldsymbol{v}_n(z) = \boldsymbol{v}_n(u) = 0$ is a solution of the initial limit problem $\{tv^+, (\frac{1}{2}x^2 + \frac{1}{2}x - tv)^{+_!}\}$. The reason is that the function $\lambda n.\ \boldsymbol{v}_n(tv)$ that maps any $n \in \mathbb{N}$ to $\boldsymbol{v}_n(tv) = \frac{1}{2}n^2 + \frac{1}{2}n - 1$ satisfies $+$, i.e., $\lim_{n\mapsto\omega}(\frac{1}{2}n^2 + \frac{1}{2}n - 1) = \omega$. Similarly, the function

$\lambda n.\ \boldsymbol{v}_n(\frac{1}{2}x^2 + \frac{1}{2}x - tv) = \lambda n.\ 1$ satisfies $+_!$. Section 5.2 will show how to infer such solutions of limit problems automatically. Thus, there is an n_0 such that $\boldsymbol{v}_n \models \mathsf{guard}(t)$ holds for all $n \geq n_0$. Hence, we get the asymptotic lower bound $\mathsf{rc}_{\mathcal{T}}(|\boldsymbol{v}_n|) \in \Omega(\boldsymbol{v}_n(\mathsf{cost}(t))) = \Omega(\frac{1}{8}n^4 + \frac{1}{4}n^3 + \frac{7}{8}n^2 + \frac{7}{4}n) = \Omega(n^4)$.

Theorem 16 (Asymptotic Bounds for Simplified Programs). *Given a transition t of a simplified program $\mathcal{T}$ with $\mathsf{guard}(t) = a_1 > 0 \wedge \cdots \wedge a_k > 0$, let the family $\boldsymbol{v}_n$ be a solution of an initial limit problem $\{a_1^{\bullet_1}, \ldots, a_k^{\bullet_k}\}$ with $\bullet_i \in \{+, +_!\}$ for all $1 \leq i \leq k$. Then $\mathsf{rc}_{\mathcal{T}}(|\boldsymbol{v}_n|) \in \Omega(\boldsymbol{v}_n(\mathsf{cost}(t)))$.*

Of course, if $\mathcal{T}$ has several transitions, then we try to take the one which results in the highest lower bound. Moreover, one should extend the initial limit problem $\{a_1^{\bullet_1}, \ldots, a_k^{\bullet_k}\}$ by $\mathsf{cost}(t)^+$. In this way, one searches for valuations $\boldsymbol{v}_n$ where $\boldsymbol{v}_n(\mathsf{cost}(t))$ depends on n, i.e., where the costs are not constant.

The costs are *unbounded* (i.e., they only depend on temporary variables) iff the initial limit problem $\{a_1^{\bullet_1}, \ldots, a_k^{\bullet_k}, \mathsf{cost}(t)^+\}$ has a solution $\boldsymbol{v}_n$ where $\boldsymbol{v}_n(x)$ is constant for all $x \in \mathcal{PV}$. Then we can even infer $\mathsf{rc}_{\mathcal{T}}(n) \in \Omega(\omega)$. For instance, after chaining the transition $\bar{t}$ of Example 9 with the transition from the start location (see Example 5), the resulting initial limit problem $\{x^{+_!}, tv^+, (tv \cdot y + 1)^+\}$ has the solution $\boldsymbol{v}_n$ with $\boldsymbol{v}_n(x) = \boldsymbol{v}_n(y) = 1$ and $\boldsymbol{v}_n(tv) = n$, which implies $\mathsf{rc}_{\mathcal{T}}(n) \in \Omega(\omega)$.

If the costs are not unbounded, we say that they *depend* on $x \in \mathcal{PV}$ iff the initial limit problem $\{a_1^{\bullet_1}, \ldots, a_k^{\bullet_k}, \mathsf{cost}(t)^+\}$ has a solution $\boldsymbol{v}_n$ where $\boldsymbol{v}_n(y)$ is constant for all $y \in \mathcal{PV} \setminus \{x\}$. If x corresponds to a "secret", then the program can be subject to side-channel attacks. For example, in Example 15 we have $\boldsymbol{v}_n(\mathsf{cost}(t)) = \frac{1}{8}n^4 + \frac{1}{4}n^3 + \frac{7}{8}n^2 + \frac{7}{4}n$. Since $\boldsymbol{v}_n$ maps all program variables except x to constants, the costs of our program depend on the program variable x. So if x is "secret", then the program is not safe from side-channel attacks.

Theorem 16 results in bounds of the form "$\mathsf{rc}_{\mathcal{T}}(|\boldsymbol{v}_n|) \in \Omega(\boldsymbol{v}_n(c))$", which depend on the *sizes* $|\boldsymbol{v}_n|$. Let $f(n) = \mathsf{rc}_{\mathcal{T}}(n)$, $g(n) = |\boldsymbol{v}_n|$, and let $\Omega(\boldsymbol{v}_n(c))$ have the form $\Omega(n^k)$ or $\Omega(k^n)$ for a $k \in \mathbb{N}$. Moreover for all $x \in \mathcal{PV}$, let $\boldsymbol{v}_n(x)$ be a polynomial of at most degree m, i.e., let $g(n) \in \mathcal{O}(n^m)$. Then the following observation from [15] allows us to infer a bound for $\mathsf{rc}_{\mathcal{T}}(n)$ instead of $\mathsf{rc}_{\mathcal{T}}(|\boldsymbol{v}_n|)$.

Lemma 17 (Bounds for Function Composition). *Let $f : \mathbb{N} \to \mathbb{R}_{\geq 0}$ and $g : \mathbb{N} \to \mathbb{N}$ where $g(n) \in \mathcal{O}(n^m)$ for some $m \in \mathbb{N} \setminus \{0\}$. Moreover, let $f(n)$ be weakly and let $g(n)$ be strictly monotonically increasing for large enough n.*

- *If $f(g(n)) \in \Omega(n^k)$ with $k \in \mathbb{N}$, then $f(n) \in \Omega(n^{\frac{k}{m}})$.*
- *If $f(g(n)) \in \Omega(k^n)$ with $k \in \mathbb{N}$, then $f(n) \in \Omega(k^{\sqrt[m]{n}})$.*

Example 18 (Bound for Fig. 6 Continued). In Example 15, we inferred $\mathsf{rc}_{\mathcal{T}}(|\boldsymbol{v}_n|) \in \Omega(n^4)$ where $\boldsymbol{v}_n(x) = n$ and $\boldsymbol{v}_n(y) = \boldsymbol{v}_n(z) = \boldsymbol{v}_n(u) = 0$. Hence, we have $|\boldsymbol{v}_n| = n \in \mathcal{O}(n^1)$. By Lemma 17, we obtain $\mathsf{rc}_{\mathcal{T}}(n) \in \Omega(n^{\frac{4}{1}}) = \Omega(n^4)$.

Example 19 (Non-Polynomial Bounds). Let $\mathcal{T} = \{(\ell_0, x = y^2, \mathsf{id}, y, \ell)\}$. By Definition 14, the family $\boldsymbol{v}_n$ with $\boldsymbol{v}_n(x) = n^2$ and $\boldsymbol{v}_n(y) = n$ is a solution of the initial limit problem $\{(x - y^2 + 1)^{+_!}, (y^2 - x + 1)^{+_!}, y^+\}$. Due to Theorem 16, this proves $\mathsf{rc}_{\mathcal{T}}(|\boldsymbol{v}_n|) \in \Omega(n)$. As $|\boldsymbol{v}_n| = n^2 + n \in \mathcal{O}(n^2)$, Lemma 17 results in $\mathsf{rc}_{\mathcal{T}}(n) \in \Omega(n^{\frac{1}{2}})$.

5.2 Transformation of Limit Problems

A limit problem S is *trivial* iff all terms in S are variables and there is no variable x with $x^{\bullet_1}, x^{\bullet_2} \in S$ and $\bullet_1 \neq \bullet_2$. For trivial limit problems S we can immediately obtain a particular solution $\boldsymbol{v}_n^S$ which instantiates variables "according to S".

Lemma 20 (Solving Trivial Limit Problems). *Let S be a trivial limit problem. Then $\boldsymbol{v}_n^S$ is a solution of S where for all $n \in \mathbb{N}$, $\boldsymbol{v}_n^S$ is defined as follows:*

$$\begin{array}{lll} \boldsymbol{v}_n^S(x) = \;\; n, \textit{ if } x^+ \in S & \boldsymbol{v}_n^S(x) = \;\; 1, \textit{ if } x^{+_!} \in S & \boldsymbol{v}_n^S(x) = 0, \textit{ otherwise} \\ \boldsymbol{v}_n^S(x) = -n, \textit{ if } x^- \in S & \boldsymbol{v}_n^S(x) = -1, \textit{ if } x^{-_!} \in S & \end{array}$$

For instance, if $\mathcal{V}_{\ell_0} = \{x, y, tv\}$ and $S = \{x^+, y^{-_!}\}$, then S is a trivial limit problem and $\boldsymbol{v}_n^S$ with $\boldsymbol{v}_n^S(x) = n$, $\boldsymbol{v}_n^S(y) = -1$, and $\boldsymbol{v}_n^S(tv) = 0$ is a solution for S.

However, in general the initial limit problem $S = \{a_1^{\bullet_1}, \ldots, a_k^{\bullet_k}, \mathsf{cost}(t)^+\}$ is not trivial. Therefore, we now define a transformation $\rightsquigarrow$ to simplify limit problems until one reaches a trivial problem. With our transformation, $S \rightsquigarrow S'$ ensures that each solution of S' also gives rise to a solution of S.

If S contains $f(a_1, a_2)^\bullet$ for some standard arithmetic operation f like addition, subtraction, multiplication, division, and exponentiation, we use a so-called *limit vector* $(\bullet_1, \bullet_2)$ with $\bullet_i \in \{+, -, +_!, -_!\}$ to characterize for which kinds of arguments the operation f is increasing (if $\bullet = +$) resp. decreasing (if $\bullet = -$) resp. a positive or negative constant (if $\bullet = +_!$ or $\bullet = -_!$).[6] Then S can be transformed into the new limit problem $S \setminus \{f(a_1, a_2)^\bullet\} \cup \{a_1^{\bullet_1}, a_2^{\bullet_2}\}$.

For example, $(+, +_!)$ is an increasing limit vector for subtraction. The reason is that $a_1 - a_2$ is increasing if a_1 is increasing and a_2 is a positive constant. Hence, our transformation $\rightsquigarrow$ allows us to replace $(a_1 - a_2)^+$ by a_1^+ and $a_2^{+_!}$.

To define limit vectors formally, we say that $(\bullet_1, \bullet_2)$ is an *increasing* (resp. *decreasing*) *limit vector* for f iff the function $\lambda n.\ f(g(n), h(n))$ satisfies $+$ (resp. $-$) for any functions g and h that satisfy $\bullet_1$ and $\bullet_2$, respectively. Here, "$\lambda n.\ f(g(n), h(n))$" is the function from $\mathbb{N} \to \mathbb{R}$ that maps any $n \in \mathbb{N}$ to $f(g(n), h(n))$. Similarly, $(\bullet_1, \bullet_2)$ is a *positive* (resp. *negative*) *limit vector* for f iff $\lambda n.\ f(g(n), h(n))$ satisfies $+_!$ (resp. $-_!$) for any functions g and h that satisfy $\bullet_1$ and $\bullet_2$, respectively.

With this definition, $(+, +_!)$ is indeed an increasing limit vector for subtraction, since $\lim_{n \mapsto \omega} g(n) = \omega$ and $\lim_{n \mapsto \omega} h(n) = m$ with $m > 0$ implies $\lim_{n \mapsto \omega}(g(n) - h(n)) = \omega$. In other words, if $g(n)$ satisfies $+$ and $h(n)$ satisfies $+_!$, then $g(n) - h(n)$ satisfies $+$ as well. In contrast, $(+, +)$ is not an increasing limit vector for subtraction. To see this, consider the functions $g(n) = h(n) = n$. Both $g(n)$ and $h(n)$ satisfy $+$, whereas $g(n) - h(n) = 0$ does not satisfy $+$. Similarly, $(+_!, +_!)$ is not a

[6] To ease the presentation, we restrict ourselves to binary operations f. For operations of arity m, one would need limit vectors of the form $(\bullet_1, \ldots, \bullet_m)$.

positive limit vector for subtraction, since for $g(n) = 1$ and $h(n) = 2$, both $g(n)$ and $h(n)$ satisfy $+_!$, but $g(n) - h(n) = -1$ does not satisfy $+_!$.

Limit vectors can be used to simplify limit problems, cf. (A) in the following definition. Moreover, for numbers $m \in \mathbb{Z}$, one can easily simplify constraints of the form $m^{+_!}$ and $m^{-_!}$ (e.g., $2^{+_!}$ is obviously satisfied since $2 > 0$), cf. (B).

Definition 21 ($\rightsquigarrow$). *Let S be a limit problem. We have:*

(A) $S \cup \{f(a_1, a_2)^\bullet\} \rightsquigarrow S \cup \{a_1^{\bullet_1}, a_2^{\bullet_2}\}$ *if* $\bullet$ *is* $+$ *(resp.* $-, +_!, -_!$*) and* $(\bullet_1, \bullet_2)$ *is an increasing (resp. decreasing, positive, negative) limit vector for* f

(B) $S \cup \{m^{+_!}\} \rightsquigarrow S$ *if* $m \in \mathbb{Z}$ *with* $m > 0$, $S \cup \{m^{-_!}\} \rightsquigarrow S$ *if* $m < 0$

Example 22 (Bound for Fig. 6 Continued). For the initial limit problem from Example 15, we have $\{tv^+, (\frac{1}{2}x^2 + \frac{1}{2}x - tv)^{+_!}\} \rightsquigarrow \{tv^+, (\frac{1}{2}x^2 + \frac{1}{2}x)^{+_!}, tv^{-_!}\} \rightsquigarrow \{tv^+, (\frac{1}{2}x^2)^{+_!}, (\frac{1}{2}x)^{+_!}, tv^{-_!}\} \rightsquigarrow^* \{tv^+, x^{+_!}, tv^{-_!}\}$ using the positive limit vector $(+_!, -_!)$ for subtraction and the positive limit vector $(+_!, +_!)$ for addition.

The resulting problem in Example 22 is not trivial as it contains tv^+ and $tv^{-_!}$, i.e., we failed to compute an asymptotic lower bound. However, if we substitute tv with its upper bound $\frac{1}{2}x^2 + \frac{1}{2}x - 1$, then we could reduce the initial limit problem to a trivial one. Hence, we now extend $\rightsquigarrow$ by allowing to apply substitutions.

Definition 23 ($\rightsquigarrow$ **Continued**). *Let S be a limit problem and let $\sigma : \mathcal{V}_{\ell_0} \to \mathcal{A}(\mathcal{V}_{\ell_0})$ be a substitution such that x does not occur in $x\sigma$ and $\boldsymbol{v}(x\sigma) \in \mathbb{Z}$ for all valuations $\boldsymbol{v} \in \mathcal{V}al_{\ell_0}$ and all $x \in \mathcal{V}_{\ell_0}$. Then we have*[7]

(C) $S \rightsquigarrow^\sigma S\sigma$

Example 24 (Bound for Fig. 6 Continued). For the initial limit problem from Example 15, we now have[8] $\{tv^+, (\frac{1}{2}x^2 + \frac{1}{2}x - tv)^{+_!}\} \rightsquigarrow^{[tv/\frac{1}{2}x^2+\frac{1}{2}x-1]} \{(\frac{1}{2}x^2 + \frac{1}{2}x - 1)^+, 1^{+_!}\} \rightsquigarrow \{(\frac{1}{2}x^2 + \frac{1}{2}x - 1)^+\} \rightsquigarrow \{(\frac{1}{2}x^2 + \frac{1}{2}x)^+, 1^{+_!}\} \rightsquigarrow^* \{x^+\}$, which is trivial.

Although Definition 23 requires that variables may only be instantiated by integer terms, it is also useful to handle limit problems that contain non-integer terms.

Example 25 (Non-Integer Metering Functions Continued). After chaining the accelerated transition of Example 8 with the transition from the start location, for the resulting initial limit problem we get $\{tv^+, (\frac{1}{2}x - tv + 1)^{+_!}, (tv + 1)^+\} \rightsquigarrow^2 \{tv^+, (\frac{1}{2}x - tv + 1)^{+_!}\} \rightsquigarrow^{[x/2tv-1]} \{tv^+, \frac{1}{2}^{+_!}\} \rightsquigarrow \{tv^+, 1^{+_!}, 2^{+_!}\} \rightsquigarrow^2 \{tv^+\}$, using the positive limit vector $(+_!, +_!)$ for division. This allows us to infer $\mathsf{rc}_{\mathcal{T}}(n) \in \Omega(n)$.

However, up to now we cannot prove that, e.g., a transition t with $\mathsf{guard}(t) = x^2 - x > 0$ and $\mathsf{cost}(t) = x$ has a linear lower bound, since $(+, +)$ is not an increasing limit vector for subtraction. To handle such cases, the following rules allow us to neglect polynomial sub-expressions if they are "dominated" by other polynomials of higher degree or by exponential sub-expressions.

[7] The other rules for $\rightsquigarrow$ are implicitly labeled with the identical substitution id.

[8] $\sigma = [tv/\frac{1}{2}x^2 + \frac{1}{2}x - 1]$ satisfies the condition $\boldsymbol{v}(y\sigma) \in \mathbb{Z}$ for all $\boldsymbol{v} \in \mathcal{V}al_{\ell_0}$ and $y \in \mathcal{V}_{\ell_0}$.

Definition 26 ($\rightsquigarrow$ Continued). *Let S be a limit problem, let $\pm \in \{+,-\}$, and let $a, b, e \in \mathcal{A}(\{x\})$ be (univariate) polynomials. Then we have:*

(D) $S \cup \{(a \pm b)^\bullet\} \rightsquigarrow S \cup \{a^\bullet\}$, if $\bullet \in \{+,-\}$, and a has a higher degree than b

(E) $S \cup \{(a^e \pm b)^+\} \rightsquigarrow S \cup \{(a-1)^\bullet, e^+\}$, if $\bullet \in \{+,+_!\}$.

Thus, $\{(x^2-x)^+\} \rightsquigarrow \{(x^2)^+\} = \{(x \cdot x)^+\} \rightsquigarrow \{x^+\}$ by the increasing limit vector $(+,+)$ for multiplication. Similarly, $\{(2^x - x^3)^+\} \rightsquigarrow \{(2-1)^{+_!}, x^+\} \rightsquigarrow \{x^+\}$. Rule (E) can also be used to handle problems like $(a^e)^+$ (by choosing $b = 0$).

Theorem 27 states that $\rightsquigarrow$ is indeed correct. When constructing the valuation from the resulting trivial limit problem, one has to take the substitutions into account which were used in the derivation. Here, $(\boldsymbol{v}_n \circ \sigma)(x)$ stands for $\boldsymbol{v}_n(\sigma(x))$.

Theorem 27 (Correctness of $\rightsquigarrow$). *If $S \rightsquigarrow^\sigma S'$ and the family $\boldsymbol{v}_n$ is a solution of S', then $\boldsymbol{v}_n \circ \sigma$ is a solution of S.*

Example 28 (Bound for Fig. 6 Continued). Example 24 leads to the solution $\boldsymbol{v}'_n \circ \sigma$ of the initial limit problem for the program from Fig. 6 where $\sigma = [tv/\frac{1}{2}x^2 + \frac{1}{2}x - 1]$, $\boldsymbol{v}'_n(x) = n$, and $\boldsymbol{v}'_n(tv) = \boldsymbol{v}'_n(y) = \boldsymbol{v}'_n(z) = \boldsymbol{v}'_n(u) = 0$. Hence, $\boldsymbol{v}'_n \circ \sigma = \boldsymbol{v}_n$ where $\boldsymbol{v}_n$ is as in Example 15. As explained in Example 18, this proves $\mathsf{rc}_\mathcal{T}(n) \in \Omega(n^4)$.

So we start with an initial limit problem $S = \{a_1^{\bullet_1}, \ldots, a_k^{\bullet_k}, \mathsf{cost}(t)^+\}$ that represents $\mathsf{guard}(t)$ and requires non-constant costs, and transform S with $\rightsquigarrow$ into a trivial S', i.e., $S \rightsquigarrow^{\sigma_1} \ldots \rightsquigarrow^{\sigma_m} S'$. For automation, one should leave the $\bullet_i$ in the initial problem S open, and only instantiate them by a value from $\{+,+_!\}$ when this is needed to apply a particular rule for the transformation $\rightsquigarrow$. Then the resulting family $\boldsymbol{v}_n^{S'}$ of valuations gives rise to a solution $\boldsymbol{v}_n^{S'} \circ \sigma_m \circ \ldots \circ \sigma_1$ of S. Thus, we have $\mathsf{rc}_\mathcal{T}(|\boldsymbol{v}_n^{S'} \circ \sigma|) \in \Omega(\boldsymbol{v}_n^{S'}(\sigma(\mathsf{cost}(t))))$, where $\sigma = \sigma_m \circ \ldots \circ \sigma_1$, which leads to a lower bound for $\mathsf{rc}_\mathcal{T}(n)$ with Lemma 17.

Our implementation uses the following strategy to apply the rules from Definitions 21, 23, 26 for $\rightsquigarrow$. Initially, we reduce the number of variables by propagating bounds implied by the guard, i.e., if $\gamma \implies x \geq a$ or $\gamma \implies x \leq a$ for some $a \in \mathcal{A}(\mathcal{V}_{\ell_0} \setminus \{x\})$, then we apply the substitution $[x/a]$ to the initial limit problem by rule (C). For example, we simplify the limit problem from Example 19 by instantiating x with y^2, as the guard of the corresponding transition implies $x = y^2$. So here, we get $\{(x - y^2 + 1)^{+_!}, (y^2 - x + 1)^{+_!}, y^+\} \rightsquigarrow^{[x/y^2]} \{1^{+_!}, y^+\} \rightsquigarrow \{y^+\}$. Afterwards, we use (B) and (D) with highest and (E) with second highest priority. The third priority is trying to apply (A) to univariate terms (since processing univariate terms helps to guide the search). As fourth priority, we apply (C) with a substitution $[x/m]$ if $x^{+_!}$ or $x^{-_!}$ in S, where we use SMT solving to find a suitable $m \in \mathbb{Z}$. Otherwise, we apply (A) to multivariate terms. Since $\rightsquigarrow$ is well founded and, except for (C), finitely branching, one may also backtrack and explore alternative applications of $\rightsquigarrow$. In particular, we backtrack if we obtain a contradictory limit problem. Moreover, if we obtain a trivial S' where $\boldsymbol{v}_n^{S'}(\sigma(\mathsf{cost}(t)))$ is a polynomial, but $\mathsf{cost}(t)$ is a polynomial of higher degree or an exponential function, then we backtrack to search for other solutions which might lead to a higher lower bound. However, our implementation can of course fail, since solvability of limit problems is undecidable (due to Hilbert's Tenth Problem).

6 Experiments and Conclusion

We presented the first technique to infer lower bounds on the worst-case runtime complexity of integer programs, based on a modular program simplification framework. The main simplification technique is *loop acceleration*, which relies on *recurrence solving* and *metering functions*, an adaptation of classical ranking functions. By eliminating loops and locations via *chaining*, we eventually obtain *simplified programs*. We presented a technique to infer *asymptotic lower bounds* from simplified programs, which can also be used to find vulnerabilities.

Our implementation LoAT ("Lower Bounds Analysis Tool") is freely available at [23]. It was inspired by KoAT [8], which alternates runtime- and size-analysis to infer *upper* bounds in a modular way. Similarly, LoAT alternates runtime-analysis and recurrence solving to transform loops to loop-free transitions independently. LoAT uses the recurrence solver PURRS [4] and the SMT solver Z3 [10].

We evaluated LoAT on the benchmarks [5] from the evaluation of [8]. We omitted 50 recursive programs, since our approach cannot yet handle recursion. As we know of no other tool to compute worst-case lower bounds for integer programs, we compared our results with the asymptotically smallest results of leading tools for upper bounds: KoAT, CoFloCo [14], Loopus [26], RanK [2]. We did not compare with PUBS [1], since the cost relations analyzed by PUBS significantly differ from the integer programs handled by LoAT. Moreover, as PUBS computes *best-case* lower bounds, such a comparison would be meaningless since the worst-case lower bounds computed by LoAT are no valid best-case lower bounds. We used a timeout of 60 s. In the following, we disregard 132 examples where $\mathsf{rc}_{\mathcal{T}}(n) \in \mathcal{O}(1)$ was proved since there is no non-trivial lower bound in these cases.

$\mathsf{rc}_{\mathcal{T}}(n)$	$\Omega(1)$	$\Omega(n)$	$\Omega(n^2)$	$\Omega(n^3)$	$\Omega(n^4)$	EXP	$\Omega(\omega)$
$\mathcal{O}(1)$	(132)	–	–	–	–	–	–
$\mathcal{O}(n)$	45	125	–	–	–	–	–
$\mathcal{O}(n^2)$	9	18	33	–	–	–	–
$\mathcal{O}(n^3)$	2	–	–	3	–	–	–
$\mathcal{O}(n^4)$	1	–	–	–	2	–	–
EXP	–	–	–	–	–	5	–
$\mathcal{O}(\omega)$	57	31	3	–	–	–	173

LoAT infers non-trivial lower bounds for 393 (78 %) of the remaining 507 examples. *Tight* bounds (i.e., the lower and the upper bound coincide) are proved in 341 cases (67 %). Whenever an exponential upper bound is proved, LoAT also proves an exponential lower bound (i.e., $\mathsf{rc}_{\mathcal{T}}(n) \in \Omega(k^n)$ for some $k > 1$). In 173 cases, LoAT infers unbounded runtime complexity. In some cases, this is due to non-termination, but for this particular goal, specialized tools are more powerful (e.g., whenever LoAT proves unbounded runtime complexity due to non-termination, the termination analyzer T2 [7] shows non-termination as well). The average runtime of LoAT was 2.4 s per example. These results could be improved further by supplementing LoAT with invariant inference as implemented in tools like APRON [20]. For a detailed experimental evaluation of our implementation as well as the sources and a pre-compiled binary of LoAT we refer to [16].

Acknowledgments. We thank S. Genaim and J. Böker for discussions and comments.

References

1. Albert, E., Genaim, S., Masud, A.N.: On the inference of resource usage upper and lower bounds. ACM Trans. Comput. Logic **14**(3), 22:1–22:35 (2013)
2. Alias, C., Darte, A., Feautrier, P., Gonnord, L.: Multi-dimensional rankings, program termination, and complexity bounds of flowchart programs. In: Cousot, R., Martel, M. (eds.) SAS 2010. LNCS, vol. 6337, pp. 117–133. Springer, Heidelberg (2010)
3. Alonso-Blas, D.E., Genaim, S.: On the limits of the classical approach to cost analysis. In: Miné, A., Schmidt, D. (eds.) SAS 2012. LNCS, vol. 7460, pp. 405–421. Springer, Heidelberg (2012)
4. Bagnara, R., Pescetti, A., Zaccagnini, A., Zaffanella, E.: PURRS: towards computer algebra support for fully automatic worst-case complexity analysis. CoRR abs/cs/0512056 (2005)
5. Benchmark examples. https://github.com/s-falke/kittel-koat/tree/master/koat-evaluation/examples
6. Bradley, A.R., Manna, Z., Sipma, H.B.: Linear ranking with reachability. In: Etessami, K., Rajamani, S.K. (eds.) CAV 2005. LNCS, vol. 3576, pp. 491–504. Springer, Heidelberg (2005)
7. Brockschmidt, M., Cook, B., Fuhs, C.: Better termination proving through cooperation. In: Sharygina, N., Veith, H. (eds.) CAV 2013. LNCS, vol. 8044, pp. 413–429. Springer, Heidelberg (2013)
8. Brockschmidt, M., Emmes, F., Falke, S., Fuhs, C., Giesl, J.: Alternating runtime and size complexity analysis of integer programs. In: Ábrahám, E., Havelund, K. (eds.) TACAS 2014 (ETAPS). LNCS, vol. 8413, pp. 140–155. Springer, Heidelberg (2014)
9. Carbonneaux, Q., Hoffmann, J., Shao, Z.: Compositional certified resource bounds. In: Grove, D., Blackburn, S. (eds.) PLDI 2015, pp. 467–478, ACM (2015)
10. de Moura, L., Bjørner, N.S.: Z3: an efficient SMT solver. In: Ramakrishnan, C.R., Rehof, J. (eds.) TACAS 2008. LNCS, vol. 4963, pp. 337–340. Springer, Heidelberg (2008)
11. Debray, S., López-García, P., Hermenegildo, M.V., Lin, N.: Lower bound cost estimation for logic programs. In: Maluszynski, J. (ed.) ILPS 1997, pp. 291–305. MIT Press (1997)
12. Falke, S., Kapur, D., Sinz, C.: Termination analysis of imperative programs using bitvector arithmetic. In: Joshi, R., Müller, P., Podelski, A. (eds.) VSTTE 2012. LNCS, vol. 7152, pp. 261–277. Springer, Heidelberg (2012)
13. Farzan, A., Kincaid, Z.: Compositional recurrence analysis. In: Kaivola, R., Wahl, T. (eds.) FMCAD 2015, pp. 57–64. IEEE (2015)
14. Flores-Montoya, A., Hähnle, R.: Resource analysis of complex programs with cost equations. In: Garrigue, J. (ed.) APLAS 2014. LNCS, vol. 8858, pp. 275–295. Springer, Heidelberg (2014)
15. Frohn, F., Giesl, J., Emmes, F., Ströder, T., Aschermann, C., Hensel, J.: Inferring lower bounds for runtime complexity. In: Fernández, M. (ed.) RTA 2015. LIPIcs, vol. 36, pp. 334–349. Dagstuhl Publishing (2015)
16. Frohn, F., Naaf, M., Hensel, J., Brockschmidt, M., Giesl, J.: Proofs and empirical evaluation of "Lower Runtime Bounds for Integer Programs" (2016). http://aprove.informatik.rwth-aachen.de/eval/integerLower/
17. Gonnord, L., Halbwachs, N.: Combining widening and acceleration in linear relation analysis. In: Yi, K. (ed.) SAS 2006. LNCS, vol. 4134, pp. 144–160. Springer, Heidelberg (2006)

18. Hofbauer, D., Lautemann, C.: Termination proofs and the length of derivations. In: Dershowitz, N. (ed.) Rewriting Techniques and Applications. LNCS, vol. 355, pp. 167–177. Springer, Heidelberg (1989)
19. Hoffmann, J., Aehlig, K., Hofmann, M.: Multivariate amortized resource analysis. ACM Trans. Program. Lang. Syst. **34**(3), 14:1–14:62 (2012)
20. Jeannet, B., Miné, A.: APRON: a library of numerical abstract domains for static analysis. In: Bouajjani, A., Maler, O. (eds.) CAV 2009. LNCS, vol. 5643, pp. 661–667. Springer, Heidelberg (2009)
21. Jeannet, B., Schrammel, P., Sankaranarayanan, S.: Abstract acceleration of general linear loops. ACM SIGPLAN Not. **49**(1), 529–540 (2014)
22. Kroening, D., Lewis, M., Weissenbacher, G.: Under-approximating loops in C programs for fast counterexample detection. Form. Meth. Sys. Des. **47**(1), 75–92 (2015)
23. LoAT. https://github.com/aprove-developers/LoAT
24. Madhukar, K., Wachter, B., Kroening, D., Lewis, M., Srivas, M.K.: Accelerating invariant generation. In: Kaivola, R., Wahl, T. (eds.) FMCAD 2015, pp. 105–111. IEEE (2015)
25. Podelski, A., Rybalchenko, A.: A complete method for the synthesis of linear ranking functions. In: Steffen, B., Levi, G. (eds.) VMCAI 2004. LNCS, vol. 2937, pp. 239–251. Springer, Heidelberg (2004)
26. Sinn, M., Zuleger, F., Veith, H.: A simple and scalable static analysis for bound analysis and amortized complexity analysis. In: Biere, A., Bloem, R. (eds.) CAV 2014. LNCS, vol. 8559, pp. 745–761. Springer, Heidelberg (2014)

Translating Scala Programs to Isabelle/HOL
System Description

Lars Hupel[1(✉)] and Viktor Kuncak[2]

[1] Technische Universität München, München, Germany
lars.hupel@tum.de
[2] École Polytechnique Fédérale de Lausanne (EPFL), Lausanne, Switzerland

Abstract. We present a trustworthy connection between the Leon verification system and the Isabelle proof assistant. Leon is a system for verifying functional Scala programs. It uses a variety of automated theorem provers (ATPs) to check verification conditions (VCs) stemming from the input program. Isabelle, on the other hand, is an interactive theorem prover used to verify mathematical specifications using its own input language Isabelle/Isar. Users specify (inductive) definitions and write proofs about them manually, albeit with the help of semi-automated tactics. The integration of these two systems allows us to exploit Isabelle's rich standard library and give greater confidence guarantees in the correctness of analysed programs.

Keywords: Isabelle · HOL · Scala · Leon · Compiler

1 Introduction

This system description presents a new tool that aims to connect two important worlds: the world of interactive proof assistant users who create a body of verified theorems, and the world of professional programmers who increasingly adopt functional programming to develop important applications. The Scala language (www.scala-lang.org) enjoys a prominent role today for its adoption in industry, a trend most recently driven by the Apache Spark data analysis framework (to which, e.g., IBM committed 3500 researchers recently [16]). We hope to introduce some of the many Scala users to formal methods by providing tools they can use directly on Scala code. Leon system (http://leon.epfl.ch) is a verification and synthesis system for a subset of Scala [2,10]. Leon reuses the Scala compiler's parsing and type-checking frontend and subsequently derives verification conditions to be solved by the automated theorem provers, such as *Z3* [13] and *CVC4* [1]. Some of these conditions arise naturally upon use of particular Scala language constructs (e.g. completeness for pattern matching), whereas others stem from Scala assertions (`require` and `ensuring`) and can naturally express universally quantified conjectures about computable functions.

Interactive proof assistants have long contained functional languages as fragments of the language they support. Isabelle/HOL [14,20] offers definitional

N. Olivetti and A. Tiwari (Eds.): IJCAR 2016, LNAI 9706, pp. 568–577, 2016.
DOI: 10.1007/978-3-319-40229-1_38

facilities for functional programming, e.g. the `datatype` command for inductive data types and `fun` for recursive functions. A notable feature of Isabelle is its code generator: certain executable specifications can be translated into source code in target languages such as ML, Haskell, Scala, OCaml [5,7]. Yet many Scala users do not know Isabelle today.

Aiming to bring the value of trustworthy formalized knowledge to many programmers familiar with Scala, we introduce a mapping in the opposite direction: instead of generating code from logic, we show how to map programs in the purely functional fragment of Scala supported by Leon into Isabelle/HOL. We use Isabelle's built-in tactics to discharge the verification conditions. Compared to use of automated solvers in Leon alone, the connection with Isabelle has two main advantages:

1. Proofs in Isabelle, even those generated from automated tactics, are justified by a minimal inference kernel. In contrast to ATPs, which are complex pieces of software, it is far less likely that a kernel-certified proof is unsound.
2. Isabelle's premier logic, HOL, has seen decades of development of rich mathematical libraries and formalizations such as Archive of Formal Proofs. Proofs carried out in Isabelle have access to this knowledge, which means that there is a greater potential for reuse of existing developments.

Establishing the formal correspondence means embedding Scala in HOL, requiring non-trivial transformations (Sect. 2). We use a *shallow embedding*, that is, we do not model Scala's syntax, but rather perform a syntactic mapping from Scala constructs to their equivalents in HOL. For our implementation we developed an idiomatic Scala API for Isabelle based on previous work by Wenzel [18,21] (Sect. 3). We implemented as much functionality as possible inside Isabelle to leverage checking by Isabelle's proof kernel. The power of Isabelle's tactics allows us to prove more conditions than what is possible with the Z3 and CVC4 backends (Sect. 4). We are able to import Leon's standard library and a large amount of its example code base into Isabelle (Sect. 5), and verify many of the underlying properties.

Contribution. We contribute a mechanism to import functional Scala code into Isabelle, featuring facilities for embedding Isabelle/Isar syntax into Scala via Leon and reusing existing constants in the HOL library without compromising soundness. This makes Isabelle available to Leon as a drop-in replacement for Z3 or CVC4 to discharge verification conditions. We show that Isabelle automation is already useful for processing such conditions.

Among related works we highlight a Haskell importer for Isabelle [6], which also uses a shallow embedding and has a custom parser for Haskell, but does not perform any verification. Breitner et al. have formalised "large parts of Haskell's standard prelude" in Isabelle [4]. They use the HOLCF logic, which is extension on HOL for domain theory, and have translated library functions manually. Mehnert [12] implemented a verification system for Java in Coq using separation logic.

In the following text, we are using the term "Pure Scala" to refer to the fragment of Scala supported by Leon [2, Sect. 3], whereas "Leon" denotes the system itself. More information about Leon and Pure Scala is available from the web deployment of Leon at http://leon.epfl.ch in the Documentation section.

2 Bridging the Gap

Isabelle is a general specification and proof toolkit with the ability of functional programming in its logic Isabelle/HOL. Properties of programs need to be stated and proved explicitly in an interactive IDE. While the system offers *proof tactics*, the order in which they are called and their parameters need to be specified by the user. Users can also write custom tactics which deal with specific classes of problems.

Leon is more specialised to verification of functional programs and runs in batch mode. The user annotates a program and then calls Leon which attempts to discharge resulting verification conditions using ATPs. If that fails, the user has to restructure the program. Leon has been originally designed to be fully automatic; consequently, there is little support for explicitly guiding the prover. However, because of its specialisation, it can leverage more automation in proofs and counterexample finding on first-order recursive functions.

Due to their differences, both systems have unique strengths. Their connection allows users to benefit from this complementarity.

```
sealed abstract class List[A]
case class Cons[A](head: A, tail: List[A]) extends List[A]
case class Nil[A]() extends List[A]

def size[A](l: List[A]): BigInt = (l match {
  case Nil => BigInt(0)
  case Cons(_, xs) => 1 + size(xs)
}) ensuring(_ >= 0)
```

(a) Pure Scala version

```
datatype 'a list = Nil | Cons 'a "'a list"

fun size :: "'a list => int" where
"size Nil = 0" |
"size (Cons _ xs) = 1 + size xs"

lemma "size xs >= 0" by (induct xs) auto
```

(b) Isabelle version

Fig. 1. Example programs: Linked lists and a size function

2.1 Language Differences

Both languages use different styles in how functional programs are expressed. Figure 1 shows a direct comparison of a simple program accompanied by a (trivial) proof illustrating the major differences:

- Pure Scala uses an object-oriented encoding of algebraic data types (*sealed classes* [15]), similar to Java or C#. Isabelle/HOL follows the ML tradition by having direct syntax support [3].
- (Pre-) and postconditions in Leon are annotated using the `ensuring` function, whereas Isabelle has a separate `lemma` command. In a sense, verification conditions in Leon are "inherent", but need to be stated manually in Isabelle.
- Pure Scala does not support top-level pattern matching (e.g. $rev\ (x : xs) = \ldots$).

The translation of data types and terms is not particularly interesting because it is mostly a cavalcade of technicalities and corner cases. However, translating functions and handling recursion poses some interesting theoretical challenges.

2.2 Translating Functions

A *theory* is an Isabelle/Isar source file comprising a sequence of definitions and proofs, roughly corresponding to the notion of a "module" in other languages. Theory developments are strictly monotonic. Cyclic dependencies between definitions are not allowed [11], however, a definition may consist of multiple constants. In Pure Scala, there are no restrictions on definition order and cyclicity.

Consequentially, the Isabelle integration has to first compute the dependency graph of the functions and along with it the set of strongly connected components. A single component contains a set of mutually-recursive functions. Collapsing the components in the graph then results in a directed acyclic graph which can be processed in any topological ordering.

The resulting function definitions are not in idiomatic Isabelle/HOL style; in particular, they are not useful for automated tactics. Consider Fig. 1: the naive translation would produce a definition $size\ xs = \texttt{case}\ xs\ \texttt{of}\ y \# ys \rightarrow \ldots size\ ys \ldots$ Isabelle offers a generic term rewriting tactic (the *simplifier*), which is able to substitute equational rules. Such a rule, however, constitutes a non-terminating simplification chain, because the right-hand side contains a subterm which matches the left-hand side.

This can be avoided by splitting the resulting definition into cases that use Haskell-style top-level pattern matching. A verified routine to perform this translation is integrated into Isabelle, producing terminating equations which can be used by automated tactics. From this, we also obtain a better induction principle which can be used in subsequent proofs.

When looking at the results of this procedure, the example in Fig. 1 is close to reality. The given Pure Scala input program produces almost exactly the Isabelle theory below, modulo renaming. Because of our implementation, the user normally does not see the resulting theory file (see Sect. 3). However, for this example, the internal constructions we perform are roughly equivalent to what Isabelle/Isar would perform (see Sect. 5).

2.3 Recursion

Leon has a separate termination checking pass, which can run along with verification and can be turned off. Leon's verification results are only meant to be valid under the assumption that its termination checker succeeded (i.e. ensuring partial correctness).

Isabelle's proof kernel does not accept recursive definitions at all. We use the *function package* by Krauss [9] to translate a set of recursive equations into a low-level, non-recursive definition. To automate this construction, the package provides a `fun` command which can be used in regular theories (see Fig. 1), but also programmatically. To justify its internal construction against the kernel, it needs to prove termination. By default, it searches for a lexicographic ordering involving some subset of the function arguments.

This also means that when Leon is run using Isabelle, termination checking is no longer independent of verification, but rather "built in". Krauss' package also supports user-specified termination proofs. In the future, we would like to give users the ability to write those in Scala.

A further issue is recursion in data types. Negative recursion can lead to unsoundness, e.g. introducing non-termination in non-recursive expressions. While Leon has not implemented a wellformedness check yet, Isabelle correctly rejects such data type definitions. Because we map Scala data types syntactically, we obtain this check for free when using Isabelle in Leon.

2.4 Cross-Language References

One of the main reasons why we chose a shallow embedding of Pure Scala into Isabelle is the prospect of reusability of Isabelle theories in proofs of imported Pure Scala programs. For example, the dominant collection data structure in functional programming – and by extension both in Pure Scala and Isabelle/HOL – are lists. Both languages offer dozens of library functions such as `map`, `take` or `drop`. Isabelle's `List` theory also contains a wealth of theorems over these functions. All of the existing theorems can be used by Isabelle's automated tactics to aid in subsequent proofs, and are typically unfolded automatically by the simplifier.

However, when importing Pure Scala programs, all its data types and functions are defined again in a runtime Isabelle theory. While the imported `List.map` function may end up having the same shape as HOL's `List.map` function, they are nonetheless distinct constants, rendering pre-existing theorems unusable.

The naive approach of annotating Pure Scala's `map` function to not be imported and instead be replaced by HOL's `map` function is unsatisfactory: The user would need to be trusted to correctly annotate Pure Scala's library, negatively impacting correctness. Hence, we implemented a hybrid approach: We first import the whole program unchanged, creating fresh constants. Later, for each annotated function, we try to prove an equivalence of the form $f' = f$ where f' is the imported definition and f is the existing Isabelle library function, and register the resulting theorem with Isabelle's automated tools. This establishes a trustworthy relationship between the imported Pure Scala program and the existing Isabelle libraries.

Depending on the size of the analysed program (including dependencies), this approach turns out to be rather inefficient.[1] According to Leon conventions, we introduced a flag which skips the equivalence proofs for Pure Scala library functions and just asserts the theorems as axioms. This also alleviates another practical problem: not all desired equivalences can be proven automatically by Isabelle. Support for specifying manual equivalence proofs would be useful, but is not yet implemented.

3 Technical Considerations

Isabelle has been smoothly integrated into Leon by providing an appropriate instance of a *solver*. In that sense, Isabelle acts as "yet another backend" which is able to check validity of a set of assertions.

3.1 Leon Integration

A solver in Leon terminology is a function checking the consistency of a set of assumptions. A pseudo-code type signature could be given as $\mathcal{P}(\mathcal{F}) \to \{\texttt{sat}, \texttt{unsat}, \texttt{unknown}\}$, where $\mathcal{F}$ is the set of supported formulas. According to program verification convention, a result of `unsat` means that no contradiction could be derived from the assumptions, i.e. that the underlying program is correct. If a solver however returns `sat`, it is expected to produce a counterexample which violates verification conditions, e.g. a value which is not matched by any clause in a pattern match.

The Isabelle integration is exactly such a function, but with the restriction that it never returns `sat`, because a failed proof attempt does not produce a suitable counterexample. Since Leon offers a sound and complete counterexample procedure for higher-order functions [17], implementing this feature for Isabelle would not be useful.

3.2 Process Communication

Communication between the JVM process running Leon and the Isabelle process works via our *libisabelle* library which extends Wenzel's PIDE framework [19,21] to cater to non-IDE applications. It introduces a remote procedure call layer on top of PIDE, reusing much of its functionality. Leon is then able to update and query state stored in the prover process. Procedure calls are typed and asynchronous, using an implementation of type classes in ML and Scala's *future* values by Haller et al. [8], respectively.

[1] Because our implementation uses Isabelle in interactive instead of in batch mode, we cannot produce pre-computed heap images to be loaded for later runs.

While being a technologically more complicated approach, it offers benefits over textual Isabelle/Isar source generation. Most importantly, because communication is typed, the implementation is much more robust. Common sources of errors, e.g. pretty printing of Isabelle terms or escaping, are completely eliminated.

4 Example

Figure 2 shows a fully-fledged example of an annotated Pure Scala program. As background, assume the `List` definition from the previous example enriched with some standard library functions, a `Nat` type, and a **`listSum`** function.[2] The functions in the example are turned into lemma statements in Isabelle. The string parameter of the `proof` annotation is an actual Isar method invocation, that is, it is interpreted by the Isabelle system. For hygienic purposes, names of Pure Scala identifiers are not preserved during translation, but suffixed with unique numbers. To allow users to refer back to syntactic entities using their original names, the `<var _>` syntax has been introduced.

Running Leon with the Isabelle solver on this example will show that all conditions hold. The first proof merely reuses a lemma which is already in the library. The other two need specific guidance, i.e. an annotation, for them to be accepted by the system. The proofs involve Isabelle library theorems, for example distributivity of $(+, *)$ on natural numbers. For comparison, Leon+Z3 cannot prove any proposition. When also instructed to perform induction, it can prove **`sumConstant`**. (Same holds for Leon+CVC4.) There is currently no way in Leon to concisely specify the use of a custom induction rule for Z3 (or CVC4) as required by the last proposition (simultaneous induction over two lists of equal length).

This example also demonstrates another instance of the general Isabelle philosophy of *nested languages:* Pure Scala identifiers may appear inside Isar text which appears inside Pure Scala code. Further nesting is possible because Isabelle text can itself contain nested elements (e.g. ML code, ...).

```
def sumReverse[A](xs: List[Nat]) =
  (listSum(xs) == listSum(xs.reverse)).holds

@proof(method = """(induct "<var xs>", auto)""")
def sumConstant[A](xs: List[A], k: Nat) =
  (listSum(xs.map(_ => k)) == length(xs) * k).holds

@proof(method = "(clarsimp, induct rule: list_induct2, auto)")
def mapFstZip[A, B](xs: List[A], ys: List[B]) = {
  require(length(xs) == length(ys))
  xs.zip(ys).map(_._1)
} ensuring { _ == xs }
```

Fig. 2. Various induction proofs about lists

[2] The full example is available at https://git.io/vznVH.

5 Evaluation

In this section, we discuss implementation coverage of Pure Scala's syntactic constructs, trustworthiness of the translation and overall performance.

Coverage. The coverage of the translation is almost complete. A small number of Leon primitives, among them array operations have not been implemented yet.[3] All other primitives are mapped as closely as possible and adaptations to Isabelle are proven correct when needed. Leon's standard library contains – as of writing – 177 functions with a total of 289 verification conditions, out of which Isabelle can prove 206 ($\approx 71\,\%$).

Trustworthiness. Our mapping uses only definitional constructs of Isabelle and thus the theorems it proves have high degree of trustworthiness. Using a shallow embedding always carries the risk of semantics mismatches. A concern is that since the translation of Pure Scala to Isabelle works through an internal API, the user has no possibility to convince themselves of the correctness of the implemented routines short of inspecting the source code. For that reason, all operations are logged in Isabelle. A user can request a textual output of an Isar theory file corresponding to the imported Pure Scala program, containing all definitions and lemma statements, but no proofs. This file can be inspected manually and re-used for other purposes, and represents faithfully the facts that Isabelle actually proved in a readable form.

Performance. On a contemporary dual-core laptop, just defining all data types from the Pure Scala library (as of writing: 13), but no functions or proofs, Leon+Isabelle takes approximately 30 s. Defining all functions adds another 70 s to the process. Using Leon+Z3, this is much faster: it takes less than 10 s. The considerable difference (factor ≈ 10) can be explained by looking at the internals of the different backends. Z3 has data types and function definitions built into its logic. Isabelle itself does not: both concepts are implemented in HOL, meaning that every definition needs to be constructed and justified against the proof kernel. The processing time of an imported Pure Scala programs is comparable to that of a hand-written, idiomatic Isabelle theory file. In fact, during processing the Pure Scala libraries, thousands of messages are passed between the JVM and the Isabelle process, but the incurred overhead is negligible compared to the internal definitional constructions.

6 Conclusion

We have implemented an extension to Leon which allows using Isabelle to discharge verification conditions of Pure Scala programs. Because it supports the

[3] In fact, while attempting to implement array support we discovered that Leon's purely functional view of immutably used arrays does not respect Scala's reference equality implementation of arrays, leading to a decision to disallow array equality in Leon's Pure Scala.

vast majority of syntax supported by Leon, we consider it to be generally usable. It is incorporated in the Leon source repository,[4] supporting the latest Isabelle version (Isabelle2016).

With this work, it becomes possible to co-develop a specification in both Pure Scala and Isabelle, use Leon to establish a formal correspondence, and prove interesting results in Leon and/or Isabelle/Isar. Because of the embedded Isar syntax, complicated correctness proofs can also be expressed concisely in Leon. To the best of our knowledge, this constitutes the first bi-directional integration between a widespread general purpose programming language and an interactive proof assistant.

An unintended consequence is that since Isabelle can export code in Haskell and now import code from Pure Scala, there is a fully-working Scala-to-Haskell cross-compilation pipeline. The transformations applied to the Pure Scala code to make it palatable to Isabelle's automation also results in moderately readable Haskell code.

Acknowledgements. We would like to thank the people who helped "making the code work": Ravi Kandhadai, Etienne Kneuss, Manos Koukoutos, Mikäel Mayer, Nicolas Voirol, Makarius Wenzel. Cornelius Diekmann, Manuel Eberl, and Tobias Nipkow suggested many textual improvements to this paper.

References

1. Barrett, C., Conway, C.L., Deters, M., Hadarean, L., Jovanović, D., King, T., Reynolds, A., Tinelli, C.: CVC4. In: Gopalakrishnan, G., Qadeer, S. (eds.) CAV 2011. LNCS, vol. 6806, pp. 171–177. Springer, Heidelberg (2011)
2. Blanc, R.W., Kneuss, E., Kuncak, V., Suter, P.: An overview of the Leon verification system: verification by translation to recursive functions. In: Scala Workshop (2013)
3. Blanchette, J.C., Hölzl, J., Lochbihler, A., Panny, L., Popescu, A., Traytel, D.: Truly modular (co)datatypes for Isabelle/HOL. In: Klein, G., Gamboa, R. (eds.) ITP 2014. LNCS, vol. 8558, pp. 93–110. Springer, Heidelberg (2014)
4. Breitner, J., Huffman, B., Mitchell, N., Sternagel, C.: Certified HLints with Isabelle/HOLCF-Prelude, Haskell and Rewriting Techniques (HART), June 2013
5. Haftmann, F.: Code generation from specifications in higher-order logic. Ph.D. thesis, Technische Universität München (2009)
6. Haftmann, F.: From higher-order logic to Haskell: there and back again. In: Proceedings of the 2010 ACM SIGPLAN Workshop on Partial Evaluation and Program Manipulation, pp. 155–158. ACM (2010)
7. Haftmann, F., Nipkow, T.: Code generation via higher-order rewrite systems. In: Blume, M., Kobayashi, N., Vidal, G. (eds.) FLOPS 2010. LNCS, vol. 6009, pp. 103–117. Springer, Heidelberg (2010)
8. Haller, P., Prokopec, A., Miller, H., Klang, V., Kuhn, R., Jovanovic, V.: Futures and promises (2012). http://docs.scala-lang.org/overviews/core/futures.html
9. Krauss, A.: Partial and nested recursive function definitions in higher-order logic. J. Autom. Reason. **44**(4), 303–336 (2009)

[4] https://github.com/epfl-lara/leon.

10. Kuncak, V.: Developing verified software using Leon. In: Havelund, K., Holzmann, G., Joshi, R. (eds.) NFM 2015. LNCS, vol. 9058, pp. 12–15. Springer, Heidelberg (2015)
11. Kunčar, O.: Correctness of Isabelle's cyclicity checker: implementability of overloading in proof assistants. In: Proceedings of the 2015 Conference on Certified Programs and Proofs, CPP 2015, pp. 85–94. ACM, New York (2015)
12. Mehnert, H.: Kopitiam: modular incremental interactive full functional static verification of java code. In: Bobaru, M., Havelund, K., Holzmann, G.J., Joshi, R. (eds.) NFM 2011. LNCS, vol. 6617, pp. 518-524. Springer, Heidelberg (2011)
13. de Moura, L., Bjørner, N.S.: Z3: an efficient SMT solver. In: Ramakrishnan, C.R., Rehof, J. (eds.) TACAS 2008. LNCS, vol. 4963, pp. 337–340. Springer, Heidelberg (2008)
14. Nipkow, T., Klein, G.: Concrete Semantics. Springer, New York (2014)
15. Odersky, M., Spoon, L., Venners, B.: Programming in Scala, 2nd edn. Artima Inc, Walnut Creek (2010)
16. Terdoslavich, W.: IBM bets on apache spark as 'the future of enterprise data'. http://www.informationweek.com/big-data/ibm-bets-on-apache-spark-as-the-future-of-enterprise-data/d/d-id/1320855
17. Voirol, N., Kneuss, E., Kuncak, V.: Counter-example complete verification for higher-order functions. In: Scala Symposium (2015)
18. Wenzel, M.: Isabelle as document-oriented proof assistant. In: Davenport, J.H., Farmer, W.M., Urban, J., Rabe, F. (eds.) MKM 2011 and Calculemus 2011. LNCS, vol. 6824, pp. 244–259. Springer, Heidelberg (2011)
19. Wenzel, M.: Isabelle/jEdit – a prover IDE within the PIDE framework. In: Jeuring, J., Campbell, J.A., Carette, J., Dos Reis, G., Sojka, P., Wenzel, M., Sorge, V. (eds.) CICM 2012. LNCS, vol. 7362, pp. 468–471. Springer, Heidelberg (2012)
20. Wenzel, M.: The Isabelle/Isar Reference Manual (2013)
21. Wenzel, M.: Asynchronous user interaction and tool integration in Isabelle/PIDE. In: Klein, G., Gamboa, R. (eds.) ITP 2014. LNCS, vol. 8558, pp. 515–530. Springer, Heidelberg (2014)

Author Index

Printed in the United States
By Bookmasters